JEPPESEN
A Bo

GUIDED FLIGHT DISCOVERY
PRIVATE PILOT

The charts, tables, and graphs used in this publication are for illustration purposes only and cannot be used for navigation or to determine actual aircraft performance.

Cover Photo: Cirrus airplane in flight courtesy of Cirrus Aircraft

ISBN 978-0-88487-700-4

CBTA Competencies and Observable Behaviors tables included in Chapter 1 Section D
Copyright notice © European Union, 1998-2023
Unless otherwise specified, you can re-use the legal documents published in
EUR-Lex for commercial or non-commercial purposes […]
('© European Union, http://eur-lex.europa.eu/, 1998-2023')

Jeppesen
55 Inverness Drive East
Englewood, CO 80112-5498
Web site: www.jeppesen.com
Email: Captain@jeppesen.com
Copyright © Jeppesen
All Rights Reserved.
Published 1997, 1998, 1999, 2000, 2001, 2002,
2004, 2007, 2010, 2011, 2013, 2015, 2018, 2024

Acknowledgments

The latest version of this textbook could not have been produced without the passion and hard work of these Guided Flight Discovery team members:

Liz Kailey—Primary Writer/Editor

Rick Patterson—Graphic Designer
Courtney Kraig—Graphic Designer
Veronica Mahr—Graphic Designer

Jennifer Bukacek—Instructional Designer

Thomas Huismann—Writer
Don Bachner—Writer
Santiago Rosell—Writer
Rebecca Albert—Writer

Michael Moffatt—Technical Editor

Welcome to Guided Flight Discovery

The Guided Flight Discovery Pilot Training System provides the finest pilot training available. Rather than just teaching facts, Guided Flight Discovery concentrates on an application-oriented approach to pilot training. The comprehensive and complete system emphasizes the why and how of aeronautical concepts when they are presented. As you progress through your training, you will find that the revolutionary Guided Flight Discovery system leads you through essential aeronautical knowledge and exposes you to a variety of interesting and useful information that will enhance and expand your understanding of the world of aviation.

Feeling comfortable with your grasp of aviation concepts and your ability to apply them is fundamental to conducting safe and enjoyable flight operations. Historically, the majority of problems that occur during flight can be traced to a pilot's judgment and decision making. Aeronautical judgment is based primarily on the pilot's ability to apply the knowledge learned during training and gained through experience. The information presented in this textbook and the related Guided Flight Discovery materials are designed to provide you with the foundation of knowledge and experience needed to exercise good judgment and make sound decisions throughout your flying experience. Although you can use each element of the Guided Flight Discovery Pilot Training System separately, the effectiveness of the materials can be maximized by using all of the individual components in a systems approach. The primary components of the Private Pilot Program are described below.

PRIVATE PILOT TEXTBOOK/E-BOOK

This *Private Pilot* textbook/e-book is your primary source for initial study and review. The text contains complete and concise explanations of the fundamental concepts and ideas that every private pilot needs to know. The subjects are organized in a logical manner to build upon previously introduced topics. Subjects are often expanded upon through the use of Discovery Insets, which are strategically placed throughout the chapters. The Summary Checklists, Key Terms, and Questions are designed to help you review and prepare for both the knowledge and practical tests. A more detailed explanation of this textbook's unique features is contained in the section entitled "How the Textbook Works" starting on page x.

Jeppesen e-books are electronic versions of traditional textbooks and reference materials that you can view on computers and other devices. Jeppesen e-books are available on iOS or Android devices and PC or Mac computers. Jeppesen e-books provide valuable features, including the ability to quickly jump to specific information, bookmark pages, take notes voice to text, and more. Direct linking to chapters in each book is provided through the table of contents.

PRIVATE PILOT MANEUVERS MANUAL

When used in conjunction with the other components of the Private Pilot Program, the *Private Pilot Maneuvers Manual* provides an effective, practical approach to your training. Maneuvers are numbered for ease of reference, are grouped into categories based on similar operational characteristics, and presented in the order in which they are typically introduced during training. This book uses colorful graphics and step-by-step procedure descriptions to help you visualize and understand each maneuver that you will perform in the airplane. Expanded instructional guidance, helpful hints, and explanations of common errors help you perform the maneuver more precisely the first time. QR codes link students to animations in the Jeppesen Private Pilot online course.

PRIVATE PILOT SYLLABUS

The syllabus provides a basic framework for your training in a logical sequence. Ground and flight lessons are coordinated to ensure that your training progresses smoothly and that you are consistently introduced to topics on the ground prior to being required to apply that knowledge in the airplane. The syllabus is available in print and e-book versions.

FAR/AIM

The Jeppesen FAR/AIM e-book includes the current Federal Aviation Regulations (FARs) and the Aeronautical Information Manual (AIM) in one publication. The FAR/AIM includes FAR Parts 1, 3, 11, 43, 48, 61, 67, 68, 71, 73, 91, 97, 103, 105, 107, 110, 119, 135, 136, 137, 141, 142, NTSB 830, and TSRs 1552 and 1562. The AIM is a reproduction of the FAA publication with full-color graphics and the Pilot/Controller Glossary. The AIM contains basic flight information and the ATC procedures to operate effectively in the U.S. National Airspace System

The FAR/AIM-FC (Flight Crew) contains the regulations used in professional pilot flight training: FAR Parts 1, 5, 25, 63, 65, 91, 110, 111, 117, 119, 120, 121, 129, 135, HMR 175 and TSA 1544.

PRIVATE PILOT AIRMAN KNOWLEDGE TEST GUIDE

The *Private Pilot Airman Knowledge Test Guide* helps you understand the learning objectives for the test questions so that you can take the FAA knowledge test with confidence. The test guide contains sample FAA Private Pilot airplane test questions, with correct answers, explanations, and study references. Explanations of why the other choices are wrong are included where appropriate. Full-color figures identical to the figures on the FAA test are also included. The test guide is intended to supplement your instructor-led flight and ground training.

PRIVATE PILOT PRACTICAL TEST STUDY GUIDE

The *Private Pilot FAA Practical Test Study Guide* provides guidance for you to pass your practical test with ease. The guide presents the information that you need to meet the knowledge, risk management, and skill requirements for each task in the Airman Certification Standards (ACS). An effective question and answer format helps you prepare for the oral portion of the test and step-by-step diagrams with helpful hints and common errors provides insight into performing the maneuvers proficiently during the flight.

PRIVATE PILOT ONLINE COURSE

Available from jeppdirect.com, the Private Pilot online course provides academic content in ground lessons with exams and interactive maneuvers lessons in a complete ground school, and outlines for every flight lesson. Ground school and maneuvers lessons are included using a combination of audio, video and graphics to clearly explain each topic. A Learning Management System (LMS) tracks your completions and test results specific to each question to assist you in identifying your strengths and weaknesses. You can use the online course together with the textbook and other Jeppesen products to enhance your learning experience.

Preface

The purpose of the *Private Pilot* textbook is to provide you with the most complete explanations of aeronautical concepts in the most effective and easy-to-use manner possible. Through the use of colorful illustrations, full-color photos, and a variety of innovative design techniques, the *Private Pilot* textbook and other Guided Flight Discovery materials are closely coordinated to make learning fun and easy. To help you organize your study, the *Private Pilot* textbook is divided into five parts:

PART I — FUNDAMENTALS OF FLIGHT

The information needed to begin your aviation journey is introduced in this part. The first chapter, Discovering Aviation, answers many of your questions about the training process. Chapter 2 introduces you to the basics of airplane systems. By exploring Chapter 3 you will gain an understanding of aerodynamic principles.

PART II — FLIGHT OPERATIONS

Part II contains information you need to know about the environment in which you will fly. You will study subjects such as airport facilities, air traffic control services, communication procedures, and sources of flight information.

PART III — AVIATION WEATHER

In Part III, you will be introduced to the characteristics of the atmosphere and the weather within it. You will explore how weather patterns and hazards affect aircraft operations and learn how to interpret aviation weather reports and forecasts that help you maximize safety by minimizing your exposure to weather-related aviation hazards.

PART IV — PERFORMANCE AND NAVIGATION

Aircraft capabilities and limitations in terms of performance parameters are covered in Part IV. You also will learn the basics of pilotage and dead reckoning, as well as VOR and GPS navigation.

PART V — INTEGRATING PILOT KNOWLEDGE AND SKILLS

The application of aeronautical decision-making principles and flight-related physiological factors is discussed in Chapter 10. A scenario in Chapter 11 provides insight into how previously learned knowledge and skills can be applied during a cross-country flight.

Table of Contents

HOW THE TEXTBOOK WORKS

The *Private Pilot* textbook is structured to highlight important topics and concepts and promote an effective and efficient study/review method of learning. To get the most out of your textbook, as well as the entire Guided Flight Discovery Pilot Training System, you will find it beneficial to review the major design elements incorporated in this textbook.

HOW THE TEXTBOOK WORKS ■ Private Pilot

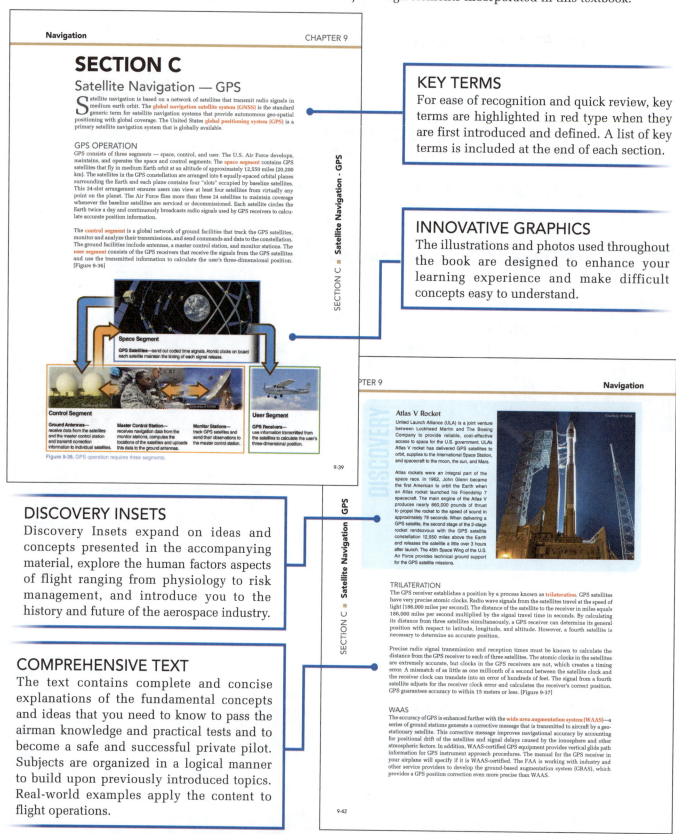

KEY TERMS
For ease of recognition and quick review, key terms are highlighted in red type when they are first introduced and defined. A list of key terms is included at the end of each section.

INNOVATIVE GRAPHICS
The illustrations and photos used throughout the book are designed to enhance your learning experience and make difficult concepts easy to understand.

DISCOVERY INSETS
Discovery Insets expand on ideas and concepts presented in the accompanying material, explore the human factors aspects of flight ranging from physiology to risk management, and introduce you to the history and future of the aerospace industry.

COMPREHENSIVE TEXT
The text contains complete and concise explanations of the fundamental concepts and ideas that you need to know to pass the airman knowledge and practical tests and to become a safe and successful private pilot. Subjects are organized in a logical manner to build upon previously introduced topics. Real-world examples apply the content to flight operations.

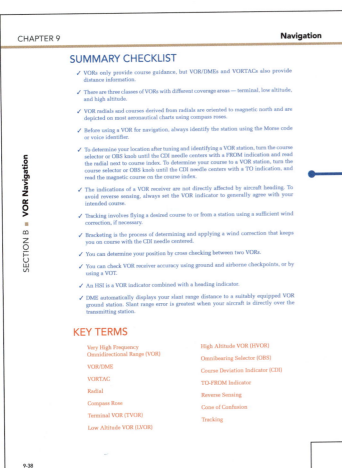

CHAPTER 9   **Navigation**

SUMMARY CHECKLIST

SECTION B ■ **VOR Navigation**

✓ VORs only provide course guidance, but VOR/DMEs and VORTACs also provide distance information.

✓ There are three classes of VORs with different coverage areas — terminal, low altitude, and high altitude.

✓ VOR radials and courses derived from radials are oriented to magnetic north and are depicted on most aeronautical charts using compass roses.

✓ Before using a VOR for navigation, always identify the station using the Morse code or voice identifier.

✓ To determine your location after tuning and identifying a VOR station, turn the course selector or OBS knob until the CDI needle centers with a FROM indication and read the radial next to course index. To determine your course to a VOR station, turn the course selector or OBS knob until the CDI needle centers with a TO indication, and read the magnetic course on the course index.

✓ The indications of a VOR receiver are not directly affected by aircraft heading. To avoid reverse sensing, always set the VOR indicator to generally agree with your intended course.

✓ Tracking involves flying a desired course to or from a station using a sufficient wind correction, if necessary.

✓ Bracketing is the process of determining and applying a wind correction that keeps you on course with the CDI needle centered.

✓ You can determine your position by cross checking between two VORs.

✓ You can check VOR receiver accuracy using ground and airborne checkpoints, or by using a VOT.

✓ An HSI is a VOR indicator combined with a heading indicator.

✓ DME automatically displays your slant range distance to a suitably equipped VOR ground station. Slant range error is greatest when your aircraft is directly over the transmitting station.

KEY TERMS

Very High Frequency Omnidirectional Range (VOR)

VOR/DME

VORTAC

Radial

Compass Rose

Terminal VOR (TVOR)

Low Altitude VOR (LVOR)

High Altitude VOR (HVOR)

Omnibearing Selector (OBS)

Course Deviation Indicator (CDI)

TO-FROM Indicator

Reverse Sensing

Cone of Confusion

Tracking

9-38

SUMMARY CHECKLISTS
Summary Checklists are included at the end of each section to help you identify and review the major points introduced in the section.

HOW THE TEXTBOOK WORKS ■ **Private Pilot**

Navigation   CHAPTER 9

Bracketing

Triangulation

VOR Orientation

VOR Checkpoint

VOR Test Facilities (VOT)

Horizontal Situation Indicator (HSI)

Distance Measuring Equipment (DME)

Slant Range Distance

QUESTIONS

1. What navigation capability does a VORTAC provide?

2. Identify the components of the VOR indicator shown in the accompanying figure.

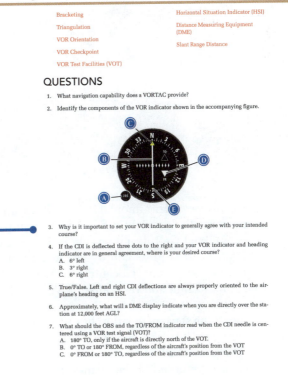

3. Why is it important to set your VOR indicator to generally agree with your intended course?

4. If the CDI is deflected three dots to the right and your VOR indicator and heading indicator are in general agreement, where is your desired course?
   A. 6° left
   B. 3° right
   C. 6° right

5. True/False. Left and right CDI deflections are always properly oriented to the airplane's heading on an HSI.

6. Approximately, what will a DME display indicate when you are directly over the station at 12,000 feet AGL?

7. What should the OBS and the TO/FROM indicator read when the CDI needle is centered using a VOR test signal (VOT)?
   A. 180° TO, only if the aircraft is directly north of the VOT.
   B. 0° TO or 180° FROM, regardless of the aircraft's position from the VOT
   C. 0° FROM or 180° TO, regardless of the aircraft's position from the VOT

SECTION B ■ **VOR Navigation**

QUESTIONS
Questions are provided at the end of each section to help you evaluate your understanding of the concepts that were presented in the accompanying section. Several question formats are provided including completion, matching, true/false, and essay.

9-39

Exploring Flight

WHAT IS FLYING ALL ABOUT?

Science, freedom, beauty, adventure—aviation offers it all.

— Charles Lindbergh

Welcome to the world of aviation. You are about to embark on a journey of adventure, exploration, and discovery. Throughout history, we have dreamed about achieving the freedom and power of flight. We have looked to the sky, marveled at the birds, and wondered what it must be like to escape the bonds of earth to join them.

For once you have tasted flight, you will walk the earth with your eyes turned skyward, for there you have been, and there you will long to return.

— Leonardo da Vinci, *On Flight of Birds*

One of history's creative geniuses was Leonardo da Vinci; an artist, scientist, and dreamer who was fascinated with flight. He spent countless hours studying the flight of birds and his 15th century manuscripts contained approximately 160 pages of descriptions and sketches of flying machines. One such machine was the ornithopter, which was designed to imitate the wing structure of birds and bats.

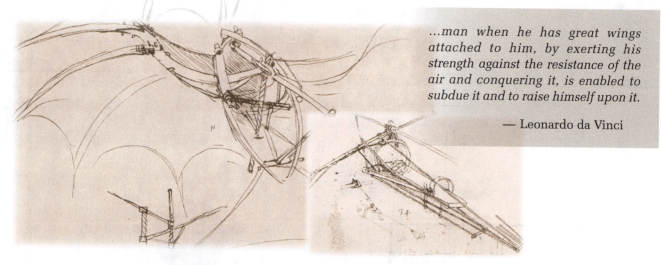

> *...man when he has great wings attached to him, by exerting his strength against the resistance of the air and conquering it, is enabled to subdue it and to raise himself upon it.*
>
> — Leonardo da Vinci

A human-powered ornithopter is virtually incapable of flight due to the dramatic difference in the strength-to-weight ratio of birds compared to humans. Da Vinci's manuscripts also contained well-developed descriptions of finned projectiles, parachutes, and the helicopter. These ideas could have advanced the course of aviation history and flight might have been achieved centuries sooner, but unfortunately, the manuscripts were not made public until 300 years after da Vinci's death.

Although the story of aviation has its share of missed opportunities, unrealized dreams, and failures, it is nonetheless a story of unparalleled success. When you learn to fly you become a part of this success story. You may never break a record or have your flying feats recorded in the history books, but as a pilot, you make your mark as one of the unique individuals who has dared to do what others only dream about. At the controls of an airplane, you can experience some of the same magic that the pioneers of aviation realized.

FIRST MANNED FLIGHT

November 21, 1783 — Launched from the garden of the Chateau La Muette near Paris, the first manned flight in history is made by Pilatre de Rozier and the Marquis d'Arlandes in a hot-air balloon designed by the brothers Joseph and Etienne Montgolfier.

The machine, say the public, rose with majesty… I was surprised at the silence and absence of movement which our departure caused among the spectators, and believed them to be astonished and perhaps awed at the strange spectacle…

— The Marquis d'Arlandes in a letter to a friend from *The Saga of Flight*, edited by Neville Duke and Edward Lanchbery

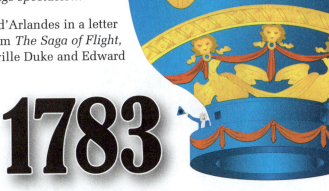

1783

FIRST MANNED FLIGHTS IN GLIDERS

1881 through 1896 — German engineer and inventor, Otto Lilienthal with the help of his brother Gustav proved to the western world that flight in a heavier-than-air machine was achievable. The Lilienthal brothers used their mechanical training to translate conclusions made about the flight of birds into practical air vehicles. From an artificial hill constructed for launching his gliders, Otto Lilienthal made over 2,000 successful glides.

1881

FIRST POWERED FLIGHT

December 17, 1903 — Near Kitty Hawk, North Carolina, Orville and Wilbur Wright achieve the first powered, sustained, and controlled airplane flights in history. Four flights are made; the first for 12 seconds, the last for 59 seconds.

Flight was generally looked upon as an impossibility, and scarcely anyone believed it until he actually saw it with his own eyes.

— Orville Wright

The flight lasted only twelve seconds, but it was nevertheless the first in the history of the world in which a machine carrying a man had raised itself by its own power into the air in full flight, had sailed forward without reduction of speed, and had finally landed at a point as high as that from which it started.

— Orville Wright

FIRST SOLO TRANSATLANTIC FLIGHT

May 21, 1927 — Charles Lindbergh lands his airplane, the *Spirit of St. Louis*, at Le Bourget field in Paris after completing the first solo nonstop transatlantic flight. His total flight time from New York to Paris was 33 hours, 30 minutes and 29.8 seconds.

The Spirit of St. Louis swings around and stops rolling, resting on the solidness of the earth, in the center of Le Bourget. I start to taxi back toward the floodlights and hangars—But the entire field ahead is covered with running figures!

— Charles Lindbergh, *The Spirit of St. Louis*

FIRST SOLO TRANSATLANTIC FLIGHT BY A WOMAN PILOT

May 21, 1932 — Amelia Earhart becomes the first woman to pilot an airplane solo across the Atlantic. Gaining fame for being the first woman passenger in a flight across the Atlantic four years earlier, Earhart was disappointed that pilot Wilmer Stultz did all the flying while she just rode along like *"a sack of potatoes"* as she phrased it. She was determined to prove that she could accomplish the flight herself, and she did when she landed in Northern Ireland after taking off from Newfoundland 14 hours and 52 minutes earlier. On August 25 of the same year, Earhart completed the first woman's solo nonstop transcontinental flight, which covered 2,448 miles from Los Angeles to Newark.

Courtesy of The Ninety-Nines Inc. International Organization of Women Pilots Archive Collection, Oklahoma City, Oklahoma

1932

FIRST TIME THE SOUND BARRIER IS BROKEN

October 14, 1947 — Captain Charles E. "Chuck" Yeager becomes the first person to fly an aircraft beyond the speed of sound. He pilots the air-launched experimental Bell X-1 rocket-propelled research airplane named *Glamorous Glennis* (after Yeager's wife) at a speed of 700 mph at 42,000 feet over Muroc Dry Lake, California.

Leveling off at 42,000 feet, I had thirty percent of my fuel, so I turned on rocket chamber three and immediately reached .96 Mach. I noticed that the faster I got, the smoother the ride. Suddenly the Mach needle began to fluctuate. It went up to 0.965 Mach—then tipped right off the scale. I thought I was seeing things! We were flying supersonic! ...I was thunderstruck. After all the anxiety, breaking the sound barrier turned out to be a perfectly paved speedway.

— *Yeager: An Autobiography* by General Chuck Yeager and Leo Janos

In those few moments, the supersonic age was born.

Courtesy of NASA Dryden Research Center

1947

FIRST MANNED MOON LANDING

July 20, 1969 — As astronaut Michael Collins maintains orbit in the Apollo 11 Command Module *Columbia*, astronauts Neil Armstrong and Edwin Aldrin land the Lunar Module *Eagle* on the moon and become the first humans to step on another celestial body.

HOUSTON: *Okay, Neil, we can see you coming down the ladder now.*
NEIL ARMSTRONG: *Okay, I just checked—getting back up to that first step. Buzz, it's not even collapsed too far, but it's adequate to get back up ... It takes a pretty good little jump ... I'm at the foot of the ladder. The LM footpads are only depressed in the surface about one or two inches. Although the surface appears to be very, very fine-grained, as you get close to it. It's almost like a powder. Now and then, it's very fine ... I'm going to step off the LM now ...That's one small step for [a] man, one giant leap for mankind.*

Courtesy of NASA

1969

FIRST SPACE SHUTTLE LAUNCH

April 12, 1981 — The United States launches the space shuttle *Columbia*, the world's first reusable manned space vehicle and the most complex flying machine built up to that time. Pilot Robert L. Crippen describes *Columbia's* landing by space shuttle commander John W. Young.

Courtesy of NASA

We made a gliding circle over our landing site, Runway 23 on Rogers Dry Lake at Edwards Air Force Base. On final approach I was reading out the airspeeds to John so he wouldn't have to scan the instruments as closely. Columbia almost floated in. John only had to make minor adjustments in pitch. We were targeted to touch down at 185 knots, and the very moment I called out 185, I felt us touch down. I have never been in any flying vehicle that landed more smoothly. If you can imagine the smoothest landing you've ever had in an airliner, ours was at least that good. John really greased it in. "Welcome home, Columbia," said Houston. "Beautiful, beautiful." "Do you want us to take it up to the hangar?" John asked.

— "Our Phenomenal First Flight," by John Young and Robert Crippen in *National Geographic*

1981

FIRST NONSTOP FLIGHT AROUND THE WORLD WITHOUT REFUELING

December 23, 1986 — Piloted by Dick Rutan and Jeana Yeager, the aircraft *Voyager* completes the historic flight in 9 days, 3 minutes, and 44 seconds. *Voyager* was designed by pilot Dick Rutan's brother Burt Rutan.

With its 7,011.5 pounds of fuel aboard at takeoff amounting to 72.3 percent of the airplane's gross weight, *Voyager* was literally a flying fuel tank. *Voyager's* takeoff from the 15,000-foot runway at Edwards Air Force Base took over two minutes as the airplane's wingtips, heavy with fuel, were dragging on the runway. After finally lifting off with only 800 feet of runway remaining, copilot Jeana Yeager radioed, *"If it were easy, it would have been done before."*

1986

Courtesy of NASA Dryden Research Center

FIRST PRIVATE MANNED SPACE FLIGHT

October 4, 2004 — *SpaceShipOne* becomes the first private manned spacecraft to fly into suborbital space (at least 100 km above the earth's surface) twice within a 14-day period. Burt Rutan's company, Scaled Composites, designed and built the spacecraft, which was lifted to approximately 46,000 feet by its carrier plane, the *White Knight,* and released. The back-to-back flights were piloted by Michael Melville and Brian Binnie to altitudes of 337,500 feet (103 km) and 367,442 feet (112 km).

Courtesy of Scaled Composites, LLC

2004

FIRST SOLO NONSTOP FLIGHT AROUND THE WORLD WITHOUT REFUELING

March 3, 2005 — Steve Fossett completes the nonstop solo flight in the *GlobalFlyer*, and also sets an around-the-world speed record by finishing the flight in 67 hours, 1 minute, and 10 seconds, more than three times as fast as *Voyager*. The following year, on February 11, 2006, Fossett again flies the *GlobalFlyer* around the world and keeps going, making a second Atlantic crossing to beat *Voyager's* distance record of 24,986 miles.

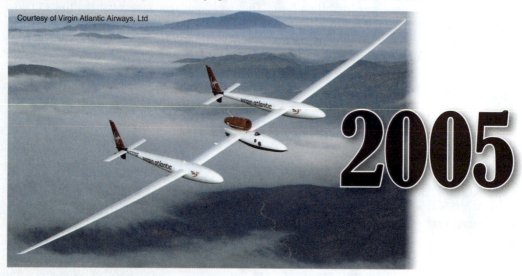

Courtesy of Virgin Atlantic Airways, Ltd

2005

FIRST FLIGHT AROUND THE WORLD BY A SOLAR-POWERED AIRPLANE

July 26, 2016 — *Solar Impulse II* lands in Abu Dhabi, UAE, after departing the same location 16 months earlier. Swiss pilots Bertrand Piccard and André Borschberg took turns flying the *Solar Impulse II* solo on each of the trip's 17 legs. The aircraft was powered entirely by solar energy from its 17,000 solar cells, storing extra energy by day to continue flying through each night.

© Revillard | Rezo.ch, Courtesy of Solar Impulse

2016

FIRST BLACK WOMAN U.S. NAVY FIGHTER PILOT

July 31, 2020—Lieutenant Madeline Swegle is the first Black woman to become a U.S. Navy fighter pilot, marking a significant milestone for U.S. Naval aviation. After earning her "Navy Wings of Gold," Swegle advanced to graduate-level pilot training at Naval Air Station Whidbey Island, Washington, where she trained to fly the Boeing EA-18G Growler, a carrier-based all-weather, electronic warfare attack aircraft. After completing pilot training in the EA-18G Growler, Lieutenant Swegle was assigned to the Electronic Attack Squadron (VAQ) 130s—the oldest electronic warfare squadron in the U.S. Navy nicknamed the "Zappers."

Photo courtesy of LT. j.g. Luke Redito

The appearance of U.S. Department of Defense (DoD) visual information does not imply or constitute DoD endorsement.

2020

FIRST FLIGHT OF AN AIRCRAFT ON ANOTHER PLANET

April 19, 2023—NASA successfully flies an unmanned helicopter on Mars. On its first flight, Ingenuity flew to a height of 3 meters and hovered in a stable holding position for 30 seconds. Ingenuity completed 72 flights, far exceeding its original planned technical demonstration of up to 5 flights. As an homage to all of its aerial predecessors, the Ingenuity helicopter carries with it a postage-stamp sized fragment of the wing of the 1903 Wright Flyer.

Photo courtesy of NASA

2023

EXPLORING FLIGHT ■ **Private Pilot**

Fundamentals of Flight

The bird has learned [his] art…so thoroughly that its skill is not apparent to our sight. We only learn to appreciate it when we try to imitate it.

— Wilbur Wright

PART I

Although we have never been able to duplicate the skill of birds, we have mastered the art of flying in our own unique way. We have built flying vehicles to transport us from town to town, coast to coast, around the world, and into space. As you explore Part I, you will begin to understand not only why we endeavor to fly, but how the goal of flight is achieved. *Discovering Aviation* answers your questions about the pilot training process and introduces you to the world of aviation. You will learn how the components of the airplane operate in *Airplane Systems*, and as you examine *Aerodynamic Principles*, you will gain knowledge of the forces acting on an airplane in flight.

CHAPTER 1

Discovering Aviation

SECTION A ■ Pilot Training

Section A
Pilot Training

In the early days of aviation, no government regulations controlled pilot training or aircraft construction. Little guidance was provided for those who wanted to fly airplanes and most would-be aviators learned to fly by trial and error. As a result, flight training was very risky and required a tremendous amount of courage and commitment. Although you might never encounter the same hazards faced by the early aviators, becoming a pilot still presents a challenge that requires hard work and dedication. Nonetheless, the time and energy you invest in flying can yield great rewards.

Individuals learn to fly for different reasons. Some want the challenge of achieving an extraordinary goal, some yearn to travel and experience the world from a new perspective, some are looking for an exciting career, and still others simply want the satisfaction and sense of accomplishment that comes from mastering a skill. You might be learning to fly for one or more of these reasons, or have an entirely different motivation. Whatever the reason, you now have the chance to spread your wings and expand your horizons.

The first step in your training process is to have your questions answered. The following information contains answers to some of the most frequently asked questions about pilot training. With this information, you will find it easier to make effective decisions about your training, and enjoy a more positive flying experience.

SAFETY STANDARDS

All countries have a national (or civil) aviation authority that governs aviation through a set of rules and regulations. Non-regulatory industry organizations provide guidance for standards and coordinate discussions promoting good practices. Government agencies, like the Federal Aviation Administration (FAA) in the U.S. and the European Union Aviation Safety Agency (EASA), and industry organizations, like the International Civil Aviation Organization (ICAO) at the United Nations, play a major role in promoting and ensuring safe skies.

WHAT IS THE FAA?

Pilot training in the U.S. is regulated by an agency called the **Federal Aviation Administration (FAA)**, which governs commercial and general aviation, private space flight, and the operation of unmanned aircraft systems (UAS)—drones. The **Federal Aviation Regulations (FARs)**, which are issued by the FAA, are rules that apply to all areas of aviation, including flight operations, the construction of aircraft, and the training requirements that you must meet to obtain pilot certificates and ratings.

The FARs are divided into numbered parts (FAR Part 61, FAR Part 91, etc.) and regulations are typically identified by the part number, followed by the specific regulation number, for example; FAR 91.106. During your training, you will become familiar with the regulations that apply to you. The FARs are contained in *Title 14—Aeronautics and Space* within the *Code of Federal Regulations* (CFR). A complete citation is 14 CFR Part 91 or 14 CFR 91.106, but pilots commonly use the shorter FAR reference. [Figure 1-1]

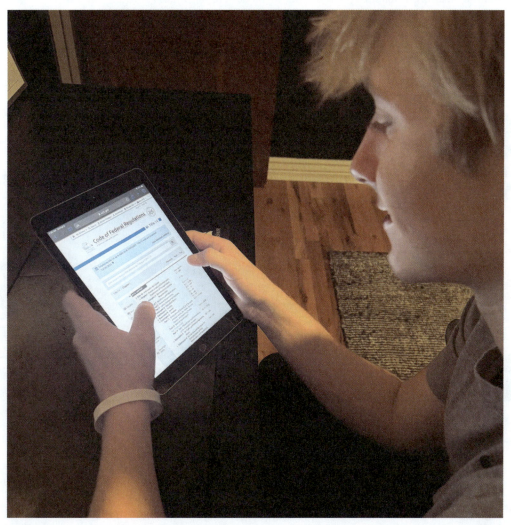

Figure 1-1. The FAA publishes the electronic Code of Federal Regulations (eCFR), which contains the rules that apply to all areas of aviation.

As outlined in the FARs, you must meet specific training requirements to obtain a private pilot certificate. During your course of training, you will take a knowledge test and at the completion of your pilot training, a practical test to obtain your pilot certificate. Although you must meet certain requirements prior to taking these tests, pilot training is generally flexible. You can usually choose your instructor, methods of learning, and lesson schedule that best suits your needs.

WHAT ARE AIRMAN CERTIFICATION STANDARDS (ACS)?

When you take your private pilot **practical test,** typically referred to as a **checkride,** you are evaluated using the FAA Private Pilot **Airman Certification Standards (ACS).** During your training, your instructor will familiarize you with the ACS knowledge areas, risk management concepts, and technical flight skills that apply to each maneuver and procedure that you must demonstrate on your checkride. The ACS assigns a unique code to each knowledge, risk management, and skills element.

The ACS clearly states what you must know, consider, and do when performing each task during both the ground and flight portion of your checkride. In addition, the ACS provides a single source of standards for both the knowledge and the practical tests. The codes designated for the ACS knowledge subjects correlate to questions on the computerized knowledge test that you must take for private pilot certification. The ACS enhances safety by making tests meaningful and relevant to actual operations and contributes to standardization in teaching and testing these concepts. [Figure 1-2]

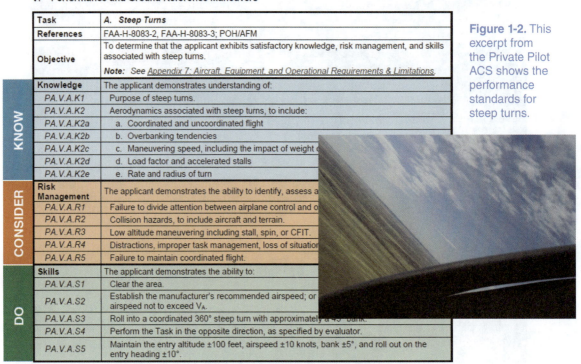

V. Performance and Ground Reference Maneuvers

Task	**A. Steep Turns**	
References	FAA-H-8083-2, FAA-H-8083-3; POH/AFM	
Objective	To determine that the applicant exhibits satisfactory knowledge, risk management, and skills associated with steep turns.	
	Note: See *Appendix 7: Aircraft, Equipment, and Operational Requirements & Limitations.*	
KNOW	**Knowledge**	The applicant demonstrates understanding of:
	PA.V.A.K1	Purpose of steep turns.
	PA.V.A.K2	Aerodynamics associated with steep turns, to include:
	PA.V.A.K2a	a. Coordinated and uncoordinated flight
	PA.V.A.K2b	b. Overbanking tendencies
	PA.V.A.K2c	c. Maneuvering speed, including the impact of weight
	PA.V.A.K2d	d. Load factor and accelerated stalls
	PA.V.A.K2e	e. Rate and radius of turn
CONSIDER	**Risk Management**	The applicant demonstrates the ability to identify, assess a
	PA.V.A.R1	Failure to divide attention between airplane control and o
	PA.V.A.R2	Collision hazards, to include aircraft and terrain.
	PA.V.A.R3	Low altitude maneuvering including stall, spin, or CFIT.
	PA.V.A.R4	Distractions, improper task management, loss of situation
	PA.V.A.R5	Failure to maintain coordinated flight.
DO	**Skills**	The applicant demonstrates the ability to:
	PA.V.A.S1	Clear the area.
	PA.V.A.S2	Establish the manufacturer's recommended airspeed; or airspeed not to exceed V$_A$.
	PA.V.A.S3	Roll into a coordinated 360° steep turn with approximately a 45° bank.
	PA.V.A.S4	Perform the Task in the opposite direction, as specified by evaluator.
	PA.V.A.S5	Maintain the entry altitude ±100 feet, airspeed ±10 knots, bank ±5°, and roll out on the entry heading ±10°.

Figure 1-2. This excerpt from the Private Pilot ACS shows the performance standards for steep turns.

WHAT IS ICAO?

The **International Civil Aviation Organization (ICAO)** is an agency of the United Nations that advances the techniques and principles of international air navigation and fosters the planning and development of international civil air transport. ICAO is not a global regulator, rather, this organization works closely with national governments and aviation industry organizations to support and promote safe air transport policy and standardized innovations. Although the FAA regulates U.S. pilot training, you will encounter many standards based on ICAO recommendations that apply to the U.S. flight environment ranging from airport identifiers to radio phraseology.

TRAINING PLAN CONSIDERATIONS

You must consider several factors before deciding on your training plan. Consider your flying goals; why you want to fly and where you want to fly. Think about your location and finances, and how those will impact the time it takes to complete your training. Below are answers to common questions about training plan considerations.

WHERE CAN I OBTAIN PILOT TRAINING?

You usually do not have to travel any further than your local airport to launch your aviation journey. Many pilot training schools are located at airport facilities called **fixed-base operators (FBOs)**. In addition to pilot training, FBOs provide a variety of services to pilots, including aircraft rental, fueling, maintenance, parking, and the sale of pilot supplies. Two types of pilot training schools are: FAA-certificated schools governed by FAR Part 141 and less-regulated schools governed by FAR Part 61. Both schools employ **FAA certificated flight instructors (CFIs)** who can provide dual instruction in the airplane. A Part 141 certificated pilot school must meet prescribed standards for equipment, facilities, personnel, and curricula.

You can make a more-informed decision about a pilot training school by conducting some research. For example: Does the school's instructional program and lesson schedule fit your needs? How long has the school been operating? What is the school's reputation and safety record? How many, and what type of aircraft are available for flight training? How are aircraft maintenance issues resolved?

Most schools offer an introductory flight lesson during which you are able to operate the controls of the airplane. This flight provides an opportunity for you to become familiar with the flight training process, evaluate the flight school, and get acquainted with a flight instructor. Probably the most important decision you will make regarding your pilot training is the selection of a flight instructor. You might speak with several CFIs and ask other pilots for instructor recommendations. If you are uncomfortable with a CFI, do not be afraid to select a different instructor. Students learn differently and another CFI might have a teaching style that you find more effective.

You can obtain flight instruction from a freelance CFI who is not employed by a school or FBO. If you plan to pursue aviation as a career, consider a large flight school, college, or university that provides highly-structured professional pilot training. [Figure 1-3]

Figure 1-3. Select a pilot training school and instructor that fits your flying goals. A Part 141 school is a good choice for career minded students. A Part 61 program can be a good fit for either a career goal or recreational flying.

SECTION A ■ **Pilot Training**

WHAT ARE THE REQUIREMENTS TO TAKE LESSONS?

You can go on an introductory flight with a CFI at any time, but in order to take flight lessons, the Transportation Security Administration (TSA) requires that you show proof of U.S. citizenship or, for non-U.S. citizens, comply with specific TSA requirements. Your CFI will guide you through the process to meet the TSA requirements, if applicable.

WHAT ARE THE REQUIREMENTS TO FLY SOLO?

Although you can start taking lessons if you meet the TSA requirements, you must meet additional requirements to solo an airplane. You must be 16 years old and receive the required endorsements in your logbook from an authorized flight instructor who has determined that you have achieved the knowledge and flight training necessary to pilot an airplane safely on your own. In addition, you must possess a student pilot certificate and a medical certificate. Depending on your training plan and how quickly you progress, the solo flight milestone might happen sooner than you think. Be sure to obtain your student pilot and medical certificates early in your training—well before your first solo. [Figure 1-4]

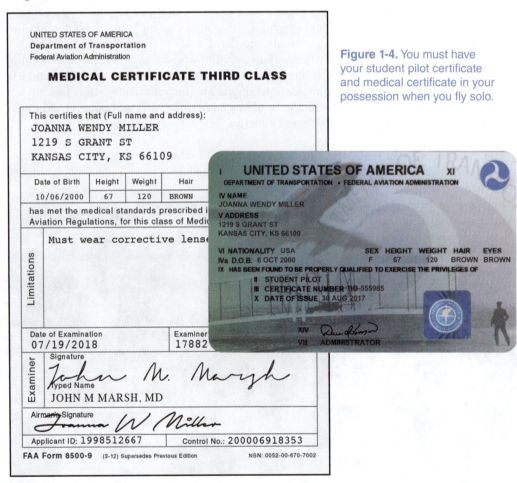

Figure 1-4. You must have your student pilot certificate and medical certificate in your possession when you fly solo.

HOW DO I OBTAIN A STUDENT PILOT CERTIFICATE?

To obtain a **student pilot certificate**, the FARs require you to be at least 16 years of age and be able to read, speak, write, and understand the English language. The student pilot certificate is a card you receive by completing an FAA electronic or paper application. Your instructor can help you fill out and submit an electronic application online at iacra.faa.gov.

Integrated Airman Certification and Rating Application (IACRA) is the web-based certification/rating application system that guides you through the FAA's process to apply for pilot certificates and ratings, including the student pilot certificate and your private pilot certificate. Not only does IACRA help ensure that you meet regulatory and policy requirements through extensive data validation, the system uses electronic signatures, eliminates paper forms, and prints temporary certificates. The FAA normally mails your student pilot certificate within a few weeks of the time you submit your application. Your student pilot certificate does not expire, but you must surrender it when you get your private pilot certificate.

HOW DO I OBTAIN A MEDICAL CERTIFICATE?

To obtain a **medical certificate,** you must pass a physical exam administered by an FAA-authorized **aviation medical examiner (AME)**. The three classes of medical certificates are: first-class, required for airline transport pilot operations; second-class, required for commercial pilot operations other than airline transport; and third-class for student, recreational, and private pilot operations. The medical standards for each class of certificate are described by FAR Part 67. First-class medical certificates have the highest physical requirements, followed by second-class, and then third-class. After you decide to begin pilot training, you should obtain a medical certificate as soon as possible to make sure you meet the required standards. Your FBO, flight school, CFI, or another pilot can recommend an AME for your exam. You can also search online for the AMEs in your area.

If you want a career as a commercial or airline pilot, consider applying for a second- or first-class medical certificate even though you do not need one yet to ensure that you are medically qualified for your desired flying career. A number of serious medical conditions, generally those that could incapacitate a pilot, require an AME to deny a medical certificate. In other cases, the FAA will review the application with information about your current treatment, and give you a special issuance medical certificate, usually with limitations.

HOW MUCH DOES PILOT TRAINING COST?

Private pilot training expenses can vary widely. Cost factors include reference materials, ground instruction, and the actual flight lessons themselves. Schools might offer a package of lessons for which you pay in advance, or you might pay by the hour for all your flight training time. Broadly, the cost factors include the type of training program, location, type of airplane, and the time you have available for training.

If you are renting an aircraft for your flight lessons, the charge is normally based on the time from engine start to engine shutdown. This period is recorded by a digital timer in the airplane that is often called a Hobbs meter. Your instructor fee is based on the time on the Hobbs meter, plus the time spent conducting preflight and postflight briefings. Additionally, the type of airplane and its onboard equipment affects the cost charged per hour. [Figure 1-5]

Figure 1-5. Your primary flight training expense is for the flight time recorded by a meter in the airplane

SECTION A ■ Pilot Training

You also must consider the expense of ground instruction. That cost varies depending on which method of ground instruction you use: ground school classes, individual lessons with your instructor, home study, or a combination of several methods. Training content, such as the Jeppesen curriculum, flight information sources, and the pilot's operating handbook for the training airplane, are important resources. You also need pilot supplies, such as a logbook, aeronautical charts, a flight computer, and a navigational plotter. You might purchase flight planning apps or an electronic flight bag (EFB) that require subscriptions to update. In addition, you will pay fees to take the knowledge and practical tests, and to get your medical exam.

The number of flight hours that you need to become proficient depends on several factors, such as your initial comfort level with the airplane, whether you have any previous flying experience, and the flight training environment. Training at a busy airport with a control tower typically requires more hours than at a non towered field. Another significant expense factor is the frequency of your flight lessons. You master the material faster if your lessons are more frequent. When significant time passes between lessons, you forget more of what you have learned, and you need to spend more time reviewing previous skills. Flying at least several times a week can make your training more cost-effective.

TRAINING PROGRAM

Pilot training is a commitment of both your time and financial resources. Managing your training progress is essential to staying on track toward your flying goals. You should know how long the training program should take, the sequence of subject matter topics, and the completion standards for each lesson. Training programs will vary, but all should have clear documentation that outlines the training sequence and completion expectations.

WHAT IS THE TRAINING COURSE SEQUENCE?

A private pilot training program can generally be divided into three phases: presolo and first solo, cross-country and night, and practical test/checkride preparation. Each phase includes both ground and flight instruction.

PRESOLO AND FIRST SOLO

From the initial lesson to your first solo flight, your ground and flight training includes all the knowledge, maneuvers, and procedures necessary to ensure you can operate the airplane safely and confidently on your own. Before endorsing your logbook for solo flight, your CFI ensures that you can operate the airplane safely and that you meet the knowledge and flight proficiency standards required by the FARs, including passing a presolo knowledge test administered by your instructor. Typically during your first solo, you complete several takeoffs and landings at your local airport while your instructor supervises from the ground. On subsequent solo flights, your instructor assigns specific maneuvers in the local practice area.

CROSS-COUNTRY AND NIGHT

During the cross-country and night phase of your training, you learn how to plan and perform flights beyond the local area by interpreting aeronautical charts, obtaining weather information, calculating airplane performance, and using various navigation methods. You complete at least one cross-country flight to several airports with your instructor and you are also introduced to local and cross-country night flying.

You perform several solo cross-country flights, including a flight with a total distance of at least 150 nautical miles. Prior to these flights, your instructor endorses your logbook for solo cross-country flight and reviews your preflight planning and preparation for each cross-country to determine if you can conduct the flight safely.

CHECKRIDE PREPARATION

The final phase is preparing for your practical test, which pilots generally refer to as the checkride. You review all the maneuvers and procedures during both solo and dual flight lessons and study the knowledge areas that have been covered throughout your training program. Prior to your practical test, your instructor endorses your logbook to indicate that you have completed the required training and that you are competent to perform each operation safely as a private pilot. Your instructor will help you submit an application for the private pilot certificate, which is typically accomplished in the IACRA system.

Your checkride is administered by an FAA-designated pilot examiner (DPE) or FAA inspector pilot. The examiner has you plan a cross-country flight and assesses your knowledge through oral quizzing. During the flight, you are evaluated on tasks that are listed in the Private Pilot Airman Certification Standards (ACS). After you have passed the checkride, the examiner issues you a temporary private pilot certificate that is valid for a specific time period. You receive your permanent certificate in the mail. [Figure 1-6]

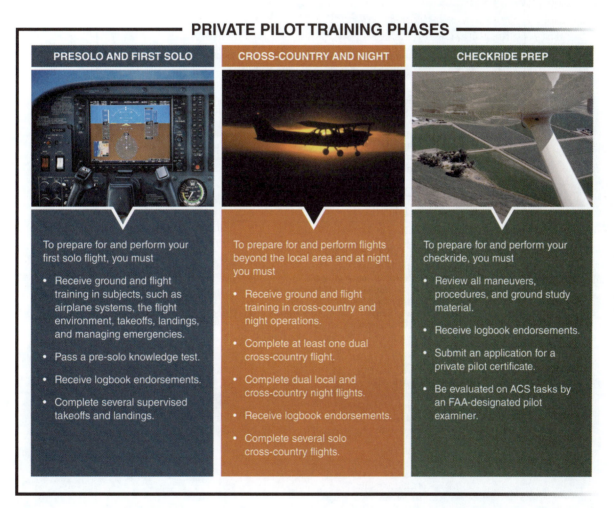

PRIVATE PILOT TRAINING PHASES

PRESOLO AND FIRST SOLO

To prepare for and perform your first solo flight, you must

- Receive ground and flight training in subjects, such as airplane systems, the flight environment, takeoffs, landings, and managing emergencies.

- Pass a pre-solo knowledge test.

- Receive logbook endorsements.

- Complete several supervised takeoffs and landings.

CROSS-COUNTRY AND NIGHT

To prepare for and perform flights beyond the local area and at night, you must

- Receive ground and flight training in cross-country and night operations.

- Complete at least one dual cross-country flight.

- Complete dual local and cross-country night flights.

- Receive logbook endorsements.

- Complete several solo cross-country flights.

CHECKRIDE PREP

To prepare for and perform your checkride, you must

- Review all maneuvers, procedures, and ground study material.

- Receive logbook endorsements.

- Submit an application for a private pilot certificate.

- Be evaluated on ACS tasks by an FAA-designated pilot examiner.

Figure 1-6. Each phase of the training course sequence includes both ground and flight instruction.

SECTION A ■ **Pilot Training**

HOW IS A SYLLABUS USED IN PILOT TRAINING?

Typically, both ground and flight lessons are organized by a syllabus that provides structure to your training and helps ensure that no procedures are overlooked. A properly designed syllabus is based upon the building-block theory of learning, which recognizes that each item taught must be presented on the basis of previously learned knowledge and skills. Knowledge support materials are coordinated with the flight lessons so that the material pertinent to a flight lesson is taught prior to the flight. [Figure 1-7]

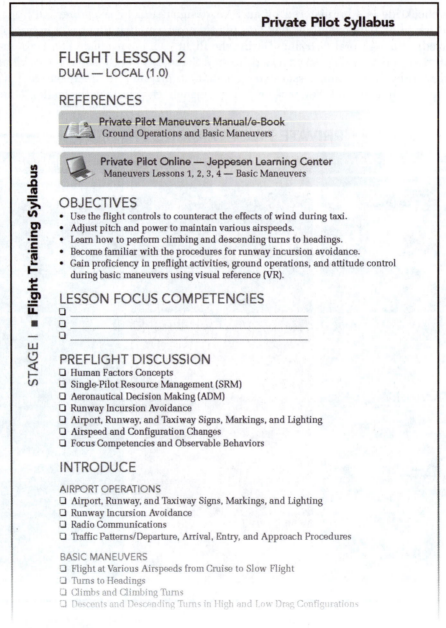

SECTION A ■ Pilot Training

STAGE I ■ Flight Training Syllabus

Private Pilot Syllabus

FLIGHT LESSON 2
DUAL — LOCAL (1.0)

REFERENCES

Private Pilot Maneuvers Manual/e-Book
Ground Operations and Basic Maneuvers

Private Pilot Online — Jeppesen Learning Center
Maneuvers Lessons 1, 2, 3, 4 — Basic Maneuvers

OBJECTIVES
- Use the flight controls to counteract the effects of wind during taxi.
- Adjust pitch and power to maintain various airspeeds.
- Learn how to perform climbing and descending turns to headings.
- Become familiar with the procedures for runway incursion avoidance.
- Gain proficiency in preflight activities, ground operations, and attitude control during basic maneuvers using visual reference (VR).

LESSON FOCUS COMPETENCIES
- ☐ _____
- ☐ _____
- ☐ _____

PREFLIGHT DISCUSSION
- ☐ Human Factors Concepts
- ☐ Single-Pilot Resource Management (SRM)
- ☐ Aeronautical Decision Making (ADM)
- ☐ Runway Incursion Avoidance
- ☐ Airport, Runway, and Taxiway Signs, Markings, and Lighting
- ☐ Airspeed and Configuration Changes
- ☐ Focus Competencies and Observable Behaviors

INTRODUCE

AIRPORT OPERATIONS
- ☐ Airport, Runway, and Taxiway Signs, Markings, and Lighting
- ☐ Runway Incursion Avoidance
- ☐ Radio Communications
- ☐ Traffic Patterns/Departure, Arrival, Entry, and Approach Procedures

BASIC MANEUVERS
- ☐ Flight at Various Airspeeds from Cruise to Slow Flight
- ☐ Turns to Headings
- ☐ Climbs and Climbing Turns
- ☐ Descents and Descending Turns in High and Low Drag Configurations

Figure 1-7. The items that are to be performed during each lesson are outlined in a training syllabus.

GROUND AND FLIGHT LESSONS

As you progress through your pilot training, your instructor helps you correlate aeronautical knowledge to practical flight skills and risk management. A pragmatic approach to ground and flight lessons enables you to apply theoretical concepts to real world problems while still on the ground and focus on flying during the flight lesson.

WHAT IS GROUND INSTRUCTION?

Ground instruction is an essential part of pilot training. To operate an aircraft safely as a private pilot, you must have knowledge in a wide variety of subject areas, including weather, aerodynamics, aircraft systems, flight planning, and regulations. You can obtain the required ground instruction individually from your flight instructor or through formal ground school classes offered by a school or FBO. In addition, self-study courses and online resources are available that effectively cover the required aeronautical knowledge. Your instructor still needs to review subject areas with you to determine that you understand the material well enough to pass the knowledge test and operate safely as a private pilot. [Figure 1-8]

Some companies offer concentrated weekend test preparation courses that focus on passing the knowledge test. Although these courses can help you with your final preparation for the knowledge test, they are not a substitute for the comprehensive ground instruction necessary to become a competent and safe pilot.

Figure 1-8. You can obtain ground instruction from your instructor, ground school classes with other students, home study courses, or a combination of these methods.

WHAT ARE FLYING LESSONS LIKE?

A typical flight lesson (excluding a cross-country flight) lasts approximately two hours. You spend 1 to 1 1/2 hours in the airplane and the remainder of the period consists of preflight and postflight discussion. [Figure 1-9]

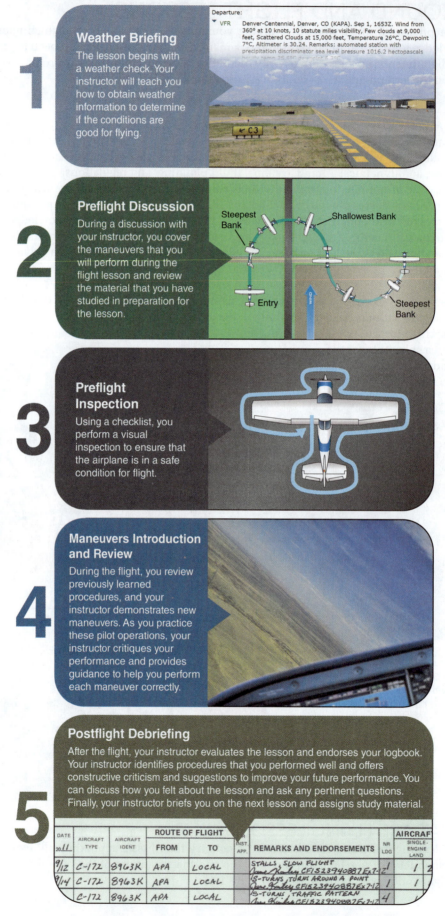

Weather Briefing

The lesson begins with a weather check. Your instructor will teach you how to obtain weather information to determine if the conditions are good for flying.

Departure:
▼ VFR   Denver-Centennial, Denver, CO (KAPA). Sep 1, 1653Z. Wind from 360° at 10 knots, 10 statute miles visibility, Few clouds at 9,000 feet, Scattered Clouds at 15,000 feet, Temperature 26°C, Dewpoint 7°C, Altimeter is 30.24. Remarks: automated station with precipitation discriminator sea level pressure 1016.2 hectopascals

1

Preflight Discussion

During a discussion with your instructor, you cover the maneuvers that you will perform during the flight lesson and review the material that you have studied in preparation for the lesson.

Steepest Bank   Shallowest Bank

Entry   WIND   Steepest Bank

2

Preflight Inspection

Using a checklist, you perform a visual inspection to ensure that the airplane is in a safe condition for flight.

3

Maneuvers Introduction and Review

During the flight, you review previously learned procedures, and your instructor demonstrates new maneuvers. As you practice these pilot operations, your instructor critiques your performance and provides guidance to help you perform each maneuver correctly.

4

Postflight Debriefing

After the flight, your instructor evaluates the lesson and endorses your logbook. Your instructor identifies procedures that you performed well and offers constructive criticism and suggestions to improve your future performance. You can discuss how you felt about the lesson and ask any pertinent questions. Finally, your instructor briefs you on the next lesson and assigns study material.

5

DATE 20__	AIRCRAFT TYPE	AIRCRAFT IDENT	ROUTE OF FLIGHT		INST. APP.	REMARKS AND ENDORSEMENTS	NR LDG	AIRCRAFT SINGLE-ENGINE LAND	
			FROM	TO					
9/12	C-172	8963K	APA	LOCAL		STALLS, SLOW FLIGHT	2	1	2
						June Finley CFI52394O887Ex7-12			
9/14	C-172	8963K	APA	LOCAL		S-TURNS, TURNS AROUND A POINT	1	1	
						June Finley CFI52394O887Ex7-12			
	C-172	8963K	APA	LOCAL		S-TURNS, TRAFFIC PATTERN	4		
						June Finley CFI52394O887Ex7-12			

Figure 1-9. Although each flight is unique, lessons normally follow a general sequence.

PRIVATE PILOT REQUIREMENTS

Your motivation for obtaining a private pilot certificate drives you to completing the requirements and ultimately enjoying the privileges of flying an airplane. Logging ground and flight training time, ensuring you meet the medical certificate requirements, and completing your lessons with a qualified flight instructor are important requirements toward a successful FAA checkride and realizing your goal.

In addition to the student pilot requirements, to be eligible for a private pilot certificate, you must be at least 17 years of age, complete specific training and flight time requirements described in the FARs, pass a knowledge test, and successfully complete a practical test that consists of an oral quiz, performing pilot operations, and performing maneuvers in the airplane.

HOW DO I TAKE THE KNOWLEDGE TEST?

Normally, you take the FAA Private **Pilot Airman Knowledge Test** after completing most of your ground and flight training, so that you have gained practical understanding of the subject matter. You complete the computerized test at an FAA-designated airman knowledge testing center. After you finish the test, the system grades your test and provides a report that shows the codes of the questions that you missed. These codes correspond to ACS knowledge areas. The passing grade is 70 percent. You must review the knowledge test report with your instructor and take it with you to the checkride. The test results are valid for 24 calendar months.

WHAT ARE THE FLIGHT REQUIREMENTS?

The FARs require that you obtain training in specific flight operations, as well as ground instruction in specific knowledge areas. In addition, you must meet minimum flight hour requirements before you apply for a private pilot certificate. FAR Part 61 requires at least 40 hours of flight time consisting of at least 20 hours of dual instruction and at least 10 hours of solo flight. If you are enrolled at an approved school under FAR Part 141, you are required to complete at least 35 hours of flight training, including 20 hours of dual instruction and at least 5 hours of solo flight training. These are *minimum* hour requirements. The average student with no prior flying experience requires approximately 65 to 75 flight hours to meet the proficiency standards necessary to pass the checkride and operate safely as a private pilot.

WHY DO I NEED A PILOT LOGBOOK?

Your logbook is a prized possession representing your accomplishments as a pilot and a legal requirement. FAA regulations require you to keep an accurate record of pilot training time and aeronautical experience used to meet the requirements for a certificate, rating, or flight review. You must also log flights required for meeting recent flight experience requirements. Instructor endorsements that are required to exercise pilot privileges also must be recorded in your logbook.

Electronic logbooks are an effective alternative to paper log books. Some electronic flight bags (EFBs) such as ForeFlight, have a logbook built into the app. One of the primary advantages of an electronic logbook is automatic entry and calculations. For example, ForeFlight's logbook will automatically calculate night time and landings from associated track logs, flight plans, and manually-entered data. [Figure 1-10]

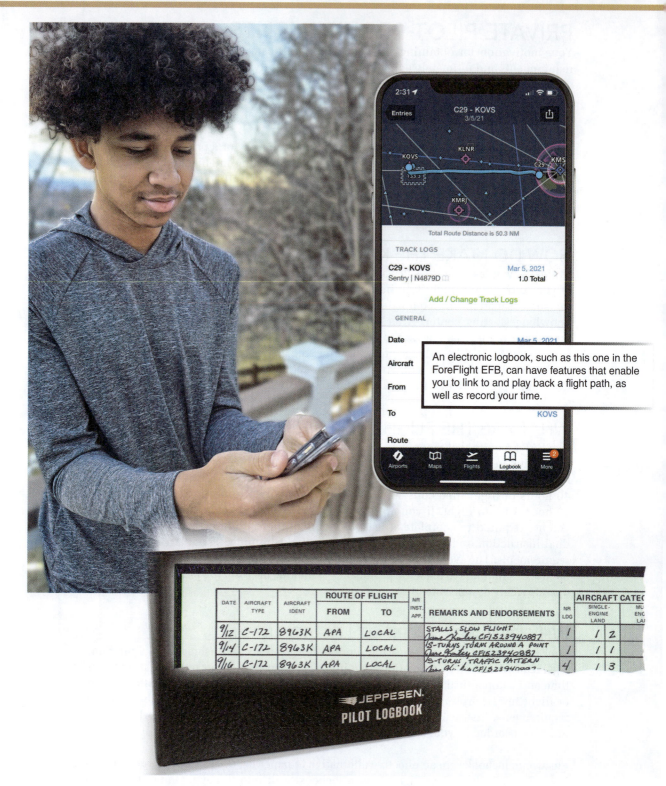

An electronic logbook, such as this one in the ForeFlight EFB, can have features that enable you to link to and play back a flight path, as well as record your time.

Figure 1-10. You might choose to use either an electronic or paper logbook or both.

WHEN DOES MY MEDICAL CERTIFICATE EXPIRE?

The three classes of medical certificates have different durations, depending on your age and the type of operation. For example, a third-class medical certificate is valid for 60 calendar months if you are less than 40 years of age, and 24 calendar months if you are 40 years of age or older. So, if you are 25 years old on the date of the exam, your third-class medical certificate will expire in 5 years on the last day of the month in which it was issued. [Figure 1-11]

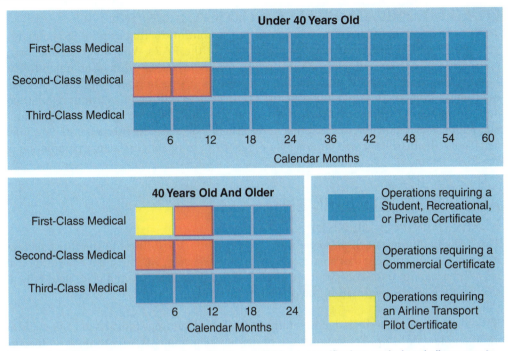

Figure 1-11. Each class of medical certificate is valid for a specific time period and allows you to exercise certain privileges as a pilot.

WHAT IS BASICMED?

To continue acting as pilot in command after your medical certificate expires, you must obtain a new exam from an AME and get a new first-, second-, or third-class medical certificate. However, if you qualify, you can continue operating as a pilot under the BasicMed rule if have previously held a medical certificate that was not suspended, revoked or withdrawn. This rule allows noncommercial flight operations with a driver's license instead of a medical certificate, by complying with certain limitations and by meeting the BasicMed requirements. [Figure 1-12]

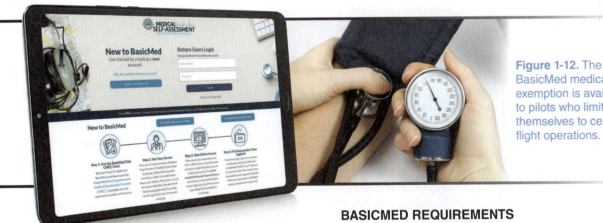

Figure 1-12. The BasicMed medical exemption is available to pilots who limit themselves to certain flight operations.

BASICMED LIMITATIONS

Under the BasicMed rule, you may conduct flight operations:

- In an aircraft certificated to carry no more than 6 occupants, including the pilot, and with a maximum certificated takeoff weight of no more than 6,000 pounds.
- Below 18,000 feet MSL.
- At a maximum airspeed of 250 knots.
- That are entirely within the United States.
- That are not for compensation or hire.

BASICMED REQUIREMENTS

To qualify for the BasicMed rule, you must:

- Possess and carry a valid U.S. driver's license.
- Have completed an approved medical education course—available online from the Aircraft Owners and Pilots Association (AOPA)—within the previous 24 calendar months.
- Have received a comprehensive medical exam from a state-licensed physician within the previous 48 months.
- Certify that are under the care and treatment of a physician for any diagnosed medical condition that could impact your ability to fly and that you are medically fit to fly.
- Agree to a National Driver Register check.

SECTION A ■ Pilot Training

DOES MY PRIVATE PILOT CERTIFICATE EXPIRE?

The private pilot certificate does not expire, but to continue acting as pilot in command, you must meet specific currency requirements in the FARs. You must satisfactorily complete a **flight review** every 24 calendar months to act as pilot in command. The flight review consists of at least one hour of ground instruction and one hour of flight instruction from an authorized instructor. The instructor will review current regulations and procedures with you and ensure that you are proficient in performing pilot operations and maneuvers. Upon determining that you can safely exercise the privileges of your pilot certificate, the instructor will endorse your logbook to indicate satisfactory completion of the flight review.

ARE THERE REQUIREMENTS FOR MAINTAINING CURRENCY?

The FARs define recent flight experience requirements. In addition to the flight review, to act as pilot in command of an aircraft carrying passengers, you must have performed at least three takeoffs and landings in an aircraft of the same category and class (and type, if a type rating is required) within the preceding 90 days.

If the flight is to be conducted at night, the takeoffs and landings must have been made at night and to a full stop. For the purpose of recency requirements, nighttime is defined as the period beginning one hour after sunset to one hour before sunrise. If the flight is to be conducted in a tailwheel airplane, the takeoffs and landings must have been in a tailwheel airplane and to a full stop.

These currency requirements are *minimums*. To maintain proficiency, you should fly regularly and accomplish frequent refresher training with an instructor. In addition to maintaining your flying skills, regularly review the latest FAA and industry publications to ensure that your aeronautical knowledge is up to date.

WHAT IS THE FAA WINGS—PILOT PROFICIENCY PROGRAM?

An effective way to maintain proficiency in all maneuvers and procedures, is to enroll in the FAA **WINGS—Pilot Proficiency Program**. The WINGS Program is designed to mitigate the primary factors that cause general aviation accidents by providing ground and flight training opportunities for pilots to apply risk management, enhance their knowledge and skills, and increase proficiency.

To enroll in the WINGS Program, create an account at FAASafety.gov. WINGS encourages an on-going training program that enables you to fly on a regular basis with an authorized flight instructor. You select the category and class of aircraft in which you wish to receive training and demonstrate your flight proficiency. In addition, you choose from a list of flight activities ranging from takeoffs and landings, to additional ratings proficiency checks. Opportunities to complete online courses, attend seminars and other events, and participate in webinars also are an integral part of the program.

If you satisfactorily complete a current WINGS phase within the previous 24 calendar months, you will not have to complete the flight review required by FAR 61.56. At FAASafety.gov, you can find detailed information regarding the WINGS Program, including links to a WINGS Program User's Guide and to AC 61-91, WINGS—Pilot Proficiency Program.

PRIVATE PILOT PRIVILEGES

You can exercise the privileges of your pilot certificate immediately after passing the checkride. As stated in the FARs, the **pilot in command (PIC)** of an aircraft is directly responsible for, and is the final authority as to, the operation of that aircraft.

WHAT OPERATIONS MAY I CONDUCT?

As a private pilot, you can carry passengers and equally share operating expenses, as long as you do not pay less than the pro rata share of the operating expenses, such as fuel, oil, airport expenditures, or rental fees. You may *not* have passengers pay the entire cost of a flight, or otherwise carry passengers or property for compensation or hire.

Under certain circumstances specified in the regulations you may operate an aircraft in connection with a business, as an aircraft salesperson, or during flights sponsored by charitable organizations in which the charitable organization receives compensation for the flight.

Once you have a minimum of 100 hours of flight time in your logbook, the opportunity to tow gliders becomes available to you. To act as pilot in command of an aircraft towing a glider, you must have made, within the preceding 12 months, at least three actual or simulated glider tows while accompanied by a qualified pilot.

WHAT DOCUMENTATION DO I NEED TO CARRY WHEN I FLY?

When you fly an airplane as a certificated private pilot, you must have in your possession, or readily accessible in the aircraft, the following documentation: your private pilot certificate, photo identification such as a driver's license or passport, and an appropriate medical certificate or BasicMed documents. Your logbook is only required to be in your possession when you are flying solo with a student pilot certificate. However, the FAA can request that you present your logbook with all required entries in a reasonable period of time.

HOW DO I KNOW MY AIRPLANE IS SAFE TO FLY?

To maintain the airworthiness of an aircraft, the FAA requires frequent inspections and maintenance. Annual inspections, and other required inspections, must be performed by FAA-certificated maintenance technicians who approve the aircraft for return to service with an entry in the aircraft's maintenance logbook.

In addition, you will perform a preflight inspection prior to every flight using an approved checklist to determine if your airplane is airworthy. If you discover a discrepancy during your preflight inspection, discuss the situation with your flight instructor or with maintenance personnel. If the airplane is unsafe to fly, it should be grounded until the problem is addressed.

WHAT ARE CATEGORIES AND CLASSES?

With respect to pilot certification, aircraft are organized into category, class, and type. **Category** is the broadest grouping of aircraft—airplane, rotorcraft, glider, lighter-than-air, and powered-lift. With the exception of gliders and powered-lift, each category is further broken down into a **class**, such as single-engine land, single-engine sea, multi-engine land, multi-engine sea, helicopter, gyroplane, airship, and balloon. Finally, the type is the make and model, such as Cessna 172, Hughes 500, or Boeing 747. The type is not listed for small airplanes. Your private pilot certificate will state the category (airplane) and class (single-engine land) of aircraft that you are authorized to fly. [Figure 1-13]

SECTION A ■ Pilot Training

SECTION A ■ Pilot Training

Figure 1-13. For pilot certification, aircraft are divided into categories and classes according to the specific knowledge and skills required to fly them.

HOW ARE AIRPLANES CATEGORIZED?

With respect to aircraft (not pilot) certification, category relates to the intended use or operating limitations. The normal and utility categories are common to most small airplanes. Depending on how they are loaded, many airplanes used in flight training are certificated in both of these categories. When loaded for the utility category, the airplane can withstand heavier stresses than it can in the normal category. Acrobatic aircraft have the fewest operating limitations because their design requirements demand more strength than those of the normal or utility category. Commuter aircraft are designed to carry passengers, but are limited to 19 seats and 19,000 pounds or less. Transport usually refers to airliners and other large aircraft above a certain weight or passenger-carrying capacity.

The restricted category is for special-purpose aircraft such as agricultural spray planes or slurry bombers used to fight forest fires. Limited category refers to military aircraft that are now allowed to be used only for limited purposes in civil aviation. The provisional category is really an interim measure for newly designed aircraft that have not met all the requirements for initial certification, but still can be operated for certain purposes. Experimental refers to a wide range of aircraft such as amateur homebuilt and racing airplanes, as well as research and development aircraft used to test new design concepts. Some small aircraft that are intended exclusively for pleasure and personal use are certificated in the primary category. [Figure 1-14]

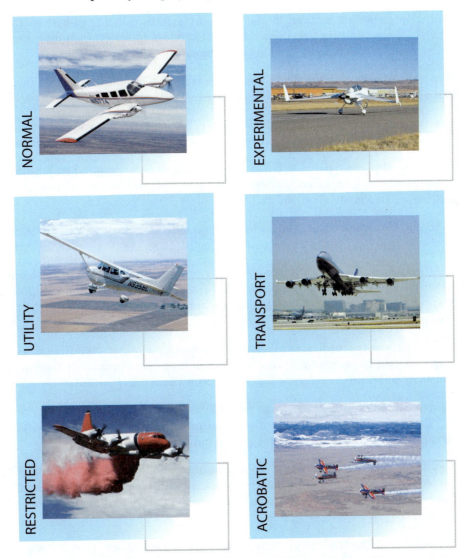

NORMAL

EXPERIMENTAL

UTILITY

TRANSPORT

RESTRICTED

ACROBATIC

Figure 1-14. For aircraft certification, category is related to the aircraft's intended use or operating limitations.

SECTION A ■ Pilot Training

SUMMARY CHECKLIST

✓ The Federal Aviation Administration (FAA) is the agency that governs commercial and general aviation, private space flight, and operation of unmanned aircraft systems (UAS).

✓ The Federal Aviation Regulations (FARs) are rules that apply to all areas of aviation, including flight operations, the construction of aircraft, and the training requirements that you must meet to obtain pilot certificates and ratings.

✓ The FAA Private Pilot Airman Certification Standards (ACS) contain the knowledge areas, risk management concepts, and technical flight skills that apply to each maneuver and procedure that you must demonstrate on your private pilot checkride.

✓ The codes designated for the ACS knowledge subject correlate to questions on the computerized FAA Private Pilot Airman Knowledge Test that you must take for private pilot certification.

✓ The International Civil Aviation Organization (ICAO) is an agency of the United Nations that advances the techniques and principles of international air navigation and fosters planning and development of international civil air transport.

✓ Many pilot training schools are located at airport facilities called fixed-base operators (FBOs). These facilities provide a variety of services to pilots, including aircraft rental, fueling, maintenance, parking, and the sale of pilot supplies.

✓ Two types of pilot training schools are: FAA-certificated schools governed by FAR Part 141 and less-regulated schools governed by FAR Part 61.

✓ The Transportation Security Administration (TSA) requires that you show proof of U.S. citizenship or, for non-U.S. citizens, comply with specific TSA requirements.

✓ To solo an airplane, you must be 16 years old, receive the required endorsements in your logbook from an authorized flight instructor, and possess a student pilot certificate and a medical certificate.

✓ To be eligible for a student pilot certificate you must be at least 16 years of age and be able to read, speak, and understand the English language.

✓ Integrated Airman Certification and Rating Application (IACRA) is the web-based certification/rating application system that guides you through the FAA's process to apply for pilot certificates and ratings, including the student pilot certificate and your private pilot certificate.

✓ The three classes of medical certificates are: first-class, required for airline transport pilots; second-class, required for commercial pilot operations other than airline transport; and third-class for student, sport, recreational, and private pilots.

✓ A private pilot training program can generally be divided into three phases: presolo and first solo; cross-country and night; and practical test/checkride preparation. Each phase includes both flight and ground training.

✓ A syllabus provides structure to your training, contains both ground and flight lessons, and is based upon the building-block theory of learning, which recognizes that each item taught must be presented on the basis of previously learned knowledge and skills.

✓ In addition to the student pilot requirements, to be eligible for a private pilot certificate you must be at least 17 years of age, complete specific training and flight time requirements described in the FARs, pass a knowledge test, and successfully complete a practical test that consists of oral quizzing, performing pilot operations, and performing maneuvers in the airplane.

✓ Minimum hours for both ground training and flight training are required by Part 61 and Part 141 for private pilot certification.

✓ You complete a knowledge test at an FAA-designated airman knowledge testing center. You must review the knowledge test report with your instructor and take it with you to the practical test.

✓ The knowledge text passing grade is 70 percent and the test results are valid for 24 calendar months.

✓ FAA regulations require you to keep an accurate record of pilot training time and aeronautical experience used to meet the requirements for a certificate, rating, or flight review.

✓ You must log flights required for meeting recent flight experience requirements and have instructor endorsements that are required to exercise pilot privileges in your logbook.

✓ Medical certificates have different durations, depending on the type of certificate and the pilot privileges being exercised.

✓ To continue acting as a pilot after your medical certificate expires, you must obtain a new exam from an aviation medical examiner or if you qualify, comply with the requirements of the BasicMed rule, which permits certain operations with a driver's license instead of a medical certificate.

✓ To act as pilot in command of an aircraft, you must satisfactorily complete a flight review every 24 calendar months.

✓ To act as pilot in command of an aircraft carrying passengers, you must have performed at least three takeoffs and landings in an aircraft of the same category and class within the preceding 90 days. If the flight is at night, the takeoffs and landings must have been to a full stop at night.

✓ An effective way to maintain proficiency is to enroll in the FAA WINGS—Pilot Proficiency Program at FAASafety.gov.

✓ The pilot in command of an aircraft is directly responsible for, and is the final authority as to, the operation of that aircraft.

✓ As a private pilot, you may not have passengers pay the entire cost of a flight, or otherwise carry passengers or property for compensation or hire.

✓ To act as pilot in command of an aircraft towing a glider, you must have logged a minimum of 100 hours of pilot flight time in powered aircraft. Additionally, you must have made, within the preceding 12 months, at least three actual or simulated glider tows while accompanied by a qualified pilot.

✓ When you fly an airplane as a certificated private pilot, you must have in your possession, or readily accessible in the aircraft, the following documentation: your private pilot certificate, photo identification such as a driver's license or passport, and an appropriate medical certificate or BasicMed documents.

✓ Your logbook is only required to be in your possession when you are flying solo with a student pilot certificate. However, the FAA can request that you present your logbook with all required entries in a reasonable period of time.

✓ In addition to verifying that the required maintenance and inspections have been accomplished, you will perform a preflight inspection prior to every flight using an approved checklist to determine if your airplane is airworthy.

SECTION A ■ **Pilot Training**

✓ For pilot certification, aircraft are organized into category, class, and type. Your private pilot certificate will state the category, class, and type (if appropriate) of aircraft that you are authorized to fly.

✓ For aircraft certification, category relates to the intended use of an aircraft and sets strict limits on its operation.

KEY TERMS

Federal Aviation Administration (FAA)

Federal Aviation Regulations (FARs)

Practical Test

Checkride

Airman Certification Standards (ACS)

International Civil Aviation Organization (ICAO)

Fixed Base Operator (FBO)

FAA Certificated Flight Instructor (CFI)

Student Pilot Certificate

Integrated Airman Certification and Rating Application (IACRA)

Medical Certificate

Aviation Medical Examiner (AME)

Syllabus

Airman Knowledge Test

BasicMed

Flight Review

WINGS—Pilot Proficiency Program

Pilot in Command (PIC)

Category

Class

QUESTIONS

1. What are the two requirements for obtaining a student pilot certificate?

2. Before solo, you must have what kind of medical certificate?
   A. BasicMed
   B. First-class
   C. First-class, second-class, or third-class

3. You were 25 years old when you obtained a second-class medical certificate on March 15, 2023. For exercising private pilot privileges, when does your medical certificate expire?
   A. March 15, 2028
   B. March 31, 2025
   C. March 31, 2028

4. The BasicMed rule allows you to use a driver's license instead of an FAA medical certificate for which flight operations?
   A. You may operate below 18,000 feet MSL at a maximum airspeed of 250 knots.
   B. You may operate outside the United States as long as the flight is not for compensation or hire.
   C. You may carry up to 6 passengers in an aircraft with a maximum certificated takeoff weight of no more than 8,000 pounds.

5. What requirements must you meet to utilize the BasicMed rule?
   A. You must have completed an approved BasicMed exam from an aviation medical examiner within the previous 48 months.
   B. You must have completed an approved medical education course within the previous 24 calendar months and a comprehensive medical exam from a physician within the previous 48 months.
   C. You must have completed an approved medical education course within the previous 48 months and a comprehensive medical exam from a physician within the previous 24 calendar months.

6. What is the required age to obtain a private pilot certificate for an airplane?
   A. 15
   B. 16
   C. 17

7. Which flight time must you record in your logbook?
   A. All flight time
   B. All solo flight time
   C. Flight time needed to meet the requirements for a certificate, rating, flight review, or recency of experience

8. When is a flight review required?
   A. Within the past 24 calendar months to act as pilot in command if carrying passengers.
   B. Within the past 24 calendar months to act as pilot in command whether flying solo or carrying passengers.
   C. Within the past 60 calendar months to act as pilot in command whether flying solo or carrying passengers.

9. What requirement must you meet before acting as pilot-in-command while carrying passengers?
   A. You must have completed three takeoffs and landings within the previous 90 days in the same category and class of aircraft.
   B. You must have completed three takeoffs and landings within the previous 180 days in the same category and class of aircraft.
   C. You must have completed three takeoffs and landings within the previous 24 calendar months in the same category and class of aircraft.

10. With respect to pilot certification, which are categories of aircraft?
    A. Gyroplane, helicopter, airship, free balloon
    B. Single-engine land and sea, multi-engine land and sea
    C. Airplane, rotorcraft, powered lift, glider, lighter than air

11. With respect to pilot certification, which are classes of aircraft?
    A. Airplane, rotorcraft, glider, lighter-than-air
    B. Single-engine land and sea, multi-engine land and sea
    C. Lighter-than-air, airship, hot air balloon, gas balloon

12. With respect to the certification of aircraft, which are categories of aircraft?
    A. Landplane, seaplane
    B. Normal, utility, acrobatic
    C. Airplane, rotorcraft, glider

SECTION A ■ **Pilot Training**

SECTION B
Aviation Opportunities

Welcome to your future. One of the unique joys of aviation is that there is always a challenge to be met; a new adventure on which to embark; one more goal to be achieved. A private pilot certificate opens a door to a future of exciting opportunities and endless possibilities. Completion of private pilot training does not signify an ending, but a beginning to an aviation journey filled with new experiences. Whether you are flying for personal pleasure or you have career aspirations, you can navigate many different courses on this voyage.

As soon as you earn your private pilot certificate, new experiences await you; new scenery, new airports, and new responsibilities. You will be able to carry passengers for the first time, and you can fly cross-country to airports that you have not yet explored. [Figure 1-15]

Figure 1-15. A new experience after you obtain your private pilot certificate is to carry a passenger for the first time

This section introduces you to a wide variety of avocational options and careers that you can pursue as a pilot. As you explore these aviation opportunities, icons for Training, Logbook, Knowledge Test, and Practical Test provide a guideline to the type of training and experience necessary to meet each goal. Before you begin any additional instruction, you should refer to the FARs for specific experience requirements and discuss the training course thoroughly with your flight instructor. [Figure 1-16]

TRAINING ONLY	Ground instruction and flight training are necessary to become competent in this skill but no specific requirements are described by the FARs.
LOGBOOK	The FARs require training in specific pilot operations and a logbook endorsement from a qualified flight instructor.
KNOWLEDGE TEST	You must pass a computerized knowledge test.
PRACTICAL TEST	Specific ground instruction and flight training are required by the FARs, and you must pass a practical test.

	TRAINING ONLY	LOGBOOK ENDORSEMENT	KNOWLEDGE TEST	PRACTICAL TEST
REFRESHER	✓			
MOUNTAIN	✓			
AEROBATIC	✓			
HIGH-PERFORMANCE/ COMPLEX		✓		
TECHNICALLY ADVANCED AIRPLANE	✓			
TAILWHEEL		✓		
HOMEBUILT	✓			
INSTRUMENT RATING			✓	✓
MULTI-ENGINE RATING				✓
SEAPLANE RATING				✓
ROTORCRAFT				✓
GLIDER				✓
BALLOON				✓
COMMERCIAL PILOT			✓	✓
CERTIFICATED FLIGHT INSTRUCTOR			✓	✓
AIRLINE TRANSPORT PILOT			✓	✓
SPORT PILOT			✓	✓

Figure 1-16. The requirements listed here assume that you have a private pilot certificate with an airplane category rating and single-engine class rating. The exception is the sport pilot certificate, which would be obtained instead of a private pilot certificate,

SECTION B ■ Aviation Opportunities

NEW AVIATION EXPERIENCES

As you gain flying experience and confidence, you might feel the need to expand your aviation horizons. The best way to sharpen your abilities, master new skills, and reenergize your enthusiasm for flight is to pursue additional training.

TRAINING ONLY REFRESHER TRAINING

Although your objectives might not include exploring a new area of aviation, frequent **refresher training** is essential to keep your skills sharp and to keep you informed on current pilot information. You might want to seek instruction in pilot operations that you do not perform frequently. In addition, by enrolling in the FAA WINGS—Pilot Proficiency Program, you can obtain ground and flight training that enables you to apply risk management, enhance your knowledge and skills, and increase your proficiency in challenging flight operations. [Figure 1-17]

You are planning a trip to an unfamiliar airport, but have not ventured out of the local area very often.

You are accustomed to operating at a small airport with very little traffic and would like to gain experience operating at a larger, busier airport.

You plan to fly frequently at night and do not have extensive night flying experience.

A substantial period of time has passed since you have practiced a series of takeoffs and landings, attitude instrument flying, or emergency procedures.

Courtesy of Piper Aircraft, Inc.

Figure 1-17. Refresher training is a good way to help you feel more comfortable in the airplane if you do not fly frequently.

SECTION B ■ **Aviation Opportunities**

▰▰ TRAINING ONLY MOUNTAIN FLYING

If all of your flight training was accomplished at low-elevation airports, learning to fly in the mountains can be a fresh, rewarding experience. **Mountain flying** is challenging and requires proficiency in all of your piloting skills. Flying in the mountains is beautiful but can be hazardous if you have not received proper training. You have to make special considerations for weather, airport operations, course selection, and aircraft performance.

Before you fly at high elevations or in mountainous terrain, obtain ground and flight training from a qualified instructor. You might have the opportunity to attend a mountain flying course, which can consist of both ground and flight instruction. In addition, to prepare for your training, many publications can provide you with information about mountain operations. [Figure 1-18]

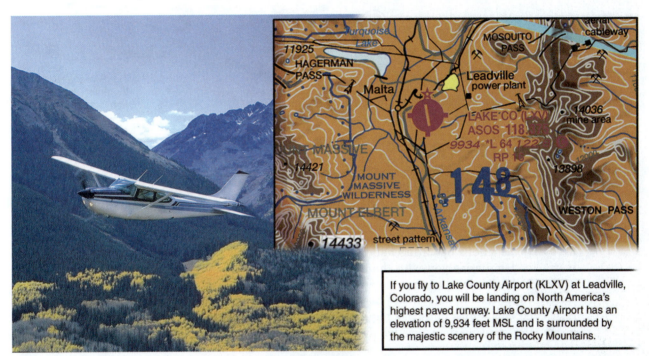

If you fly to Lake County Airport (KLXV) at Leadville, Colorado, you will be landing on North America's highest paved runway. Lake County Airport has an elevation of 9,934 feet MSL and is surrounded by the majestic scenery of the Rocky Mountains.

Figure 1-18. Mastery of mountain flying skills allows you to operate at some unique airports.

▰▰ TRAINING ONLY AEROBATIC FLIGHT

The aileron roll, snap roll, barrel roll, Cuban eight, loop, and Immelmann turn are all maneuvers that you might perform during **aerobatic flight training.** In addition to being just plain fun, aerobatics instruction increases your proficiency as a pilot. As you master each aerobatic maneuver, your timing, coordination, and reactions improve, your confidence level as a pilot increases, and you gain a better understanding of the capabilities and limitations of the airplane. Aerobatic training also presents aerodynamics in a unique way that can not be duplicated in a classroom. Although no specific flight hour training requirements apply to operate an aircraft in aerobatic flight, the FARs do place certain restrictions on aerobatic maneuvers. For example, the aircraft must be certificated in the acrobatic category and parachutes are required when carrying passengers. [Figure 1-19]

Copyright Corel

Figure 1-19. Aerobatic competitors are judged on how smoothly and precisely they perform a series of predetermined maneuvers in a box of air with lateral and vertical parameters.

SECTION B ■ **Aviation Opportunities**

DISCOVERY

Immelmann Turn

The Immelmann turn is an aerobatic maneuver named after World War I German ace, Lieutenant Max "the Eagle" Immelmann. Lieutenant Immelmann was renowned for outwitting his opponents by performing aerobatic maneuvers in his Fokker monoplane.

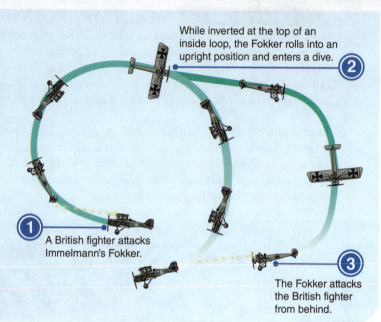

While inverted at the top of an inside loop, the Fokker rolls into an upright position and enters a dive. **2**

1 A British fighter attacks Immelmann's Fokker.

3 The Fokker attacks the British fighter from behind.

AIRPLANE TRANSITIONS

An enormous variety of airplanes exist to explore, and the transition to another make and model of airplane can be one of the most fascinating aspects of flying. To safely pilot an unfamiliar airplane, you must receive transition training, often referred to as an **aircraft checkout.**

Studying the pilot's operating handbook and discussing the new aircraft with your instructor familiarize you with the airplane's systems, performance, and limitations. During flight training, you practice normal and emergency procedures and learn how to safely control the airplane in all phases of flight. Whether you own or rent an airplane, insurance companies normally require that you have a specific number of hours of instruction in the make and model of airplane before you can fly it as pilot in command.

LOGBOOK

HIGH-PERFORMANCE AND COMPLEX AIRPLANES

If you want faster cruising speeds, increased performance, and the opportunity to master the operation of new systems, a checkout in a high-performance, or complex airplane is the solution. A **high-performance airplane** has an engine with more than 200 horsepower. A **complex airplane** has retractable landing gear, flaps, and a controllable pitch propeller, including airplanes equipped with an engine control system consisting of a digital computer and associated accessories for controlling the engine and propeller, such as a full authority digital engine control (FADEC). The transition to these airplanes can be challenging—requiring additional planning, judgment, and pilot skill. To act as pilot in command of a high-performance or complex airplane, you must receive instruction and a logbook endorsement stating that you are competent to pilot such an airplane. [Figure 1-20]

Figure 1-20. Training in high-performance or complex airplanes will focus on the operation of systems that are new to you, such as the retractable landing gear.

TRAINING ONLY TECHNICALLY ADVANCED AIRPLANES

If you accomplish your private pilot training in an airplane with analog instruments with minimal navigation equipment, you should obtain additional training to operate a **technically advanced airplane (TAA)**. A TAA has:

- A primary flight display (PFD) with an airspeed indicator, turn coordinator, attitude indicator, heading indicator, altimeter, and vertical speed indicator;

- A multifunction display (MFD) with a moving map using GPS navigation to display the aircraft position;

- A two-axis autopilot integrated with the navigation and heading guidance system.

The FAA does not require an endorsement to fly a TAA, but you should treat the transition to any airplane with unfamiliar equipment and systems as an important training endeavor that is essential to flight safety. [Figure 1-21]

Figure 1-21. You should obtain training on the specific equipment in each airplane you fly.

LOGBOOK TAILWHEEL AIRPLANES

Learning to fly a **tailwheel airplane** presents a new challenge to many pilots trained in airplanes with tricycle landing gear. Due to their design and structure, tailwheel airplanes, also known as **conventional landing gear airplanes**, exhibit operational and handling characteristics that are different from airplanes with tricycle landing gear. To act as pilot in command of a tailwheel airplane, you are required to obtain training in specific pilot operations outlined in the FARs. A logbook endorsement from a qualified flight instructor must state that you have achieved competency in normal and crosswind takeoffs and landings, wheel landings (unless the manufacturer has recommended against such landings), and go-around procedures. [Figure 1-22]

Figure 1-22.
A tailwheel checkout is a perfect way for you to expand your airplane options, and can be the first step toward aerobatic training.

Courtesy of Jeffrey Miller

TRAINING ONLY | HOMEBUILT AIRCRAFT

You can construct a **homebuilt aircraft** from scratch using wood, fabric, metal, or composite material, or you can purchase a kit that requires assembly of larger components. Although hundreds of completed aircraft of a particular design might exist, no two homebuilts are ever the same. Builders add individual touches to their aircraft that make each one unique. The spectrum of homebuilt aircraft ranges from very light airplanes that cruise at 50 mph to composite airplanes that reach speeds of 350 mph.

You can request FAA certification of your homebuilt airplane as an experimental, amateur-built aircraft. The FAA certification process requires documentation of the construction process, aircraft inspections, and specific test flights. Because flying characteristics of homebuilt aircraft can vary widely, effective training requires an instructor who has considerable flight experience in the same type of aircraft. To obtain information on building and flying homebuilt aircraft, contact the local chapter of the Experimental Aircraft Association or your local FAA Flight Standards District Office. [Figure 1-23]

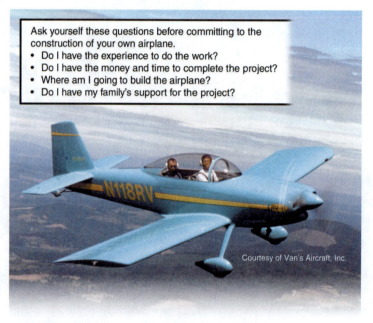

Ask yourself these questions before committing to the construction of your own airplane.
• Do I have the experience to do the work?
• Do I have the money and time to complete the project?
• Where am I going to build the airplane?
• Do I have my family's support for the project?

Courtesy of Van's Aircraft, Inc.

Figure 1-23. Although building your own aircraft can be an extremely satisfying experience, it also requires hard work and a substantial commitment of time and resources.

ADDITIONAL PILOT RATINGS

The FARs describe several ratings that might be added to your private pilot certificate; the instrument rating introduces you to new procedures and flight operations; and class ratings expand your airplane options. Type ratings are required for turbojet aircraft, aircraft that weigh more than 12,500 pounds, certain helicopter operations, and other aircraft that are specified by the FAA.

KNOWLEDGE TEST | INSTRUMENT RATING
PRACTICAL TEST | The addition of an **instrument rating** is an option that allows you to fly in a wider range of weather conditions and increases your skill at precisely controlling the aircraft. As a private pilot without an instrument rating, you must fly under visual flight rules (VFR) conditions. To remain in VFR conditions, the FARs require that you maintain a specific flight visibility and that you operate the aircraft a certain distance from clouds. Operating under instrument flight rules (IFR) allows you to fly in the clouds with no reference to the ground or horizon.

Among other requirements, you must have at least 40 hours of instrument flight time to obtain an instrument rating. To accumulate the required flight time, you train with a view-limiting device that restricts your view outside the aircraft so that you maintain reference only to the flight deck instruments. Because instrument procedures are much different than VFR operations, you are required to pass a knowledge test and practical test to add the instrument rating to your certificate. Some of your instrument training might be conducted in a flight simulation training device (FSTD) or aviation training device (ATD). [Figure 1-24]

Courtesy of Redbird Flight Simulations

Figure 1-24. Instrument training time with a qualified flight instructor can be acquired in actual instrument weather conditions, by using a view-limiting device, or while operating an aviation training device.

AIRPLANE CLASS RATINGS

Adding a multi-engine rating or a seaplane rating to your certificate can be an exciting way to explore new airplanes without requiring a substantial amount of training time.

PRACTICAL TEST

MULTI-ENGINE RATING

If you are pursuing a career in aviation, the **multi-engine rating** is an essential requirement for most flying jobs. No specific ground or flight instruction hours are required for the addition of the multi-engine rating, but you will have to pass a practical test. Typically, the training can be completed in a short period of time, but most aircraft insurance policies require that you obtain a substantial amount of multi-engine flight time before operating the airplane as pilot in command. [Figure 1-25]

Figure 1-25. To accumulate the necessary experience, you might be able to share flight time and expenses with a qualified pilot who meets the insurance requirements.

PRACTICAL TEST　　### SEAPLANE RATING

Seaplanes conjure up images of flying to exotic and distant locations that other humans have yet to encounter. Although you might never be the first individual to set foot on the shores of a remote jungle lagoon, as a seaplane pilot, you might rediscover some of the romance of flight when you touch down on the glassy surface of a clear mountain lake.

Seaplanes can have twin floats or can be designed with a single main float plus small wing floats or similar lateral supports. The single-main-float seaplane is frequently referred to as a flying boat. A twin float seaplane is known as a floatplane. In addition, when a seaplane is equipped with wheels, it becomes an amphibian, which is capable of operating from land or water.

No minimum number of flight hours are required to add a **single-engine sea rating** or **multi-engine sea rating** to your private pilot certificate, but you must pass a practical test. The Seaplane Pilot's Association (SPA) publishes a Water Landing Directory to help you determine what lakes or other water surfaces are legal for seaplane operations. [Figure 1-26]

Copyright Corel

Figure 1-26. Seaplane designs include aircraft with twin floats and those with a main float and small wing floats.

CATEGORY AND CLASS RATINGS

After you experience the thrill of flight behind the controls of an airplane, you might want to spread your wings and explore a new category and class of aircraft.

PRACTICAL TEST   ROTORCRAFT — HELICOPTER RATING

The principal difference between an airplane and a **helicopter** is how each aircraft develops lift. The airplane's lift is produced by a fixed wing, while the helicopter generates lift from a rotating airfoil called the rotor. Aircraft are classified as either fixed-wing or rotating-wing. Lift generation by a rotating wing gives the helicopter its unique ability to hover.

Helicopters are one of the most maneuverable types of aircraft, and while hovering, the helicopter can be moved in every possible direction or combination of directions. Flying a helicopter requires precise control inputs and coordination. To add a rotorcraft category rating with a helicopter class rating to your private pilot certificate, you must have a minimum of 40 hours of flight time that includes at least 20 hours of dual instruction in helicopters and at least 10 hours of solo flight in helicopters, and you must pass a practical test.

Copyright Corel

Helicopters are used in a variety of commercial operations such as corporate transport, pipeline laying and patrolling, news and traffic reporting, aerial photography, sightseeing tours, and general police work. In addition, helicopters play an important role in the oil, timber, and agriculture industries. [Figure 1-27]

Figure 1-27. An advantage to the helicopter is its ability to take off and land in a confined or restricted area.

DISCOVERY

Da Vinci Helicopter Design

In the 15th century, Leonardo da Vinci made sketches of a helicopter design, which was to be composed of wood, reeds, and tafetta. The vehicle had a screw-type thread on a vertical shaft, which if properly shaped and powered, would be able to takeoff vertically, hover, and land. Nearly 500 years later, on November 13, 1907, a two-rotor helicopter built by Paul Cornu lifted a man off the ground during a flight test in France, but it would be another 30 years until practical helicopters were actually in use.

PRACTICAL TEST

ROTORCRAFT — GYROPLANE RATING

Gyroplanes are rotating-wing aircraft that generate lift as the aircraft's forward movement drives air up through the rotor blades. Most gyroplanes are homebuilt from kits, have only one or two seats, and few have radios or lights. To add a rotorcraft category with a gyroplane class rating to your private pilot certificate, you must pass a practical test after obtaining a minimum of 20 hours of dual instruction in gyroplanes and at least 10 hours of solo flight in gyroplanes. [Figure 1-28]

Courtesy NASA Langley Research Center

Figure 1-28. The rotating wings of the gyroplane are free-spinning and while it can't hover, the gyroplane needs very little forward speed to stay airborne.

PRACTICAL TEST

GLIDER RATING

If you are seeking a new challenge, yet want to remain with a fixed-wing aircraft, a **glider rating** is an excellent choice. Gliders are unpowered, and typically an airplane tows them to an altitude that enables them to search for atmospheric lift. Gliders also can be launched using an automobile or winch tow.

To add a glider rating to your private pilot certificate, the FARs require a total of 40 hours of flight time as a pilot in a heavier-than-air aircraft with at least 3 hours of flight training in a glider and 10 solo flights. The glider rating does not require a knowledge test, provided you hold a powered category rating. [Figure 1-29]

Figure 1-29. A glider allows you to soar with the hawks, and share thermals with the eagles.

SECTION B ■ Aviation Opportunities

LIGHTER-THAN-AIR — BALLOON RATING

The **hot air balloon** is the most common of the three types of balloons; gas, hot air, and Rozier (combination gas/hot air). The hot air balloon rises because the heated air inside the balloon becomes less dense than the outside air.

To obtain a lighter-than-air certificate with a balloon rating, you must accumulate a total of 10 hours of flight training that includes at least 6 training flights. A knowledge test is not required for a fixed-wing pilot, but you must pass a practical test. To find a qualified balloon instructor, you can contact the Balloon Federation of America (BFA), the world's largest organization of balloonists. The BFA can provide you with a list of balloonists who participate in the Master Instruction program. [Figure 1-30]

Many balloon students get their start by being part of the balloon crew. A two or three person crew helps the pilot rig the balloon, holds open the envelope as it fills with air, and applies weight to the outside of the basket before the launch. The crew then follows by car, and after the balloon lands, helps the pilot pack up.

Courtesy of coffeebreakwithlizandkate.com

Balloons usually fly within two to three hours of sunrise or sunset when the air is cool, the winds are calm, and conditions are most stable.

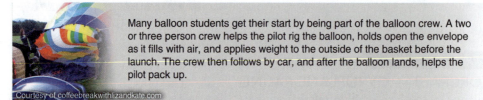

To share the experience with the passengers, crew, and spectators, balloonists traditionally end each flight with a toast of champagne, a practice dating back to the 1700s.

Figure 1-30. From the first manned flight in 1783 to the present, pilots have relished the quiet, tranquil sensation of floating aloft in a balloon.

DISCOVERY

Powered-Lift

The powered-lift category aircraft combines the hovering capability and vertical takeoffs and landings of helicopters with the horizontal flight of airplanes. The role of rotor equipment in the aviation industry might be considerably expanded with the use of the tilt-rotor aircraft. Some tilt-rotor aircraft can reach cruise speeds of close to 300 mph.

Courtesy NASA Dryden Research Center

ADDITIONAL PILOT CERTIFICATES

KNOWLEDGE TEST

PRACTICAL TEST

COMMERCIAL PILOT

Obtaining the **commercial pilot certificate** is the first step toward a professional pilot career. As stated in the FARs, the privileges of a commercial pilot certificate include the ability to "act as pilot in command of an aircraft carrying persons or property for hire." However, the types of commercial operations that you can perform are limited unless you receive additional training beyond the commercial certificate.

To apply for a commercial pilot certificate, you must meet fairly substantial flight time requirements. Depending on the type of pilot school you attend, a total of 190 to 250 hours of flight time is required, which normally includes a minimum of 100 hours of pilot-in-command time and 50 hours of cross-country time. In addition, you must have 10 hours of flight training in an airplane that has retractable landing gear, flaps, and a controllable pitch propeller or FADEC or 10 hours of flight training in a technically advanced airplane (TAA). To become a commercial pilot, you must pass a knowledge test and complete a practical test on precision flight maneuvers.

KNOWLEDGE TEST

PRACTICAL TEST

CERTIFICATED FLIGHT INSTRUCTOR

To become a certificated flight instructor (CFI), you must continue training beyond the commercial certificate. Your instruction will focus on aspects of teaching that include the learning process, student evaluation, and lesson planning. No specific number of flight hours is required for CFI training, but you must pass two knowledge exams and a practical test.

KNOWLEDGE TEST

PRACTICAL TEST

AIRLINE TRANSPORT PILOT

The **airline transport pilot (ATP) certificate** is required for pilots flying passengers or cargo under FAR Part 121 and for some FAR Part 135 operations. To apply for an ATP certificate, you must be at least 23 years of age. The flight time

Copyright Boeing

Figure 1-31. You must hold an ATP certificate to fly for the airlines under FAR Part 121.

requirements to obtain an ATP certificate are demanding: a total of 1,500 hours of flight time including 250 hours of pilot-in-command time, 500 hours of cross-country time, 100 hours of night flight, and 75 hours of instrument experience. The ATP knowledge test emphasizes subjects such as navigation, meteorology, aircraft performance, and air carrier flight procedures. During the practical test, your instrument skills will be evaluated, as well as your ability to correctly perform emergency procedures. [Figure 1-31]

SECTION B ■ **Aviation Opportunities**

The FAA requires additional training prior to taking your ATP knowledge test in the form of an authorized **Airline Transport Pilot Certification Training Program (ATP CTP)**. This training requirement ensures that you have completed a flight training program that integrates academic training and aeronautical experience in a flight simulation training device (FSTD) to prepare you to function effectively in a multipilot (multicrew) environment; manage adverse weather conditions; perform high-altitude operations; and adhere to the highest professional standards in an air carrier environment. If you are a pilot with less than 1,500 hours of flight time and a graduates of specific aviation degree programs authorized by the FAA, you may obtain a restricted privileges ATP (R-ATP) certificate. This certificate allows you to serve as a co-pilot until you obtain the necessary 1,500 hours. [Figure 1-32]

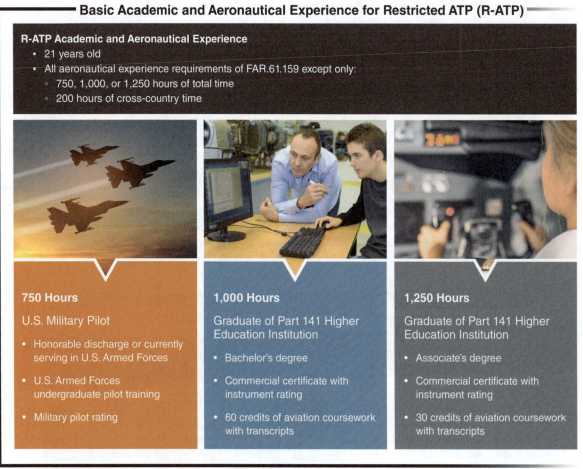

Basic Academic and Aeronautical Experience for Restricted ATP (R-ATP)

R-ATP Academic and Aeronautical Experience
- 21 years old
- All aeronautical experience requirements of FAR.61.159 except only:
  - 750, 1,000, or 1,250 hours of total time
  - 200 hours of cross-country time

750 Hours

U.S. Military Pilot

- Honorable discharge or currently serving in U.S. Armed Forces
- U.S. Armed Forces undergraduate pilot training
- Military pilot rating

1,000 Hours

Graduate of Part 141 Higher Education Institution

- Bachelor's degree
- Commercial certificate with instrument rating
- 60 credits of aviation coursework with transcripts

1,250 Hours

Graduate of Part 141 Higher Education Institution

- Associate's degree
- Commercial certificate with instrument rating
- 30 credits of aviation coursework with transcripts

Figure 1-32. If you meet specific criteria, you can obtain an R-ATP with reduced age and flight time requirements.

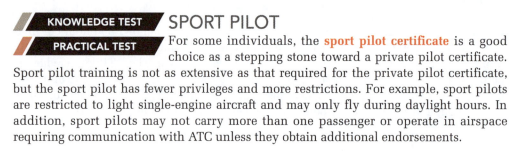

KNOWLEDGE TEST

PRACTICAL TEST

SPORT PILOT

For some individuals, the **sport pilot certificate** is a good choice as a stepping stone toward a private pilot certificate. Sport pilot training is not as extensive as that required for the private pilot certificate, but the sport pilot has fewer privileges and more restrictions. For example, sport pilots are restricted to light single-engine aircraft and may only fly during daylight hours. In addition, sport pilots may not carry more than one passenger or operate in airspace requiring communication with ATC unless they obtain additional endorsements.

SECTION B ■ **Aviation Opportunities**

Light Sport Aircraft

A light sport aircraft, or LSA, is an aircraft intended primarily for fun. The definition includes airplanes, balloons, airships, gliders, gyroplanes, powered parachutes, and weight-shift-controlled aircraft. Light sport airplanes can carry one or two people. They have a stall speed of 45 knots or less, a maximum level flight speed of 120 knots or less, fixed landing gear, and a fixed-pitch or ground-adjustable propeller. Landplanes can have a maximum weight of 1,320 pounds, and seaplanes a maximum weight of 1,430 pounds. Pilots with a sport pilot certificate or higher may fly light sport aircraft.

AVIATION CAREERS

Some pilots transform flying as a hobby into flying as a career. To choose between sitting behind the controls of an airplane and sitting behind a desk might not be as easy as it sounds. The pursuit of an aviation career requires a solid commitment of time, energy, and financial resources. However, as most professional pilots will tell you, the rewards of a flying career are worth the dedication and hard work. [Figure 1-33]

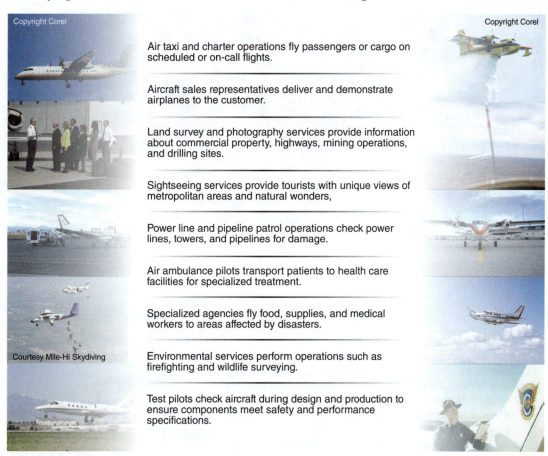

Copyright Corel

Copyright Corel

Courtesy Mile-Hi Skydiving

Air taxi and charter operations fly passengers or cargo on scheduled or on-call flights.

Aircraft sales representatives deliver and demonstrate airplanes to the customer.

Land survey and photography services provide information about commercial property, highways, mining operations, and drilling sites.

Sightseeing services provide tourists with unique views of metropolitan areas and natural wonders,

Power line and pipeline patrol operations check power lines, towers, and pipelines for damage.

Air ambulance pilots transport patients to health care facilities for specialized treatment.

Specialized agencies fly food, supplies, and medical workers to areas affected by disasters.

Environmental services perform operations such as firefighting and wildlife surveying.

Test pilots check aircraft during design and production to ensure components meet safety and performance specifications.

Figure 1-33. Highly trained professional pilots use their skills in a wide variety of fields.

FLIGHT INSTRUCTING

Many career-oriented pilots use flight instruction as a stepping stone to enhance their professional qualifications. For example, a substantial amount of flight experience must be gained to apply for most flying jobs, and as a flight instructor, you can accumulate

SECTION B ■ Aviation Opportunities

flight time as you earn money. This does not mean that flight instructors are not dedicated professionals who take the job of teaching very seriously. Some instructors focus on instruction for beginning students, while others expand their instructing privileges to include teaching instrument or multi-engine students. Many pilots have established careers in other fields, but find part-time flight instructing an extremely satisfying part of their lives.

As a flight instructor you can be self-employed or apply for a position at a pilot training school. Although many flight instructors are paid per flight hour, you might earn a salary and receive benefits if employed at a larger pilot training facility. One advantage to many flight instructor positions is your ability to determine your own lesson schedule. [Figure 1-34]

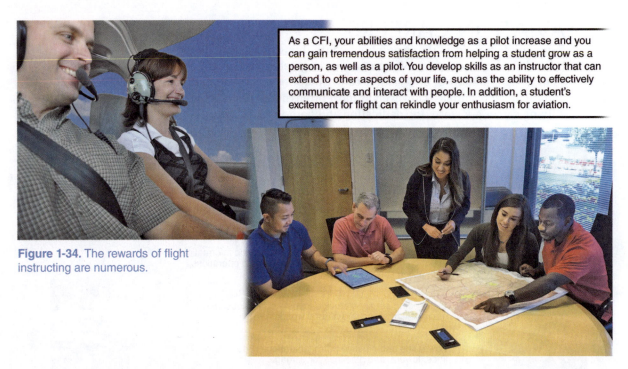

As a CFI, your abilities and knowledge as a pilot increase and you can gain tremendous satisfaction from helping a student grow as a person, as well as a pilot. You develop skills as an instructor that can extend to other aspects of your life, such as the ability to effectively communicate and interact with people. In addition, a student's excitement for flight can rekindle your enthusiasm for aviation.

Figure 1-34. The rewards of flight instructing are numerous.

Airmail Pilots

In the early days of aviation, American pilots often had to choose between the unstable life of a barnstormer and the dangerous career of flying the mail. The United States Airmail Service was founded in 1918 and was operated by the government for 9 years. An airmail pilot's life expectancy was 4 years, and 31 of the first 40 pilots were killed before the service was turned over to private industry. While the Airmail Service was in operation, 1 in every 6 pilots was killed in a flying accident while delivering the mail.

It was the spring of 1920. There has not been a great deal of airmail flying, and the pilots were getting killed almost as fast as the Post Office Department could employ them . . . So the airmail was considered pretty much a suicide club, and only pilots desperate to fly would join it. But the pay was excellent. We were soon making from $800 to $1,000 a month, and in the early twenties this was really a tremendous amount of money.

— Dean Smith, United States airmail pilot in 1920, as quoted in *The American Heritage History of Flight*

SECTION B ■ Aviation Opportunities

DISCOVERY

REGIONAL AIRLINES

Many pilots view flying for a regional, or commuter, airline as a way to enhance their qualifications while working toward an airline position. As a pilot for a **regional airline** you fly advanced turboprop or small jet aircraft during scheduled passenger-carrying flights. You must have an ATP or restricted ATP certificate to fly for a regional airline. [Figure 1-35]

Figure 1-35. Flying for a regional airline is an excellent training ground for the airlines.

MAJOR AIRLINES

Flying for a **major airline** is a rewarding career that requires a serious commitment and hard work. Hiring qualifications for major airlines vary as the job market shifts. Although each airline has specific minimum requirements, you must achieve competitive qualifications for the market at the time. [Figure 1-36]

Copyright Boeing

Flight Experience
Most pilots hired by major airlines have regional airline, corporate, or military flight experience. As the number of qualified pilots grows, the average total flight time expected by the airlines increases.

Certificates and Ratings
The FARs require you to hold an ATP certificate with a multi-engine rating. However, you may act as a first officer if you hold a restricted ATP (R-ATP). In addition, you must hold a first-class medical certificate.

Education
At a minimum, your credentials should include a four-year college degree in any area of study. However, to be eligible for a R-ATP, you must complete flight training under Part 141 and graduate from an FAA-approved university aviation program.

Interview Skills
During an airline interview you are evaluated on how well you communicate, your leadership skills, and your ability to perform as a crewmember. Typically, you also must pass a stringent medical exam and your flying skills are assessed during a simulator flight.

Figure 1-36. The qualifications for an airline pilot position under FAR Part 121 can be divided into four categories.

SECTION B ■ **Aviation Opportunities**

First U.S. Commercial Passenger Service

One of the first commercial passenger services in the U.S. used a combination of airline and railroad transportation. Passengers left New York City's Pennsylvania Station by train and traveled to Columbus, Ohio where they boarded a Ford Trimotor transport for a flight to Waynoka, Oklahoma. Upon arrival in Waynoka, the passengers boarded a train that took them to Clovis, New Mexico. A Ford Trimotor delivered them to their final destination on the west coast. This service reduced travel time from the east coast to the west coast from 72 hours to 48 hours. Today, a Boeing 757 makes the journey nonstop in 6 hours.

DISCOVERY

SECTION B ■ **Aviation Opportunities**

Courtesy of Embraer Aircraft Holdings, Inc.

Figure 1-37. Corporate airplanes can range from small single- and multi-engine airplanes to large jets.

Figure 1-38. Agricultural aircraft operations provide timely and economical pesticide application over large and often remote areas.

CORPORATE FLYING

Corporate flying offers some unique benefits. For example, you might have the opportunity to bring a spouse along for any extended layovers. Although some corporate flight departments operate regularly scheduled flights, most have very few prescheduled trips and pilots are on-call.

Many corporate jobs are not advertised and pilots are hired upon referral by another pilot. Although most corporations prefer that you have experience in the type of aircraft operated by the company, your lack of experience might not be a factor if you have been recommended by a pilot within the corporation. Typically, the minimum pilot qualifications for corporate flying include a commercial pilot certificate with an instrument rating and a multi-engine rating. An ATP certificate and type rating in a jet or turboprop airplane are preferred by many corporations. [Figure 1-37]

AERIAL APPLICATION

Aerial application, or crop dusting, provides farmers with seeding, pollinating, and control over damaging insects and weeds. You should have a knowledge of chemistry, agriculture, and aviation. Many aerial applicators have a degree in agriculture or chemical engineering. To become employed as an aerial applicator you must hold a commercial pilot certificate and receive additional training in agricultural aircraft operations. [Figure 1-38]

MILITARY AVIATION

The United States military offers a wide variety of aviation career opportunities. As a candidate for military pilot training, you must meet specific academic qualifications, demonstrate certain physical abilities, and possess strong leadership skills. In addition, you must be willing to commit to a specified service obligation at the completion of your training. Although many pilots choose a career in the military, others find that the flight training and experience gained during military service are excellent preparation for similar jobs in civilian life. [Figure 1-39]

Courtesy of Lockheed Martin. Lockheedmartin.com

The Lockheed/Martin F-35B Joint Strike Fighter is the world's only supersonic short takeoff/vertical landing (STOVL) fighter aircraft with radar-evading stealth technology. The F-35B has the unique ability to operate from a variety of small naval ships, roads, and unimproved airstrips near frontline combat zones. The F-35B STOVL capabilities are due to a single Pratt & Whitney F135 afterburner turbofan engine with a rotating exhaust nozzle and a shaft-driven Rolls Royce LiftFan propulsion system—mounted aft of the cockpit—that produces more than 20,000 pounds of thrust.

The Boeing T-7A Red Hawk is an Advanced Pilot Training System (APTS) for the U.S. Air Force, with flexibility to evolve as technologies, missions, and training needs change. The T-7A was named Red Hawk to honor the famous Red Tails of the 332nd Fighter Group in World War II—the Tuskegee Airmen fighter pilots, who painted their P-51 Mustangs with the distinctive red tail.

"…the T-7A Red Hawk is a game changer, providing advanced mission systems, a glass touchscreen cockpit, stadium seating, and embedded training capability."
– Col. Kirt Cassell, U.S. Air Force T-7A Program Manager

Copyright Boeing

Figure 1-39. The Army, Navy, Air Force, and Marines provide extensive and demanding pilot training in some of the most technologically advanced aircraft in the world.

THE COAST GUARD

Courtesy of U.S. Coast Guard

Figure 1-40. Flying Hercules HC-130s, the Coast Guard's International Ice Patrol provides a service to monitor the extent of iceberg danger in the North Atlantic.

The U.S. Coast Guard conducts extensive flight operations as the principal federal agency with maritime authority for the United States. Part of the Department of Homeland Security during peacetime, the Coast Guard operates under the Department of Defense during wartime. The four primary missions of the Coast Guard are maritime law enforcement, maritime safety, marine environmental protection, and national security. [Figure 1-40]

SECTION B ■ **Aviation Opportunities**

MORE AVIATION CAREERS

Your interest in flight might lead to a career in the aviation industry that does not require you to pilot an aircraft. For example, airlines employ flight dispatchers, flight attendants, ticket agents, and ramp service personnel. Airport managers, linepersons, and fixed base operators are just some of the jobs that can be explored at an airport. Aircraft manufacturing provides jobs for scientists, engineers and technicians who specialize in aircraft design and construction. You might be interested in pursuing a career as an aircraft maintenance or avionics technician. In addition, becoming an air traffic controller, meteorologist, or Flight Service specialist might appeal to you. [Figure 1-41]

Figure 1-41. These are just a few of the occupations that you can pursue in the field of aviation, and obtaining a pilot certificate greatly enhances your job qualifications.

AVIATION ORGANIZATIONS

One way to become involved as a pilot is by joining an aviation organization or club. Numerous associations furnish pilots with information, sponsor flying activities, and promote safety. You might want to ask other pilots in your area about local flying clubs. Many national and international groups have local chapters. Several examples of unique aviation organizations are described here.

AIRCRAFT OWNERS AND PILOTS ASSOCIATION

Aircraft Owners and Pilots Association (AOPA) is a nonprofit organization dedicated to general aviation. AOPA provides a wide variety of benefits to its members including pilot information, legal services, and loan programs, as well as pilot and aircraft insurance. In addition, the AOPA Air Safety Institute offers services ranging from online courses, safety quizzes and videos, webinars, and podcasts to in-person seminars, accident case studies, and real pilot stories. [Figure 1-42]

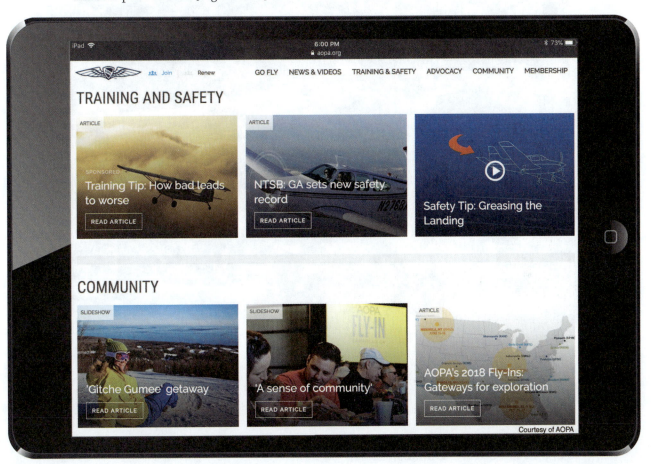

Figure 1-42. On the AOPA website, you find out about pilot services, enroll in training, obtain safety materials, and explore information about AOPA community activities.

SECTION B ■ **Aviation Opportunities**

SECTION B ■ **Aviation Opportunities**

EXPERIMENTAL AIRCRAFT ASSOCIATION

The **Experimental Aircraft Association (EAA),** which was founded in 1953 by a dozen aircraft homebuilders, has grown to be one of the most significant forces in general aviation. In addition to homebuilt aircraft, there are EAA divisions for vintage aircraft, warbirds, and aerobatics. The EAA is a leader in aviation education, and provides inspiration for innovative ideas in aircraft design, construction, and flight technique. One of the lasting symbols of the magnificence of flight is the airshow and the EAA sponsors one of the world's greatest—AirVenture Oshkosh. [Figure 1-43]

Courtesy of the EAA and Mike Husar

Figure 1-43. Each year, hundreds of thousands of aviation enthusiasts converge on the town of Oshkosh, Wisconsin to experience the best aviation has to offer.

DISCOVERY

EAA AirVenture Oshkosh

Burt Rutan designed the *Voyager* aircraft, several popular homebuilt aircraft, and the first private manned spacecraft, *SpaceShipOne*. As quoted in the EAA publication *Sport Aviation*, Rutan expressed his thoughts on aviation, Oshkosh, and the EAA.

I hope in the not too distant future ... that my Oshkosh trip will be about a half an hour, and half of that time will be spent outside the atmosphere, just coasting along ... getting ready for re-entry over Oshkosh. That sounds a little crazy, but it's not. It's much sooner than you think. We're at a period when the barnstormer, entrepreneur and businessman are gonna get into space... So I'm looking forward to it and I can see an EAA moving forever upward doing more and more exciting things ... we are nowhere near our limits.

THE NINETY-NINES, INC.

The Ninety-Nines, Inc. is an international organization of women pilots with members from over 40 countries. The mission of The Ninety-Nines is to promote fellowship through flight, to provide networking and scholarship opportunities for women, sponsor aviation education in the community, and to preserve the unique history of women in aviation. [Figure 1-44]

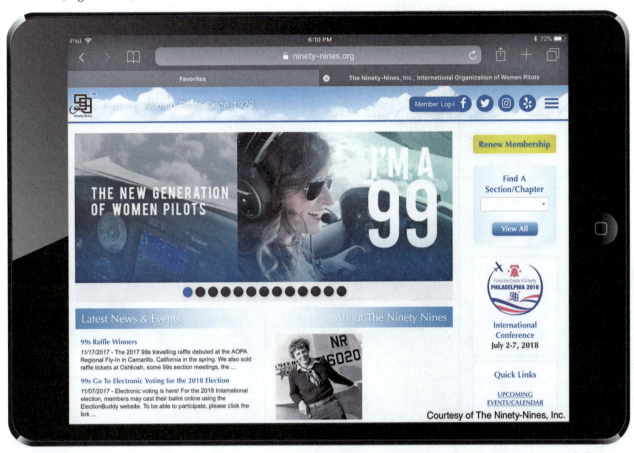

Figure 1-44. The Ninety-Nines, Inc. offers a wide variety of activities and opportunities for "women who love to fly!"

History of The Ninety-Nines

On November 2, 1929, twenty-six women gathered in a hangar at Curtiss Airport in New York to form an organization for women pilots. The club's goals included promoting fellowship, locating aviation jobs, and maintaining a central office with files on women in aviation. Some suggestions for a club name included The Climbing Vines, Noisy Birdwomen, Homing Pigeons, and Gadflies. Amelia Earhart and Jean Davis Hoyt proposed that the name reflect the number of charter members. The group was first called The Eighty-Sixes, then The Ninety-Sevens and finally The Ninety-Nines. In 1931, Amelia Earhart became the organization's first elected president.

Courtesy of The Ninety-Nines Inc. International Organization of Women Pilots

CIVIL AIR PATROL

The **Civil Air Patrol (CAP)** was founded on December 1st, 1941, to mobilize the nation's civilian aviation resources for national defense in World War II. In 1946, President Truman declared the CAP a benevolent, nonprofit organization with three primary missions: aerospace education, cadet programs, and emergency services. In 1948, the CAP became an all-volunteer auxiliary of the U.S. Air Force. Approximately 90 percent of all search and rescue missions issued by the Air Force Rescue Coordination Center are performed by CAP volunteers. The CAP also plays a vital role in national disaster relief, and CAP pilots fly drug interdiction reconnaissance missions on behalf of several government agencies.

The CAP Cadet Program provides young people the opportunity to develop leadership skills through their interest in flight, and many CAP cadets pursue aviation careers. For example, each year approximately 10 percent of the U.S. Air Force Academy appointees are former CAP cadets. In addition, CAP provides aerospace education materials and workshops to thousands of teachers at universities throughout the nation. [Figure 1-45]

Courtesy of Civil Air Patrol

Figure 1-45. The Civil Air Patrol operates one of the world's largest fleets of civil aircraft and one-third of all CAP members are FAA-certificated pilots.

SUMMARY CHECKLIST

✓ Frequent refresher training is essential to keep your flying skills sharp and to keep you informed on current pilot information.

✓ Receiving proper training is essential to fly in the mountains because you must make special considerations for weather, airport operations, course selection, and aircraft performance.

✓ Aerobatics instruction increases your proficiency as a pilot. Although there are no specific flight hour training requirements to operate an aircraft in aerobatic flight, the FARs do place certain restrictions on aerobatic maneuvers.

✓ The FARs require that you receive training and a logbook endorsement stating competency before you can operate as pilot in command of a high-performance, complex, or tailwheel airplane.

✓ If you accomplish your private pilot training in an airplane with analog instruments with minimal navigation equipment, you should obtain additional training to operate a technically advanced airplane (TAA).

✓ A TAA has: a primary flight display (PFD) with electronic instruments , multifunction flight display (MFD with a GPS moving map), and a two-axis autopilot.

✓ To obtain an instrument rating, you are required to have a least 40 hours of instrument flight time as well as pass a knowledge test and practical test.

✓ No specific ground or flight instruction hours are required for the addition of a multi-engine rating to your certificate, but you will have to pass a practical test.

✓ No minimum number of flight hours are required to add a single-engine sea rating or multi-engine sea rating to your private pilot certificate, but you must pass a practical test.

✓ To pilot aircraft such as helicopters, gliders, or hot air balloons, you will need to obtain an appropriate category and class rating.

✓ To apply for a commercial pilot certificate, you must accumulate a total of 190 to 250 hours of flight time (depending on the type of pilot school that you attend) that typically include a minimum of 100 hours of pilot-in-command time and 50 hours of cross-country time.

✓ No specific number of flight hours are required for CFI training, but you must pass two knowledge exams and a practical test.

✓ The airline transport pilot (ATP) certificate is required for pilots flying passengers or cargo under FAR Part 121 and for some FAR Part 135 operations.

✓ To obtain an ATP certificate, you must have a total of 1,500 hours of flight time including 250 hours of pilot-in-command time, 500 hours of cross-country time, 100 hours of night flight, and 75 hours of instrument experience.

✓ The FAA requires additional training prior to taking your ATP knowledge test in the form of an authorized Airline Transport Pilot Certification Training Program (ATP CTP).

✓ If you are a pilot with less than 1,500 hours of flight time and a graduate of a specific aviation degree program authorized by the FAA, you may obtain a restricted privileges ATP certificate (R-ATP). This certificate allows you to serve as a co-pilot until you obtain the necessary 1,500 hours.

✓ Typically, the minimum pilot qualifications to fly as a corporate pilot include a commercial pilot certificate with an instrument rating and a multi-engine rating.

✓ To become employed as an aerial applicator you must hold a commercial pilot certificate and receive additional training in agricultural aircraft operations.

✓ The Army, Navy, Air Force, and Marines provide extensive and demanding pilot training in some of the most technologically advanced aircraft in the world.

✓ Aviation organizations such as AOPA, the EAA, The Ninety-Nines Inc., and the CAP sponsor flying activities, promote safety, and furnish pilot information.

SECTION B ■ **Aviation Opportunities**

KEY TERMS

Refresher Training

Mountain Flying

Aerobatic Flight Training

Aircraft Checkout

High-Performance Airplane

Complex Airplane

Technically Advanced Airplane (TAA)

Tailwheel Airplane

Conventional Landing Gear Airplane

Homebuilt Aircraft

Instrument Rating

Multi-Engine Rating

Single-Engine Sea Rating

Multi-Engine Sea Rating

Helicopter

Glider Rating

Hot Air Balloon

Commercial Pilot Certificate

Airline Transport Pilot (ATP) Certificate

Airline Transport Pilot Certification Training Program (ATP CTP)

Sport Pilot Certificate

Regional Airline

Major Airline

Corporate Flying

Aerial Application

Aircraft Owners and Pilots Association (AOPA)

Experimental Aircraft Association (EAA)

The Ninety-Nines, Inc.

Civil Air Patrol (CAP)

QUESTIONS

1. Which of these requires a logbook endorsement?
   A. Aerobatic flight
   B. Flying in the mountains
   C. Operating as pilot in command of a tailwheel airplane

2. How do the FARs define a high-performance airplane?

3. How do the FARs define a complex airplane?

4. Which rating requires both a computerized knowledge test and practical test?
   A. Seaplane
   B. Instrument
   C. Multi-engine

5. What rating allows you to operate in the clouds without a reference to the ground or horizon?

6. True/False. The glider rating does not require a knowledge test, provided you hold a powered category rating.

7. Which is an example of a category and class rating?
   A. Multi-engine land
   B. Rotorcraft helicopter
   C. Powered-lift gyroplane

8. True/False. You must have an ATP certificate to operate as pilot-in-command for a major airline under Part 121.

SECTION C
Introduction to Human Factors

There is more to pilot training than acquiring technical knowledge and gaining proficiency in airplane control. Understanding how your mind and body function when you fly is as important as knowing the operation of your airplane's systems and equipment.

In the early years of aviation, the majority of accidents were the result of mechanical difficulties or severe weather conditions. Aviation's safety record steadily improved as technology progressed. Today, aircraft have sophisticated equipment and systems, pilots have access to detailed weather information, an extensive air traffic control system manages flights, and pilot skills have increased through advanced training methods. Yet accidents still occur. Why? Despite all the changes in aviation, one factor has remained the same; the human factor. It is estimated that approximately 75% of all aviation accidents are human-factors related.

The goal of human factors training is to increase aviation safety by optimizing pilot performance and reducing human error. Instruction in human factors principles focuses on two primary subject areas: single-pilot resource management and aviation physiology.

- **Single-pilot resource management (SRM)** is the art and science of managing all available resources (both onboard the airplane and from outside sources) prior and during flight to ensure the successful completion of a flight.

- Aviation physiology is the study of the performance and limitations of the body in the flight environment. [Figure 1-46]

Figure 1-46. You must master SRM skills and understand how your body functions in flight to be an effective pilot in command.

This section defines SRM concepts and introduces some physiological factors that you should be aware of as you begin flight training. The Human Element Insets located throughout this textbook help you to correlate human factors concepts to specific pilot operations and expand upon the fundamental SRM principles introduced in this section. Chapter 10 provides a more extensive examination of SRM and aviation physiology as they apply to private pilot operations.

SINGLE-PILOT RESOURCE MANAGEMENT

Human factors-related accidents motivated the airline industry to implement **crew resource management (CRM)** training for flight crews. The training helped crews recognize hazards and provided tools for them to eliminate the hazard or minimize its impact. CRM training provided the foundation for SRM training. Applying SRM means using hardware, information, and human resources, such as dispatchers, weather briefers, maintenance personnel, and air traffic controllers, to gather information, analyze your situation, and make effective decisions about the current and future status of your flight. SRM includes these six concepts:

- Aeronautical decision making
- Risk management
- Task management
- Situational awareness
- Controlled flight into terrain awareness
- Automation management

AERONAUTICAL DECISION MAKING

Aeronautical decision making (ADM) is a systematic approach for aircraft pilots to consistently determine the best course of action in response to specific circumstances. Your ability to make effective decisions as a pilot depends on a number of factors. Some factors, such as the time available to make a decision, might be beyond your control. However, you can learn to recognize the factors that you can manage, and learn skills to improve your decision-making ability and judgment.

ADM PROCESS

Some situations, such as emergencies, require you to respond immediately using established procedures, with little time for detailed analysis. This reflexive type of decision making—anchored in training and experience—is often referred to as naturalistic or automatic decision making. However, typically during a flight, you have time to recognize changes that occur, gather information, examine options, and assess risk before reaching a decision. Then, after implementing a course of action, you determine how your decision could affect other phases of the flight. To make an analytical decision, you use the **ADM process**. The ADM process consists of the steps that you use to make effective decisions as pilot in command. [Figure 1-47]

The ADM Process
Recognize a change. Identify changes in your situation and be alert for sudden changes that can lead to abnormal and emergency situations.
Define the problem. Use experience and resources to determine the exact nature of the problem.
Choose a course of action. Consider the expected outcome of each possible action and assess the risk involved with each before making a decision.
Implement your decision. Take the necessary action to solve the problem.
Evaluate the outcome. Think ahead and keep track of the situation to ensure that your actions are producing the desired outcome.

Figure 1-47. The ADM process includes defining the problem and monitoring the outcome after you implement a decision.

Although the basic steps are the same, A variety of mnemonics are used by pilots to remember the steps in the decision-making process—you might hear of FOR-DEC, NMATE, DODAR, SAFE, or the FAA's DECIDE model:

1. **D**etect the fact that a change has occurred.
2. **E**stimate the need to counter or react to the change.
3. **C**hoose a desirable outcome for the success of the flight.
4. **I**dentify actions which could successfully control the change.
5. **D**o the necessary action to adapt to the change.
6. **E**valuate the effect of the action.

SELF ASSESSMENT

As pilot in command, you are the ultimate decision maker and your choices determine the outcome of the flight. Just as you must thoroughly check your aircraft to determine if it is airworthy, you must evaluate your own fitness for flight. Your general health, level of stress or fatigue, attitude, knowledge, skill level, and recency of experience are several factors that affect your performance as pilot in command. Establish personal limitations for flight and create a checklist to help you determine if you are prepared for a particular flight. For example, based on your experience, determine your own weather minimums and set limitations for the maximum amount of crosswind that you are comfortable with. After you have reviewed your personal limitations, you can use the **I'M SAFE checklist** to further evaluate your fitness for flight. [Figure 1-48]

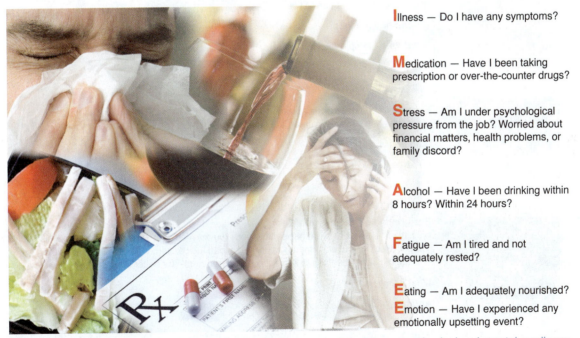

Illness — Do I have any symptoms?

Medication — Have I been taking prescription or over-the-counter drugs?

Stress — Am I under psychological pressure from the job? Worried about financial matters, health problems, or family discord?

Alcohol — Have I been drinking within 8 hours? Within 24 hours?

Fatigue — Am I tired and not adequately rested?

Eating — Am I adequately nourished?
Emotion — Have I experienced any emotionally upsetting event?

Figure 1-48. Using the I'M SAFE checklist is an effective way to determine your physical and mental readiness for flight.

SECTION C ■ Introduction to Human Factors

SECTION C ■ Introduction to Human Factors

HAZARDOUS ATTITUDES

Whether you are fit to fly depends on more than your experience and physical condition. Your attitude also affects the quality of your decisions. Studies have identified five **hazardous attitudes** among pilots that can interfere with a pilot's ability to make effective decisions. [Figure 1-49]

Anti-authority — You display this attitude if you resent having someone tell you what to do, or you regard rules and procedures as unnecessary.

Don't tell me.

Antidote — *Follow the rules. They are usually right.*

Impulsivity — If you feel the need to act immediately and do the first thing that comes to mind without considering the best solution to a problem, then you are exhibiting impulsivity.

Do it quickly.

Antidote — *Not so fast. Think first.*

Invulnerability — You are more likely to take chances and increase risk if you think accidents will not happen to you.

It won't happen to me.

Antidote — *It could happen to me.*

Macho — If you have this attitude, you might take risks trying to prove that you are better than anyone else. Women are just as likely to have this characteristic as men.

I can do it.

Antidote — *Taking chances is foolish.*

Resignation — You are experiencing resignation if you feel that no matter what you do it will have little effect on what happens to you. You may feel that when things go well, it is just good luck and when things go poorly, it is bad luck or someone else is responsible. This feeling can cause you to leave the action to others—for better or worse.

What's the use?

Antidote — *I'm not helpless. I can make a difference.*

Figure 1-49. As pilot in command, you must examine your decisions carefully to ensure that your choices have not been influenced by hazardous attitudes.

SELF-CRITIQUES

In addition to assessing your condition prior to and during flight, you should perform **self-critiques** after each flight to evaluate your performance, determine the skills that need improvement, and create a plan for increasing your proficiency. If you feel you need to improve your skills, review aircraft manuals, practice procedures using an aviation training device, or obtain refresher training. During flight lessons, both you and your instructor should evaluate your performance and resolve any differences in your assessments before creating a plan for improvement. This is referred to as **learner-centered grading**.

RISK MANAGEMENT

Risk management is critical to making effective decisions. During each flight, you are required to make decisions that involve four fundamental risk elements: the pilot, the aircraft, the environment, and the type of operation. Pilots use a variety of tools to identify, assess, and mitigate risks associated with the risk elements. Two frequently-used tools are **PAVE** and the **5Ps**. [Figure 1-50]

PAVE

Pilot – Evaluate your training, experience, and fitness.
Aircraft – Determine airworthiness, performance, and proper configuration. Check avionics airworthiness.
en**V**ironment – Assess items such as airport conditions, terrain and airspace, and weather.
External Pressures – Evaluate the purpose of the flight and how critical it is to maintain the schedule.

5Ps

Pilot – Evaluate your training, experience, and fitness.
Passengers – Consider your passengers' experience, flexibility, and fitness.
Plane – Determine airworthiness, performance, and proper configuration.
Programming – Check avionics airworthiness, operation, and configuration.
Plan – Assess items such as airport conditions, terrain and airspace, and weather. Evaluate the mission—the purpose of the flight—and how critical it is to maintain the schedule.

Figure 1-50. Both PAVE and the 5Ps remind you of the risk factors that you must manage when planning and implementing flights.

USING THE 5PS DURING FLIGHT PLANNING

Whether you use PAVE or the 5Ps, a risk management tool helps you make an effective Go/No-Go decision during flight planning. For example, 5P checklists provide guidelines on the risk factors to consider as you prepare for a flight. Create your own or make copies of the checklists at the end of this section to use for your flights. [Figure 1-51]

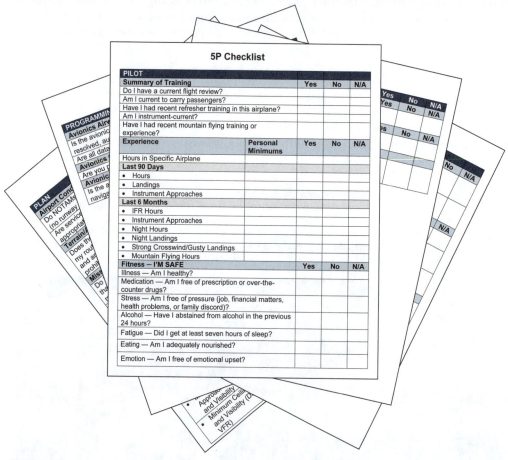

Figure 1-51. Use the 5P checklists to identify and mitigate risks prior to flight. The I'M SAFE checklist is part of the Pilot checklist.

USING THE 5PS IN FLIGHT

Managing risk does not end with a Go decision; you must continue to assess risk to make effective decisions during the flight. The risk management process continues as you evaluate the situation using the 5P check at decision points that correspond to the phases of flight. [Figure 1-52]

At each decision point, consider each of the 5Ps and ask these questions:
- What is the situation?
- What has changed since my Go decision?
- Is the risk associated with a change acceptable?
- What can I do to mitigate risk?

| Before Takeoff | Climb and Initial Cruise | Enroute Cruise | Descent | Before Approach and Landing |

Figure 1-52. You must reevaluate each of the 5Ps during the flight to recognize any changes that might increase your risk.

TASK MANAGEMENT

Task management involves planning and prioritizing tasks to avoid work overload, identifying and using resources to accomplish tasks, and managing distractions. When you are effectively managing tasks, you avoid fixating on one task at the expense of others and maintain positive control of the airplane.

PLANNING AND PRIORITIZING

When flying an airplane, your tasks are not evenly distributed over time. By planning ahead and prioritizing tasks, you can prepare for high workload periods during times of low workload. As you gain experience, you realize which tasks you can accomplish ahead of time, and which tasks you need to leave until the moment. Tasks such as organizing charts in the order of use, setting radio frequencies, and planning a descent to an airport help you prepare for what comes next.

RESOURCE USE

Because tools and sources of information are not always readily apparent, you must learn to recognize all the resources available to you and use them effectively. A wide variety of resources both inside and outside the airplane can help you manage tasks and make effective decisions. [Figure 1-53]

External Resources
- Air Traffic Controllers
- Maintenance Technicians
- Flight Service Briefers

Internal Resources
- Your Own Knowledge and Skills
- Your Instructor and Other Pilots
- Passengers
- Aircraft Equipment
- Aeronautical Charts
- Pilot's Operating Handbook
- Checklists

Figure 1-53. You use these resources continually during routine flights and they become even more critical in emergency situations.

SECTION C ■ Introduction to Human Factors

CHECKLISTS

Checklists are valuable resources that help you manage distractions while you perform procedures. You typically use one of two methods for following checklists. With a **do-list**, you read the checklist item and the associated action and then perform the action. Use a do-list when you have time and completing each step in the correct order is critical. A **flow pattern** guides you through the flight deck in a logical order as you perform each step without the written checklist. After completing the flow pattern, refer to the checklist and verify that you have accomplished each item. Use a flow pattern when the checklist item sequence is not critical. Emergency checklists are unique because many have items that you must perform immediately from memory before referring to the checklist. [Figure 1-54]

4A.3.3 STARTING ENGINE

(a) Cold engine

1. Strobe light (ACL) . . . . ON
2. Electrical fuel pump . . . . ON, note pump noise
   (≈ functional check of pump)
3. Throttle . . . . 3 cm (1.2 in) forward from
   IDLE (measured from rear of
   slot)
4. Mixture control lever . . . . RICH for 3 - 5 sec, then
   LEAN
5. Throttle . . . . 1 cm (0.4 in) forward from
   IDLE (measured from rear of
   slot)

WARNING

Before starting the

Do-Lists
Use do-lists for abnormal procedures, such as addressing an electrical malfunction.

Flow Patterns
Use flow patterns to perform normal procedures, such as configuring the airplane and the avionics for specific phases of flight.

Emergency Checklists
Perform critical tasks from memory and then refer to the checklist to manage specific emergencies, such as an engine failure.

Figure 1-54. Use do-lists and flow patterns based on the procedure.

SITUATIONAL AWARENESS

Situational awareness is the accurate perception of all the operational and environmental factors that affect flight safety before, during, and after the flight. At any period of time, you should be able to accurately assess the current and future status of the flight. This includes the status of operational conditions, such as airplane systems, fuel, autopilot, and passengers, as well as the status of environmental conditions, such as your relationship to terrain, traffic, weather, and airspace. Using SRM, including risk management tools such as the 5Ps, task management, and available resources enables you to maintain situational awareness. Resources, such as navigation, traffic, terrain, and weather displays are particularly valuable for maintaining situational awareness if you understand how to use them properly. [Figure 1-55]

Figure 1-55. You are maintaining situational awareness when you have a solid mental picture of the condition of the pilot, passengers, plane, programming, and plan.

BRIEFINGS

Briefings are an effective tool to help you maintain situational awareness by preparing you for critical phases of flight. Standard briefings include a passenger briefing, a takeoff briefing, and a before-landing briefing. Regulations require that you explain to your passengers how to fasten and unfasten the safety belts and shoulder harnesses and when the safety belts must be fastened. The FAA also recommends that you cover certain safety considerations with passengers before flight. You can remember the elements of a passenger briefing by using the acronym SAFETY. [Figure 1-56]

Safety Belts
- How to fasten and unfasten the safety belts and shoulder harnesses.
- When safety belts must be fastened—prior to movement on the surface, takeoff, and landing

Air Vents
- Location and operation
- Operation of heating or air conditioning controls

Fire Extinguisher
Location and operation

Egress and Emergency
- Operation of doors and windows
- Location of the survival kit
- Use of onboard emergency equipment

Traffic and Talking
- Pointing out traffic
- Use of headsets
- Avoiding unnecessary conversation during critical phases of flight

Your Questions
Solicit questions from your passengers.

Figure 1-56. Perform the passenger briefing prior to starting the engine.

SECTION C ■ Introduction to Human Factors

SECTION C ■ Introduction to Human Factors

The takeoff briefing enables you to mentally rehearse what is about to happen during and after takeoff, and it prepares any other crewmembers or passengers for takeoff. You normally perform the before-landing briefing 15 to 20 miles from the destination airport, after you've obtained airport information. [Figure 1-57]

Takeoff Briefing
- Wind direction and velocity
- Runway length
- Takeoff distance
- Initial heading
- Initial altitude
- Takeoff and climb speeds
- Departure procedures
- Emergency plan in case of an engine failure after takeoff

Before-Landing Briefing
- Airport information and weather conditions
- Active runway
- Terrain and obstacles
- Airport elevation and pattern altitude
- Traffic pattern entry

Figure 1-57. The takeoff briefing and before-landing briefing help you maintain situational awareness of the airport environment.

OBSTACLES TO MAINTAINING SITUATIONAL AWARENESS

Fatigue, stress, and work overload can cause you to fixate on one aspect of the flight and omit others from your attention. A contributing factor in many accidents is a distraction that diverts the pilot's attention from monitoring the instruments or scanning outside the aircraft. A minor problem, such as a gauge that is not reading correctly, has the potential to become a major problem if you divert your attention to the perceived problem and neglect to properly control the airplane.

Complacency presents another obstacle to maintaining situational awareness. When activities become routine, you can have a tendency to relax and put less effort into your performance. Flight deck automation can lead to complacency—you could assume that the autopilot is doing what you expect, and neglect to cross check the instruments or the airplane's position.

SITUATIONAL AWARENESS DURING GROUND OPERATIONS

In addition to keeping track of your status while in flight, you must maintain situational awareness during ground operations. As you gain pilot experience, you will learn techniques to correctly follow taxi instructions, to know your position on the airport in relation to runways and other aircraft, and to minimize your workload.

CONTROLLED FLIGHT INTO TERRAIN AWARENESS

Controlled flight into terrain (CFIT) occurs when an aircraft is flown into terrain or water with no prior awareness on the part of the crew that the crash is imminent. Air carriers and professional flight departments have significantly reduced the number of CFIT accidents in the United States by implementing extensive training programs and installing specialized aircraft equipment. CFIT is more prevalent in general aviation because pilots do not have the same training and equipment. CFIT normally results from a combination of factors including weather, unfamiliar environment, nonstandard procedures, breakdown or loss of communication, loss of situational awareness, lack of perception of hazards, and lack of sound risk management techniques. Throughout your flight training, you will learn strategies for preventing CFIT during each phase of flight. [Figure 1-58]

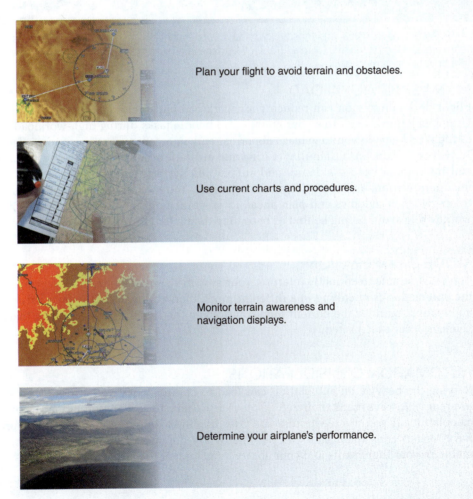

Plan your flight to avoid terrain and obstacles.

Use current charts and procedures.

Monitor terrain awareness and navigation displays.

Determine your airplane's performance.

Figure 1-58. To prevent a CFIT accident, you must maintain positional awareness.

SECTION C ■ **Introduction to Human Factors**

AUTOMATION MANAGEMENT

Flight deck automation has the potential to increase or decrease the flight safety, depending on how well you use the equipment. Although automation can consist of a simple autopilot that maintains heading and altitude combined with traditional analog instruments and navigation equipment, the concept of **automation management** typically applies to an airplane with an advanced avionics system that includes digital displays, GPS equipment, a moving map, and an integrated autopilot. [Figure 1-59]

Figure 1-59. Advanced avionics airplanes typically have a primary flight display (PFD) with digital instrumentation and a multifunction display (MFD) that depicts a moving map, terrain, traffic, weather, and other flight environment information.

MANAGING WORKLOAD

Flight deck automation can reduce your workload and increase situational awareness. Use of an autopilot can free your attention to handle tasks during high-workload phases of flight and enable you to manage abnormal and emergency situations more effectively. However, if you are unfamiliar with your airplane's equipment, trying to program and interpret advanced avionics and automation systems might be distracting, lead to misinterpretation, and cause programming errors. You must thoroughly understand how to operate your avionics and plan ahead to program equipment during periods of lower workload to avoid falling behind or becoming distracted during high-workload periods.

MODE OF OPERATION

You must be able to correctly interpret your system's annunciations and recognize when the automation is operating in a different mode than you expect. To effectively manage automation, monitor the current mode, anticipate the next mode, and verify that mode changes occur as expected.

AUTOMATION CONSIDERATIONS

Relying too heavily on automation can lead to complacency and a loss of situational awareness. Always monitor aircraft displays, use charts to verify information, and confirm calculations if you use electronic databases for flight planning. Equipment failure can have serious consequences if you become overly dependent on automation. You must maintain your flight skills and your ability to maneuver the airplane manually.

You also need to recognize when automation is increasing your workload and switch to a simpler mode or turn equipment off. For example, if trying to program the GPS equipment or engage the autopilot starts to overwhelm you, flying the procedure manually might be safer. To help you manage automation and other avionics equipment, you can consider using one of three **equipment operating levels** during flight operations. [Figure 1-60]

LEVEL 1

Level 1 — Control the airplane manually and use the minimum equipment necessary to perform procedures.

LEVEL 2

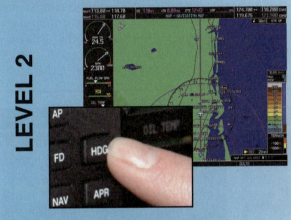

Level 2 — Use the autopilot to help manage workload, but manually control the airplane at times. In addition, use the flight environment avionics information to enhance situational awareness and to make effective decisions.

LEVEL 3

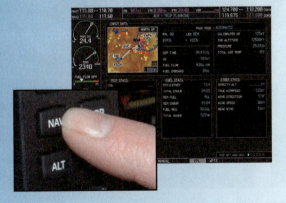

Level 3 — Control the airplane primarily by the autopilot. Use a wide variety of avionics tools, including navigation and flight planning information to manage workload and maintain an increased level of situational awareness.

Figure 1-60. As you gain experience with your airplane's specific equipment, you will learn how to determine which equipment operating level to use in specific situations.

SECTION C ■ Introduction to Human Factors

THREAT AND ERROR MANAGEMENT

Threat and error management (TEM) has its origins in aviation as risk management. TEM training teaches you to detect and respond to threats and errors so that a situation does not progress into an **undesired aircraft state (UAS)**—a pilot-induced airplane position/speed deviation, misapplied flight controls, or an incorrect airplane configuration. TEM strategies enable you to prevent or to mitigate a potentially unsafe outcome.

THREATS

A **threat** is an expected or unexpected risk, or hazard. Threats:

- Occur outside the influence of, and are not controlled by you as the pilot.

- Increase the operational complexity of a flight.

- Can appear suddenly and limit the time available for you to analyze.

- Must be effectively managed to contain risk within acceptable levels.

Threats can involve a number of hazards. A **hazard** is a condition that can cause, or contribute to, an aircraft incident or accident. Examples of hazards include adverse weather, challenging airports surrounded by high mountains, or congested airspace. You can anticipate some hazards, such as forecast adverse weather, while other hazards are unforeseen, such as a sudden in-flight mechanical malfunction. [Figure 1-61]

Figure 1-61. Examples of threats include, but are not limited to, issues with the airplane, airport, weather, and flight environment.

ERRORS

Errors are caused by pilot actions or inactions, and lead to deviations from expected outcomes. Errors absorb your attention and increase your workload. Errors can cause confusion, increase risk, and reduce safety margins. Errors can be spontaneous or can be linked to threats. If you fail to manage or mismanage an error, you increase the probability of an adverse outcome, including a UAS. Sometimes, a sequence of errors create a chain that ultimately leads to an incident or accident. Errors can be skill-based, decision-based, perceptual, or involve SRM/CRM. [Figure 1-62]

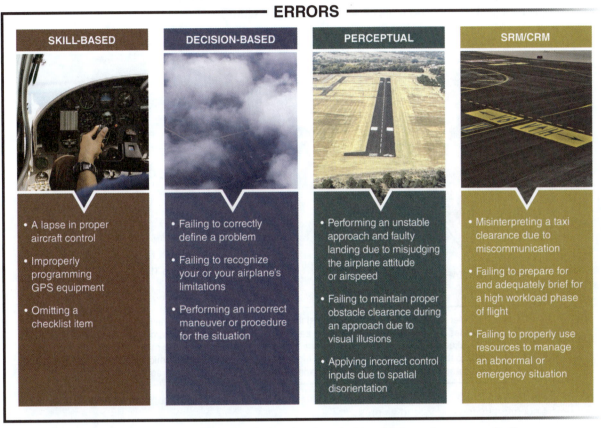

ERRORS

SKILL-BASED	DECISION-BASED	PERCEPTUAL	SRM/CRM
• A lapse in proper aircraft control • Improperly programming GPS equipment • Omitting a checklist item	• Failing to correctly define a problem • Failing to recognize your or your airplane's limitations • Performing an incorrect maneuver or procedure for the situation	• Performing an unstable approach and faulty landing due to misjudging the airplane attitude or airspeed • Failing to maintain proper obstacle clearance during an approach due to visual illusions • Applying incorrect control inputs due to spatial disorientation	• Misinterpreting a taxi clearance due to miscommunication • Failing to prepare for and adequately brief for a high workload phase of flight • Failing to properly use resources to manage an abnormal or emergency situation

Figure 1-62. Examples of pilot errors range from faulty aircraft control to misinterpreting communication.

When a threat appears, or an error occurs, you must make a decision. You must guard against a hazardous reaction that will impair your ability to mitigate a threat or error, such as:

- Failing to carefully consider choices and reacting quickly and impulsively.
- Failing to respond in a timely manner.
- Reacting with resignation.

Becoming skilled at single-pilot resource management by using tools to effectively assess and mitigate risks, manage workload, maintain situational awareness, and make effective decisions will help you perform TEM. In addition, briefing potential threats and mitigation strategies prior to a flight will make you better prepared to handle situations during flight.

AVIATION PHYSIOLOGY

An essential component of human factors training is aviation physiology, which is the study of the performance and limitations of the body in the flight environment. Most healthy people do not experience any physical difficulties as a result of flying. However, there are some physiological factors that you should be aware of as you begin flight training. An expanded description of how your body functions in flight is contained in Chapter 10.

SECTION C ■ Introduction to Human Factors

PRESSURE EFFECTS

As the airplane climbs and descends, variations in atmospheric pressure affect many parts of your body. As outside air pressure changes, air trapped in the ears, teeth, sinus cavities, and gastrointestinal tract can cause pain and discomfort.

EAR AND SINUS BLOCK

Ear pain is normally the result of a difference between air pressure in the middle ear and outside air pressure. When the air pressure in the middle ear is equal to the pressure in the ear canal, there is no blocked feeling or pain. [Figure 1-63]

The ear is composed of three sections—the outer ear, the middle ear, and the inner ear.

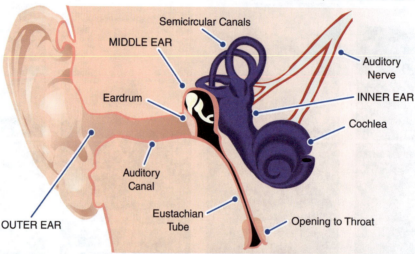

The outer ear includes the auditory canal that extends to the eardrum. The eardrum separates the outer ear from the middle ear, which is located within the temporal bone of the skull. A short slit-like tube called the Eustachian tube connects the middle ear cavity to the back wall of the throat. The inner ear, which contains the semicircular canals and cochlea, is used for both hearing and maintaining a sense of equilibrium.

Figure 1-63. If the small passageways in the ear become blocked, passengers can experience discomfort during changes in air pressure.

As you ascend during flight, the pressure in the auditory canal decreases and usually the higher pressure in the middle ear will open the eustachian tube to equalize the pressure. If the tube does not open, you may feel a fullness in the ear, notice a slight hearing loss and experience discomfort because the eardrum is distended and cannot vibrate as freely. During a descent, the outside air pressure in the auditory canal will become higher than the pressure in the middle ear. This situation is harder to correct, since the eustachian opens more easily to let positive pressure out than it does to allow air back into the middle ear. Slow descent rates can help prevent or reduce the severity of ear problems and the eustachian tube can sometimes be opened by yawning, swallowing, or chewing.

In addition, pressure can be equalized by holding the nose and mouth shut and forcibly exhaling. This procedure, which is called the Valsalva maneuver, forces air up the eustachian tube into the middle ear. If you have a cold, an ear infection, or sore throat, you may not be able to equalize the pressure in your ears. A flight in this condition can be extremely painful, as well as damaging to your eardrums. If you are experiencing minor congestion, nose drops or nasal sprays may reduce the chance of a painful ear blockage. Before you use any medication, though, check with an aviation medical examiner to ensure that it will not affect your ability to fly.

If you have an inflammation of the sinuses from an allergy or a cold, you may experience pain from trapped air in your sinus cavities. As with the ears, slow descent rates, use of nasal sprays, or employing the Valsalva maneuver can help equalize the pressure.

TOOTHACHE

Expansion of trapped air in the cavities caused by imperfect fillings, damaged root canals, and dental abscesses can produce pain at altitude. If you experience a toothache while flying, a descent to a lower altitude may bring relief, but it is recommended that you visit a dentist for examination and treatment.

GASTROINTESTINAL PAIN

At any given time, your gastrointestinal tract contains about one quart of gas. Most of this is swallowed air, and the rest is gas caused by the digestion process. As altitude increases, this gas expands and can cause abdominal pain. You are less likely to have this problem if you maintain good eating habits and avoid foods that produce excess gas prior to flying.

SCUBA DIVING

The reduction of atmospheric pressure that accompanies flying can produce physical problems for scuba divers. **Decompression sickness**, more commonly referred to as "the bends," occurs when nitrogen absorbed during a scuba dive comes out of solution and forms bubbles in the tissues and bloodstream, much like uncapping a bottle of soda. This condition is very serious and can produce extreme pain, paralysis, and, if severe enough, death.

Even though you may finish a dive well within the no-decompression limits, the reduced atmospheric pressure of flying can cause the onset of decompression sickness. If you or a passenger plan to fly after scuba diving, it is important that enough time is allowed for the body to rid itself of excess nitrogen absorbed during diving. The recommended waiting time before ascending to 8,000 feet MSL is at least 12 hours after a dive that has not required a controlled ascent (nondecompression stop diving), and at least 24 hours after a dive that has required a controlled ascent (decompression stop diving). The waiting time before going to flight altitudes above 8,000 feet MSL should be at least 24 hours after any scuba dive.

MOTION SICKNESS

Motion sickness, or airsickness, is caused by the brain receiving conflicting messages about the state of the body. You may experience motion sickness during initial flights, but it generally goes away within the first 10 lessons. Anxiety and stress, which you may feel as you begin flight training, can contribute to motion sickness. Symptoms of motion sickness include general discomfort, nausea, dizziness, paleness, sweating, and vomiting.

It is important to remember that experiencing airsickness is no reflection on your ability as a pilot. Let your flight instructor know if you are prone to motion sickness since there are techniques that can be used to overcome this problem. For example, you may want to avoid lessons in turbulent conditions until you are more comfortable in the airplane or start with shorter flights and graduate to longer instruction periods. If you experience symptoms of motion sickness during a lesson, you can alleviate some of the discomfort by opening fresh air vents or by focusing on objects outside the airplane. Although medication like Dramamine can prevent airsickness in passengers, it is not recommended while you are flying since it can cause drowsiness.

SECTION C ■ Introduction to Human Factors

STRESS

Stress can be defined as the body's response to physical and psychological demands placed upon it. Reactions of your body to stress include the release of chemical hormones (such as adrenaline) into the blood and the speeding of the metabolism to provide energy to the muscles. In addition, blood sugar, heart rate, respiration, blood pressure, and perspiration all increase. The term stressor is used to describe an element that causes you to experience stress.

Stressors include conditions associated with the environment, such as temperature and humidity extremes, noise, vibration, and lack of oxygen. Your physical condition can contribute to stress. For example, lack of physical fitness, fatigue, sleep loss, missed meals (leading to low blood sugar levels), and illness are sources of stress. Psychological stressors include social or emotional factors, such as a death in the family, a divorce, a sick child, a demotion at work, or the mental workload of in-flight conditions.

A certain amount of stress is good for you because it keeps you alert and prevents complacency. However, stress effects are cumulative and if not coped with adequately, they eventually add up to an intolerable burden. Performance generally increases with the onset of stress, peaks, and then begins to fall off rapidly as stress levels exceed your ability to cope.

There are several techniques that can be applied to manage the accumulation of life stresses and help prevent stress overload. Including relaxation time in your schedule and maintaining a program of physical fitness can help reduce stress levels. Learning to manage your time more efficiently can help you avoid the heavy pressures imposed by getting behind schedule and not meeting deadlines. By taking an assessment of yourself, you can determine your capabilities and limitations, which will enable you to set realistic goals. Whenever possible, avoid stressful situations and encounters, and finally, be aware of specialized techniques, such as meditation, which can help you cope with stress.

FATIGUE

Fatigue is frequently associated with pilot error. Some of the effects of fatigue include degradation of attention and concentration, impaired coordination, and decreased ability to communicate. These factors can seriously influence your ability to make effective decisions. Physical fatigue can result from sleep loss, exercise, or physical work. Factors such as stress and prolonged performance of cognitive work can result in mental fatigue.

If you become fatigued on the flight deck, no amount of training or experience can overcome the detrimental effects. Getting adequate rest is the only way to prevent fatigue from occurring. You should avoid flying when you have not had a full night's rest, when you have been working excessive hours, or have had an especially exhausting or stressful day.

NOISE

Flight deck noise can contribute to fatigue, stress, and even airsickness. The understanding of speech also can be severely impeded by flight deck noise, which can impair communication between persons on board the aircraft and radio exchanges with air traffic controllers. Tests have shown that in certain airplanes under full takeoff power conditions, the intelligibility of a communication from an air traffic controller can sometimes drop from 100 percent to zero.

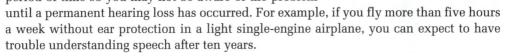

The long-term effects of flight deck noise can be serious. Although some hearing loss will be experienced by every person during an average lifetime, the problem may occur sooner for pilots who fly frequently without ear protection. Hearing loss generally occurs very slowly over an extended period of time so you may not be aware of the problem until a permanent hearing loss has occurred. For example, if you fly more than five hours a week without ear protection in a light single-engine airplane, you can expect to have trouble understanding speech after ten years.

Almost all of the problems associated with noise on the flight deck can be eliminated with the use of earplugs. Earplugs are devices that are inserted in, or pressed against, the external ear canal to reduce the effect of ambient sound on the auditory system. Wearing noise-attenuating headsets or earphones also can reduce the impact of aircraft noise, but you should verify that the specific device provides the necessary noise reduction. Both headsets and earplugs must fit snugly to work properly.

ALCOHOL, DRUGS, AND PERFORMANCE

Illness and disease also can affect the functioning and performance of your body, as can the drugs that are meant to fight these illnesses. There are two things you should consider before flying while using a drug. First, what is the condition you are treating, and second, what are the side effects of the drug used to treat the condition? Some conditions are serious enough to prohibit flying, even if the illness is being treated successfully with drugs. Always let your physician know you are a pilot and ask about the side effects of prescription medication. You should consult an aviation medical examiner about any medication, including over-the-counter drugs that you suspect will adversely affect your ability to pilot an aircraft.

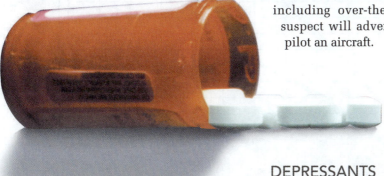

DEPRESSANTS

Depressants are drugs that reduce the body's functioning in many areas. These drugs lower blood pressure, reduce mental processing, and slow motor and reaction responses. There are several types of drugs that can cause a depressing effect on the body, including tranquilizers, motion sickness medication, some types of stomach medication, decongestants, and antihistamines. The most common depressant is alcohol.

SECTION C ■ **Introduction to Human Factors**

SECTION C ■ **Introduction to Human Factors**

Courtesy Rick Patterson

ALCOHOL

Ethyl alcohol is the most widely used and abused drug. Although some alcohol is used for medicinal purposes, the majority of it is consumed as a beverage. Alcohol requires no digestion and can be absorbed into the bloodstream unchanged from the stomach and small intestine. It then passes almost immediately through the liver. This produces a depressing effect on the nervous system and a dulling of the senses is experienced. The rate at which alcohol is absorbed into your body varies with the percentage of alcohol in the drink, the rate at which it is consumed, the amount and type of foods you have eaten, and the length of time you have been drinking.

Intoxication is determined by the amount of alcohol in the bloodstream. This is usually measured as a percentage by weight in the blood. [Figure 1-64] The FARs require that your blood alcohol level be less than .04 percent and that 8 hours pass between drinking alcohol and piloting an aircraft. If you have a blood alcohol level of .04 percent or greater after 8 hours, you cannot fly until your blood alcohol falls below that amount. Even though your blood alcohol may be well below .04 percent, you cannot fly sooner than 8 hours after drinking alcohol. Although the regulations are quite specific, it is a good idea to be more conservative than the FARs. Most pilots allow a minimum of 12 hours to pass after the last drink before flying; commercial airlines generally require their pilots to wait 24 hours.

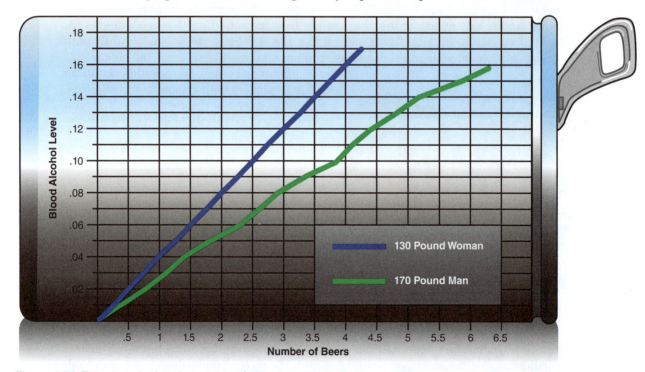

Figure 1-64. This chart depicts the number of 12 ounce, 5% alcohol beers consumed and the corresponding blood alcohol level for a 130 pound woman and a 170 pound man.

Hundreds of decisions, some of them time-critical, must be made during the course of a flight. The safe outcome of any flight depends on your ability to make the correct decisions and take the appropriate actions during routine occurrences, as well as abnormal situations. The influence of alcohol drastically reduces the chances of completing your flight without incident. Even in small amounts, alcohol can impair your judgment, decrease your sense of responsibility, affect your coordination, constrict your visual field, diminish your memory, reduce your reasoning power, and lower your attention span. As little as one ounce of alcohol can decrease the speed and strength of your muscular reflexes, lessen the efficiency of your eye movements while reading, and increase the frequency at which you commit errors. Impairments in vision and hearing occur at alcohol blood levels as low as .01 percent.

There are several regulations in FAR Parts 61 and 91 that apply to drug and alcohol violations and testing requirements, including motor vehicle offenses involving alcohol. Any violation or refusal to submit to an alcohol test may result in the denial of an application for a pilot certificate. If you already hold a pilot certificate, it can be suspended or revoked. If you have any questions about how a violation may affect your pilot training, you should discuss these regulations with your instructor.

HANGOVER

When you have a hangover, you are still under the influence of alcohol. Although you may think that you are functioning normally, the impairment of motor and mental responses still remains. Considerable amounts of alcohol can remain in the body for over 16 hours, so you should be cautious about flying too soon after drinking.

PAIN KILLERS

Pain killers can be grouped into two broad categories: analgesics and anesthetics. Over-the-counter analgesics, such as acetylsalicylic acid (Aspirin), acetaminophen (Tylenol), and ibuprofen (Advil) have few side effects when taken in the correct dosage. Although some people are allergic to certain analgesics or may suffer from stomach irritation, flying usually is not restricted when 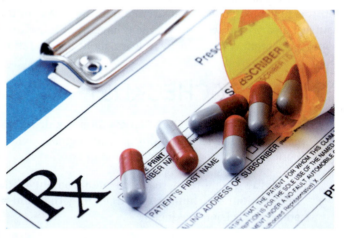 taking these drugs. However, flying is almost always precluded while using prescription analgesics, such as Darvon, Percodan, Demerol, and codeine since these drugs may cause side effects such as mental confusion, dizziness, headaches, nausea, and vision problems.

Anesthetics are drugs that deaden pain or cause a loss of consciousness. These drugs are commonly used for dental and surgical procedures. Most local anesthetics used for minor dental and outpatient procedures wear off within a relatively short period of time. The anesthetic itself may not limit flying so much as the actual procedure and subsequent pain.

STIMULANTS

Stimulants are drugs that excite the central nervous system and produce an increase in alertness and activity. Amphetamines, caffeine, and nicotine are all forms of stimulants. Common uses of these drugs include appetite suppression, fatigue reduction, and mood elevation. Some of these drugs may cause a stimulant reaction, even though this reaction is not their primary function. In some cases, stimulants can produce anxiety and drastic mood swings, both of which are dangerous when you fly.

SECTION C ■ Introduction to Human Factors

OTHER PROBLEM DRUGS

Some drugs that can neither be classified as stimulants nor depressants, have adverse effects on flying. For example, some forms of antibiotics can produce dangerous side effects, such as balance disorders, hearing loss, nausea, and vomiting. While many antibiotics are safe for use while flying, the infection requiring the antibiotic may prohibit flight. In addition, unless specifically prescribed by a physician, you should not take more than one drug at a time nor mix drugs with alcohol. The effects are often unpredictable.

The dangers of illegal drugs are well documented. Certain illegal drugs can have hallucinatory effects that occur days or weeks after the drug is taken. Obviously, these drugs have no place in the aviation community.

FITNESS FOR FLIGHT

Prior to operating an aircraft, you should ask yourself several key questions to determine your physical suitability for flight. If I have an illness, does the condition present a hazard to safe flight? If I am taking a drug for an illness and it wears off during a flight, will it cause an unsafe condition? Can the drug that I am taking produce any side effect that would influence my motor, perceptual, or psychological condition? Am I fatigued? Am I experiencing excessive stress from work or home? If the answer to any of these questions is "Yes" or "I don't know," you may not be fit to operate an aircraft. In addition, if you have a known medical deficiency that would make you unable to meet the qualifications of your current medical certificate, the FARs prohibit you from acting as pilot in command. If you are not sure of your physical suitability for flight, consult an aviation medical examiner.

Your ability to assess your mental and physical fitness for flight, and your skill at making effective decisions are essential to flight safety. As you explore the Human Element insets in this textbook, your human factors training will continue and your knowledge in this subject area will grow. Now that you have been introduced to how your mind and body operate in flight, you are ready to learn about the operation of the airplane in Chapter 2.

SUMMARY CHECKLIST

✓ It is estimated that 75% of all aviation accidents are human-factors related so the goal of human factors training is to increase aviation safety by optimizing pilot performance through reducing human error.

✓ Applying single-pilot resource management (SRM) involves using hardware, information, and human resources to gather and analyze flight information for effective decision making.

✓ The airline industry implemented crew resource management (CRM) training to provide tools for flight crews to eliminate hazards or minimize the impact of hazards.

✓ Aeronautical decision making (ADM) is a systematic approach for aircraft pilots to consistently determine the best course of action in response to specific circumstances.

✓ The ADM process consists of the steps that you use to make effective decisions as pilot in command.

✓ Your general health, stress or fatigue level, attitude, knowledge, skill level, and recency of experience are factors that affect your performance as pilot in command so it is imperative to determine if you are prepared for a particular flight.

✓ Establish personal limitations for flight and use the I'M SAFE checklist to evaluate your fitness for flight.

✓ Five hazardous attitudes that can interfere with a pilot's ability to make effective decisions are: anti-authority, macho, impulsivity, invulnerability, and resignation.

✓ Perform self-critiques after each flight to evaluate your performance, determine the skills that need improvement, and create a plan for increasing your proficiency.

✓ Ideally, both you and your instructor should evaluate your performance and resolve any differences in your assessments before creating a plan for improvement. This is called learned-centered grading.

✓ Risk management involves making decision about four fundamental risk elements related to the flight: the pilot, the aircraft, the environment, and the type of operation.

✓ PAVE and the 5Ps remind you of the risk factors that you must manage when planning and implementing flights.

✓ 5P checklists provide guidelines on the risk factors to consider during flight planning.

✓ During flight, use the 5P check to evaluate the situation at decision points that correspond to the phases of flight.

✓ Task management involves planning and prioritizing tasks to avoid work overload, identifying and using resources to accomplish tasks, and managing distractions.

✓ When using a do-list, read the checklist item and the associated action and then perform the action.

✓ A flow pattern guides you through the flight deck in a logical order. Following completion of the flow pattern, refer to the checklist and verify that you have accomplished each item.

✓ Situational awareness is the accurate perception of all the operational and environmental factors that affect flight safety before, during, and after the flight.

✓ Passenger briefings, takeoff briefings, and before-landing briefings are effective tools to help you maintain situational awareness by preparing you for critical phases of flight.

✓ Fatigue, stress, and work overload can cause you to fixate on one aspect of the flight and omit others from your attention.

✓ Complacency is an obstacle to maintaining situational awareness; when activities are routine, you might relax and put less effort into your performance.

✓ Controlled flight into terrain (CFIT) occurs when an aircraft is flown into terrain or water with no prior awareness on the part of the crew that the crash is imminent.

✓ The concept of automation management typically applies to an airplane with an advanced avionics system that includes digital displays, GPS equipment, a moving map, and an integrated autopilot.

✓ To effectively manage automation, you must be able to correctly interpret your system's annunciations and recognize when the automation is operating in a different mode than you expect.

✓ Relying too heavily on automation can lead to serious consequences in the event of an equipment failure so it is important to maintain your ability to maneuver the airplane manually.

✓ Threat and error management (TEM) training, developed from risk management, focuses on detecting and responding to threats to prevent an undesired aircraft state (UAS).

✓ Undesired aircraft states are pilot-induced airplane position/speed deviations, misapplied flight controls, or incorrect airplane configurations.

SECTION C ■ **Introduction to Human Factors**

✓ Threats are expected or unexpected risks, or hazards, beyond your influence as a pilot. Threats increase operational complexity, can appear suddenly, and must be effectively managed to contain risk within acceptable levels.

✓ A hazard is a condition that could cause, or contribute to, an aircraft incident or accident. Examples of hazards include adverse weather, challenging airports surrounded by high mountains, or congested airspace.

✓ Errors are caused by pilot actions or inactions and cause confusion, increase risk, and lead to deviations from expected outcomes.

✓ An essential component of human factors training is aviation physiology, which is the study of the performance and limitations of the body in the flight environment.

✓ Ear pain is normally the result of a difference between air pressure in the middle ear and outside air pressure.

✓ Slow descent rates can help prevent or reduce the severity of ear pain and to equalize pressure, the eustachian tube can sometimes be opened by yawning, swallowing, chewing, or employing the Valsalva maneuver.

✓ The reduction of atmospheric pressure during flight can cause scuba divers to experience decompression sickness.

✓ The recommended waiting time before ascending above 8,000 feet MSL after scuba diving is 24 hours unless the dive did not require a controlled ascent, in which case the waiting period is only 12 hours.

✓ Motion sickness is caused by the brain receiving conflicting messages about the state of the body.

✓ Stress can be defined as the body's response to physical and psychological demands placed upon it.

✓ Frequently associated with pilot error, some of the effects of fatigue include degradation of attention and concentration, impaired coordination, and decreased ability to communicate.

✓ Flight deck noise can contribute to excessive fatigue, stress, and airsickness, as well as severely impede the understanding of speech.

✓ Pilots who fly frequently may experience serious hearing loss over a period of time unless ear protection is used.

✓ Depressants are drugs that reduce the body's functioning in many areas. The most common depressant is alcohol.

✓ Intoxication is determined by the amount of alcohol in the bloodstream, which is usually measured as a percentage by weight in the blood. The FARs require that your blood alcohol level be less than .04 percent and that 8 hours pass between drinking alcohol and piloting an aircraft.

✓ Stimulants are drugs that excite the central nervous system and produce an increase in alertness and activity. Amphetamines, caffeine, and nicotine are all forms of stimulants.

✓ Your ability to assess your mental and physical fitness for flight, and your skill at making effective decisions are essential to flight safety.

KEY TERMS

Single-Pilot Resource Management (SRM)

Crew Resource Management (CRM)

Aeronautical Decision Making (ADM)

ADM Process

I'M SAFE Checklist

Hazardous Attitudes

Self-Critiques

Learner-Centered Grading

Risk Management

PAVE

5Ps

Task Management

Checklists

Do-List

Flow Pattern

Situational Awareness

Briefings

Controlled Flight into Terrain (CFIT)

Automation Management

Equipment Operating Levels

Threat and Error Management (TEM)

Undesired Aircraft State (UAS)

Threat

Hazard

Error

Decompression Sickness

Motion Sickness

Stress

Fatigue

Depressants

Alcohol

Stimulants

QUESTIONS

1. Select the true statement regarding the ADM process.
   A. The ADM process is used when you need to respond immediately using established procedures.
   B. Part of the ADM process is to evaluate the outcome of your decision to ensure your actions are producing the desired result.
   C. The first step of the ADM process is to choose a course of action after considering the risk involved with each possible action.

2. List each of the factors that you should assess when you follow the I'M SAFE checklist prior to a flight.

3. List the five hazardous attitudes and their antidotes.

4. List the risk factors associated with the 5Ps.

5. Select the true statement regarding the use of checklists.
   A. With a do-list, perform all of the actions then verify the actions with the do-list.
   B. Use flow patterns for abnormal procedures, such as an equipment malfunctions, in place of a checklist.
   C. Use flow patterns to perform normal procedures, such as configuring the airplane and avionics, then verify with a checklist.

6. List the items included in a passenger briefing using the SAFETY acronym.

SECTION C ■ Introduction to Human Factors

7. Which is an item to include the takeoff briefing?
   A. Traffic pattern entry
   B. Wind direction and velocity
   C. Taxi instructions to the runway

8. What is controlled flight into terrain (CFIT)?
   A. A controller issues an instruction that causes a pilot to fly into terrain.
   B. An aircraft is flown into terrain or water when the crew mismanages an engine failure.
   C. An aircraft is flown into terrain or water with no prior awareness on the part of the crew that the crash is imminent.

9. What is an undesired aircraft state (UAS) state?

10. Name at least 3 characteristics of a threat.

11. What are four types of errors? Provide an example of each.

12. When a pilot is faced with an unusual situation, what potentially hazardous reactions might occur?

13. List three examples of hazards that could potentially contribute to an aircraft accident or incident.

14. Select the true statement regarding ear and sinus block during flight.
   A. During descent, high pressure in the middle ear typically opens the eustachian tube to equalize the pressure.
   B. Pressure in the ear can be equalized by performing the Valsalva maneuver—holding the nose and mouth shut and forcibly exhaling.
   C. Rapid descent rates can help increase the pressure differential between the auditory canal and the middle ear helping to open the eustachian tube.

15. Name at least two ways to help prevent or alleviate motion sickness.

16. In addition to pressure effects and motion sickness, name at least three factors that can impair your fitness for flight.

17. The FARs require that your blood alcohol level be less than what percent?
   A. .04%
   B. .06%
   C. .08%

18. How many hours must pass between drinking alcohol and piloting an airplane according to the FARs?

5P CHECKLISTS

Use these 5P checklists to identify and manage risks when you plan flights.

Pilot				
Summary of Training		**Yes**	**No**	**N/A**
Do I have a current flight review?				
Am I current to carry passengers?				
Have I had recent refresher training in this airplane?				
Am I instrument-current?				
Have I had recent mountain flying training or experience?				
Experience	**Personal Minimums**	**Yes**	**No**	**N/A**
Hours in Specific Airplane				
Last 90 Days				
• Hours				
• Landings				
• Instrument Approaches				
Last 6 Months				
• IFR Hours				
• Instrument Approaches				
• Night Hours				
• Night Landings				
• Strong Crosswind/Gusty Landings				
• Mountain Flying Hours				
Fitness — I'M SAFE		**Yes**	**No**	**N/A**
Illness — Am I healthy?				
Medication — Am I free of prescription or over-the-counter drugs?				
Stress — Am I free of pressure (job, financial matters, health problems, or family discord)?				
Alcohol — Have I abstained from alcohol in the previous 24 hours?				
Fatigue — Did I get at least seven hours of sleep?				
Eating — Am I adequately nourished?				
Emotion — Am I free of emotional upset?				

Passengers				
Experience	**Yes**	**No**	**N/A**	
Are my passengers comfortable flying? (spent time in small aircraft, certificated pilots, etc.)				
Fitness	**Yes**	**No**	**N/A**	
Are my passengers feeling well? (sickness, likely to experience airsickness, etc.)				
Flexibility	**Yes**	**No**	**N/A**	
Are my passengers flexible and well-informed about the changeable nature of flying? (arriving late, diverting to an alternate, etc.)				

SECTION C ■ **Introduction to Human Factors**

Plane			
Airworthiness	Yes	No	N/A
Are the aircraft inspections current and appropriate to the type of flight? (annual and 100-hour inspections, VOR check, etc.)			
Is the required equipment on board and working for the type of flight? (lights for night flight, onboard oxygen, survival gear, etc.)			
Have all prior maintenance issues been taken care of? (squawks resolved, inoperative equipment placarded, etc.)			
Performance	Yes	No	N/A
Can the aircraft carry the planned load within weight and CG limits?			
Is the aircraft performance (takeoff, climb, enroute, and landing) adequate for the available runways, density altitude, and terrain conditions? — Both engines operating			
Is the aircraft performance (takeoff, climb, enroute, and landing) adequate for the available runways, density altitude, and terrain conditions? — One engine inoperative			
Is the fuel capacity adequate for the proposed flight legs, including to an alternate airport if required?			

Programming			
Avionics Airworthiness	Yes	No	N/A
Is the avionics equipment working properly? (squawks resolved, autopilot functional)			
Are all databases current? (GPS navdata, terrain, etc.)			
Avionics Operation	Yes	No	N/A
Are you proficient at operating the avionics equipment?			
Avionics Configuration	Yes	No	N/A
Is the avionics configuration appropriate for the navigation required?			

Plan					
Airport Conditions			Yes	No	N/A
Do NOTAMs indicate my flight can proceed as planned? (no runway or navaid closures, and so on)					
Are services available at the airport during the appropriate time? (fuel, ATC, Unicom, etc.)					
Terrain/Airspace			Yes	No	N/A
Does the airspace and terrain in the area allow me to fly my route as planned? (Check for mountainous terrain, and areas to avoid, such as TFRs, restricted or prohibited areas).					
Mission			Yes	No	N/A
Do I have alternate plans to manage any commitments that exist at my destination? (reschedule meeting, airline reservations, etc.)					
Did I tell the people whom I'm meeting at my destination that I might be late?					
Do I have an overnight kit containing any necessary prescriptions and toiletries?					
Weather		Location	Yes	No	N/A
Are the weather conditions acceptable? (no hazards such as thunderstorms, icing, turbulence, etc.)		departure			
		enroute			
		destination			
Is there a suitable airport that meets the regulatory requirements for an alternate if the forecast at my destination requires an alternate airport?					
Weather Limitations	**Personal Limitations**	**Location**	Yes	No	N/A
Are the weather conditions for my flight within my personal limitations?					
• Minimum IFR Approach Ceiling and Visibility		departure			
		destination			
• Minimum Ceiling and Visibility (Day VFR)		departure			
		enroute			
		destination			
• Minimum Ceiling and Visibility (Night VFR)		departure			
		enroute			
		destination			
• Maximum Surface Wind Speed and Gusts		departure			
		destination			
• Maximum Direct Crosswind		departure			
		destination			

SECTION C ■ Introduction to Human Factors

SECTION D
Developing Pilot Resilience with CBTA

The single-pilot resource management (SRM) concepts (to which you were previously introduced) provide the foundation for understanding the abilities that you will need to become a resilient pilot. **Competency-Based Training and Assessment (CBTA)** is a framework that builds on that foundation. CBTA focuses on developing **pilot competencies** that help you to become a resilient pilot who is ready to meet the many challenges of flight. Aviation training industry leaders have identified nine competencies that pilots need to operate safely, efficiently, and effectively in today's flight environment. [Figure 1-65]

Application of Knowledge (KNO)

Application of Procedures and Compliance with Regulations (PRO)

Airplane Flight Path Management — Manual Control (FPM)

Airplane Flight Path Management — Automation (FPA)

Technical Competencies

Communication (COM)

Leadership and Teamwork (LTW)

Workload Management (WLM)

Situation Awareness and Management of Information (SAW)

Problem Solving — Decision Making (PSD)

Human Factors Competencies

Figure 1-65. Both technical and human factors competencies are taught during CBTA. Note: you will see the term "situation awareness" used instead of "situational awareness" when referring to the competencies.

BECOMING A RESILIENT PILOT

Resilient pilots are present and aware, highly informed, able to cope with challenges, have stress management strategies, and can quickly recover from adversity. When driving a car, if you have a flat tire, a mechanical problem, or are lost, you can pull over to the side of the road and take time out to resolve the issue. When flying an airplane, you must directly confront and resolve any issues you face, where you are. Although you might not have the luxury of "pulling over to the side of the road," you will learn how to effectively fly the airplane while you manage any situation that arises. For example, to continue on your planned flight or find a safe place to land, you might need to determine options for avoiding adverse weather, contact ATC for assistance, or perform a checklist to resolve a mechanical issue. Your passengers and people on the ground are depending on you to be a resilient private pilot.

A notable example of resiliency during an in-flight emergency is USAir Flight 1549. Despite multiple catastrophic bird strikes that caused the failure of both engines, Captain Chesley Sullenberger and First Officer Jeffrey Skiles demonstrated proficiency in the pilot competencies in a matter of seconds. Their ability to draw on a combination of the competencies while under duress in a critical situation illustrates the capabilities of resilient pilots. [Figure 1-66]

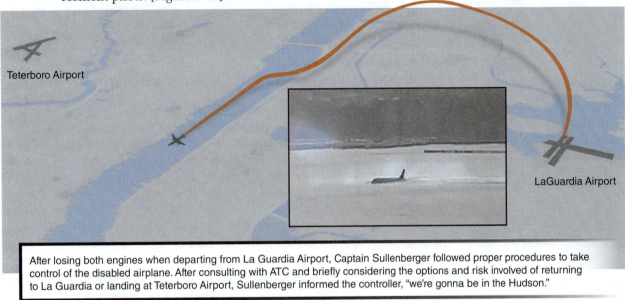

Teterboro Airport

LaGuardia Airport

After losing both engines when departing from La Guardia Airport, Captain Sullenberger followed proper procedures to take control of the disabled airplane. After consulting with ATC and briefly considering the options and risk involved of returning to La Guardia or landing at Teterboro Airport, Sullenberger informed the controller, "we're gonna be in the Hudson."

Figure 1-66. After multiple bird strikes caused a dual engine failure, the resilient pilots of USAir Flight 1549 safely landed on the Hudson River. All passengers and crew survived.

COMPETENCIES AND OBSERVABLE BEHAVIORS

During your training, you should be able to demonstrate specific **observable behaviors (OBs)** to indicate that you have mastered each competency. All the competencies and most of the observable behaviors are applicable to you as a student pilot from your very first flight. To apply CBTA to flight training, your instructor selects competencies and associated observable behaviors on which to focus during each flight lesson. These focus competencies are aligned with the lesson objectives and represent the essential technical and/or human factors emphasis areas that apply to the flight. You and your instructor evaluate your performance on tasks *and* the pilot competencies.

APPLICATION OF KNOWLEDGE

The **application of knowledge (KNO)** competency provides the foundation for mastering the other technical and human factors competencies. Throughout your pilot training, you acquire knowledge through self-study, individual ground lessons with your instructor, classroom instruction, or a combination of methods. You apply this knowledge to perform safe ground and flight operations and to make effective decisions as a pilot. [Figure 1-67]

From the beginning of your training, and as you gain experience, the knowledge you accumulate must be robust and well-learned. During normal flying operations, you must be able to retrieve the necessary information from your long-term memory to effectively perform tasks. When faced with challenging circumstances, your ability to retrieve information and apply knowledge will make you less susceptible to stress and enable you to effectively manage the situation. [Figure 1-68]

APPLICATION OF PROCEDURES AND COMPLIANCE WITH REGULATIONS

Mastering the **application of procedures and compliance with regulations (PRO)** competency promotes safety by ensuring you fly the airplane in accordance with the FARs, aircraft manufacturer recommendations, and flight school guidelines. [Figure 1-69]

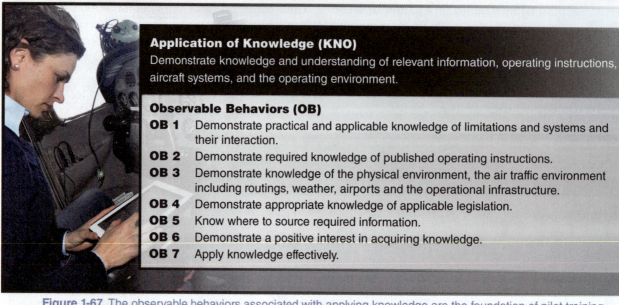

Application of Knowledge (KNO)

Demonstrate knowledge and understanding of relevant information, operating instructions, aircraft systems, and the operating environment.

Observable Behaviors (OB)

OB 1 Demonstrate practical and applicable knowledge of limitations and systems and their interaction.

OB 2 Demonstrate required knowledge of published operating instructions.

OB 3 Demonstrate knowledge of the physical environment, the air traffic environment including routings, weather, airports and the operational infrastructure.

OB 4 Demonstrate appropriate knowledge of applicable legislation.

OB 5 Know where to source required information.

OB 6 Demonstrate a positive interest in acquiring knowledge.

OB 7 Apply knowledge effectively.

Figure 1-67. The observable behaviors associated with applying knowledge are the foundation of pilot training.

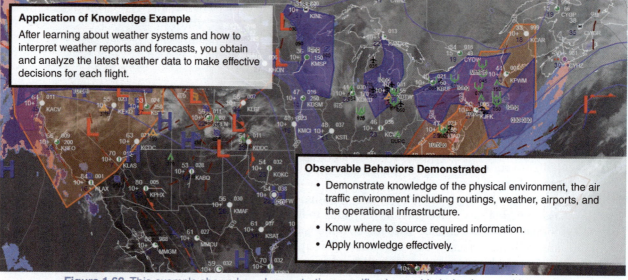

Application of Knowledge Example

After learning about weather systems and how to interpret weather reports and forecasts, you obtain and analyze the latest weather data to make effective decisions for each flight.

Observable Behaviors Demonstrated

- Demonstrate knowledge of the physical environment, the air traffic environment including routings, weather, airports, and the operational infrastructure.
- Know where to source required information.
- Apply knowledge effectively.

Figure 1-68. This example shows how demonstrating specific observable behaviors enables you to effectively apply knowledge.

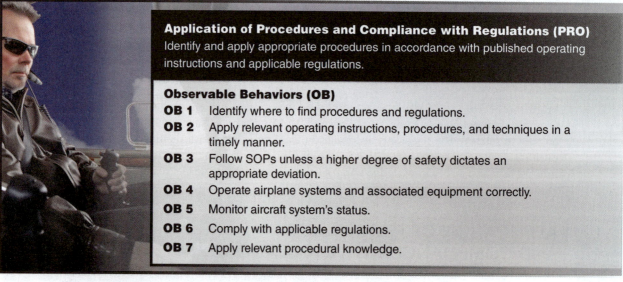

Application of Procedures and Compliance with Regulations (PRO)

Identify and apply appropriate procedures in accordance with published operating instructions and applicable regulations.

Observable Behaviors (OB)

OB 1 Identify where to find procedures and regulations.

OB 2 Apply relevant operating instructions, procedures, and techniques in a timely manner.

OB 3 Follow SOPs unless a higher degree of safety dictates an appropriate deviation.

OB 4 Operate airplane systems and associated equipment correctly.

OB 5 Monitor aircraft system's status.

OB 6 Comply with applicable regulations.

OB 7 Apply relevant procedural knowledge.

Figure 1-69. The observable behaviors associated with applying procedures include knowing where to access the proper procedures and applicable regulations.

The aviation industry has long emphasized standard operating procedures (SOPs), which are defined uniform standards and methodical instructions that set expectations on how to perform routine operations accurately, efficiently, and safely. In addition, you must understand and comply with regulations during ground and flight operations. [Figure 1-70]

Application of Procedures Example
Use checklists and follow the appropriate procedures to configure the airplane for approach and landing.

Observable Behaviors Demonstrated
- Apply relevant operating instructions, procedures, and techniques in a timely manner.
- Operate aircraft systems and associated equipment correctly.
- Apply relevant procedural knowledge.

Figure 1-70. This example shows how demonstrating specific observable behaviors enables you to effectively apply procedures.

AIRPLANE FLIGHT PATH MANAGEMENT – MANUAL CONTROL (FPM)

You learn to master the technical competency of **airplane flight path management – manual control (FPM)** during each lesson as you practice a variety of maneuvers ranging from steep turns to landings. [Figure 1-71]

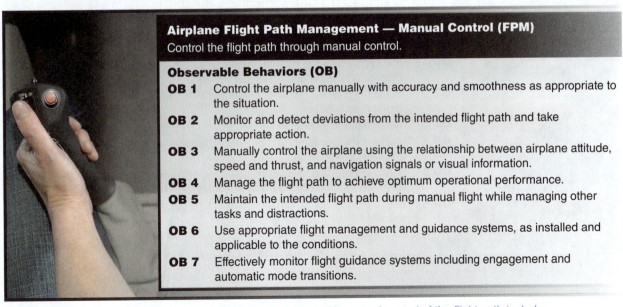

Airplane Flight Path Management — Manual Control (FPM)
Control the flight path through manual control.

Observable Behaviors (OB)

OB 1 Control the airplane manually with accuracy and smoothness as appropriate to the situation.

OB 2 Monitor and detect deviations from the intended flight path and take appropriate action.

OB 3 Manually control the airplane using the relationship between airplane attitude, speed and thrust, and navigation signals or visual information.

OB 4 Manage the flight path to achieve optimum operational performance.

OB 5 Maintain the intended flight path during manual flight while managing other tasks and distractions.

OB 6 Use appropriate flight management and guidance systems, as installed and applicable to the conditions.

OB 7 Effectively monitor flight guidance systems including engagement and automatic mode transitions.

Figure 1-71. The observable behaviors associated with manual control of the flight path include being able to detect deviations and take appropriate action.

SECTION D ■ **Developing Pilot Resilience with CBTA**

One of the aviation industry's greatest concerns is the erosion of manual flying skills, due to overreliance on automation. You must be able to turn off automation and still retain full control of the airplane in all flight situations. [Figure 1-72]

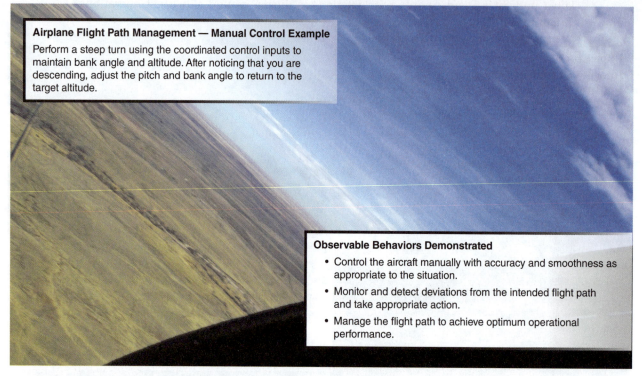

Airplane Flight Path Management — Manual Control Example

Perform a steep turn using the coordinated control inputs to maintain bank angle and altitude. After noticing that you are descending, adjust the pitch and bank angle to return to the target altitude.

Observable Behaviors Demonstrated

- Control the aircraft manually with accuracy and smoothness as appropriate to the situation.
- Monitor and detect deviations from the intended flight path and take appropriate action.
- Manage the flight path to achieve optimum operational performance.

Figure 1-72. This example shows how demonstrating specific observable behaviors enables you to manually control your flight path.

AIRPLANE FLIGHT PATH MANAGEMENT — AUTOMATION

After you have learned to fly the airplane manually, you will learn to master the technical competency of **airplane flight path management — automation (FPA)** in airplanes that are equipped with autopilots. [Figure 1-73]

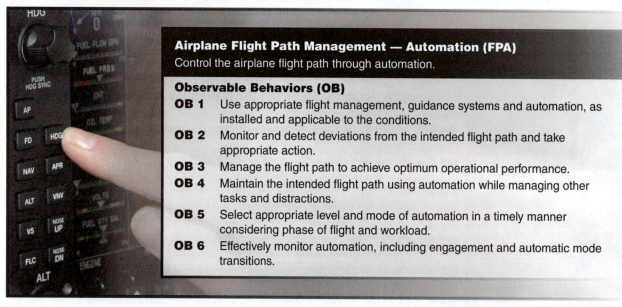

Airplane Flight Path Management — Automation (FPA)

Control the airplane flight path through automation.

Observable Behaviors (OB)

OB 1 Use appropriate flight management, guidance systems and automation, as installed and applicable to the conditions.

OB 2 Monitor and detect deviations from the intended flight path and take appropriate action.

OB 3 Manage the flight path to achieve optimum operational performance.

OB 4 Maintain the intended flight path using automation while managing other tasks and distractions.

OB 5 Select appropriate level and mode of automation in a timely manner considering phase of flight and workload.

OB 6 Effectively monitor automation, including engagement and automatic mode transitions.

Figure 1-73. The observable behaviors associated with automated control of the flight path include selecting the appropriate level and mode of automation.

To manage your flight path, you must have knowledge of how the specific autopilot in your airplane is integrated with other systems, including the flight controls and navigation systems. You must be able to correctly engage the autopilot to maintain the correct flight path for the phase of flight by selecting the appropriate level and mode. After automation is engaged, you must actively monitor your flight path and the interpret the autopilot annunciations so that you can take action when deviations occur. [Figure 1-74]

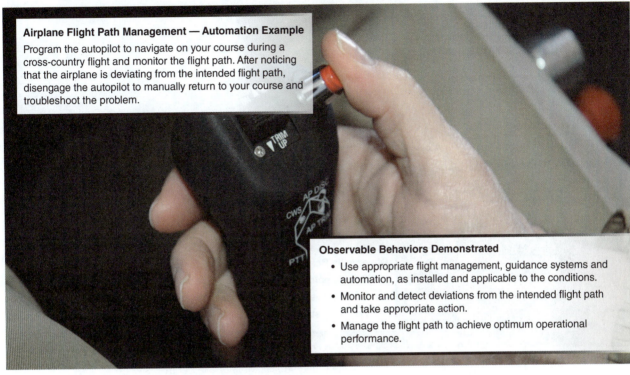

Airplane Flight Path Management — Automation Example
Program the autopilot to navigate on your course during a cross-country flight and monitor the flight path. After noticing that the airplane is deviating from the intended flight path, disengage the autopilot to manually return to your course and troubleshoot the problem.

Observable Behaviors Demonstrated
- Use appropriate flight management, guidance systems and automation, as installed and applicable to the conditions.
- Monitor and detect deviations from the intended flight path and take appropriate action.
- Manage the flight path to achieve optimum operational performance.

Figure 1-74. This example shows how demonstrating specific observable behaviors enables you to effectively use automation to control your flight path.

COMMUNICATION

You will apply the **communication (COM)** competency when interacting with a variety of individuals, such as your instructor, air traffic control (ATC), passengers, co-pilots, other aircraft, and aircraft maintenance technicians (AMTs). [Figure 1-75]

Communication (COM)
Communicate through appropriate means in the operational environment, in both normal and non-normal situations.

Observable Behaviors (OB)

OB 1	Determine that the recipient is ready and able to receive information.
OB 2	Select appropriately what, when, how and with whom to communicate.
OB 3	Convey messages clearly, accurately and concisely.
OB 4	Confirm that the recipient demonstrates understanding of important information.
OB 5	Listen actively and demonstrate understanding when receiving information.
OB 6	Ask relevant and effective questions.
OB 7	Use appropriate escalation in communication to resolve identified deviations.
OB 8	Use and interpret non-verbal communication in a manner appropriate to the organizational and social structure.
OB 9	Adhere to standard radio phraseology and procedures.
OB 10	Accurately read, interpret, construct, and respond to datalink messages.

Figure 1-75. The observable behaviors associated with communication include being able to both convey messages and actively listen.

SECTION D ■ **Developing Pilot Resilience with CBTA**

You must master proper radio phraseology and methods for communication with ATC, In addition, you must be proficient in performing briefings, conveying information to co-pilots and passengers, and actively listening. Because English is the international language of civil aviation, you must be proficient in reading, speaking, and comprehending the English language. [Figure 1-76]

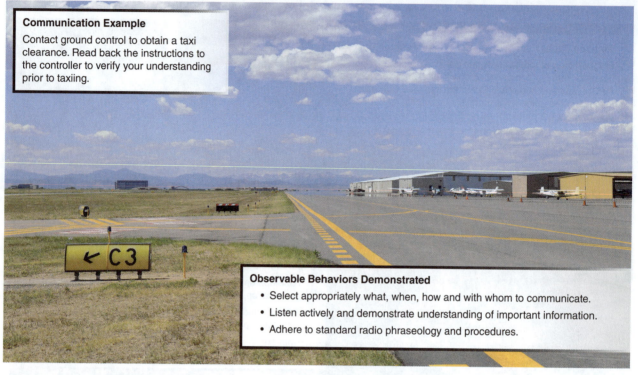

Communication Example

Contact ground control to obtain a taxi clearance. Read back the instructions to the controller to verify your understanding prior to taxiing.

Observable Behaviors Demonstrated

• Select appropriately what, when, how and with whom to communicate.
• Listen actively and demonstrate understanding of important information.
• Adhere to standard radio phraseology and procedures.

Figure 1-76 This example shows how demonstrating specific observable behaviors enables you to effectively communicate.

LEADERSHIP AND TEAMWORK

The **leadership and teamwork (LTW)** competency does not just apply to working with a professional flight crew. During training, you perform leadership and teamwork behaviors with your flight instructor and as a private pilot, you must be proficient in this competency to successfully engage with other pilots and passengers. [Figure 1-77]

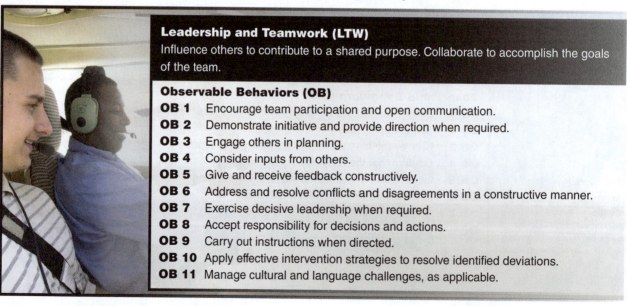

Leadership and Teamwork (LTW)

Influence others to contribute to a shared purpose. Collaborate to accomplish the goals of the team.

Observable Behaviors (OB)

OB 1   Encourage team participation and open communication.
OB 2   Demonstrate initiative and provide direction when required.
OB 3   Engage others in planning.
OB 4   Consider inputs from others.
OB 5   Give and receive feedback constructively.
OB 6   Address and resolve conflicts and disagreements in a constructive manner.
OB 7   Exercise decisive leadership when required.
OB 8   Accept responsibility for decisions and actions.
OB 9   Carry out instructions when directed.
OB 10   Apply effective intervention strategies to resolve identified deviations.
OB 11   Manage cultural and language challenges, as applicable.

Figure 1-77. The observable behaviors associated with leadership and teamwork include addressing and resolving conflicts.

As a student pilot, focus on fostering an effective relationship with your flight instructor, mentoring fellow student pilots who might be following in your footsteps, and utilizing flight school staff who can provide you with resources, assistance, or expertise. [Figure 1-78]

Leadership and Teamwork Example

After obtaining your private pilot certificate, when flying with other pilots, determine roles and responsibilities prior to departure to help prevent conflict. Consider the other pilot as a resource to provide input during the flight.

Observable Behaviors Demonstrated

- Demonstrate initiative and provide direction when required.
- Consider input from others.
- Address and resolve conflicts and disagreements in a constructive manner.

Figure 1-78. This example shows how demonstrating specific observable behaviors enables you to lead and work as a team

WORKLOAD MANAGEMENT

Workload management (WLM) requires you to plan, prioritize, schedule, and delegate tasks to avoid work overload. [Figure 1-79]

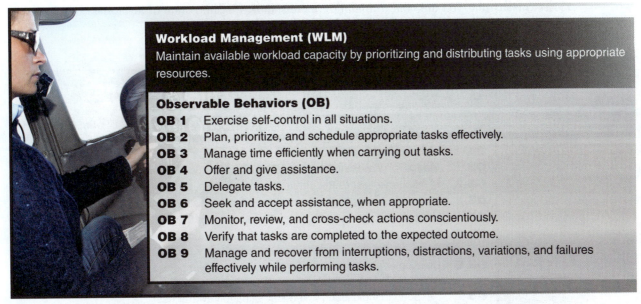

Workload Management (WLM)

Maintain available workload capacity by prioritizing and distributing tasks using appropriate resources.

Observable Behaviors (OB)

OB 1 Exercise self-control in all situations.
OB 2 Plan, prioritize, and schedule appropriate tasks effectively.
OB 3 Manage time efficiently when carrying out tasks.
OB 4 Offer and give assistance.
OB 5 Delegate tasks.
OB 6 Seek and accept assistance, when appropriate.
OB 7 Monitor, review, and cross-check actions conscientiously.
OB 8 Verify that tasks are completed to the expected outcome.
OB 9 Manage and recover from interruptions, distractions, variations, and failures effectively while performing tasks.

Figure 1-79. The observable behaviors associated with workload management include planning, prioritizing, and delegating tasks.

You should plan to complete tasks during times of low workload in preparation for the high-workload phases of flight. To help you manage workload, use both internal

SECTION D ■ Developing Pilot Resilience with CBTA

resources, such as checklists, and external resources, such as ATC assistance. If you feel overburdened, your performance can decline so do not hesitate to speak up to your flight instructor during training. As a private pilot, divide roles and responsibilities when flying with other pilots to decrease your workload. [Figure 1-80]

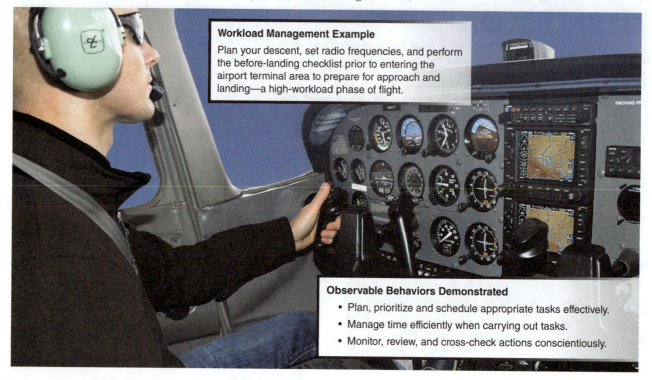

Workload Management Example

Plan your descent, set radio frequencies, and perform the before-landing checklist prior to entering the airport terminal area to prepare for approach and landing—a high-workload phase of flight.

Observable Behaviors Demonstrated
- Plan, prioritize and schedule appropriate tasks effectively.
- Manage time efficiently when carrying out tasks.
- Monitor, review, and cross-check actions conscientiously.

Figure 1-80. This example shows how demonstrating specific observable behaviors enables you to manage your workload.

SITUATION AWARENESS AND MANAGEMENT OF INFORMATION

Mastering the competency of **situation awareness and management of information (SAW)** means that you are able to maintain a conscious, elevated state of awareness of your surroundings. [Figure 1-81]

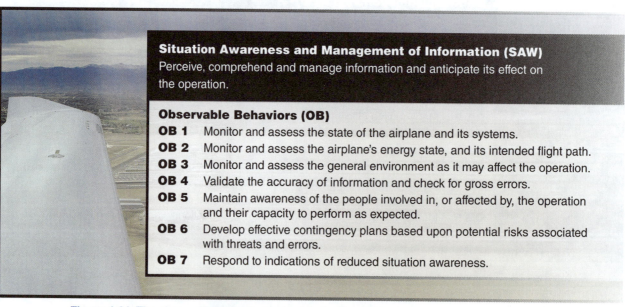

Situation Awareness and Management of Information (SAW)

Perceive, comprehend and manage information and anticipate its effect on the operation.

Observable Behaviors (OB)

OB 1 Monitor and assess the state of the airplane and its systems.

OB 2 Monitor and assess the airplane's energy state, and its intended flight path.

OB 3 Monitor and assess the general environment as it may affect the operation.

OB 4 Validate the accuracy of information and check for gross errors.

OB 5 Maintain awareness of the people involved in, or affected by, the operation and their capacity to perform as expected.

OB 6 Develop effective contingency plans based upon potential risks associated with threats and errors.

OB 7 Respond to indications of reduced situation awareness.

Figure 1-81. The observable behaviors associated with situation awareness include being able to monitor and assess the state of the airplane and the flight environment.

To become proficient in the situation awareness competency requires you to practice making accurate perceptions of operational and environmental factors that affect your flight. You must be able to analyze the status of factors, such as the airplane, the flight environment, and yourself to have a clear picture of what is happening now and predict future events. You will learn to use monitoring techniques and cross-check information with various sources to maintain situation awareness. Developing plans to identify and mitigate threats and errors is also key to ensuring that you do not lose situation awareness. [Figure 1-82]

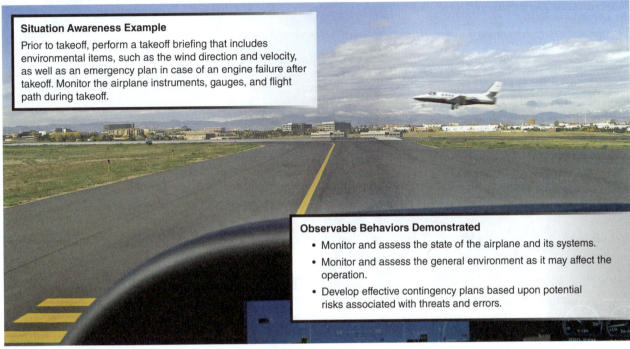

Situation Awareness Example

Prior to takeoff, perform a takeoff briefing that includes environmental items, such as the wind direction and velocity, as well as an emergency plan in case of an engine failure after takeoff. Monitor the airplane instruments, gauges, and flight path during takeoff.

Observable Behaviors Demonstrated

- Monitor and assess the state of the airplane and its systems.
- Monitor and assess the general environment as it may affect the operation.
- Develop effective contingency plans based upon potential risks associated with threats and errors.

Figure 1-82. This example shows how demonstrating specific observable behaviors enables you to maintain situation awareness.

PROBLEM SOLVING — DECISION MAKING

Mastering other competencies, such as workload management and situation awareness enhances your ability to effectively manage threats and errors and become proficient in **problem solving — decision making (PSD)**. [Figure 1-83]

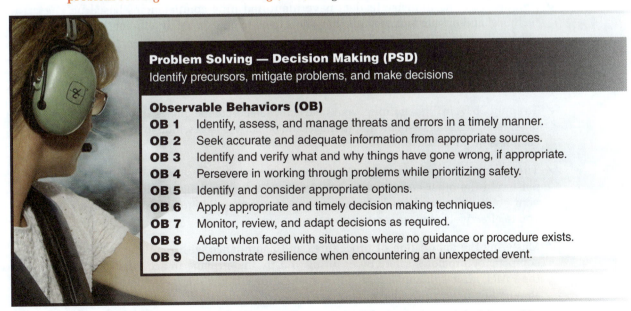

Problem Solving — Decision Making (PSD)

Identify precursors, mitigate problems, and make decisions

Observable Behaviors (OB)

OB 1	Identify, assess, and manage threats and errors in a timely manner.
OB 2	Seek accurate and adequate information from appropriate sources.
OB 3	Identify and verify what and why things have gone wrong, if appropriate.
OB 4	Persevere in working through problems while prioritizing safety.
OB 5	Identify and consider appropriate options.
OB 6	Apply appropriate and timely decision making techniques.
OB 7	Monitor, review, and adapt decisions as required.
OB 8	Adapt when faced with situations where no guidance or procedure exists.
OB 9	Demonstrate resilience when encountering an unexpected event.

Figure 1-83. The observable behaviors associated with problem solving and decision making include being able to identify and manage threats.

SECTION D ■ Developing Pilot Resilience with CBTA

Being skilled at threat and error management and using the aeronautical decision making process (ADM) to define problems and implement solutions are key to solving problems and making effective decisions. [Figure 1-84]

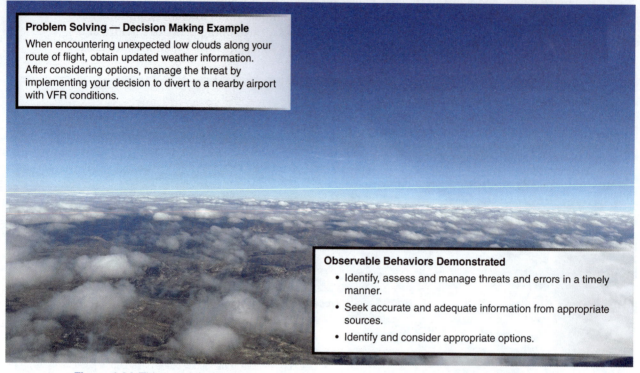

Problem Solving — Decision Making Example
When encountering unexpected low clouds along your route of flight, obtain updated weather information. After considering options, manage the threat by implementing your decision to divert to a nearby airport with VFR conditions.

Observable Behaviors Demonstrated
- Identify, assess and manage threats and errors in a timely manner.
- Seek accurate and adequate information from appropriate sources.
- Identify and consider appropriate options.

Figure 1-84. This example shows how demonstrating specific observable behaviors enables you to solve problems and make effective decisions.

SUMMARY CHECKLIST

✓ Competency-Based Training and Assessment (CBTA) is a framework that builds on single-pilot resource management concepts to focus on developing pilot competencies that help you to become a resilient pilot.

✓ Aviation training industry leaders have identified nine competencies based on technical and human factors skills that pilots need to operate safely, efficiently, and effectively in today's flight environment.

✓ During your training, you should be able to demonstrate specific observable behaviors to indicate that you have mastered each competency.

✓ To apply CBTA to flight training, your instructor selects competencies and associated observable behaviors—aligned with the lesson objectives—on which to focus during each lesson.

✓ The application of knowledge (KNO) competency provides the foundation for mastering the other competencies. When faced with challenging circumstances, your ability to apply knowledge makes you less susceptible to stress and enables you to effectively manage the situation.

✓ Mastering the application of procedures and compliance with regulations (PRO) competency promotes safety by ensuring you fly the airplane in accordance with the FARs, aircraft manufacturer recommendations, flight school guidelines, and standard operating procedures (SOPs).

✓ To master the airplane flight path management – manual control (FPM), you must be able to turn off automation and still retain full control of the airplane in all flight situations.

✓ To become proficient in the competency of airplane flight path management — automation (FPA), you must actively monitor your flight path and interpret autopilot annunciations so that you can take action when deviations occur.

✓ You apply the communication (COM) competency when interacting with a variety of individuals, such as your instructor, air traffic control (ATC), passengers, co-pilots, other aircraft, and aircraft maintenance technicians (AMTs).

✓ During training, you perform the behaviors associated with the leadership and teamwork (LTW) competency with your flight instructor. As a private pilot, you must successfully engage with other pilots and passengers.

✓ The workload management (WLM) competency requires you to plan, prioritize, schedule, and delegate tasks to avoid work overload.

✓ Mastering the competency of situation awareness and management of information (SAW) means that you use monitoring techniques, cross-check information, and mitigate threats and errors to maintain a conscious, elevated state of awareness of your surroundings.

✓ Being skilled at threat and error management and using the aeronautical decision making process (ADM) to define problems and implement solutions are key to becoming proficient in the problem solving — decision making (PSD) competency.

KEY TERMS

Competency-Based Training and
Assessment (CBTA)

Pilot Competencies

Resilient Pilot

Observable Behaviors (OBs)

Application of Knowledge (KNO)

Application of Procedures and
Compliance with Regulations (PRO)

Airplane Flight Path Management —
Manual Control (FPM)

Airplane Flight Path Management —
Automation (FPA)

Communication (COM)

Leadership and Teamwork (LTW)

Workload Management (WLM)

Situation Awareness and Management
of Information (SAW)

Problem Solving — Decision Making
(PSD)

SECTION D ■ **Developing Pilot Resilience with CBTA**

QUESTIONS

1. Select the true statement regarding CBTA.
   A. CBTA only addresses technical pilot operations, while SRM focuses on human factors skills.
   B. Each competency applies to a specific maneuver or procedure that you perform during flight.
   C. Mastering competencies and their associated observable behaviors can help you to become a resilient pilot.

Match the following observable behavior example to the applicable competency.

2. Accept responsibility for an error and collaborate with your instructor to determine an effective way to prevent similar deviations in the future.

3. After becoming distracted by a traffic advisory, return to your checklist tasks to resume preparing descent and landing.

4. Use the pilot's operating handbook to learn how your training airplane's equipment and systems operate.

5. Ask the controller to repeat an instruction that you did not understand.

6. Follow a standard operating procedure to provide a standard safety briefing prior to flight.

7. Perform a crosswind approach and landing to a short runway.

8. When your fuel consumption is greater than originally planned, gather data and select an appropriate airport for a fuel stop along your route.

9. Engage the autopilot to navigate on your flight planned course.

10. Monitor the airplane systems, flight path, traffic situation, and airport conditions as you approach a busy airport for landing.

A. Application of Knowledge

B. Application of Procedures and Compliance with Regulations

C. Airplane Flight Path Management — Manual Control

D. Airplane Flight Path Management — Automation

E. Communication

F. Situation Awareness and Information Management

G. Workload Management

H. Leadership and Teamwork

I. Problem Solving — Decision Making

CHAPTER 2

Airplane Systems

SECTION A
Airplanes

Although airplanes are designed for a variety of purposes, the basic components of most airplanes are essentially the same. After the practical aspects of building an airworthy craft are resolved, what ultimately becomes the final model is largely a matter of the original design objectives and aesthetics. Therefore, airplane design is a combination of art and science. Although the artistic possibilities are limitless, the relatively inflexible scientific requirements for manned flight dictate that most airplane structures include, at a minimum, a fuselage, wings, an empennage, landing gear, and a powerplant. [Figure 2-1]

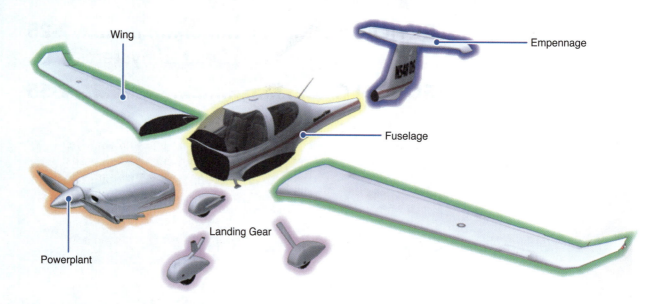

Figure 2-1. The central component of an airplane is the fuselage, which contains the passenger cabin.

SECTION A ■ Airplanes

Wright Brothers —
First Powered Flight

Orville and Wilbur Wright made the first powered flight on December 17, 1903 at Kill Devil Hill on a North Carolina beach. The first flight in this aircraft, named the *Wright Flyer*, covered a distance of 120 feet. The Wright brothers built the aircraft in a nearby workshop. Its essential design features appear in some form on most modern aircraft. The Wright Flyer's wings could be slightly warped to work with a rudder and make coordinated turns. The aircraft had elevators, skid-type landing gear, and a small powerplant and propeller.

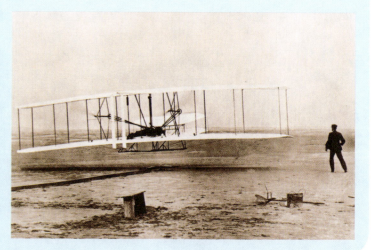

FUSELAGE

The **fuselage** houses the cabin and cockpit, or flight deck, which contains the controls for the airplane, seats for the occupants, and room for cargo. The fuselage also provides attachment points for the other major airplane components. Although uncommon today, most early aircraft utilized an **open truss** structure, which can be identified by the clearly visible struts and wire bracing. [Figure 2-2]

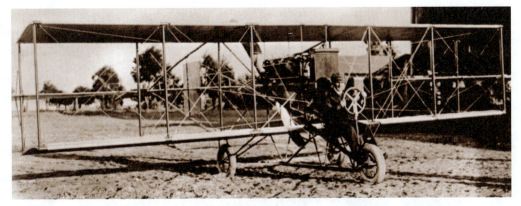

Figure 2-2. Open truss construction was common to many early aircraft.

Soon, aircraft designers enclosed the truss members to streamline the airplane and improve performance. This was originally accomplished with cloth fabric, but later, more rigid materials added strength. A **stressed skin** design uses skin composed of sheet metal, plywood, or composite (fiberglass or carbon fiber) to carry some of the flight loads.

Monocoque is a French word meaning single shell. The **monocoque** design uses the skin to support almost all imposed loads. This design also uses formers and bulkheads as attachment points for the skin, which help shape and strengthen the fuselage. The monocoque structure can be very strong, but depending upon the construction material, this design might not tolerate any dents or deformation of the surface. This characteristic is easily demonstrated using a thin aluminum beverage can. You can apply a large force to the ends of the can without it giving way. However, if the side of the can is dented only slightly, the can easily collapses.

Because of limitations of the monocoque design, many aircraft use a **semi-monocoque** structure. Like the monocoque structure, the semi-monocoque design uses formers and bulkheads. However, an additional strength increase is accomplished by adding stringers that reinforce the skin making it more resistant against dents. [Figure 2-3]

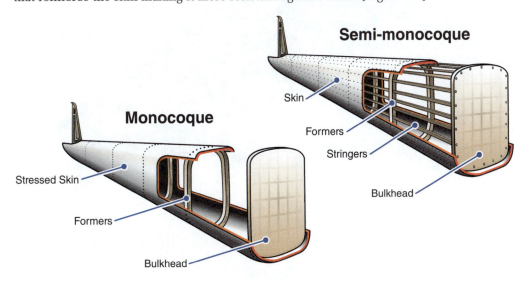

Figure 2-3. Semi-monocoque construction incorporates a substructure for added strength.

SECTION A ■ Airplanes

2-3

WINGS

When air flows around the wings of an airplane, it generates lift, which is essential for flight. **Wings** are attached at the top, middle, or lower portion of the fuselage. These designs are referred to as high-, mid-, and low-wing, respectively. The number of wings can also vary. Airplanes with a single set of wings are referred to as **monoplanes**, while those with two sets are called **biplanes**. [Figure 2-4]

Copyright Corel

Copyright Corel

Figure 2-4. Monoplanes have a single set of wings while biplanes use two sets of wings stacked vertically.

Airplane wings consist of several components, each playing an essential role in the structure and function of the wing. The wing spar, which extends from the root to the tip, is the main structural component of the wing. Ribs give the wing its shape to support lift and help distribute the loads and forces acting on the wing in flight. Stringers add to the wing's strength and rigidity. Fuel tanks in most airplanes are located in the wings to use the space that might otherwise be empty and aid in maintaining airplane balance and stability. Attached to the rear, or trailing edge, of the wings are two types of control surfaces—ailerons and flaps. [Figure 2-5]

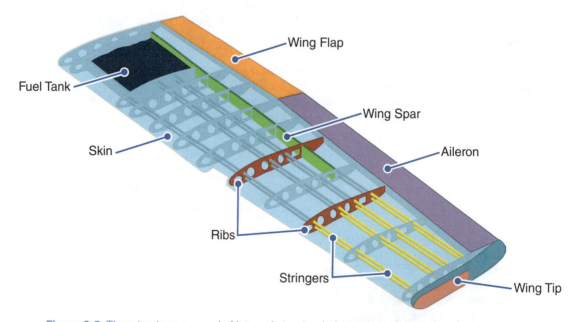

Figure 2-5. The wing is composed of internal structural elements and control surfaces.

Ailerons are located on the outside portion of each wing. They are normally connected to the flight deck control—a control wheel or control stick—by a series of chains, cables, bellcranks, and pulleys. When you move the control wheel, the ailerons move in opposite directions to create aerodynamic forces that roll the airplane in and out of turns. **Flaps** are located on the inner portion of each wing. They are normally flush with the wing's surface during cruising flight. When extended, the flaps move simultaneously downward to increase the wing's lift for takeoffs and landings. [Figure 2-6]

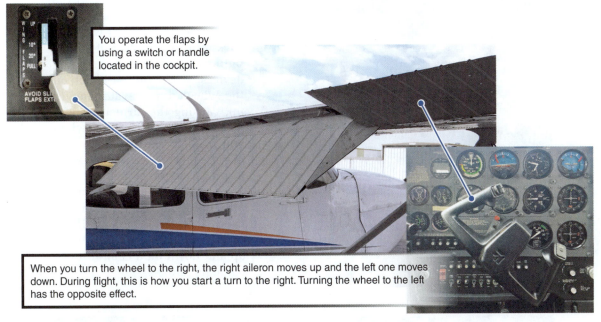

You operate the flaps by using a switch or handle located in the cockpit.

When you turn the wheel to the right, the right aileron moves up and the left one moves down. During flight, this is how you start a turn to the right. Turning the wheel to the left has the opposite effect.

Figure 2-6. The ailerons and flaps are the control surfaces on the trailing edge of each wing.

EMPENNAGE

The **empennage** typically consists of the **vertical stabilizer**, or fin, the rudder, the **horizontal stabilizer**, and the elevator. These surfaces act like the feathers on an arrow to steady the airplane and help maintain a straight path through the air. [Figure 2-7]

Vertical Stabilizer

N20JA

Horizontal Stabilizer

Figure 2-7. The vertical and horizontal stabilizers are located on the empennage.

The **rudder** is attached to the back of the vertical stabilizer. It creates aerodynamic forces that move the nose of the airplane left or right, much like a rudder is used to turn a ship. You operate the rudder with your feet, using pedals. When you press the left pedal, connecting cables move the rudder to the left, which causes the airplane's nose to move to the left. Pressing the right pedal moves the rudder and the nose to the right. You use the rudder in combination with the ailerons to turn the airplane. [Figure 2-8]

Figure 2-8 The rudder deflects the airflow over the vertical stabilizer to the left or right, causing the nose of airplane to move left or right.

The **elevator** is attached to the back of the horizontal stabilizer. It is adjusted by a control wheel or stick through a system of cables, pulleys, and other connecting devices. When you pull back on the wheel, the elevator moves up, causing the airplane's nose to move, or pitch, up; when you push forward, the elevator moves down, which causes the airplane's nose to pitch down. Some empennage designs use a one-piece horizontal stabilizer called a **stabilator**. Used in lieu of an elevator, the stabilator pivots up and down on a central hinge point. [Figure 2-9]

Figure 2-9. The elevator or the stabilator deflects the airflow over the horizontal stabilizer up or down, causing the nose of airplane to pitch up or down.

TRIM TABS

Some airplanes may include small hinged trim devices attached to the trailing edge of one or more of the control surfaces. These mechanisms help minimize your workload by aerodynamically helping you move a control surface or maintain the surface in a desired position. One such device commonly used on training airplanes is called a **trim tab**. The trim tab counteracts the resistance you feel on the flight controls due to airflow over the associated control surface. Although your airplane might have both rudder and elevator trim, if an airplane uses only one trim tab, it is usually located on the elevator. [Figure 2-10]

Figure 2-10. This elevator trim tab is controlled by a wheel on the flight deck (inset).

ANTI-SERVO TAB

A stabilator normally uses an **anti-servo tab** to provide you with a control "feel" similar to an elevator. Without the anti-servo tab, control forces from the stabilator would be so light that you might "over control" the airplane, moving the control wheel too far to obtain the desired result. The anti-servo tab also functions as a trim tab to maintain the stabilator in the desired position. Another very similar device is a servo tab that works in a similar way to an anti-servo tab except it reduces the control force required to move the controls. [Figure 2-11]

Courtesy of Piper Aircraft Inc.

Figure 2-11 Some airplanes, such as Piper, use a stabilator instead of a horizontal stabilizer and elevator combination.

GROUND ADJUSTABLE TRIM TABS

Ground adjustable tabs typically are thin pieces of metal located on the rudder and/or ailerons that are create a trim force to aid in maintaining coordinated flight. The correct displacement is determined by trial and error. You should consult with an aviation maintenance technician (AMT) and the manufacturer's recommendations before adjusting these tabs. [Figure 2-12]

SECTION A ■ **Airplanes**

Figure 2-12. Ground adjustable trim tabs—often located on the ailerons or rudder—can be adjusted to help the airplane maintain coordinated flight.

Figure 2-13. Because of the added clearance between the propeller and the ground, tailwheel airplanes are desirable for operations on unimproved fields.

LANDING GEAR

The landing gear supports the airplane while it is on the ground, and also absorbs landing loads. On landplanes, the landing gear normally consists of three wheels—two **main wheels**, one on each side of the fuselage, and a third wheel positioned either at the front or rear of the airplane. Landing gear with a rear-mounted wheel is called **conventional landing gear.** Airplanes with conventional landing gear are commonly called **tailwheel** airplanes. [Figure 2-13]

When the third wheel is located on the nose, it is called a **nosewheel**, and the design is referred to as **tricycle gear**. [Figure 2-14]

Nosewheels can be either steerable or castering. Steerable nosewheels are linked to the rudder pedals by cables or rods while castering nosewheels are not. In both cases, you steer the airplane using the rudder pedals, however, airplanes with a castering nosewheel might require more use of the brakes, particularly in tight turns.

Landing gear can also be classified as either fixed or retractable. **Fixed gear** always remains extended and has the advantage of simplicity combined with low cost. **Retractable gear** streamlines the airplane by stowing the landing gear inside the structure during cruising flight. The increased weight and cost of retractable gear systems normally limits their use to high-performance aircraft. [Figure 2-15]

Courtesy of Boeing

Figure 2-14. Tricycle landing gear is used on most currently-manufactured airplanes, including almost all modern transport category airplanes.

Seaplanes, Amphibious Aircraft, and Flying Boats

Airplanes that can take off and land on water are practical for flights to remote areas where landing strips are not available. Many landplanes can be fitted with twin floats that support them on water. Normally, these types of airplanes are referred to as seaplanes. A plane that can operate from both land and water is known as an amphibian. Some amphibian airplanes use a hull like that of a boat and others have retractable landing gear built into their floats. You must have an appropriate FAA rating to fly a seaplane or amphibian.

Copyright Corel

SECTION A ■ Airplanes

Courtesy of Piper Aircraft Inc.

Figure 2-15. Higher-performance airplanes have retractable landing gear, while most trainers have fixed landing gear.

Courtesy of Piper Aircraft Inc.

LANDING GEAR STRUTS

The wheels of an airplane are attached to the aircraft structure by struts that absorb the shock of landing and taxiing over rough ground. The spring steel and bungee cord struts transmit shock to the aircraft structure at a rate that reduces stress, and also reduces the tendency for the airplane to bounce into the air after a hard landing. [Figure 2-16]

Photos Copyright of Corel

Figure 2-16. Spring steel (left) and bungee cord struts (right) transmit shock to the airplane at an acceptable rate.

The most widely used type of strut is an air-oil shock absorber that is normally an integral part of the wheel attachment assembly. Referred to as an **oleo strut**, this type of shock uses a piston enclosed in a cylinder with oil and compressed air to absorb the bumps and jolts encountered during landing and taxi operations. [Figure 2-17]

Figure 2-17. The oleo strut consists of an enclosed cylinder, which houses a piston, oil, and air. It absorbs pressure rapidly and then slowly releases it.

BRAKES

The typical training airplane uses **disc brakes** located on the main wheels. [Figure 2-18] You can apply the brakes by pressing on the top of each rudder pedal. To stop or slow the airplane in a straight line, you apply equal pressure on both the left and right brakes. You can also apply **differential braking** to help steer the airplane while taxiing. Most airplanes also provide a hand-operated parking brake that holds pressure on both brakes to keep the airplane from rolling when your feet are off the pedals.

Figure 2-18. Most light general aviation airplanes use a hydraulically actuated disc brake on each main wheel.

POWERPLANT

In small airplanes, the **powerplant** includes both the engine and the propeller. The primary function of the **engine** is to provide the power to turn the propeller. Accessories connected to the engine generate electrical power, create vacuum power for some of the flight instruments and, in most single-engine airplanes, provide heat for the cabin. A **firewall**, which is located between the engine compartment and the flight deck, protects aircraft occupants and serves as a mounting point for the engine. The engine compartment is enclosed by a **cowling**, which streamlines the airplane and increases engine cooling effectiveness by ducting air around the cylinders. [Figure 2-19]

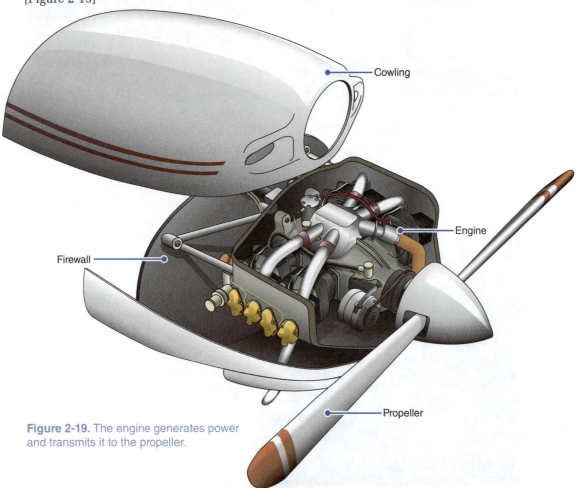

Cowling

Engine

Firewall

Propeller

Figure 2-19. The engine generates power and transmits it to the propeller.

The **propeller**, mounted on the front of the engine, translates the rotational force of the engine into a forward-acting force called thrust, which moves the airplane through the air. Although the number of blades on a propeller can vary, most general aviation training airplanes use a two-bladed propeller. [Figure 2-20]

Figure 2-20. The propeller receives power from the engine and provides thrust to move the airplane through the air.

AIRCRAFT CONSTRUCTION MATERIALS

Airplane construction has evolved throughout the years. In the early days, airplane designers used a truss structure that was made of mostly wood and fabric. As airplanes evolved into monocoque and semi monocoque designs, designers used metal materials, like titanium, steel, and aluminum. Today, modern airplane engineers favor composite materials for strength and sustainability.

EARLY CONSTRUCTION

Many early airplanes were built with wood and fabric, which were abundant and easy to access materials. [Figure 2-21] Today, wood is not commonly used to construct airplanes due to a variety of disadvantages when compared to modern metal and composite construction, including greater weight, reduced strength, and susceptibility to deterioration in moist conditions. However, some manufacturers still utilize wood and fabric for wings for ultra-light, sport, and homebuilt kits.

Figure 2-21. The Wright Brothers used wood as their primary construction material due to its good weight-to-strength ratio and its ability to be molded into spars and beams for the fuselage, wings, and other components.

METAL CONSTRUCTION

Metal advancements in the 1920s led to the development of airplanes with full monocoque metal fuselages. The first metal airplanes used corrugated steel. As metal construction evolved, aluminum was used as the demand for lighter, more efficient airplanes increased. Today, airplane manufacturers continue to use steel, aluminum, and titanium because metal is strong and durable, easy to manufacture and form into components, and is weather resistant. But, while metal is not easy to crack; depending on the airplanes design, it can be easy to dent. Although most modern airplanes use a semi-monocoque structure that provides strength and support, dents and corrosion leading to costly maintenance repairs still remain a concern. [Figure 2-22]

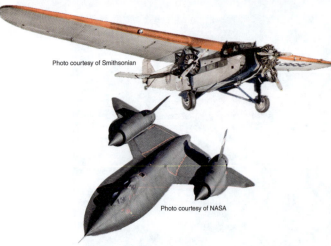

Photo courtesy of Smithsonian

Photo courtesy of NASA

Figure 2-22. The Ford Trimotor (Tin Goose), which first took flight in 1925 was constructed of an aluminum alloy while the Lockheed SR-71 (Blackbird) reconnaissance airplane is primarily composed of titanium.

COMPOSITE CONSTRUCTION

Composite construction combines two or more materials to make a structure that is much stronger than the individual components. Unlike metal, materials used in composites like carbon fiber, Kevlar, and fiberglass are not vulnerable to corrosion so composite structures have a longer lifespan with less maintenance.

Composite technology gives airplane engineers the flexibility to improve on complex airplane designs. Composite materials provide a stronger, smoother skin, with a lighter weight structure. These characteristics create performance benefits such as the reduction of drag and greater fuel efficiency. [Figure 2-23]

Figure 2-23 This Boeing 787 fuselage is manufactured from composite materials as a one piece barrel.

Composite construction has some disadvantages. Because composite materials do not break easily or dent like aluminum, if an impact occurs, damage to the interior structure of the airplane can be difficult to detect. If you think an impact might have occurred to the surface of your airplane, ask an inspector familiar with composites to examine the structure to determine if underlying damage exists. In addition, composite materials can be damaged by excessive heat and are expensive to manufacture and repair. [Figure 2-24]

Figure 2-24. A discrepancy noticed on the airplane skin during preflight can alert you to possible interior structural damage.

AIRWORTHINESS REQUIREMENTS

Before every flight, you use a checklist to perform a preflight inspection to ensure that your airplane is safe to fly. For example, you check for damage, verify that the tires are properly inflated, and confirm the proper levels of fuel and oil. As part of this inspection, you must ensure that airworthiness requirements are met and determine the actions to take if you discover a discrepancy.

REQUIRED DOCUMENTS

Your airplane is required to have specific documentation on board for every flight. An **airworthiness certificate** must be displayed in a location where passengers can see it. This certificate means that at the time of issuance by the FAA, the aircraft met its design and manufacturing standards and was in condition for safe flight. The airworthiness certificate is valid as long as the required maintenance and inspections are performed on the aircraft within the specified time periods and the aircraft remains in airworthy condition. A **registration certificate** issued to the owner must be on board and aircraft operating outside of the U.S. require a **radio station class license**. In addition, the airplane flight manual/pilot's operating handbook (AFM/POH) contains **operating limitations** and **weight and balance data.** Use the acronym **ARROW** to help you remember these required documents during your preflight inspection. [Figure 2-25]

irworthiness Certificate (91.203)

- Issued by the FAA.
- Must be displayed in a location visible to passengers.
- Means that the airplane meets its design and manufacturing standards.
- Valid if the required maintenance and inspections are performed.

egistration Certificate (91.203)

- Issued to the owner.
- Indicates that the airplane is listed in the FAA Aircraft Registry
- Valid for 7 years or for 90 days as a temporary certificate, which is pink.

adio Station Class License

Required by the Federal Communications Commission (FCC) when transmitting to ground stations outside the United States.

perating Limitations (91.9)

- Included in the in the AFM/POH and on markings and placards in the airplane.
- Established by the airplane manufacturer and the FAA to ensure safe airplane operation.

eight and Balance Data

- Contained in the AFM/POH to provide the necessary procedures, tables, and graphs for determining if your airplane is within weight and balance limits.
- Includes information specific to your airplane, such as the current empty weight and the equipment list that are updated if a modification is made to the airplane.

Figure 2-25. Ensuring your airplane is airworthy includes verifying that the required documents are on board.

PILOT'S OPERATING HANDBOOK

You can find most of the pertinent information about a particular make and model of airplane in the **pilot's operating handbook (POH)**. The format and content of the POH is standardized to make it easy to find information and to transition between different makes and models of airplanes. The FAA requires all currently-manufactured airplanes to be equipped with an **FAA approved airplane flight manual (AFM)**, which is specifically assigned to the individual airplane and must be accessible by the pilot during flight. To satisfy the regulatory requirement, the POH for most of these aircraft is also designated as the AFM. [Figure 2-26]

(1) GENERAL – Presents basic information such as loading, handling, and preflight of the aircraft. Also includes definitions, abbreviations, symbology, and terminology explanations.

(2) LIMITATIONS – Includes operating limitations, instrument markings, color coding, and basic placards necessary for the safe operation of the airplane.

(3) EMERGENCY PROCEDURES – Provides checklists followed by amplified procedures for coping with various types of emergencies or critical situations. Related recommended airspeeds are also included. At the manufacturer's option, a section on abnormal procedures may be included to describe recommendations for handling equipment malfunctions or other abnormalities that are not of an emergency nature.

(4) NORMAL PROCEDURES – Includes checklists followed by amplified procedures for conducting normal operations. Related recommended airspeeds are also provided.

(5) PERFORMANCE – Gives performance information appropriate to the airplane, plus optional information presented in the most likely order for use in flight.

(6) WEIGHT AND BALANCE – Includes weighing procedures, weight and balance records, computation instructions, and the equipment list.

(7) AIRPLANE AND SYSTEMS DESCRIPTION – Describes the airplane and its systems in a format considered by the manufacturer to be most informative.

(8) HANDLING, SERVICE, AND MAINTENANCE – Includes information on airplane inspection periods, preventative maintenance that can be performed by the pilot, ground handling procedures, servicing, cleaning, and care instructions.

(9) SUPPLEMENTS – Contains information necessary to safely and efficiently operate the airplane's various optional systems and equipment.

(10) SAFETY AND OPERATIONAL TIPS – Includes optional information from the manufacturer of a general nature addressing safety practices and procedures.

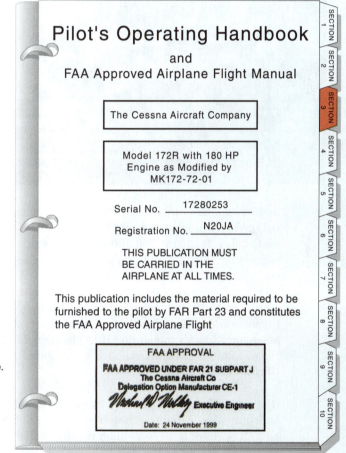

Figure 2-26. The FAA approved airplane flight manual can contain as many as ten sections as well as an optional alphabetical index.

REQUIRED MAINTENANCE AND INSPECTIONS

The FAA requires specific inspections and maintenance to be performed on aircraft at regular intervals to maintain airworthiness. An appropriately certificated aviation maintenance technician (AMT) must perform the **annual inspection**, the **100-hour inspection**, and additional required inspections and make logbook entries approving the aircraft for return to service. [Figure 2-27]

Required Inspections

Every 12 Calendar Months (all aircraft)
- Annual
- Emergency locator transmitter (ELT)

Every 24 Calendar Months
- Transponder (all aircraft)
- Altimeter and static system (aircraft operated under IFR)

Every 100 Hours
- Aircraft used for flight instruction for hire and provided by the flight instructor
- Aircraft that carry any person, other than a crewmember, for hire

Figure 2-27. To remain airworthy, an aircraft must have all of its required inspections completed on schedule and documented in the appropriate maintenance records.

Inspections are normally scheduled based on calendar months. For example, if an annual inspection is completed on October 6, 2017, the next annual inspection will be due on October 31, 2018. The 100-hour inspection is based on time in service and must be completed within 100 hours after the previous inspection, even if that inspection was completed earlier than required. However, if you are enroute to an airport where the 100-hour inspection is to be performed, you may overfly the 100-hour time by up to 10 hours, but the next inspection is still due within 100 hours of the time the original inspection was due. Another requirement that is not based on calendar months is the emergency locator transmitter (ELT) battery. The battery must be replaced or recharged if the transmitter has been operated for a total of 1 hour or after 50 percent of useful battery life (or charge) has expired.

Aircraft maintenance records normally consist of logbooks for the airframe, engine, and propeller. Although these records are not required to be kept in the airplane, the aircraft operator must have them readily available. Before you fly an airplane, you should personally examine these records to ensure that the airplane has the required inspections. In addition, expect to take the logbooks to your checkride and show the examiner the entries that document the airworthiness of the airplane. [Figure 2-28]

Most aircraft maintenance and repairs are required to be performed by FAA-certificated aviation mechanics, but FAR 43.7 allows **preventive maintenance** to be performed by private pilots. Examples of preventive maintenance are: replacing and servicing batteries, replacing spark plugs, and servicing wheel bearings and struts.

Although most required maintenance involves minor alterations, if an alteration or repair substantially affects the operation of an aircraft in flight, that aircraft must be test flown by at least a private pilot and approved for return to service before you may fly passengers.

Figure 2-28. Before flying the airplane, check the aircraft logbooks to verify that the required maintenance and inspections have been completed.

AIRWORTHINESS DIRECTIVES

An aircraft is considered airworthy at the time the FAA issues an airworthiness certificate for that aircraft. However, unsafe conditions can later be found that make an aircraft, and other aircraft of the same design, unsafe to operate. When that occurs, the FAA imposes additional requirements to maintain airworthiness. When an unsafe condition might exist or develop in an aircraft because of a design defect, maintenance, or another cause, the FAA publishes an **airworthiness directive (AD)**. These ADs are legally enforceable rules governed by FAR Part 39. An AD could be a one-time fix for a defect, recurring maintenance or inspections to address a specific issue, or limitations on the operation of an aircraft. ADs are divided into two categories. An emergency AD requires immediate compliance before the aircraft may be flown again. If the AD is not an emergency AD, it normally requires compliance within a specified period of time.

The aircraft owner or operator must maintain records in the maintenance logbooks that show compliance with all pertinent ADs. An operator might schedule some recurring ADs to be addressed during 100-hour inspections or other events such as oil changes. However, unlike regular inspections, which might include provisions for going over the time limit, ADs are not flexible—their time interval may not be "overflown" unless a provision is included in the AD allowing it. You can keep track of ADs that are published for your aircraft by searching for *Airworthiness Directives* on faa.gov. or by signing up for email updates. [Figure 2-29]

SECTION A ■ **Airplanes**

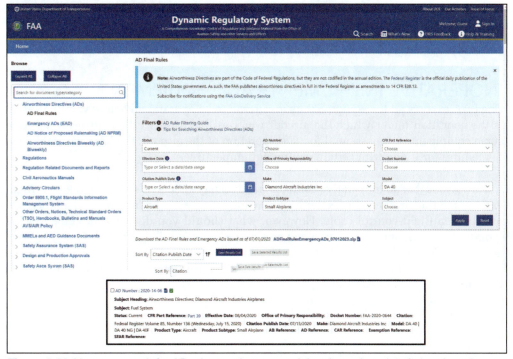

Figure 2-29. You can search for ADs that apply to a special make and model of aircraft on the FAA website.

AIRCRAFT EQUIPMENT REQUIRED FOR VFR

FAR 91.205 requires specific equipment to be installed and operational for VFR flight. Your airplane most likely has equipment beyond that which is required by this regulation. However, you must know the minimum equipment required for VFR day and night operations, which includes flight instruments, engine and system monitoring indicators, and safety equipment. [Figure 2-30]

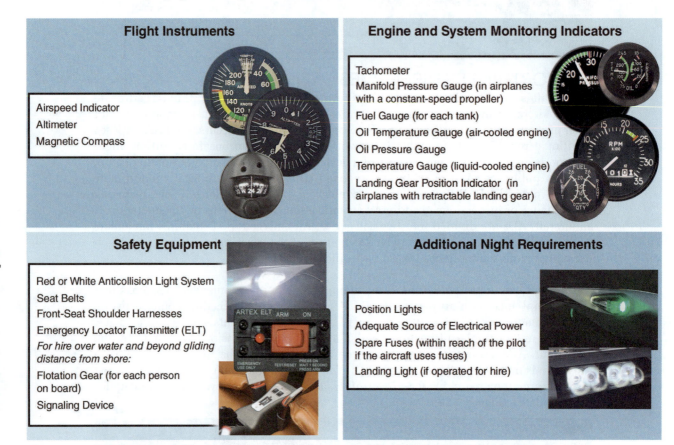

Flight Instruments

Airspeed Indicator
Altimeter
Magnetic Compass

Engine and System Monitoring Indicators

Tachometer
Manifold Pressure Gauge (in airplanes with a constant-speed propeller)
Fuel Gauge (for each tank)
Oil Temperature Gauge (air-cooled engine)
Oil Pressure Gauge
Temperature Gauge (liquid-cooled engine)
Landing Gear Position Indicator (in airplanes with retractable landing gear)

Safety Equipment

Red or White Anticollision Light System
Seat Belts
Front-Seat Shoulder Harnesses
Emergency Locator Transmitter (ELT)
For hire over water and beyond gliding distance from shore:
Flotation Gear (for each person on board)
Signaling Device

Additional Night Requirements

Position Lights
Adequate Source of Electrical Power
Spare Fuses (within reach of the pilot if the aircraft uses fuses)
Landing Light (if operated for hire)

Figure 2-30. FAR 91.205 specifies the minimum equipment required for VFR day and night operations.

INOPERATIVE INSTRUMENTS AND EQUIPMENT

The FAA expects all the equipment on an aircraft to be operational before you fly it—that is the condition under which the aircraft was certified as airworthy. However, because some equipment is nonessential for safe operation of the aircraft, the FAA has published procedures for flying with certain equipment inoperative.

If it is obvious that an inoperative component adversely affects safety, postpone your flight and ground the airplane. For example, you would not fly an airplane with a bad magneto or a broken flight control cable. However, what should you do when the effect of the inoperative equipment on safety is not as obvious? Can you fly VFR with an inoperative attitude indicator? What if one of the radios is inoperative or the landing light does not work? For these situations, you need to utilize an appropriate procedure for determining whether the aircraft is safe and legal to fly.

If a minimum equipment list (MEL) exists for your aircraft, you must use it. The MEL takes into consideration the regulations and the specific requirements for your aircraft and flight operation and indicates the equipment that is allowed to be inoperative for a particular flight. Because MELs are not common for light single-engine piston-powered airplanes, you will most likely use the procedure described in FAR 91.213. [Figure 2-31]

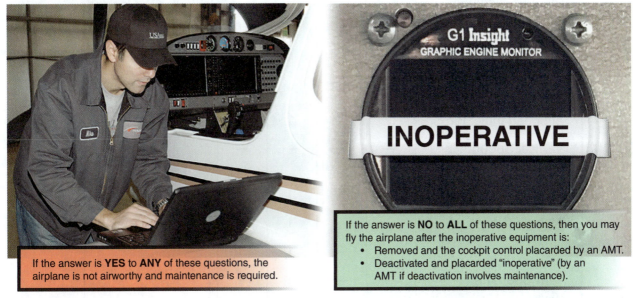

Answer yes or no—Do any of the following require the inoperative equipment?
• The VFR-day type certificate requirements prescribed in the airworthiness certification regulations
• FAR 91.205 for the specific kind of flight operation (e.g. day or night VFR) or by other flight rules for the specific kind of flight to be conducted
• The aircraft's equipment list or the kinds of operations equipment list (KOEL)
• An airworthiness directive (AD)

If the answer is **YES** to **ANY** of these questions, the airplane is not airworthy and maintenance is required.

If the answer is **NO** to **ALL** of these questions, then you may fly the airplane after the inoperative equipment is:
• Removed and the cockpit control placarded by an AMT.
• Deactivated and placarded "inoperative" (by an AMT if deactivation involves maintenance).

Figure 2-31. Before flying with inoperative equipment, you must ask questions to determine if you may operate the aircraft safely and legally without that equipment.

The first question is whether the inoperative equipment is part of the list of instruments and equipment prescribed in the airworthiness regulations under which the aircraft was type-certificated. For small airplanes other than light sport, this is normally FAR 23. This equipment list is different than the basic instruments and equipment required for VFR flight, and could include items such as a stall-warning device. You can often find an equipment list in the AFM with each item of equipment labeled as to whether it is required for flight. If you are not sure where this information is located or how to interpret it, ask an AMT.

Next, determine if the inoperative equipment is required by the regulations for day or night VFR specified in FAR 91.205, or by other rules under which the flight is to be conducted. For example, a transponder is required in Class B airspace. Without ATC authorization, you may not fly the airplane in Class B airspace areas until the transponder repairs are complete. The aircraft manufacturer may include additional required items and specify them in a **kinds of operations equipment list (KOEL)**, which is normally in the limitations section of the AFM. The KOEL specifies the required equipment for conditions such as DAY VFR, NIGHT VFR, or IFR. [Figure 2-32] Finally, check whether an airworthiness directive requires the equipment to be operational. Look in the aircraft maintenance records for a list of ADs and their status.

SECTION A ■ Airplanes

SECTION A ■ Airplanes

System, Instrument, and/or Equipment	Kinds of Operation				Remarks, Notes, and/or Exceptions
	VFR Day	VFR Nt.	IFR Day	IFR Nt.	
Standby Airspeed Indicator	1	1	1	1	Serials 0435 & subs w/ PFD only
PFD Altimeter	-	-	1	1	Serials 0435 & subs w/ PFD only.
Standby Altimeter	1	1	1	1	Serials 0435 & subs w/ PFD only.
PFD Heading	-	-	1	1	Serials 0435 & subs w/ PFD only.
Altimeter	1	1	1	1	
Airspeed Indicator	1	1	1	1	
Vertical Speed Indicator	—	—	—	—	
Magnetic Compass	1	1	1	1	
Attitude Gyro	—	—	1	1	
HSI	—	—	1	1	
Turn Coordinator (Gyro)	—	—	1	1	
Clock	—	—	1	1	
Nav Radio	—	—	1	1	
Pitot System	1	1	1	1	
Static System, Normal	1	1	1	1	

Figure 2-32. Aircraft manufacturers may specify required equipment beyond what is listed in FAR 91.205.

SPECIAL FLIGHT PERMIT

If you determine that the airworthiness requirements are not met, the FAA may permit the airplane to be flown to a location where the needed repairs can be made by issuing a **special flight permit**, sometimes called a **ferry permit**. To obtain this permit, an application must be submitted to the nearest FAA flight standards district office (FSDO). However, if an airworthiness directive prohibits further flight until the AD is satisfied, the FSDO may not issue a ferry permit. In this case, an AMT might have to be transported to the aircraft's location to resolve the AD.

SUMMARY CHECKLIST

✓ The fuselage houses the cabin and flight deck , and provides attachment points for the other major airplane components.

✓ Wings are attached at the top, middle, or lower portion of the fuselage and are contoured to take maximum advantage of the lifting force created by the passing airflow.

✓ The empennage typically consists of the vertical stabilizer, rudder, horizontal stabilizer, and elevator, which act to steady the airplane and maintain a straight path through the air.

✓ The elevator, which is attached to the back of the horizontal stabilizer, is adjusted by a control wheel or stick through a system of cables, pulleys, and other connecting devices. When you pull back on the wheel or stick, the elevator moves up, causing the airplane's nose to pitch up.

✓ Ailerons are located on the outside portion of each wing and create aerodynamic forces to roll the airplane in and out of turns.

✓ Trim devices help minimize your workload by aerodynamically helping you move a control surface, or maintain the surface in a desired position.

✓ Landing gear with a rear-mounted wheel is called conventional landing gear and airplanes with this type of gear are called tailwheel airplanes. When the third wheel is located on the nose, the design is referred to as tricycle gear.

✓ You apply the brakes by pressing on the top of each rudder pedal. You can stop or slow the airplane in a straight line by applying equal pressure on the left and right brakes, or use differential braking to help steer the airplane while taxiing.

✓ The engine provides power to turn the propeller, generate electrical energy, create vacuum power for some flight instruments, and in most single-engine airplanes, provides a source of heat for the pilot and passengers.

✓ The propeller translates the rotational force of the engine into thrust, which moves the airplane forward through the air.

✓ Airplane construction has evolved throughout the years. Modern airplane engineers favor composite materials that are much stronger than individual materials.

✓ Unlike metal, materials used in composites like carbon fiber, Kevlar, and fiberglass are not vulnerable to corrosion so composite structures have a longer lifespan with less maintenance.

✓ Many early airplanes were built with wood and fabric, which were abundant and easy to access materials.

✓ Airplane manufacturers continue to use steel, aluminum, and titanium because metal is strong and durable, easy to manufacture and form into components, and is weather resistant. But, while metal is not easy to crack, depending on the airplanes design, it can be easy to dent.

SECTION A ■ **Airplanes**

✓ Prior to flying, you must verify all required inspections and documentation requirements have been met. Use the acronym ARROW to remember the required documentation: **A**irworthiness certificate; **R**egistration; **R**adio station class license; **O**perating limitations; **W**eight and balance data.

✓ The FAA requires all currently-manufactured airplanes to be equipped with an FAA approved airplane flight manual (AFM), which is specifically assigned to the individual airplane and that must be accessible during flight. To satisfy the regulatory requirement, the pilot's operating handbook (POH) may be designated as the AFM.

✓ You can find most of the pertinent information about a particular make and model of airplane in the POH or AFM.

✓ You may not fly an aircraft unless it has received an annual inspection and an ELT inspection within the previous 12 calendar months.

✓ Transponder inspections are required within the previous 24 calendar months for all flights. Altimeter and static system inspections are required every 24 calendar months if the aircraft is flown under IFR.

✓ 100-hour inspections are required on aircraft that are used for hire, including flight instructing.

✓ Required inspections must be conducted by an appropriately certificated aviation maintenance technician (AMT) and documented in the aircraft's maintenance records.

✓ If an alteration or repair substantially affects the operation of an aircraft in flight, that aircraft must be test flown by at least a private pilot and approved for return to service before you may fly passengers.

✓ When an unsafe condition might exist or develop in an aircraft or other aircraft of the same design, the FAA publishes an airworthiness directive (AD). ADs are legally enforceable rules and compliance is mandatory.

✓ FAR 91.205 requires specific equipment to be installed and operational for day and night VFR flight, which includes flight instruments, engine and system monitoring indicators, and safety equipment.

✓ If an airplane has inoperative equipment but does not have an MEL, you must determine whether the equipment affects flight safety and if it is required by the FARs.

✓ If inoperative equipment is not required, you may fly the airplane if the equipment is removed and the flight deck control for the equipment is placarded "inoperative."

✓ To fly an airplane that is not airworthy to a location where repairs can be made, you can submit an application to the nearest FAA flight standards district office (FSDO) for a special flight permit.

KEY TERMS

Fuselage

Open Truss

Stressed Skin

Monocoque

Semi-Monocoque

Wings

Monoplanes

Biplanes

Ailerons

Flaps

Empennage

Vertical Stabilizer

Horizontal Stabilizer

Rudder

Elevator

Stabilator

Trim Tab

Anti-Servo Tab

Main Wheels

Conventional Landing Gear

Tailwheel

Nosewheel

Tricycle Gear

Fixed Gear

Retractable Gear

Oleo Strut

Disc Brakes

Differential Braking

Powerplant

Engine

Firewall

Cowling

Propeller

Composite

Airworthiness Certificate

Registration Certificate

Radio Station Class License

Operating Limitations

Weight and Balance Data

ARROW

Pilot's Operating Handbook (POH)

FAA Approved Airplane Flight Manual (AFM)

Annual Inspection

100-Hour Inspection

Preventive Maintenance

Airworthiness Directive (AD)

Kinds of Operations Equipment List (KOEL)

Special Flight Permit (Ferry Permit)

SECTION A ■ **Airplanes**

QUESTIONS

1. Identify the major components of the airplane that are depicted in the accompanying illustration.

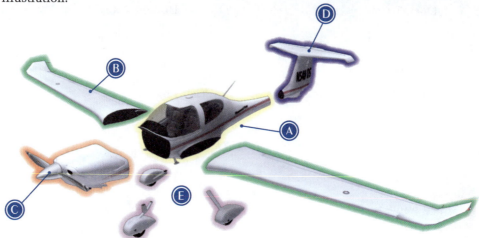

2. What is the primary difference between monocoque and semi-monocoque aircraft construction?

3. When you move the control wheel to the left, which way do the ailerons move?

4. True/False. The rudder is located on the horizontal stabilizer.

5. What is the purpose of trim devices?

6. An airplane with a wheel mounted on the tail is equipped with what type of landing gear?

7. In addition to providing power to turn the propeller, what other functions does the engine in a typical training airplane perform?

8. Why do modern airplane engineers favor composite materials over individual materials?

9. Which of the following is specific to the airplane and must be accessible during flight?
   A. Aircraft maintenance logbooks
   B. Airworthiness Directives
   C. FAA approved airplane flight manual (AFM)

10. What are the documents that are required to be on board the airplane for each flight?

11. An aircraft's annual inspection was performed on June 13, this year. When is the next annual inspection due?
    A. June 1, next year
    B. June 13, next year
    C. June 30, next year

12. A 100-hour inspection was due at 2202.5 hours on the tachometer. The 100-hour inspection was performed at 2209.5 hours. When is the next 100-hour inspection due?
    A. 2302.5 hours
    B. 2312.5 hours
    C. 2309.5 hours

13. You discover inoperative equipment on an airplane that you are planning to fly. What four requirements must you check to determine if the equipment must be operational for this flight?

SECTION B
The Powerplant and Related Systems

As early as 1483, an aerial craft propelled by a screw-type device was conceived by Leonardo da Vinci. However, without a practical means of sustained power to drive the screw, the concept remained unachievable for centuries. It wasn't until 1860 that the first practical piston engine was developed. Although adequate for some land-based functions, further development was necessary before the engine could be used in airborne applications. By 1903, the Wright brothers and their mechanic, Charles Taylor, were able to refine previous designs and build a powerplant that was lightweight, yet powerful enough to meet the requirements of sustained flight. Although woefully inefficient by today's standards, it provided a starting point for continued development. The resulting modern aircraft powerplant still maintains several similarities with its predecessor, including the requirement for precise interaction of the engine, propeller, and other related systems.

ENGINES

Engines in widespread use today can be divided into two categories—reciprocating and turbine. Most large passenger carrying airplanes use a form of the **turbine engine**, which is relatively costly but extremely powerful. Because the large power output of the turbine engine is not required in most general aviation training airplanes, a form of the more economical, but still very reliable, **reciprocating engine** is used. [Figure 2-33]

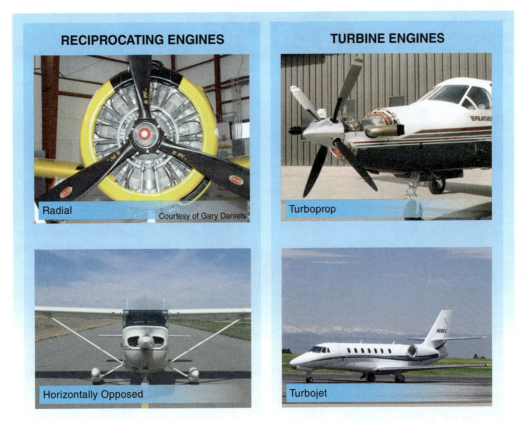

RECIPROCATING ENGINES

Radial
Courtesy of Gary Daniels

Horizontally Opposed

TURBINE ENGINES

Turboprop

Turbojet

Figure 2-33. A variety of reciprocating and turbine engines are in use today.

RECIPROCATING ENGINE OPERATION

Although engine design differs between manufacturers, the principles of reciprocating engine operation are essentially the same. The basic process of converting the chemical energy in fuel into mechanical energy occurs within the cylinders. A fuel/air mixture within the cylinder is compressed by a piston and then ignited. The combustion forces the piston down and that force is transferred through a crankshaft to a propeller. [Figure 2-34]

Figure 2-34. Connecting rods and a crankshaft, like pedals and a bicycle crank, change the back-and-forth motion of the pistons to rotary motion that turns a propeller.

THE FOUR-STROKE OPERATING CYCLE

In typical reciprocating airplane engines, the continuous energy-creating process is referred to as the **four-stroke operating cycle**. The steps in this cycle are: the intake of the fuel/air mixture, the compression by the piston, the ignition and expansion of the gases, and the venting of the burned gases. [Figure 2-35]

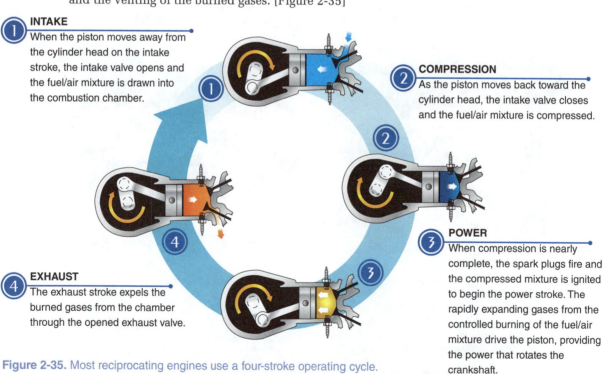

1 INTAKE
When the piston moves away from the cylinder head on the intake stroke, the intake valve opens and the fuel/air mixture is drawn into the combustion chamber.

2 COMPRESSION
As the piston moves back toward the cylinder head, the intake valve closes and the fuel/air mixture is compressed.

3 POWER
When compression is nearly complete, the spark plugs fire and the compressed mixture is ignited to begin the power stroke. The rapidly expanding gases from the controlled burning of the fuel/air mixture drive the piston, providing the power that rotates the crankshaft.

4 EXHAUST
The exhaust stroke expels the burned gases from the chamber through the opened exhaust valve.

Figure 2-35. Most reciprocating engines use a four-stroke operating cycle.

Even when the engine is operated at a fairly slow speed, the four-stroke cycle takes place several hundred times each minute. In a four-cylinder engine, each cylinder operates on a different stroke. Continuous rotation of the crankshaft is maintained by the precise timing of the power strokes in each cylinder.

Continuous operation of the engine as a whole is dependent on the simultaneous function of ancillary systems including the induction, ignition, fuel, oil, cooling, and exhaust systems.

Turbine Engines

At first glance, the turbine engine and reciprocating engine look decidedly different. The surprising fact is however, that the events that take place in converting fuel to thrust-producing energy are essentially the same—intake, compression, combustion, and exhaust. Although the steps occur sequentially within the cylinder of the reciprocating engine, the turbine engine differs in that all four events occur simultaneously in sections of the engine specifically designed for each function. Thrust from a turbojet engine is obtained from the reaction to the rapidly expanding gases as they leave the engine.

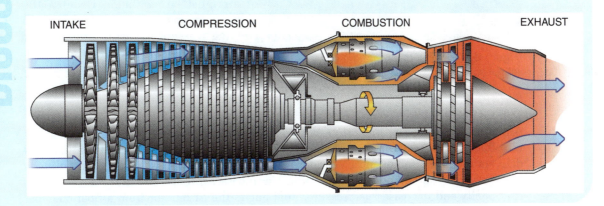

INTAKE   COMPRESSION   COMBUSTION   EXHAUST

INDUCTION SYSTEMS

The purpose of the induction system is to bring outside air into the engine, mix it with fuel in the proper proportion, and deliver it to the cylinders where combustion occurs. Your control over the amount of fuel and air that is introduced into the engine cylinders is maintained by two controls in the cockpit—the **throttle** and the **mixture**. The throttle controls engine power by regulating the amount of fuel and air mixture that flows into the cylinders, while the mixture controls the fuel/air ratio. [Figure 2-36]

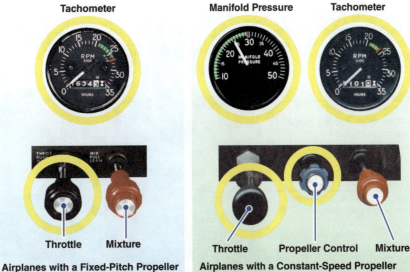

Figure 2-36. You monitor engine power differently on an airplane with a constant-speed propeller than on an airplane with a fixed-pitch propeller.

Airplanes with a Fixed-Pitch Propeller
Many training airplanes have fixed-pitch propellers. You control engine power with the throttle and monitor engine speed, in revolutions per minute (RPM), on a tachometer. Use the mixture control to maintain the proper blend of fuel and air.

Airplanes with a Constant-Speed Propeller
This type of propeller is more efficient because you can adjust it from the cockpit to obtain the optimum blade angle. An airplane with a constant-speed propeller has a propeller control in addition to the throttle and mixture controls. The throttle sets the engine power output, which you read on a manifold pressure gauge, in inches of mercury (Hg). The propeller control sets the engine RPM, which you read on the tachometer, independent of the engine power output.

SECTION B ■ The Powerplant and Related Systems

Figure 2-37. Engine intake air normally enters an intake port through an air filter.

Outside air enters the induction system through an **intake port** at the front of the engine compartment. This port normally contains an air filter that inhibits the entry of dust and other foreign objects. [Figure 2-37] Because the filter can occasionally become clogged, an alternate source of air must be available. Usually, the alternate air comes from inside the engine cowling where it bypasses a clogged air filter. Some alternate air sources function automatically, while others must be operated manually.

THE CARBURETOR

After the air enters the induction system, it moves through a system of ducts and is introduced into the carburetor. The **carburetor** mixes the incoming air with fuel and delivers it to the combustion chamber. A float-type carburetor system is used on many light aircraft. When the air enters the carburetor it passes through a venturi. This increases its velocity and decreases its pressure. Fuel enters the carburetor from a float chamber where it is maintained at a nearly constant level by a float device. The float chamber is vented to the outside so that pressure inside remains equal to the atmospheric pressure, even during climbs and descents. Because the discharge nozzle is located in an area of low pressure created by the venturi, the fuel is forced through the discharge nozzle by the higher atmospheric pressure in the float chamber. [Figure 2-38]

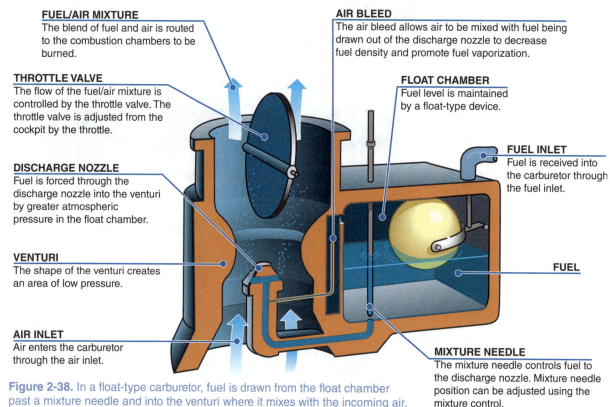

FUEL/AIR MIXTURE
The blend of fuel and air is routed to the combustion chambers to be burned.

THROTTLE VALVE
The flow of the fuel/air mixture is controlled by the throttle valve. The throttle valve is adjusted from the cockpit by the throttle.

DISCHARGE NOZZLE
Fuel is forced through the discharge nozzle into the venturi by greater atmospheric pressure in the float chamber.

VENTURI
The shape of the venturi creates an area of low pressure.

AIR INLET
Air enters the carburetor through the air inlet.

AIR BLEED
The air bleed allows air to be mixed with fuel being drawn out of the discharge nozzle to decrease fuel density and promote fuel vaporization.

FLOAT CHAMBER
Fuel level is maintained by a float-type device.

FUEL INLET
Fuel is received into the carburetor through the fuel inlet.

FUEL

MIXTURE NEEDLE
The mixture needle controls fuel to the discharge nozzle. Mixture needle position can be adjusted using the mixture control.

Figure 2-38. In a float-type carburetor, fuel is drawn from the float chamber past a mixture needle and into the venturi where it mixes with the incoming air.

Carburetors are calibrated at sea level, and the correct fuel-to-air mixture ratio is established at that altitude with the mixture control set in the FULL RICH position. However, as altitude increases, the density of air entering the carburetor decreases while the density of the fuel remains the same. This creates a progressively richer mixture, which can result in engine roughness. The roughness normally is due to spark plug fouling from excessive carbon buildup on the plugs. This occurs because the excessively rich mixture lowers the temperature inside the cylinder, inhibiting complete combustion of the fuel. This condition can occur during the pretakeoff runup at high-elevation airports and during climbs or cruise flight at high altitudes. To maintain the correct fuel/air mixture, you must lean the mixture using the mixture control.

During a descent from high altitude, you must remember to enrich the mixture or it can become too lean. An overly lean mixture can result in high engine temperatures, which can cause excessive engine wear or even failure over a period of time. The best way to maintain the proper mixture is to monitor the engine temperature and enrich the mixture as needed. Because the process of adjusting the mixture can vary from one airplane to another, it's important that you refer to the pilot's operating handbook (POH) to determine the specific procedures for your airplane.

One disadvantage of a float-type carburetor is its icing tendency. **Carburetor ice** occurs due to the sharp temperature drop caused by fuel vaporization and decreasing air pressure in the venturi of the carburetor. If water vapor in the air condenses when the carburetor temperature is at or below freezing, ice can form on internal surfaces of the carburetor, including the throttle valve. Carburetor ice reduces the size of the air passage to the engine, restricting the flow of the fuel/air mixture and reducing power. If enough ice builds up, the engine can stop operating. [Figure 2-39]

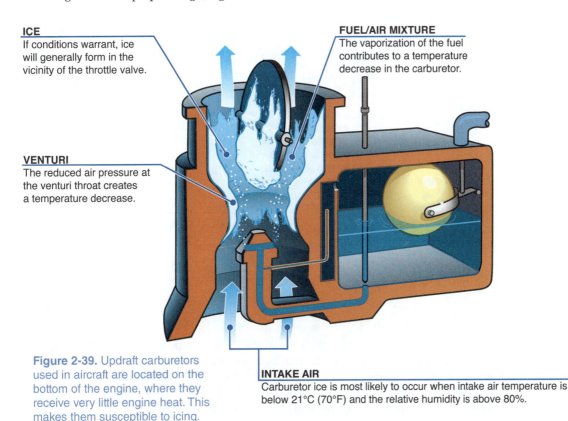

ICE
If conditions warrant, ice will generally form in the vicinity of the throttle valve.

FUEL/AIR MIXTURE
The vaporization of the fuel contributes to a temperature decrease in the carburetor.

VENTURI
The reduced air pressure at the venturi throat creates a temperature decrease.

INTAKE AIR
Carburetor ice is most likely to occur when intake air temperature is below 21°C (70°F) and the relative humidity is above 80%.

Figure 2-39. Updraft carburetors used in aircraft are located on the bottom of the engine, where they receive very little engine heat. This makes them susceptible to icing.

Carburetor ice is more likely to occur when temperatures are below 21°C (70°F) and relative humidity is above 80 percent. However, due to the sudden cooling that takes place in the carburetor, icing can occur even with temperatures as high as 38°C (100°F) and humidity as low as 50 percent. [Figure 2-40]

The first indication of carburetor icing in an airplane with a fixed-pitch propeller is a decrease in engine RPM, followed by engine roughness and possible stoppage. An airplane with a constant-speed propeller will experience a decrease in manifold pressure instead of RPM loss. Although carburetor ice can occur during any phase of flight, it is most likely to occur during reduced power operations, like during descent. One hazard is that carburetor ice can accumulate unnoticed until you try to add power, and then the engine

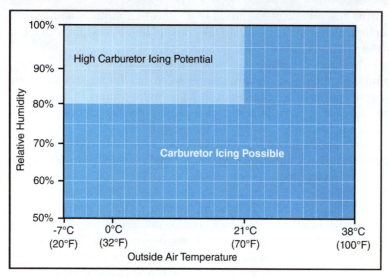

Figure 2-40.
Carburetor ice is likely when the temperature and humidity are in the ranges shown on this chart, but it can also form under other conditions.

fails to respond. To combat ice, engines with carburetors employ a **carburetor heat** system designed to eliminate ice by routing air across a hot surface before it enters the carburetor. Because the heat source is the engine's exhaust system, it is critically important to turn on carburetor heat early, before the engine has already lost power and the exhaust system has cooled down. [Figure 2-41]

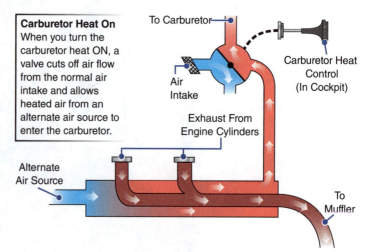

Carburetor Heat On
When you turn the carburetor heat ON, a valve cuts off air flow from the normal air intake and allows heated air from an alternate air source to enter the carburetor.

Figure 2-41. The exhaust system provides heat to prevent and melt carburetor ice.

The use of carburetor heat causes a slight decrease in engine power, because the heated air is less dense than the cold outside air that had been entering the engine. This enriches the mixture. When ice is present in an airplane with a fixed-pitch propeller and you use carburetor heat, there is a slight decrease in RPM, followed by a gradual increase in RPM as the ice melts. If ice is not present, the RPM will decrease slightly, then remain constant. In an airplane with a constant-speed propeller, these effects appear as changes in manifold pressure.

Generally, you should use full carburetor heat whenever you reduce engine RPM below the normal operating range for your airplane, or when you suspect the presence of carburetor ice. Normally, you do not use carburetor heat continuously when full power is required, such as during a takeoff, because of the reduced power available when the carburetor heat is on. Check the POH for specific recommendations.

FUEL INJECTION

One way to eliminate the threat of carburetor ice is to remove the carburetor altogether. Instead of carburetors, many aircraft have fuel injection systems. With these systems, the fuel is not vaporized until it is sprayed directly into the hot engine intakes—that means no risk of ice from chilled fuel. In addition, fuel metering is more precise and distribution of the fuel/air mixture is more consistent than in engines equipped with carburetors.

Fuel injection increases engine efficiency, offering lower fuel consumption, increased horsepower, lower operating temperatures, and longer engine life. Although fuel-injected engines are generally superior to engines equipped with carburetors, they do have some disadvantages, including increased sensitivity to fuel contaminants and more complex starting procedures, particularly when the engine is hot.

The basic components of a fuel injection system are shown in Figure 2-42. The electric fuel pump (sometimes referred to as the auxiliary or boost pump) provides fuel under pressure to the fuel control unit for engine starting and/or emergency use. After starting, the engine-driven fuel pump provides fuel under pressure from the fuel tank to the fuel control unit. The fuel control unit, which replaces the carburetor, meters the fuel based upon the mixture control setting and sends it to the fuel manifold valve at the rate set by the throttle. The fuel then flows to the fuel discharge nozzles located in each cylinder head. Air is mixed with the fuel and the mixture is injected into each intake port where it is drawn into the cylinders for combustion when the intake valve opens.

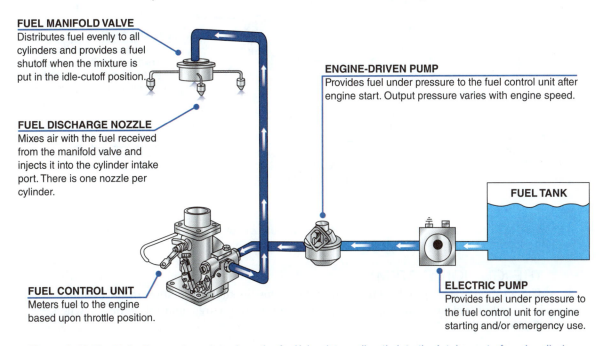

FUEL MANIFOLD VALVE
Distributes fuel evenly to all cylinders and provides a fuel shutoff when the mixture is put in the idle-cutoff position.

ENGINE-DRIVEN PUMP
Provides fuel under pressure to the fuel control unit after engine start. Output pressure varies with engine speed.

FUEL DISCHARGE NOZZLE
Mixes air with the fuel received from the manifold valve and injects it into the cylinder intake port. There is one nozzle per cylinder.

FUEL TANK

FUEL CONTROL UNIT
Meters fuel to the engine based upon throttle position.

ELECTRIC PUMP
Provides fuel under pressure to the fuel control unit for engine starting and/or emergency use.

Figure 2-42. Fuel injection systems introduce the fuel/air mixture directly into the intake port of each cylinder.

SUPERCHARGING AND TURBOCHARGING

When you operate a reciprocating engine at high altitudes, engine power decreases because of lower air density, even though the volume of airflow into the engine remains the same. This is a characteristic of any normally aspirated engine, where the fuel/air mixture is not compressed. If intake air could be compressed, more fuel could be added to the mixture, resulting in more engine power.

A supercharging or a turbocharging system often enables an engine to produce sea-level performance even at high altitudes. A **supercharger** compresses the incoming air using a pump driven by the engine. Although effective, some engine power must be used to drive the supercharger, reducing the net power increase. On the other hand, a **turbocharger** is more efficient because it pressurizes the air using a mechanism driven by engine exhaust gases that would otherwise be vented overboard. In some airplanes, the turbocharger also supplies air for cabin pressurization in addition to compressed air for the engine induction system. [Figure 2-43] Both supercharged and turbocharged engines usually are fuel injected and have tight operating parameters that require you to make careful mixture adjustments and monitor engine operation closely.

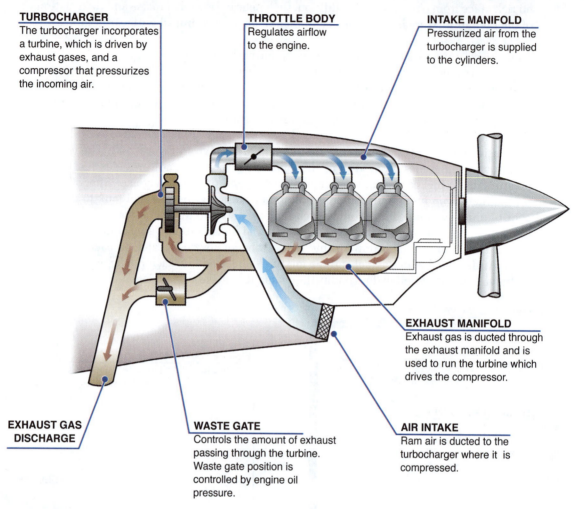

TURBOCHARGER
The turbocharger incorporates a turbine, which is driven by exhaust gases, and a compressor that pressurizes the incoming air.

THROTTLE BODY
Regulates airflow to the engine.

INTAKE MANIFOLD
Pressurized air from the turbocharger is supplied to the cylinders.

EXHAUST MANIFOLD
Exhaust gas is ducted through the exhaust manifold and is used to run the turbine which drives the compressor.

EXHAUST GAS DISCHARGE

WASTE GATE
Controls the amount of exhaust passing through the turbine. Waste gate position is controlled by engine oil pressure.

AIR INTAKE
Ram air is ducted to the turbocharger where it is compressed.

Figure 2-43. Turbocharging systems increase the power output of an engine, especially at high altitude, as well as supplying air for cabin pressurization.

THE IGNITION SYSTEM

The ignition system provides the spark that ignites the fuel/air mixture in the cylinders. It is made up of magnetos, spark plugs, interconnecting wires, and the ignition switch. [Figure 2-44]

A **magneto** is a self-contained, engine-driven unit that supplies electrical current to the spark plugs. It uses a permanent magnet to generate an electrical current completely independent of the aircraft's electrical system. The magneto generates sufficiently high voltage to jump a spark across the spark plug gap in each cylinder. The system begins to fire when the crankshaft begins to turn and continues to operate whenever the crankshaft is rotating.

Most airplanes incorporate a dual ignition system with two individual magnetos, separate sets of wires, spark plugs, and other components to enhance safety and increase reliability of the ignition system. Each magneto operates independently to fire one of the two spark plugs in each cylinder. The firing of two spark plugs improves combustion of the fuel/air mixture and results in a slightly higher power output. If one of the magnetos fails, the other is unaffected. The engine continues to operate normally, although you can expect a slight decrease in engine power. The same is true if one of the two spark plugs in a cylinder fails.

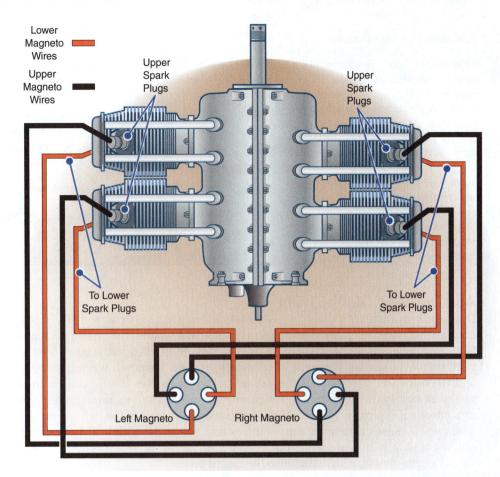

Lower Magneto Wires

Upper Magneto Wires

Upper Spark Plugs

Upper Spark Plugs

To Lower Spark Plugs

To Lower Spark Plugs

Left Magneto

Right Magneto

Figure 2-44. Each magneto is connected to one of the two spark plugs in each cylinder.

You control the operation of the magnetos from the cockpit with the **ignition switch**, which is normally labeled: OFF, RIGHT, LEFT, BOTH, and START. When you select the LEFT or RIGHT position, only the associated magneto is activated. When you select BOTH, the system operates on both magnetos. [Figure 2-45]

You can identify a malfunctioning ignition system during the pretakeoff check by observing the decrease in RPM that occurs when you first move the ignition switch from BOTH to RIGHT, and then from BOTH to LEFT. A small decrease in engine RPM is normal during this check, however, you should consult your POH for the permissible decrease. An excessive decrease could signal the presence of fouled plugs, broken or shorted wires between the magneto and the plugs, or improperly timed firing of the plugs.

Figure 2-45. The magnetos are controlled by the ignition switch on the flight deck.

Following engine shutdown, turn the ignition switch to the OFF position. If you leave the ignition switch ON and someone moves the propeller, the engine could fire and turn over with the potential for serious injury. The engine can run even with the battery and master switches are OFF because the magneto requires no outside source of electrical power.

Loose or broken wires in the ignition system also can cause problems. For example, if the ignition switch is OFF, the magneto will continue to fire if the ignition switch ground wire is disconnected. If this occurs, the only way to stop the engine is to move the mixture lever to the idle cutoff position, then have the system checked by a qualified mechanic.

SECTION B ■ The Powerplant and Related Systems

Turbine Engine Power

A reciprocating engine can convert only about 30 percent of the heat energy it creates into useful work, and propellers convert only about 90 percent of that engine torque into actual thrust. Even with technological advancements, reciprocating engines are inherently limited in power output. That is why large transport and military aircraft use turbine engines to meet their demanding power requirements. These engines produce large amounts of power for their weight and are fuel efficient at the high altitudes they are capable of reaching. Some high-performance military fighter airplanes can extract additional thrust from their jet engines using a technique called afterburning. Afterburners use an enormous amount of fuel and can only operate for short periods of time.

Copyright Boeing

ABNORMAL COMBUSTION

During normal combustion, the fuel/air mixture burns in a very controlled and predictable manner. Although the process occurs in a fraction of a second, the mixture actually begins to burn at the point where it is ignited by the spark plugs, then burns away from the plugs until it's all consumed. This type of combustion causes a smooth buildup of temperature and pressure and ensures that the expanding gases deliver the maximum force to the piston at exactly the right time in the power stroke. **Detonation**, on the other hand, is an uncontrolled, explosive ignition of the fuel/air mixture within the cylinder's combustion chamber. It causes excessive temperatures and pressures which, if not corrected, can quickly lead to failure of the piston, cylinder, or valves. In less severe cases, detonation causes engine overheating, roughness, or loss of power.

Detonation can happen anytime you allow the engine to overheat or if you use a lower than recommended fuel grade. The potential for engine overheating is greatest under the following conditions: takeoff with an engine that is very near the maximum allowable temperature, operation at high RPM and low airspeed, and extended operations above 75% power with an extremely lean mixture. If detonation is suspected on climbout, you can help cool the engine by retarding the throttle and by climbing at a slower rate.

Detonation can often lead to another problem, known as **preignition**. As the name implies, preignition occurs when the fuel/air mixture is ignited in advance of the normal timed ignition. Preignition is caused by a residual hot spot in the cylinder such as a small carbon deposit on a spark plug, a cracked ceramic spark plug insulator, or almost any damage around the combustion chamber. [Figure 2-46]

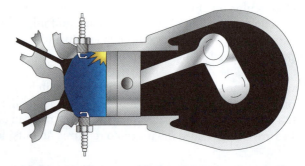

Figure 2-46. Preignition can be caused by an area roughened and heated by detonation.

Preignition and detonation can occur simultaneously. Because both are likely to cause engine roughness and high engine temperatures, you might find it difficult to distinguish between the two. If problems occur, you should attempt to lower cylinder temperature by retarding the throttle, enriching the fuel mixture, and/or lowering the nose attitude.

FUEL SYSTEMS

Fuel systems include a number of individual components such as tanks, lines, vents, valves, drains, and gauges. There are two general types of fuel systems found in light airplanes— those that require a fuel pump and those that operate on gravity feed. A **fuel-pump system** is used in airplanes with fuel injection systems to provide sufficient pressure to the injector nozzles. [Figure 2-47]

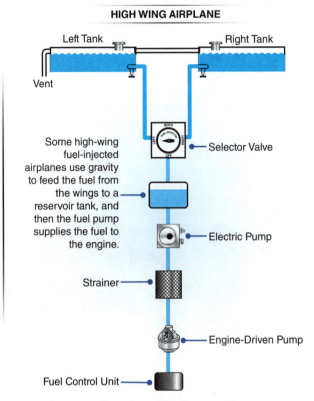

LOW-WING AIRPLANE

- Fuel Control Unit
- Engine-Driven Pump
- Electric Pump
- Strainer
- Left Tank
- Right Tank
- Selector Valve

HIGH WING AIRPLANE

- Left Tank
- Right Tank
- Vent
- Some high-wing fuel-injected airplanes use gravity to feed the fuel from the wings to a reservoir tank, and then the fuel pump supplies the fuel to the engine.
- Selector Valve
- Electric Pump
- Strainer
- Engine-Driven Pump
- Fuel Control Unit

Figure 2-47. Fuel pump systems are used in both low- and high-wing airplanes with fuel injection systems.

Fuel pump systems are also found in low-wing airplanes with carburetors because the fuel tanks are located below the engine. In a fuel pump system an engine-driven pump provides fuel under pressure from the fuel tanks to the engine. Because the engine-driven pump operates only when the engine is running, an electric fuel pump, which is controlled by a switch in the cockpit, provides fuel under pressure for engine starting and as a backup, should the engine-driven pump malfunction.

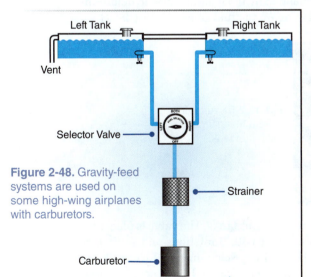

- Left Tank
- Right Tank
- Vent
- Selector Valve
- Strainer
- Carburetor

Figure 2-48. Gravity-feed systems are used on some high-wing airplanes with carburetors.

The fuel pump system usually includes a **fuel pressure gauge** that can be helpful in detecting fuel pump malfunctions. If fuel pressure drops below the normal operating range, turning on the electric pump will ensure a steady flow of fuel to the engine. Any interruption of fuel flow in a fuel pump system can cause problems. In a fuel pump equipped airplane, running a tank completely dry can allow air to enter the fuel system. When this situation develops, it might be difficult, or impossible, to restart the engine. **Vapor lock** is a related situation that can occur when high engine temperatures cause bubbles of fuel vapor to form in the fuel lines or fuel pump.

High-wing airplanes with a carburetor can have a **gravity-feed system**—fuel flows by the force of gravity from the fuel tanks to the engine. The difference in height between the wing-mounted fuel tanks and the engine enables the fuel to flow under sufficient pressure to meet the requirements of the engine. [Figure 2-48]

Most gravity-feed, and some fuel-pump systems incorporate a manually operated pump called a primer. It is used to pump fuel directly into the intake system prior to engine start. The primer is useful in cold weather when fuel in the carburetor is difficult to vaporize. [Figure 2-49]

Figure 2-49. This manual pump-type primer can be used to help start a cold engine.

FUEL SYSTEM COMPONENTS

The **fuel tanks**, which are usually located in the wings, contain vents that enable air pressure inside the tanks to remain the same as that outside the tanks. This prevents the formation of a vacuum that would restrict fuel from flowing out of the tank. The vents can be located in the filler caps, or the tank might be vented through a small tube extending through the wing surface. The tanks also contain an overflow drain that prevents the rupture of the tank due to fuel expansion. The overflow drain can be combined with the fuel tank vent, or it can have a separate opening. On hot days, it is not unusual to see a small amount of fuel coming from the overflow drain. [Figure 2-50]

Figure 2-50. On this Cessna 172, the left tank is vented through a tube (shown) while the right tank is vented through the filler cap.

The **fuel quantity gauges** are located on the instrument panel, where they are usually grouped with engine monitoring gauges. The amount of fuel is measured by a sensing unit in each fuel tank and is usually displayed on the gauge in gallons, pounds or metric units if needed. Do not depend solely on your fuel gauges for determining fuel quantity. Visually check the fuel level in each tank during the preflight inspection or use other methods recommended by the manufacturer to determine fuel quantity and then compare it with the indication on the fuel gauge. [Figure 2-51]

Figure 2-51. Aircraft have a fuel gauge for each fuel tank.

The **fuel selector valve** enables you to select fuel from various tanks. The valve is placarded to show the amount of usable fuel in each tank, together with limitations on its use. One type of selector valve has LEFT, RIGHT, and BOTH positions. Selecting the LEFT or RIGHT position allows fuel to feed only from that tank, while selecting the BOTH position feeds fuel from both tanks. Normally, you place this type of selector in the BOTH position. You can use the LEFT or RIGHT position to balance the fuel amount in each tank in level flight during a cross-country, if necessary. [Figure 2-52]

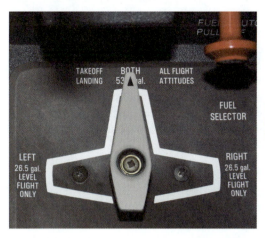

Figure 2-52. This fuel selector valve does not have an OFF position—you pull back the red fuel shutoff knob to stop the flow of fuel.

Another common type of fuel selector switch has LEFT, RIGHT, and OFF, but no BOTH position. With this type of system, you must switch tanks at regular intervals to keep the fuel load balanced. Do not run a tank completely out of fuel before switching tanks. [Figure 2-53]

Figure 2-53. The three position fuel selector demands that you pay close attention to fuel tank quantities.

The fuel passes through a strainer before it enters the engine. The **fuel strainer** removes moisture and other sediments that might be in the system. Because these contaminants are heavier than aviation fuel, they settle in a sump at the bottom of the strainer assembly. It is generally recommended that you drain the fuel strainer before each flight, and check the fuel visually to ensure that no moisture is present. In addition, drain and inspect samples of fuel from each tank. [Figure 2-54]

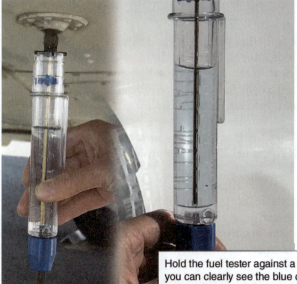

Figure 2-54. Check the fuel for contaminants by draining a sample into a fuel tester.

Hold the fuel tester against a white background so you can clearly see the blue color of the fuel. Any water in the fuel sinks to the bottom of the tester.

SECTION B ■ **The Powerplant and Related Systems**

If moisture is present in the sump, it probably means there is more water in the fuel tanks, and you should continue to drain them until there is no evidence of contamination. If you are in doubt, have the system inspected by a qualified mechanic. In any event, never take off until you are certain that all moisture has been removed from the engine fuel system. Moisture is hazardous because in cold weather it can freeze and block fuel lines. In warm weather, it can flow into the induction system and stop the engine. To help prevent moisture buildup it is a good practice to ensure that an airplane's tanks are refueled following the last flight of the day because moisture can condense in partially filled fuel tanks.

REFUELING

The fuel tanks are replenished with fuel through a filler cap on top of the wing using a fuel nozzle and hose. Whether you fuel the airplane yourself or request fueling from an FBO, refueling can be safely accomplished if a few simple procedures are accomplished.

The major refueling hazard is the possible combustion of the fuel by a spark that causes fumes to ignite. The most probable cause of a spark is from static electricity that discharges between refueling equipment and the airplane. To reduce the risk of sparking, ensure a ground wire is attached to the airplane before the fuel cap is removed from the tank. The airplane should be grounded throughout the refueling procedure and the fuel truck should be grounded to the airport surface. [Figure 2-55] If you are refueling, be careful not to allow the fuel nozzle spout to project very far into the filler opening because it can damage the tank. After refueling, ensure the fuel caps are secure.

Figure 2-55. A ground wire should be attached to the airplane during all refueling operations.

In addition to using the proper refueling technique, you must ensure that you are using the proper grade of fuel. The recommended fuel grade for your airplane is listed in the POH and shown on a placard next to the fuel cap. Fuel grades are identified by octane, or performance number, and different grades are dyed different colors so that you can identify them. The most common fuel for piston-powered aircraft is avgas 100 low lead (100LL), which is dyed blue. You might also find avgas 82 unleaded (82UL), which is designed for low-compression engines that do not require the high octane of Avgas 100 and that can use unleaded fuel. Avgas 82UL is dyed purple. [Figure 2-56]

In an emergency, using the next higher grade of fuel for a short period of time is not considered harmful, provided it's authorized by the manufacturer. On the other hand, using a lower rated fuel can be extremely harmful. In addition, aviation fuel can lose its characteristic color and become clear if various fuel grades are mixed together. This can be particularly dangerous because turbine fuel, which reciprocating engines cannot tolerate, is nearly colorless. Normally, it is not possible to accidently put jet fuel in your aircraft because jet fuel nozzles do not fit in fuel tanks designed for avgas.

Figure 2-56. Color coding of fuel and placement of decals near fuel tank filler caps help ensure that you use the proper grade of fuel.

DISCOVERY

Fuel Tanker Aircraft

Because military aircraft are often optimized for performance at the expense of range, they sometimes need to be refueled during flight. The Boeing KC-46A Pegasus is a sixth-generation fuel tanker based on the B767 airframe. The Pegasus fuel tanker can carry 212,299 pounds (96,265 kg) of fuel at speeds up to 0.86 Mach (565 knots). This Pegasus shown here is refueling two F/A-18 Super Hornets.

Copyright Boeing

OIL SYSTEMS

The engine oil system performs several important functions including lubrication of the engine's moving parts and cooling of the engine by reducing friction and removing some of the heat from the cylinders. Additionally, engine oil improves engine efficiency by providing a seal between the cylinder walls and pistons. During circulation, engine oil also carries away contaminants, which are removed as the oil passes through a filter. Reciprocating engines generally use either a wet-sump or dry-sump oil system. In a **dry-sump system**, the oil is contained in a separate tank and circulated through the engine by pumps. Most light airplanes use a **wet-sump system** in which all of the oil is carried in a sump that is an integral part of the engine. This system is simple, reliable, and suitable for most small reciprocating engines. The oil pump draws oil from the sump and routes it throughout the engine. After the oil passes through the engine, it returns to the sump. In some airplanes, additional lubrication is supplied by the rotating crankshaft, which splashes oil onto portions of the engine. [Figure 2-57]

SECTION B ■ **The Powerplant and Related Systems**

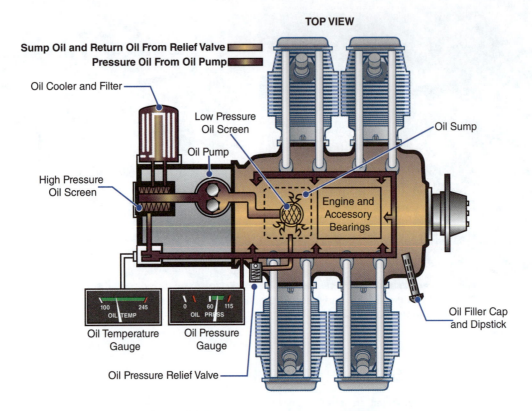

TOP VIEW

Sump Oil and Return Oil From Relief Valve
Pressure Oil From Oil Pump

Oil Cooler and Filter

Low Pressure
Oil Screen

Oil Pump

Oil Sump

High Pressure
Oil Screen

Engine and
Accessory
Bearings

Oil Filler Cap
and Dipstick

100   OIL TEMP   245   0   OIL   PRESS   115
                              60

Oil Temperature   Oil Pressure
Gauge   Gauge

Oil Pressure Relief Valve

Figure 2-57 In a wet-sump system the oil pump draws oil from the sump—the crankcase—and pumps it to the engine.

The oil filler cap and dipstick for measuring the oil quantity are accessible through a panel in the engine cowling. You should check the oil quantity before each flight. The pilot's operating handbook and placards near the access panel indicate the minimum operating quantity, as well as the recommended oil type and weight for your airplane. [Figure 2-58]

Figure 2-58. When checking the oil during your preflight inspection, you can verify the oil type and quantity by referring to the placard on the inside of the oil door.

After engine start, you should monitor the oil pressure and oil temperature gauges. [Figure 2-59] The **oil pressure gauge** provides a direct indication of the oil system operation. It can indicate pressure in pounds per square inch (psi), or bars. A below-normal pressure can mean that the oil pump is not putting out enough pressure to circulate oil throughout the engine, while an above-normal pressure might indicate a clogged oil line. Consider any abnormal indication to mean that vital engine parts are not receiving the necessary lubrication. If that happens, follow your POH or abnormal procedures checklist. Most manufacturers recommend that you shut down the engine if the oil pressure does not begin to rise within 30 seconds after an engine start in warm weather, or within 60 seconds in cold weather.

Figure 2-59. The green band indicates the normal operating range, while red markings indicate the minimum and maximum pressures.

The **oil temperature gauge** is usually located next to the oil pressure gauge, so that you can check both at the same time. It can indicate temperature in degrees Fahrenheit or Celsius. Changes in oil temperature occur more slowly than changes in oil pressure. After starting a cold engine, it can take several minutes or longer for the gauge to show any increase in oil temperature.

You should check the oil temperature periodically to ensure that it remains within the normal operating range. This is particularly important when you are using a high power setting and slow airspeed, such as in a climb, because this increases oil temperature. Abnormally high indications could also indicate a plugged oil line or a low oil quantity.

COOLING SYSTEMS

The combustion process that takes place within the engine's cylinders produces intense heat. Excessively high engine temperatures can result in a loss of power, high oil consumption, and engine damage. Some of the combustion heat exits through exhaust gases. The oil system transfers additional heat away from the internal components of the engine to an oil cooler outside the engine. Outside air flowing around the engine dissipates heat from the oil cooler and the engine cylinders into the outside air. Engine cowlings are carefully designed to provide optimal flow of cooling air with minimum drag. Outside air usually enters the engine compartment through an inlet behind the propeller hub. Baffles direct it to the hottest parts of the engine, primarily the cylinders, which have fins that increase the area exposed to the airflow. [Figure 2-60]

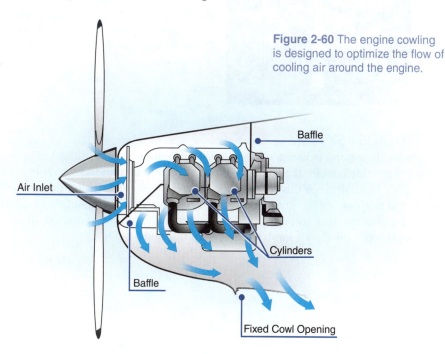

Figure 2-60 The engine cowling is designed to optimize the flow of cooling air around the engine.

Air Inlet

Baffle

Baffle

Cylinders

Fixed Cowl Opening

SECTION B ■ The Powerplant and Related Systems

Air cooling becomes less effective during takeoffs, go-arounds, or any other flight maneuver that combines low airspeed with high power. Conversely, during high-speed descents, the increased airflow, combined with reduced engine power, can result in excessive cooling. To provide more control over engine cooling, some airplanes equipped with cowl flaps, normally placed on the underside of the engine cowling. Opening the **cowl flaps** creates a larger opening for air to exit the engine compartment, which increases the cooling airflow. The proper use of cowl flaps can reduce temperature fluctuations and prolong engine life. [Figure 2-61]

Figure 2-61. You might find cowl flaps on airplanes with high-performance engines that generate a lot of heat.

Airplanes equipped with cowl flaps normally have a **cylinder head temperature gauge** that provides a direct temperature reading from one of the cylinders. You should compare the readings from the cylinder head temperature and oil temperature gauges. A disparity between the two could indicate a malfunction in one of the instruments. Oil and cylinder head temperatures above the normal operating range can indicate that you are using too much power with the mixture set too lean. [Figure 2-62] By monitoring the cylinder head temperature, you can regulate the flow of cooling air by adjusting the position of the cowl flaps using a control in the cockpit. Other methods for reducing engine temperatures include enriching the mixture, reducing the rate of climb, increasing airspeed, and when conditions permit, decreasing the power setting.

Figure 2-62. Like the other engine gauges, the normal range for the cylinder head temperature gauge is marked in green.

THE EXHAUST SYSTEM

In many airplanes, the exhaust system is used not only to vent burned gases overboard, but also to provide heat for the cabin and for defrosting the windscreen. A typical exhaust system for a light airplane directs exhaust out below the engine compartment through a muffler and tailpipe. As the hot exhaust gases heat the muffler, metal shrouds around the muffler capture the heat and duct it to the cabin. The amount of heated air entering the cabin can be controlled by a knob in the cockpit. [Figure 2-63]

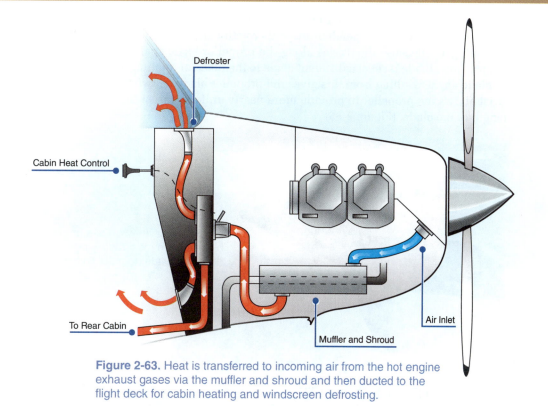

Figure 2-63. Heat is transferred to incoming air from the hot engine exhaust gases via the muffler and shroud and then ducted to the flight deck for cabin heating and windscreen defrosting.

DISCOVERY

Cracked Muffler and Carbon Monoxide

NTSB Narrative: The student solo pilot returned from a cross country flight [in a Cessna 150] complaining of headache, nausea, and difficulty walking. The pilot stated that she had smelled exhaust fumes, was not taught of a potential connection between exhaust smells and carbon monoxide poisoning, and continued flight. Medical tests revealed elevated carbon monoxide, which required 5 1/2 hours of 100% oxygen to reduce to normal levels. Post-flight inspection revealed a crack in the repaired muffler which had been installed 18 hours earlier.

The symptoms experienced by the student pilot are common with carbon monoxide poisoning, which in severe cases, can result in death, or can incapacitate a pilot leading to a deadly crash. Although carbon monoxide itself is colorless and odorless, if you smell exhaust gases, you can assume that carbon monoxide is present. If you experience a headache, dizziness, drowsiness, or a loss of muscle power, you must act immediately before you lose consciousness. Turn off the heater, open the fresh air vents, and use supplemental oxygen, if it is available.

PROPELLERS

The propeller converts engine rotational power into thrust that propels the airplane through the air. It is normally connected directly to the crankshaft, although some propellers are driven through a gear mechanism. The propeller consists of a central hub with two or more blades attached. Each blade is an airfoil that acts like a rotating wing, producing thrust. Because the tip of the propeller blade travels faster in the rotational plane than the portion near the hub, a low-speed airfoil is used near the hub, transitioning to a high-speed airfoil at the tip. [Figure 2-64]

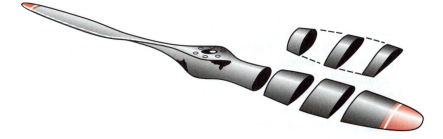

Figure 2-64. The cross sections of this propeller blade illustrate the changing airfoil shapes along the length of the propeller blade.

SECTION B ■ **The Powerplant and Related Systems**

The amount of thrust produced by each section of the propeller is not only a function of its shape, but is also dependent upon its rotational speed and position relative to the oncoming air. Because this varies along the propeller blade span, each small section of the propeller blade is set at a different angle to the plane of rotation. The gradual decrease in blade angle resulting from this gives the propeller blade its twisted appearance. Blade twist allows the propeller to provide more nearly uniform thrust throughout most of the length of the blade. [Figure 2-65]

Figure 2-65. The twist of a propeller blade can be most readily seen by looking down the blade from the tip to the hub.

FIXED-PITCH PROPELLER

The overall blade angle is the average of the blade pitch along the length of the propeller. With a **fixed-pitch propeller**, blade angle is selected on the basis of what is best for the primary function of the airplane, and cannot be changed by the pilot. A propeller with a low blade angle, known as a **climb propeller**, provides the best performance for takeoff and climb, while one with a high blade angle, known as a **cruise propeller**, is optimized for high-speed cruise and high-altitude flight. The only power control for a fixed-pitch propeller is the throttle, and the only power indicator is the tachometer. [Figure 2-66]

Figure 2-66. A fixed-pitch propeller is simple, but cannot deliver optimal performance under varying conditions of speed, power, and pitch attitude.

CONSTANT-SPEED PROPELLER

Compared to a fixed-pitch propeller, a **constant-speed propeller** is more efficient. Found on higher-performance airplanes, they are sometimes referred to as variable-pitch or controllable-pitch propellers because you can adjust the blade angle for the most efficient operation. As a result, a constant-speed propeller converts a high percentage of the engine's power into thrust over a wide range of RPM and airspeed combinations. The engine power on airplanes equipped with a constant-speed propeller is controlled directly by the throttle and indirectly by the propeller control. The power output of the engine is indicated on the manifold pressure gauge and adjusted using the throttle. The **propeller control** is used to change the pitch of the propeller blades. The resulting engine RPM is indicated on the tachometer. [Figure 2-67]

Figure 2-67. The propeller control sets the RPM, which is indicated on the tachometer.

The propeller control enables you to select a low blade angle and high RPM setting for maximum thrust on takeoff. As you reach cruising flight conditions, you can use a higher pitch and a lower RPM setting to maintain the required thrust for the desired airspeed. This is like using a low gear in your car to accelerate, then using a high gear for cruising speed. Most hydraulic pitch-change mechanisms on single-engine airplanes use high-pressure oil to push against the aerodynamic twisting forces acting on the blades. [Figure 2-68]

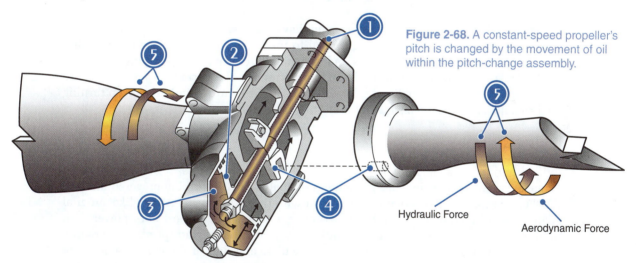

Figure 2-68. A constant-speed propeller's pitch is changed by the movement of oil within the pitch-change assembly.

Hydraulic Force

Aerodynamic Force

① High-pressure oil enters the cylinder through the center of the propeller shaft and piston rod. The propeller control regulates the flow of high-pressure oil through a governor.

② A hydraulic piston in the hub of the propeller is connected to each blade by a piston rod. This rod is attached to forks that slide over the pitch-change pin mounted in the root of each blade.

③ The oil pressure moves the piston toward the rear of the cylinder, moving the piston rod and forks aft.

④ The forks push the pitch-change pin of each blade toward the rear of the hub, causing the blades to twist toward the high-pitch position.

⑤ The governor regulates the oil pressure sent to the pitch-change mechanism to maintain an equilibrium between aerodynamic and hydraulic pitch-changing forces at the selected RPM.

For a given RPM setting, there is usually a maximum allowable manifold pressure. Setting the throttle above this level can cause internal engine stress. Refer to your POH for the allowable settings for your engine. As a general rule, you should avoid high manifold pressures with low RPM settings.

DISCOVERY

Multi-Blade Airplane Propellers

Higher engine power requires greater propeller blade area to efficiently convert the torque to thrust. However, if the blades are too wide or too long, propeller efficiency is degraded. Blade length is limited by ground clearance and blade tip speed. On longer propellers, the tip speed can approach the speed of sound, resulting in increased drag and other adverse performance effects. One way to handle more engine power is to increase the number of blades, which also enables the propeller to turn more slowly, reducing noise. Many higher-performance general aviation airplanes use three-bladed propellers, and larger propeller airplanes, like this Lockheed C-130 Hercules military transport can use four or more blades.

Copyright Corel

PROPELLER HAZARDS

The propeller is extremely dangerous to persons outside of the airplane, and when the engine is running, it is mostly invisible. Warn your passengers about propeller hazards, stressing the need to stay well clear of the propeller area at all times. Generally, you should shut down the engine any time people are near the airplane. Avoid having passengers approach or board the airplane while the engine is running.

If you are flying an old airplane that does not have a starter, or you get stuck at a remote airport with a dead battery, you might need to turn the propeller by hand to start the engine. Hand-propping is dangerous and is generally discouraged by flying clubs and FBOs. Obtain a demonstration and get instruction on the correct procedure before attempting it yourself. A qualified pilot should be at the controls, and the person turning the propeller should be in charge of the starting procedure.

FADEC

Full authority digital engine control (FADEC) is a computer with associated systems that manage an aircraft's engine and propeller. Originally used in turbine-powered aircraft, FADEC is becoming more common in aircraft with piston engines. FADEC, like a digital engine control module on a car, optimizes engine parameters for best horsepower with minimum fuel consumption. FADEC continuously monitors engine speed, temperature, and pressure through a series of sensors to calculate and deliver the correct amount of fuel for each injector. FADEC systems do not have magnetos or mixture controls, and automatically handle engine priming. FADEC-equipped aircraft have a single power lever for the engine—just set the lever to the desired power, and the computer adjusts the engine and propeller. [Figure 2-69]

Courtesy of Diamond Aircraft Industries GmbH

Figure 2-69. This diesel-powered Diamond DA40 NG has FADEC.

Although operation of an engine with FADEC is simpler than one with magnetos, the system itself is more complex and the engine will stop running if the FADEC fails. Therefore, FADEC engines must have redundant engine control units (ECUs) as well as backup electrical power for each ECU, and you must check these systems for proper operation before takeoff.

ELECTRICAL SYSTEMS

The electrical systems in light airplanes power sophisticated avionics and important accessories. Because you depend on the airplane's electrical system for most flight operations, you should understand the system and its basic components. [Figure 2-70]

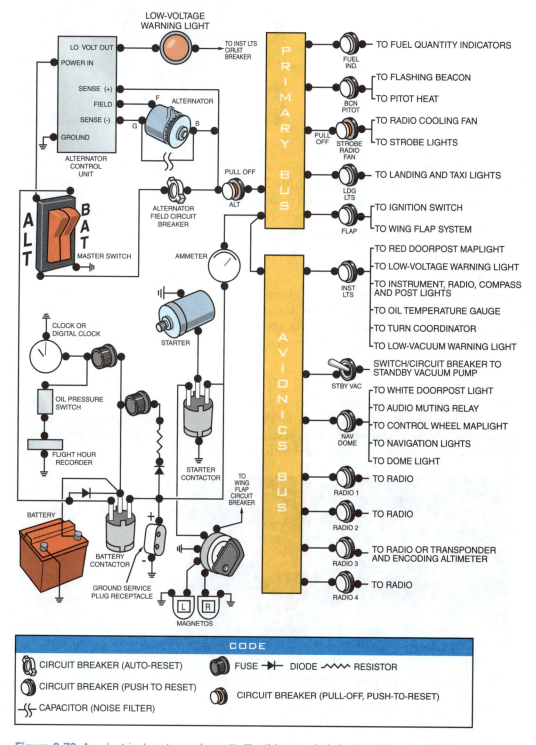

Figure 2-70. An electrical system schematic like this sample is included in most POHs.

SECTION B ■ **The Powerplant and Related Systems**

ALTERNATOR

On light airplanes, electrical energy is supplied by a 14- or 28-volt direct-current system powered by an engine-driven **alternator**. Some older airplanes are equipped with generators, but alternators have many advantages, including as light weight, lower maintenance, and uniform output, even at low engine RPM. [Figure 2-71]

Figure 2-71. Alternators are mounted to and driven by the engine.

Alternators produce **alternating current (AC)** that a built-in rectifier converts to **direct current (DC)**. The current is delivered to a bus bar where it is distributed to various electrical components on the aircraft. Although the components are protected by circuit breakers, you normally should turn off most equipment before starting the engine. This protects sensitive components, particularly the avionics, from damage that could be caused by voltage surges from the starter.

BATTERY

Another essential part of the electrical system is the storage battery. Its main purpose is to provide a means of starting the engine, but it also enables limited operation of electrical components, such as the radios, without starting the engine. In addition, the battery is a source of standby or emergency electrical power in case of alternator malfunction.

AMMETER AND LOADMETER

An **ammeter** shows if the alternator is producing an adequate supply of electrical power and also indicates whether or not the battery is receiving an electrical charge. A **loadmeter** reflects the total percentage of the load placed on the generating capacity of the electrical system by the electrical accessories and battery. [Figure 2-72]

Ammeter

When the pointer is on the plus side, it shows the charging rate of the battery. On the minus side, it shows the discharge rate of the battery.

Loadmeter

This instrument indicates the total load placed on the alternator by the electrical accessories and the battery (if charging). It does not indicate whether the battery is charging or discharging.

Figure 2-72. Ammeter and loadmeters are used to monitor the performance of the aircraft electrical system.

A charging indication on the ammeter (needle on the plus side) is normal following an engine start because the battery power used to start the engine is being replaced. After the battery is charged, the ammeter should stabilize near zero as the alternator supplies only the electrical needs of the equipment. A discharge indication means the electrical load exceeds the output of the alternator, and the battery is making up the difference. This could mean the alternator is malfunctioning, or that electrical load is excessive. If this occurs, turn off equipment to conserve battery power and land as soon as practical.

With a loadmeter, you can tell immediately if the alternator is operating normally because it indicates the amount of current being drawn from the alternator. Check the POH for the normal load to expect. Loss of the alternator will cause the loadmeter to indicate zero.

DISCOVERY

The Pipistrel Alpha Electro Trainer

Pipistrel has developed an electric light sport trainer to serve alongside its gasoline-engine-powered aircraft at flight schools. The electric version is suitable for traffic pattern practice, with similar performance to its gasoline-powered trainers at a fraction of the operating cost. For cross-country and other flights that require increased endurance, students can switch to the Rotax gasoline-powered airplanes that have similar cockpit instruments and controls.

Courtesy of Pipistrel Aircraft

MASTER SWITCH

The **master switch** controls the entire electrical system, except for the ignition system, which is independent of the main electrical system. However, the engine's starter won't operate unless the master switch is ON because power for the starter comes from the electrical system, not the magnetos. Most airplanes use a split-rocker type master switch. The right half is labeled BAT and controls all power to the electrical system. The left half is labeled ALT and controls the alternator. During normal operations, both sides of the master switch are on. If the alternator malfunctions, it can be turned off to isolate it from the system. If you want to check equipment on the ground before starting the engine, you can turn on only the battery side. [Figure 2-73]

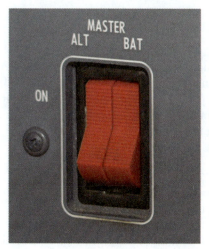

Figure 2-73. Most airplanes use a split-rocker type master switch that enable separate operation of the alternator.

SECTION B ■ **The Powerplant and Related Systems**

CIRCUIT BREAKERS AND FUSES

As the electrical system schematic shows, circuit breakers or fuses are used to protect various components from overloads. With circuit breakers, resetting the breaker usually will reactivate the circuit, unless an overload or short exists. If this is the case, the circuit breaker will continue to pop, indicating an electrical problem. [Figure 2-74]

On an airplane equipped with fuses, manufacturers usually provide a holder for spare fuses in case you need to replace one in flight. The FARs require extra fuses for night flight.

In some airplanes, certain electrical system problems are signaled by the illumination of a low-voltage warning light. Refer to the POH for normal and emergency operations related to the electrical system.

Figure 2-74. Each circuit protection device is labeled with the name of the protected component or system.

BACKUP ELECTRICAL SYSTEMS

The proliferation of airplanes with integrated digital displays has required more electrical system redundancy to ensure that essential instruments continue operating long enough to land safely if an electrical failure occurs at a critical time. Loss of electric power results in the displays going dark, which means that most flight and engine instruments are lost. Although these systems are more critical when flying under instrument flight rules (IFR), as a VFR pilot you should be aware of the safeguards built into the electrical systems of airplanes with digital displays.

DUAL ALTERNATORS

Dual alternators are standard on multi-engine airplanes, but not so typical on single-engine airplanes. However, when all the instruments are electrically powered, a second alternator might be used to provide a backup electrical source on a single-engine airplane. The alternator can be smaller and connected to an essential bus that powers only the equipment that is required for the airplane to be airworthy for IFR operations.

STANDBY BATTERY

A battery that is isolated from the main electrical system by a relay or switch can power essential electrical equipment for a limited time after failure of the main electrical system. Some airplanes use a secondary battery that is kept charged by the main power bus. If the main bus voltage drops below a specific amount, the **standby battery** system disconnects from the main bus and powers the essential bus, which keeps the primary flight instruments, radios and indicator lights operating for a limited time. Another system uses a battery pack that can power the backup attitude indicator and flood lights. [Figure 2-75]

SECTION B ■ **The Powerplant and Related Systems**

Figure 2-75. Standby batteries can be activated by switches on the flight deck in the event of an electrical system failure.

This emergency switch on the Diamond DA40 provides standby power to the electric backup attitude indicator if the main electrical system fails.

The standby battery on this Cessna 172 can power essential avionics for a short time if the main electrical system fails.

FADEC BACKUP POWER

Another design feature that requires backup electric power is FADEC. Loss of power means that the ECUs stop working and the engine stops. Therefore, these systems must incorporate backup power sources to ensure that the engine continues running after failure of the main electrical system. During normal operation, backup batteries are charged from the main power system. However, if power to an ECU bus is lost, a relay is de-energized, disconnecting the backup battery from the main power system and connecting it to the backup engine control unit for that engine. The engine keeps running.

SUMMARY CHECKLIST

✓ The continuous energy-creating process in typical reciprocating airplane engines is referred to as the four-stroke operating cycle. The steps in this cycle are: the intake of the fuel/air mixture, the compression by the piston, the ignition and expansion of the gases, and the venting of the burned gases.

✓ Engine speed for aircraft equipped with a fixed-pitch propeller is displayed on a tachometer in revolutions per minute (RPM).

✓ A constant-speed propeller is adjustable from the cockpit. A manifold pressure gauge is used on these types of airplanes to monitor engine output by displaying the pressure inside the engine in inches of mercury (Hg).

✓ The carburetor mixes incoming air with fuel and delivers it to the combustion chamber.

✓ The operating principle of float-type carburetors is based on the difference in pressure at the venturi throat and the air inlet.

✓ The fuel/air mixture can be adjusted from the cockpit with the mixture control.

✓ Carburetor ice can be caused by fuel vaporization and decreasing air pressure in the venturi, which causes a sharp temperature drop in the carburetor.

✓ Carburetor ice is more likely to occur when temperatures are below 21°C (70°F) and relative humidity is above 80 percent. To combat the effects of carburetor ice, engines with float-type carburetors employ a carburetor heat system that is designed to eliminate ice by routing air across a heat source before it enters the carburetor.

SECTION B ■ **The Powerplant and Related Systems**

✓ One of the most significant advantages of the fuel injection system is the relative freedom from the formation of induction icing.

✓ Sea-level performance can be obtained even at high altitudes using either a supercharging or a turbocharging system.

✓ The ignition system is made up of magnetos, spark plugs, interconnecting wires, and the ignition switch.

✓ Detonation occurs when fuel in the cylinders explodes instead of burning smoothly.

✓ Preignition is a result of the fuel/air mixture being ignited in advance of the normal timed ignition.

✓ A fuel-pump system is used in airplanes with fuel injection systems to provide sufficient pressure to the injector nozzles. Fuel pump systems are also found in low-wing airplanes with carburetors because the fuel tanks are located below the engine.

✓ Because the engine-driven pump operates only when the engine is running, an electric fuel pump, which is controlled by a switch in the cockpit, provides fuel under pressure for engine starting and as a backup, should the engine-driven pump malfunction.

✓ High-wing airplanes with a carburetor can have a gravity-feed system—fuel flows by the force of gravity from the fuel tanks to the engine.

✓ To help prevent moisture buildup it is a good practice to ensure that an airplane's tanks are refueled following the last flight of the day.

✓ A wet-sump system uses an oil pump to draw oil from the sump and route it to the engine. Oil system operation can be monitored by referring to the oil pressure and temperature gauges.

✓ Cooling air enters the engine compartment through an inlet behind the propeller hub where it is further directed to the hottest part of the engine by baffles.

✓ Exhaust is normally directed out below the engine compartment through a muffler and tailpipe. Metal shrouds around the muffler capture heat that is used to defrost the windscreen and heat the cabin.

✓ A fixed-pitch propeller uses a single blade angle that is selected on the basis of what is best for the primary function of the airplane.

✓ A constant-speed propeller control enables you to select a blade angle that is the most appropriate for the flight operation being conducted. The propeller control regulates engine RPM as shown on the tachometer, while the throttle controls engine power output, as indicated on the manifold pressure gauge.

✓ With a constant-speed propeller, you should avoid low RPM settings with high manifold pressure.

✓ When hand-propping an airplane, always ensure that you have received instruction on the correct procedure, and a qualified pilot is at the controls.

✓ Full authority digital engine control (FADEC) is a computer with associated systems that manage an aircraft's engine and propeller.

✓ Alternators produce alternating current (AC) first, and then convert it to direct current (DC) for use in the airplane electrical system.

✓ An ammeter shows if the alternator is producing an adequate supply of electrical power and whether or not the battery is receiving an electrical charge.

✓ A loadmeter reflects the total percentage of the load placed on the generating capacity of the electrical system by the electrical accessories and battery.

✓ Backup electrical systems include dual alternators and backup batteries to power equipment, such as integrated digital displays and FADEC, in the event of an electrical system failure.

KEY TERMS

Turbine Engine

Reciprocating Engine

Four-Stroke Operating Cycle

Throttle

Mixture

Intake Port

Carburetor

Carburetor Ice

Carburetor Heat

Fuel Injection

Supercharger

Turbocharger

Magneto

Ignition Switch

Detonation

Preignition

Fuel-Pump System

Fuel Pressure Gauge

Vapor Lock

Gravity-Feed System

Fuel Tanks

Fuel Quantity Gauges

Fuel Selector Valve

Fuel Strainer

Dry-Sump System

Wet-Sump System

Oil Pressure Gauge

Oil Temperature Gauge

Cowl Flaps

Cylinder Head Temperature Gauge

Fixed-Pitch Propeller

Climb Propeller

Cruise Propeller

Constant-Speed Propeller

Propeller Control

Full Authority Digital Engine Control (FADEC)

Alternator

Alternating Current (AC)

Direct Current (DC)

Ammeter

Loadmeter

Master Switch

Standby Battery

SECTION B ■ The Powerplant and Related Systems

QUESTIONS

1. Identify the four-stroke operating cycle step shown in each of the following illustrations.

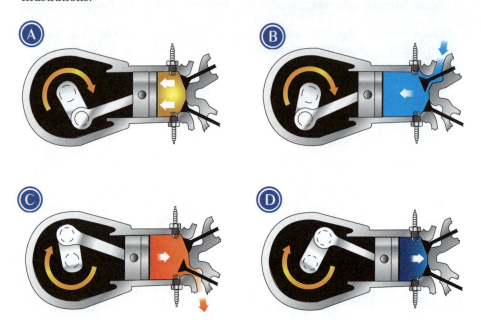

2. As an airplane climbs, do you enrich or lean the mixture to maintain an optimum fuel/air ratio?

3. What is your first indication of carburetor ice in an airplane equipped with a fixed-pitch propeller?

4. Explain why an engine equipped with a fuel injection system is less susceptible to induction icing than one equipped with a float-type carburetor.

5. What is a magneto?

6. The uncontrolled, explosive ignition of the fuel/air mixture within the cylinder's combustion chamber describes which type of abnormal combustion? What actions can you take while airborne to help correct for this problem?

7. Select the true statement regarding fuel systems.
   A. High- and low-wing airplanes with a carburetor typically have gravity-feed systems.
   B. An electric fuel pump provides fuel under pressure to the fuel control unit after engine start.
   C. A fuel-pump system is used in airplanes with fuel injection systems to provide sufficient pressure to the injector nozzles.

8. True/False. If the fuel grade specified for your airplane is not available, you can use a higher grade of fuel.

9. Describe at least two functions performed by the engine oil system.

10. If a constant-speed propeller is set to a high RPM, will the blade pitch (angle) be high or low?

11. True/False. To prevent internal engine damage in an airplane equipped with a constant-speed propeller, you should avoid low RPM settings with a high manifold pressure setting.

12. Immediately after engine start you notice that the ammeter shows a discharge. Is this normal? Why or why not?

SECTION C
Flight Instruments

Of all the instruments located on the airplane flight deck, the indicators that provide you information regarding the airplane's attitude, direction, altitude, and speed are collectively referred to as the flight instruments. Traditionally, the flight instruments are sub-divided into categories according to their method of operation. The instruments that reflect your speed, rate of climb or descent, and altitude operate on air pressure differentials and are called pitot-static instruments. A pictorial view of the airplane's attitude and rate of turn is provided by the attitude indicator and turn coordinator, which operate on gyroscopic principles. The airplane's heading indicator, which also operates using a gyroscope, is usually set by using information from another flight instrument, the magnetic compass. [Figure 2-76]

THE FLIGHT INSTRUMENTS

MAGNETIC
Magnetic Compass

GYROSCOPIC
Attitude Indicator
Turn Coordinator
Heading Indicator

PITOT-STATIC
Airspeed Indicator
Altimeter
Vertical Speed Indicator

Figure 2-76. The flight instruments can be grouped according to method of operation — gyroscopic, magnetic, and pitot-static.

PITOT-STATIC INSTRUMENTS

Pitot-static instruments rely on air pressure differences to measure speed and altitude. The airspeed indicator, altimeter, and vertical speed indicator all use surrounding, or static, air pressure. Pitot pressure, which is only used by the airspeed indicator, is the combination of the static pressure plus the pressure generated as the aircraft moves through the air.

EFFECTS OF ATMOSPHERIC CONDITIONS

Because changes in static pressure can affect pitot-static instrument operation, you should understand some basic atmospheric principles. Atmospheric pressure can be defined as the weight of a single column of air. The pressure decreases from the point of measurement at sea level to the top of the atmosphere. Typically, air exerts about 14.7 pounds per square inch (lb/in²) at sea level. [Figure 2-77] As altitude increases, pressure steadily decreases. For example, at 18,000 feet, atmospheric pressure decreases to approximately one-half of sea level pressure.

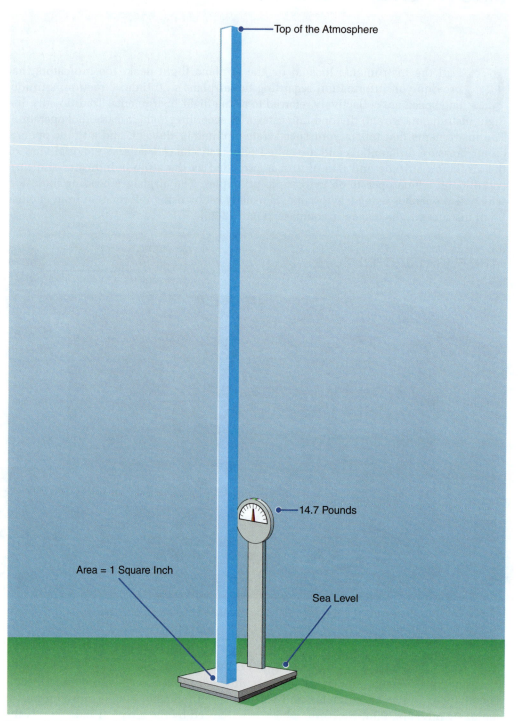

- Top of the Atmosphere
- 14.7 Pounds
- Area = 1 Square Inch
- Sea Level

Figure 2-77. You can think of atmospheric pressure as the weight of a column of air, decreasing as you go higher in that column from sea level.

In addition to changes in altitude, atmospheric pressure can be affected by changes in temperature. For instance, assuming that all other variables remain constant, a decrease in temperature will result in a lower atmospheric pressure. This occurs because a cooler temperature slows the movement of the air molecules, thereby lowering the pressure they exert on the surrounding atmosphere. On the other hand, a warmer temperature increases atmospheric pressure, all else being equal.

Digital Flight Instruments in Airliners

The advanced digital instrumentation that appears in most new general aviation aircraft was first introduced in military and airline flight decks. As this equipment becomes less expensive, small airplanes enjoy more safety benefits from this technology. Meanwhile, transport category aircraft manufacturers continue to raise the standard. This Boeing 787 Dreamliner flight deck is an example of the latest enhancements in airline electronic flight deck instrumentation.

Copyright Boeing

To provide a common reference for temperature and pressure, the **International Standard Atmosphere (ISA)** was established. These standard conditions are the basis for certain flight instruments and most airplane performance data. At sea level, when standard atmospheric conditions exist, the air exerts a pressure 14.7 lb/in² and has a temperature of 15°C (59°F). Atmospheric pressure can be measured with mercury in an inverted tube that is closed on one end. The weight of mercury in the column is balanced by the pressure (weight) of the atmosphere over the mercury reservoir. [Figure 2-78] It turns out that the standard pressure of 14.7 lb/in² supports a column of mercury 29.92 inches high, so standard sea level is defined as 29.92 inches Hg. The metric equivalent is 1013.2 millibars.

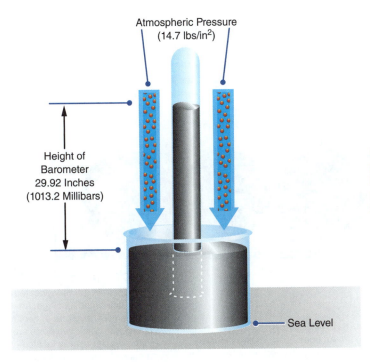

Atmospheric Pressure
(14.7 lbs/in²)

Height of
Barometer
29.92 Inches
(1013.2 Millibars)

Sea Level

Figure 2-78. Atmospheric pressure can be measured by how far it pushes mercury (Hg) up in an inverted tube.

Because both pressure and temperature normally decrease with altitude, **standard lapse rates** can help you calculate the temperatures and pressures you can anticipate at various altitudes. In the lower atmosphere, the standard pressure lapse rate for each 1,000 feet of altitude is approximately 1.00 in. Hg, and the standard temperature lapse rate is 2°C (3.5°F). Although ISA is useful for planning purposes, remember that there can be large variations from standard conditions in the real atmosphere. [Figure 2-79]

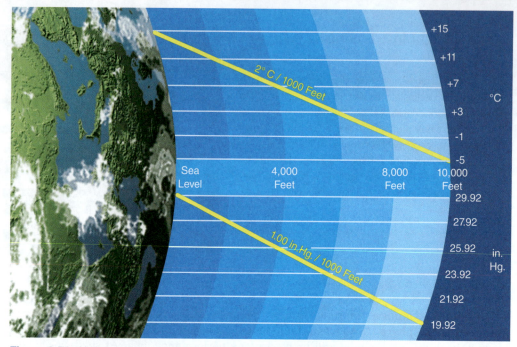

Figure 2-79. Altimetry is based on the observation that atmospheric pressure and temperature decrease by a predictable amount as altitude increases.

PITOT-STATIC SYSTEM

Pitot-static instruments use pressure-sensitive devices to convert pressure supplied by the pitot-static system to instrument indications on the flight deck. Pitot pressure, also called impact or ram air pressure, is supplied by the pitot tube or head. The **pitot tube** is usually mounted on the wing or on the nose section, so the opening is exposed to the relative wind. This arrangement allows ram air pressure to enter the pitot tube before it is affected by the airplane's structure. Because the pitot tube opening faces forward, an increase in speed increases ram air pressure. Static pressure enters the pitot-static system through a **static port,** which is normally flush-mounted on the side of the fuselage in an area of relatively undisturbed air. [Figure 2-80]

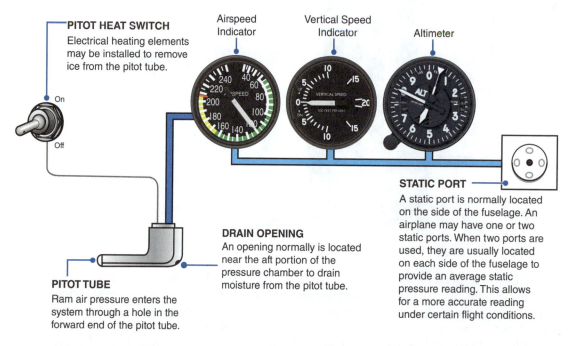

PITOT HEAT SWITCH
Electrical heating elements may be installed to remove ice from the pitot tube.

DRAIN OPENING
An opening normally is located near the aft portion of the pressure chamber to drain moisture from the pitot tube.

STATIC PORT
A static port is normally located on the side of the fuselage. An airplane may have one or two static ports. When two ports are used, they are usually located on each side of the fuselage to provide an average static pressure reading. This allows for a more accurate reading under certain flight conditions.

PITOT TUBE
Ram air pressure enters the system through a hole in the forward end of the pitot tube.

Figure 2-80. A typical pitot-static system uses a single pitot tube and one or two static ports. Some designs combine the static port with the pitot tube.

How Altimetry was Discovered

The altimeter in your airplane is essentially a barometer that is calibrated to measure pressure changes in feet rather than inches of mercury (or hectoPascals). However, before an altimeter could be invented, someone had to figure out the relationship between air pressure and altitude, a relationship that was not always understood. Blaise Pascal (1623-1662) and Evangelista Torricelli (1608-1647) conjectured that if the atmosphere exerted pressure (weight), it would have to weigh less at a higher altitude. Carrying a mercury barometer up a mountain and recording the height of the mercury column at several stops on the way to the top, they determined that the atmosphere does have weight, and that the weight decreases with altitude. This is the basis of altimetry.

Blaise Pascal   Evangelista Torricelli

AIRSPEED INDICATOR

The airspeed indicator is the only instrument to operate using both pitot and static pressure. The speed of your airplane through the air is determined by comparing ram air pressure with static air pressure — the greater the differential, the greater the speed. The airspeed indicator is divided into color-coded arcs that define speed ranges for different phases of flight. The upper and lower limits of the colored arcs correspond to some airspeed limitations, called **V-speeds**. [Figure 2-81]

SECTION C ■ Flight Instruments

Figure 2-81. In addition to delineating various speed ranges, the boundaries of the color-coded arcs also identify airspeed limitations.

V_{S0}—the lower limit of the white arc; the stalling speed or the minimum steady flight speed in the landing configuration. In small airplanes, this is the power-off stall speed at the maximum landing weight in the landing configuration (gear and flaps down).

V_{S1}—the lower limit of the green arc; the stalling speed or the minimum steady flight speed obtained in a specified configuration. For small airplanes, this is the power-off stall speed at the maximum takeoff weight in the clean configuration (gear up, if retractable, and flaps up). You should check the POH for specific information on your airplane.

V_{FE}—the upper limit of the white arc; the maximum speed with the flaps extended.

V_{NO}—the upper limit of the green arc; the maximum structural cruising speed. Do not exceed this speed except in smooth air.

V_{NE} **(Red Line)**—the never-exceed speed. Operating above this speed is prohibited because it can result in damage or structural failure.

White Arc—commonly referred to as the flap operating range because its lower limit represents the full flap stall speed and its upper limit provides the maximum flap speed. You usually fly approaches and landings at speeds within the white arc.

Green Arc—the normal operating range. Most of your flying occurs within this range.

Yellow Arc—the caution range. Fly within this range only in smooth air, and then only with caution.

Although the airspeed indicator presents important airspeed limitations, not all V-speeds are shown. For example, one very important speed not displayed is V_A, or maneuvering speed. This represents the maximum speed at which you may apply full and abrupt control movement without the possibility of causing structural damage. It also represents the maximum speed that you can safely use during turbulent flight conditions. V_A is listed in the POH and also can be found on a placard on the flight deck. It's important to check the corresponding aircraft weight when referencing these speeds because V_A changes with weight. For instance, V_A can be 100 knots when an airplane is heavily loaded and 90 knots when the load is light.

When you are flying a retractable-gear airplane, two other important speeds, which are not specifically depicted on the airspeed indicator, should be taken into consideration. One speed, known as V_{LE}, should not be exceeded when the gear is extended. The other speed, called V_{LO}, is the maximum speed at which you can raise or lower the landing gear. You should reference the airplane's POH for the corresponding airspeeds.

TYPES OF AIRSPEED

The airspeed you are likely to become most familiar with is that read directly from the airspeed indicator, aptly referred to as indicated airspeed. Important performance airspeeds such as takeoff, landing, and stall speeds are always the same indicated airspeed, regardless of altitude. Other important speeds are shown in Figure 2-82.

Figure 2-82. Speed is measured in different ways, depending on its purpose.

INDICATED AIRSPEED (IAS)
Indicated airspeed is the reading you get from the airspeed indicator. Since the airspeed indicator is designed to indicate true airspeed under standard sea level conditions, IAS does not reflect variations in air density as you climb to higher altitudes. IAS is also uncorrected for installation (position) and instrument errors.

CALIBRATED AIRSPEED (CAS)
Calibrated airspeed is indicated airspeed corrected for installation and instrument errors. Although attempts are made to minimize these errors, it is not possible to eliminate them entirely throughout the full range of operating speeds, weights, and flap settings. To determine calibrated airspeed, read indicated airspeed and then correct it by using the chart or table in the POH.

TRUE AIRSPEED (TAS)
True airspeed represents the true speed of your airplane through the air. It is calibrated airspeed corrected for altitude and nonstandard temperature. As altitude or air temperature increases, the density of the air decreases. For a given IAS, TAS increases with altitude.

GROUNDSPEED (GS)
Groundspeed represents the actual speed of your airplane over the ground. It is true airspeed adjusted for wind. Groundspeed decreases with a headwind and increases with a tailwind.

SECTION C ■ Flight Instruments

North American X-15

Most airspeed indicators are calibrated in knots or mph, but those that fly near or above the speed of sound also have a Machmeter. This device measures speed in relation to the speed of sound. The Machmeter indicates Mach 1.0 at the speed of sound, Mach 0.75 at 75% of the speed of sound, and so on. On October 3, 1967, U.S. Air Force test pilot Major William J. "Pete" Knight attained the fastest speed ever recorded in a manned airplane, a top speed of 4,534 mph (Mach 6.72) in the North American X-15A-2 rocket-powered research airplane. This is nearly 5 times as fast as a a .357 Magnum bullet.

Courtesy of NASA Dryden Research Center

ALTIMETER

The altimeter senses pressure changes and displays altitude in feet. It usually has three pointers, or hands, to indicate the altitude—one for hundreds of feet, one for thousands of feet, and one for tens of thousands of feet. Because changes in air pressure directly affect the accuracy of the altitude readout, the altimeter is equipped with an adjustable barometric scale. [Figure 2-83]

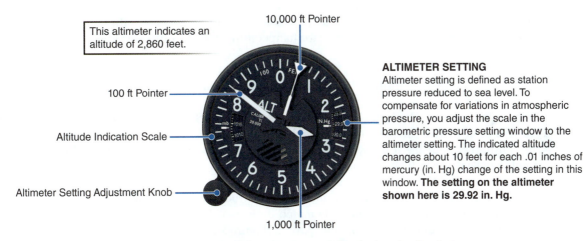

This altimeter indicates an altitude of 2,860 feet.

10,000 ft Pointer

100 ft Pointer

Altitude Indication Scale

Altimeter Setting Adjustment Knob

1,000 ft Pointer

ALTIMETER SETTING
Altimeter setting is defined as station pressure reduced to sea level. To compensate for variations in atmospheric pressure, you adjust the scale in the barometric pressure setting window to the altimeter setting. The indicated altitude changes about 10 feet for each .01 inches of mercury (in. Hg) change of the setting in this window. **The setting on the altimeter shown here is 29.92 in. Hg.**

Figure 2-83. The hands on an altimeter are read like the hands of a clock.

TYPES OF ALTITUDE

The altimeter measures the vertical elevation of an object above a given reference point. The reference can be the surface of the earth, mean sea level (MSL), or some other point. There are several different types of altitude, depending on the reference point used. The six most common types are: indicated, pressure, density, true, calibrated, and absolute. **Indicated altitude** is the altitude measured by the altimeter, and the one you use most often during flight. If you set the altimeter to the standard sea level atmospheric pressure of 29.92 in. Hg, your indicated altitude will be equivalent to **pressure altitude**. This is the vertical distance above the theoretical plane where atmospheric pressure is equal to 29.92 in. Hg. The theoretical pressure line is referred to as the standard datum plane. [Figure 2-84]

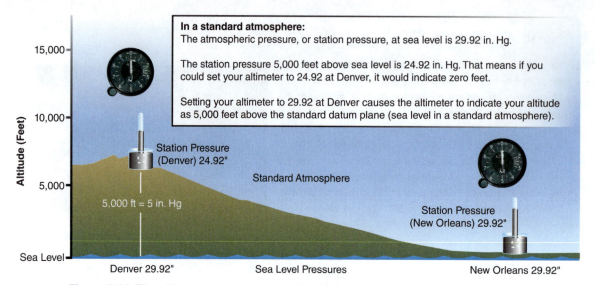

Figure 2-84. The altimeter is set to indicate altitude above a datum plane—normally sea level.

Pressure altitude can be adjusted for temperature to obtain a theoretical value called **density altitude,** an important factor in determining airplane performance. At standard temperature, density altitude is equal to pressure altitude. However, when the ambient air temperature is above standard, the air density is reduced and aircraft performance degrades. Density altitude is altitude at which the aircraft acts like it is performing. Under very warm conditions, the air density is lower and the density altitude is higher. Conversely, when it is colder than standard temperature, density altitude is lower and the aircraft performs better.

Calibrated altitude is indicated altitude corrected to compensate for instrument error.

True altitude is the actual height above mean sea level (MSL). On aeronautical charts, the elevation figures for fixed objects, such as airports, towers, and TV antennas, are true altitudes. Your try to set your altimeter so that it indicates true altitude during flight. However, even when properly set, it indicates true altitude only when standard atmospheric conditions exist. In practice, conditions always vary from standard, which causes the indicated altitude to have some error from true altitude. Only when the airplane is on the airport ramp will true altitude match indicated altitude (field elevation), when the altimeter is set to the local altimeter setting. [Figure 2-85]

Figure 2-85. The actual vertical distance of the airplane above mean sea level is called true altitude.

SECTION C ■ Flight Instruments

The actual height of the airplane above the earth's surface over which it is flying is referred to as **absolute altitude.** This altitude varies with the height of the airplane, as well as the height of the surface. Absolute altitude is commonly referred to as height above ground level (AGL). [Figure 2-86]

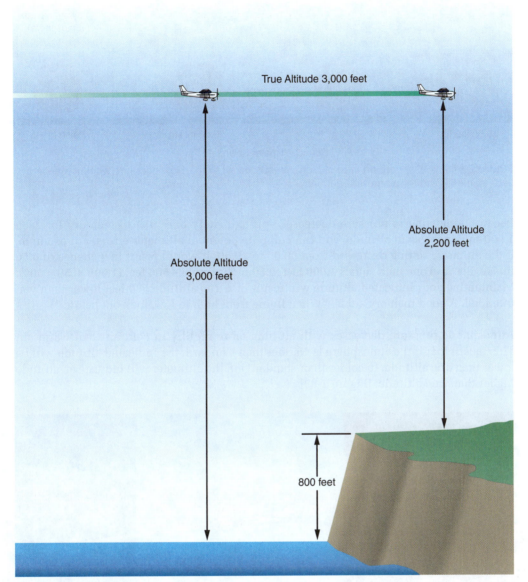

Figure 2-86. Your absolute altitude can change rapidly as you pass over varying terrain elevations.

ALTIMETER ERRORS

Although the altimeter is calibrated based on the International Standard Atmosphere, actual atmospheric conditions seldom match standard values. In addition, local pressure readings within a given area normally change over a period of time, and pressure frequently changes as you fly from one airport to another. As a result, altimeter indications are subject to errors. The extent of the errors depends on how much the pressure, temperature, and lapse rates deviate from standard, as well as how recently you have reset your altimeter.

The most common altimeter error is also the easiest to correct. It occurs when you fail to keep the altimeter set to the local altimeter setting. For example, assume your altimeter is set to 30.00 in. Hg and you are flying at a constant altitude of 3,500 feet MSL. If you fly into an area where atmospheric pressure is 29.50 in. Hg, the altimeter will sense this decrease in pressure as an increase in altitude and will indicate higher. To maintain your "desired" altitude, you will be inclined to lower the nose and descend. [Figure 2-87]

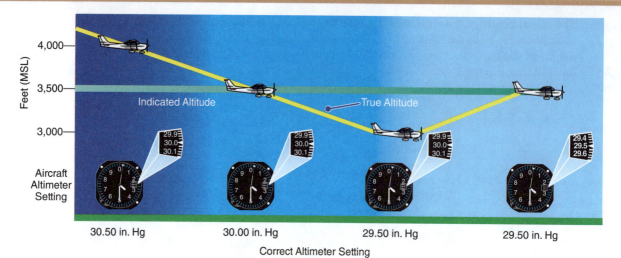

Figure 2-87. If you fly from an area of high pressure to an area of low pressure without resetting your altimeter, your airplane will be too low.

Because atmospheric pressure decreases approximately one inch of mercury for each 1,000-foot increase in altitude, you can compute potential altimeter errors. For example, if the altimeter setting decreased from 30.00 to 29.50, it would result in a change of 0.50 in. Hg. Because one inch equals 1,000 feet, 0.50 inches equals 500 feet (1,000 × .50 = 500). Maintaining your indicated altitude will result in a true altitude 500 feet lower than you intended. A good memory aid is, "When flying from high to low, look out below."

Atmospheric pressure decreases with altitude more rapidly in cold air than warm air. This means that if the atmosphere is warmer than standard, the indicated altitude will be lower than true altitude. In colder-than-standard air, the altimeter will indicate an altitude higher than true altitude. [Figure 2-88]

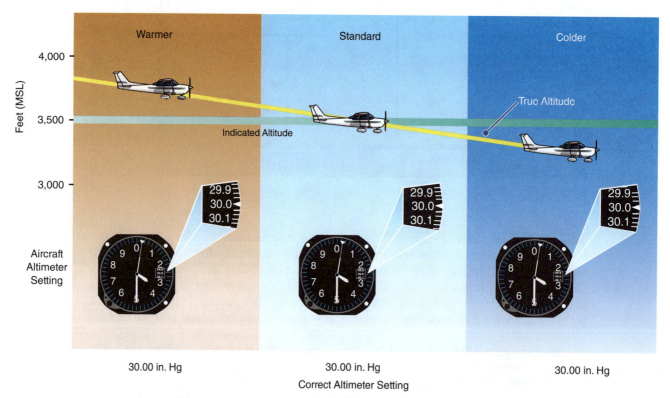

Figure 2-88. On warm days, true altitude is higher than indicated altitude. Aircraft true altitude is lower than indicated in colder air.

Temperature errors are generally smaller than those associated with sea level pressure variations. For example, if the actual temperature is 10°C warmer than standard, true altitude is about 4% higher. This is only 40 feet at 1,000 feet, but the error increases with increased altitude. At 12,000 feet MSL, a 500-foot error would exist. This discrepancy could be critical while flying over mountainous terrain.

The best way to minimize altimeter errors resulting from pressure changes is to update your altimeter setting frequently. In most cases, you use the current altimeter setting of the nearest reporting station along your route of flight. If you encounter areas of extremely low pressure (below 28.00 in. Hg) or unusually high pressure (above 31.00 in. Hg), you should exercise caution because large variations from standard atmospheric pressure conditions can have an increasingly detrimental effect on altimeter accuracy.

VERTICAL SPEED INDICATOR

The vertical speed indicator (VSI), which sometimes is called a vertical velocity indicator (VVI), uses static pressure to display a rate of climb or descent in feet per minute. As the airplane climbs or descends, the VSI determines the vertical speed by measuring how fast the ambient air pressure is increasing or decreasing. [Figure 2-89]

The VSI displays two different types of information. One is called trend, and the other is called rate. **Trend information** shows an immediate indication of an increase or decrease in the airplane's rate of climb or descent. **Rate information** shows a stabilized rate of change. For example, if you are maintaining a steady 500 ft/min climb, and you lower the nose slightly, the VSI will immediately sense this change and display a decrease in the rate of climb. This first indication is called the trend. After a short period of time, the VSI will stabilize and display the new rate of climb, which, in this example, would be something less than 500 ft/min.

BLOCKAGE OF THE PITOT-STATIC SYSTEM

Although pitot-static instruments have some limitations, usually they are very reliable. Gross errors almost always indicate blockage of the pitot tube, the static port(s), or both. Blockage can be caused by moisture (including ice), dirt, or even insects. During preflight, you should make sure the pitot tube cover is removed. Then, check the pitot and static port openings. If they are clogged, the openings should be cleaned by a certificated mechanic. A clogged pitot tube only affects the accuracy of the airspeed indicator. However, a blockage of the static system not only affects the airspeed indicator, but can also cause errors in the altimeter and vertical speed indicator.

300 ft/min descent

300 ft/min climb

Figure 2-89. The vertical speed indicator tells you how fast you are climbing or descending.

BLOCKED PITOT SYSTEM

The pitot system can become blocked completely, or only partially if the pitot tube drain hole remains open. If the pitot tube becomes clogged and its associated drain hole remains clear, ram air will no longer be able to enter the pitot system. Air already in the system will vent through the drain hole, and the remaining pressure will drop to ambient (outside) air pressure. Under these circumstances, the airspeed indicator reading decreases to zero, because the airspeed indicator senses no difference between ram and static air pressure. In other words, the airspeed indicator acts as if the airplane is stationary on the ramp. The apparent loss of airspeed is not usually instantaneous. Instead, the airspeed generally drops slowly to zero. [Figure 2-90]

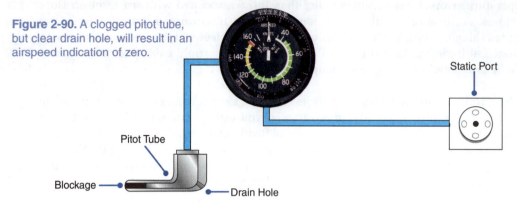

Figure 2-90. A clogged pitot tube, but clear drain hole, will result in an airspeed indication of zero.

If the pitot tube, drain hole, and static system all become clogged in flight, changes in airspeed will not be indicated due to the trapped pressures. However, if the static system remains clear, the airspeed indication will increase as you climb and decrease as you descend—opposite from the way it normally behaves. An apparent increase in the ram air pressure relative to static pressure will occur as altitude increases above the level where the pitot tube and drain hole became clogged. This pressure differential causes the airspeed indicator to show an increase in speed. A decrease of indicated airspeed will occur as the airplane descends below the altitude where the pitot system became obstructed. [Figure 2-91]

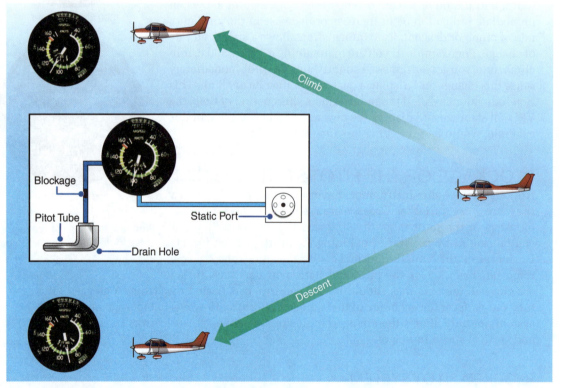

Figure 2-91. If the pitot system becomes clogged while the static system remains clear, the airspeed indicator will act like an altimeter, indicating an increase as altitude is increased.

It is possible for the pitot tube to become obstructed during flight through visible moisture when temperatures are near the freezing level. If your airplane is equipped with pitot heat, you should turn it on to prevent the pitot tube from becoming clogged with ice. Consult your POH for specific procedures regarding the use of pitot heat.

BLOCKED STATIC SYSTEM

If the static system becomes blocked but the pitot tube remains clear, the airspeed indicator will continue to operate, however it is inaccurate. Airspeed indications will be slower than the actual speed when the airplane is operated above the altitude where the static ports became clogged because the trapped static pressure is higher than normal for that altitude. Conversely, when you operate at a lower altitude, a faster than actual airspeed will be displayed due to the relatively low static pressure trapped in the system.

A blockage of the static system also affects the altimeter and VSI. Trapped static pressure will cause the altimeter to freeze at the altitude at which the blockage occurred. In the case of the VSI, a blocked static system will produce a continuous zero indication. [Figure 2-92]

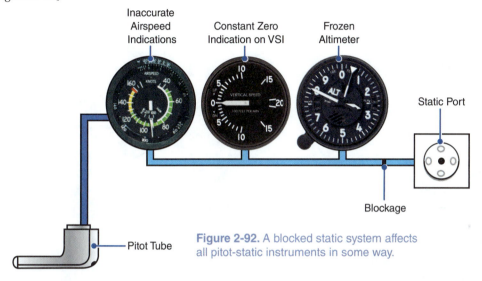

Figure 2-92. A blocked static system affects all pitot-static instruments in some way.

In some airplanes, you can bypass a blocked static system by using an alternate static source. In most cases, the alternate source is vented inside the flight deck where ambient air pressure is lower than outside static pressure. As a result, minor pitot-static instrument errors can occur, such as slightly higher than normal airspeed and altimeter indications. In addition, the VSI might display a momentary climb immediately after the alternate static source is opened. You should check your POH for information on the use of alternate air.

GYROSCOPIC INSTRUMENTS

The primary gyroscopic instruments used on most training airplanes include the turn coordinator, attitude indicator, and heading indicator. [Figure 2-93] Gyroscopic instrument operation is based on two fundamental concepts that apply to gyroscopes—rigidity in space and precession.

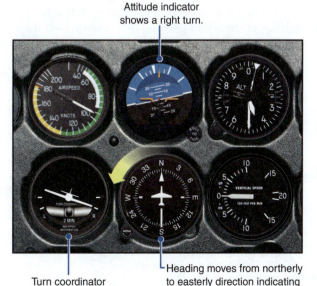

Figure 2-93. As you climb, descend, or turn, you can cross-check the gyroscopic instruments to confirm your attitude and direction.

Attitude indicator shows a right turn.

Turn coordinator indicates a right turn.

Heading moves from northerly to easterly direction indicating a right turn.

RIGIDITY IN SPACE

Rigidity in space refers to the principle that a wheel spun rapidly will remain in a fixed position in the plane in which it is spinning. By mounting this wheel, or gyroscope, on a set of gimbal rings, the gyro is able to rotate freely in any direction. Thus, if the gimbal rings are tilted, twisted, or otherwise moved, the gyro remains in the plane in which it was originally spinning. [Figure 2-94]

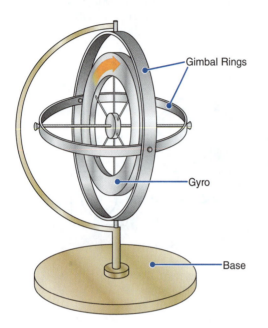

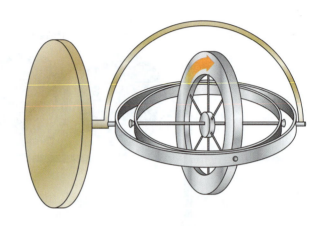

Figure 2-94. Regardless of the position of its base, a gyro tends to remain rigid in space, with its axis of rotation pointed in a constant direction.

PRECESSION

Precession is the tilting or turning of a gyro in response to pressure. Unfortunately, it is not possible to mount a gyro in a frictionless environment. A small force is applied to the gyro whenever the airplane changes direction. The reaction to this force occurs in the direction of rotation, approximately 90° ahead of the point where the force was applied. This causes slow drifting and minor erroneous indications in the gyroscopic instruments.

SOURCES OF POWER

A source of power is required to keep the gyros spinning. Most small airplanes use two different sources of power to ensure that you have at least one reliable indication of aircraft bank attitude if a system failure occurs. The turn coordinator is typically an electrically-powered instrument while the attitude and heading indicators normally receive power from a **vacuum (suction) system**. Air is first drawn into the vacuum system through a filter assembly, and then moves through the attitude and heading indicators where it spins the gyros. The air exits the system through the engine-driven vacuum pump. A relief valve prevents the vacuum pressure from exceeding prescribed limits. [Figure 2-95]

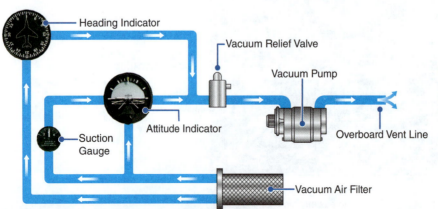

Figure 2-95. The vacuum pump pulls air through the heading indicator and attitude indicator, to spin the gyros.

It is important that you monitor vacuum pressure during flight, because the attitude and heading indicators might not work properly when the suction pressure is low. The vacuum or suction gauge generally is marked to indicate the normal range. [Figure 2-96]

Figure 2-96. Reliable operation of the attitude and heading indicators requires that the suction is in the proper range.

Some airplanes are equipped with a warning light that illuminates when the vacuum pressure drops below the acceptable level. In addition, some airplanes are equipped with an electrically-driven backup vacuum pump.

TURN COORDINATOR

The turn coordinator is a newer version of the old turn-and-slip indicator. Both types of indicators provide an indication of turn direction and quality as well as a backup source of bank information if the attitude indicator fails. The primary difference between the two is that the turn coordinator has a tilted gyro that also indirectly indicates the rate of roll. The turn-and-slip indicator uses a pointer, called a turn needle, and the turn coordinator uses a miniature airplane. Both indicators have a ball in a tube, called an inclinometer, to provide information about the quality (coordination) of the turn. [Figure 2-97]

Turn-and-Slip Indicator

Figure 2-97. These two types of gyroscopic turn indicators provide similar information.

Turn Coordinator

Turn coordinators are more prevalent in training aircraft. As you roll into or out of a turn, the miniature airplane banks in the direction the airplane is rolled. A rapid roll rate causes the miniature airplane to bank more steeply than a slow roll rate. You can use the turn coordinator to establish and maintain a **standard-rate turn** by aligning the wing of the miniature airplane with the turn index. At this rate, you will turn 3 degrees per second and complete a 360-degree turn in 2 minutes. The turn coordinator indicates only the rate of turn and does not display a specific angle of bank. [Figure 2-98]

Figure 2-98. When you roll into a bank, the miniature airplane on a turn coordinator indicates a bank in the same direction.

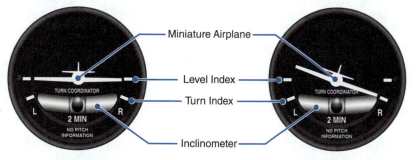

Level Flight

Standard-Rate Right Turn

The inclinometer depicts airplane yaw, which is the side-to-side movement of the airplane's nose. It consists of a liquid-filled, curved tube with a ball inside. During coordinated straight-and-level flight, the force of gravity causes the ball to rest in the lowest part of the tube, centered between the reference lines. You maintain coordinated flight by keeping the ball centered. If the ball is not centered, you can center it using the rudder. To do this, you push on the rudder on the side where the ball is deflected.

If aileron and rudder are coordinated during a turn, the ball will remain centered in the tube. If aerodynamic forces are unbalanced, the ball moves away from the center of the tube. In a **slip**, the rate of turn is too slow for the angle of bank, and the ball moves to the inside of the turn. In a **skid**, the rate of turn is too great for the angle of bank, and the ball moves to the outside of the turn. To correct for these conditions and improve the quality of the turn, you should "Step on the ball." [Figure 2-99] You also can vary the angle of bank to help restore coordinated flight from a slip or skid. To correct for a slip, decrease bank or apply more rudder pressure in the direction of the turn to increase the rate of turn. To correct for a skid, increase the bank or reduce rudder pressure in the direction of the turn to decrease the rate of turn.

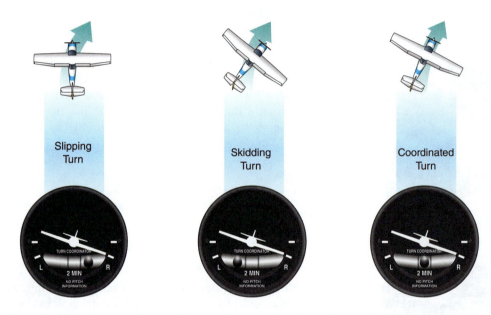

Figure 2-99. In this example of a right turn, inadequate right rudder results in a slip and excessive right rudder results in a skid through the turn.

ATTITUDE INDICATOR

The attitude indicator senses pitch—the up and down position of the airplane's nose—as well as roll. It uses an artificial horizon and miniature airplane to depict the position of your airplane in relation to the true horizon. This is especially useful when the natural horizon is obscured by clouds, reduced visibility, or darkness.

The attitude indicator presents you with a view of the airplane as it would appear from behind. The angle of bank is shown both pictorially by the relationship of the miniature aircraft to the deflected horizon bar and by the alignment of the pointer with the bank scale at the top of the instrument. Pitch is indicated by the position of the "nose," or center, of the miniature airplane with respect to the horizon bar. [Figure 2-100]

Prior to flight, you should set the miniature airplane symbol so that it is level with the horizon bar. Once properly adjusted, modern attitude indicators normally are very reliable as long as the correct vacuum pressure is maintained. Occasionally, however, attitude indicators fail gradually without providing obvious warning signals. Therefore, you should remember to periodically cross-check it with outside visual references and other flight instruments.

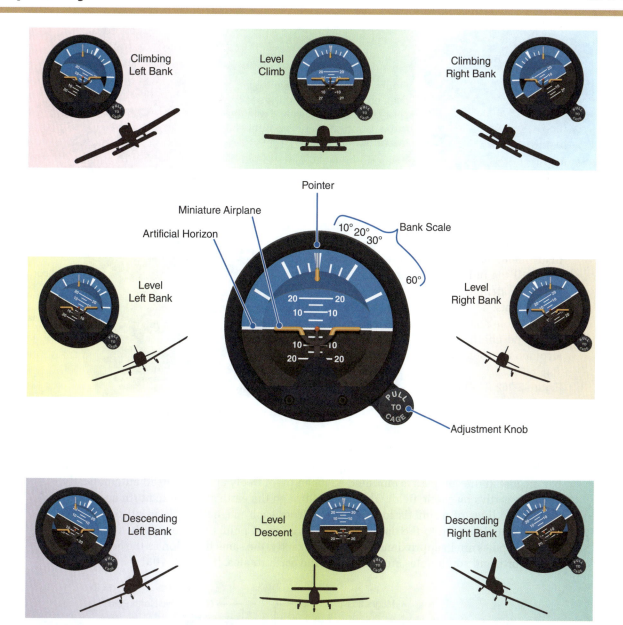

Figure 2-100. The attitude indicator, or artificial horizon, shows both pitch and bank attitude.

HEADING INDICATOR

The heading indicator, also called a directional gyro (DG), senses airplane movement around the vertical axis and displays heading on a 360° azimuth card, with the final zero omitted. For example, 6 indicates 60°, 21 indicates 210°, and so on. When properly set, it is your primary source of heading information. Heading indicators in most training airplanes are referred to as "free" gyros. This means they have no automatic, north-seeking system built into them. For the heading indicator to display the correct heading, you must align it with the magnetic compass before flight. However, precession can cause the selected heading to drift from the set value. For this reason, you should check the heading indicator every 15 minutes against the magnetic compass. When aligning the heading indicator while airborne, be certain you are in straight-and-level, unaccelerated flight, with the magnetic compass showing a steady indication. [Figure 2-101]

Like most gyroscopic instruments, the heading indicator can "tumble" during excessive pitch and roll conditions. If the indicator has tumbled, you must realign it with a known heading or with a stabilized indication from the magnetic compass.

SECTION C ■ **Flight Instruments**

Figure 2-101. A basic heading indicator does not have any direction-seeking ability and must be aligned with the magnetic compass before flight and at regular intervals during flight.

MAGNETIC COMPASS

The magnetic compass was one of the first instruments to be installed in an airplane, and it is still the only direction-seeking instrument in most airplanes that do not have digital instruments. The magnetic compass is a simple and reliable source of heading information but it can be challenging to interpret unless the airplane is in steady flight. [Figure 2-102]

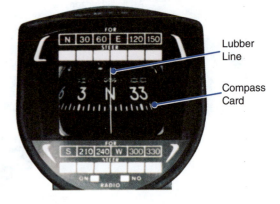

Figure 2-102. The compass is your primary source of direction information, but is not as easy to read as a gyroscopic heading indicator.

The magnetic compass is a self-contained unit that does not require electrical or suction power. To determine direction, the compass uses a simple bar magnet with two poles. The bar magnet in the compass is mounted so it can pivot freely and align itself automatically with the earth's magnetic field, which goes through the earth's magnetic north and south poles. The lines of magnetic force flow out from each magnetic pole in all directions, and eventually return to the opposite pole. From most points on the earth's surface, the magnetic poles are in approximately, but not exactly, the same direction as the geographic (true) north and south poles around which the earth rotates. [Figure 2-103]

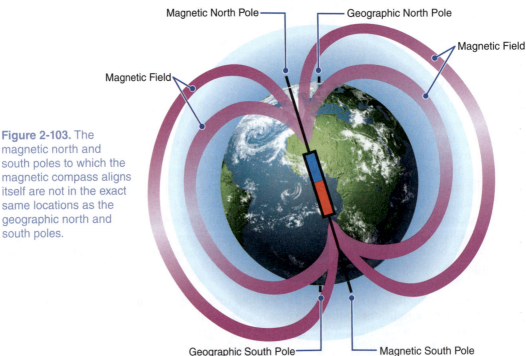

Figure 2-103. The magnetic north and south poles to which the magnetic compass aligns itself are not in the exact same locations as the geographic north and south poles.

VARIATION

The angular difference between the true and magnetic poles at a given point is referred to as **variation**. Because most aviation charts are oriented to true north and the aircraft compass is oriented to magnetic north, you must convert a true direction to a magnetic direction by correcting for the variation. The amount of correction you need to apply depends upon your location on the earth's surface. [Figure 2-104]

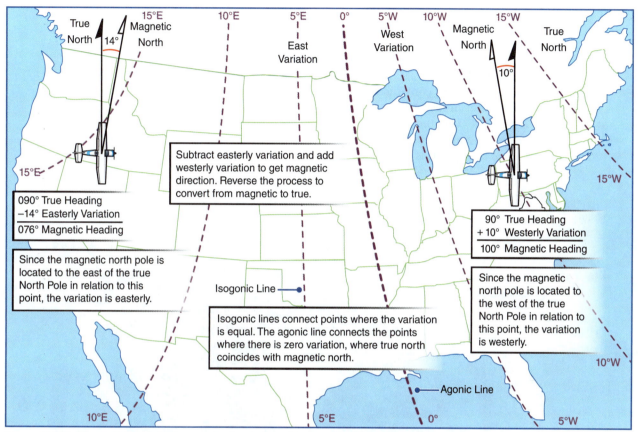

Subtract easterly variation and add westerly variation to get magnetic direction. Reverse the process to convert from magnetic to true.

090° True Heading
−14° Easterly Variation
076° Magnetic Heading

Since the magnetic north pole is located to the east of the true North Pole in relation to this point, the variation is easterly.

Isogonic Line

Isogonic lines connect points where the variation is equal. The agonic line connects the points where there is zero variation, where true north coincides with magnetic north.

90° True Heading
+ 10° Westerly Variation
100° Magnetic Heading

Since the magnetic north pole is located to the west of the true North Pole in relation to this point, the variation is westerly.

Agonic Line

Figure 2-104. The variation between magnetic north and true north differs depending on your location.

DEVIATION

Deviation is a compass error that occurs due to disturbances from magnetic fields produced by metals and electrical accessories within the airplane. Although it cannot be completely eliminated, deviation error can be decreased by adjusting compensating magnets within the compass housing. The remaining error is recorded on a compass correction card, which is mounted near the compass. For example, on the card shown in Figure 2-105, if you want to fly a magnetic heading (MH) of 060°, the compass heading (CH) also is 060°; to fly 180° magnetic, you must fly a compass heading of 183°.

FOR (MH)	0°	30°	60°	90°	120°	150°	180°	210°	240°	270°	300°	330°
STEER (CH)	359°	30°	60°	88°	120°	152°	183°	212°	240°	268°	300°	329°
RADIO ON ☑						**RADIO OFF** ☐						

Figure 2-105. This compass correction card shows headings you must steer to on the magnetic compass (with the radio on) to get actual magnetic headings.

SECTION C ■ **Flight Instruments**

COMPASS ERRORS

Although you can correct for variation and deviation, the compass is susceptible to other types of errors which, although predictable, can make it difficult to use. For example, the freedom of movement necessary for the compass to orient itself to magnetic north makes it sensitive to in-flight turbulence. In light turbulence, you might be able to use the compass by averaging the readings. For instance, if the compass swings between 030° to 060°, you can estimate an approximate heading of 045°. In heavier turbulence, however, the compass can be of very little use. Even in smooth air, additional errors can occur while you are turning or changing speed due to a phenomenon known as magnetic dip.

MAGNETIC DIP

When the bar magnet contained in the compass is pulled by the earth's magnetic field, it tends to point north and somewhat downward. The downward pull, called **magnetic dip**, is greatest near the poles and diminishes as you approach the equator. Within approximately 300 miles of either magnetic pole, these errors are so great that use of the compass for navigation is impractical. [Figure 2-106]

Figure 2-106. Magnetic dip occurs because the compass magnet tilts downward, as well as pointing horizontally, toward the pole.

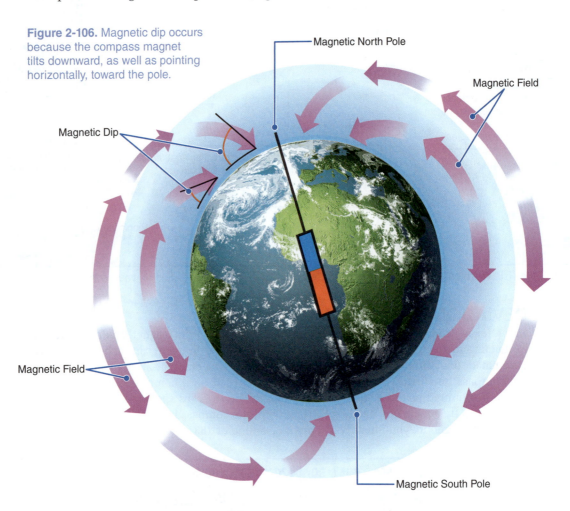

Magnetic North Pole

Magnetic Field

Magnetic Dip

Magnetic Field

Magnetic South Pole

The internal pivot point that suspends the compass card is designed to minimize the tilting force on the bar magnet caused by magnetic dip. Unfortunately, this system, as well as magnetic dip itself, both contribute to acceleration and turning errors.

ACCELERATION ERROR

If you accelerate or decelerate an airplane on an easterly or westerly heading, an erroneous indication will occur. As you accelerate an airplane, the internal arrangement of the compass components causes the card to turn toward the north. During a deceleration, the same factors turn the compass card toward a southerly heading even though no change of direction has taken place. The compass will return to its previous, and proper, heading once the acceleration or deceleration subsides.

Acceleration error is more pronounced as you move closer to due east or west. The error doe not occur when you are flying on a directly north or south heading because the bar magnet is in line with the direction of travel. In addition, these acceleration errors are valid only for the northern hemisphere. The effects are reversed in the southern hemisphere. The memory aid, ANDS (accelerate north, decelerate south), can help you recall how acceleration error affects the compass in the northern hemisphere. [Figure 2-107]

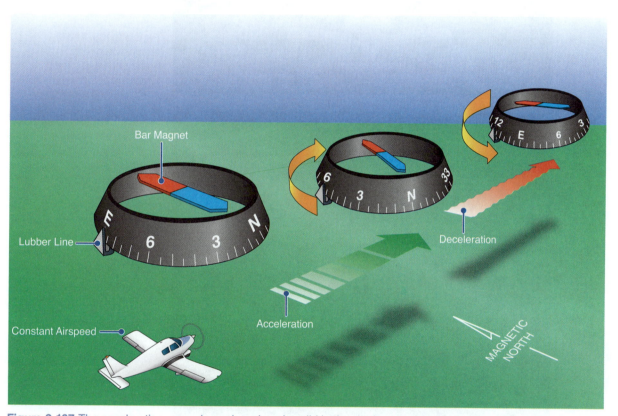

Figure 2-107. The acceleration error shown here is only valid in the northern hemisphere.

TURNING ERROR

Turning error is directly related to magnetic dip; the greater the dip, the greater the turning error. It is most pronounced when you are turning to or from headings of north or south. When you begin a turn from a heading of north, the compass initially indicates a turn in the opposite direction. When the turn is established, the compass begins to turn in the correct direction, but it lags behind the actual heading. The amount of lag decreases as the turn continues, then disappears as the airplane reaches a heading of east or west.

When turning from a heading of east or west to a heading of north, there is no error as you begin the turn. However, as the heading approaches north, the compass increasingly lags behind the airplane's actual heading. When you turn from a heading of south, the compass initially indicates a turn in the proper direction but leads the airplane's actual heading. This error also cancels out as the airplane reaches a heading of east or west. Turning from east or west to a heading of south causes the compass to move correctly at the start of a turn, but then it increasingly leads the actual heading as the airplane nears a southerly direction. [Figure 2-108]

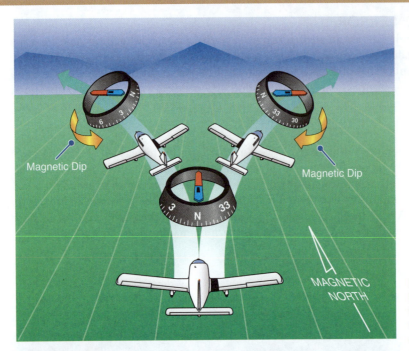

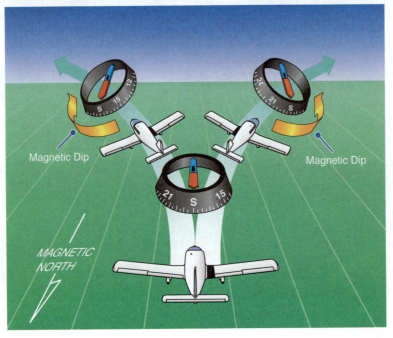

Figure 2-108. The left portion of the figure shows the initial tendency of the magnetic compass in a turn from a northerly heading. The initial turning error that occurs during turns from a southerly heading is shown on the right.

The amount of lead or lag is approximately equal to the latitude of the airplane. For example, if you are turning from a heading of south to a heading of west while flying at 40° north latitude, the compass will rapidly turn to a heading of 220° (180° + 40°). At the midpoint of the turn, the lead will decrease to approximately half (20°), and upon reaching a heading of west, it will be zero. As in acceleration errors, these lead and lag errors are only valid for flight in the northern hemisphere. Lead and lag errors in the southern hemisphere act in the opposite directions.

COPING WITH COMPASS ERRORS

Due to its potential for errors, the magnetic compass is normally used as a backup source of heading information while the gyroscopic heading indicator is used as the primary heading reference. If your heading indicator fails and you understand the limitations of the magnetic compass, you should still be able to navigate properly. When you are referring to the compass for heading information, remember that it is accurate only when your airplane is in smooth air and in straight-and-level, unaccelerated flight.

ELECTRONIC FLIGHT DISPLAYS

Many of today's training airplanes have digital flight instruments that are integrated into a single display, as part of a glass flight deck. Digital flight instruments provide essentially the same information as analog instruments, but in a different format. Because a variety of different **electronic flight display** systems exist, make sure that you are proficient in interpreting and operating the specific system in your airplane. The systems used in smaller general aviation airplanes typically have two screens: a **primary flight display (PFD)** and a **multifunction display (MFD)**. [Figure 2-109]

Primary Flight Display
The PFD contains the primary flight instruments positioned directly in front of you.

Multifunction Display
The MFD contains a variety of information on different pages, such as a moving map display, airport, terrain, and weather data, aircraft systems indications, checklists, and instrument charts.

Backup Instruments
Selected backup analog instruments enable you to control the airplane if digital displays fail.

Figure 2-109. Although some of the features of electronic flight displays vary based on the manufacturer's design, primary and multifunction displays typically contain the same basic information.

PRIMARY FLIGHT DISPLAY

The PFD contains digital versions of traditional analog flight instruments, including the airspeed indicator, altimeter, vertical speed indicator, and attitude indicator. In addition to pitch and bank information, the attitude indicator includes a slip/skid indicator. A horizontal situation indicator (HSI) displays heading and navigation information and includes a turn rate indicator. [Figure 2-110]

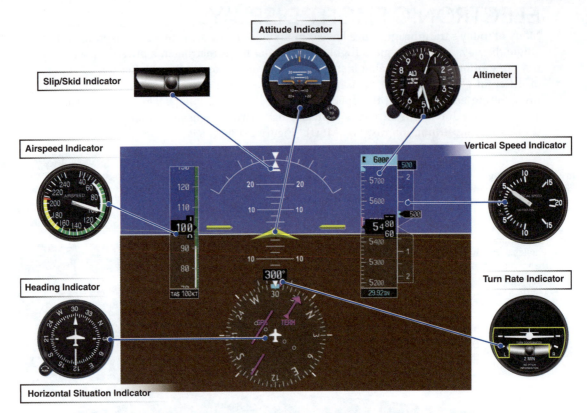

Figure 2-110. The PFD conveys the same information as the six traditional flight instruments.

ATTITUDE AND HEADING REFERENCE SYSTEM

To provide attitude, heading, rate of turn, and slip/skid information, electronic flight displays use an **attitude and heading reference system (AHRS)**. The AHRS uses inertial sensors such as electronic gyroscopes and accelerometers to determine the aircraft's attitude relative to the horizon. An electronic magnetometer provides magnetic heading data. In some systems, GPS equipment also provides data to the AHRS—a general indication of the airplane's attitude is determined by comparing signals received from three antennas located in different parts of the airplane.

MAGNETOMETER

A **magnetometer** used in electronic avionics displays senses the earth's magnetic field to function as a magnetic compass, but without some of the errors associated with a conventional compass. Instead of a suspended bar magnet, the magnetometer uses a flux valve or flux gate, which is an electronic means of sensing magnetic lines of force. The magnetometer is usually located as far as possible from sources of magnetic fields within the aircraft, and connected by a cable to the AHRS. Because the AHRS automatically updates the heading display, you do not have to set the heading to match the compass. The magnetometer does not lead or lag during turns, and does not experience errors due to acceleration.

MEMS GYROS

Some AHRS use MEMS gyros, which are technically not gyroscopes, but angular rate sensors. MEMS gyros take advantage of Coriolis force, which causes a moving object to experience a linear acceleration when it is rotated. In a typical MEMS gyro, tiny quartz fingers or discs move back and forth in one plane, and when the aircraft rotates, that plane is slightly displaced. Sensitive electronic circuitry detects the displacements as small changes in capacitance between the moving and non-moving parts of the sensor, translating the changes into signals that the AHRS uses to determine attitude and rates of rotation.

RING LASER GYROS

The ring laser gyro, which uses light to detect rotation and changes in attitude, is another type of AHRS gyro. Lasers emit light that is all of the same wavelength and moving in the same direction. In a ring laser gyro, a beam of laser light is split in half, and each half is directed by mirrors around a triangular track in opposite directions. If the aircraft changes attitude and the ring laser gyro rotates, changes occur in the wavelengths of the light beams. The AHRS uses this data to determine aircraft attitude. By arranging three ring laser gyros perpendicular to each other, rotation in any direction can be measured. [Figure 2-111]

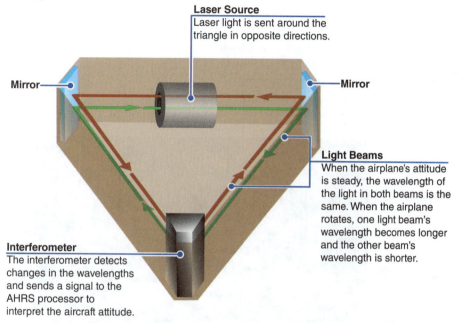

Laser Source
Laser light is sent around the triangle in opposite directions.

Mirror

Mirror

Light Beams
When the airplane's attitude is steady, the wavelength of the light in both beams is the same. When the airplane rotates, one light beam's wavelength becomes longer and the other beam's wavelength is shorter.

Interferometer
The interferometer detects changes in the wavelengths and sends a signal to the AHRS processor to interpret the aircraft attitude.

Figure 2-111. Ring laser gyros bounce laser beams around a triangular path and measure shifts in wavelengths of the light.

ATTITUDE INDICATOR

A digital attitude indicator can have a virtual blue sky and brown ground with a white horizon line or can be displayed over a realistic representation of terrain features. The horizon line can extend the width of the PFD. Near the center of the display, a miniature aircraft symbol shows whether the nose is above or below the horizon and a pitch scale shows pitch angles of 5, 10, 15, and 20 degrees. Bank angle is shown with a roll scale with reference marks at 10, 20, 30, 45, and 60 degrees. Some systems have optional settings to let you choose whether the roll scale moves and the pointer remains stationary, or the pointer moves while the roll scale remains stationary. A **slip/skid indicator** below the roll pointer helps you maintain coordinated flight. [Figure 2-112]

Figure 2-112. The digital attitude indicator shows pitch, bank, and slip/skid information.

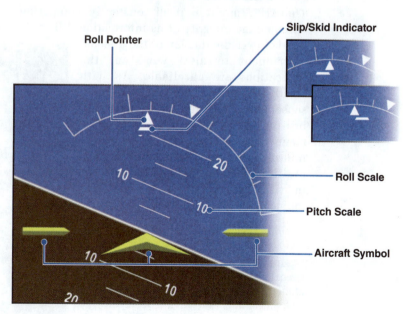

Roll Pointer

Slip/Skid Indicator

Roll Scale

Pitch Scale

Aircraft Symbol

SECTION C ■ **Flight Instruments**

HORIZONTAL SITUATION INDICATOR

In addition to a compass card, the digital **horizontal situation indicator (HSI)** displays the current airplane heading in a window. The course indicator arrow and the course deviation indicator (CDI) change color based on the navigation source you select—magenta for a GPS source, and green if the source is a VOR or localizer. Many HSIs have a heading bug to set for navigation systems or autopilot functions or to use as a reference mark when you are hand-flying. A **turn rate indicator** with index marks and a **trend vector** helps you make standard-rate turns. [Figure 2-113]

You might be able to display additional bearing pointers to navaids and GPS waypoints on the HSI for increased situational awareness. Optional windows associated with the HSI can display information such as distance to navaids and their frequencies.

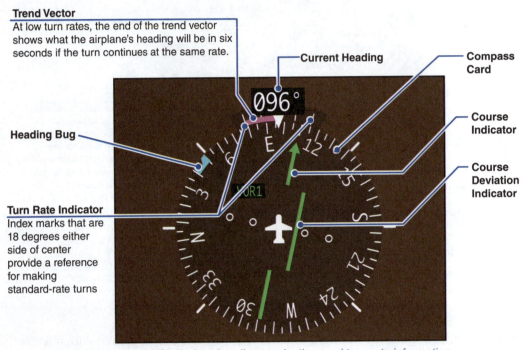

Trend Vector
At low turn rates, the end of the trend vector shows what the airplane's heading will be in six seconds if the turn continues at the same rate.

Current Heading

Compass Card

Heading Bug

Course Indicator

Course Deviation Indicator

Turn Rate Indicator
Index marks that are 18 degrees either side of center provide a reference for making standard-rate turns

Figure 2-113. The digital HSI displays heading, navigation, and turn-rate information.

AHRS ERRORS

The AHRS monitors itself constantly, comparing the data from different inputs and checking the integrity of its information. When the system detects a problem, it places a red X over the display of the affected instrument to alert you that the indications are unreliable. As with analog instruments, failures of different sensors affect different combinations of instruments. For example, failure of the magnetometer affects the HSI heading information. Some sensor failures affect multiple instruments—failure of the inertial sensors affects both the attitude indicator and the HSI. [Figure 2-114]

Figure 2-114. A red X on the attitude indicator and heading window and removal of the compass card values means the attitude indicator and HSI are not receiving attitude information from the AHRS.

AHRS INSTRUMENT CHECK

During taxi, check the operation of the attitude and heading instruments. Verify that the heading indicator agrees with the compass. Ensure that the heading indicator and turn indicator show turns in the correct direction and the slip/skid indicator moves to the outside of the turn. In addition, check and set the backup instruments.

AIR DATA COMPUTER

In an electronic flight display system, the pitot tube, static source, and outside air temperature probe provide information to the **air data computer (ADC)**. The ADC uses these pressure and temperature inputs to determine the appropriate readings for the airspeed indicator, altimeter, and vertical speed indicator. In addition, the ADC provides information to display true airspeed and outside air temperature on the PFD.

AIRSPEED INDICATOR

On the digital airspeed indicator, a central window and a pointer on a moving vertical scale show the indicated airspeed. The vertical scale is called the airspeed tape. The tape has colored bars to indicate airspeed operating ranges. Many digital airspeed displays have a trend vector that indicates how much the airplane is accelerating or decelerating. The length of the trend vector is proportional to the rate of change, and the tip of the trend vector shows what the airspeed will be in six seconds if the acceleration or deceleration continues at the same rate. [Figure 2-115]

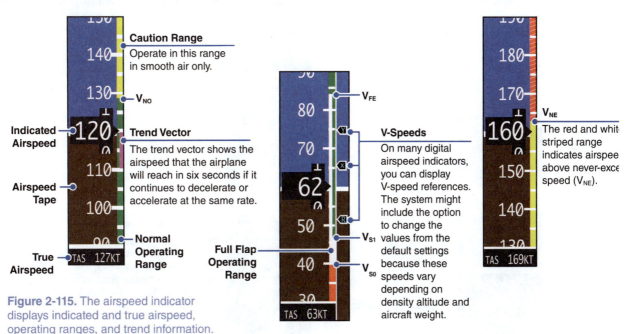

Figure 2-115. The airspeed indicator displays indicated and true airspeed, operating ranges, and trend information.

ALTIMETER

Like the airspeed indicator, the digital altimeter provides a central window and a pointer on a moving tape to display the indicated altitude. The window at the bottom of the display shows the altimeter setting. When the airplane is climbing or descending, the altimeter displays a trend vector. The trend vector shows the altitude that the airplane will reach in six seconds if it continues to climb or descend at the same rate. The altimeter might have a bug to select the reference altitude for the autopilot. You can also set the bug as a reminder to level off at an assigned altitude or at a minimum descent altitude or decision altitude on an instrument approach. The altitude that you select appears in a window above the altimeter tape. [Figure 2-116]

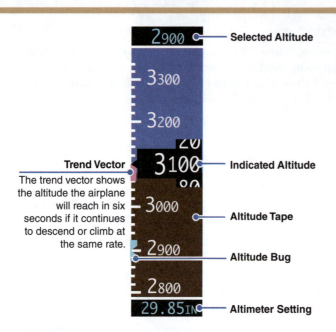

Selected Altitude

Trend Vector

The trend vector shows the altitude the airplane will reach in six seconds if it continues to descend or climb at the same rate.

Indicated Altitude

Altitude Tape

Altitude Bug

Altimeter Setting

Figure 2-116. The digital altimeter includes a trend vector and a bug that you can set as an altitude reference.

VERTICAL SPEED INDICATOR

The vertical speed indicator (VSI) is adjacent to the altimeter. As with the airspeed indicator and altimeter, you read the vertical speed in feet per minute in a window that also serves as a pointer, but on the vertical speed display, the tape remains motionless as the window moves up or down over the scale. [Figure 2-117]

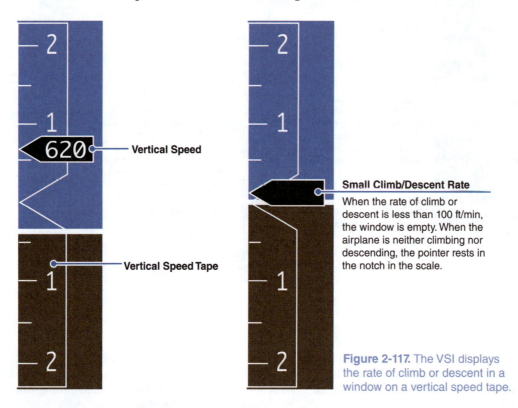

Vertical Speed

Vertical Speed Tape

Small Climb/Descent Rate

When the rate of climb or descent is less than 100 ft/min, the window is empty. When the airplane is neither climbing nor descending, the pointer rests in the notch in the scale.

Figure 2-117. The VSI displays the rate of climb or descent in a window on a vertical speed tape.

ADC SYSTEM ERRORS

The ADC relies on inputs from sensors, such as the pitot tube, static source, and outside air temperature probe, to provide reliable instrument indications. If one or more of these sources stops providing input, or if the computer determines that its own internal operations are not correct, it places a red X over the display of the affected instrument. The failure of a single sensor might affect only one instrument. For example, a blocked pitot tube can disable the airspeed indicator without affecting the other instruments. [Figure 2-118]

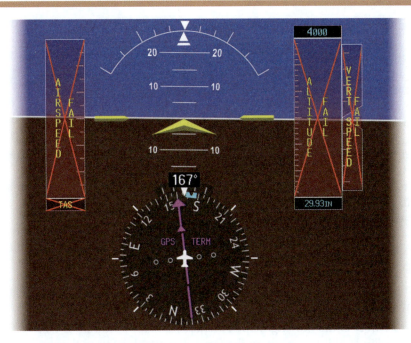

Figure 2-118. A red X on the airspeed indicator, altimeter, and vertical speed indicator means that these instruments are not receiving valid information from the ADC.

ADC INSTRUMENT CHECK

Before takeoff, verify that the airspeed indicator reads zero when the aircraft is at rest and indicates airspeed appropriately during the takeoff roll. Ensure that the altimeter indicates within 75 feet of a known elevation when set to the current altimeter setting. Verify that the VSI reads zero. In addition, check and set the backup instruments.

PFD SCREEN FAILURE

In the most commonly used electronic flight display systems, the screens have a white light source behind an electronic matrix of tiny liquid crystal pixels that can change the white light coming through them into thousands of colors. If the back-light in the display fails, the screen can go black although all the information is still there. The electronic flight display is configured so that the functions of the PFD can be transferred to the MFD screen, and vice versa. If the PFD screen turns black, the PFD instruments should automatically display on the MFD in reversionary mode. In the event that the PFD does not appear on the MFD, most systems enable you to manually switch to reversionary mode. [Figure 2-119]

Reversionary Mode
In the event of a PFD screen failure, the PFD should automatically appear on the MFD screen. The flight instruments replace the information normally shown on the MFD except for the engine indicators.

Display Backup
Manually switch to reversionary mode by pushing the Display Backup button.

Figure 2-119. The Garmin G1000 electronic flight display enables you to manually switch to reversionary mode using a Display Backup button.

SECTION C ■ Flight Instruments

ELECTRICAL SYSTEM FAILURE

The components of an aircraft electrical system are designed, built, and tested to be extremely reliable, but they can and do occasionally fail. If your aircraft experiences an electrical failure in flight, redundant power sources exist for the instrument systems. Some aircraft have two alternators, and others have a separate battery to power critical systems. Be sure that you understand how your electrical system works, and how to make the best use of your remaining energy resources if an electrical failure occurs.

Even if you have a catastrophic electrical failure and lose all electric power, in many light general aviation aircraft, the backup instruments might not require electric power. The backup attitude indicator can be vacuum-powered and the backup altimeter, airspeed indicator, and magnetic compass usually need no electrical power to function. Other aircraft have electric backup attitude indicators powered by a separate battery that is isolated from the main electrical system. The key to safely handling a failure is knowing your system and practicing for emergency situations. Periodically, practice emergency scenarios with a certificated flight instructor.

MULTIFUNCTION DISPLAY

The primary feature of the MFD is the moving map. The moving map on the MFD can usually display many other kinds of information in the form of overlays, such as terrain, instrument procedures, graphical weather, lightning strikes, and traffic information. In addition to the moving map, the MFD provides separate pages that can include airport information, flight plan data, systems indications, checklists, and instrument charts. [Figure 2-120]

Figure 2-120. The MFD can include items such as the aircraft's system's indications, flight plan progress, and terrain awareness displays.

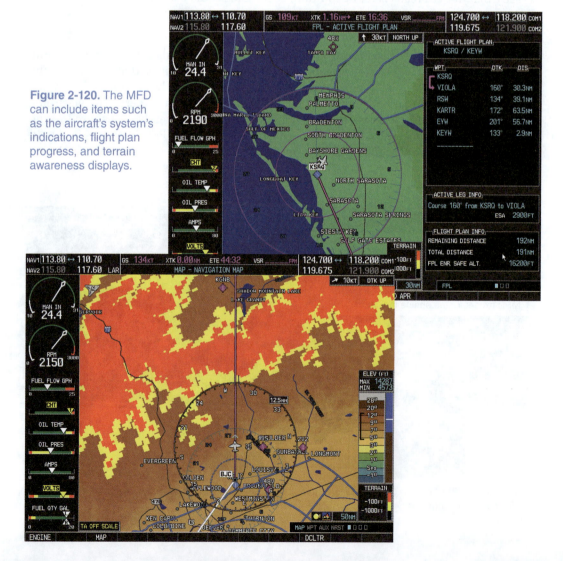

VISION SYSTEMS

Vision systems technology can improve your situational awareness when flying with an electronic flight display. Vision systems fall under four categories based on their features and how you use them during flight operations. Enhanced vision systems, synthetic vision systems, and combined (enhanced and synthetic) vision systems can be displayed on PFDs and navigation displays. The enhanced flight vision system must be displayed on a conformal head-up display (HUD).

An enhanced vision system (EVS) provides a display of the forward external scene topography through the use of imaging sensors, such as forward looking infrared, millimeter wave radiometry or radar, and low-light-level image intensifying. A synthetic vision system (SVS) is a computer-generated image of the surrounding topography and airport environment created from a database of terrain, obstacles, and cultural features and a navigation source for the aircraft's position, altitude, heading, and track. The image is displayed from the flight crew's perspective or as a plan-view moving map. [Figure 2-121]

Figure 2-121. SVS real-time, color 3-D imagery of the flight environment enhances situational awareness.

An enhanced flight vision system (EFVS) uses a real-time imaging sensor to provide highly accurate vision performance in low visibility conditions for pilots flying under IFR. The EFVS projects an image onto a HUD. Required visual references for instrument approaches become visible in the image before they are visible naturally out the window. In addition, depending on atmospheric conditions and the strength of energy emitted and/or reflected from the scene, you can see these visual references on the display in more detail than you can by looking through the window without enhanced vision. [Figure 2-122]

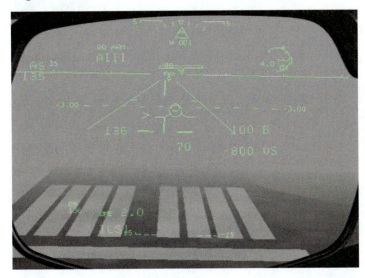

Figure 2-122. Primarily used in airliners and military aircraft, EFVS displays increase situational awareness during instrument approaches in low visibility conditions.

SECTION C ■ **Flight Instruments**

SUMMARY CHECKLIST

✓ The airspeed indicator, altimeter, and vertical speed indicator all use static pressure. The airspeed indicator is the only instrument that uses pitot pressure.

✓ At sea level, the standard atmosphere consists of a barometric pressure of 29.92 in. Hg (1013.2 millibars) and a temperature of 15°C (59°F).

✓ In the lower atmosphere (below 36,000 feet), the standard pressure lapse rate for each 1,000 feet of altitude is approximately 1.00 in. Hg, and the standard temperature lapse rate is 2°C (3.5°F).

✓ The airspeed indicator is divided into color-coded arcs that define speed ranges for different phases of flight. The upper and lower limits of the arcs correspond to specific airspeed limitations, called V-speeds.

✓ V_A, or maneuvering speed, is the maximum speed at which you may apply full and abrupt control movement without the possibility of causing structural damage. Since V_A changes with aircraft weight, it is not depicted on the airspeed indicator.

✓ Regardless of altitude, the indicated airspeed at which a particular airplane stalls in a specific configuration remains the same.

✓ Indicated altitude is the altitude measured, and displayed, by your altimeter. Pressure altitude is the vertical distance above the standard datum plane, while density altitude corrects pressure altitude for nonstandard temperature. True altitude is the actual height of an object above mean sea level. Absolute altitude is the actual height of the airplane above the earth's surface over which it is flying.

✓ If you fly from an area of high pressure to an area of lower pressure without resetting your altimeter, the altimeter will indicate higher than the true altitude. True altitude will be higher than indicated altitude if you do not reset your altimeter when flying from a low pressure area to an area of high pressure.

✓ A one-inch change in the setting on your altimeter results in a 1,000-foot indicated altitude change in the same direction.

✓ If atmospheric temperature is higher than standard, true altitude will be higher than your indicated altitude. In colder than standard temperatures, true altitude will be lower than indicated altitude.

✓ Trend information shows an immediate indication of an increase or decrease in the airplane's rate of climb or descent, while rate information shows you a stabilized rate of change.

✓ Blockage of the pitot tube only affects the airspeed indicator, but a clogged static system affects all three pitot-static instruments.

✓ Rigidity in space refers to the principle that a wheel spun rapidly tends to remain fixed in the plane in which it is spinning.

✓ Precession causes slow drifting and minor erroneous indications in the gyroscopic instruments.

✓ The turn coordinator typically uses electrical power while an engine-driven vacuum pump is used to power the attitude and heading indicators.

✓ The turn coordinator provides an indication of turn direction and quality as well as a backup source of bank information in the event of attitude indicator failure.

✓ The attitude indicator uses an artificial horizon and miniature airplane to depict the position of your airplane in relation to the true horizon.

✓ Due to precession, the heading indicator must be aligned periodically with the magnetic compass. When aligning the heading indicator, be certain you are in straight-and-level, unaccelerated flight with the magnetic compass showing a steady indication.

✓ The magnetic compass shows a turn to the north if you accelerate an airplane in the northern hemisphere; if you decelerate, it indicates a turn to the south. The error does not occur when you are flying on a north or south heading.

✓ Turning error causes the magnetic compass to lead or lag the actual magnetic heading of the airplane during turns.

✓ The PFD contains digital versions of traditional analog flight instruments, including the attitude indicator, airspeed indicator, altimeter, vertical speed indicator, and HSI.

✓ The AHRS uses inertial sensors such as electronic gyroscopes, accelerometers, and a magnetometer to determine the aircraft's attitude relative to the horizon and heading.

✓ A magnetometer senses the earth's magnetic field to function as a magnetic compass, but without some of the errors associated with a conventional compass.

✓ A slip/skid indicator below the roll pointer of the attitude indicator helps you maintain coordinated flight.

✓ In addition to a compass card, the digital HSI displays the current airplane heading in a window, a course indicator arrow and CDI, and a turn rate indicator with index marks and a trend vector.

✓ When the AHRS system detects a problem, it places a red X over the display of the affected instrument to alert you that the indications are unreliable.

✓ The pitot tube, static source, and outside air temperature probe provide information to the air data computer (ADC), which uses these pressure and temperature inputs to determine the appropriate readings for the airspeed indicator, altimeter, and VSI.

✓ On the digital airspeed indicator, a central window and a pointer on the airspeed tape shows the indicated airspeed.

✓ A trend vector on the digital airspeed indicator shows what the airspeed will be in six seconds if the acceleration or deceleration continues at the same rate.

✓ The trend vector on the digital altimeter shows the altitude that the airplane will reach in six seconds if it continues to climb or descend at the same rate.

✓ The digital altimeter might have a bug that you set to a reference altitude for the autopilot or use as a reminder to level off at a specific altitude.

✓ On the digital VSI, you read vertical speed in feet per minute in a window that also serves as a pointer. The VSI tape remains motionless as the window moves up or down over the scale.

✓ If one or more sensors stops providing input, or if the ADC determines that its own internal operations are not correct, it places a red X over the display of the affected instrument.

✓ If the PFD goes black, electronic flight displays typically enable you to display PFD information on the MFD display in reversionary mode.

✓ If you have a catastrophic electrical failure and lose all electric power, you must use your backup instruments. The backup instruments are not electrically powered or are powered by a separate battery that is isolated from the main electrical system.

SECTION C ■ **Flight Instruments**

2-87

✓ The multifunction display (MFD) provides separate pages that can include a moving map, airport information, flight plan data, systems indications, checklists, and instrument charts.

✓ Enhanced vision systems (EVS), synthetic vision systems (SVS), and combined vision systems can improve situational awareness of the flight environment on PFDs.

✓ An enhanced flight vision system (EFVS) displayed on a conformal head-up display (HUD) uses a real-time imaging sensor that provides highly accurate vision performance in low visibility conditions.

KEY TERMS

International Standard Atmosphere (ISA)

Standard Lapse Rates

Pitot Tube

Static Port

V-Speeds

Indicated Altitude

Pressure Altitude

Density Altitude

Calibrated Altitude

True Altitude

Absolute Altitude

Trend Information

Rate Information

Rigidity In Space

Precession

Vacuum (Suction) System

Standard-Rate Turn

Slip

Skid

Variation

Deviation

Magnetic Dip

Electronic Flight Display

Primary Flight Display (PFD)

Multifunction Display (MFD)

Attitude and Heading Reference System (AHRS)

Magnetometer

Slip/Skid Indicator

Horizontal Situation Indicator (HSI)

Turn Rate indicator

Trend Vector

Air Data Computer (ADC)

QUESTIONS

1. What is the atmospheric pressure and temperature at sea level in a standard atmosphere?

2. Pitot pressure is used by which flight instrument(s)?

3. Referring to the airspeed indicator below, identify the V-speeds associated with the colored arcs.

4. Which important airspeed limitation changes with aircraft weight and is not depicted on the airspeed indicator?

Match the following types of altitude with the corresponding description.

5. Pressure Altitude

6. Density Altitude

7. True Altitude

8. Absolute Altitude

    A. The height of the airplane above the earth's surface

    B. The actual height of an object above mean sea level

    C. The vertical distance above the standard datum plane

    D. Pressure altitude corrected for non-standard temperature

9. You fly from an area of high pressure to an area of low pressure but do not reset your altimeter. If you maintain a consistent indicated altitude, will you be at your desired altitude? Why?

10. What will the effect be on the airspeed indicator if the static system becomes clogged, but the pitot system remains unobstructed? Why?

11. What type of movement is shown by the attitude indicator, but not by the turn coordinator?

12. True/False. If you accelerate an airplane in the northern hemisphere on a heading of east, your compass will indicate a turn to the south.

13. Describe the function of the AHRS.

14. Select the true statement regarding the digital attitude indicator.
    A. The roll scale reference marks are at 10, 25, 45, and 60 degrees.
    B. The turn-rate vector located on the roll scale indicates standard-rate turns.
    C. In a slip, the trapezoid of the slip/skid indicator located beneath the roll pointer moves to the inside of the turn.

15. What information is provided by the trend vector on the HSI?

16. If the AHRS detects a problem with the integrity of the sensor information, what occurs?
    A. The system reverts to reversionary mode and PFD information is displayed on the MFD.
    B. A red X is placed over the display of the affected instrument (attitude indicator or HSI).
    C. After an alert message appears, you must determine the affected instrument by comparing the indications of all instruments.

17. Select the true statement about the ADC.
    A. The pitot tube, static source, and outside air temperature probe provide information to the ADC.
    B. The ADC determines the readings for the airspeed indicator, attitude indicator, and altimeter.
    C. The failure of a single sensor affects every instrument that receives information from the ADC.

18. What is true about the indications on the altimeter?
    A. In six seconds, the airplane will reach an altitude of 8,500 feet MSL if it continues to climb at the same rate.
    B. In ten seconds, the airplane will reach an altitude of 8,460 feet MSL if it continues to climb at the same rate.
    C. In six seconds, the airplane will reach an altitude of 8,460 feet MSL if it continues to climb at the same rate.

19. Describe how the electronic flight display system compensates for a PFD screen failure.

20. True/False. An enhanced flight vision system (EFVS) can be displayed on a PFD.

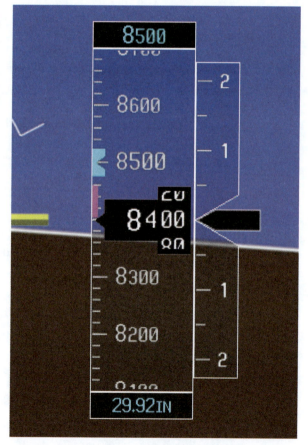

CHAPTER 3

Aerodynamic Principles

SECTION A
Four Forces of Flight

The science of aerodynamics deals with the motion of air and the forces acting on bodies moving relative to the air. When you study aerodynamics, you are learning about why and how an airplane flies. Although aerodynamics is a complex subject, exploring the fundamental principles that govern flight can be an exciting and rewarding experience. The challenge to understand what makes an airplane fly begins with learning the four forces of flight.

During flight, the four forces acting on the airplane are lift, weight, thrust, and drag. **Lift** is the force created by the effect of airflow as it passes over and under the wing. The airplane is typically supported in flight by lift. **Weight** is caused by the downward pull of gravity. **Thrust** is the forward force that propels the airplane through the air. It varies with the amount of engine power being used. **Drag** is a backward, or retarding, force that limits the speed of the airplane. In unaccelerated flight, the four forces are in equilibrium. Unaccelerated flight means that the airplane is maintaining a constant airspeed and is neither accelerating nor decelerating. [Figure 3-1]

In straight-and-level, unaccelerated flight:
- Lift is equal to and directly opposite weight.
- Thrust is equal to and directly opposite drag.

Figure 3-1. Four forces act on an airplane during flight.

The arrows that show the forces acting on an airplane are often called **vectors**. The magnitude of a vector is indicated by the arrow's length, while the direction is shown by the arrow's orientation. When two or more forces act on an object at the same time, they combine to create a resultant. [Figure 3-2]

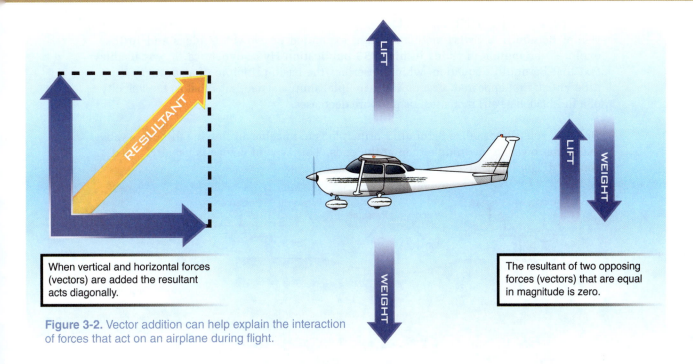

When vertical and horizontal forces (vectors) are added the resultant acts diagonally.

The resultant of two opposing forces (vectors) that are equal in magnitude is zero.

Figure 3-2. Vector addition can help explain the interaction of forces that act on an airplane during flight.

LIFT

Lift is the key aerodynamic force. It is the force that opposes weight in straight-and-level, unaccelerated flight. When weight and lift are equal, an airplane is in a state of equilibrium.

When an airplane is stationary on the ramp, it is also in equilibrium, but the aerodynamic forces are not a factor. In calm wind conditions, the atmosphere exerts equal pressure on the upper and lower surfaces of the wing. Movement of air about the airplane, particularly the wing, is necessary before the aerodynamic force of lift becomes effective. Knowledge of some of the basic principles of motion will help you to understand the force of lift.

NEWTON'S LAWS OF FORCE AND MOTION

In the 17th century, Sir Isaac Newton, a physicist and mathematician presented principles of motion which, today, help to explain the creation of lift by an airfoil. **Newton's three laws of motion** are as follows:

> Newton's first law: A body at rest tends to remain at rest, and a body in motion tends to remain moving at the same speed and in the same direction. For example, an airplane at rest on the ramp will remain at rest unless a force is applied that is strong enough to overcome the airplane's inertia.

> Newton's second law: When a body is acted upon by a constant force, its resulting acceleration is inversely proportional to the mass of the body and is directly proportional to the applied force. This law may be expressed by the formula: Force = mass × acceleration (**F = ma**).

> Newton's third law: Every force applied to an object is opposed by an equal force in the opposite direction, or as it is often stated, for every action there is an equal and opposite reaction. This principle applies whenever two things act upon each other, such as the air and the propeller, or the air and the wing of an airplane.

SECTION A ■ **Four Forces of Flight**

BERNOULLI'S PRINCIPLE

Daniel Bernoulli, a Swiss mathematician, expanded on Newton's ideas and further explored the motion of fluids in his 1738 publication Hydrodynamica. It was in this text that Bernoulli's equation, which describes the basic principle of airflow pressure differential, first appeared. **Bernoulli's principle**, simply stated, says that as the velocity of a fluid (such as air) increases, its pressure decreases.

One way you can visualize Bernoulli's principle is to imagine air flowing through a tube that is narrower in the middle than at the ends. This type of device is usually called a **venturi**. [Figure 3-3]

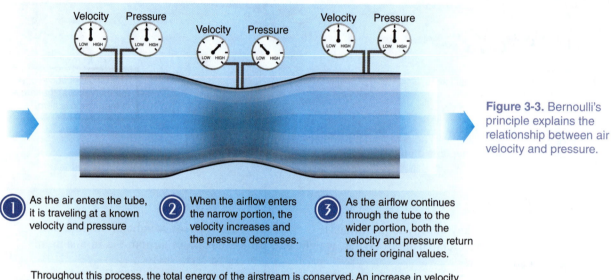

Figure 3-3. Bernoulli's principle explains the relationship between air velocity and pressure.

① As the air enters the tube, it is traveling at a known velocity and pressure

② When the airflow enters the narrow portion, the velocity increases and the pressure decreases.

③ As the airflow continues through the tube to the wider portion, both the velocity and pressure return to their original values.

Throughout this process, the total energy of the airstream is conserved. An increase in velocity (kinetic energy) is accompanied by a decrease in static pressure (potential energy).

AIRFOILS

An **airfoil** is any surface, such as a wing, that provides aerodynamic force when it interacts with a moving stream of air. Some of the terms used to describe the wing, and the interaction of the airflow about it, are defined in Figures 3-4 and 3-5.

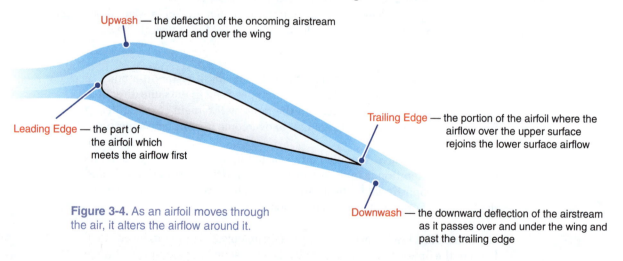

Upwash — the deflection of the oncoming airstream upward and over the wing

Leading Edge — the part of the airfoil which meets the airflow first

Trailing Edge — the portion of the airfoil where the airflow over the upper surface rejoins the lower surface airflow

Figure 3-4. As an airfoil moves through the air, it alters the airflow around it.

Downwash — the downward deflection of the airstream as it passes over and under the wing and past the trailing edge

Chord line and camber are terms that help define the wing's shape, while flight path and relative wind describe the movement of the wing with respect to the surrounding air.

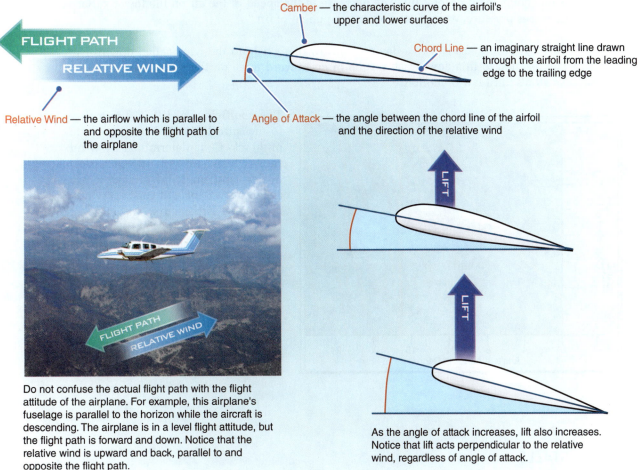

Camber — the characteristic curve of the airfoil's upper and lower surfaces

Chord Line — an imaginary straight line drawn through the airfoil from the leading edge to the trailing edge

Relative Wind — the airflow which is parallel to and opposite the flight path of the airplane

Angle of Attack — the angle between the chord line of the airfoil and the direction of the relative wind

Do not confuse the actual flight path with the flight attitude of the airplane. For example, this airplane's fuselage is parallel to the horizon while the aircraft is descending. The airplane is in a level flight attitude, but the flight path is forward and down. Notice that the relative wind is upward and back, parallel to and opposite the flight path.

As the angle of attack increases, lift also increases. Notice that lift acts perpendicular to the relative wind, regardless of angle of attack.

Figure 3-5. The angle of attack is one factor that determines the amount of lift generated by the wing.

Lift occurs when an airstream circulates about an airfoil and affects the pressure distribution on that airfoil. As air circulates around a wing's surface, some regions experience lower-than-atmospheric pressure and others experience higher-than-atmospheric pressure. The exact pressure distribution varies with angle of attack. The net force from these pressure differentials is the total lift generated by the airfoil. [Figure 3-6]

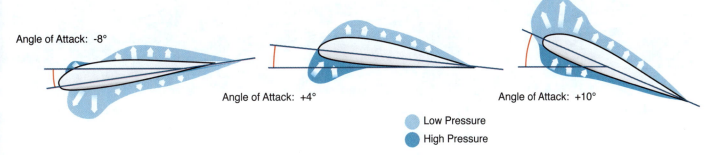

Angle of Attack: -8°

Angle of Attack: +4°

Angle of Attack: +10°

⬤ Low Pressure
⬤ High Pressure

Figure 3-6. Newton's laws and Bernoulli's principle explain the circulation of air around a wing and the pressure distribution on the wing's surface.

SECTION A ■ **Four Forces of Flight**

The airplane wing's shape is designed to take advantage of Newton's laws and Bernoulli's principle. The shape of an airfoil causes air to accelerate as it passes over the wing, and to decelerate as it passes under the wing. According to Bernoulli's theorem, the increased speed of the air on the top of an airfoil produces a pressure drop and this lowered pressure is one component of total lift. The decrease in the speed of the air on the lower surface increases pressure, which is the other component of lift.

This airflow pattern causes a downward flow of air behind the wing called downwash. The reaction to this downwash results in an upward force on the wing, demonstrating Newton's third law of motion.

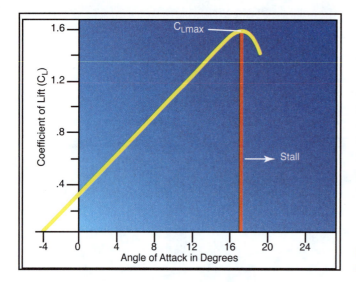

The **coefficient of lift (C_L)** is a way to measure lift as it relates to angle of attack. C_L is determined by wind tunnel tests and is based on airfoil design and angle of attack. Every airfoil has an angle of attack at which maximum lift occurs. As angle of attack increases up to this point, C_L also increases. This point of maximum lift is called C_{Lmax}. If the maximum lift angle is exceeded, lift decreases rapidly and the wing stalls. [Figure 3-7]

Figure 3-7. In this example, C_{Lmax} occurs at an angle of attack of about 17 degrees.

STALLS

A **stall** is caused by the separation of airflow from the wing's upper surface. This results in a rapid decrease in lift. For a given airplane, a stall always occurs at the same angle, regardless of airspeed, flight attitude, or weight. This angle is the stalling or **critical angle of attack**. [Figure 3-8]

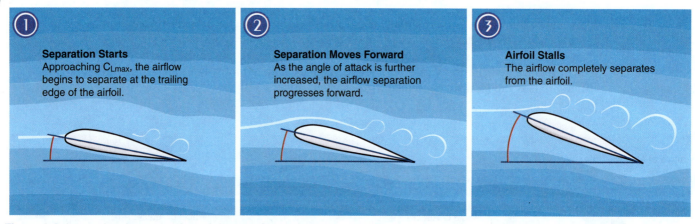

① **Separation Starts**
Approaching C_{Lmax}, the airflow begins to separate at the trailing edge of the airfoil.

② **Separation Moves Forward**
As the angle of attack is further increased, the airflow separation progresses forward.

③ **Airfoil Stalls**
The airflow completely separates from the airfoil.

Figure 3-8. Increasing the angle of attack beyond C_{Lmax} causes progressive disruption of airflow from the upper surface of the wing.

Stall characteristics vary with different airplanes. However, in training airplanes during most normal maneuvers, the onset of a stall is gradual. The first indications might be a mushy feeling in the flight controls, a stall warning device, or a slight buffeting of the airplane. To recover from a stall, you must restore the smooth airflow by decreasing the angle of attack to a point below the critical angle of attack.

WING DESIGN FACTORS

Wing design is based on the anticipated use of the airplane, cost, and other factors. The main design considerations are wing planform, camber, aspect ratio, and total wing area.

Camber affects the difference in the velocity of the airflow between the upper and lower surfaces of the wing. If the upper camber increases and the lower camber remains the same, the velocity differential increases.

DISCOVERY

The Boundary Layer

Examining the boundary layer can lead to a better understanding of the cause of airflow separation from the wing. The boundary layer is a thin layer of air next to the surface of an airfoil that shows a reduction in speed due to the air's viscosity or stickiness. The boundary layer can be described as either laminar or turbulent based on the type of airflow. Laminar flow begins near the leading edge and consists of smooth laminations of air sliding over one another. At some point along the airfoil, this laminar layer transitions to a thicker turbulent flow with higher velocities.

Figure A depicts the development of the boundary layer on a flat plate. The velocity profiles can help you visualize the local velocity of the airstream in the boundary layer and provide a comparison between the laminar and turbulent airflow.

Proceeding back from the leading edge of the airfoil, pressure decreases with distance. This favorable pressure gradient (high to low) assists the flow of the boundary layer. At the point where the local velocity of the air at the surface is zero, the pressure gradient reverses and an adverse pressure gradient exists (low to high). As the angle of attack increases, the unfavorable pressure gradient grows longer, and the airflow begins to separate from the wing. [Figure B] When the airflow does not adhere to the surface near the leading edge, a stall occurs. The high velocity airflow of the turbulent boundary layer helps to prevent the airflow separation, which can cause a stall.

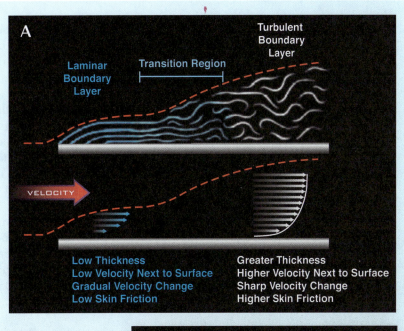

A

Laminar Boundary Layer

Transition Region

Turbulent Boundary Layer

VELOCITY

Low Thickness
Low Velocity Next to Surface
Gradual Velocity Change
Low Skin Friction

Greater Thickness
Higher Velocity Next to Surface
Sharp Velocity Change
Higher Skin Friction

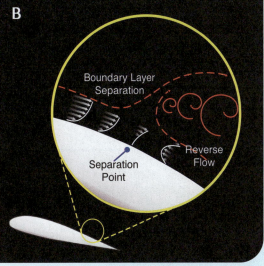

B

Boundary Layer Separation

Separation Point

Reverse Flow

SECTION A ■ Four Forces of Flight

DISCOVERY

Airfoil Design and Wind Tunnels

Orville and Wilbur Wright constructed a wind tunnel in 1901 and tested several hundred airfoil shapes to determine optimum performance before their aircraft was built. Today, aircraft designers use wind tunnels to test specific designs and organizations such as the National Aeronautics and Space Administration (NASA) use wind tunnels to perform research on the development of airfoil and aircraft shapes.

The photo shows an F/A-18 fighter aircraft in the giant wind tunnel (80 feet by 120 feet) at NASA's Ames Research Center. The U.S. Navy, through the Naval Air Systems Command, supplied the F/A-18 aircraft to NASA. It was the first full-scale aircraft to undergo tests in the world's largest wind tunnel.

There is, of course, a limit to the amount of camber that can be used. After a certain point, air will no longer flow smoothly over the airfoil. Once this happens, the lifting capacity diminishes. The ideal camber varies with the airplane's performance specifications, especially the speed range and the load-carrying requirements.

Aspect ratio is the relationship between the length and width of a wing—it is the span of the wing, wingtip to wingtip, divided by its average chord. It is one of the primary factors in determining lift/drag characteristics. In general, the higher the aspect ratio, the higher the lifting efficiency of the wing. For example, gliders can have an aspect ratio of 10-30, while typical training aircraft have an aspect ratio of about 6-10. At a given angle of attack, a higher aspect ratio produces less drag for the same amount of lift. [Figure 3-9]

Wing area is the total surface area of the wings. Most wings don't produce a great amount of lift per square foot, so wing area must be sufficient to support the weight of the airplane. For example, in a training aircraft at normal operating speed, the wings produce only about 14 pounds of lift for each square foot of wing area. This means a wing area of 175 square feet is required to support an airplane weight of 2,450 pounds during straight-and-level flight.

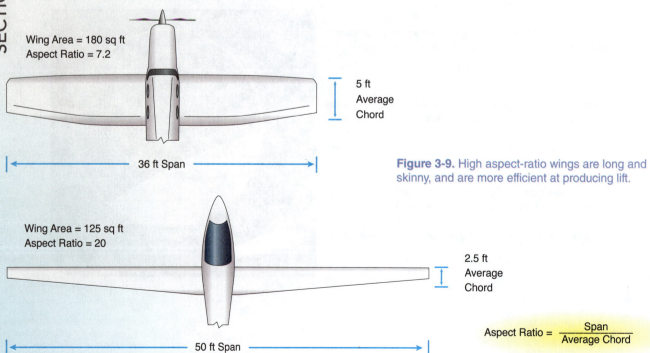

Figure 3-9. High aspect-ratio wings are long and skinny, and are more efficient at producing lift.

$$\text{Aspect Ratio} = \frac{\text{Span}}{\text{Average Chord}}$$

Planform refers to the shape of the airplane's wing when viewed from above or below. Each planform design has advantages and disadvantages. [Figure 3-10]

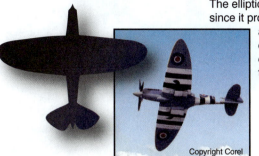

The elliptical wing is ideal for flight at slow speeds since it provides a minimum of drag for a given aspect ratio. This type of planform is difficult to construct, though, and its stall characteristics are not as favorable as those of the rectangular wing.

The rectangular wing is not as efficient as the elliptical wing, but it has a tendency to stall first at the wing root which provides adequate stall warning and aileron effectiveness.

Tapering provides a decrease in drag and increase in lift which is most effective at high speeds. A highly tapered wing has a tendency to stall first slightly inboard of the wingtip. A good compromise on planform for low-speed aircraft is a combination of both rectangular and tapered configurations. The rectangular inboard section exhibits good stall characteristics and is cost effective. The tapered outboard portion allows for a reduction in weight and an increase in aspect ratio.

Sweptback wings, including delta wings, are efficient at high speeds but low-speed performance is degraded by this design.

Figure 3-10. Each planform design has its own specific aerodynamic characteristics.

SECTION A ■ Four Forces of Flight

The Swept Wing

Sweepback is an important design feature of high-speed airplanes. This characteristic allows the airplane to fly at higher speeds without reaching the critical Mach number—the speed at which the wing experiences supersonic airflow. While sweepback enhances high-speed performance, it degrades performance at low speeds. A significant part of the air velocity is flowing spanwise, not contributing to lift. This raises the stall speed and also causes the wingtips to stall first.

By employing variable-sweep wings, the B1 Bomber can change its flying configuration to meet different aerodynamic and performance requirements such as takeoff and landing or high-speed flight. [Figure A]

A

Copyright © 3daddict | Dreamstime.com

A wing with forward sweep has the same effect on the airflow as a sweptback wing. Forward sweep also reduces the critical Mach number over the wing. A forward-swept wing has the advantage of a spanwise flow directed inboard and does not have the problem of the wingtips stalling first as with aft-swept wings. Therefore, the forward-swept wing is more efficient at slow speeds. One drawback with this design is the tendency of the wing to twist more than a sweptback wing when high flight loads are applied, which can cause structural failure.

The Grumman Corporation's forward-swept wing research airplane, the X-29, used high-tech composite construction, which made the wing lightweight and rigid to prevent twisting in flight. [Figure B]. The X-29 was approximately 30% to 40% more efficient at producing lift than most conventional fighters with aft-swept wings.

B

Courtesy of NASA Dryden Research Center

Once the design of the wing is determined, the wing must be mounted on the airplane. Usually it is attached to the fuselage with the chord line inclined upward at a slight angle, which is called the **angle of incidence**. [Figure 3-11]

Longitudinal Axis

Angle of Incidence

Figure 3-11. Angle of incidence refers to the angle between the wing chord line and a line parallel to the longitudinal axis of the airplane.

When wing twist, or washout, is incorporated into the wing design, the wingtip has a lower angle of incidence than the wing root. This results in the wingtip having a lower angle of attack than the root during the approach to a stall. Wing twist is used by airplane designers to prevent the wingtips from stalling before the roots—an undesirable characteristic, because the disrupted airflow near the wingtip can reduce aileron effectiveness so much that it could be impossible to level the wings. With an appropriate amount of washout, the wingtip and ailerons will continue flying and provide positive bank control after the wing root has stalled. [Figure 3-12]

Another method sometimes used to ensure positive control during the stall is installation of **stall strips**, which consist of two metal strips attached to the leading edge of each wing near the fuselage. These strips disrupt the airflow at high angles of attack, causing the wing area directly behind them to stall before the wingtips stall.

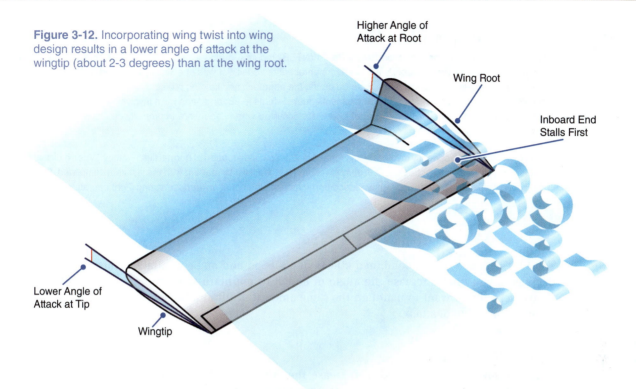

Figure 3-12. Incorporating wing twist into wing design results in a lower angle of attack at the wingtip (about 2-3 degrees) than at the wing root.

Higher Angle of Attack at Root

Wing Root

Inboard End Stalls First

Lower Angle of Attack at Tip

Wingtip

PILOT CONTROL OF LIFT

The amount of lift generated by an airplane is controlled by the pilot as well as determined by aircraft design factors. For example, you can change the angle of attack and the airspeed or you can change the shape of the wing by lowering the flaps. Anytime you do something to increase lift, drag also increases. Drag is always a by-product of lift.

CHANGING ANGLE OF ATTACK

You have direct control over angle of attack. During flight at normal operating speeds, if you increase the angle of attack, you increase lift. Anytime you change the pitch of the airplane during flight, you change the angle of attack of the wings. At the same time, you are changing the coefficient of lift.

CHANGING AIRSPEED

The faster the wing moves through the air, the greater the lift. Actually, lift is proportional to the square of the airplane's speed. For example, at 200 knots, an airplane has four times the lift of the same airplane traveling at 100 knots, if the angle of attack and other factors are constant. On the other hand, if the speed is reduced by one-half, lift is decreased to one-quarter of the previous value.

Although airspeed is an important factor in the production of lift, it is only one of several factors. The airspeed required to sustain an aircraft in flight depends on the flap position, the angle of attack, and the weight.

ANGLE OF ATTACK AND AIRSPEED

The relationship between angle of attack and airspeed in the production of lift is not as complex as it might seem. Angle of attack establishes the coefficient of lift for the airfoil. At the same time, lift is proportional to the square of the airplane's speed. Because you can control both angle of attack and airspeed, you can control lift.

SECTION A ■ **Four Forces of Flight**

DISCOVERY

Failure to Properly Use Checklists

According to National Transportation Safety Board (NTSB) records, during a certain 4 1/2-year period, there were 87 accidents in which failure to properly use an aircraft procedures checklist was specifically attributed as a cause or factor. The records indicate that 43 accidents occurred during the approach and landing phase of flight, while 35 occurred during takeoff. Landing gear was indicated in 36 accidents, fuel systems in 28, and flaps were involved in 11 of the reports. The most serious accidents involving misuse of checklists occurred when flaps were improperly configured, usually during takeoff.

This type of accident frequently involves some form of distraction or disruption of normal flight routine. When a distraction occurs while using a written checklist, it is easy to resume the checklist by starting with the last item completed. Properly using checklists can help you manage workload and maintain situational awareness during flight operations.

Total lift depends on the combined effects of airspeed and angle of attack. When speed decreases, you must increase the angle of attack to maintain the same amount of lift. Conversely, if you want to maintain the same amount of lift at a higher speed, you must decrease the angle of attack.

HIGH-LIFT DEVICES

High-lift devices are designed to increase the efficiency of the airfoil at low speeds. The most common high-lift device is the trailing-edge flap. When properly used, **flaps** increase the lifting efficiency of the wing and decrease stall speed. This allows you to fly at a reduced speed while maintaining sufficient control and lift for sustained flight. Remember, though, that when you retract the flaps, the stall speed increases.

The ability to fly at slow speeds is particularly important during the approach and landing phases. For example, an approach with full flaps permits you to fly at a fairly steep descent angle without gaining airspeed, which allows the airplane to touch down at a slower speed. In addition, you can land near the approach end of the runway, even when there are obstacles along the approach path.

In training airplanes, **configuration** normally refers to the position of the landing gear and flaps. When the gear and flaps are up, an airplane is in a clean configuration. If the gear is fixed rather than retractable, the airplane is considered to be in a clean configuration when the flaps are in the up position. During flight, you can change configuration by raising or lowering the gear, or by moving the flaps.

Lowering the flaps increases lift (and drag) by increasing the wing's effective camber and changing the average chord line, which increases the average angle of attack. In some cases, flaps also increase the area of the wing. [Figure 3-13]

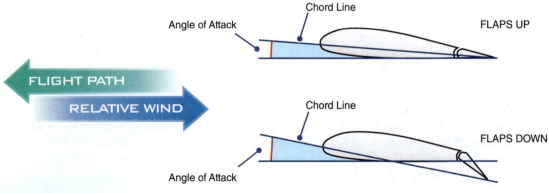

Figure 3-13. Most flaps, when fully extended, form an angle of 30° to 40° relative to the wing.

Vortex Generators

Vortex generators are small airfoil-like surfaces on the wing that project vertically into the airstream. [Figure A] Vortices are formed at the tip of these generators just as they are on ordinary wingtips. These vortices add energy to the boundary layer (the layer of air next to the surface of the wing) to prevent airflow separation. This reduces stall speeds and can increase takeoff and landing performance. Although most commonly seen on high-speed aircraft, vortex generators also are used on some light general aviation aircraft.

Many animals, including bats, owls, beetles, flies, moths and even dolphins, have mechanisms for controlling lift and drag through control of the boundary layer. Blood vessels in the wings of a worker bee stabilize the membranes and increase the energy of the turbulent boundary layer flow. [Figure B]

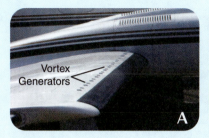

Vortex Generators

Copyright Corel

There are several common types of flaps. The **plain flap** is attached to the wing by a hinge. When deflected downward, it increases the effective camber and changes the wing's chord line. Both of these factors increase the lifting capacity of the wing. [Figure 3-14]

The **split flap** is hinged only to the lower portion of the wing. This type of flap also increases lift, but it produces greater drag than the plain flap because of the turbulent wake it causes. [Figure 3-15]

The **slotted flap** is similar to the plain flap. In addition to changing the wing's camber and chord line, it also allows a portion of higher pressure air beneath the wing to travel through a slot. This increases the velocity of the airflow over the flap and provides additional lift. The high energy air from the slot accelerates the upper surface airflow and delays airflow separation to a higher angle of attack. [Figure 3-16]

Another type of flap is the **Fowler flap**. It is attached to the wing by a track and roller system. When extended, the Fowler flap moves rearward and down. This rearward motion increases the total wing area, as well as the camber and chord line. [Figure 3-17]

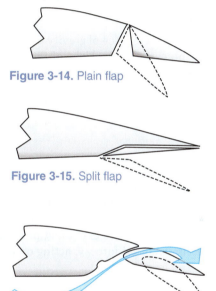

Figure 3-14. Plain flap

Figure 3-15. Split flap

Figure 3-16. Slotted flap

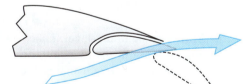

Figure 3-17. Fowler flap

Although the amount of lift and drag created by a specific flap system varies, a few general observations can be made. As the flaps are extended, at first they will produce a relatively large amount of lift for a small increase in drag. However, once the flap extension reaches approximately the midpoint, this relationship reverses. Now, a significant increase in drag will occur for a relatively small increase in lift. Because of the large increase in drag beyond the half-flap position, most manufacturers limit the takeoff setting to half flaps or less.

High-speed airplanes often employ high-lift devices on the leading edge of the wing to increase lift at slow speeds and spoilers on top to *reduce* lift for rapid descents. Leading-edge flaps increase wing camber, which provides additional lift. Fixed slots and movable slats conduct the flow of high energy air beneath the wing into the airflow on the wing's upper surface, which delays airflow separation to a higher angle of attack. [Figure 3-18]

SECTION A ■ **Four Forces of Flight**

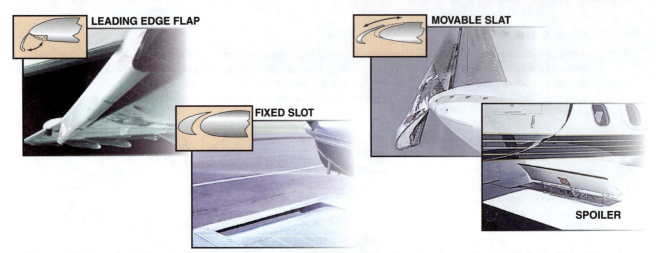

Figure 3-18. Leading edge devices operate in concert with trailing edge flaps to change the coefficient of lift of the wing. Spoilers reduce lift and increase drag, to allow higher descent rates and to prevent becoming airborne again after touchdown.

WEIGHT

Weight is the force of gravity that acts through the center of the airplane toward the center of the earth. The weight of the airplane is not a constant. It varies with the equipment installed, passengers, cargo, and fuel load. During the course of a flight, the total weight of the airplane decreases as fuel is consumed. Additional weight reduction may also occur during some specialized flight activities, such as crop dusting, fire fighting, or sky diving flights.

THRUST

Thrust is the forward-acting force that opposes drag and propels the airplane. In most general aviation airplanes, this force is provided when the engine turns the propeller. The same physical principles involved in the generation of lift also apply when describing the force of thrust. As explained previously in this chapter, Newton's second law states that an unbalanced force, **F**, acting on a mass, m, will accelerate, **a**, the mass in the direction of the force ($\mathbf{F} = m\mathbf{a}$).

In the case of airplane thrust, the force is provided by the expansion of the burning gases in the engine that turn the propeller. A mass of air moves through the propeller, a rotating airfoil, and is accelerated opposite to the direction of the flight path. The equal and opposite reaction illustrated by Newton's third law is thrust, a force on the airplane in the direction of flight.

During straight-and-level, unaccelerated flight, the forces of thrust and drag are equal. You increase thrust by using the throttle to increase power. When you increase power, thrust exceeds drag, causing the airplane to accelerate. This acceleration, however, is accompanied by a corresponding increase in drag. The airplane continues to accelerate only while the force of thrust exceeds the force of drag. When drag again equals thrust, the airplane ceases to accelerate and maintains a constant airspeed. However, the new airspeed is higher than the previous one.

When you reduce thrust, the force of drag causes the airplane to decelerate. But as the airplane slows, drag diminishes. When drag has decreased enough to equal thrust, the airplane no longer decelerates. Once again, it maintains a constant airspeed. Now, however, the airspeed is slower than the one previously flown.

DRAG

Drag acts opposite the direction of flight, opposes the forward-acting component of thrust, and limits the forward speed of the airplane. Drag is broadly classified as either parasite or induced.

PARASITE DRAG

Parasite drag is caused by any aircraft surface that deflects or interferes with the smooth airflow around the airplane. Parasite drag normally is divided into three types: form drag, interference drag, and skin friction drag.

Form drag results from the turbulent wake caused by the separation of airflow from the surface of a structure. The amount of drag is related to both the size and shape of the structure that protrudes into the relative wind. [Figure 3-19]

Interference drag occurs when the airflow around one part of the airplane interacts with the airflow around an adjacent part. Drag results when flows of air moving at different speeds or in different directions combine. For example, interference drag occurs where the wing joins the fuselage of a low wing airplane because air traveling across the top of the wing moves at a higher speed than the air flowing along the fuselage. [Figure 3-20]

Courtesy of Diamond Aircraft Industries

Figure 3-20. Design features such as wheel fairings and retractable landing gear can reduce both form and interference drag.

Skin friction drag is caused by the roughness of the airplane's surfaces. Even though these surfaces may appear smooth, under a microscope, they may be quite rough. A thin layer of air clings to these rough surfaces and creates small eddies that contribute to drag. [Figure 3-21]

Figure 3-21. Skin friction drag can be minimized by applying a glossy, smooth finish to surfaces, and by eliminating protruding rivet heads, roughness, and other irregularities.

Figure 3-19. It is easy to visualize the creation of form drag by examining the airflow around a flat plate. Streamlining decreases form drag by reducing the airflow separation.

Each type of parasite drag varies with the speed of the airplane. The combined effect of all parasite drag varies proportionately to the square of the airspeed. If airspeed is doubled, parasite drag increases fourfold. Therefore, parasite drag is predominant at high speeds and negligible at low speeds, near a stall. [Figure 3-22]

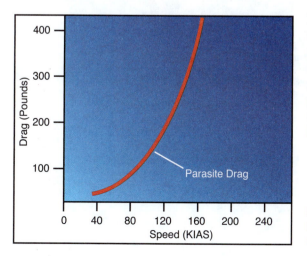

Figure 3-22. Parasite drag increases with the square of the airspeed. For example, an airplane has four times as much parasite drag at 160 knots as it does at 80 knots.

SECTION A ■ Four Forces of Flight

INDUCED DRAG

Induced drag is generated by the airflow circulation around the wing as it creates lift. The high pressure air beneath the wing joins the low pressure air above the wing at the trailing edge and wingtips. This causes a spiral or vortex that trails behind each wingtip whenever lift is being produced. These **wingtip vortices** have the effect of deflecting the airstream downward in the vicinity of the wingtip, creating an increase in downwash. Therefore, the wing operates in an *average* relative wind that is inclined downward and rearward near the wing. Because the lift produced by the wing is perpendicular to the relative wind, the lift is inclined aft by the same amount. The component of lift acting in a rearward direction is induced drag. [Figure 3-23]

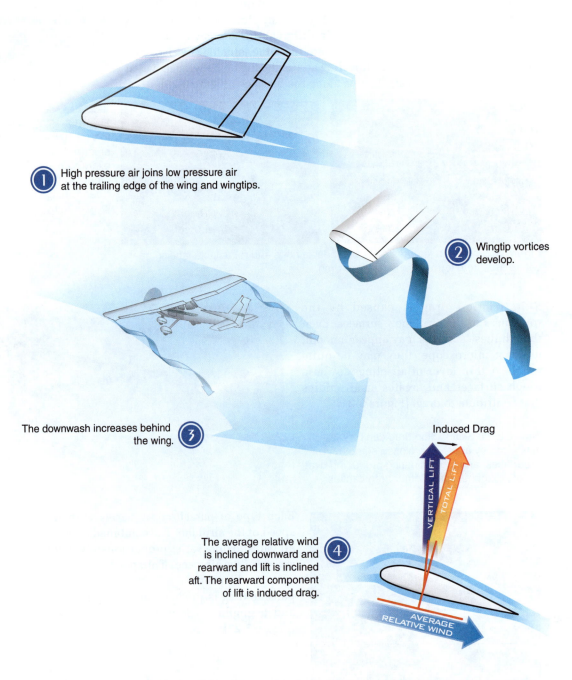

1 High pressure air joins low pressure air at the trailing edge of the wing and wingtips.

2 Wingtip vortices develop.

3 The downwash increases behind the wing.

4 The average relative wind is inclined downward and rearward and lift is inclined aft. The rearward component of lift is induced drag.

Induced Drag

VERTICAL LIFT

TOTAL LIFT

AVERAGE RELATIVE WIND

Figure 3-23. The formation of induced drag is associated with the downward deflection of the airstream near the wing.

As the air pressure differential increases with an increase in angle of attack, stronger vortices form and induced drag increases. Because the wing usually is at a low angle of attack at high speed, and a high angle of attack at low speed, induced drag is *inversely* proportional to the square of the speed. If speed is decreased by half, induced drag increases fourfold. It is the major cause of drag at reduced speeds near a stall, but, as speed increases, induced drag decreases dramatically. [Figure 3-24]

Figure 3-24. Induced drag is inversely proportional to the square of the speed. For example, an airplane has four times as much induced drag at 60 knots than it does at 120 knots.

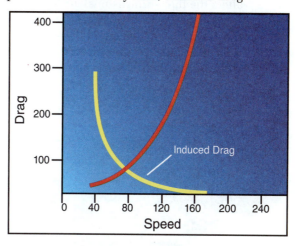

TOTAL DRAG

Total drag for an airplane is the sum of parasite and induced drag. The total drag curve represents these combined forces and is plotted against airspeed. The low point on the total drag curve shows the airspeed at which drag is at its minimum. This point, where the lift-to-drag ratio is greatest, is referred to as L/D_{max}. At this speed, the total lift capacity of the airplane, when compared to the total drag of the airplane, is most favorable. This is important in airplane performance. [Figure 3-25]

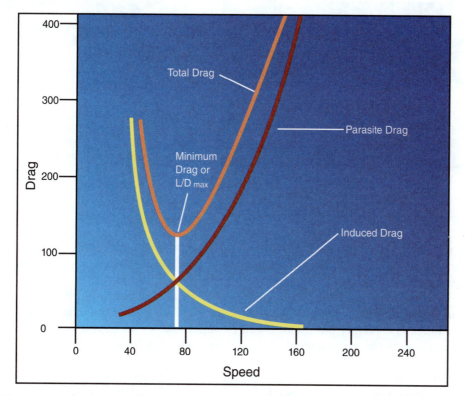

Figure 3-25. The orange total drag curve shows the sum of induced drag and parasite drag.

SECTION A ■ **Four Forces of Flight**

GROUND EFFECT

The phenomenon of **ground effect** is associated with the reduction of induced drag. During takeoffs or landings, when you are flying close to the ground, the earth's surface alters the three-dimensional airflow pattern around the airplane, causing a reduction in wingtip vortices and a decrease in upwash and downwash. Because ground effect restricts the downward deflection of the airstream, induced drag decreases. Ground effect is most noticeable near the surface; it rapidly diminishes as the airplane climbs and is negligible when the airplane is about one wingspan above the surface. [Figure 3-26]

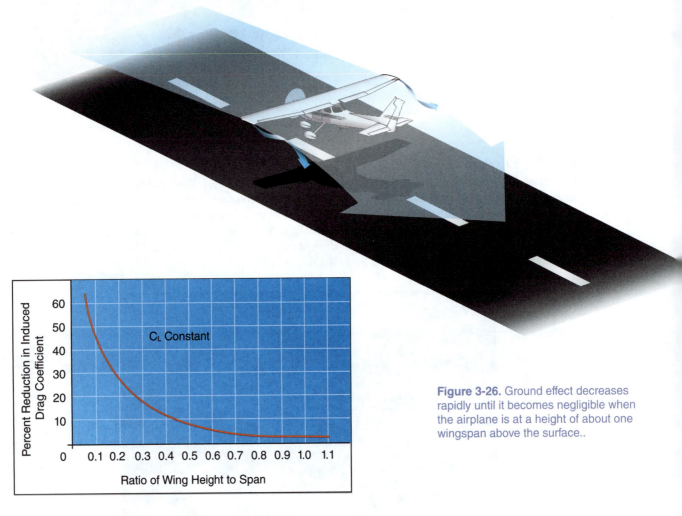

Figure 3-26. Ground effect decreases rapidly until it becomes negligible when the airplane is at a height of about one wingspan above the surface..

With the reduction of induced drag in ground effect, the amount of thrust required to produce lift is reduced. What this means is that your airplane is capable of lifting off at lower-than-normal speed. Although you might initially think that this is desirable, consider what happens as you climb out of ground effect. The power (thrust) required to sustain flight increases significantly as the normal airflow around the wing returns and induced drag is suddenly increased. If you attempt to climb out of ground effect before reaching the speed for normal climb, the airplane might sink back to the surface.

Ground effect is noticeable in the landing phase of flight, too, just before touchdown. Within one wingspan above the ground, the decrease in induced drag makes your airplane seem to float on the cushion of air beneath it. Because of this, power reduction usually is required during the flare to help the airplane land. Although all airplanes may experience ground effect, it is more noticeable in low-wing airplanes, simply because the wings are closer to the ground.

SUMMARY CHECKLIST

✓ During flight, the four forces acting on the airplane are lift, weight, thrust, and drag.

✓ The four forces are in equilibrium during unaccelerated flight.

✓ Lift is the upward force created by the effect of airflow as it passes over and under the wing.

✓ The airplane wing's shape is designed to take advantage of both Newton's laws and Bernoulli's principle.

✓ According to Bernoulli's principle, the increase in speed of air on the top of an airfoil produces a drop in pressure and this lowered pressure is a component of total lift. The decrease in speed of air on the bottom of an airfoil produces an increase in pressure, providing the other main component of total lift.

✓ The reaction to downwash causes an upward reaction according to Newton's third law of motion.

✓ Planform, camber, aspect ratio, and wing area are some of the design factors that affect a wing's lifting capability.

✓ A stall is caused by the separation of airflow from the wing's upper surface. For a given airplane, a stall always occurs at the critical angle of attack, regardless of airspeed, flight attitude, or weight.

✓ Total lift depends on the combined effects of airspeed and angle of attack. When speed decreases, you must increase the angle of attack to maintain the same amount of lift.

✓ Flaps increase lift (and drag) by increasing the wing's effective camber and changing the chord line that increases the angle of attack. Flap types include plain, split, slotted, and Fowler.

✓ Weight is the force of gravity that acts vertically through the center of gravity of the airplane toward the center of the earth.

✓ Thrust is the forward-acting force that opposes drag and propels the airplane.

✓ Drag acts in opposition to the direction of flight, opposes the forward-acting force of thrust, and limits the forward speed of the airplane.

✓ Parasite drag is caused by any aircraft surface that deflects or interferes with the smooth airflow around the airplane. Parasite drag normally is divided into three types: form drag, interference drag, and skin friction drag. If airspeed is doubled, parasite drag increases fourfold.

✓ Induced drag is generated by the airflow circulation around the wing as it creates lift. Induced drag increases with flight at slow airspeeds as the angle of attack increases.

✓ The phenomenon of ground effect occurs close to the ground where the earth's surface restricts the downward deflection of the airstream from the wing, decreasing induced drag.

SECTION A ■ Four Forces of Flight

KEY TERMS

Lift

Weight

Thrust

Drag

Vectors

Newton's Three Laws of Motion

Bernoulli's Principle

Venturi

Airfoil

Leading Edge

Trailing Edge

Upwash

Downwash

Relative Wind

Camber

Chord Line

Angle of Attack

Coefficient of Lift (C_L)

Stall

Critical Angle of Attack

Aspect Ratio

Wing Area

Planform

Angle of Incidence

Stall Strips

Flaps

Configuration

Plain Flap

Split Flap

Slotted Flap

Fowler Flap

Parasite Drag

Form Drag

Interference Drag

Skin Friction Drag

Induced Drag

Wingtip Vortices

Ground Effect

SECTION A ■ **Four Forces of Flight**

QUESTIONS

1. Select the true statement regarding the four forces of flight.
   A. During accelerated flight, thrust and drag are equal.
   B. The four forces are in equilibrium during unaccelerated flight.
   C. In straight-and-level unaccelerated flight, all four forces are equal in magnitude.

2. Refer to the following illustration and identify the aerodynamic terms associated with the airfoil.

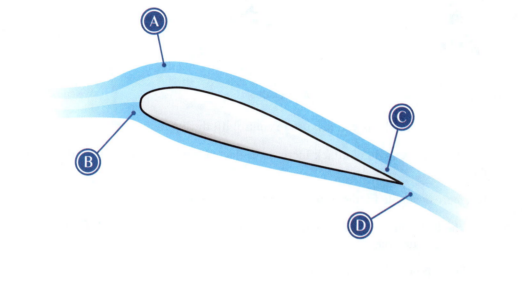

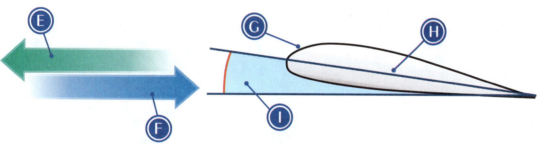

3. Describe how Newton's laws of motion and Bernoulli's principle explain the generation of lift by an airfoil.

4. True/False. As airspeed increases, the angle of attack at which an airfoil stalls also increases.

5. Determine the aspect ratio of the following planforms.

Wing Span = 196 ft
Average Chord = 28 ft

Wing Span = 35 ft
Average Chord = 5 ft

Wing Span = 37 ft
Average Chord = 11 ft

6. Identify three methods you can use to control lift during flight.

7. Will the wing's angle of attack increase or decrease when you lower trailing edge flaps?

8. Is it more desirable for the wing root or wingtips to stall first and why?

9. List the three forms of parasite drag and provide examples of aircraft features that reduce parasite drag.

10. Explain why induced drag increases as airspeed decreases.

11. The reduction in induced drag due to ground effect is most noticeable when the airplane is within what distance from the earth's surface?

SECTION A ■ Four Forces of Flight

SECTION B
Stability

Although no airplane is completely stable, all airplanes must have desirable stability and handling characteristics. An inherently stable airplane is easy to fly and reduces pilot fatigue. This quality is essential throughout a wide range of flight conditions—during climbs, descents, turns, and at both high and low airspeeds. An aircraft's inherent stability also affects its ability to recover from stalls and spins. In fact, stability, maneuverability, and controllability are all interrelated design characteristics.

Stability is the characteristic of an airplane in flight that causes it to return to a condition of equilibrium, or steady flight, after it is disturbed. For example, if you are flying a stable airplane and a wind gust disrupts it while in straight-and-level flight, it has a tendency to return to the same attitude. The initial tendency to return to the position from which it was displaced is called **positive static stability**. However, the airplane does not immediately return to the original position, but instead does so over a period of time through a series of successively smaller oscillations—this behavior is called **positive dynamic stability**. [Figure 3-27]

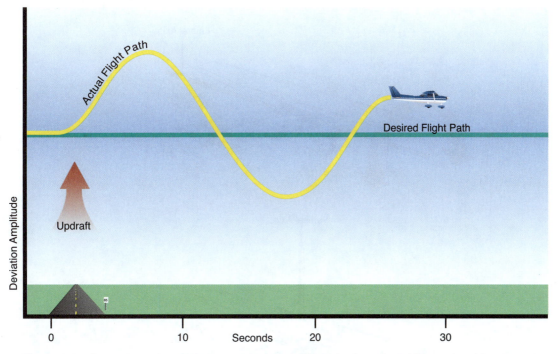

Figure 3-27. General aviation airplanes are designed to have both static stability and positive dynamic stability.

The amount of time that it takes for the oscillations to cease is a measure of the degree of stability. After a significant disturbance, oscillations for typical light airplanes normally damp to half of the original deviation in 20 to 30 seconds. Because an inherently stable platform is highly desirable in training aircraft, they are normally designed to possess both positive static and positive dynamic stability.

Maneuverability is the characteristic of an airplane that permits you to maneuver it easily and allows it to withstand the stress resulting from the maneuvers. An airplane's size, weight, flight control system, structural strength, and thrust determine its maneuverability.

Controllability is the capability of an airplane to respond to your control inputs, especially with regard to attitude and flight path. Stability, maneuverability, and controllability all refer to movement of the aircraft about one or more of three axes of rotation.

THREE AXES OF FLIGHT

Because an aircraft operates in a three dimensional environment, aircraft movement takes place around one or more of three axes of rotation. They are called the **longitudinal**, **lateral**, and **vertical axes** of flight. The common reference point for the three axes is the airplane's **center of gravity (CG)**, which is the theoretical point where the entire weight of the airplane is considered to be concentrated. Because all three axes pass through this point, you can say that the airplane always moves about its CG, regardless of which axis is involved. The ailerons, elevator (or stabilator), and rudder create aerodynamic forces that cause the airplane to rotate about the three axes. [Figure 3-28]

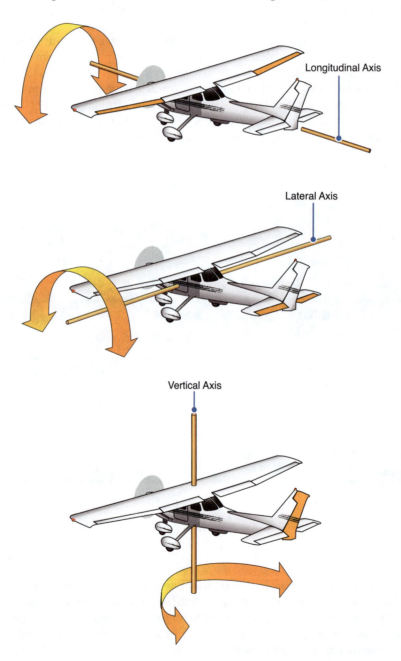

Figure 3-28. Ailerons control roll movement about the longitudinal axis; the elevator controls pitch movement about the lateral axis; and the rudder controls yaw movement about the vertical axis.

LONGITUDINAL AXIS

When you deflect the ailerons to begin a turn, they create an immediate rolling movement about the longitudinal axis. Deflected ailerons alter the chord line and change the effective camber of the outboard section of each wing. Because the ailerons move in opposite directions, they have an opposite effect on the aerodynamic shape of each wing and the associated production of lift. In Figure 3-29, the down aileron on the right wing increases its angle of attack, increasing its lift. At the same time, the up aileron on left wing decreases its angle of attack, decreasing its lift. The airplane rolls to the left as long as the ailerons are deflected. To stop the roll, center the control wheel or stick to neutralize the ailerons.

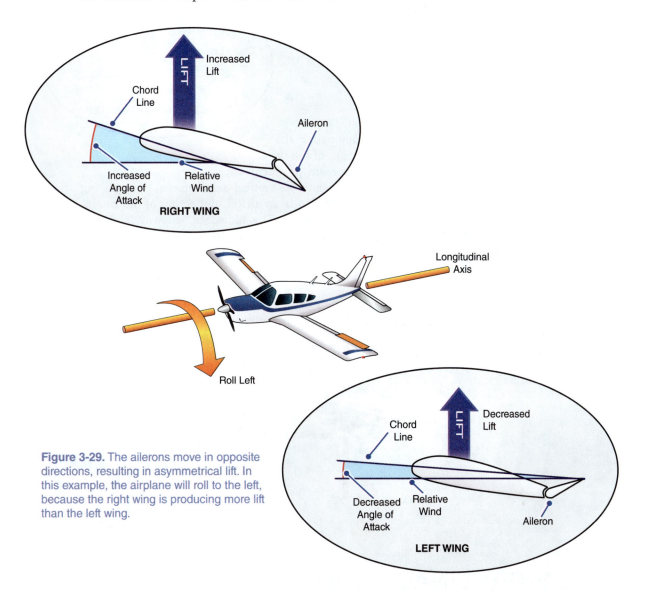

Figure 3-29. The ailerons move in opposite directions, resulting in asymmetrical lift. In this example, the airplane will roll to the left, because the right wing is producing more lift than the left wing.

LATERAL AXIS

The elevator (or stabilator) controls movement about the lateral axis, normally referred to as an adjustment to pitch attitude. The chord line and effective camber of the stabilizer are changed by deflection of the elevator/stabilator. Movement of the control wheel or stick fore or aft moves the elevator/stabilator and changes the pitch. When you push the elevator/stabilator control forward, it decreases the airplane's pitch attitude, which decreases the angle of attack on the wings. Conversely pulling back on the elevator/stabilator control increases pitch attitude, which increases the angle of attack. [Figure 3-30]

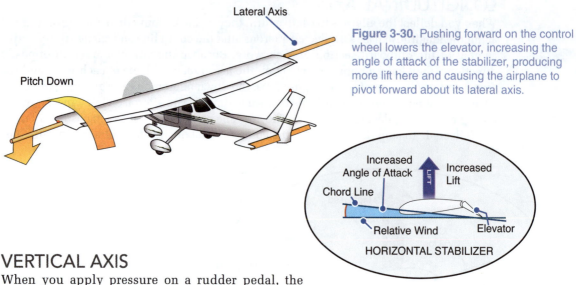

Figure 3-30. Pushing forward on the control wheel lowers the elevator, increasing the angle of attack of the stabilizer, producing more lift here and causing the airplane to pivot forward about its lateral axis.

VERTICAL AXIS

When you apply pressure on a rudder pedal, the rudder deflects left or right in the direction of the pedal you push. This deflection into the airstream produces an aerodynamic force that rotates the airplane about its vertical axis, referred to as yawing the airplane. Because the vertical stabilizer also is an airfoil, deflection of the rudder alters the stabilizer's effective camber and chord line. This increases "sideways" lift—for example, left rudder deflection increases the angle of attack on the left side of the vertical fin, which pushes the tail to the right, causing the nose of the airplane to yaw to the left. [Figure 3-31]

Figure 3-31. In this example, left rudder pressure moves the rudder to the left, which pushes the tail to the right and causes the nose of the airplane to yaw to the left.

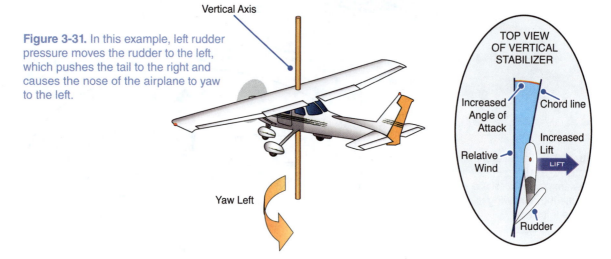

LONGITUDINAL STABILITY

The **longitudinal stability** of an airplane involves the pitching motion or tendency of the aircraft to move about its lateral axis. An airplane that is longitudinally stable will tend to return to its trimmed angle of attack after displacement. This is desirable because an airplane with this characteristic tends to resist either excessively nose-high or nose-low pitch attitudes. If an airplane is longitudinally unstable, it has the tendency to climb or dive until a stall or a steep dive develops. As a result, a longitudinally unstable airplane is very dangerous to fly.

BALANCE

An important consideration when designing a longitudinally stable airplane is the balance between the center of gravity (CG), the center of pressure of the wing, and tail-down force. The **center of pressure** is a point along the wing chord line where lift is considered to be concentrated—sometimes referred to as the **center of lift**. On a typical cambered wing, the center of lift changes position along the chord line at different flight attitudes. It moves forward at high angles of attack and aft as angle of attack decreases. As a result, the airplane tends to pitch down when the angle of attack is decreased and the center of lift moves farther aft of the center of gravity. To maintain balance and aid longitudinal stability, most airplanes are designed so that during normal operations, the center of pressure remains aft of the center of gravity. [Figure 3-32]

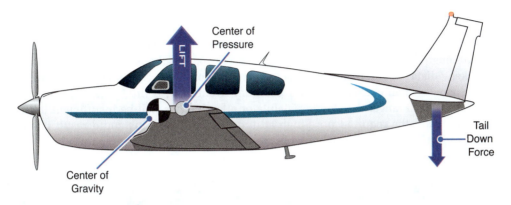

Figure 3-32. To maintain stability, most airplanes operate with some tail-down force and with the center of lift aft of the center of gravity.

CENTER OF GRAVITY POSITION

The position of the center of gravity, which is determined by the distribution of weight either by design or by the pilot, can also affect the longitudinal stability of an airplane. If the CG is too far forward, the airplane is very nose heavy; if the CG is too far aft, the airplane may become tail heavy. To achieve longitudinal stability, most airplanes are designed so they're slightly nose heavy. This is accomplished during the engineering and development phase by placing the center of gravity slightly forward of the center of pressure.

You control the CG location by what you load into the airplane, and where you put it. This includes the weight of fuel, passengers, and baggage. For example, if you load heavy baggage into an aft baggage compartment, it might shift the CG too far back, which could result in severe control problems. As you might expect, for an airplane to be controllable during flight, the CG must be located within a reasonable distance forward or aft of an optimum position. All airplanes have forward and aft limits for the position of the CG published in the pilot's operating handbook (POH). The area between these limits is the **CG range**. An airplane must be loaded so that the weight distribution does not adversely affect longitudinal balance. [Figure 3-33]

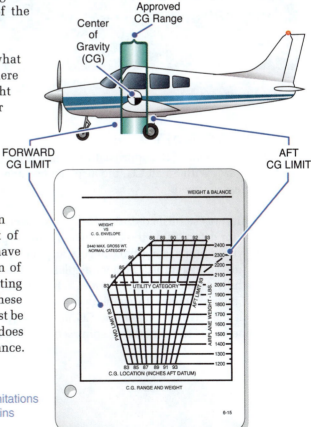

Figure 3-33. You must observe loading limitations to ensure that the position of the CG remains within the approved range.

SECTION B ■ **Stability**

When the CG is within the approved CG range, the airplane not only is controllable, but its longitudinal stability also is satisfactory. If the CG is located near the forward or aft limit of the approved CG range, a slight loss of longitudinal stability may be noticeable, but elevator/stabilator effectiveness is still adequate to control the airplane during all approved maneuvers. However, loading an aircraft in such a way as to move the CG too far forward or aft could result in a situation in which the capability of the elevator/stabilator to control the aircraft is exceeded.

CG TOO FAR FORWARD

If you load your airplane so that the CG is forward of the forward CG limit, it will be too nose heavy. Although this tends to make the airplane seem stable, adverse side effects include longer takeoff distance and higher stalling speeds. The condition gets progressively worse as the CG moves to an extreme forward position. Eventually, if the CG is well forward of the approved CG range, the elevator/stabilator effectiveness will be insufficient to lift the nose to obtain the required nose-high attitude for landing. This could result in the nosewheel to striking the runway before the main gear touches down. [Figure 3-34]

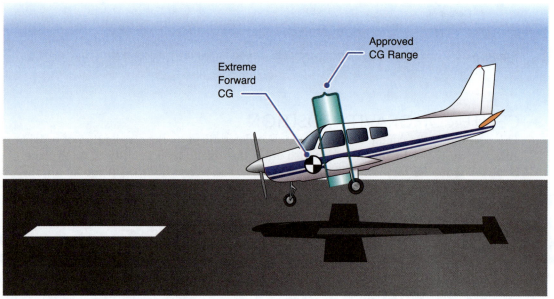

Figure 3-34. If the CG is too far forward, the airplane will be nose heavy and it might not be possible to lift the nose for the landing flare.

CG TOO FAR AFT

A CG located aft of the approved CG range is even more dangerous than a CG that is too far forward. With an aft CG, the airplane becomes tail heavy and very unstable in pitch, regardless of speed.

CG limits are established during initial testing and airworthiness certification. One of the criteria for determining the CG range in light airplanes is spin recovery capability. If the CG is within limits, a normal category airplane must demonstrate that it can recover from a one-turn spin; a utility category airplane that is approved for spins must be recoverable from a fully developed spin. The aft CG limit is the most critical factor in spin recovery. As the CG moves aft, elevator/stabilator effectiveness decreases to its minimum permissible value at the aft limit. When the CG moves beyond the aft limit, the elevator/stabilator could be ineffective for stall or spin recovery. [Figure 3-35]

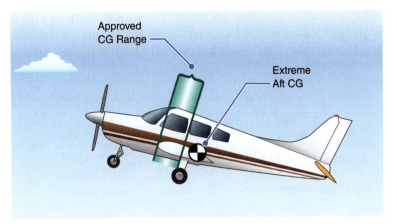

Figure 3-35. If the CG is too far aft, you might be unable to recover from a stall or spin.

As a pilot, there are certain actions you can take to prevent an aft CG position. You can make sure the heaviest passengers and baggage, or cargo, are loaded as far forward as practical. Lighter passengers and baggage normally should be loaded in aft seats or compartments. The main thing you must do is follow the airplane manufacturer's loading recommendations in the POH. If you do this, your airplane will be loaded so the CG is within the approved range where longitudinal stability is adequate and, at the same time, where you can control the airplane during all approved maneuvers. Two important points to remember are that a CG beyond acceptable limits adversely affects longitudinal stability, and the most hazardous condition is an extreme aft CG position. You will learn more about the effects of adverse loading in the section on weight and balance in Chapter 8.

HORIZONTAL STABILIZER

When the airplane is properly loaded, the CG remains forward of the center of pressure and the airplane is slightly nose heavy. The nose-heavy tendency is offset by the horizontal stabilizer, which is designed with a negative angle of attack. This produces a downward force, or negative lift on the tail, to counteract the nose heaviness. The downward force is called the **tail-down force**—the balancing force during most flight conditions that aids in longitudinal stability. [Figure 3-36]

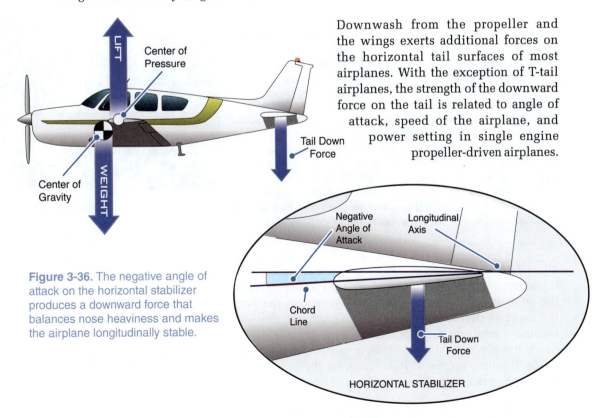

Downwash from the propeller and the wings exerts additional forces on the horizontal tail surfaces of most airplanes. With the exception of T-tail airplanes, the strength of the downward force on the tail is related to angle of attack, speed of the airplane, and power setting in single engine propeller-driven airplanes.

Figure 3-36. The negative angle of attack on the horizontal stabilizer produces a downward force that balances nose heaviness and makes the airplane longitudinally stable.

SECTION B ■ **Stability**

3-29

Any variance in the strength of the downwash, such as a power change, affects the horizontal tail's contribution to longitudinal stability. T-tail designs are not subject to the same downwash effect, simply because the horizontal tail surface is above most, or all, of the downwash. [Figure 3-37]

Wing Downwash

Prop Downwash

Figure 3-37. The downwash from the propeller and the wings passing over the horizontal stabilizer influences the longitudinal stability of the airplane.

The Canard Design

Although the tail-down force created by the horizontal stabilizer is excellent for longitudinal stability and balance, it is aerodynamically inefficient. The wings must support the negative lift created by the tail, and the negative angle of attack on the stabilizer increases drag. If an airplane design permitted two lifting surfaces, aerodynamic efficiency would be much greater.

A canard is a stabilizer that is located in front of the main wings. Canards are something like miniature forward wings. They were used in the pioneering days of aviation, most notably on the Wright Flyer, and are now reappearing on several original designs. The Rutan VariEze (see photo) is a typical canard design. The canard provides longitudinal stability about the lateral axis by lifting the nose of the airplane.

Because both the main wings and the canard produce positive lift, the design is aerodynamically efficient. A properly designed canard is also stall/spin resistant. The canard stalls at a lower angle of attack than the main wings. In doing so, the canard's angle of attack immediately decreases after it stalls. This breaks the stall and effectively returns the canard to a normal lift-producing angle of attack before the main wings have a chance to stall. Ailerons remain effective throughout the stall because they are attached to the main wings. In spite of its advantages, the canard design has limitations in total lift capability. Critical design conditions also must be met to maintain adequate longitudinal stability throughout the flight envelope.

Copyright Corel

POWER EFFECTS

If you reduce power during flight, a definite nose-down pitching tendency occurs due to the reduction of downwash from the wings and the propeller, which reduces elevator/stabilator effectiveness. Although this is a destabilizing factor, it is a desirable characteristic because it tends to result in a nose-down attitude during power reductions. The nose-down attitude helps you maintain, or regain, airspeed. Increasing power has the opposite effect. It causes increased downwash on the horizontal stabilizer, which decreases its contribution to longitudinal stability and causes the nose of the airplane to rise.

The influence of power on longitudinal stability also depends on the overall design of the airplane. Because power provides thrust, the alignment of thrust in relation to the longitudinal axis, the CG, the wings, and the stabilizer are all factors. The **thrustline** is determined by where the propeller is mounted and by the general direction in which thrust acts. In most light general aviation airplanes, the thrustline is parallel to the longitudinal axis and above the CG. This creates a slight pitching moment around the CG that counteracts the effects of downwash. Decreasing the thrust reduces the pitching moment and decreases nose heaviness. Increasing the thrust increases the pitching moment and increases nose heaviness. This thrustline design arrangement minimizes the destabilizing effects of power changes and improves longitudinal stability. [Figure 3-38]

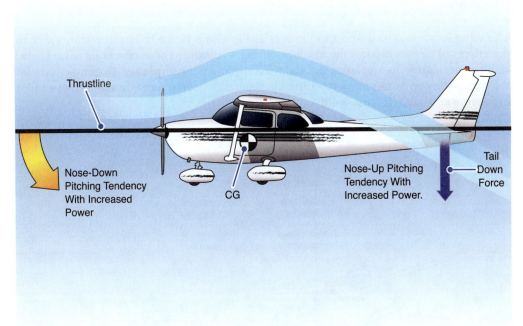

Figure 3-38. Airplanes with the thrustline above the CG produce a pitching moment about the CG that partially counteracts downwash effects.

High power settings combined with low airspeed produce a situation in which increased downwash and decreased airspeed reduce the overall stabilizing effect of the horizontal stabilizer. Additionally, the extension of high-lift devices, such as flaps, can increase downwash and its debilitating effects on longitudinal stability. Therefore, it is particularly important to maintain precise aircraft control during power-on approaches or go-arounds because longitudinal stability can be reduced.

LATERAL STABILITY

Stability about an airplane's longitudinal axis, which extends nose to tail, is called **lateral stability**. If one wing is lower than the opposite wing, lateral stability helps return the wings to a level attitude. This tendency to resist lateral, or roll, movement is aided by specific design characteristics. Four of the most common design features that influence lateral stability are weight distribution, dihedral, sweepback, and keel effect. Two of these, sweepback and keel effect, also help provide directional stability about the vertical axis.

You have no control over the design features that help maintain lateral stability, but you can control the distribution of weight and improve lateral stability. For example, most training airplanes have two fuel tanks, one inside each wing. Before you takeoff on a long flight, you normally fill both tanks. If you use fuel from only one tank, you will soon notice that the airplane wants to roll toward the wing with the full tank. The distribution of weight is uneven and lateral stability is affected. You can prevent the imbalance by switching tanks before a significant difference in weight can occur.

DIHEDRAL

The most common design for lateral stability is known as wing dihedral. **Dihedral** is the upward angle of the airplane's wings with respect to the horizontal. When you look at an airplane, dihedral makes the wings appear to form a shallow V. Dihedral usually is just a few degrees.

If an airplane with dihedral enters an uncoordinated roll during gusty wind conditions, one wing will be elevated and the opposite wing will drop. This causes an immediate sideslip downward toward the low wing. Because the relative wind is now coming from the side, the low wing experiences an increased angle of attack while the high wing's angle of attack is reduced. The increased angle of attack on the low wing produces more lift for that wing and tends to roll the aircraft back toward a level flight attitude. [Figure 3-39]

The wing location with respect to the fuselage is also a factor in lateral stability due to the effects of airflow around the fuselage during a sideslip. These effects increase lateral stability on high-wing airplanes and reduce it on low-wing airplanes. To compensate for this factor, low-wing airplanes typically have about 5 degrees more dihedral than similar high-wing airplanes. [Figure 3-40]

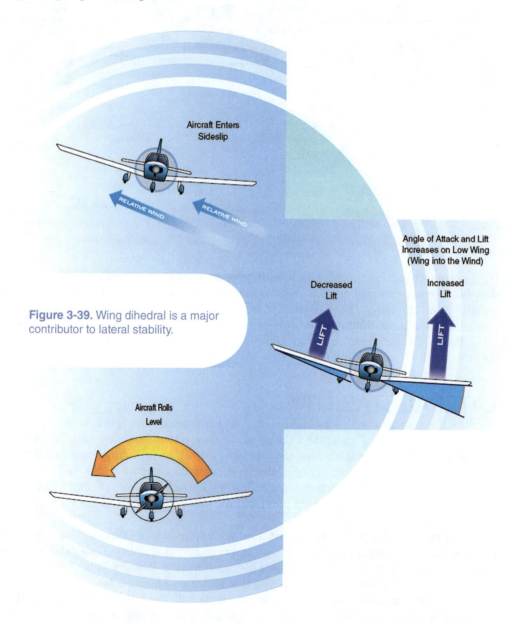

Aircraft Enters Sideslip

RELATIVE WIND

RELATIVE WIND

Angle of Attack and Lift Increases on Low Wing (Wing into the Wind)

Decreased Lift

Increased Lift

LIFT

LIFT

Aircraft Rolls Level

Figure 3-39. Wing dihedral is a major contributor to lateral stability.

Figure 3-40. Because they are inherently more laterally stable, high-wing aircraft such as the Cessna 172 on the left are designed with less dihedral than typical low-wing aircraft, such as the Beechcraft Bonanza on the right.

From an operational standpoint, it's important to note that, in certain situations, the propeller slipstream can reduce the lateral stability of the airplane by reducing the effect of wing dihedral. At high power settings and low airspeeds, propwash increases the effectiveness of the inboard sections of the wings, which decreases the effect of dihedral, thereby reducing lateral stability. This warrants particular attention because high-power, low-airspeed conditions also contribute to a degradation of longitudinal stability.

SECTION B ■ **Stability**

SWEEPBACK

In many airplanes, the leading edges of the wings do not form right angles with the longitudinal axis. Instead, the wings are angled backward from the wing root to the wingtips. This design characteristic is referred to as wing sweep or **sweepback**. In high-performance airplanes with pronounced sweepback, the design is used primarily to maintain the center of lift aft of the CG and reduce wave drag when operating at airspeeds near the speed of sound. Sweepback design improves lateral stability by generating forces that tend to level the wings and straighten the airplane if it enters a sideslip. [Figure 3-41]

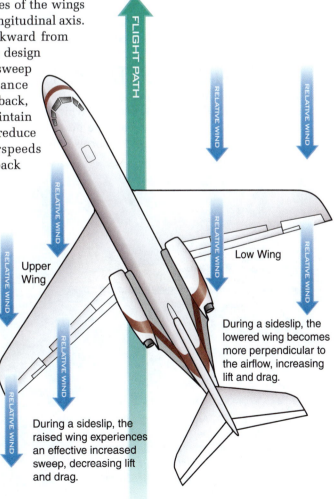

Figure 3-41. Sweepback helps counteract a sideslip.

FLIGHT PATH

RELATIVE WIND

Upper Wing

Low Wing

During a sideslip, the lowered wing becomes more perpendicular to the airflow, increasing lift and drag.

During a sideslip, the raised wing experiences an effective increased sweep, decreasing lift and drag.

Sweepback can also aid slightly in directional stability. If an airplane rotates about its vertical axis or yaws to the left, the right wing has less sweep and a slight increase in drag. The left wing has more sweep and less drag. This tends to force the airplane back into alignment with the relative wind.

KEEL EFFECT

Lateral stability also is provided by the vertical fin and side area of the fuselage reacting to the airflow much like the keel of a ship. **Keel effect** is the steadying influence exerted by the side area of the fuselage and vertical stabilizer. In the example in Figure 3-42, as the airplane rolls to the right, a side force is applied to the right side of the fuselage. Because the majority of the surface area lies above the CG, the keel effect tends to roll the airplane back toward an upright position.

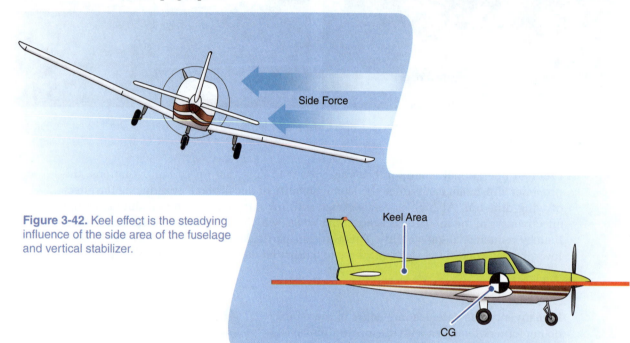

Figure 3-42. Keel effect is the steadying influence of the side area of the fuselage and vertical stabilizer.

DIRECTIONAL STABILITY

Stability about the vertical axis is called **directional stability**. The primary contributor to directional stability is the vertical tail, which causes an airplane in flight to act much like a weather vane. You can compare the pivot point on the weather vane to the CG of the airplane. The nose of the airplane corresponds to the weather vane's arrowhead, and the vertical fin on the airplane acts like the tail of the weather vane. [Figure 3-43]

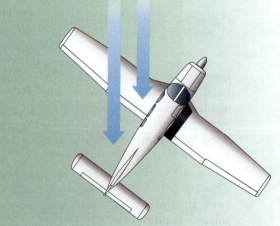

Figure 3-43. An airplane must have more surface area behind the CG than it has in front of it. When an airplane enters a sideslip, the greater surface area behind the CG helps keep the airplane aligned with the relative wind.

B-2 Directional Stability

In most aircraft designs, the primary source of directional stability is the vertical tail. Aircraft designed without tail assemblies, such as flying wings, must somehow still maintain directional stability and control, but how is this possible?

One of the most notable flying wing designs is employed by the U.S. Air Force B-2 *Stealth* bomber manufactured by Northrop Corporation. The B-2 uses a four-times redundant fly-by-wire flight control system that is controlled by approximately 200 computer processors. The actual flight controls are primarily located on the trailing edge of wing (fuselage). Yaw control is accomplished using "drag rudders" located near each wingtip. The drag rudders extend or retract independently to control the direction of nose movement. Note the extended drag rudder on the left wing. If both drag rudders are extended simultaneously, they act as speed brakes. Pitch and roll control is accomplished through the use of "elevons" located inboard of the drag rudders.

INTERACTION OF LATERAL AND DIRECTIONAL STABILITY

For ease-of-understanding, lateral and directional stability have been discussed separately up to this point. However, it is impossible to yaw an aircraft without also creating a rolling motion. This interaction between the lateral and directional stabilizing design elements can sometimes uncover some potentially undesirable side effects. Two of the most common are Dutch roll and spiral instability.

Dutch roll is a combination of rolling/yawing oscillations caused either by your control input or by wind gusts. Dutch roll will normally occur when the dihedral effects of an aircraft are more powerful than the directional stability. After a disturbance resulting in a yawing motion and sideslip, the dihedral effect will tend to roll the aircraft away from the direction of the initial yaw. However, due to weak directional stability, the rolling movement may overshoot the level position and reverse the sideslip. This motion continues to repeat, creating an oscillation that can be felt by the pilot as side-to-side wagging of the aircraft's tail. If Dutch roll tendency is not effectively dampened, it is considered objectionable.

The alternative to an airplane that exhibits Dutch roll tendencies is a design that has better directional stability than lateral stability. If directional stability is increased and lateral stability is decreased, the Dutch roll motion is adequately suppressed. However, this design arrangement tends to cause spiral instability.

Spiral instability is associated with airplanes that have strong directional stability in comparison with lateral stability. When an airplane susceptible to spiral instability is disturbed from a condition of equilibrium and a sideslip is introduced, the strong directional stability tends to yaw the airplane back into alignment with the relative wind. Due to the yaw back into the relative wind, the outside wing travels faster than the inside wing and, as a result, more lift is generated by the outside wing. The rolling moment increases the angle of bank, which increases the sideslip. The comparatively weak dihedral effect lags in restoring lateral stability and the yaw forces the nose of the airplane down while the angle of bank continues to increase, tightening the spiral. Spiral instability is normally easily overcome by the pilot. However, if left uncorrected, the motion could increase into a tight spiral dive, sometimes referred to as a graveyard spiral.

As you can see, even a well-designed airplane may have some undesirable characteristics. Generally, increased dihedral reduces spiral instability while an increased vertical tail surface increases spiral instability. Because Dutch roll is considered less tolerable than spiral instability, designers attempt to minimize the Dutch roll tendency. The compromise results in a small degree of spiral instability, which generally is considered acceptable.

SECTION B ■ **Stability**

STALLS

The inherent stability of an airplane is particularly important as it relates to the aircraft's ability to recovery from stalls and spins (which can result from aggravated stalls). Familiarization with the causes and effects of stalls is especially important during flight at slow airspeeds, such as during takeoff and landing, where the margin above the stall speed is small.

It's important to understand the variables that affect stall development. As indicated earlier in this chapter, a stall will always occur when the maximum lift, or critical angle of attack (C_{Lmax}) is exceeded. If an airplane's speed is too slow, the required angle of attack to maintain lift may be exceeded, causing a stall. It's important to note, however, that the airspeed at which an aircraft may be stalled is not fixed. For example, although the extension of flaps increases drag, it also increases the wing's ability to produce lift, thereby reducing the stall speed.

Stall speed also can be affected by a number of other factors such as weight and environmental conditions. As aircraft weight increases, a higher angle of attack is required to maintain the same airspeed because some of the lift must be used to support the increased weight. This causes an increase in the aircraft's stall speed. The distribution of weight also affects the stall speed of an aircraft. For example, a forward CG creates a situation that requires the tail to produce more downforce to balance the aircraft. This, in turn, causes the wings to produce more lift than if the CG was located more rearward. So, you can see that a more forward CG also increases stall speed.

Any modification to the wing surface also can affect the stall speed of the aircraft. Although man-made high-lift devices can decrease stall speed, the opposite can occur due to natural factors. Snow, ice or frost accumulation on the wing's surface not only changes the shape of the wing, disrupting the airflow, but also increases weight and drag, all of which will increase stall speed.

Another environmental factor that can affect stall speed is turbulence. The unpredictable nature of turbulence encounters can significantly and suddenly cause an aircraft to stall at a higher airspeed than the same aircraft in stable conditions. This occurs when a vertical gust changes the direction of the relative wind and abruptly increases the angle of attack. During takeoff and landing operations in gusty conditions, an increase in airspeed usually is necessary in order to maintain a wide margin above stall speed.

TYPES OF STALLS

There are three basic types of stalls that will normally be practiced during training to familiarize you with stall recognition and recovery in particular flight regimes. **Power-off stalls** are practiced to simulate the conditions and aircraft configuration you will most likely encounter during a normal landing approach. **Power-on stalls** are normally encountered during takeoff, climb-out, and go-arounds when the pilot fails to maintain proper control due to premature flap retraction or excessive nose-high trim. To help you understand how stalls may occur at higher than normal stall speed, your instructor may demonstrate **accelerated stalls** and show you associated recovery techniques.

Most stalls are practiced while maintaining coordinated flight. However, uncoordinated, or crossed-control, inputs can be very dangerous when operating near a stall. One type of stall, sometimes referred to as the **crossed-control stall** is most likely to occur when a pilot tries to compensate for overshooting a runway during a turn from base to final while on landing approach. [Figure 3-44]

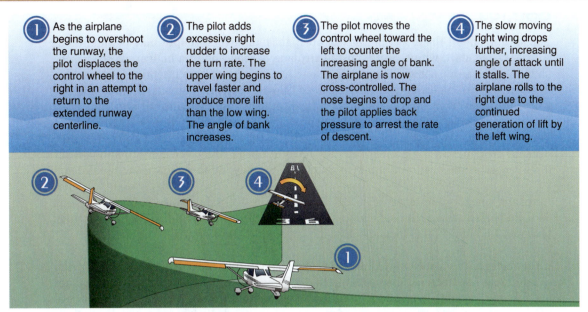

① As the airplane begins to overshoot the runway, the pilot displaces the control wheel to the right in an attempt to return to the extended runway centerline.

② The pilot adds excessive right rudder to increase the turn rate. The upper wing begins to travel faster and produce more lift than the low wing. The angle of bank increases.

③ The pilot moves the control wheel toward the left to counter the increasing angle of bank. The airplane is now cross-controlled. The nose begins to drop and the pilot applies back pressure to arrest the rate of descent.

④ The slow moving right wing drops further, increasing angle of attack until it stalls. The airplane rolls to the right due to the continued generation of lift by the left wing.

Figure 3-44. If a stall occurs during a skidding turn close to the ground, such as during the turn to final, there might not be sufficient altitude for recovery.

STALL RECOGNITION

There are a number of ways to recognize that a stall is imminent. Ideally, you should be able to detect the first signs of an impending stall and make appropriate corrections before it actually occurs. If you have a good understanding of the types of stalls, recognition is much easier. Recovery at the first indication of a stall is quite simple; but, if you allow the stalled condition to progress, recovery becomes more difficult.

A typical indication of a stall is a mushy feeling in the flight controls and less control effect as the aircraft's speed decreases. The reduction in control effectiveness is primarily due to reduced airflow over the flight control surfaces. In fixed-pitch propeller airplanes, a loss of revolutions per minute (RPM) may be noticeable as you approach a stall in power-on conditions. Also, a reduction in the sound of air flowing along the fuselage is usually evident. Just before the stall occurs, buffeting, uncontrollable pitching, or vibrations may begin. Finally, your kinesthetic sense (ability to recognize changes in direction or speed) may also provide a warning of decreased speed or the beginning of a sinking feeling.

STALL RECOVERY

Your primary consideration after a stall occurs should be to regain positive control of the aircraft. If you do not recover promptly by reducing the angle of attack, a secondary stall and/or spin may result. A **secondary stall** is normally caused by poor stall recovery technique, such as attempting flight prior to attaining sufficient flying speed. If you encounter a secondary stall, you should apply normal stall recovery procedures. The following basic guidelines should be used to effect a proper stall recovery.

1. Decrease the angle of attack. Depending on of the type of aircraft, you might find that a different amount of forward pressure on the control wheel is required. Too little forward movement may not be enough to regain lift; too much may impose a negative load on the wing, hindering recovery.

2. Smoothly apply maximum allowable power. If you are not already at maximum allowable power, increase the throttle to minimize altitude loss and increase airspeed.

3. Adjust the power as required. As the airplane recovers, maintain coordinated flight while adjusting the power to a normal level.

You can usually prevent an accidental stall by knowing when you are most susceptible to a stall and recognizing the indicators of an impending stall. Unless you are practicing stalls and stall recoveries, don't wait for the stall to fully develop—apply stall recovery techniques at the first indication of an impending stall.

SECTION B ■ **Stability**

3-37

SPINS

The spin is one of the most complex of all flight maneuvers. A **spin** may be defined as an aggravated stall, which results in the airplane descending in a helical, or corkscrew, path. Single-engine, normal category airplanes are prohibited from intentional spins. This is indicated by a placard with words such as "No acrobatic maneuvers, including spins, approved." However, during aircraft certification tests, normal category airplanes must demonstrate recovery from a one-turn spin or a three-second spin, whichever takes longer. The recovery must take place within one additional turn with normal control inputs. Because airplanes in the normal category have not been tested for more than one-turn/three second spins, their performance characteristics beyond these limits are unknown.

Acrobatic category airplanes must fully recover from fully developed spins within one and one-half additional turns. Certification in this category also requires six turns or three seconds, whichever takes longer, before the recovery control inputs are applied.

Utility category airplanes may be tested under the one-turn (normal) criteria or they may satisfy the six-turn (acrobatic) spin requirements. However, spins in utility category airplanes may be approved only with specific loading, such as a reduced weight and with a forward CG position. It is extremely important for you to understand all of the operating limitations for your airplane. Applicable limitations are placarded in the aircraft and/or included in the POH.

PRIMARY CAUSES

A stalled aircraft is a prerequisite for a spin. However, a properly executed stall is essentially a coordinated maneuver where both wings are equally or nearly equally stalled. In contrast, a spin is an uncoordinated maneuver with the wings unequally stalled. In this case, the wing that is more completely stalled drops before the other, and the nose of the aircraft yaws in the direction of the low wing. The upgoing, lesser-stalled wing experiences more lift and less drag. The opposite, more-stalled wing is forced down and back due to less lift and increased drag. [Figure 3-45]

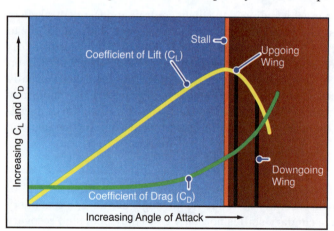

Figure 3-45. During a spin, the wings are unequally stalled, which sustains the rolling and yawing motions of the spin.

Typically, the cause of an inadvertent spin is exceeding the critical angle of attack while performing an uncoordinated maneuver. The lack of coordination is normally caused by either too much or not enough rudder control for the amount of aileron being used. The result is a crossed-control condition. If you do not initiate the stall recovery promptly, the airplane is more likely to enter a full stall that may develop into a spin. The spin that occurs from crossed-controlling usually results in rotation in the direction of the rudder being applied, regardless of which wing is raised. In a skidding turn, where both aileron and rudder are applied in the same direction, rotation will be in that direction. However, in a slipping turn, where opposite aileron is held against the rudder, the resultant spin will usually occur in the direction opposite the aileron that is being applied.

Coordinated use of the flight controls is important, especially during flight at slow airspeeds. Although most pilots are able to maintain coordination of the flight controls during routine maneuvers, this ability often deteriorates when distractions occur and their attention is divided between important tasks. Distractions that have caused problems include preoccupation with situations inside or outside the cockpit, maneuvering to avoid other aircraft, and maneuvering to clear obstacles during takeoffs, climbs, approaches, or landings. Because of this, you will be required to learn how to recognize and cope with these distractions by practicing "flight at slow airspeeds with realistic distractions" during your flight training. In addition, although you are not required to demonstrate flight proficiency in spins during private pilot training, you will need to exhibit knowledge of the situations where unintentional spins may occur, as well as the general spin recognition and recovery procedures for the airplane you use for your practical test.

TYPES OF SPINS

In general, spins are divided into three primary types. The most common type is the upright, or **erect spin**, which is characterized by a slightly nose down rolling and yawing motion in the same direction. In an inverted spin, the aircraft is spinning upside down with yaw and roll occurring in opposite directions. **Inverted spins** are most likely to occur during aerobatic maneuvers. The third type of spin can be the most deadly. In a **flat spin**, the aircraft simply yaws about its vertical axis with a pitch attitude approximately level with the horizon. Although it sounds fairly benign, recovery is usually very difficult or impossible except in specialized aerobatic aircraft. Most general aviation airplanes are designed to prevent entry into flat spins provided the loading and CG location are within approved limits. [Figure 3-46]

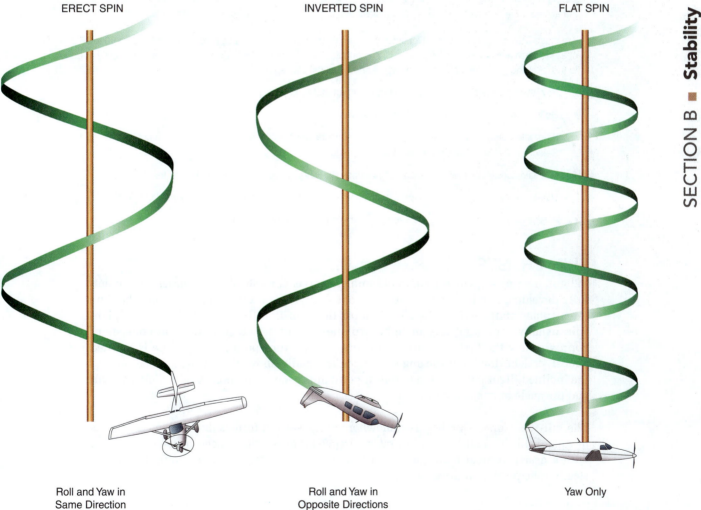

ERECT SPIN	INVERTED SPIN	FLAT SPIN
Roll and Yaw in Same Direction	Roll and Yaw in Opposite Directions	Yaw Only

Figure 3-46. A spin may be characterized as erect, inverted, or flat, depending on the roll and yaw motion of the aircraft.

SECTION B ■ **Stability**

3-39

WEIGHT AND BALANCE CONSIDERATIONS

Even minor weight and balance changes can affect an aircraft's spin characteristics. Heavier weights generally result in slow spin rates initially; but, as the spin progresses, heavier weights tend to cause an increasing spin rate and longer recovery time. Distribution of weight is even more significant. Forward center of gravity positions usually inhibit the high angles of attack necessary for a stall. Thus, an airplane with a forward CG tends to be more stall and spin resistant than an aircraft with an aft CG. In addition, spins with aft CG positions are more likely to become flat.

In a training airplane, the addition of a back seat passenger or a single suitcase to an aft baggage compartment can affect the CG enough to change the characteristics of a spin. In addition, any concentration of weight, or unbalanced weight distribution, that is particularly far from the CG is undesirable. This type of loading may occur with tip tanks or outboard wing tanks. If the fuel in these tanks becomes unbalanced, an asymmetrical condition exists. The worst asymmetric condition is full fuel in the wing on the outside of the spin and no fuel in the tanks on the inside of the turn. Once the spin is developed, the momentum (inertial force) makes recovery unlikely.

DISCOVERY

STALL/SPIN ACCIDENT PREVENTION

During a recent four-year period, 171 stall/spin accidents occurred, 73 percent of which were fatal. By far the majority of the accidents were a result of unintentional stalls and spins that occurred close to the ground.

Are these accidents preventable? Consider that human error has been cited as a contributing factor in 90 to 95 percent of all stall/spin accidents, and the answer is a resounding yes. A review of related accident reports suggest that adhering to the following guidelines can help you avoid an accidental stall/spin.

- Pay attention to aircraft loading. An aircraft with an aft CG is more prone to stall/spin entry.

- Do not take off with snow, ice, or frost on the wings.

- If an emergency that requires a forced landing occurs immediately after takeoff, don't attempt to return to the runway. Select a suitable landing site straight ahead or slightly off to the side.

- Maintain coordinated flight as much as possible. Particularly avoid skidding turns near the ground.

- Use a somewhat higher than normal airspeed during takeoffs and landings in gusty winds.

- Always concentrate on flying the aircraft and avoid prolonged distractions.

SPIN PHASES

In light, training airplanes, a complete spin maneuver consists of three phases—incipient, fully developed, and recovery. The **incipient spin** is that portion of a spin from the time the airplane stalls and rotation starts until the spin is fully developed. The incipient spin usually occurs rapidly in light airplanes (about 4 to 6 seconds) and consists of approximately the first two turns. At about the half-turn point, the airplane is pointed almost straight down but the angle of attack is usually above that of the stall because of the inclined flight path. As the one-turn point approaches, the nose might come back up and the angle of attack continues to increase.

The **fully developed spin** begins somewhere in the second turn as the airplane continues to rotate and the angular rotation rates, airspeed, and vertical speed become stabilized with a nearly vertical flight path. The fully developed stage is often referred to as the steady-state portion of the spin.

Spin recovery is the final stage of the spin that occurs when anti-spin forces overcome pro-spin forces. Anti-spin control inputs result in a slowing and eventual cessation of rotation coupled with a decrease in angle of attack below C_{Lmax}. This phase can range from one-quarter of a turn to several turns. [Figure 3-47]

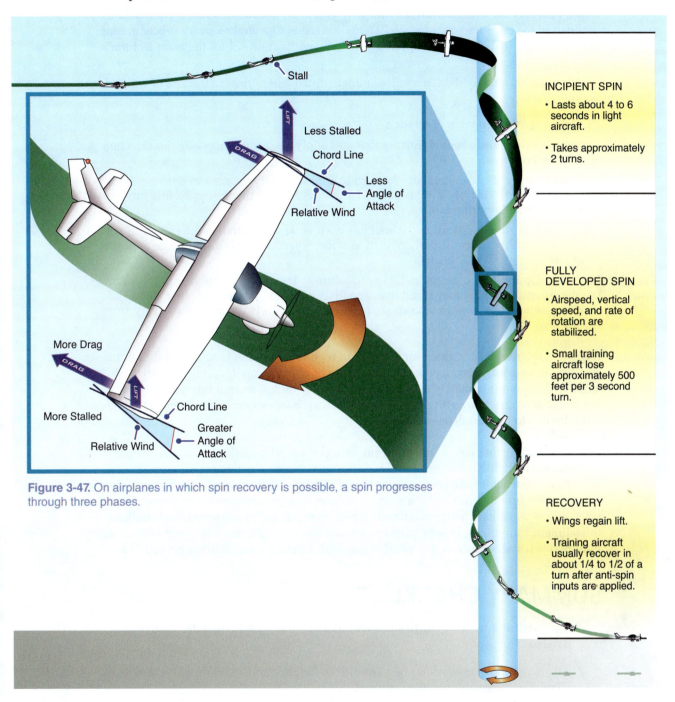

Figure 3-47. On airplanes in which spin recovery is possible, a spin progresses through three phases.

Stall

Less Stalled

Chord Line

Less Angle of Attack

Relative Wind

More Drag

More Stalled

Chord Line

Relative Wind

Greater Angle of Attack

INCIPIENT SPIN
- Lasts about 4 to 6 seconds in light aircraft.
- Takes approximately 2 turns.

FULLY DEVELOPED SPIN
- Airspeed, vertical speed, and rate of rotation are stabilized.
- Small training aircraft lose approximately 500 feet per 3 second turn.

RECOVERY
- Wings regain lift.
- Training aircraft usually recover in about 1/4 to 1/2 of a turn after anti-spin inputs are applied.

SECTION B ■ Stability

SPIN RECOVERY

While some characteristics of a spin are predictable, every airplane spins differently. In addition, the same airplane's spin behavior changes with variations in configuration, loading, and several other factors. Therefore, it's easy to understand why spin recovery techniques vary for different aircraft and why you must follow the recovery procedures outlined in the POH for your airplane. The following is a general recovery procedure for erect spins, but it should not be applied arbitrarily without regard for the manufacturer's recommendations.

1. Move the throttle to idle. This will eliminate thrust and minimize the loss of altitude.

2. Neutralize the ailerons.

3. Determine the direction of rotation.

4. Apply full opposite rudder. Ensure that you apply the rudder opposite the direction of rotation.

5. Briskly apply elevator/stabilator forward to approximately the neutral position. Some aircraft require merely a relaxation of back pressure; others require full forward elevator/stabilator pressure.

6. As rotation stops (indicating the stall has been broken), neutralize the rudder. If you don't neutralize the rudder when rotation stops, you could enter a spin in the opposite direction.

7. Gradually apply aft elevator/stabilator to return to level flight. Applying the elevator/stabilator too quickly may result in a secondary stall, and possibly another spin. Also, make sure you adhere to aircraft airspeed and load limits during the recovery from the dive.

Normally, the recovery from an incipient spin requires less time (and altitude) than the recovery from a fully developed spin. As a rule of thumb, small aircraft authorized for spins will lose approximately 500 feet of altitude per each 3-second turn and recover in 1/4 to 1/2 of a turn. More altitude will be lost and recoveries may be prolonged at higher altitudes due to the less dense air.

It should be clear that you, as an applicant for a private pilot certificate, are not required to demonstrate flight proficiency in spin entries or spin recovery techniques. Even though your flight instructor may demonstrate a spin at some point during your training, you should never intentionally enter a spin, even after you become certificated as a private pilot, unless you obtain additional training from an experienced instructor. The emphasis in stall/spin training for private pilots is awareness of conditions that could lead to an unintentional stall or spin and to provide you with some general recovery procedures.

SUMMARY CHECKLIST

✓ Most training aircraft are designed to display both positive static and positive dynamic stability.

✓ All aircraft movement takes place around the longitudinal, lateral, and vertical axes, all of which pass through the center of gravity.

✓ Longitudinal stability relates to movement about the airplane's lateral axis. Longitudinal stability is influenced by the relationship between the center of pressure and the center of gravity as well as the effects of power changes and the design of the horizontal stabilizer.

✓ Stability around the aircraft's longitudinal axis is referred to as lateral stability. Wing dihedral, sweepback, keel effect, and weight distribution are design features that affect an airplane's lateral stability.

✓ Directional stability, or stability about the vertical axis, of most aircraft is maintained by the vertical tail.

✓ Dutch roll is most likely to occur on aircraft with weak directional stability and strong lateral stability.

✓ Aircraft with strong directional stability and weak lateral stability are susceptible to spiral instability.

✓ A stall will always occur when the critical angle of attack, or C_{Lmax}, is exceeded. This can occur at any airspeed and in any configuration or attitude.

✓ A spin will not develop unless both wings are stalled. A normal, erect spin results in the airplane entering a nose-low autorotative descent with one wing stalled more than the other.

KEY TERMS

Stability	Sweepback
Positive Static Stability	Keel Effect
Positive Dynamic Stability	Directional Stability
Maneuverability	Dutch Roll
Controllability	Spiral Instability
Longitudinal Axis	Power-Off Stalls
Lateral Axis	Power-On Stalls
Vertical Axis	Accelerated Stalls
Center of Gravity (CG)	Crossed-Control Stall
Longitudinal Stability	Secondary Stall
Center of Pressure	Spin
Center of Lift	Erect Spin
CG Range	Inverted Spin
Tail-Down Force	Flat Spin
Thrustline	Incipient Spin
Lateral Stability	Fully Developed Spin
Dihedral	Spin Recovery

SECTION B ■ **Stability**

QUESTIONS

1. Referring to the airplane diagram below, identify the three axes of flight and the type of movement associated with each axis.

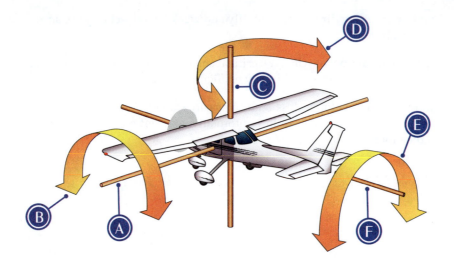

Match the following control surface with the associated aircraft movement.

2. Roll movement   A. Elevator/stabilator

3. Pitch movement   B. Ailerons

4. Yaw movement   C. Rudder

5. In relation to the center of gravity, in which direction would the center of pressure normally move as angle of attack is increased on a cambered wing?

6. What factors can affect the longitudinal stability of an airplane at high power settings and low airspeed?

7. Does the propwash resulting from high power settings increase or decrease the contribution of wing dihedral to the lateral stability of an airplane?

8. An aircraft with strong directional stability and weak lateral stability is prone to what type of undesirable side effect?

9. True/False. When landing in gusty winds, airspeed should be increased above normal to help guard against a stall.

10. List the basic guidelines for stall recovery.

11. List the basic guidelines for spin recovery.

Section C
Aerodynamics of Maneuvering Flight

The extent to which an airplane can perform a variety of maneuvers is primarily a matter of design and a measure of its overall performance. Although aircraft design and performance may differ, the aerodynamic forces acting on any maneuvering aircraft are essentially the same. Understanding the aerodynamics of maneuvering flight can help you perform precise maneuvers while maintaining your airplane within its design limitations.

CLIMBING FLIGHT

The aerodynamic forces acting on an airplane established in a stabilized climb are in equilibrium; however, because the flight path is inclined, the relationship between these forces is altered. For example, the total force of weight no longer acts perpendicular to the flight path, but is comprised of two components. Although one component still acts 90° to the flight path, a rearward component of weight acts in the same direction as drag, opposing thrust. [Figure 3-48]

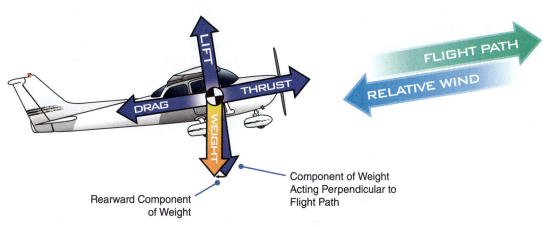

Figure 3-48. In a climb, the rearward component of weight is opposed by thrust, while the component of weight acting perpendicular to the flight path is supported by lift.

A transition from level flight into a climb normally combines a change in pitch attitude with an increase in power. If you attempt to climb just by pulling back on the control wheel to raise the nose of the airplane, momentum will cause a brief increase in altitude, but airspeed will soon decrease. The amount of thrust generated by the propeller for cruising flight at a given airspeed is not enough to maintain the same airspeed in a climb. Excess thrust, not excess lift, is necessary for a sustained climb. In fact, as the angle of climb steepens, thrust will not only oppose drag, but also will increasingly replace lift as the force opposing weight. At the point where the climb becomes exactly vertical, weight and drag are opposed solely by thrust, and lift no longer acts to support the aircraft in flight.

DISCOVERY

When Is No Lift Required To Fly?

During a normal sustained climb, a component of weight is opposed by lift. However, in a true sustained vertical climb, such as the one performed by the General Dynamics F-16C pictured in Figure A, the wings supply no vertical lift, and thrust is the only force opposing weight. Depending on an aircraft's thrust-to-weight ratio, a sustained vertical climb may be maintained under certain conditions. The thrust-to-weight ratio of the F-16 is approximately 1.1 to 1. In contrast, the thrust-to-weight ratio of the Boeing 747 [Figure B] is about 0.26 to 1.

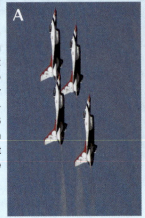

LEFT-TURNING TENDENCIES

In addition to the basic aerodynamic forces present in a climb, a combination of physical and aerodynamic forces can contribute to a left-turning tendency in propeller-driven airplanes. The forces of torque, gyroscopic precession, asymmetrical thrust, and spiraling slipstream all work to create a left-turning tendency during high-power, low-airspeed flight conditions. A thorough understanding of left-turning tendencies will help you anticipate and correct for their effects.

TORQUE

You can understand **torque** most easily by remembering Newton's third law of motion: "For every action there is an equal and opposite reaction." In most airplanes with a single engine mounted on the front of the aircraft, the propeller rotates clockwise when viewed from the pilot's seat. The clockwise action of a spinning propeller causes a torque reaction, which tends to rotate the airplane counterclockwise about its longitudinal axis. [Figure 3-49]

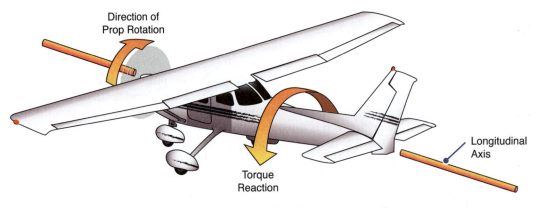

Figure 3-49. Torque is the opposite reaction created by the turning propeller.

GYROSCOPIC PRECESSION

The turning propeller of an airplane also exhibits characteristics of a gyroscope — rigidity in space and precession. The characteristic that produces a left-turning tendency is precession. **Gyroscopic precession** is the resultant reaction when a force is applied to the rim of a rotating disc. The reaction to a force applied to a gyro acts in the direction of rotation and approximately 90° ahead of the point where force is applied. You will experience the effects of precession only when the attitude of the aircraft is changed. [Figure 3-50]

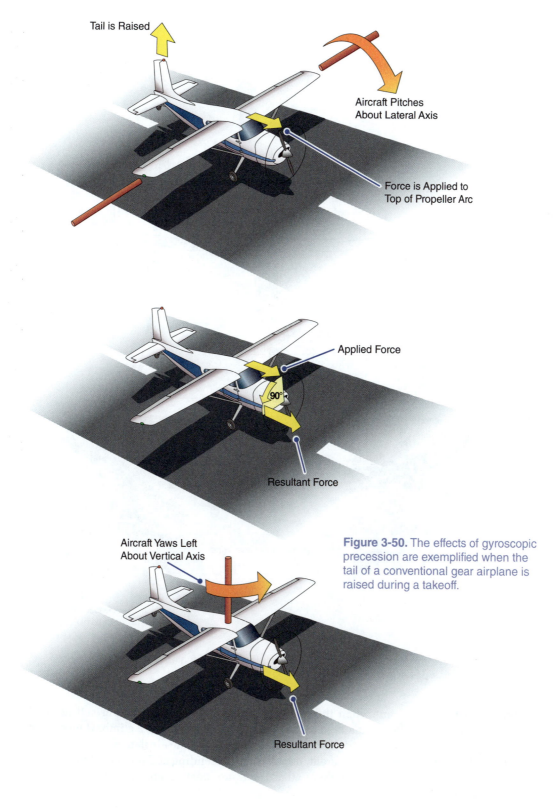

Tail is Raised

Aircraft Pitches About Lateral Axis

Force is Applied to Top of Propeller Arc

Applied Force

90°

Resultant Force

Aircraft Yaws Left About Vertical Axis

Figure 3-50. The effects of gyroscopic precession are exemplified when the tail of a conventional gear airplane is raised during a takeoff.

Resultant Force

SECTION C ■ **Aerodynamics of Maneuvering Flight**

ASYMMETRICAL THRUST

When you are flying a propeller-driven airplane at a high angle of attack, the descending blade of the propeller takes a greater "bite" of air than the ascending blade on the other side. The greater bite is caused by a higher angle of attack for the descending blade, compared to the ascending blade. This creates the uneven, or **asymmetrical thrust**, which is known as P-factor. Because the descending blade is normally on the right side of the airplane (as viewed from the flight deck), **P-factor** makes an airplane yaw to the left about its vertical axis. [Figure 3-51]

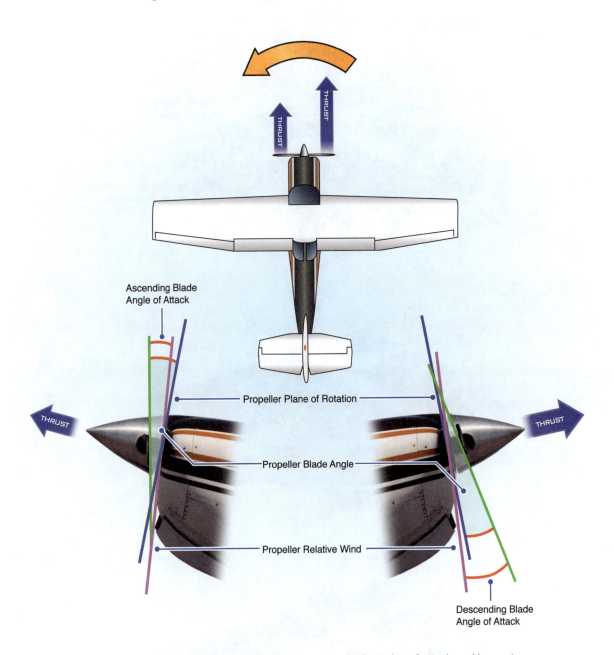

Figure 3-51. Asymmetrical thrust from P-factor occurs at a high angles of attack and is most noticeable at high power settings.

P-factor is most apparent when the engine is operating at a high power setting, and when the airplane is flown at a high angle of attack, such as during a climb. Under these conditions, you need to apply right rudder pressure to keep the airplane straight. In level cruising flight, P-factor is not apparent, because both ascending and descending propeller blades are at nearly the same angle of attack, and are creating approximately the same amount of thrust.

SPIRALING SLIPSTREAM

As the propeller rotates, it produces a backward flow of air, or slipstream, which wraps around the airplane. This **spiraling slipstream** causes a change in airflow around the vertical stabilizer. Due to the direction of propeller rotation, the resultant slipstream strikes the left side of the vertical fin. The resulting sideward force yaws the airplane about its vertical axis, moving the nose to the left. [Figure 3-52]

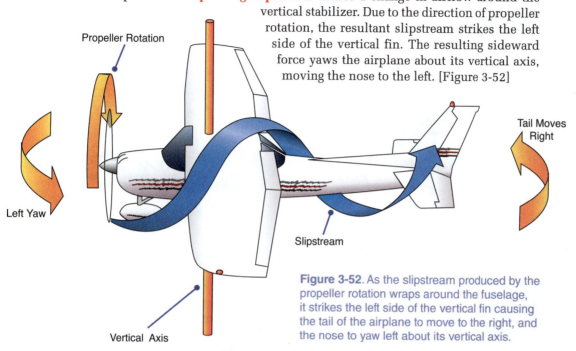

Propeller Rotation

Tail Moves Right

Left Yaw

Slipstream

Vertical Axis

Figure 3-52. As the slipstream produced by the propeller rotation wraps around the fuselage, it strikes the left side of the vertical fin causing the tail of the airplane to move to the right, and the nose to yaw left about its vertical axis.

AIRCRAFT DESIGN CONSIDERATIONS

Some aircraft manufacturers include design elements that help counteract left-turning tendencies and make the airplane easier to control. One such design places a small metal tab on the trailing edge of the rudder to help combat the effects of spiraling slipstream. The tab is bent to the left so that pressure from the passing airflow will push on the tab and force the rudder slightly to the right. The slight right-hand rudder displacement creates a yawing moment that opposes the left-turning tendency caused by spiraling slipstream. [Figure 3-53]

A horizontally canted engine also may be used to help counteract the left-turning tendency caused by spiraling slipstream. In this arrangement, the engine is turned slightly toward the right, effectively offsetting the airplane's thrustline and compensating for left yaw.

Rudder Trim Tab

N50

Figure 3-53. An adjustable metal trim tab is attached to the trailing edge of the rudder on this Diamond DA40.

DESCENDING FLIGHT

In stabilized descending flight, aerodynamic forces are in equilibrium with the force of weight comprised of two components. One component of weight acts perpendicular to the flight path, while the other component of weight acts forward along the flight path. [Figure 3-54]

SECTION C ■ Aerodynamics of Maneuvering Flight

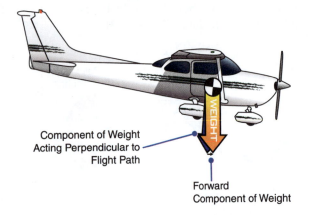

Component of Weight Acting Perpendicular to Flight Path

WEIGHT

Forward Component of Weight

FLIGHT PATH
RELATIVE WIND

Figure 3-54. The force of weight is comprised of two components during a descent.

As the nose of the aircraft is lowered in the descent, the component of weight acting forward along the flight path increases and, assuming that power remains the same, an increase in speed occurs. Increasing airspeed results in an increase in parasite drag that works to balance the force of weight. Once speed is stabilized, the four forces of flight are once again in equilibrium. If the power is at idle, the force of thrust is removed and a larger component of weight must be allocated to counteract drag and maintain a constant airspeed. This is accomplished by lowering the nose of the airplane further. [Figure 3-55]

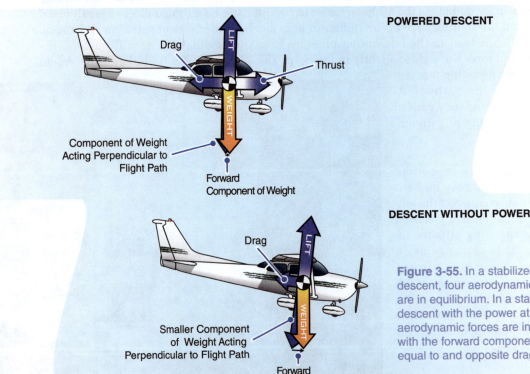

POWERED DESCENT

Drag

LIFT

Thrust

WEIGHT

Component of Weight Acting Perpendicular to Flight Path

Forward Component of Weight

DESCENT WITHOUT POWER

Drag

LIFT

WEIGHT

Smaller Component of Weight Acting Perpendicular to Flight Path

Forward Component of Weight

Figure 3-55. In a stabilized powered descent, four aerodynamic forces are in equilibrium. In a stabilized descent with the power at idle, three aerodynamic forces are in equilibrium with the forward component of weight equal to and opposite drag.

Without the aid of power, your ability to control your rate of descent and flight path is somewhat limited. However, if you understand the glide characteristics of your airplane, you can maximize your chances of making a successful landing if a power failure should occur.

LIFT-TO-DRAG RATIO

The lift-to-drag ratio (L/D) can be used to measure the gliding efficiency of your airplane. The angle of attack resulting in the least drag on your airplane will give the **maximum lift-to-drag ratio (L/D$_{max}$)**, the best glide angle, and the maximum gliding distance. [Figure 3-56]

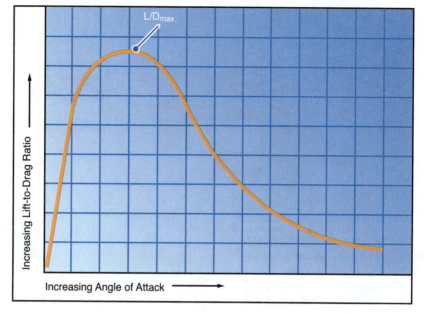

Figure 3-56. L/D$_{max}$ is the specific angle of attack that generates the greatest lift with the least amount of corresponding drag.

GLIDE SPEED

At a given weight, L/D$_{max}$ will correspond to a certain airspeed. This important performance speed is called the **best glide speed**. In most cases, it is the only speed that will give you the maximum gliding distance. In the event of an engine failure, maintaining the best glide speed is extremely important. Any speed other than the best glide speed creates more drag. If your airspeed is too high, parasite drag increases; and if you descend with too slow of an airspeed, induced drag increases. [Figure 3-57]

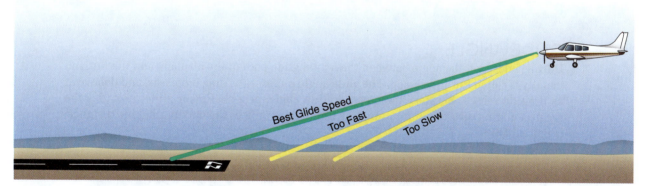

Figure 3-57. The best glide speed is normally achieved at L/D$_{max}$. Any deviation from best glide speed will increase drag and reduce the distance you can glide.

GLIDE RATIO

The **glide ratio** represents the distance an airplane will travel forward, without power, in relation to altitude loss. For example, a glide ratio of 10:1 means that an airplane will travel 10,000 feet of horizontal distance (approximately 1.6 NM) for every 1,000 feet of altitude lost in the descent. The best glide ratio of an aircraft is available only at the optimum angle of attack associated with L/D$_{max}$.

SECTION C ■ **Aerodynamics of Maneuvering Flight**

SECTION C ■ Aerodynamics of Maneuvering Flight

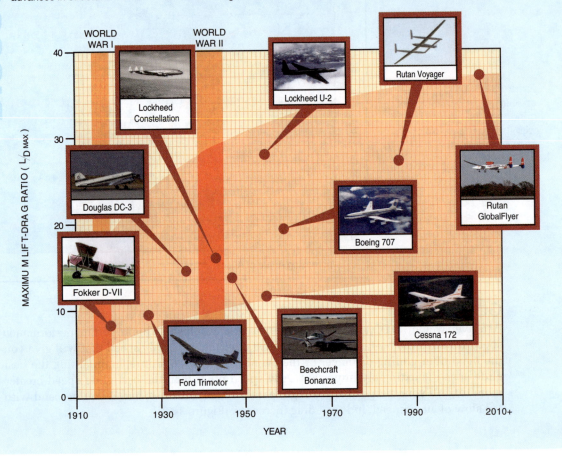

Trends in Lift-to-Drag Ratio

Improvements in aircraft design resulted in a general increase in maximum lift-to-drag ratios (L/D$_{max}$) between the years of 1920 and 1960. The higher aspect ratios that accompanied the emergence of the monoplane led to the relatively sharp rise in lift-to-drag ratios between World War I and World War II. Except for specialized one-of-a-kind aircraft, L/D ratios have stayed about the same since the 1950s. Further increases in L/D$_{max}$ values will probably result from advances in structural materials that will allow significant reductions in drag and/or increased aspect ratio.

GLIDE ANGLE

During a descent, the angle between the actual glide path of the airplane and the horizon usually is called the **glide angle**. The glide angle increases as drag increases, and decreases as drag decreases. Because a decreased glide angle, or a shallower glide, provides the greatest gliding distance, minimum drag normally produces the maximum glide distance. [Figure 3-58]

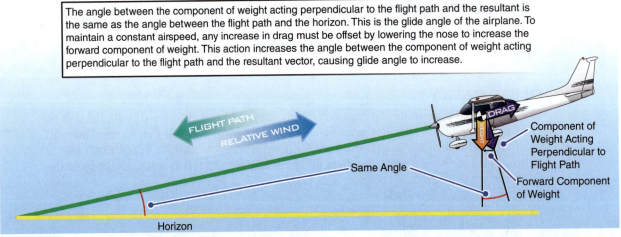

The angle between the component of weight acting perpendicular to the flight path and the resultant is the same as the angle between the flight path and the horizon. This is the glide angle of the airplane. To maintain a constant airspeed, any increase in drag must be offset by lowering the nose to increase the forward component of weight. This action increases the angle between the component of weight acting perpendicular to the flight path and the resultant vector, causing glide angle to increase.

Figure 3-58. Glide angle increases with drag during the glide.

FACTORS AFFECTING THE GLIDE

In general, maintaining the best glide speed published by the manufacturer of your aircraft will assure an optimum glide. However, some factors may affect the efficiency of your glide. These factors include airplane weight, configuration, and wind.

WEIGHT

Variations in weight do not affect the glide ratio of an airplane, however, there is a specific airspeed that is optimum for each weight. Two aerodynamically identical aircraft with different weights can glide the same distance from the same altitude. This can only be accomplished, however, if the heavier airplane flies at a higher airspeed than the airplane with the lighter load. Although the heavier airplane sinks faster and will reach the ground sooner, it will travel the same distance as the lighter airplane as long as the appropriate glide speed is maintained.

CONFIGURATION

Glide distance will be reduced if you increase drag. For example, if you extend the landing gear, drag increases, which reduces the airplane's maximum lift-to-drag ratio and glide ratio. To maintain airspeed, you must lower the nose of the airplane to increase the component of weight acting along your flight path to overcome drag. [Figure 3-59]

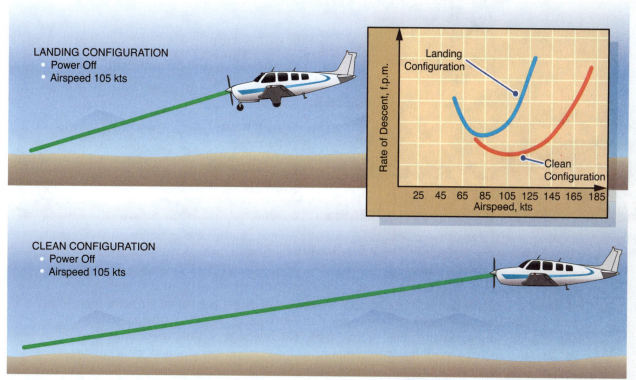

Figure 3-59. Any change in configuration that increases drag reduces the glide distance and requires that you lower the nose of the airplane to maintain the same airspeed.

WIND

A headwind reduces your glide distance and a tailwind increases it. In light wind conditions you can normally use the best glide speed and obtain close to the best possible glide range for the wind conditions. However, with a strong headwind or tailwind (greater than 25% of glide speed), adjusting your glide speed can maximize your travel over the ground. Increase your glide speed if you have a strong headwind, and decrease it if you have a strong tailwind. Glider pilots use graphs to figure out the best speeds to fly, but as a rule of thumb, you could try a speed adjustment of about half of the headwind component and come close to obtaining the best possible distance with wind. The adjustment for a tailwind is smaller—avoid slowing down enough to increase your sink rate. [Figure 3-60]

SECTION C ■ Aerodynamics of Maneuvering Flight

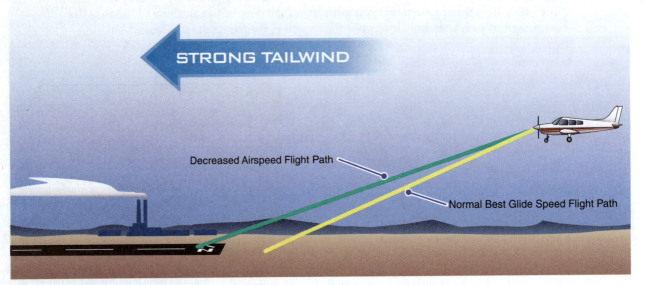

Figure 3-60. With a strong headwind, increasing glide speed will provide best distance over the ground. With a strong tailwind, decreasing glide speed slightly will increase glide distance.

TURNING FLIGHT

In order for an airplane to turn, it must overcome inertia—its tendency to continue in a straight line. You create the necessary turning force by using the ailerons to bank the airplane so that the direction of total lift is tilted sideways. This is accomplished by dividing the force of lift into two components; one component still acts vertically to oppose weight, while the other acts horizontally. [Figure 3-61]

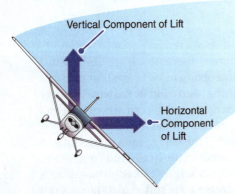

Figure 3-61. To make an airplane turn, part of the lift must act horizontally.

To maintain altitude during a turn, you must increase lift until the vertical component of lift equals weight. The horizontal component of lift creates a force toward the center of the turn known as **centripetal force**—which causes the airplane to turn. Because centripetal force works against the tendency of the aircraft to continue in a straight line, inertia tends to oppose centripetal force toward the outside of the turn. This opposing impetus is called **centrifugal force**, which is not a true force but an apparent force that results from the effect of inertia in the turn. [Figure 3-62]

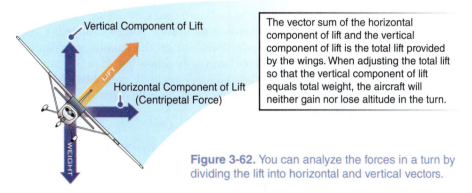

The vector sum of the horizontal component of lift and the vertical component of lift is the total lift provided by the wings. When adjusting the total lift so that the vertical component of lift equals total weight, the aircraft will neither gain nor lose altitude in the turn.

Figure 3-62. You can analyze the forces in a turn by dividing the lift into horizontal and vertical vectors.

ADVERSE YAW

When you roll into a turn, the aileron on the inside of the turn is raised, and the aileron on the outside of the turn is lowered. The lowered aileron on the outside increases the angle of attack and produces more lift for that wing. Because induced drag is a by-product of lift, the outside wing also produces more drag than the inside wing. This creates a tendency for the airplane to yaw toward the outside of the turn, which is called **adverse yaw**. [Figure 3-63]

Figure 3-63. Because the left wing produces more lift in this right turn, it also has more drag—causing the nose to yaw to the left, opposite the turn.

SECTION C ■ **Aerodynamics of Maneuvering Flight**

The coordinated use of aileron and rudder helps compensate for adverse yaw. For example, when you enter a turn to the left, you should press the left rudder pedal slightly. After you are established in the turn and neutralize the ailerons to prevent further roll, adverse yaw is removed and you can relax the rudder pressure. When you roll out of the turn, you should apply coordinated right aileron and rudder pressure to return to a wings-level attitude.

OVERBANKING TENDENCY

As you enter a turn and increase the angle of bank, you may notice the tendency of the airplane to continue rolling into a steeper bank, even though you neutralize the ailerons. This **overbanking tendency** is caused by additional lift on the outside, or raised, wing. Because the outside wing is traveling faster than the inside wing, it produces more lift and the airplane tends to roll beyond the desired bank angle. [Figure 3-64] To correct for overbanking tendency, you can use a small amount of opposite aileron, away from the turn, to maintain your desired angle of bank.

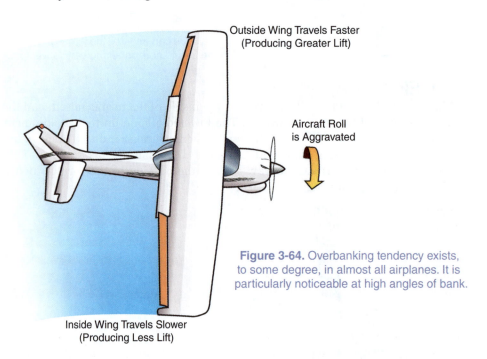

Outside Wing Travels Faster
(Producing Greater Lift)

Aircraft Roll
is Aggravated

Figure 3-64. Overbanking tendency exists, to some degree, in almost all airplanes. It is particularly noticeable at high angles of bank.

Inside Wing Travels Slower
(Producing Less Lift)

RATE AND RADIUS OF TURN

Two major components that define airplane performance during turning flight are rate and radius of turn. **Rate of turn** refers to the amount of time it takes for an airplane to turn a specified number of degrees. If flown at the same airspeed and angle of bank, every aircraft will turn at the same rate. If airspeed increases and the angle of bank remains the same, the rate of turn will decrease. Conversely, a constant airspeed coupled with an angle of bank increase will result in a faster rate of turn.

The amount of horizontal distance an aircraft uses to complete a turn is referred to as the **radius of turn**. Although the radius of turn is also dependent on an airplane's airspeed and angle of bank, the relationship is the opposite of rate of turn. For example, as an airplane's airspeed is increased with the angle of bank held constant, the radius of turn increases. On the other hand, if the angle of bank is increased and the airspeed remains the same, the radius of turn is decreased. [Figure 3-65]

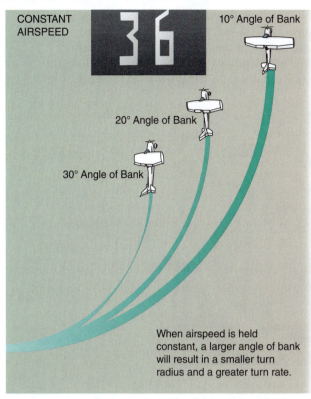

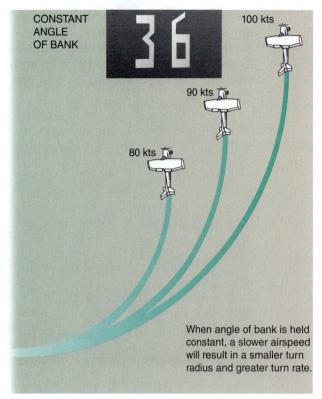

CONSTANT AIRSPEED — 10° Angle of Bank — 20° Angle of Bank — 30° Angle of Bank

When airspeed is held constant, a larger angle of bank will result in a smaller turn radius and a greater turn rate.

CONSTANT ANGLE OF BANK — 100 kts — 90 kts — 80 kts

When angle of bank is held constant, a slower airspeed will result in a smaller turn radius and greater turn rate.

Figure 3-65. Angle of bank and airspeed regulate the rate and radius of a turn.

LOAD FACTOR

If you attempt to improve turn performance by increasing angle of bank while maintaining airspeed, you should pay close attention to airplane limitations due to the effects of increasing load factor. **Load factor** is the ratio of the load supported by the airplane's wings to the actual weight of the aircraft and its contents. An airplane in cruising flight, while not accelerating in any direction, has a load factor of one. This one-G condition means the wings are supporting only the actual weight of the airplane and its contents. If the wings are supporting twice as much weight as the weight of the airplane and its contents, the load factor is two. You may be more familiar with the term "G-forces" as a way to describe flight loads caused by aircraft maneuvering. Because G-force increases commonly occur when pulling back on the control wheel to increase back pressure, the term "pulling G's" is sometimes used when referring to an increase in load factor. [Figure 3-66]

Figure 3-66. When a pilot pulls G's during aerobatic maneuvers in warm, humid conditions, the increased wing loading and decreased pressure above the wings can contribute to the formation of a distinctive condensation cloud.

SECTION C ■ **Aerodynamics of Maneuvering Flight**

As you enter a banked turn while riding on a roller coaster, the forces you will experience are very similar to the forces that act on a turning airplane. On a roller coaster, you can feel the resultant force created by the combination of centripetal force and inertia as an increase in seat pressure. This pressure is an increased positive load factor that causes you to feel heavier in the turn than when you are on a flat portion of the track. [Figure 3-67]

Figure 3-67. Riders on this rollercoaster experience a maximum load factor of 3.85G's.

If you abruptly push the control wheel forward while flying, you could experience a sensation of weightlessness. This is caused by inertia attempting to keep your body moving forward while the aircraft is diving away from the original flight path. If the effects of inertia and centripetal force cancel each other out, you will experience a weightless sensation of zero G's. If inertia exceeds centripetal force, you could experience negative G-loading.

It is important to note that a change in load factor can occur at any time due to either pilot input, or environmental conditions, such as turbulence. In rare instances, you may experience a rapid change in G-forces. For example, in extremely turbulent air, you might be subjected to a variety of G-forces including positive G's, negative G's, and sideward, or transverse, G's.

LOAD FACTOR IN TURNS

When an airplane is turning, you must compensate for the apparent increase in weight and loss of vertical lift, or you will lose altitude. You do this by increasing the angle of attack with back pressure on the control wheel. This increases the total lift of the airplane and imposes additional loads that must be supported by the wings. For example, during a constant-altitude turn with a 60° bank, two G's are required to maintain level flight. This means the airplane's wings must support twice the weight of the airplane and its contents, although the actual mass of the airplane does not increase. [Figure 3-68]

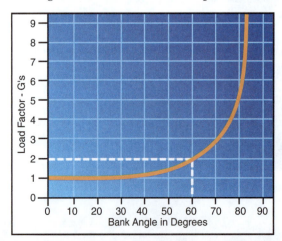

Figure 3-68. If maintaining constant altitude during a turn, you will feel pushed down in your seat by additional G-force as the angle of bank increases.

LOAD FACTOR AND STALL SPEED

When you maintain altitude during a turn by increasing the angle of attack, the stall speed increases as the angle of bank increases. The percent of increase in stall speed is fairly moderate with bank angles less than 45°. However, beyond 45° the stall speed rises rapidly. For example, in a 60° turn the stall speed increases by approximately 41 percent; a 75° bank increases stall speed by about 100 percent. [Figure 3-69]

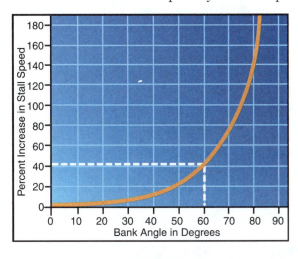

Figure 3-69. Stall speed increases with the square root of the load factor in a constant-altitude turn.

The stall speed increases in proportion to the square root of the load factor. For example, an airplane with a one-G stalling speed of 55 knots will stall at twice that speed (110 knots) with a load factor of four G's. Stalls that occur from increased G-forces on an airplane are called **accelerated stalls**. These stalls occur at higher airspeeds than the normal one-G stall speed. Accelerated stalls demonstrate that exceeding the critical angle of attack—not slow airspeed—is what causes a stall.

HOW MANY G'S ARE TOO MANY?

From the files of the NTSB: *The witness stated the pilot was performing aerobatic maneuvers. At the top of the loop (approx. 3,000 ft AGL), the aircraft remained inverted, power was reduced, and an inverted spin was entered. The aircraft remained in the inverted spin to water impact. The pilot does not recall the accident flight, but stated he had been having problems with G-loads and low blood pressure.*

It is possible, even likely, that some aerobatic aircraft may be able to withstand more G's than the pilot. A particular pilot's G-tolerance is a function of many factors, including the intensity, duration, and direction of the G-forces.

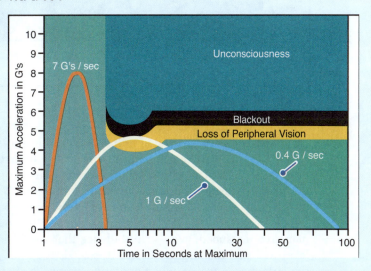

The main physical problems associated with G-forces are caused by basic changes within the cardiovascular system. Positive G's create a pooling of blood in the lower extremities of the body, impairing circulation and reducing blood pressure at head level. Continued or increased G-loading will result in a decrease of visual acuity, ultimately followed by unconsciousness, or blackout.

The human body is less tolerant of negative G's, which force blood into the head. Large amounts of sustained negative G's can result in uncomfortable symptoms such as facial pain and redout. Although some experienced aerobatic pilots may be able to withstand 7 or 8 positive G's before blackout occurs, most will be incapacitated by only -3G's. You can improve your G-tolerance by maintaining good physical conditioning and avoiding smoking, hyperventilation, and hypoxia. Most civil pilots, however, will not encounter G-forces of sufficient strength during normal flight to cause any major problems.

LIMIT LOAD FACTOR

The amount of stress, or load factor, that an airplane can withstand before structural damage or failure occurs is expressed as the airplane's **limit load factor**. Primarily a function of airplane design, an individual airplane's limit load factor is published in the pilot's operating handbook in terms of maximum positive or negative G's.

Most small general aviation airplanes weighing 12,500 pounds or less, and nine passenger seats or less, are certificated in the normal, utility, or acrobatic categories. The maximum limit load factor in the normal category is 3.8 positive G's, and 1.52 negative G's, which is sufficient for basic training maneuvers. An airplane certificated in the utility category may be used for several maneuvers requiring additional stress on the airframe. A limit of 4.4 positive G's or 1.76 negative G's is permitted in the utility category. An acrobatic category airplane may be flown in any flight attitude as long as its limit load factor does not exceed 6 positive G's or 3 negative G's. By adhering to proper loading techniques and flying within the limits listed in the pilot's operating handbook, you will avoid excessive loads on the airplane and possible structural damage.

The POH for the airplane you are flying is your best source of load limit information. Some pilot operating handbooks publish a **V-g diagram**, which graphically depicts the limit load factors for the associated airplane at a variety of airspeeds. [Figure 3-70]

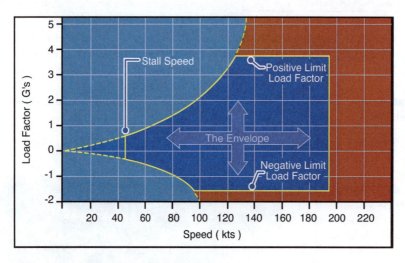

Figure 3-70. The V-g diagram depicts a specific airplane's envelope, and is valid at a particular weight, configuration, and altitude with symmetrical wing loading.

MANEUVERING SPEED

An important airspeed related to load factors and stall speed is the **design maneuvering speed (V_A)**. This is the maximum speed at which you can use full, abrupt control movement without overstressing the airframe. Any airspeed in excess of V_A can overstress the airframe during abrupt maneuvers or turbulence. The higher the airspeed, the greater the amount of excess load that can be imposed before a stall occurs. V_A is depicted on the V-g diagram where the stall curve meets the positive load limit. If you operate at or below maneuvering speed, gusts or turbulence that would increase G-loading to a dangerous level will stall the airplane before it exceeds the limit load factor, preventing damage to the airframe. However, when operating at speeds above V_A, even if you are below V_{NE}, structural damage or failure could occur if limit load factors are exceeded. [Figure 3-71]

V_A normally is not marked on the airspeed indicator, because it varies with total weight. V_A decreases as weight decreases because an aircraft operating at lighter weights is subject to more rapid acceleration from gusts and turbulence. The POH and/or a placard in the airplane are the best sources for determining V_A.

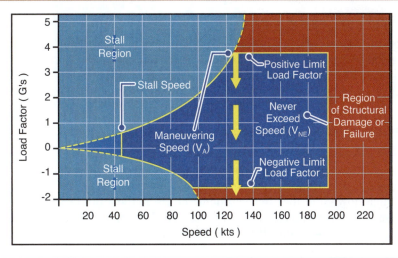

Figure 3-71. The V-g diagram shows that excessive G-loading at speeds less than V_A will result in a stall before structural damage occurs.

SECTION C ■ Aerodynamics of Maneuvering Flight

Calculating Maneuvering Speed

Maneuvering speed may only be provided by the airplane manufacturer at maximum weight. For other weights, you can mathematically determine the maneuvering speed using one of two methods.

METHOD 1

One method uses a formula that is based on limit load factor (n) and stall speed (V_S):

$$V_A = V_S \times \sqrt{n \text{ limit}}$$

The above formula requires you to know the stall speed for the associated weight. This can be determined from the POH or by using the formula,

$$V_{S2} \div V_S = \sqrt{W_2 \div W_1}$$

OR

$$V_{S2} = V_S \times \sqrt{W_2 \div W_1}$$

Using these formulas and the following information

(obtained from the POH), the V_A for a Cessna 172 at 1,800 pounds can be determined.

Given:

Limit load factor (n): 3.8G's

Maximum weight (W_1): 2,400 lbs.

Stall speed at maximum weight (V_S): 51 KCAS

First, determine the stall speed (V_{S2}) at 1,800 lbs.

$$V_{S2} = 51 \times \sqrt{1,800 \div 2,400}$$

$$V_{S2} = 51 \times \sqrt{.75}$$

$$V_{S2} = 51 \times 0.866 = 44.2 \text{ KCAS}$$

Next, use the maneuvering speed formula to determine V_{A2}.

$$V_{A2} = 44.2 \times \sqrt{3.8}$$

$$V_{A2} = 44.2 \times 1.95 = 86.2 \text{ KCAS}$$

METHOD 2

Another method allows you to determine V_A using the formula, V_A at maximum weight multiplied by the square root of the ratio of the actual weight (W_2) divided by the

maximum weight (W_1):

$$V_{A2} = V_A \times \sqrt{W_2 \div W_1}$$

Using this method, and given the following data, you can calculate the approximate maneuvering speed for a Cirrus SR20 and compare it to the value published in the POH.

Weight	Calculated V_A	POH V_A
3,000 lb	—	131 KIAS
2,600 lb	121.95 KIAS	122 KIAS
2200 lb	112.18 KIAS	111 KIAS

SECTION C ■ **Aerodynamics of Maneuvering Flight**

ENERGY MANAGEMENT

Energy management is the process of using power and pitch to manage an airplane's energy state of altitude and airspeed. Mismanagement of altitude and/or airspeed is a contributing factor in three of the most common types of fatal accidents in aviation: loss of control in-flight (LOC-I), controlled flight into terrain (CFIT), and approach-and-landing accidents.

MANAGING THE ENERGY STATE OF THE AIRPLANE

Altitude and airspeed are interdependent, exchangeable, and continuously changing during flight. The **total mechanical energy** of an airplane in flight is the sum of its **potential energy** from altitude, and **kinetic energy** from airspeed. Altitude is potential energy that you can convert to airspeed. Pitch down and the airspeed increases while the altitude decreases. Airspeed is kinetic energy that you can convert to altitude. While maintaining level flight at a cruise power setting, pitch up—as the altitude increases, airspeed decreases.

When energy is exchanged, altitude and airspeed change in opposite directions (absent any other energy or control inputs). As altitude increases, airspeed decreases and vice versa. However, this relationship is dependent on the quantity of either airspeed or altitude. For example, near a stall speed, pitching up and decreasing airspeed will not increase altitude. Neither pitch nor power independently controls altitude or airspeed—the two are connected through the total mechanical energy of the airplane. To control altitude and airspeed effectively, you must coordinate the use of pitch and power to manage the energy state of the airplane.

You adjust power to increase or decrease thrust and drag, which changes total mechanical energy. Although total mechanical energy is a function of both thrust and drag, drag can change due to airspeed changes or by using high lift/drag devices, such as wing flaps. Therefore, changes in total energy are normally initiated by changing thrust, not drag.

A flying airplane is an open energy system, constantly gaining and losing energy while in flight. An airplane gains energy from the force of engine thrust, and loses energy due to aerodynamic drag. The difference between thrust and drag is the net change, which determines whether total mechanical energy—stored as altitude and airspeed—increases, decreases, or remains the same. [Figure 3-72]

DISCOVERY

Bob Hoover — Energy Management Master

One of the world's greatest pilots, Robert A. "Bob" Hoover was the master of managing the energy state of an airplane. Hoover served as a fighter pilot in World War II and after leaving the military, he joined North American Aviation as a test pilot for many programs. Later in his career, Hoover was a popular airshow performer—flying the P-51D Mustang and the Shrike Commander 500S shown here. Hoover demonstrated a mastery of energy management when performing precision aerobatics in the Shrike Commander, constantly exchanging kinetic and potential energy. For example, to end his performance, Hoover would conduct a series of maneuvers without power:

1. First, Hoover enters a steep descent, shuts down both engines and feathers the propellers, which stops the blades from windmilling.

2. The airspeed gained in the high-speed descent is exchanged for altitude while performing a loop.

3. Then, the altitude gained at the top of the loop is exchanged for airspeed while performing an eight-point roll.

4. Finally, the altitude gained at the top of the roll is exchanged once again for airspeed while performing a 180 degree turn to final, landing, and stopping the airplane at airshow center in front of the spectators.

Hoover's Shrike Commander 500 now resides in the National Air and Space Museum, at the Udvar-Hazy Center in Virginia.

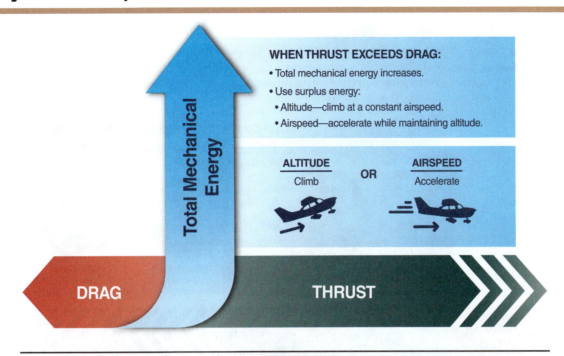

WHEN THRUST EXCEEDS DRAG:
- Total mechanical energy increases.
- Use surplus energy:
  - Altitude—climb at a constant airspeed.
  - Airspeed—accelerate while maintaining altitude.

ALTITUDE Climb   OR   **AIRSPEED** Accelerate

DRAG   THRUST

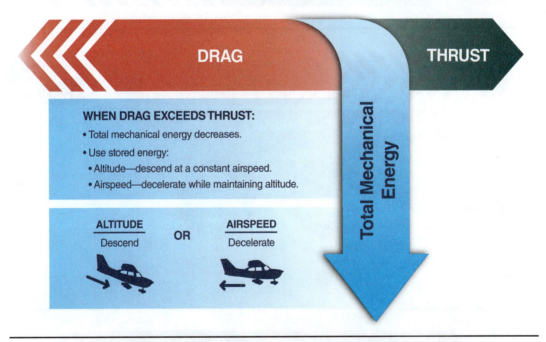

DRAG   THRUST

WHEN DRAG EXCEEDS THRUST:
- Total mechanical energy decreases.
- Use stored energy:
  - Altitude—descend at a constant airspeed.
  - Airspeed—decelerate while maintaining altitude.

ALTITUDE Descend   OR   **AIRSPEED** Decelerate

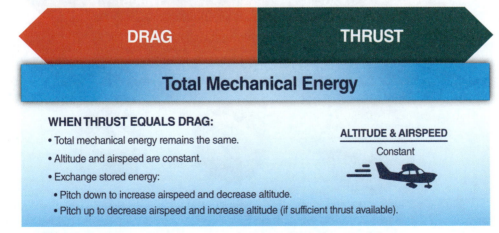

DRAG   THRUST

Total Mechanical Energy

WHEN THRUST EQUALS DRAG:
- Total mechanical energy remains the same.
- Altitude and airspeed are constant.
- Exchange stored energy:
  - Pitch down to increase airspeed and decrease altitude.
  - Pitch up to decrease airspeed and increase altitude (if sufficient thrust available).

ALTITUDE & AIRSPEED Constant

Figure 3-72. Total mechanical energy is a function of both thrust and drag.

SECTION C ■ **Aerodynamics of Maneuvering Flight**

RULES OF ENERGY CONTROL

You change altitude and airspeed by adjusting power with the throttle/power lever and adjusting pitch with the flight controls, which move the elevator/stabilator. Three rules of energy control govern how you coordinate pitch and power to move an airplane from one energy state to another. [Figure 3-73]

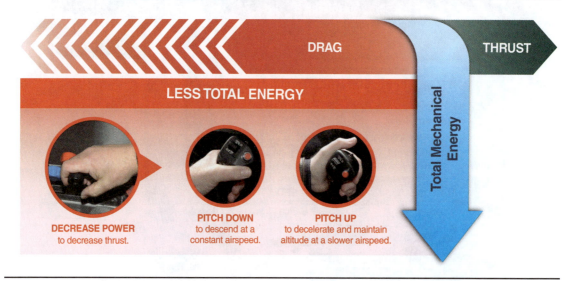

Figure 3-73. Upon reaching a new desired energy state, adjust the pitch attitude and the power setting as needed to maintain the new altitude/airspeed profile, and retrim the airplane.

ENERGY MANAGEMENT ERRORS

Monitoring your altimeter and airspeed indicator enables you to distinguish between two types of energy errors: **total energy errors**, and **energy distribution errors**. A total energy error occurs when your airplane has too much or too little energy. Altitude and airspeed deviate in the same direction (low and slow, or high and fast). To correct total energy errors, increase or decrease energy by adding or reducing power. On the other hand, with an energy distribution error, your airplane has the right amount of total energy, but the distribution over altitude and airspeed is incorrect. In this case, altitude and airspeed deviate in opposite directions (high and slow, or low and fast). Correct an energy distribution error by adjusting pitch to exchange energy between altitude and airspeed.

To correct a combination of total energy and distribution errors, you must simultaneously adjust both pitch and power. You must be able to readily identify, assess, and mitigate two major risks associated with energy management errors:

- A deviation from the desired energy state in the form of unintended altitude and/or airspeed changes.
- An unintended excessive deceleration and/or sink rate with little or no available excess power resulting in continuous airspeed and/or altitude loss.

THE POWER CURVE

You can easily visualize energy management concepts using a graph showing an airplane's power-required curve (commonly called the **power curve**). This graph shows the amount of power necessary to maintain level flight at airspeeds throughout the aircraft performance envelope. [Figure 3-74]

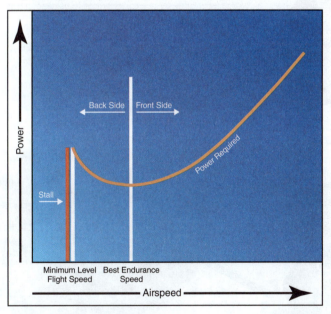

Figure 3-74. Flying on the back side of the power curve limits your available energy and options.

On the back side of the power curve, flight at low airspeeds requires a high angle of attack, and a great amount of power to overcome the resulting induced drag. Any reduction in airspeed requires an increase in power to maintain level flight. Flying on the back side of the curve is discouraged, because reducing speed can demand more power than the engine can supply or an unplanned reduction of power could result in an involuntary descent. Because these speeds are relatively close to the stall, even minor engine trouble could leave you with two choices: descend or stall. Avoid situations where you are dependent on engine power to prevent a stall.

As airspeed increases, the wing generates more lift with less induced drag, and less power is needed for level flight until, at the low point of the graph, the highest efficiency is reached. This is the point where the airplane will maintain level flight with the least amount of power. As speed increases past this point, you are on the front side of the power curve—your airplane is at a relatively high speed and a low angle of attack. On the front side of the power curve, pitching to change your airspeed has positive energy results:

- Pitching down—increases airspeed; reduces induced drag; and increases available energy.
- Pitching up—decreases airspeed; reduces parasite drag; and increases available energy.

With this excess energy, the airplane can now climb at a constant airspeed or turn in level flight while maintaining a constant airspeed at an increased load factor.

BELOW THE GLIDE PATH

Although, you must manage the airplane's energy state during all phases of flight, this skill is critical when you are at low altitude during a high workload phase of flight. Proficiency in energy management enables you to safely fly a stabilized approach to landing. A stabilized approach is one in which you establish and maintain a constant angle glide path towards a predetermined point on the landing runway. Flying a stabilized approach is based on your judgment of certain visual clues and depends on maintaining a constant final descent airspeed and configuration. You must carefully monitor your flight path and airspeed and adjust pitch and power as needed. [Figure 3-75]

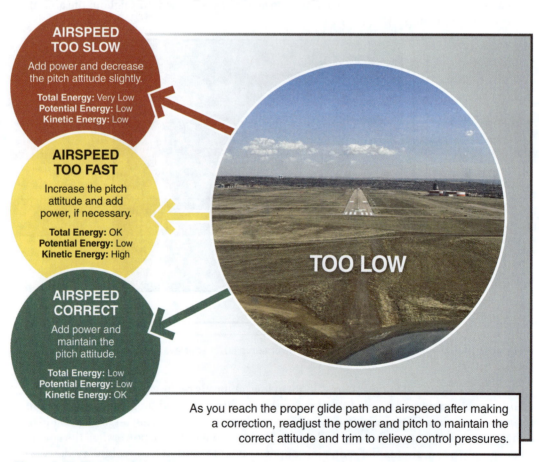

AIRSPEED TOO SLOW

Add power and decrease the pitch attitude slightly.

Total Energy: Very Low
Potential Energy: Low
Kinetic Energy: Low

AIRSPEED TOO FAST

Increase the pitch attitude and add power, if necessary.

Total Energy: OK
Potential Energy: Low
Kinetic Energy: High

AIRSPEED CORRECT

Add power and maintain the pitch attitude.

Total Energy: Low
Potential Energy: Low
Kinetic Energy: OK

TOO LOW

As you reach the proper glide path and airspeed after making a correction, readjust the power and pitch to maintain the correct attitude and trim to relieve control pressures.

Figure 3-75. These techniques are recommended to correct for deviations below the desired glide path on final approach.

THE DANGER OF LOW AND SLOW

Descending below the proper glide path and allowing the airspeed to decay below the approach speed is especially hazardous. Increase the power anytime the airplane is low and slow. Do not try to stretch a glide by pitching up without adding power. If you mismanage the airplane's energy state, and encounter an unintended, excessive deceleration and/or sink rate, coupled with little or no excess power with insufficient altitude (potential energy), you can deplete the airplane's total mechanical energy, and regardless of what you do, the airplane will impact terrain.

Applying full power to increase energy only succeeds if excess thrust is available. If not enough thrust is available to overcome drag because you have slowed to a speed where induced drag is so high, applying full power yields no surplus energy—you are on the back side of the power curve.

In this case, the only recourse might be to first trade altitude for airspeed, by slightly pitching down to reduce the angle of attack and induced drag, and only then, applying power to regain total energy. However, if the airplane is too low, you might not have enough altitude to reverse the negative energy rate, and prevent the airplane from impacting the ground.

ABOVE THE GLIDE PATH

If you are above the desired glide path, you have surplus energy in the form of altitude. If you are slow, you can pitch down to exchange altitude for airspeed. However, excess altitude is not your only consideration—you must be careful to manage your airspeed. If you are at the correct approach speed, pitching to descend without reducing power will cause your airplane to accelerate. An excessive approach speed can lead to overshooting your aiming point on the runway and a faulty landing. [Figure 3-76]

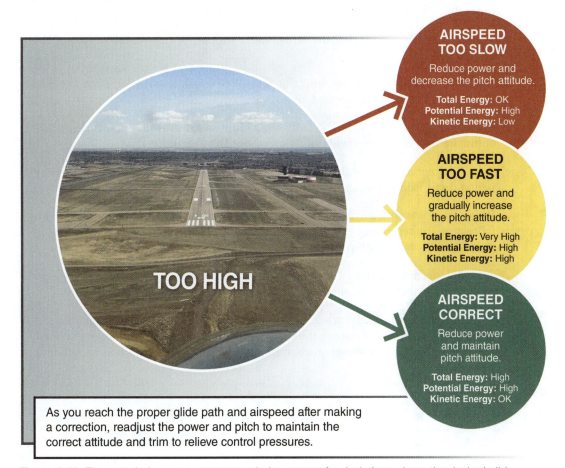

TOO HIGH

AIRSPEED TOO SLOW
Reduce power and decrease the pitch attitude.

Total Energy: OK
Potential Energy: High
Kinetic Energy: Low

AIRSPEED TOO FAST
Reduce power and gradually increase the pitch attitude.

Total Energy: Very High
Potential Energy: High
Kinetic Energy: High

AIRSPEED CORRECT
Reduce power and maintain pitch attitude.

Total Energy: High
Potential Energy: High
Kinetic Energy: OK

As you reach the proper glide path and airspeed after making a correction, readjust the power and pitch to maintain the correct attitude and trim to relieve control pressures.

Figure 3-76. These techniques are recommended to correct for deviations above the desired glide path on final approach.

SECTION C ■ Aerodynamics of Maneuvering Flight

SUMMARY CHECKLIST

✓ Excess thrust is necessary for a sustained climb. As the angle of climb steepens, thrust will not only oppose drag, but also will increasingly replace lift as the force opposing weight.

✓ Four left-turning tendencies associated with propeller-driven airplanes are torque, gyroscopic precession, asymmetrical thrust, and spiraling slipstream.

✓ During descending flight, one component of weight acts forward along the flight path, while another component acts perpendicular to the flight path.

✓ The least drag, best glide angle, and maximum gliding distance can be obtained by maintaining the angle of attack that corresponds to L/D_{max}.

✓ Changes in aircraft weight will not affect glide ratio, but a higher airspeed will have to be maintained in a heavier aircraft in order to cover the same distance over the ground.

✓ The horizontal component of lift creates a force toward the center of the turn, known as centripetal force, causes the airplane to turn.

✓ The effects of adverse yaw can be countered by maintaining a coordinated turn using rudder.

✓ To correct for overbanking tendency, you can use a small amount of opposite aileron, away from the turn, to maintain your desired angle of bank.

✓ Rate of turn increases and radius of turn decreases as angle of bank is increased in a constant airspeed turn. If angle of bank is held constant and airspeed is increased, turn rate will decrease and turn radius will increase.

✓ The ratio of the weight that the wings must support to the actual weight of the aircraft is termed load factor.

✓ Accelerated stalls occur when the critical angle of attack is exceeded at an airspeed higher than the one-G stall speed.

✓ Design maneuvering speed (V_A) is the maximum speed at which you can use full, abrupt control movement without overstressing the airframe.

✓ The V-g diagram defines the airplane's envelope, which is bounded by the stall region, limit load factor, and V_{NE}.

✓ When energy is exchanged, altitude and airspeed change in opposite directions (absent any other energy or control inputs). Energy management is the process of managing this relationship.

✓ The total mechanical energy of an airplane in flight is the sum of its potential energy from altitude, and kinetic energy from airspeed.

✓ A total energy error occurs when your airplane has too much or too little energy— altitude and airspeed deviate in the same direction (low and slow, or high and fast). To correct total energy errors, increase or decrease energy by adding or reducing power.

✓ An energy distribution error occurs when your airplane has the right amount of total energy, but the distribution over altitude and airspeed is incorrect—altitude and airspeed deviate in opposite directions (high and slow, or low and fast). Correct an energy distribution error by adjusting pitch to exchange energy between altitude and airspeed.

✓ You can easily visualize energy management concepts using a graph showing an airplane's power-required curve (commonly called the power curve).

✓ Understanding energy management concepts helps you adjust pitch and power to safely fly a stabilized approach to land.

✓ Descending below the proper glide path and allowing the airspeed to decay below the approach speed is especially hazardous. Increase the power anytime the airplane is low and slow.

KEY TERMS

Torque

Gyroscopic Precession

Asymmetrical Thrust

P-Factor

Spiraling Slipstream

Maximum Lift-to-Drag Ratio

Best Glide Speed

Glide Ratio

Glide Angle

Centripetal Force

Centrifugal Force

Adverse Yaw

Overbanking Tendency

Rate of Turn

Radius of Turn

Load Factor

Accelerated Stalls

Limit Load Factor

V-g Diagram

Design Maneuvering Speed (V_A)

Energy Management

Total Mechanical Energy

Potential Energy

Kinetic Energy

Total Energy Errors

Energy Distribution Errors

Power Curve

SECTION C ■ Aerodynamics of Maneuvering Flight

QUESTIONS

1. Identify the aerodynamic force that opposes the rearward component of weight in a climb.

2. What relative airspeed, power, and angle of attack conditions produce the most noticeable left-turning tendencies that are common to single-engine, propeller-driven aircraft?

3. Name two design elements that can be used to help offset left-turning tendencies.

4. All else being equal, will two aerodynamically identical aircraft with different weights be able to glide the same distance over the ground? If so, how can this be accomplished and why?

5. What causes an airplane to turn?

6. If angle of bank and altitude are held constant, what can be done to increase the rate of turn?

Given a wings-level, 1G stall speed of 55 knots, use the chart provided to determine the stall speed under the following conditions:

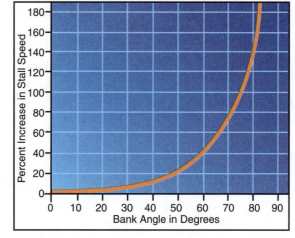

7. Bank angle, 30°

8. Bank angle, 45°

9. Bank angle, 75°

10. True/False. Maneuvering speed increases with a decrease in weight.

11. The total mechanical energy of an airplane in flight is the sum of what two types of energy?

12. Why is flying on the backside of the power curve discouraged?

13. Which is the correct action to take if you are too low on an approach to landing and your airspeed is too slow?
    A. Increase the pitch attitude and add power.
    B. Add power and decrease the pitch attitude slightly.
    C. Increase the pitch attitude and maintain the power setting.

14. When below the glide path and flying at an excessive airspeed, what is the status of the three energy states?

SECTION C ■ **Aerodynamics of Maneuvering Flight**

PART II

Flight Operations

There is a feeling about an airport that no other piece of ground can have. No matter what the name of the country on whose land it lies, an airport is a place you can see and touch that leads to a reality that can only be thought and felt.

— Richard Bach

PART II

In the early days of aviation, there were few airplanes and even fewer airports. At the busiest airports, the amount of air traffic was negligible compared to today. As air traffic grew, pilots became aware of the increased potential for midair collisions; airports evolved to manage many aircraft at once; airspace designations were created to govern the operation of aircraft; and common radio procedures were established to enhance communication. Part II contains a broad range of information that you need to operate safely in today's complex flight environment. The rules and procedures that make it possible for thousands of aircraft to efficiently take off and land each day are examined in *The Flight Environment*. As you explore *Communication and Flight Information* you will discover how to effectively communicate with air traffic control and will learn about the various sources that provide you with information essential to flight operations.

CHAPTER 4

The Flight Environment

SECTION A
Safety of Flight

Maintaining the safety of flight is your number one priority as a pilot. You must be aware of some safety issues during every flight, such as collision avoidance and maintaining minimum safe altitudes. Other safety considerations only apply in certain situations; for example, taxiing in wind, flight over hazardous terrain, and effective exchange of flight controls with your instructor. Every flight is different and, as pilot in command, you need to consider the factors that can affect your flight and take the appropriate actions to ensure safety.

PILOT IN COMMAND

As **pilot in command (PIC)**, you have the final authority and responsibility for the safe operation of the flight. Pilot-in-command responsibility starts with preflight actions , such as becoming familiar with all information concerning the flight and conducting passenger safety briefings prior to flight, then continues throughout the duration of the flight. As the PIC, you are required to determine runway lengths at airports of intended use and the takeoff and landing distance data for your aircraft. Additionally, for any flight beyond the vicinity of your home airport, you must also check weather reports and forecasts, fuel requirements, and available alternates if the flight cannot be completed as planned.

In flight, when a clearance is obtained from **air traffic control (ATC)**, you may not deviate from that clearance unless a new or amended clearance is received. An exception exists for emergency situations, where you, as the PIC, have the authority to deviate from any rule as required to handle the emergency. If a deviation is necessary, you are required to submit a written report upon request of the **Administrator**. By definition of the FAA: "Administrator means the Federal Aviation Administrator or any person to whom he/she has delegated his/her authority in the matter concerned."

COLLISION AVOIDANCE

Learning **collision avoidance** procedures begins with your first flight. The risk of an in-flight collision exists for all pilots, but you can take action to avoid this type of accident. Studies show that the majority of midair collisions occur within five miles of an airport, during daylight hours, and in VFR conditions. You will hear the terms VFR and IFR used frequently in several different ways. If you are operating under **visual flight rules (VFR)**, you are governed by specific FARs that include minimum cloud clearance and visibility requirements, also referred to as weather minimums. In comparison, **instrument flight rules (IFR)** are rules that are established to govern flight operations in weather conditions below VFR weather minimums. The terms VFR and IFR also are used to define weather conditions. For example, visual meteorological conditions (VMC) are often referred to as VFR conditions, and instrument meteorological conditions (IMC) are sometimes called IFR conditions. In addition, the terms VFR and IFR can define the type of flight plan under which an aircraft is operating. An aircraft can be on an IFR flight plan in VFR weather conditions.

Many resources are available to help you avoid midair collisions in VFR conditions, including use of exterior lights, radio transmissions, air traffic control services, and traffic information systems on the flight deck. However, the most important collision avoidance tool is your ability to effectively see and avoid other aircraft. Early detection of aircraft is crucial to avoiding a collision. For example, if your aircraft is flying at a speed of 150 knots on a head-on collision course with another aircraft also traveling at 150 knots, the closure rate is one quarter of a mile in the 3 seconds that is usually required for you or the other pilot to take action.

VISUAL SCANNING

To see and avoid other aircraft, you must develop an effective **visual scanning** pattern that is compatible with the function of your eyes. Two normal healthy eyes provide the average person with a field of vision of approximately 200 degrees. However, the area in which the eye can focus sharply and perceive detail is a relatively narrow cone (usually only about 10 degrees wide) directly in the center of the field of vision. Beyond this area, visual acuity decreases sharply in all directions. Because your eyes require time to focus on this narrow viewing area, scanning is most effective when you use a series of short, regularly-spaced eye movements. This method helps to bring successive areas of the sky into your central visual field. The FAA recommends that your eye movements not exceed 10 degrees, and that you focus for at least one second on each segment of the sky. Be sure that the scan pattern you develop covers all of the sky that you can see from the flight deck, both horizontally and vertically. [Figure 4-1] Nighttime conditions require a different scanning technique. Refer to the *Aviation Physiology* section of Chapter 10.

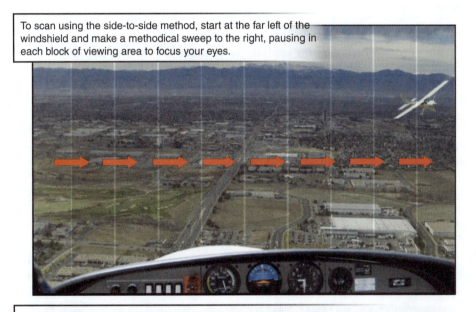

To scan using the side-to-side method, start at the far left of the windshield and make a methodical sweep to the right, pausing in each block of viewing area to focus your eyes.

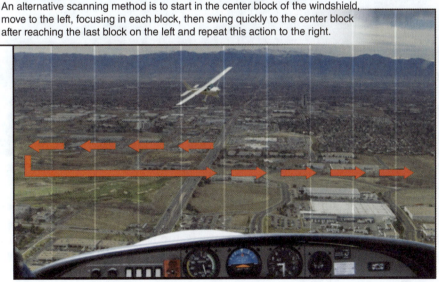

An alternative scanning method is to start in the center block of the windshield, move to the left, focusing in each block, then swing quickly to the center block after reaching the last block on the left and repeat this action to the right.

Figure 4-1. The block system of scanning involves dividing your viewing area (windshield) into segments, and then methodically scanning for traffic in each block of airspace in sequential order.

SECTION A ■ **Safety of Flight**

SECTION A ■ Safety of Flight

Visual Field

To demonstrate the limitations of your visual field, look at a wall calendar from a distance of about 5 feet. From that distance, a typical calendar is about 10 degrees wide, and represents the relatively small area in which your eyes can focus sharply. Focus on one part of the calendar, such as the picture or the name of the month, keep your eyes there, and notice that other areas of the calendar, although visible, are not sharp enough to be seen clearly.

To test your peripheral vision, close your eyes and have a friend place a hand beside your ear and about two feet away. When you open your eyes, look straight ahead. Can you see your friend's hand when it is stationary? How about when it is moving up and down? You will find that it is difficult to perceive objects in your peripheral vision unless they are moving.

The further objects are from your central visual field, the smaller the amount of detail you can discern. You might not notice objects in your peripheral vision unless there is some relative motion. An airplane on a converging course from one side does not have any relative motion unless there is a significant speed difference, and therefore, might not catch your attention. When scanning for traffic, be especially alert for an aircraft that shows no movement, because this indicates that it is most likely on a collision course with your airplane. If the other aircraft shows no lateral or vertical motion, but is increasing in size, take immediate evasive action.

As you scan, it can take several seconds for your eyes to refocus when switching from the instrument panel to distant objects outside the airplane. To counter eye fatigue, focus on exterior parts of the airplane as a transition from inside to outside the flight deck.

Sky conditions also affect the contrast an aircraft has with its background and therefore your ability to see traffic. In bright sunlight, a clean windshield and sunglasses can help you see objects more clearly. [Figure 4-2]

Figure 4-2. An aircraft below the horizon, silhouetted against a uniform landscape, has greater contrast with its environment than a similar aircraft viewed against a cluttered background.

Haze and reduced visibility make traffic and terrain features appear farther away than their actual distance. Another phenomenon that can occur in reduced visibility is called **empty field myopia**. When you are looking at a featureless sky that is devoid of objects, contrasting colors, or patterns, your eyes tend to focus at only 10 to 30 feet ahead. This means that spots on the windshield that are out of focus could appear to be aircraft, and distant traffic can go undetected.

Night myopia, or night-induced nearsightedness, is similar to empty field myopia, but is more pronounced because of the lack of visual references, With nothing to focus on, your eyes automatically focus on a point three to six feet in front of you. Searching out and focusing on distant light sources, no matter how dim, helps prevent night myopia.

When you encounter traffic, remember that the other pilot might be occupied with tasks other than scanning and might not react in the manner that you anticipate. As you take evasive action, watch the other aircraft to see if it makes any unusual maneuvers and, if so, plan your reaction accordingly. You also should be aware that more than one potential collision hazard could be in an area at the same time. The more time you spend on developing your scan in the early part of flight training, the more natural it will become later. The scanning techniques described here will help you see potential conflicts earlier and enable you to take the proper action to avoid a collision. Also, being familiar with the blind spots in your aircraft's design will improve your awareness of potential hazards.

BLIND SPOTS AND AIRCRAFT DESIGN

Airplanes, like automobiles, have **blind spots**. In both high-wing and low-wing aircraft designs, portions of your view are blocked by the fuselage and wings. [Figure 4-3] This can make it difficult to see conflicting traffic. For example, in a high-wing airplane, your view is blocked as soon as you lower the wing to start a turn. Prior to beginning the turn, you can check the area for other aircraft by lifting the wing and looking in the direction of the turn.

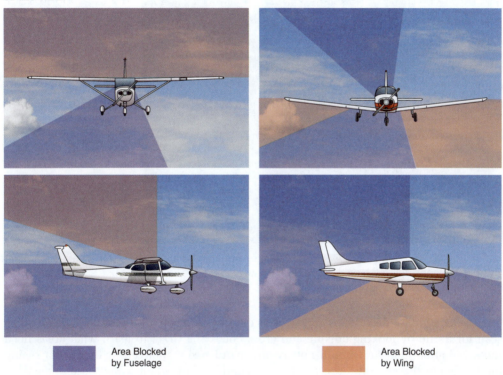

Area Blocked by Fuselage

Area Blocked by Wing

Figure 4-3. The portions of your view that are restricted depend on the design of the airplane.

Blind Spot

In a high-wing airplane it is easy to see below the airplane, but difficult to see the area above the airplane. The reverse is true for low-wing airplanes. These blind spots can develop into a serious problem, particularly during the approach and landing phases of flight. A good way to reduce the possibility of a collision during extended climbs or descents is to make shallow S-turns and avoid climbing or descending at steep angles. [Figure 4-4]

Figure 4-4. When a high-wing airplane is below a low-wing airplane on approach to landing, both airplanes can easily remain out of sight of each other. Similar problems can occur during departures.

FLIGHT DECK TRAFFIC DISPLAYS

Flight deck technology can help increase your situational awareness of other traffic. In an aircraft that is equipped with automatic dependent surveillance-broadcast (ADS-B) In and Out, you can monitor other transponder-equipped aircraft on a display that is often referred to as a **cockpit display of traffic information (CDTI)**. A CDTI can be a dedicated screen or traffic can be included on a multi-function display (MFD) in an integrated flight deck or on a GPS moving map. [Figure 4-5]

The transparent target is not considered a threat. It is 3,600 ft above your aircraft's altitude and descending.

The solid target indicates proximate traffic—nearby, but not close enough to be an imminent threat. It is 300 ft below your airplane's altitude.

The target in a yellow circle is a traffic alert, which is usually accompanied by an aural "TRAFFIC" warning, This target is about 2 miles away and within 300 ft above your airplane's altitude. Look for the aircraft and when you see it, take evasive action as needed.

The solid diamond is a proximate target without directional information.

A target's direction and projected location after a set time are depicted with a traffic motion vector.

Figure 4-5. A CDTI incorporated into a GPS navigation system can be a dedicated traffic page or integrated into other moving map pages. Although operation varies between manufacturers, most use similar symbols to depict traffic targets, with your airplane symbol (ownship) in the center.

General aviation traffic systems are advisory only, to help you locate traffic. Continuously scan for traffic by looking outside and cross check the CDTI to learn what areas need increased attention. You should not fly any avoidance maneuvers without first seeing the traffic out the window. You must avoid complacency when using these systems. A traffic display is not a replacement for deliberate scanning—it is a tool that can help you scan more effectively. And some threats, such as aircraft without operating transponders, might not show up on your display or generate a traffic alert.

AIRPORT OPERATIONS

Any operation in the vicinity of an airport warrants extra caution. Even towered airports can be hazardous because of the large amount of air traffic. A control tower does not relieve you of the responsibility to see and avoid other aircraft. At nontowered airports, sometimes referred to as uncontrolled airports, sequencing to the airport is determined by the individual pilots, traffic advisories normally are not provided by air traffic control, and aircraft without radios may operate in the area. These factors increase the risk of a collision if you are not vigilant.

To increase safety at airports, a voluntary program called **Operation Lights On** has been established by the FAA. Operation Lights On encourages you to use your landing lights during departures and approaches, both day and night, especially when operating within 10 miles of an airport, or in conditions of reduced visibility.

In addition, the anticollision lights on your airplane are required to be on whenever the engine is running, day or night. However, anticollision lights need not be turned on when they might interfere with safety. For example, strobe lights should be turned off when their brightness might impair the vision of others. In order to easily identify aircraft position at night, the FARs require that you have your position lights on from sunset to sunrise.

FORMATION FLIGHT

For collision prevention, the FARs require maintaining a safe distance from other aircraft to prevent potential collisions. However, formation flight is permissible if arranged prior to the flight by the pilot in command of each aircraft in the formation. To safely fly in formation is challenging so you should not attempt this type of flying without comprehensive instruction by a flight instructor with formation training experience.

MANEUVERS IN THE TRAINING AREA

In addition to maintaining your scan, you must always clear an area for traffic before and during your practice of maneuvers. Your instructor will teach you to make **clearing turns** before performing maneuvers in the practice area. Clearing turns, which usually consist of at least a 180-degree change in direction, enable you to see areas blocked by blind spots and make it easier to maintain visual contact with other aircraft in the area.

RIGHT-OF-WAY RULES

The FARs state that the pilot in command is responsible for seeing and avoiding all traffic in visual flight conditions. To help avoid conflicting traffic, the FAA has established **right-of-way rules**. An aircraft in distress requires immediate assistance due to a serious problem or emergency and therefore has the right-of-way over all other air traffic. Otherwise, right-of-way rules apply in three situations; overtaking another aircraft, approaching another aircraft head-on, or converging with another aircraft. An overtaking aircraft must pass the slower aircraft on the right and stay well clear. If two aircraft are approaching each other head-on or nearly so, both aircraft must give way to the right. When aircraft of the same category are converging, the aircraft to the right has the right-of-way. [Figure 4-6]

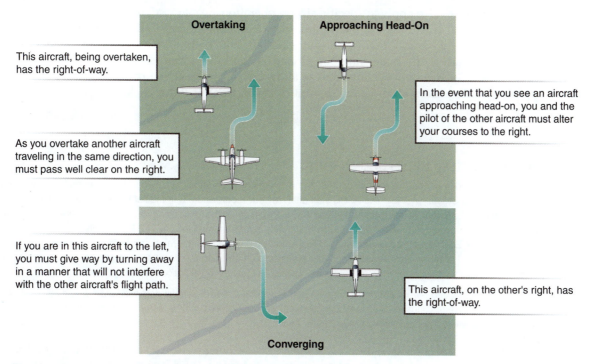

Overtaking

This aircraft, being overtaken, has the right-of-way.

As you overtake another aircraft traveling in the same direction, you must pass well clear on the right.

Approaching Head-On

In the event that you see an aircraft approaching head-on, you and the pilot of the other aircraft must alter your courses to the right.

If you are in this aircraft to the left, you must give way by turning away in a manner that will not interfere with the other aircraft's flight path.

This aircraft, on the other's right, has the right-of-way.

Converging

Figure 4-6. If your aircraft and another aircraft are on a collision course, both you and the other pilot should take the proper action to avoid each other.

SECTION A ■ **Safety of Flight**

The general rule regarding converging aircraft of different categories is that the least maneuverable aircraft usually has the right-of-way over all other air traffic. Figure 4-7 shows the order of aircraft by right of way. An aircraft that is towing or refueling another aircraft has the right-of-way over all other engine-driven aircraft.

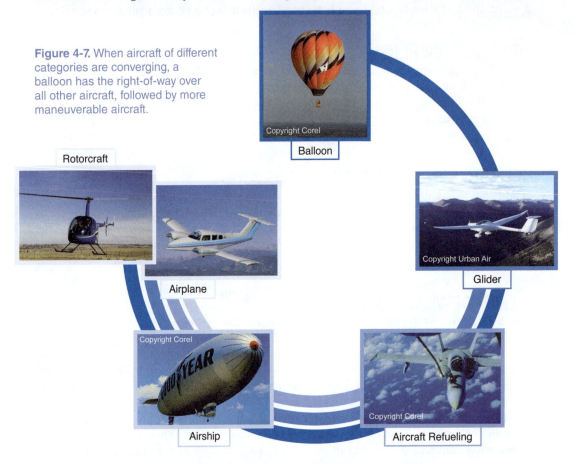

Figure 4-7. When aircraft of different categories are converging, a balloon has the right-of-way over all other aircraft, followed by more maneuverable aircraft.

Balloon

Rotorcraft

Glider

Airplane

Airship

Aircraft Refueling

Right-of-way rules also apply to specific airport operations. An aircraft on final approach or an aircraft that is landing has the right-of-way over other aircraft in the traffic pattern and those on the ground. When two or more aircraft, which are preparing to land, enter the traffic pattern at the same time, the aircraft at the lowest altitude has the right-of-way. However, this rule is not intended to allow aircraft to enter the pattern below the specified traffic pattern altitude and disrupt or cut in front of those already established in the pattern.

Aviation Safety Reporting System

If you are involved in or observe an incident that compromises aviation safety, you can submit a report to the Aviation Safety Reporting System (ASRS). The ASRS staff analyze your report and if a immediate action is needed, notify the appropriate FAA office or aviation authority. The ASRS investigates the underlying causes of a reported event and adds each report into a database for research on aviation safety and human factors. The ASRS collects nearly 100,000 reports per year. Each report is confidential and the FAA may not use ASRS information in enforcement actions against those who submit reports. The following accounts are excerpts from reports submitted to the ASRS that illustrate the importance of flying at a safe altitude and airspeed, as well as maintaining situational awareness.

Flying over uninhabited terrain, I suddenly came upon a golf course, with houses and golfers on the course. Before I was able to analyze the situation, I passed low overhead of some golfers, maybe within 500 feet. Increased familiarity with the locale would have prevented this particular event.

I came upon a softball game taking place...I became too focused on the players and failed to realize how low and close to the field I was. Suddenly there was a row of trees ahead of me. I tried to climb but due to low airspeed, I struck one of the trees. I proceeded directly back to the airstrip with reduced elevator control.

AEROBATIC FLIGHT

To perform **aerobatic flight** you must consider numerous safety factors. Aerobatic flight is an intentional maneuver involving an abrupt change in an aircraft's attitude, an abnormal attitude, or abnormal acceleration, not necessary for normal flight. You should never conduct aerobatic flight without the proper training from a qualified instructor.

To perform aerobatics, you must be at least 1,500 feet above ground level (AGL) with at least 3 statute miles of visibility. You may not perform aerobatics over a congested area or open-air assembly of persons or within the lateral boundaries of any controlled airspace (Class B, C, D, and E) extending to the surface around an airport or in the airspace designated for federal airways. In addition to aerobatic flight, operating an experimental or restricted aircraft is prohibited over densely populated areas and on congested federal airways. [Figure 4-8]

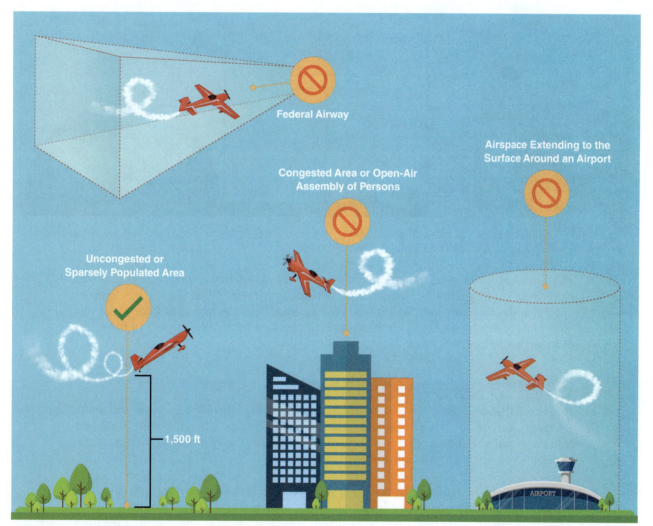

Figure 4-8. The FARs prohibit all aerobatic flight over densely populated areas, in airspace extending to the surface around an airport, or on a federal airway.

SECTION A ■ **Safety of Flight**

If carrying passengers, you may not perform any intentional maneuver that exceeds a nose-up or nose-down attitude of 30 degrees relative to the horizon (typical for aerobatics) unless each occupant on-board is wearing an approved parachute. Parachutes must be repacked periodically to satisfy the FARs—the frequency depends on the materials of the parachutes. If your parachute has a natural canopy, shroud, and harness it must have been packed by a certificated and appropriately rated parachute rigger within the preceding 60 days. Otherwise, if the canopy, shroud, and harness are made of synthetic materials, the parachute only needs to be packed within the previous 180 days.

MINIMUM SAFE ALTITUDES

The FARs specify minimum altitudes that you must maintain during flight. These **minimum safe altitudes** apply at all times except during take off and landing. Over a congested area, such as a city or metropolitan area, you are required to fly at least 1,000 feet above any obstacle within a horizontal radius of 2,000 feet of your aircraft. When flying over an uncongested area, you must fly at least 500 feet above the surface. Over sparsely populated or open water areas you may not fly within 500 feet of any person, vessel, vehicle, or structure. [Figure 4-9]

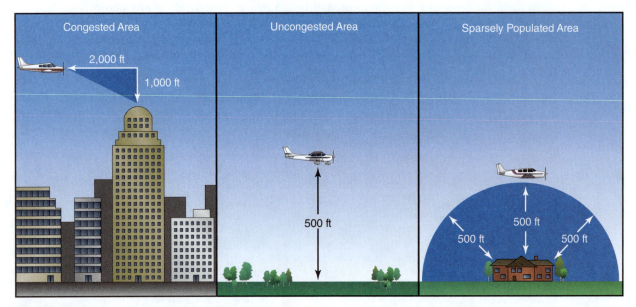

Figure 4-9. The FARs require that you fly high enough so that you do not endanger people or property on the ground and specify minimum safe altitudes over various areas.

The lowest altitude at which you may fly anywhere is one that allows you to make an emergency landing without undue hazard to persons or property on the surface. Keep in mind that the altitudes specified in the FARs are minimums. A higher altitude gives you more time to troubleshoot any problems and to choose a better landing site in the event of an engine failure.

WIRE STRIKE AVOIDANCE

An aircraft that collides with a power line or an antenna guy wire often becomes unflyable, which typically results in an accident with serious injury or death. Although you might think that **wire-strike accidents** mostly affect helicopters, any aircraft flying near the ground is susceptible to striking a wire.

Many airplane wire-strike accidents are associated with agricultural operations, pipeline patrol, and other operations that legitimately require low flying. If you are preparing for a career in one of these fields you must obtain special training in the specific operation you are conducting, and ensure that you know and consistently follow sound safety procedures. For example, military low-flying missions require pilots to study their charts carefully and fly over the top of towers that support power lines because the power lines themselves are often invisible in flight. [Figure 4-10]

Figure 4-10. Maintaining safe clearance above powerlines is particularly important because the wires can be difficult to see.

Typical wire strikes normally occur below 100 feet AGL. Because most pilots typically do not need to fly that low, flying at least 1,000 feet AGL (except during takeoff and landing) dramatically reduces your chances of a wire strike. Follow the FAA guideline to maintain a 2,000-foot horizontal distance from any tower (unless you are safely above it), to reduce the risk of colliding with an antenna guy wire.

Even if you are careful to maintain a safe altitude during enroute flight, you should be aware that some wire-strike accidents still occur when pilots fly too low during final approach, sometimes due to the lack of adequate visual guidance to the runway. Avoiding these types of accidents during night and low-visibility conditions requires extra attention to maintaining an appropriate glide path.

DROPPING OBJECTS

Although rare, certain circumstances might require dropping objects from an aircraft. The FAA permits dropping objects from an aircraft if you take precautions to avoid injury or damage to persons or property on the surface.

FLIGHT OVER HAZARDOUS TERRAIN

Depending on where you learn to fly, mountain flying and flight over open water are typically not part of private pilot training. To safely operate over these areas, you need specialized training from an instructor who knows the area over which you will fly. Mountain flying can be a rewarding experience, but it introduces numerous hazards for the inexperienced pilot. Decreased aircraft performance at high altitudes, turbulence, rapidly changing weather, and difficulty in locating a forced landing site are just a few of the challenges of a mountain flight. If you do make a successful emergency landing in a remote area, can you survive the night until help arrives? Having warm clothing, water, a first aid kit, and other survival gear onboard could easily make a life or death difference for you and your passengers.

TAXIING IN WIND

Another safety concern that you will be introduced to early in your flight training is the effect of wind during taxi. Strong winds passing over and around the wings and horizontal stabilizer can actually lift the airplane. In the most severe conditions, the airplane could flip over.

Proper use of the aileron and elevator controls normally counteracts the wind and helps you maintain control of your airplane on the ground. For example, if the wind is blowing from the left front quarter, you should turn the yoke to the left, which will hold the left aileron up and counteract the lifting tendency of the wind. When you turn while taxiing, you have to change the position of the controls for the new direction of the wind relative to the airplane. In a tricycle-gear airplane, the elevator should be in a neutral position when the wind is from the front to prevent the wind from exerting any lifting force on the tail. [Figure 4-11]

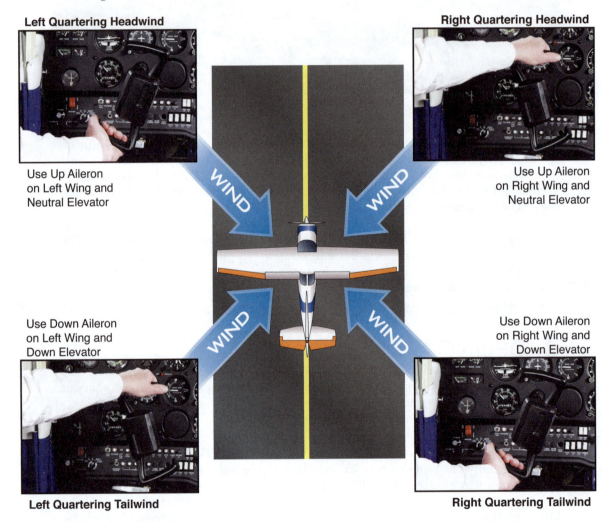

Left Quartering Headwind

Use Up Aileron on Left Wing and Neutral Elevator

Right Quartering Headwind

Use Up Aileron on Right Wing and Neutral Elevator

Use Down Aileron on Left Wing and Down Elevator

Use Down Aileron on Right Wing and Down Elevator

Left Quartering Tailwind

Right Quartering Tailwind

Figure 4-11. Your knowledge of proper control use during windy conditions will help you control the airplane while taxiing.

Differences exist in taxiing tricycle-gear airplanes versus conventional-gear, or tailwheel, airplanes. Compared with tricycle-gear airplanes in gusty and crosswind conditions, a tailwheel airplane has an increased tendency to weather vane into the wind due to the greater surface area behind the main gear. In addition, the location of the center of gravity behind the main wheels can cause the airplane to swerve further and further out of alignment once it is displaced. [Figure 4-12]

Figure 4-12. Tailwheel airplanes are more difficult to taxi in strong wind conditions than tricycle gear airplanes.

The most critical situation exists when you are taxiing a high-wing tricycle-gear airplane in a strong quartering tailwind. The high wing is susceptible to being lifted by the wind. In extreme conditions, a quartering tailwind can cause the airplane to nose over and flip on its back.

When taxiing a tailwheel airplane, position the ailerons the same as you do for a tricycle-gear airplane. However, to help keep the tailwheel on the ground, hold the elevator control aft (elevator up) in a headwind, and in a tailwind, hold the elevator control forward (elevator down).

SAFETY BELTS

As a flight crewmember, you are required to keep your safety belt and shoulder harness fastened during takeoffs and landings. Safety belts must stay fastened while enroute. As pilot in command, you must brief your passengers on the use of safety belts and notify them to fasten their safety belts during taxi, takeoff, and landing.

POSITIVE EXCHANGE OF FLIGHT CONTROLS

Frequently during your flight training, it will be necessary to exchange the flight controls with your instructor. For example, your instructor normally demonstrates a maneuver first, before passing the controls to you. To ensure that it is clear as to who has control of the aircraft, the FAA strongly recommends the use of a three-step process when exchanging the flight controls. During the preflight briefing, you should review the procedure for the **positive exchange of flight controls**:

PILOT PASSING CONTROL: *"You have the flight controls."*

PILOT TAKING CONTROL: *"I have the flight controls."*

PILOT PASSING CONTROL: *"You have the flight controls."*

The pilot passing the controls should continue to fly until the pilot taking the controls acknowledges the exchange by saying, *"I have the flight controls."* A visual check also is recommended to ensure that the other pilot actually has the controls. At times, your instructor might need to assume control of the aircraft from you. In this case, your instructor should take the controls while informing you, *"I have the flight controls."*

SECTION A ■ **Safety of Flight**

NTSB PART 830

In addition to enhancing safety during routine flights, you also learn how to handle abnormal and emergency situations. After managing an emergency, you must determine if the circumstances require you to report the event to the National Transportation Safety Board (NTSB). You should refer to **NTSB Part 830** regulations for detailed definitions and reporting requirements. General definitions in NTSB 830.2 include:

- Aircraft accident—an occurrence that takes place between the time any person boards the aircraft with the intention of flight and all persons have disembarked in which any person suffers death or serious injury, or in which the aircraft receives substantial damage.

- Substantial damage—damage or failure that adversely affects the structural strength, performance, or flight characteristics of the aircraft, and which would normally require major repair or replacement of the affected component. The definition specifically states that, among other things, engine failure and damage limited to one engine or damage to landing gear are not considered substantial damage.

- Serious injury—requires hospitalization for more than 48 hours within 7 days from the date of injury; a fracture of any bone (except simple fractures of fingers, toes, or nose); severe hemorrhages, nerve, muscle, or tendon damage; involves any internal organ; second- or third-degree burns, or any burns affecting more than 5 percent of the body surface. [Figure 4-13]

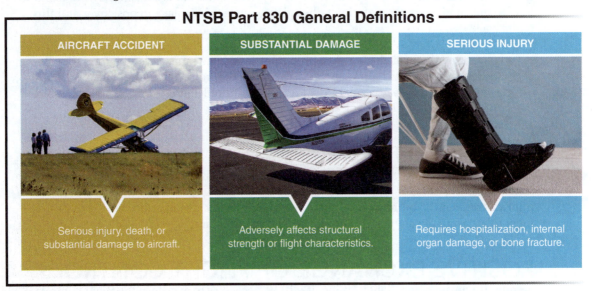

NTSB Part 830 General Definitions

AIRCRAFT ACCIDENT	SUBSTANTIAL DAMAGE	SERIOUS INJURY
Serious injury, death, or substantial damage to aircraft.	Adversely affects structural strength or flight characteristics.	Requires hospitalization, internal organ damage, or bone fracture.

Figure 4-13. General definitions in NTSB 830.2 include: aircraft accident, substantial damage, and serious injury.

NTSB 830.5 states that as the aircraft operator, you must immediately notify the NTSB if an aircraft accident or serious incident occurs. In general, the regulation defines serious incidents as:

- Flight control system malfunction or failure;

- Inability of any required flight crewmember to perform normal flight duties as a result of injury or illness;

- Failure of any internal turbine engine component that results in the escape of debris other than out the exhaust path;

- In-flight fire;

- Aircraft collision in flight;

- Damage to property, other than the aircraft, estimated to exceed $25,000 for repair;

- Release of all or a portion of a propeller blade from an aircraft;

- A complete loss of information from more than 50 percent of an aircraft's electronic flight deck displays;

- Airborne collision and avoidance system (ACAS) resolution advisories issued either when an aircraft is being operated under IFR and compliance with the advisory is necessary to avert a substantial risk of collision between two or more aircraft;

- Additional incidents that are specified for large multi-engine aircraft (more than 12,500 pounds maximum certificated takeoff weight).

The operator of an aircraft that has been involved in an accident is required to file a report with the NTSB within 10 days or after 7 days if an overdue aircraft is still missing. The operator of an aircraft that is involved in an incident that requires immediate notification is required to file a report only as requested by an authorized NTSB representative. Aircraft wreckage may be moved prior to the time the NTSB takes custody only to protect the wreckage from further damage.

SUMMARY CHECKLIST

✓ As the pilot in command, you are the final authority as to the operation of an aircraft.

✓ The PIC is directly responsible for the passenger safety briefing prior to flight.

✓ In addition to other preflight actions, you are required to determine runway lengths at airports of intended use and the takeoff and landing distance data for that aircraft.

✓ When an ATC clearance has been obtained, you may not deviate from that clearance, unless you obtain an amended clearance. The exception to this regulation is in an emergency.

✓ If an in-flight emergency requires immediate action, you may deviate from the FARs to the extent required to meet that emergency. A written report is not required unless requested by the FAA.

✓ The majority of midair collisions occur during daylight hours, in VFR conditions, and within five miles of an airport.

✓ In daylight conditions, the most effective way to scan is through a series of short, regularly-spaced eye movements in 10° sectors.

✓ You might not notice objects in your peripheral vision unless there is some relative motion.

✓ If there is no apparent relative motion between another aircraft and yours, you are probably on a collision course.

✓ Empty field myopia occurs when you are looking at a featureless sky that is devoid of objects, contrasting colors, or patterns and your eyes tend to focus at only 10 to 0 feet.

✓ Blind spots make it difficult to see conflicting traffic. In both high-wing and low-wing designs, portions of your view are blocked by the fuselage and wings.

✓ A cockpit display of traffic information (CDTI) is a dedicated screen, MFD, or GPS moving map that depicts traffic threats.

✓ General aviation traffic systems are advisory only, to help you locate traffic. Continuously scan for traffic by looking outside and cross check the CDTI to learn what areas need increased attention.

SECTION A ■ **Safety of Flight**

✓ Operation Lights On encourages you to use your landing lights during departures and approaches, both day and night, especially when operating within 10 miles of an airport, or in conditions of reduced visibility.

✓ During sunset to sunrise, except in Alaska, lighted position lights must be displayed on an aircraft.

✓ You may not operate an aircraft in formation flight except by prior arrangement with the pilot in command of each aircraft.

✓ Clearing turns allow you to see areas blocked by blind spots and make it easier to maintain visual contact with other aircraft in the practice area.

✓ An aircraft in distress has the right-of-way over all other aircraft.

✓ Primarily, right-of-way rules apply in three situations; converging with another aircraft, approaching another aircraft head-on, or overtaking another aircraft.

✓ To perform aerobatics, you must be at least 1,500 feet above ground level (AGL) with at least 3 statute miles of visibility.

✓ You must not perform aerobatics over a congested area or open-air assembly of persons or within the lateral boundaries of any controlled airspace (Class B, C, D, and E) extending to the surface around an airport or in the airspace designated for federal airways.

✓ Operating an experimental or restricted aircraft over densely populated areas and on congested federal airways is prohibited

✓ You must maintain FAA-designated minimum safe altitudes at all times except during takeoffs and landings. Complying with safe altitude rules also minimizes your risk of a wire strike accident.

✓ You may only drop objects from an aircraft if precautions are taken to avoid injury or damage to persons or property on the surface.

✓ Mountain flying and flight over open water require specialized training from experienced instructors who are familiar with the area over which the flights will be conducted.

✓ While taxiing in wind, proper use of the aileron and elevator controls will help you maintain control of the airplane.

✓ As a flight crewmember, you are required to keep their safety belts and shoulder harnesses fastened during takeoffs and landings. Safety belts must stay fastened while enroute.

✓ You must brief your passengers on the use of safety belts and notify them to fasten their safety belts during taxi, takeoff, and landing.

✓ To ensure that it is clear as to who has control of the aircraft, the FAA strongly recommends the use of a three-step process when exchanging the flight controls.

KEY TERMS

Pilot in Command (PIC)

Air Traffic Control (ATC)

Administrator

Collision Avoidance

Visual Flight Rules (VFR)

Instrument Flight Rules (IFR)

Visual Scanning

Empty Field Myopia

Blind Spots

Cockpit Display of Traffic Information (CDTI)

Operation Lights On

Clearing Turns

Right-of-Way Rules

Aerobatic Flight

Minimum Safe Altitudes

Wire-Strike Accidents

Positive Exchange of Flight Controls

NTSB Part 830

QUESTIONS

1. When is it appropriate to deviate from an ATC clearance?

2. When must the pilot in command submit a written report to the FAA after a deviation from a clearance?

3. What is the most effective method to scan for other aircraft during the day and why?

4. True/False. When looking through haze, air traffic and terrain features are not as close as they appear.

5. What is required before you can operate an aircraft in formation flight?

6. Select the true statement regarding collision avoidance.
   A. Operating at an airport with a control tower relieves you of the responsibility to see and avoid other traffic.
   B. If there is no apparent relative motion between another aircraft and yours, you are probably on a collision course.
   C. Studies show that the majority of midair collisions occur during daylight hours, in IFR conditions, and within five miles of an airport.

7. What is the appropriate way to use a cockpit display of traffic information (CDTI) to avoid a collision?
   A. To avoid complacency, use the CDTI only after you are instrument rated and flying IFR.
   B. Monitor the CDTI and if you receive a traffic alert, turn away from target shown on the display.
   C. Continuously scan for traffic by looking outside and cross check the CDTI to learn what areas need increased attention.

8. Aerobatic flight is prohibited under which circumstance?
   A. Below 1,500 feet AGL
   B. Over a congested area or open air assembly of persons.
   C. On federal airways and in controlled airspace (Class B, C, D, and E)

SECTION A ■ **Safety of Flight**

Select the aircraft that has the right-of-way in each of the following illustrations.

9.

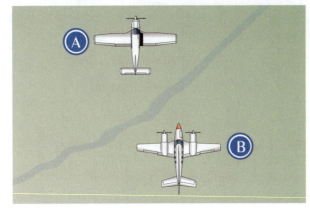

10.

11.

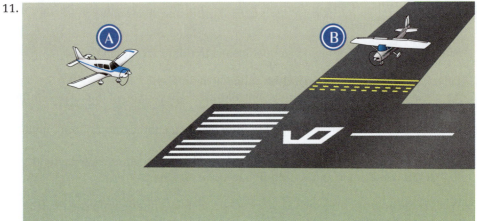

Match the minimum safe altitudes with the appropriate areas.

12. 1,000 feet above any obstacle within a horizontal radius of 2,000 feet of the aircraft

A. Uncongested areas

B. Sparsely populated or open water area

13. 500 feet above the surface

14. 500 feet from any person, vessel, vehicle, or structure

C. Congested areas

15. Select the proper control positions for taxiing in the wind condition shown.
    A. Control wheel to the right, elevator control aft
    B. Control wheel to the left, elevator control neutral
    C. Control wheel to the right, elevator control neutral

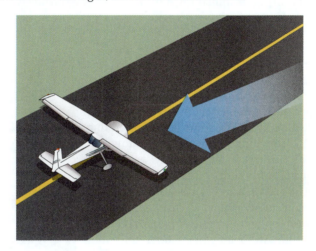

16. When are you, as PIC, typically required to have your safety belt and harness fastened?

17. Describe the recommended procedure to be used when it is necessary for you to exchange the flight controls with your instructor.

18. When should the NTSB be notified if an airplane is involved in an accident that results in substantial damage?
    A. Immediately
    B. Within 7 days
    C. Within 48 hours

SECTION A ■ **Safety of Flight**

SECTION B
Airports

Each day, aircraft take off and land at private grass strips, busy international airports, and every type of field in between. Whether most of your flying is out of your local airport or you frequently journey to new destinations, an airport will never be unfamiliar territory after you learn the basic procedures for operating in the terminal environment. Knowing how to determine the correct runway to use and understanding the markings on taxiways and runways is essential every time you fly. Night flying offers a new challenge as the airport is transformed by a myriad of lights designed to make aircraft operations safe and efficient in the darkness.

CONTROLLED AND UNCONTROLLED AIRPORTS

There are two types of airport environments—controlled and uncontrolled. A **controlled airport** has an operating control tower and is sometimes referred to as a towered airport. Because all aircraft in the vicinity, as well as those on the ground, must follow instructions issued by **air traffic control (ATC)**, a two-way radio is required to operate in the controlled airport environment. At an **uncontrolled airport**, or nontowered airport, ATC does not control VFR traffic. Although you are not required to have a two-way radio, most pilots use radios to transmit their intentions to other pilots. You also are responsible for determining the active runway and correctly entering and exiting the traffic pattern. [Figure 4-14]

Figure 4-14. Air traffic controllers direct operations at controlled airports from the tower. At uncontrolled airports, you are responsible for determining the active or favored runway and following local procedures.

RUNWAY LAYOUT

Because airplanes are directly affected by wind during takeoffs and landings, **runways** are aligned, if possible, to minimize unfavorable winds. If there is a single runway, it is aligned with the prevailing wind and any additional runways are aligned with other common wind directions.

The numbers on runways correspond to the magnetic direction of the runway, rounded off to the nearest 10 degrees, with the last zero omitted. A runway with a magnetic heading of 268 degrees is rounded off to 270 degrees and, with the zero dropped, becomes Runway 27. A runway with a magnetic heading of 88 degrees becomes Runway 9. The number at the end of the runway corresponds to the direction that you are heading when taking off or landing on that runway. So, the numbers at each end of a runway are different because the runway designators are 180 degrees apart. [Figure 4-15]

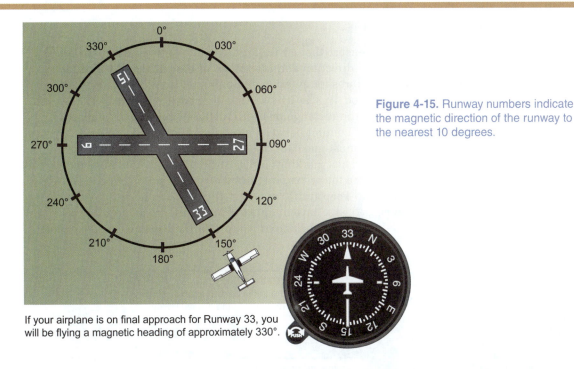

Figure 4-15. Runway numbers indicate the magnetic direction of the runway to the nearest 10 degrees.

If your airplane is on final approach for Runway 33, you will be flying a magnetic heading of approximately 330°.

At some airports, there may be two or three parallel runways with the same runway number. If there are two parallel runways, one is labeled the left runway and the other is the right; for example, 36L and 36R. If there is a third parallel runway, the one in the middle is the center runway and the respective runways are marked 36L, 36C, and 36R. [Figure 4-16]

Figure 4-16. At large airports with heavy air traffic, it is common to have parallel runways. When assigning runways for takeoffs and landings, air traffic controllers will refer to these runways as *"one-seven-left,"* and *"one-seven-right."*

SECTION B ■ Airports

DISCOVERY

Early Airports

Today, pilots take airports for granted. If you want to fly to Chicago for the weekend, you will find many airports available to you day or night, in and around the city. It wasn't always so simple. For example, the Wright Brothers flew for miles in some of their early aircraft but never went anywhere except in circles over crowds of onlookers. The first airports were

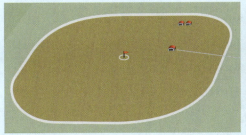

circular fields where pilots could take off or land in any direction. A white gravel circle with a windsock marked the middle of the airfield and distinguished it from fields used for livestock. The name of the nearest town was often painted atop a nearby building to help orient pilots.

TRAFFIC PATTERN

Traffic patterns are established to ensure that air traffic flows into and out of an airport in an orderly manner. Although the direction and placement of the pattern, the altitude at which it is flown, and the procedures for entering and exiting it may vary, a standard rectangular pattern with five named legs is used at most airports. [Figure 4-17] At uncontrolled airports, adhering to the rectangular traffic pattern procedures increases safety by reducing the possibility of conflict between aircraft. As specified in FAR Part 91, when approaching to land, you must make all traffic pattern turns to the left unless the airport displays light signals or visual markings indicating that turns should be made to the right. At controlled airports, the tower may instruct you to fly a pattern that differs from a standard pattern in order to keep traffic moving efficiently. In addition, some terminal areas, such as Anchorage, Alaska, have unique traffic pattern procedures that are covered in FAR Part 93, *Special Air Traffic Rules and Airport Traffic Patterns*.

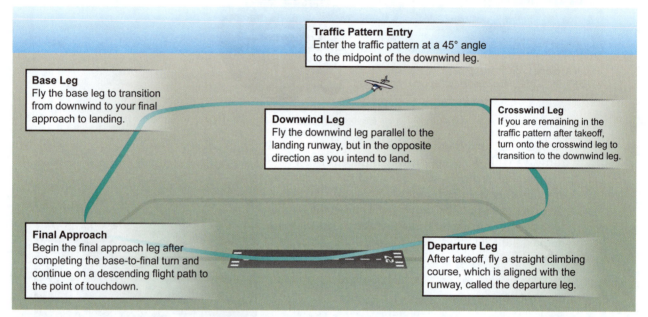

Traffic Pattern Entry
Enter the traffic pattern at a 45° angle to the midpoint of the downwind leg.

Base Leg
Fly the base leg to transition from downwind to your final approach to landing.

Downwind Leg
Fly the downwind leg parallel to the landing runway, but in the opposite direction as you intend to land.

Crosswind Leg
If you are remaining in the traffic pattern after takeoff, turn onto the crosswind leg to transition to the downwind leg.

Final Approach
Begin the final approach leg after completing the base-to-final turn and continue on a descending flight path to the point of touchdown.

Departure Leg
After takeoff, fly a straight climbing course, which is aligned with the runway, called the departure leg.

Figure 4-17. The standard traffic pattern has five named legs; downwind, base, final, departure, and crosswind.

AirVenture Oshkosh Procedures

Every year, more than 10,000 aircraft fly into Wittman Airport in Oshkosh, Wisconsin, for the Experimental Aircraft Association (EAA) AirVenture event, making Wittman the world's busiest airport during that time. To enhance safety and minimize air traffic delays, special flight procedures apply to both radio-equipped and no-radio VFR traffic, IFR traffic, and warbirds. If you have a radio and are flying under VFR, you generally head for the town of Ripon, 15 miles southwest of Oshkosh, fly northeast up the railroad tracks toward Oshkosh, wait for ATC to call out your aircraft by type and color, and follow the airplane in front of you. At Wittman airport, the tower may clear you to land on a specific portion of your assigned runway where marshalls quickly usher your airplane into the grass to make room for the next arrival.

Courtesy Paul Neuman

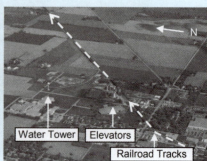

Water Tower   Elevators   Railroad Tracks   N

WIND DIRECTION INDICATORS

In most cases, you want to take off and land into the wind. At a controlled airport, the tower or ATIS normally gives you the current airport information, including surface winds and the active runway or runways. The tower controller assigns you a runway for takeoff or landing. At an uncontrolled airport, you decide which runway to use. Some uncontrolled airports have radio operators who can tell you the wind direction and speed, and advise you of the active or favored runway. And, most airports have automated weather systems that transmit wind information on a frequency shown on the chart. Airports normally have a designated calm-wind runway recommended whenever the wind is 5 knots or less. However, you must still watch carefully for other traffic that could be using other runways. When approaching an unfamiliar uncontrolled airport, it is recommended that you overfly the airport at 500 to 1,000 feet above the traffic pattern altitude to observe the flow of traffic and locate the wind direction indicator.

Of the two types of airport wind direction indicators, the most common is the **windsock**. Used at both controlled and uncontrolled airports, the windsock provides you with the present wind conditions near the touchdown zone of the runway. The stronger the wind, the straighter the extension of the windsock. Gusty conditions are indicated by movement of the windsock.

A **wind tee** also provides wind direction, but does not show intensity or gusty conditions. The tail of the tee aligns itself like a weather vane into the wind, so you need to take off or land on the runway that most closely parallels the direction of the tee. The wind tee is not as common as the windsock and in some cases, a windsock and tee may be at the same location. If so, the tee may be manually aligned to show which runway is active.

A **tetrahedron** is also a landing direction indicator, usually located near a wind direction indicator. The tetrahedron may swing around with the small end pointing into the wind, or it can be manually positioned to show the landing direction. Be careful not to use it as the sole wind direction indicator, but use it in conjunction with a windsock. [Figure 4-18]

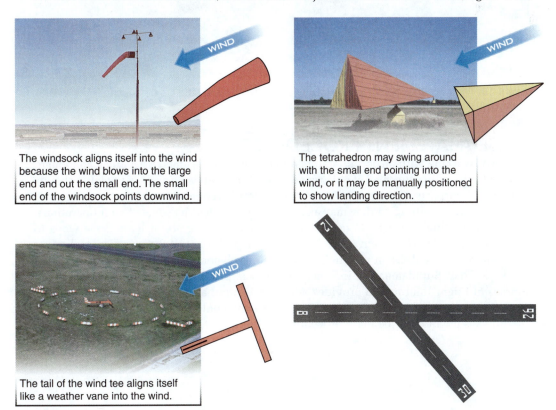

The windsock aligns itself into the wind because the wind blows into the large end and out the small end. The small end of the windsock points downwind.

The tetrahedron may swing around with the small end pointing into the wind, or it may be manually positioned to show landing direction.

The tail of the wind tee aligns itself like a weather vane into the wind.

Figure 4-18. When possible, you should select the runway that is the closest to paralleling the direction of the wind. In this case, Runway 8 would be the logical choice for takeoff and landing into the wind.

SECTION B ■ Airports

SEGMENTED CIRCLE

A wind or landing direction indicator usually is placed in the middle of a **segmented circle** at a central location on the airport. The segmented circle has two purposes. First, it helps to identify the location of the wind direction indicator. Second, extensions on the segmented circle show whether the traffic pattern is other than left-hand traffic.

To indicate the direction you should turn in the traffic pattern for a given runway, the segmented circle uses L-shaped extensions that diagram the base-to-final turn to each runway. For example, in Figure 4-19, a left hand traffic pattern is shown for approach to Runway 6 and a right-hand pattern is shown for approach to Runway 24. If right-hand turns are required for the approach, they normally are used for departure as well. If there are no traffic pattern indicators, the runways all have left-hand traffic patterns. At night, overhead lights normally illuminate both the wind direction indicator and the segmented circle.

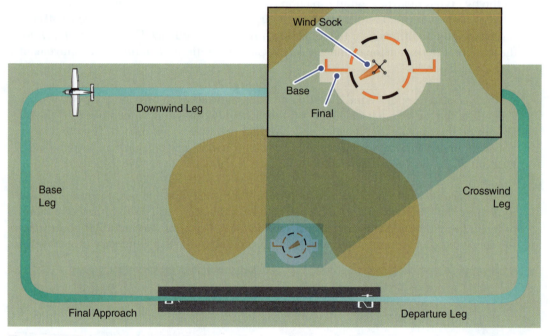

Figure 4-19. The wind sock points to Runway 6, the active runway. The L extension to Runway 6 indicates left-hand traffic.

NOISE ABATEMENT PROCEDURES

The FAA, working with airport operators and community leaders, uses **noise abatement procedures** to reduce the level of noise generated by aircraft departing over neighborhoods that are near airports. The airport authority can simply request that you use a designated runway, wind permitting, or they may also require you to restrict some of your operations, such as practicing landings, to certain time periods. There are at least three ways to determine the noise abatement procedures at an airport. First, if there is a control tower on the field, they will assign you the preferred noise abatement runway. Second, you can check the Chart Supplement for information on local procedures. Chapter 5, Section C—Sources of Flight Information provides additional information about the features of the Chart Supplement. Third, FBOs often post noise abatement information and signs next to the runways might describe these procedures. [Figure 4-20]

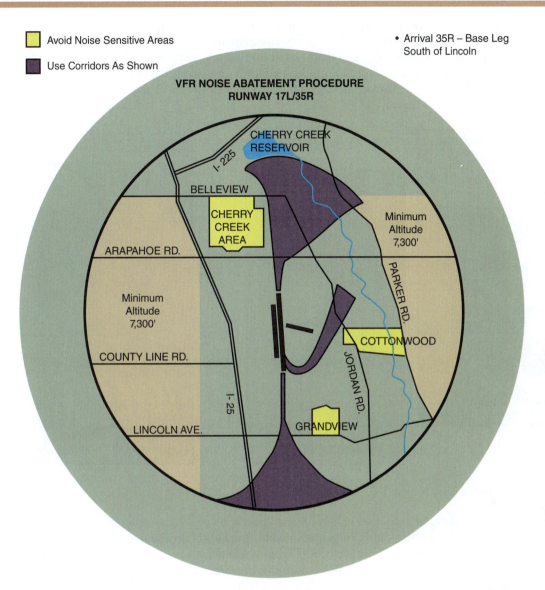

Figure 4-20. You can find diagrams that outline recommended noise abatement procedures at many airport FBOs.

AIRPORT VISUAL AIDS

When you begin your flight training, the airport environment can seem confusing. However, just as you learn how to interpret traffic signals, road signs, and highway markings, you will soon become familiar with the visual aids at an airport that help you maintain orientation and keep traffic flowing efficiently.

RUNWAY MARKINGS

Runway markings vary between runways used solely for VFR operations and those that are also used for IFR operations. A visual runway usually is marked with only the runway number and a dashed white centerline. When flying instrument approaches, pilots can use the additional markings on IFR runways as references for landing. Instrument approach procedures enable pilots to navigate to the runway using only the flight instruments.

Instrument approaches that use an electronic glide path for guidance to the landing runway, such as the instrument landing system (ILS), are called precision approaches. Nonprecision approaches do not incorporate an electronic glide path, and the corresponding runway markings vary accordingly.

When a visual runway is used with a nonprecision instrument approach, threshold and aiming point markings are added. A precision instrument runway also includes touchdown zone markings. Occasionally, you might see threshold or aiming point markings on visual runways. Runway threshold markings help identify the beginning of the runway that is available for landing. Runway aiming point markings serve as a visual aiming point for a landing aircraft. [Figure 4-21]

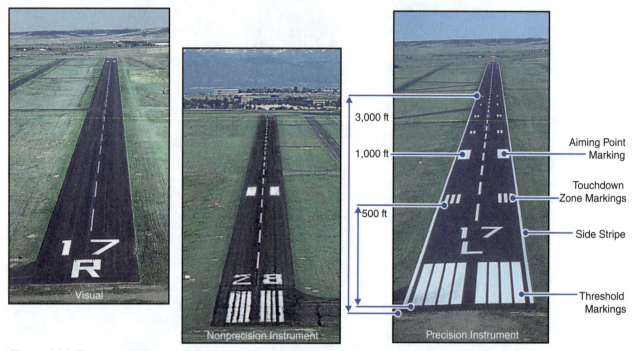

Figure 4-21. The common types of runway markings for visual, precision, and nonprecision runways are shown here.

It is not uncommon to fly into an airport that has a runway with another type of marking called a **displaced threshold**. Thresholds can be displaced because of obstructions, such as trees, powerlines, or buildings near the end of the runway that would prevent you from making a normal descent and landing to the beginning portion of the pavement. [Figure 4-22]

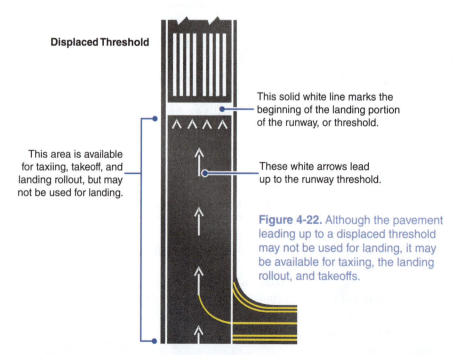

Displaced Threshold

This solid white line marks the beginning of the landing portion of the runway, or threshold.

This area is available for taxiing, takeoff, and landing rollout, but may not be used for landing.

These white arrows lead up to the runway threshold.

Figure 4-22. Although the pavement leading up to a displaced threshold may not be used for landing, it may be available for taxiing, the landing rollout, and takeoffs.

A **blast pad/stopway area**, sometimes referred to as an overrun, is different from the area preceding a displaced threshold because it cannot be used for landing, takeoff, or taxiing. The pavement strength can support the weight of an airplane in an emergency, but might not be strong enough for continuous operations. The blast pad is an area where propeller or jet blast can dissipate without creating a hazard to others. The stopway area is paved so that in the event of a rejected takeoff, an airplane can decelerate and come to a stop. [Figure 4-23]

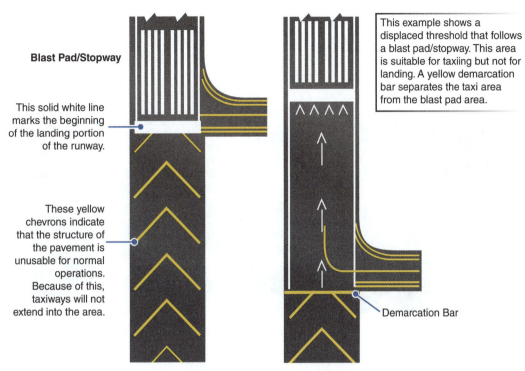

Blast Pad/Stopway

This solid white line marks the beginning of the landing portion of the runway.

These yellow chevrons indicate that the structure of the pavement is unusable for normal operations. Because of this, taxiways will not extend into the area.

This example shows a displaced threshold that follows a blast pad/stopway. This area is suitable for taxiing but not for landing. A yellow demarcation bar separates the taxi area from the blast pad area.

Demarcation Bar

Figure 4-23. Due to its reduced weight-bearing capability, a blast pad/stopway area is not suitable for taxiing, takeoff, or landing.

Sometimes construction, maintenance, or other activities require the threshold to be relocated temporarily. Methods of identifying the relocated threshold vary but one common practice is to use a ten-foot wide white threshold bar across the width of the runway. Yellow Xs indicate that an entire runway or taxiway is closed. [Figure 4-24]

Figure 4-24. Although the surface may appear to be usable, operations on a closed runway cannot be conducted safely.

A Notice to Air Missions (NOTAM) may be issued to inform pilots of a runway closure. NOTAMs contain time-critical information that is of either a temporary nature or is not known far enough in advance to permit publication on aeronautical charts or other operational publications. In addition to runway closures or construction, NOTAMs can include changes in the status of navigational aids or instrument approach facilities, radar service availability, and other information essential to planned enroute, terminal, or landing operations.

SECTION B ■ Airports

TAXIWAY MARKINGS

The links between the airport parking areas and the runways are the **taxiways**. They are easily identified by a continuous yellow centerline stripe. At some airports, taxiway edge markings define the edge of the taxiway and separate the taxiway from pavement that is not intended for aircraft use. Runway holding position markings, or **hold lines**, which are located wherever the taxiway intersects a runway, keep aircraft clear of the runway in use.

At an uncontrolled airport, you should stop at a hold line and check for traffic, crossing the line only after ensuring that no one is on an approach to land. At a controlled airport, the controller may ask you to hold short of the runway for landing traffic. In this case, you should stop before the hold line and proceed only after the controller clears you onto the runway and you have checked for traffic. [Figure 4-25]

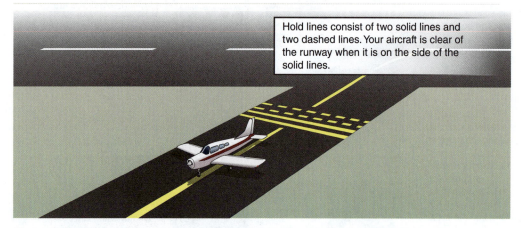

Hold lines consist of two solid lines and two dashed lines. Your aircraft is clear of the runway when it is on the side of the solid lines.

Figure 4-25. When exiting the runway, do not stop until you have cleared the hold line.

At airports equipped with an instrument landing system (ILS), it is possible for aircraft near the runway to interfere with the ILS signal. If this is the case, the hold line may be placed farther from the runway to prevent interference, or a second line, an ILS hold line, may be added. The nonmovement area boundary markings delineate the ramp area, where contact with ATC is not normally required. [Figure 4-26]

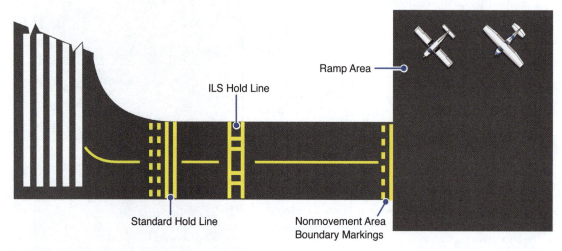

Ramp Area

ILS Hold Line

Standard Hold Line

Nonmovement Area
Boundary Markings

Figure 4-26. When ILS approaches are in progress, the controller may request that you *"hold short of the ILS critical area,"* at the ILS hold line.

RAMP AREA

The area where aircraft are parked and tied down is called the apron, or **ramp area**. The airport terminal and maintenance facilities are often located near the ramp area. You should be alert for fuel trucks driving on the ramp or in the process of refueling aircraft. Vehicle roadway markings are used to define pathways for vehicles operating on, or crossing areas that also are intended for aircraft. In addition, you should be familiar with the standard **hand signals** used by ramp personnel to direct you during ground operations. [Figure 4-27]

Figure 4-27. At many FBOs, ramp personnel will direct you to parking.

AIRPORT SIGNS

Major airports can have complex taxi routes, multiple runways, and widely dispersed parking areas. In addition, vehicular traffic in some areas may be quite heavy. As a result, most airfield signs are standardized to make it easy for you to identify taxi routes, mandatory holding positions, and boundaries for critical areas. [Figure 4-28] Another benefit, if you fly outside the United States, is that the U.S. standards are very similar to ICAO specifications. The **International Civil Aviation Organization (ICAO)** is a specialized agency of the United Nations whose objective is to develop standard principles and techniques of international air navigation and to promote development of civil aviation. Specifications for airport signs include size, height, location, and illumination requirements. [Figure 4-29] Sometimes the installation of a sign is not practical so a surface-painted sign may be used. Surface painted signs can include directional guidance or location information. For example, the runway number might be painted on the taxiway pavement near the taxiway hold line.

Direction Signs indicate directions of taxiways leading out of an intersection. They have black inscriptions on a yellow background and always contain arrows which show the approximate direction of turn.

Mandatory Instruction Signs denote an entrance to a runway, a critical area, or an area prohibited to aircraft. These signs are red with white letters or numbers. An example of a mandatory instruction sign is a runway holding position sign which is located at the holding position on taxiways that intersect a runway or on runways that intersect other runways.

Location Signs identify either the taxiway or runway where your aircraft is located. These signs are black with yellow inscriptions and a yellow border. Location signs also identify the runway boundary or ILS critical area for aircraft exiting the runway.

Runway Distance Remaining Signs provide distance remaining information to pilots during takeoff and landing operations. The signs are located along the sides of the runway, and the inscription consists of a white numeral on a black background. The signs indicate the distance remaining in thousands of feet. Runway distance remaining signs are recommended for runways used by turbojet aircraft.

Information Signs advise you of such things as areas that cannot be seen from the control tower, applicable radio frequencies, and noise abatement procedures. These signs use yellow backgrounds with black inscriptions.

Destination Signs indicate the general direction to a location on the airport, such as civil aviation areas, military areas, international areas, or FBOs. They have black inscriptions on a yellow background and always contain an arrow.

Figure 4-28. There are six basic types of airport signs — direction, mandatory, location, runway distance remaining, information, and destination.

SECTION B ■ Airports

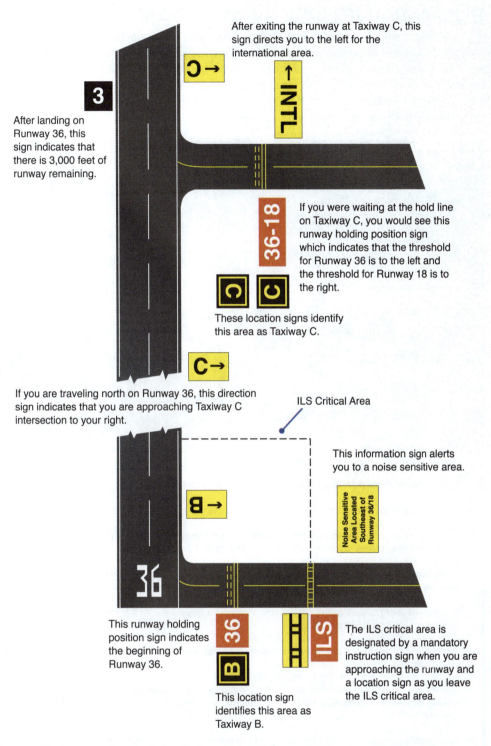

After exiting the runway at Taxiway C, this sign directs you to the left for the international area.

After landing on Runway 36, this sign indicates that there is 3,000 feet of runway remaining.

If you were waiting at the hold line on Taxiway C, you would see this runway holding position sign which indicates that the threshold for Runway 36 is to the left and the threshold for Runway 18 is to the right.

These location signs identify this area as Taxiway C.

If you are traveling north on Runway 36, this direction sign indicates that you are approaching Taxiway C intersection to your right.

ILS Critical Area

This information sign alerts you to a noise sensitive area.

Noise Sensitive Area Located Southeast of Runway 36/18

This runway holding position sign indicates the beginning of Runway 36.

This location sign identifies this area as Taxiway B.

The ILS critical area is designated by a mandatory instruction sign when you are approaching the runway and a location sign as you leave the ILS critical area.

Figure 4-29. As you gain experience in airport operations, you will become familiar with the standard airport signs.

SECTION B ■ Airports

RUNWAY INCURSION AVOIDANCE

A **runway incursion** is any occurrence at an aerodrome involving the incorrect presence of an aircraft, vehicle or person on the protected area of a surface designated for the landing and takeoff of aircraft. Runway incursions can be very hazardous, and are primarily caused by errors associated with clearances, communication, airport surface movement, and positional awareness. The FAA recommends procedures that you can follow and precautions that you can take to avoid a runway incursion. [Figure 4-30]

Take these actions to help prevent a runway incursion:

• During your preflight planning, study the airport layout by reviewing the airport diagram and taxi routes.

• Complete as many checklist items as possible before taxi or while holding short.

• Strive for clear and unambiguous pilot-controller communication. Read back (in full) all clearances involving active runway crossing, hold short, or line up and wait instructions.

• While taxiing, know your precise location and concentrate on your primary responsibilities. Do not become absorbed in other tasks, or conversation, while your airplane is moving.

• If unsure of your position on the airport, stop and ask for assistance. At a controlled airport, you can request progressive taxi instructions.

• When possible, while in a runup area or waiting for a clearance, position your airplane so you can see landing aircraft.

• Monitor the appropriate radio frequencies for information regarding other aircraft cleared onto your runway for takeoff or landing. Be alert for aircraft that may be on other frequencies or without radio communication.

• After landing, stay on the tower frequency until instructed to change frequencies.

• To help others see your airplane during periods of reduced visibility or at night, use your exterior taxi/landing lights when practical.

• Report deteriorating or confusing airport markings, signs, and lighting to the airport operator or FAA officials. Also report confusing or erroneous airport diagrams and instructions.

• Make sure you understand the required procedures if you fly into or out of an airport where LAHSO is in effect.

Figure 4-30. Following these recommendations can minimize your risk of a runway incursion.

Runway Incursion Fatal Accident

NTSB Narrative: *During the takeoff roll on runway 30R, the MD-82, N954U, collided with the Cessna 441, N441KM, which was positioned on the runway waiting for takeoff clearance. The pilot of the Cessna acted on an apparently preconceived idea that he would use his arrival runway, Runway 30R, for departure. After receiving taxi clearance to back-taxi into position and hold on runway 31, the pilot taxied into a position at an intersection of Runway 30R, which was the assigned departure runway for the MD-82. . . Air traffic control personnel were not able to maintain visual contact with the Cessna after it taxied from the well-lighted ramp area into the runway/taxiway environment of the northeast portion of the airport.*

The NTSB narrative suggested that the pilot of the light twin had incorrectly anticipated taking off on Runway 30R, and taxied to that runway even though that was not the clearance received. A thorough understanding of airport markings, signs, and lighting, as well as proper radio procedures are essential, especially when it is dark and two runways intersect each other and have similar headings. Look at the airport diagram before you taxi and if you are unfamiliar with the taxiway and runway layout, request progressive taxi instructions from ATC.

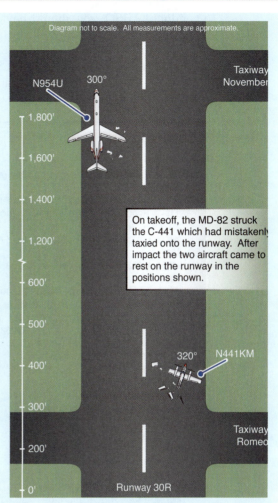

On takeoff, the MD-82 struck the C-441 which had mistakenly taxied onto the runway. After impact the two aircraft came to rest on the runway in the positions shown.

LAND AND HOLD SHORT OPERATIONS

During **land and hold short operations (LAHSO)**, an aircraft is cleared to land and stop on the runway, holding short of an intersecting runway, intersecting taxiway, or some other designated point on the runway. Used only at selected airports, LAHSO is an air traffic control tool that is used to increase airport capacity and maintain system efficiency. [Figure 4-31]

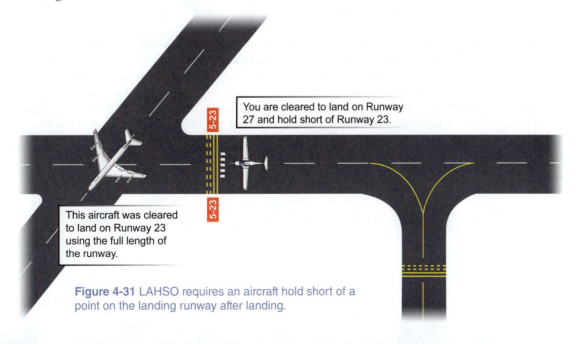

You are cleared to land on Runway 27 and hold short of Runway 23.

This aircraft was cleared to land on Runway 23 using the full length of the runway.

Figure 4-31 LAHSO requires an aircraft hold short of a point on the landing runway after landing.

SECTION B ■ **Airports**

To conduct LAHSO, you should have at least a private pilot certificate and you must understand your responsibilities. During preflight planning, you should become familiar with all available information concerning LAHSO at the destination airport. The Airport/ Facility Directory section of the Chart Supplement indicates the presence of LAHSO and includes the **available landing distance (ALD)** and runway slope, if it applies, for each LAHSO runway. [Figure 4-32]

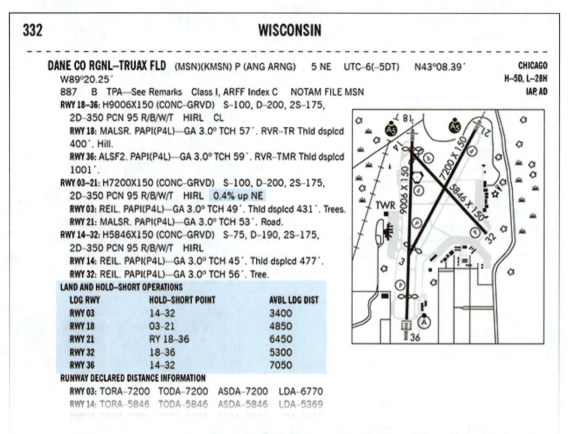

Figure 4-32. Make sure you have the available landing distance (ALD) and slope for any runway on which you might receive a LAHSO clearance. Controllers may provide the ALD on request.

Determine which runway LAHSO combinations are acceptable for your airplane's landing performance and for your personal minimums. During flight, have the published ALD and runway slope readily available and ensure that you can safely land and stop within the ALD with the existing conditions upon arrival.

When ATIS announces that LAHSO is in effect, be prepared for a LAHSO clearance from the tower—*"Diamond 77 Xray Foxtrot, cleared to land Runway 27, hold short of Runway 23 for landing traffic, Learjet."* As pilot in command, you should decline the LAHSO clearance if you consider it unsafe—*"Diamond 77 Xray Foxtrot, unable to hold short of Runway 23,"* and expect a revised or delayed landing clearance. If you need the full length of the runway, or a different runway, it is preferable that you inform the tower even before they issue a LAHSO clearance.

If you accept a LAHSO clearance, read back the full clearance including the words, "hold short of runway, taxiway, or point"—*"Diamond 77 Xray Foxtrot, cleared to land Runway 27 to hold short of Runway 23."* After you accept a LAHSO clearance, you must adhere to it, just like any other ATC clearance. Visual aids on the LAHSO runway help you identify the hold-short point. [Figure 4-33] Acceptance of a LAHSO clearance does not preclude a rejected landing. If you must go around, tell ATC immediately and maintain safe separation from other aircraft or vehicles.

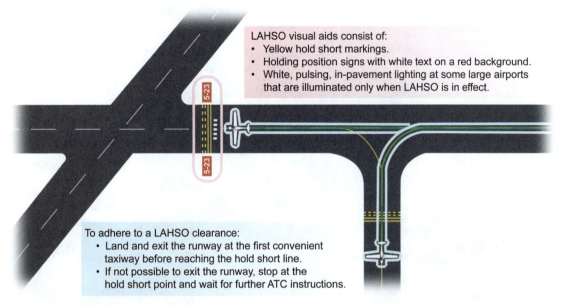

LAHSO visual aids consist of:
• Yellow hold short markings.
• Holding position signs with white text on a red background.
• White, pulsing, in-pavement lighting at some large airports that are illuminated only when LAHSO is in effect.

To adhere to a LAHSO clearance:
• Land and exit the runway at the first convenient taxiway before reaching the hold short line.
• If not possible to exit the runway, stop at the hold short point and wait for further ATC instructions.

Figure 4-33. After you accept a LAHSO clearance you must adhere to the clearance by exiting the runway or stopping before the assigned hold-short point on the landing runway.

AIRPORT LIGHTING

Your flying experiences will soon take you from flying only in the daytime to the new challenge of night flying. You will notice that airport lighting is similar from one airport to the next. To maintain continuity, airports that are lighted for nighttime operations use FAA-approved lighting systems and colors.

AIRPORT BEACON

At night, **airport beacons** are used to guide pilots to lighted airports. Airport beacons might be of the older rotating type, or the newer flashing variety, which produces the same effect. These airport (and heliport) beacons are most effective from one to ten degrees above the horizon; however, you normally can see them at altitudes well above the ten degree angle. If you maintain sufficient altitude, beacons can be seen at great distances in good visibility conditions. [Figure 4-34]

Generally, you will find that an airport's beacon is on from dusk until dawn. The beacon usually is not operating during the day unless the ceiling is less than 1,000 feet and/or the ground visibility is less than 3 statute miles, the normal VFR weather minimums. You should not rely solely on the operation of the airport beacon to indicate if weather conditions are below VFR minimums. These minimums will be discussed in greater detail in Section D of this chapter.

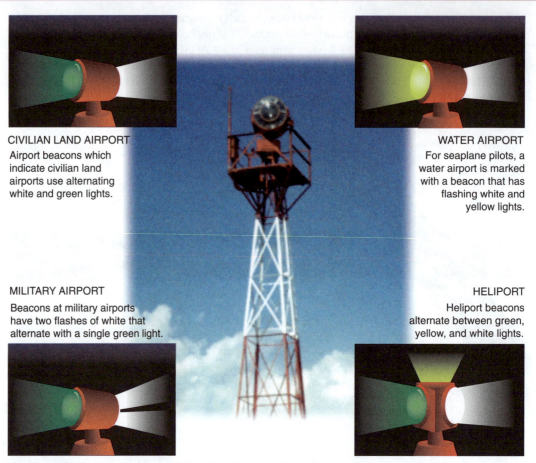

CIVILIAN LAND AIRPORT
Airport beacons which indicate civilian land airports use alternating white and green lights.

WATER AIRPORT
For seaplane pilots, a water airport is marked with a beacon that has flashing white and yellow lights.

MILITARY AIRPORT
Beacons at military airports have two flashes of white that alternate with a single green light.

HELIPORT
Heliport beacons alternate between green, yellow, and white lights.

Figure 4-34. The combination of light colors from an airport beacon indicates the type of airport. As a routine measure, you are not permitted to operate at military airports.

VISUAL GLIDESLOPE INDICATORS

Visual glideslope indicators are light systems that indicate your position in relation to the desired glide path to the runway. The indicator lights are located on the side of a basic or instrument runway and can be used for day or night approaches. One of the most frequently used installations is the **visual approach slope indicator (VASI)**. The two-bar VASI shows whether or not you are on a glide path that will take you safely to the touchdown zone of the runway. The lights are either white or red, depending on the angle of your glide path, and can be visible up to 20 miles at night. [Figure 4-35]

Above Glide Path

If both light bars are white, you are too high.

Below Glide Path

If you see red over red, you are below the glide path.

On Glide Path

If the far bar is red and the near bar is white, you are on the glide path. The memory aid "red over white

Figure 4-35. The VASI consists of light bars that change color between white and red depending on whether your approach angle is above or below the glide path.

VASI configurations vary and may have either 2 or 3 bars. Two-bar systems have near and far bars and may include 2, 4, or 12 light units. The VASI glide path provides safe obstruction clearance within 10 degrees of the extended runway centerline out to 4 nautical miles from the threshold. You should not begin a descent using VASI until your aircraft is aligned with the runway. When landing at a controlled airport that has a VASI, regulations require you to remain on or above the glide path until a lower altitude is necessary for a safe landing.

Larger airports might have a three-bar VASI system that incorporates two different glide paths. The lower glide path normally is set at three degrees, while the higher one usually is one-fourth of a degree above it. The higher glide path is used by certain transport category aircraft with high cockpits. This ensures that these aircraft will have sufficient altitude when crossing the threshold. If you encounter a three-bar VASI system, use the two lower bars as if it were a standard two-bar VASI.

Some airports have a **pulsating visual approach slope indicator (PVASI)**, which projects a two-color visual approach path into the final approach area. A pulsating red light indicates below glide path; above glide path is usually pulsating white; and the on-glide path indication is a steady white light. The useful range is about 4 miles during the day and up to 10 miles at night.

Another system is the **precision approach path indicator (PAPI)**. It has two or four lights installed in a single row instead of far and near bars. [Figure 4-36]

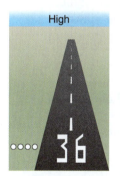    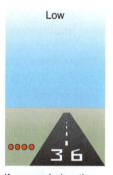

High	Slightly High	On Glide Path	Slightly Low	Low
If all of the PAPI system lights are white you are too high.	If only the light on the far right is red and the other three are white, you are slightly high.	When you are on the glide path, the two lights on the left are white and the two lights on the right are red.	If you are slightly low, only the light on the far left is white.	If you are below the glide path, all four of the lights are red.

Figure 4-36. The PAPI is normally on the left side of the runway and can be seen up to 5 miles during the day and 20 miles at night.

APPROACH LIGHT SYSTEMS

Some airports have **approach lighting systems (ALS)** to help instrument pilots transition to visual references at the completion of an instrument approach. These light systems can begin as far away as 3,000 feet along the extended runway centerline, and normally include a combination of steady and flashing lights. The most complex systems are for precision instrument runways and usually have sequenced flashing lights that look like a ball of light traveling toward the runway at high speed. For nonprecision instrument runways, the approach lighting is simpler and, for VFR runways, the system may consist only of visual glideslope indicators. [Figure 4-37]

Figure 4-37. Approach light systems can aid you in locating the runway at night.

RUNWAY EDGE LIGHTS

Runway edge lights consist of a single row of white lights bordering each side of the runway and lights identifying the runway threshold. Runway edge lights can be classified according to three intensity levels. High intensity runway lights (HIRLs) are the brightest runway lights available. Medium intensity runway lights (MIRLs) and low intensity runway lights (LIRLs) are, as their names indicate, dimmer in intensity. At some airports, you will be able to adjust the intensity of the runway lights from your cockpit by using your radio transmitter. At others, the lights are preset or are adjusted by air traffic controllers.

Some runway edge lights incorporate yellow runway remaining lights on the last half of the runway (or last 2,000 feet of runway, whichever distance is less) to inform you of the amount of runway left. These lights are two-sided, so they appear white when viewed from the opposite end of the runway.

At night, there are three ways to determine where the runway begins. If the runway has a displaced threshold, there is a set of green lights on each side of the white threshold line to indicate the beginning of the landing portion of the runway. If the threshold is not displaced, the beginning of the runway pavement has a row of green lights across it. These lights are two-sided. If you were taking off or landing on the opposite end, they would appear red to mark the end of the usable portion of the runway.

Sometimes high intensity white strobe lights are placed on each side of the runway to mark the threshold. These are called **runway end identifier lights (REILs)** and can be used in conjunction with the green threshold lights.

IN-RUNWAY LIGHTING

Some precision approach runways have flush-mounted centerline, touchdown zone, and taxiway turnoff lighting. Viewed from the threshold, the runway centerline lighting system (RCLS) is white until the last 3,000 feet of the runway. From the 3,000-foot point to the 1,000-foot point, alternating red and white lights appear, with the last 1,000 feet of lights changing to red only. This system helps instrument pilots determine the amount of runway remaining in very low visibility situations.

Touchdown zone lighting (TDZL) consists of two rows of transverse light bars on either side of the runway centerline starting at 100 feet from the threshold and extending 3,000 feet or to the midpoint of the runway, whichever is less. Taxiway lead-off lights are alternating green and yellow lights that define the curved path of aircraft travel from the runway centerline to a point on the taxiway (normally the runway holding position or ILS critical area boundary).

TAXIWAY LIGHTING

As you taxi off the active runway, blue lights, lining both edges of the taxiway, guide you from the runway to the ramp area. Because they can be seen from any direction, they are said to be omnidirectional lights. At some airports, green taxiway centerline lights also may be installed. These lights are located along the taxiway centerline in both straight and curved portions of the taxiway. They also may be located along designated taxiing paths in portions of runways and ramp areas. [Figure 4-38]

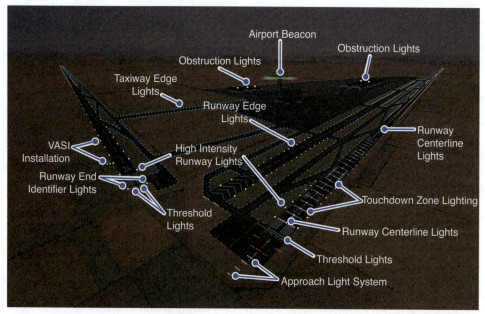

Figure 4-38. This pictorial summary shows the various types of airport marking and lighting typically found at large, controlled airports.

PILOT-CONTROLLED LIGHTING

Pilot-controlled lighting is the term used to describe systems that you can activate by keying the aircraft's microphone, or mic, on a specified radio frequency. For practical and economic reasons, the approach, runway, and taxiway lights at some unattended airports might be on a timer that turns off the lights 15 minutes after they have been activated. Keep in mind that other types of airport lighting can be pilot controlled, not just approach and runway lighting. For example, VASI and REIL lights are pilot controlled at some locations.

To activate three-step pilot-controlled lighting, key your mic seven times on the specified frequency to turn all the lights on at maximum intensity. If conditions dictate a lower intensity, key your mic five times for medium-intensity lighting and three times for the lowest intensity. For each adjustment, you must key the mic the required number of times within a period of five seconds. Remember though, using the lower intensity lighting on some installations may turn the REILs completely off. The Airport/Facility Directory section of the Chart Supplement contains a description of the type of pilot-controlled lighting available at individual airports.

OBSTRUCTION LIGHTING

Obstruction lighting is used both on and off the airport, during the day and at night. The purpose of obstruction lighting is to give you advance warning of prominent structures such as towers, buildings and, sometimes, even powerlines. Bright red and high intensity white lights typically are used and flashing lights may be employed. Remember, guy-wires might extend from the top of a tower to the ground, so be sure that you are well clear of the obstruction.

AIRPORT SECURITY

As a pilot, you are responsible for following best practices to keep your airport and aircraft secure. Always comply with security procedures that limit access to airport ramps and prevent unauthorized use of aircraft. [Figure 4-39]

Airport Gate Security
Comply with airport gate security procedures:
- Keep gate codes and lock combinations confidential.
- Close and lock access gates.
- Ensure that other vehicles do not follow you through gates—stop and wait until the gate closes before proceeding.
- Discuss airport security-related issues with the FBO or airport manager.

Securing Aircraft
Secure aircraft from unauthorized use:
- Do not leave keys in the aircraft.
- Lock aircraft doors.
- Use auxiliary locks.
- Limit access to aircraft keys after normal business hours.

Figure 4-39. Pilots must diligently protect airport ramp areas and aircraft from unauthorized users.

You also need to be able to recognize suspicious activity and know how to alert authorities if necessary. In addition to obvious criminal activities, such as someone breaking into an aircraft, other activities can alert you that a crime is likely to occur. [Figure 4-40]

Authorities take reports 24 hours a day at the phone number, **1-866-GA-SECURE**, but they do not dispatch law enforcement. Call 911 if you witness criminal activity—do not confront the person—and then call 1-866-GA-Secure, as well as your airport, FBO, or flight school manager.

Recognize Suspicious Activity

Be alert for individuals who:

- Suspiciously approach multiple aircraft.
- Appear to be breaking into an aircraft.
- Try to avoid others.
- Loiter without a clear purpose.
- Videotape or take photos of aircraft and hangars.
- Load unusual cargo—typically not suitable for GA operations.

Report Suspicious Activity

Call:

- 1-866-GA-Secure (1-866-427-3287) to report suspicious activity.
- 911 and then 1-866-GA-Secure (1-866-427-3287) to report criminal activity—an immediate threat to persons or property.
- Your airport, FBO, or flight school manager to explain the situation.

Figure 4-40. You can help keep the airport secure by recognizing and reporting suspicious and criminal activities.

DISCOVERY

Emirates Flight Training Academy

Emirates Flight Training Academy (EFTA) is one of the most advanced flight training academies in the world. Located in the critical global aviation hub of Dubai, EFTA has the unique feature of having its own exclusive airport. EFTA boasts a private control tower that manages the training airplanes taking off and landing on the 1,800-meter runway. The airport is capable of accommodating over 400 flights a day. EFTA's fleet of training airplanes include the Cirrus SR22 and the twin-jet Embraer Phenom 100E, which are serviced in the maintenance facility on the field.

Ground instruction takes place in classrooms with state-of the art interactive technology and students live in apartments at the academy. EFTA cadets begin their training with no flight experience and complete the comprehensive ground and flight training program prepared for careers as airline pilots.

Courtesy of Emirates Flight Training Academy

SUMMARY CHECKLIST

✓ A two-way radio is required to operate in the controlled airport environment because all aircraft in the vicinity, as well as those on the ground, must follow instructions issued from the control tower.

✓ ATC does not control VFR traffic at an uncontrolled airport.

✓ The number at the approach end of a runway corresponds to your magnetic heading when taking off or landing on that runway.

✓ A standard rectangular pattern with five named legs is used at most airports to ensure that air traffic flows in an orderly manner.

✓ The most common wind direction indicator is the windsock, which is used at both controlled and uncontrolled airports. It provides you with the present wind conditions near the touchdown zone of the runway.

✓ Wind tees and tetrahedrons are landing direction indicators that can swing around to point into the wind, or that can be manually positioned to show landing direction.

✓ The segmented circle helps to identify the location of the wind direction indicator and employs L-shaped extensions that show the traffic pattern turn directions for depicted runways.

✓ Adhering to noise abatement procedures reduces the level of noise over neighborhoods that are near airports.

✓ A visual runway normally is marked only with the runway number and a dashed white centerline. When flying instrument approaches, pilots can use the additional markings on IFR runways, such as threshold markings, touchdown zone markings, and aiming point markings.

✓ Usually, a runway has a displaced threshold because of an obstruction off the end of the runway that can prevent a normal descent and landing on the beginning portion of the pavement.

✓ A blast pad/stopway can serve as an emergency overrun, but is not strong enough to be used for normal landing, takeoff, or taxiing operations. The blast pad is an area where propeller or jet blast can dissipate without creating a hazard to others.

✓ Taxiways normally have yellow centerline markings, and hold lines wherever they intersect with a runway.

✓ There are six basic types of airport signs — direction, mandatory, location, runway distance remaining, information, and destination.

✓ To conduct LAHSO, you should have at least a private pilot certificate, have the published ALDs and runway slopes available, and know which runway LAHSO combinations provide acceptable landing distances with the existing conditions upon arrival.

✓ If you believe a LAHSO clearance is unsafe, you are expected to decline it. If you accept a LAHSO clearance, you must adhere to it, exiting the runway at the first convenient taxiway before the hold-short point or stopping at the hold-short point.

✓ Acceptance of a LAHSO clearance does not preclude a go-around, but you must tell ATC immediately and maintain safe separation from other aircraft and vehicles.

✓ Airport beacons guide pilots to lighted airports at night and can indicate when weather conditions are below VFR minimums during the day.

✓ The two-bar visual approach slope indicator (VASI) shows whether or not you are on a glide path that will take you safely to the touchdown zone of the runway.

✓ A variety of lighting systems, including approach light systems, runway edge lights, runway end identifier lights (REILs), in-runway lighting, and taxiway lighting are used at airports to aid pilots in identifying the airport environment at night and in low visibility conditions.

✓ Pilot-controlled lighting is the term used to describe systems that you can activate by keying the aircraft's microphone on a specified radio frequency.

✓ You are responsible for following best practices to keep your airport and aircraft secure—complying with security procedures that limit access to airport ramps and prevent unauthorized use of aircraft.

✓ You must be able to recognize suspicious activity and know how to alert authorities. If you observe suspicious activity, call 1-866-GA-SECURE (1-866-427-3287). If you observe criminal activity, call 911 and then call 1-866-GA-SECURE.

KEY TERMS

Controlled Airport

Air Traffic Control (ATC)

Uncontrolled Airport

Runways

Traffic Patterns

Windsock

Wind Tee

Tetrahedron

Segmented Circle

Noise Abatement Procedures

Displaced Threshold

Blast Pad/Stopway Area

Taxiways

Hold Lines

Ramp Area

Hand Signals

International Civil Aviation Organization (ICAO)

Runway Incursion

Land and Hold Short Operations (LAHSO)

Available Landing Distance (ALD)

Airport Beacons

Visual Approach Slope Indicator (VASI)

Pulsating Visual Approach Slope Indicator (PVASI)

Precision Approach Path Indicator (PAPI)

Approach Lighting Systems (ALS)

Runway Edge Lights

Runway End Identifier Lights (REILs)

Pilot-Controlled Lighting

1-866-GA SECURE

QUESTIONS

1. Describe how runway numbers are determined.

2. Determine the proper runway and traffic pattern for landing.
   A. Left-hand traffic for Runway 4
   B. Left-hand traffic for Runway 36
   C. Right-hand traffic for Runway 22

3. Explain the purpose of a displaced threshold and the operating limitations associated with it.

4. What marking indicates a closed runway?

5. Which airplane is on the correct side of the hold line to be clear of the runway?

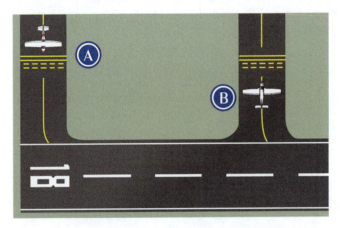

Match the signs in the illustrations to their descriptions.

6. Direction Sign

7. Location Sign

8. Mandatory Instruction Sign

9. List at least five actions you can take to help prevent a runway incursion.

Match the following airport beacon light patterns to the appropriate airport.

10. White/White/Green   A. Civilian land airport

11. White/Green   B. Military airport

12. White/Yellow   C. Water airport

13. What is expected of you in order to conduct land and hold-short operations (LAHSO)?

14. True/False. After you accept a LAHSO clearance, you must adhere to it and you may not go around.

Match each illustration to the correct glide path description. (Questions 15-18)

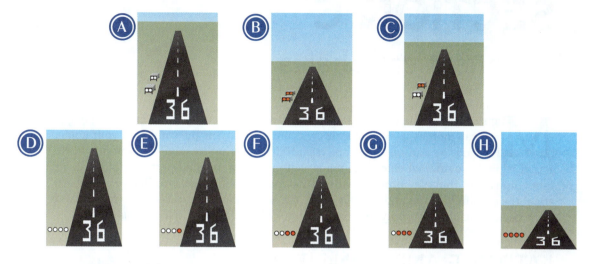

15. VASI, on glide path

16. PAPI, slightly high

17. PAPI, on glide path

18. VASI, low

19. Describe the procedure for activating three-step pilot-controlled lighting.

20. What should you do if you observe criminal activity, such as someone breaking into an airplane?
     A. Call 1-866-GA SECURE and then call 911.
     B. Call 911 and then call 1-866-GA SECURE.
     C. Call 1-866-GA SECURE and then try to detain the suspected criminal.

SECTION B ■ **Airports**

SECTION C
Aeronautical Charts

Maps conjure up images of travel, adventure, and discovery. By exploring maps, you can journey to exotic locales without ever leaving the comfort of your home. For you, as a pilot, maps are essential in turning imaginary excursions into actual trips. **Aeronautical charts** are maps that provide a detailed portrayal of an area's topography and include aeronautical and navigational information. Before you learn about the specific features and symbology of aeronautical charts, you need to understand some basic concepts that apply to representations of the earth's surface on maps.

LATITUDE AND LONGITUDE

The largest circle that can be drawn on the surface of the earth, or any sphere, is referred to as a **great circle**. A great circle's plane must pass through the center of the earth dividing it into two equal parts. A **small circle** is formed on the surface of the earth by the intersection of a plane that does not pass through the center of the earth.

Reference lines based on small and great circles are used to define locations on the earth's surface. For example, the equator forms a great circle. The equator is the imaginary line that circles the earth midway between the north and south poles. You can locate a position north or south of the equator by using **parallels**, or lines of **latitude**, which form small circles. As a reference point, the equator is labeled as 0° of latitude. The parallel lines north of the equator are numbered from 0° to 90°, with 90° north latitude positioned at the north pole. Parallels in the southern hemisphere also are numbered from 0° to 90°, with 90° south latitude representing the south pole.

Meridians, or lines of **longitude**, are imaginary lines that extend from the north to the south pole. [Figure 4-41] Because they connect the poles, lines of longitude always are given in a direction of true north and south. Just as the equator is designated 0° of latitude, the **Prime Meridian**, which passes through Greenwich, England, is labeled 0° of longitude.

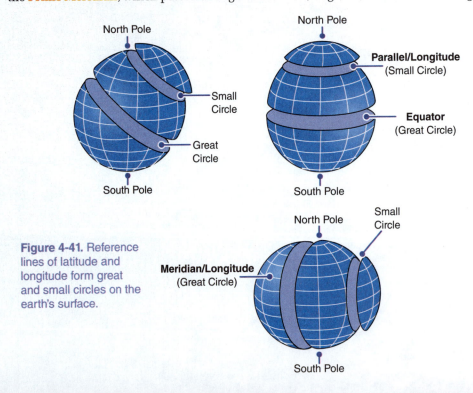

Figure 4-41. Reference lines of latitude and longitude form great and small circles on the earth's surface.

A total of 360° of longitude encompasses the earth, with 180° on the east side and 180° on the west side of the Prime Meridian. The line of 180° of longitude is on the opposite side of the earth from the Prime Meridian. The International Date Line approximately corresponds with the 180° line of longitude, although segments of the Date Line actually vary as much as 20°. When you locate a position east or west of the Prime Meridian, you are determining a position in reference to a line of longitude. The lines of latitude and longitude are printed on aeronautical charts with each degree subdivided into 60 equal segments called minutes. You can use latitude and longitude coordinates to reference the exact location of a point on the earth. [Figure 4-42]

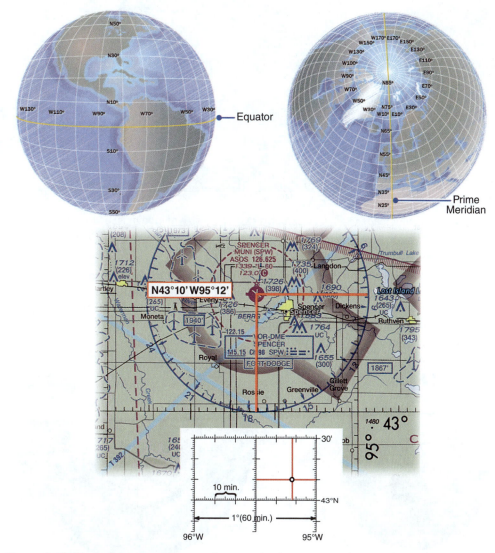

Figure 4-42. Given the geographic coordinates—the intersection of the latitude and longitude lines—you can locate any point on a globe or map.

PROJECTIONS

A globe is the most accurate reduced representation of earth, but obviously is not the most convenient navigation tool. To create a useful map or chart, a picture of the reduced-earth globe must be projected onto a flat surface. Because this is like pressing a section of orange peel on a flat surface, some distortion ultimately occurs in this process.

SECTION C ■ Aeronautical Charts

Projections are used for transfering a section of the earth's surface onto a flat chart. Two of the most common types of projections are the Mercator and the Lambert Conformal Conic. **Mercator projections** normally are used as wall charts. Distortion of landmasses on a Mercator chart increases with distance from the equator. [Figure 4-43]

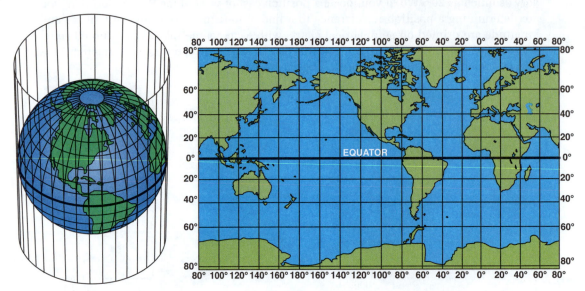

Figure 4-43. The Mercator maps used in classrooms distort the size of land masses to the extreme north and south. Greenland looks as almost as big as South America but is actually only 1/8 as large.

The **Lambert Conformal Conic projection** is frequently used to create aeronautical charts because it offers minimal distortion on the comparatively small area covered by an individual chart. When you compare miles on a Lambert chart to actual miles on earth, the overall scale inaccuracies are negligible within a single chart. [Figure 4-44]

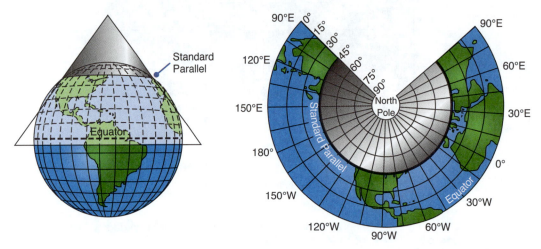

Figure 4-44. Lambert Conformal Conic charts do not have significant distortion as long as they do not cover a large area of the earth's surface.

SECTIONAL CHARTS

Sectional charts, or sectionals, cover all of the 48 mainland states, plus Alaska, Hawaii, Puerto Rico, and the Virgin Islands and are your primary VFR navigation reference, depicting topographic information, visual landmarks, and airport data. Updated every 56 days, you can download these charts from faa.gov or purchase printed versions. In addition, electronic flight bags (EFBs), such as ForeFlight, display sectional charts. Changes in aeronautical data that could affect your flight might occur between chart revisions. For this reason, you should consult the appropriate Chart Supplement prior to flight. You can find more information about sources of flight information in Chapter 5.

Each chart covers 6° to 8° of longitude and approximately 4° of latitude and is given the name of a primary city within its coverage. The scale of a paper sectional chart is 1:500,000, which means that each inch on the sectional chart represents 500,000 actual inches. This translates to one inch on the sectional equaling approximately 7 nautical, or 8 statute, miles on the earth's surface.

To choose a safe maneuvering or cruising altitude, you can refer to the contour lines, spot elevations, and color tints used to show terrain elevation on sectional charts. In addition, topographical information includes cities, towns, rivers, highways, railroads, and other distinctive landmarks that you can use as visual checkpoints. Along with airport depictions, sectionals also contain aeronautical information pertaining to navigation and communication facilities, as well as airspace and obstructions. In flight, you can display a digital chart on a tablet or use a paper sectional chart, which is folded for easy handling, storage, and identification. Use the legend on the chart panel and download the FAA *Aeronautical Chart User's Guide* at faa.gov to help you interpret chart symbols. [Figure 4-45]

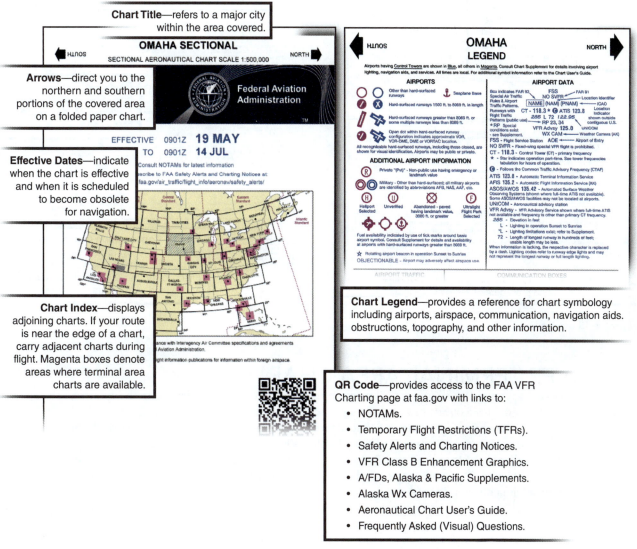

SECTION C ■ Aeronautical Charts

Figure 4-45. You will find information about chart features by referring to the front and back panels, as well as to the inside chart panels. [This figure is continued on the next page.]

SECTION C ■ **Aeronautical Charts**

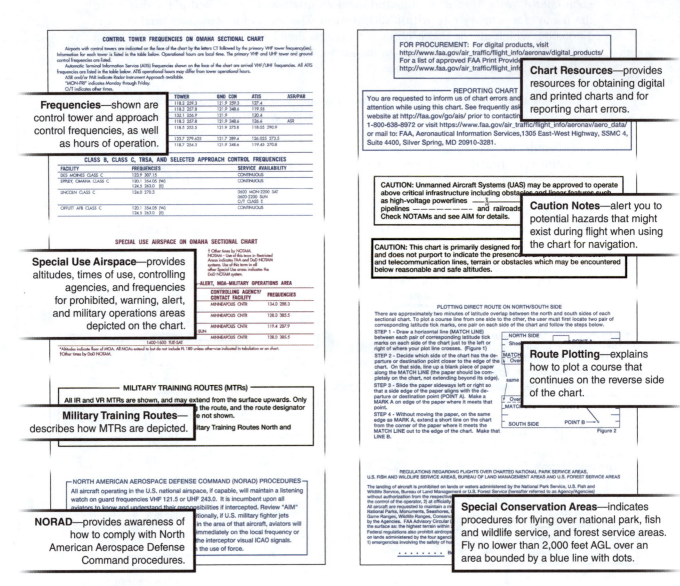

Figure 4-45 continued.

VFR TERMINAL AREA CHARTS

Whenever you are flying under VFR in or around some of the busiest airports in the country, **VFR terminal area charts (TACs)** help significantly with orientation and navigation. Most terminal area charts cover airports that have Class B airspace. You will learn more about the airspace classes in Section D of this chapter. VFR terminal area charts provide a more detailed display of topographical features and airspace on a larger scale (1:250,000) than sectional charts. Sectional charts display a white border to indicate the area covered by a terminal area chart. [Figure 4-46]

Terminal Area Charts

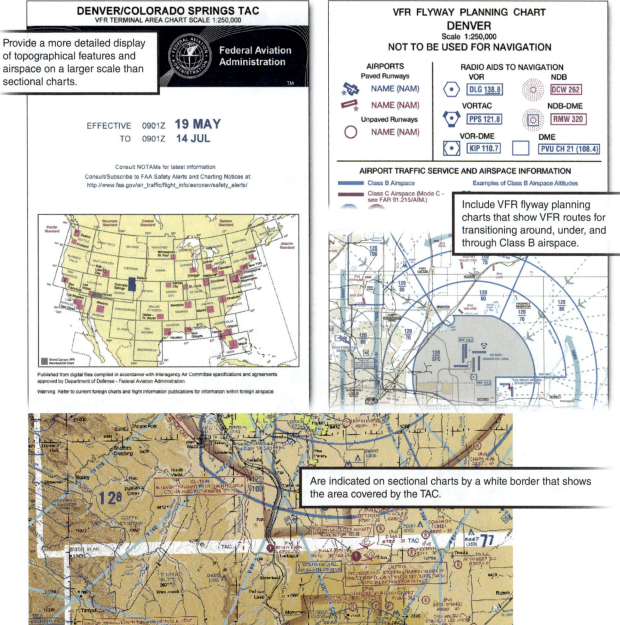

Figure 4-46. VFR terminal area charts have a larger scale with more detail to facilitate VFR navigation in and around busy airports.

IFR Charts

Before aeronautical charts were available, many pilots flying airmail and cargo used road maps for navigation and when visibility was limited, they often followed the railroad tracks, which they called "hugging the UP," or Union Pacific. If weather conditions deteriorated drastically, pilots had to land their airplanes in fields and wait until conditions improved.

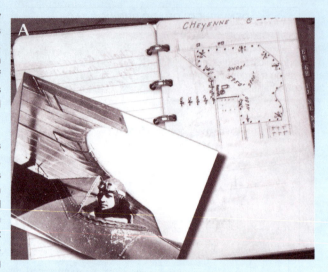

Today, instrument-rated pilots use charts specifically designed for flight in the clouds, thanks to aviation pioneer Elrey B. Jeppesen. As an airmail pilot in the 1930s, Jeppesen began recording information about terrain heights, field lengths, airport layouts, lights, and obstacles in a little black notebook. [Figure A] The notes that he took eventually turned into a thriving business that provided pilots with enroute charts depicting airways and navigation aids, as well as terminal charts with instrument approach procedures.

Although some of the symbols used on instrument charts are the same as those shown on VFR charts, instrument pilots operating in IFR conditions seldom have use for the visual landmarks that are featured on VFR charts. In Figure B, you can see that the same area depicted on a VFR chart appears much differently on a digital IFR chart.

CHART SYMBOLOGY

The **legend** is your tool for deciphering symbols and decoding aeronautical chart information. Divided into seven categories, the legend describes symbology for airports, airport data, radio aids to navigation and communication boxes, airport traffic service and airspace information, obstructions, topographic information, and miscellaneous data.

TERRAIN AND ELEVATION

When you are planning and implementing a flight, knowing the height of the terrain along your route is critical to flight safety. VFR aeronautical charts show the terrain elevation information that you need to select a safe maneuvering or cruising altitude. **Color tints** depict bands of elevation relative to sea level. Colors range from light green for the lower elevations to dark brown for the higher elevations. **Contour lines** join points of equal elevation. The pattern of these lines and their spacing help you visualize the terrain. Widely spaced contours represent gentle slopes, while closely spaced contours represent steep slopes. **Spot elevations** provide additional essential information about the height of the terrain. **Shaded relief** shows how terrain might appear from the air. [Figure 4-47]

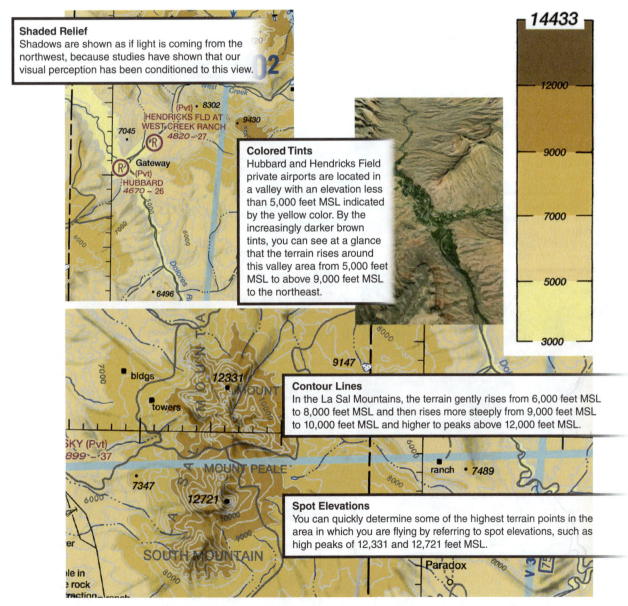

Shaded Relief
Shadows are shown as if light is coming from the northwest, because studies have shown that our visual perception has been conditioned to this view.

Colored Tints
Hubbard and Hendricks Field private airports are located in a valley with an elevation less than 5,000 feet MSL indicated by the yellow color. By the increasingly darker brown tints, you can see at a glance that the terrain rises around this valley area from 5,000 feet MSL to above 9,000 feet MSL to the northeast.

Contour Lines
In the La Sal Mountains, the terrain gently rises from 6,000 feet MSL to 8,000 feet MSL and then rises more steeply from 9,000 feet MSL to 10,000 feet MSL and higher to peaks above 12,000 feet MSL.

Spot Elevations
You can quickly determine some of the highest terrain points in the area in which you are flying by referring to spot elevations, such as high peaks of 12,331 and 12,721 feet MSL.

Figure 4-47. The ability to interpret color tints, contour lines, spot elevations, and shaded relief is a fundamental piloting skill. A legend for the color tints appears on the chart panels.

SECTION C ■ **Aeronautical Charts**

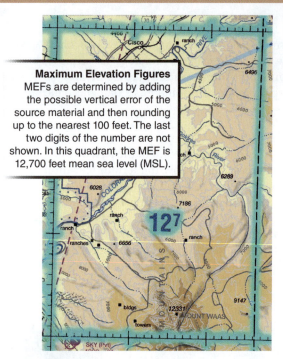

Maximum Elevation Figures
MEFs are determined by adding the possible vertical error of the source material and then rounding up to the nearest 100 feet. The last two digits of the number are not shown. In this quadrant, the MEF is 12,700 feet mean sea level (MSL).

Figure 4-48. Use maximum elevation figures to help you determine safe flying altitudes.

The **maximum elevation figure (MEF)** represents the highest elevation of terrain and other obstacles (for example; towers, trees, etc.) within a quadrant—the area bounded by ticked lines dividing each 30 minutes of latitude and each 30 minutes of longitude. [Figure 4-48]

LAND FEATURES

The locations of natural and man-made features, such as lakes, rivers, railroads, roads, and highways are shown on sectional charts as reference points for navigation. Not every symbol is shown on the chart legend so you refer to the FAA *Aeronautical Chart User's Guide* at faa.gov to help you interpret land feature symbols. [Figure 4-49]

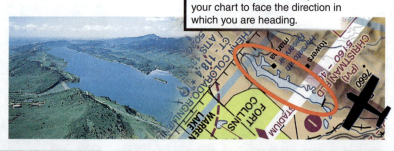

Orienting yourself in relationship to a landmark is easier if you turn your chart to face the direction in which you are heading.

With experience, you will become skilled at associating the symbol on the chart with the landmark (such as this mine) as viewed from the airplane.

Small black circles with text indicate water, oil, or gas tanks, while open circles depict oil, gas, or mineral wells. The scale of these landmarks might be increased to make them easier to read and to use for navigational reference.

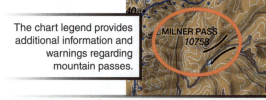

The chart legend provides additional information and warnings regarding mountain passes.

TOPOGRAPHIC INFORMATION

Power Transmission Line	Mountain Pass
Aerial Cable	11823 (Elevation of Pass)
Lookout Tower	Pass symbol does not indicate a recommended
618 (Elevation Base of Tower)	route or direction of flight and pass elevation does not indicate a recommended clearance altitude. Hazardous flight conditions may exist within and near mountain passes.

Yellow tinted areas indicate populated places. However, towns such as Milliken, which are too small to depict using the yellow tint, are shown as a circle.

Use additional landmarks to identify Milliken. For example, the town is located southwest of the intersection of a primary road and a railroad track.

Figure 4-49. With experience, you will become skilled at associating the symbol on the chart with the landmark as viewed from the airplane.

OBSTRUCTIONS

Typically, charts depict man-made obstructions (also referred to as obstacles) that are higher than 200 feet above ground level (AGL). Obstructions can impose hazards to low-level flight and you also can use them to identify your position. Charted obstacles include smokestacks, tanks, factories, lookout towers, and antennas. In high-density areas like cities, only the highest obstacle is represented by the group obstacle symbol. High-intensity strobe lighting on obstacles might operate part-time or by proximity activation. [Figure 4-50]

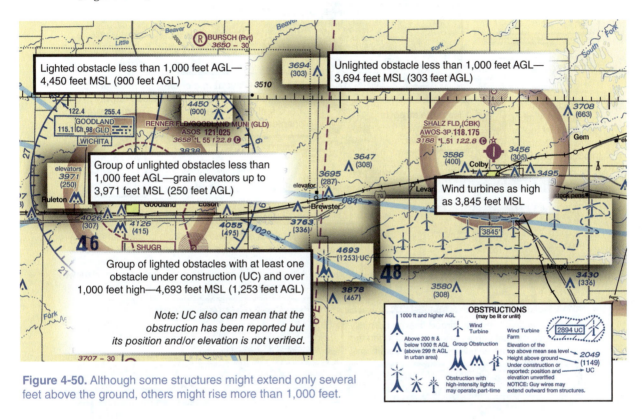

Figure 4-50. Although some structures might extend only several feet above the ground, others might rise more than 1,000 feet.

AIRPORT SYMBOLS

Because a wide variety of airport types, shapes, and sizes exist, sectional charts show several different airport diagrams to help you picture the actual airport being illustrated. Using unique symbology, charts depict civil, military, and private airports, as well as seaplane bases, heliports, and ultralight flight parks. A variety of symbols provide information regarding the type and length of runways, and the airport services available at a particular airport. In addition, you can quickly identify airports with control towers as they are shown in blue, while non-towered airports are magenta in color. [Figure 4-51]

SECTION C ■ Aeronautical Charts

SECTION C ■ Aeronautical Charts

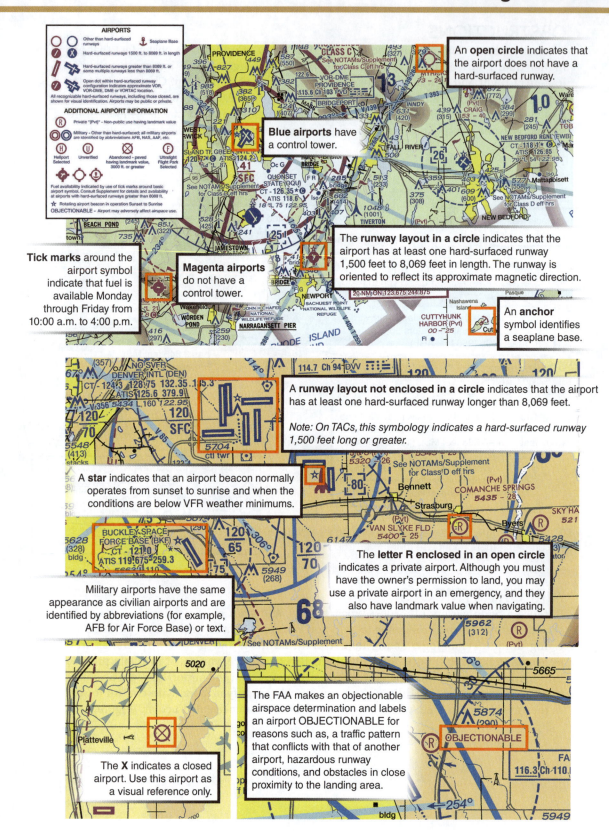

An **open circle** indicates that the airport does not have a hard-surfaced runway.

Blue airports have a control tower.

The **runway layout in a circle** indicates that the airport has at least one hard-surfaced runway 1,500 feet to 8,069 feet in length. The runway is oriented to reflect its approximate magnetic direction.

An **anchor** symbol identifies a seaplane base.

Tick marks around the airport symbol indicate that fuel is available Monday through Friday from 10:00 a.m. to 4:00 p.m.

Magenta airports do not have a control tower.

A **runway layout not enclosed in a circle** indicates that the airport has at least one hard-surfaced runway longer than 8,069 feet.

Note: On TACs, this symbology indicates a hard-surfaced runway 1,500 feet long or greater.

A **star** indicates that an airport beacon normally operates from sunset to sunrise and when the conditions are below VFR weather minimums.

The **letter R enclosed in an open circle** indicates a private airport. Although you must have the owner's permission to land, you may use a private airport in an emergency, and they also have landmark value when navigating.

Military airports have the same appearance as civilian airports and are identified by abbreviations (for example, AFB for Air Force Base) or text.

The **X** indicates a closed airport. Use this airport as a visual reference only.

The FAA makes an objectionable airspace determination and labels an airport OBJECTIONABLE for reasons such as, a traffic pattern that conflicts with that of another airport, hazardous runway conditions, and obstacles in close proximity to the landing area.

Figure 4-51. Airport symbols help you quickly determine basic airport characteristics in the event you might need to divert.

AIRPORT DATA

Airport data contains communication frequencies, including the common traffic advisory frequency (CTAF), and approximate length of the longest runway, as well as the availability of lighting and weather services. Chapter 5 covers the use of CTAF and other frequencies shown on the chart. In addition, the airport elevation is included as part of the airport data. The official **airport elevation** is defined as the highest part of usable runway surface, measured in feet above mean sea level.

A three-letter FAA location identifier is shown next to the airport name on the chart. When retrieving the airport from a navigation database or when filing a flight plan, use the 4-letter International Civilian Aviation Organization (ICAO) identifier beginning with the letter K. For example, Wittman Regional Airport located in Oshkosh, Wisconsin has an ICAO identifier of KOSH. When verbalizing the identifier to air traffic control, use the phonetic alphabet: Kilo-Oscar Sierra Hotel. [Figure 4-52]

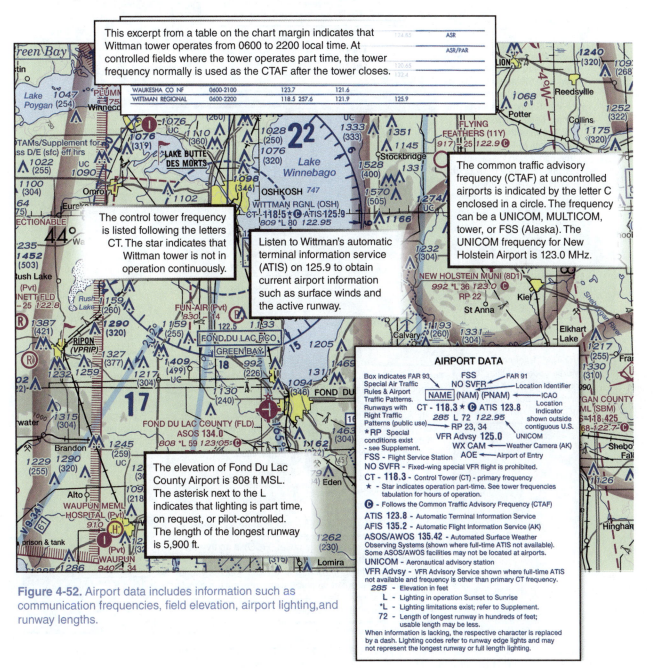

Figure 4-52. Airport data includes information such as communication frequencies, field elevation, airport lighting, and runway lengths.

NAVIGATION AIDS

For cross-country planning and flight, you can refer to navigation and communication boxes for information concerning radio aids to navigation, or **navaids**, and Flight Service frequencies in the area. You will communicate with Flight Service enroute to open and close flight plans, obtain current weather information, or for assistance in emergency situations. Chapter 5 describes the services available from Flight Service in greater detail and Chapter 9 covers the operation of navaids represented by chart symbols. [Figure 4-53]

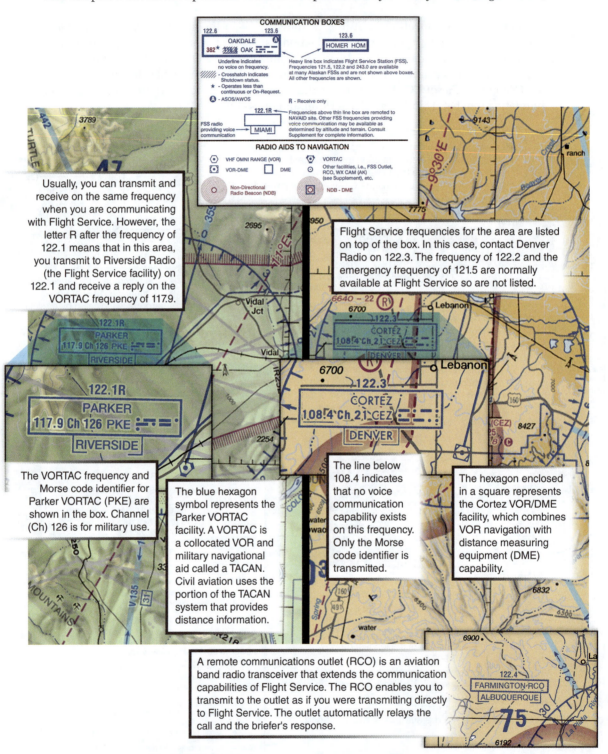

Usually, you can transmit and receive on the same frequency when you are communicating with Flight Service. However, the letter R after the frequency of 122.1 means that in this area, you transmit to Riverside Radio (the Flight Service facility) on 122.1 and receive a reply on the VORTAC frequency of 117.9.

Flight Service frequencies for the area are listed on top of the box. In this case, contact Denver Radio on 122.3. The frequency of 122.2 and the emergency frequency of 121.5 are normally available at Flight Service so are not listed.

The VORTAC frequency and Morse code identifier for Parker VORTAC (PKE) are shown in the box. Channel (Ch) 126 is for military use.

The blue hexagon symbol represents the Parker VORTAC facility. A VORTAC is a collocated VOR and military navigational aid called a TACAN. Civil aviation uses the portion of the TACAN system that provides distance information.

The line below 108.4 indicates that no voice communication capability exists on this frequency. Only the Morse code identifier is transmitted.

The hexagon enclosed in a square represents the Cortez VOR/DME facility, which combines VOR navigation with distance measuring equipment (DME) capability.

A remote communications outlet (RCO) is an aviation band radio transceiver that extends the communication capabilities of Flight Service. The RCO enables you to transmit to the outlet as if you were transmitting directly to Flight Service. The outlet automatically relays the call and the briefer's response.

Figure 4-53. Boxes placed near the appropriate navaid include the name, frequency, and Morse code identifier of the navaid. Flight Service frequencies are printed above the boxes.

Although, the FAA is gradually decommissioning a type of navaid referred to as a nondirectional beacon (NDBs), these navaids are still prevalent in Alaska. In addition, in the contiguous U.S., Flight Service is operated by a contract service provider, Leidos. However, individual Flight Service Stations (FSSs) located at airports still provide services in Alaska. [Figure 4-54]

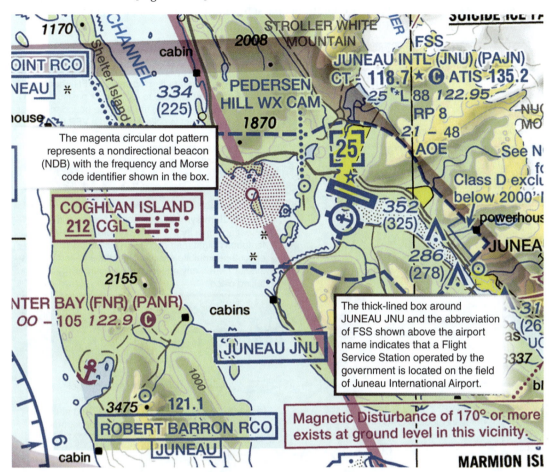

Figure 4-54. If you fly in Alaska, you might encounter some unique navaid and Flight Service facilities.

MISCELLANEOUS SYMBOLS

VFR chart symbols also represent miscellaneous information that affects the flight environment. For example, you normally navigate by magnetic reference, so during flight planning, you must correct for the difference between true and magnetic direction, which is called variation. Local magnetic variation is shown on aeronautical charts by a dashed magenta isogonic line.

Sectional charts also indicate locations of space launch activity. A horizontal spaceport includes runways for aircraft that support spaceflight operations, such as winged launch vehicles. A vertical spaceport, such as NASA's Cape Canaveral in Florida, typically is surrounded by a large safety area (range) over which launched rockets are expected to fly, and within which some components of the rockets might land. Other miscellaneous symbols include areas of glider operations, parachute jumping areas, large sports venues, VFR checkpoints, and VFR waypoints. [Figure 4-55]

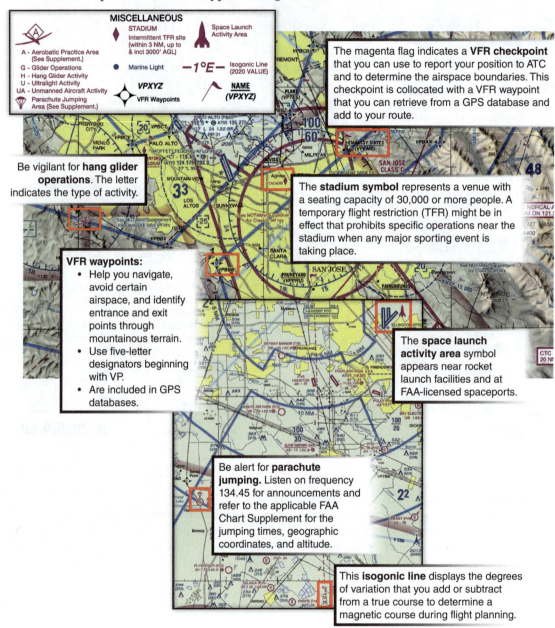

Figure 4-55. These sectional chart excerpts depict a variety of miscellaneous symbols that help you maintain situational awareness during flight.

DISCOVERY

Aeronautical Chart User's Guide

A wide variety of symbols and markings shown on aeronautical charts are not identified on the chart legend. These symbols are defined in the *Aeronautical Chart User's Guide,* which can be downloaded at faa.gov. The guide is useful as a learning aid and a quick reference. Can you correctly identify some of these unique symbols?

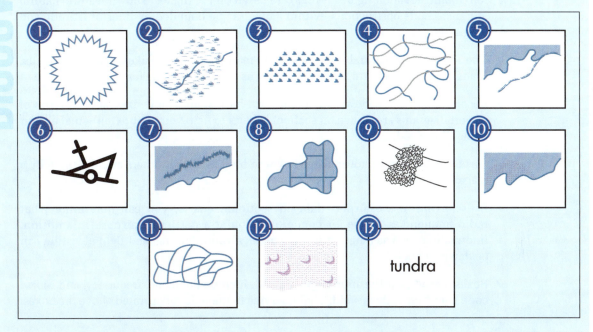

Answers: 1. high energy radiation areas; 2. swamps, marshes, and bogs; 3. rice paddies; 4. glaciers; 5. ice cliffs; 6. exposed ship wreck; 7. rocky or coral reefs; 8. fish ponds and hatcheries; 9. lava flows; 10. tidal flats exposed at low tide; 11. cranberry bog; 12. sand dunes; 13. tundra (an easy one)

SUMMARY CHECKLIST

✓ Aeronautical charts are maps that provide a detailed portrayal of an area's topography and include aeronautical and navigational information.

✓ Reference lines based on great and small circles are used to define locations on the earth's surface.

✓ You can locate a position on an aeronautical chart by knowing its coordinates of latitude and longitude.

✓ Each sectional chart covers 6° to 8° of longitude and approximately 4° of latitude and is given the name of a primary city within its coverage.

✓ VFR terminal area charts provide a more detailed display of topographical features and airspace on a larger scale (1:250,000) than sectional charts (1:500,000). Sectional charts display a white border to indicate the area covered by a terminal area chart.

✓ The chart legend describes symbology for airports, airport data, radio aids to navigation and communication boxes, airport traffic service and airspace information, obstructions, topographic information, and miscellaneous data.

✓ VFR aeronautical charts show the terrain elevation information that you need to select a safe maneuvering or cruising altitude.

✓ With the help of contour lines, spot elevations, and the elevations of obstructions, you can choose a safe cruising altitude.

✓ Maximum elevation figures (MEFs) are based on the highest known feature, natural and man-made obstructions, within a quadrangle bounded by lines of latitude and longitude.

✓ The locations of natural and man-made features, such as lakes, rivers, railroads, roads, and highways are shown on charts as reference points for navigation.

✓ Because there is a wide variety of airport types, shapes, and sizes, several types of airport diagrams are shown on sectional charts to help you picture the actual airport being illustrated.

✓ Airports with control towers are shown in blue, while all others are identified by a magenta color.

✓ By referring to the airport data on sectional charts, you can determine what radio frequencies to use for communication at a particular airport. In addition, information such as longest runway length, airport lighting, and field elevation can be determined.

✓ Boxes placed near the appropriate navaid, include the name, frequency, and Morse code identifier of the navaid. Flight Service frequencies are printed above the boxes.

KEY TERMS

Aeronautical Charts	Sectional Charts
Great Circle	Terminal Area Charts (TAC)
Small Circle	Legend
Parallels	Color Tints
Latitude	Contour Lines
Meridians	Spot Elevations
Longitude	Shaded Relief
Prime Meridian	Maximum Elevation Figures (MEFs)
Projections	Airport Elevation
Mercator Projection	Navaids
Lambert Conformal Conic Projection	

QUESTIONS

1. Determine the approximate latitude and longitude of Red Bluff Airport.

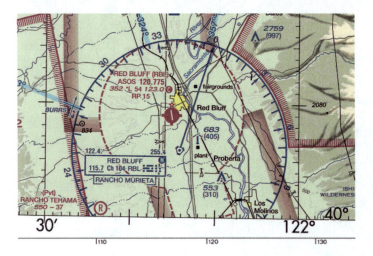

2. What is the minimum MSL altitude that you should fly over the area depicted in this chart excerpt?

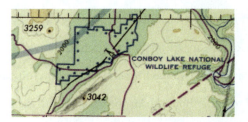

Match the airport diagrams with their descriptions.

3. Hard-surfaced runways 1,500 feet to 8,069 feet in length

4. Private airport

5. Hard-surface runways greater that 8,069 feet in length

6. Seaplane base

7. Closed airport

8. True/False. Airports with control towers are magenta on sectional charts.

Refer to this chart excerpt for questions 9 through 15.

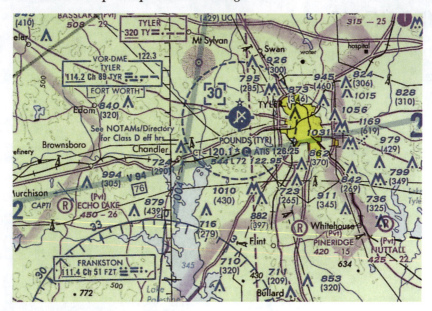

9. What is the control tower frequency for Pounds Airport?

10. What is the elevation of Pounds Airport?

11. True/False. Pounds Airport has full-time lighting.

12. What do the tick marks on the airport diagram indicate?

13. What does the star symbol above the airport diagram indicate?

14. Does the control tower at Pounds Airport operate full time?

15. You can use the frequency 122.3 located on top of the navaid box to communicate with what facility?

16. What is the height of the lighted obstruction?
    A. 1,125 feet MSL
    B. 1,467 feet AGL
    C. 2,049 feet AGL

17. How often are VFR sectional and terminal area charts updated?

SECTION D

Airspace

Webster's dictionary defines **airspace** as "the portion of the atmosphere above a particular land area, especially above a nation." To efficiently manage the large amount of air traffic that traverses the sky each day, the atmosphere above the United States is divided into several sectors, or classes. In each airspace class, specific rules apply. For example, there are **VFR weather minimums** (minimum flight visibilities and distances from clouds) that you must maintain in each airspace class. In some areas, you are required to communicate with ATC and comply with pilot certification and aircraft equipment requirements. In addition to the primary classes, the airspace over the United States includes special use and other airspace areas where certain restrictions apply or specific ATC services are provided.

Compared to the ground-based rules of driving, airspace regulations might seem very unusual. For example, when you drive across the country, you do not enter regions where the rules of the road change significantly. You need to comply with the regulations as you fly, but how do you know when you are entering a different class of airspace? There are no signposts in the sky to alert you to the fact that you are crossing over an invisible boundary into another airspace sector. The signposts that you do have are the lateral and vertical airspace dimensions that are depicted on aeronautical charts. [Figure 4-56]

Figure 4-56. The legend helps you identify the boundaries of airspace below 18,000 feet MSL that are depicted on sectional charts.

CONTROLLED AND UNCONTROLLED AIRSPACE

Airspace is divided into classes and categorized as either controlled or uncontrolled. **Controlled airspace** is a general term that describes five of the six airspace classes of airspace; Class A, Class B, Class C, Class D, and Class E. While operating in controlled airspace, you are subject to certain operating rules, as well as pilot qualification and aircraft equipment requirements. Class G airspace is referred to as **uncontrolled airspace**. There is no airspace designation for Class F in the United States.

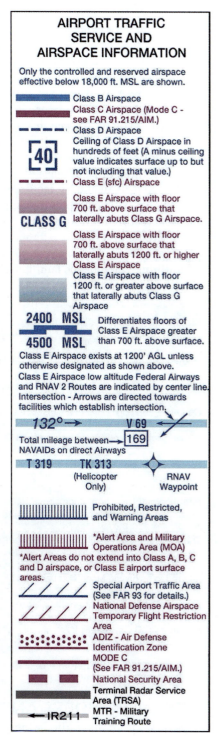

AIRPORT TRAFFIC SERVICE AND AIRSPACE INFORMATION

Only the controlled and reserved airspace effective below 18,000 ft. MSL are shown.

Class B Airspace

Class C Airspace (Mode C - see FAR 91.215/AIM.)

Class D Airspace

Ceiling of Class D Airspace in hundreds of feet (A minus ceiling value indicates surface up to but not including that value.)

Class E (sfc) Airspace

Class E Airspace with floor 700 ft. above surface that laterally abuts Class G Airspace.

Class E Airspace with floor 700 ft. above surface that laterally abuts 1200 ft. or higher Class E Airspace

Class E Airspace with floor 1200 ft. or greater above surface that laterally abuts Class G Airspace

2400 MSL / 4500 MSL — Differentiates floors of Class E Airspace greater than 700 ft. above surface.

Class E Airspace exists at 1200' AGL unless otherwise designated as shown above. Class E Airspace low altitude Federal Airways and RNAV 2 Routes are indicated by center line. Intersection - Arrows are directed towards facilities which establish intersection.

132° →   V 69

Total mileage between NAVAIDs on direct Airways → 169

T 319   TK 313 (Helicopter Only)   RNAV Waypoint

Prohibited, Restricted, and Warning Areas

*Alert Area and Military Operations Area (MOA)
*Alert Areas do not extend into Class A, B, C and D airspace, or Class E airport surface areas.

Special Airport Traffic Area (See FAR 93 for details.)

National Defense Airspace Temporary Flight Restriction Area

ADIZ - Air Defense Identification Zone

MODE C (See FAR 91.215/AIM.)

National Security Area

Terminal Radar Service Area (TRSA)

←IR211   MTR - Military Training Route

One of the primary functions of airspace classification is the separation of IFR and VFR traffic. The FARs prohibit noninstrument-rated pilots from flying when conditions are below the basic VFR weather minimums specified for each class of airspace. VFR cloud clearance and visibility requirements are designed to help you avoid flying into clouds, as well as to allow you to maintain adequate forward visibility to see and avoid other aircraft and navigate in flight. Keep in mind that, these values are legal minimums. You should establish your own personal minimums that are higher than the regulatory minimums based on your experience level. Your instructor can advise you in this area.

When the weather conditions deteriorate below VFR minimums, all aircraft in controlled airspace must be flown by instrument-rated pilots in accordance with IFR clearances issued by ATC. If you are an instrument-rated pilot on an IFR flight plan, you are not subject to any visibility or cloud clearance minimums, but ATC must issue a clearance allowing you to proceed through controlled airspace.

The following discussion follows a standard format that makes it easy to understand the features of each class of airspace. Following a general description of the airspace class and its operating requirements, a sectional or terminal area chart (TAC) depiction is shown with examples that apply to operating in and near the airspace. A table lists the VFR weather minimums that apply to the airspace class and summarizes the operating requirements. For easy reference at the end of this section, a chart combines the individual airspace class tables, and a diagram depicts the relationships between each airspace. In addition, you must be familiar with several terms to understand airspace dimensions and operating requirements. [Figure 4-57]

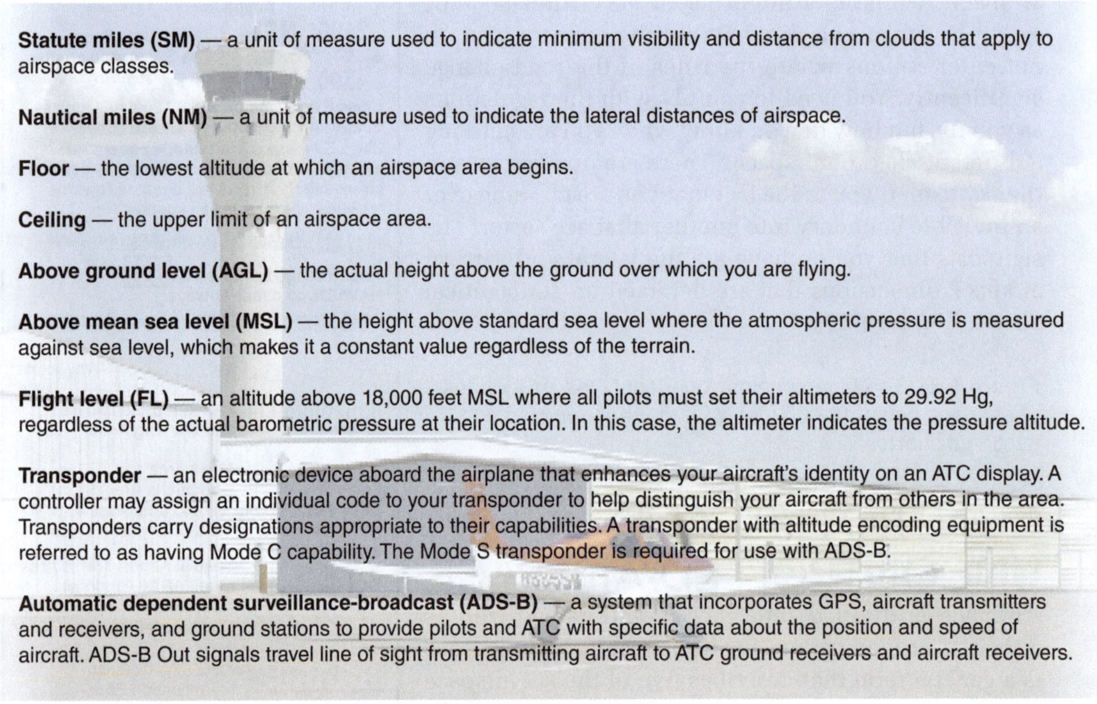

Statute miles (SM) — a unit of measure used to indicate minimum visibility and distance from clouds that apply to airspace classes.

Nautical miles (NM) — a unit of measure used to indicate the lateral distances of airspace.

Floor — the lowest altitude at which an airspace area begins.

Ceiling — the upper limit of an airspace area.

Above ground level (AGL) — the actual height above the ground over which you are flying.

Above mean sea level (MSL) — the height above standard sea level where the atmospheric pressure is measured against sea level, which makes it a constant value regardless of the terrain.

Flight level (FL) — an altitude above 18,000 feet MSL where all pilots must set their altimeters to 29.92 Hg, regardless of the actual barometric pressure at their location. In this case, the altimeter indicates the pressure altitude.

Transponder — an electronic device aboard the airplane that enhances your aircraft's identity on an ATC display. A controller may assign an individual code to your transponder to help distinguish your aircraft from others in the area. Transponders carry designations appropriate to their capabilities. A transponder with altitude encoding equipment is referred to as having Mode C capability. The Mode S transponder is required for use with ADS-B.

Automatic dependent surveillance-broadcast (ADS-B) — a system that incorporates GPS, aircraft transmitters and receivers, and ground stations to provide pilots and ATC with specific data about the position and speed of aircraft. ADS-B Out signals travel line of sight from transmitting aircraft to ATC ground receivers and aircraft receivers.

Figure 4-57. Refer to these definitions as you explore airspace.

TRANSPONDERS AND ADS-B

Separation of air traffic is the primary function of ATC, and the automatic dependent surveillance-broadcast (ADS-B) system and radar are the controller's principal tools. The FARs require that you have an operating transponder with Mode C capability and ADS-B Out equipment in Class A airspace, Class B airspace, within 30 nautical miles of Class B primary airports, and in and above Class C airspace. In addition, you must have a Mode C **transponder** and ADS-B Out equipment at or above 10,000 feet MSL (except at or below 2,500 feet AGL) in controlled airspace. You also must have a Mode C transponder if you are in uncontrolled airspace above 10,000 feet MSL.

Even if you are not in airspace where transponders are required, the FARs require you to have your transponder turned on (if your aircraft is so equipped) while operating in controlled airspace. You will learn more about ADS-B, radar, and transponders in Chapter 5, Section A—ATC Services. [Figure 4-58]

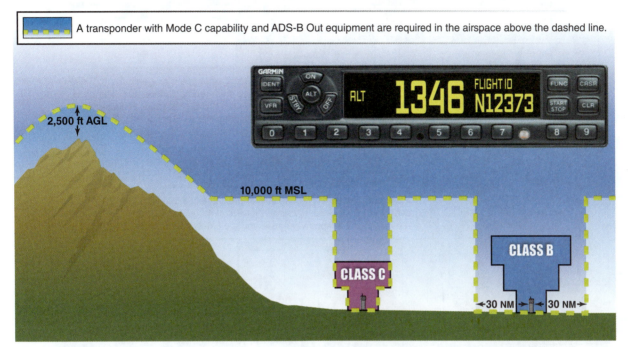

A transponder with Mode C capability and ADS-B Out equipment are required in the airspace above the dashed line.

Figure 4-58. You must have a Mode C transponder and ADS-B Out equipment to operate in the airspace shown here.

CLASS G AIRSPACE (UNCONTROLLED)

ATC does not exercise control of traffic and you are not required to communicate with controllers when operating in **Class G airspace**, unless a temporary control tower exists. Class G airspace typically starts at the surface—think G for Ground—and extends up to the base of the overlying controlled airspace (Class E), which is normally 700 or 1,200 feet AGL. In a few remote areas of the western U.S. and Alaska, Class G airspace can extend all the way up to 14,500 feet MSL, or to 1,500 feet AGL, whichever is higher. [Figure 4-59]

CONTROLLED AIRSPACE

When operating in controlled airspace, you might be subject to air traffic control. As a routine measure, IFR flights are controlled from takeoff to touchdown because they are permitted to operate in all kinds of weather. As a VFR pilot, your contact with ATC typically is limited to terminal areas. For example, when you take off or land at controlled airports, you must contact the control tower, and you often will use approach and departure control services.

SECTION D ■ **Airspace**

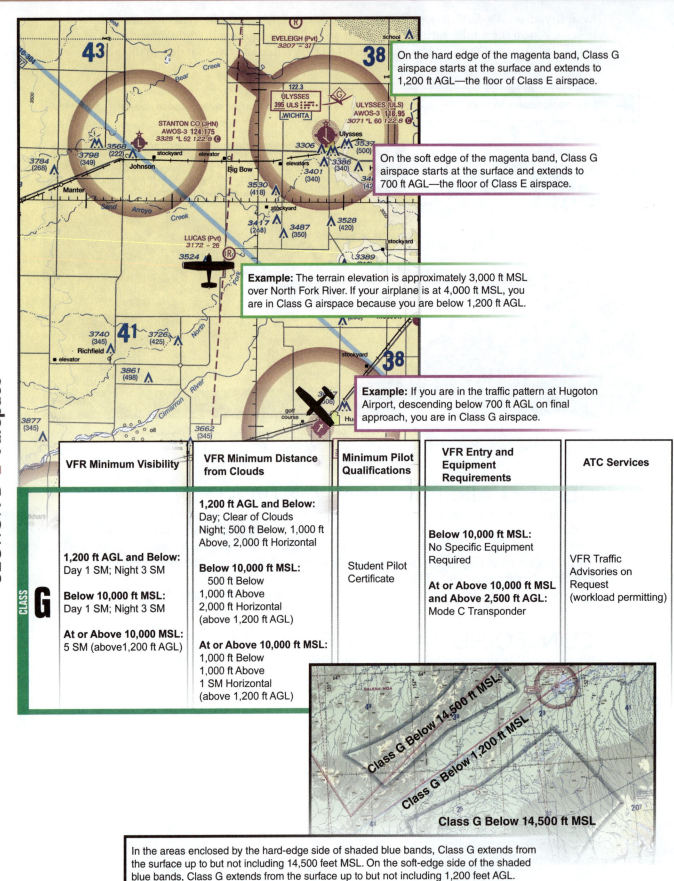

On the hard edge of the magenta band, Class G airspace starts at the surface and extends to 1,200 ft AGL—the floor of Class E airspace.

On the soft edge of the magenta band, Class G airspace starts at the surface and extends to 700 ft AGL—the floor of Class E airspace.

Example: The terrain elevation is approximately 3,000 ft MSL over North Fork River. If your airplane is at 4,000 ft MSL, you are in Class G airspace because you are below 1,200 ft AGL.

Example: If you are in the traffic pattern at Hugoton Airport, descending below 700 ft AGL on final approach, you are in Class G airspace.

	VFR Minimum Visibility	VFR Minimum Distance from Clouds	Minimum Pilot Qualifications	VFR Entry and Equipment Requirements	ATC Services
CLASS **G**	**1,200 ft AGL and Below:** Day 1 SM; Night 3 SM **Below 10,000 ft MSL:** Day 1 SM; Night 3 SM **At or Above 10,000 MSL:** 5 SM (above 1,200 ft AGL)	**1,200 ft AGL and Below:** Day; Clear of Clouds Night; 500 ft Below, 1,000 ft Above, 2,000 ft Horizontal **Below 10,000 ft MSL:** 500 ft Below 1,000 ft Above 2,000 ft Horizontal (above 1,200 ft AGL) **At or Above 10,000 ft MSL:** 1,000 ft Below 1,000 ft Above 1 SM Horizontal (above 1,200 ft AGL)	Student Pilot Certificate	**Below 10,000 ft MSL:** No Specific Equipment Required **At or Above 10,000 ft MSL and Above 2,500 ft AGL:** Mode C Transponder	VFR Traffic Advisories on Request (workload permitting)

Class G Below 14,500 ft MSL
Class G Below 1,200 ft MSL
Class G Below 14,500 ft MSL

In the areas enclosed by the hard-edge side of shaded blue bands, Class G extends from the surface up to but not including 14,500 feet MSL. On the soft-edge side of the shaded blue bands, Class G extends from the surface up to but not including 1,200 feet AGL.

Figure 4-59. Class G airspace is the least restrictive of all airspace.

CLASS E AIRSPACE

The majority of your flying time will probably be spent in the controlled airspace designated as **Class E airspace**. Unless designated otherwise, Class E begins at 1,200 feet AGL and extends to the floor of Class A airspace at 18,000 feet MSL. Class E airspace also extends upward from FL600. Class E is the most common airspace in the United States—think E for Everywhere. [Figure 4-60]

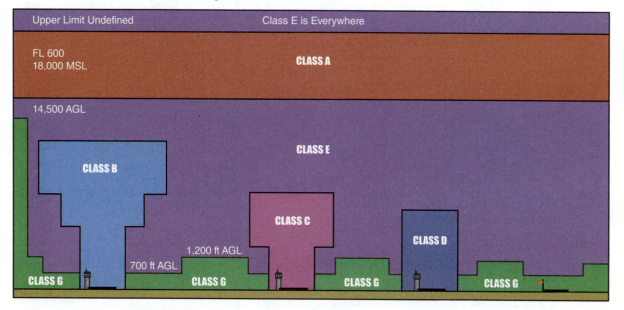

Figure 4-60. Class E airspace extends from the ceilings of Class G, D, C, and B airspace to the base of Class A airspace.

No communication requirements apply to operating within Class E airspace, but you can request traffic advisory services that ATC provides on a workload-permitting basis. In Class E airspace, you cannot fly when the weather is below VFR minimums unless you are instrument rated, have filed an IFR flight plan, and have received a clearance from ATC.

To allow IFR traffic to remain in controlled airspace while transitioning from the enroute to the terminal environment, the base of Class E starts closer to the ground near many airports. At airports without control towers that have approved instrument approach procedures, **Class E transition areas** and **Class E surface areas** are depicted on VFR aeronautical charts. Shaded magenta bands represent Class E transition areas that lower the floor of Class E airspace to 700 feet AGL. Dashed magenta lines indicate Class E surface areas that lower the floor of Class E airspace to the surface. Typically, Class E transition and surface areas are depicted as circles, but they can also contain rectangular extensions in the direction of an instrument approach. At airports where Class E begins at the surface, weather reporting services are provided by a weather observer or automatic weather observation equipment (ASOS/AWOS). [Figure 4-61]

Federal airways, or **Victor airways**, are based on VOR or VORTAC navaids and help to expedite enroute air navigation between airports or terminals. Victor airways are depicted on VFR aeronautical charts as shaded blue lines identified by a V and the airway number.

T-routes are similar to Victor airways, but are based on GPS navigation. T-routes are depicted on VFR aeronautical charts as shaded blue lines identified with a T and the route number. Both T-routes and Victor airways are typically 8 nautical miles wide—4 nautical miles on either side of the airway centerline—and are Class E airspace that starts at 1,200 feet AGL and extends up to but not including 18,000 feet MSL. [Figure 4-62]

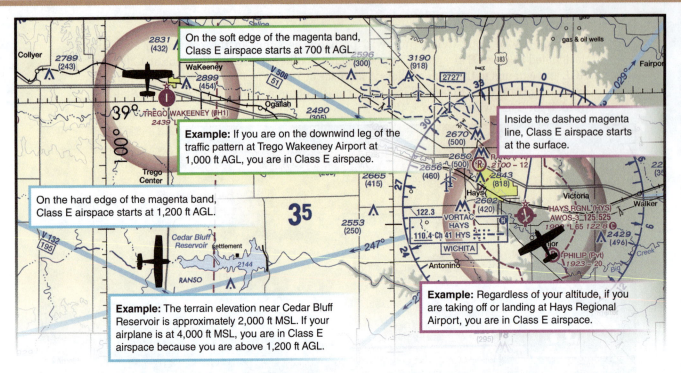

On the soft edge of the magenta band, Class E airspace starts at 700 ft AGL.

Inside the dashed magenta line, Class E airspace starts at the surface.

Example: If you are on the downwind leg of the traffic pattern at Trego Wakeeney Airport at 1,000 ft AGL, you are in Class E airspace.

On the hard edge of the magenta band, Class E airspace starts at 1,200 ft AGL.

Example: The terrain elevation near Cedar Bluff Reservoir is approximately 2,000 ft MSL. If your airplane is at 4,000 ft MSL, you are in Class E airspace because you are above 1,200 ft AGL.

Example: Regardless of your altitude, if you are taking off or landing at Hays Regional Airport, you are in Class E airspace.

Figure 4-61. Class E airspace encircling an airport extends up to the base of the overlying or adjacent controlled airspace.

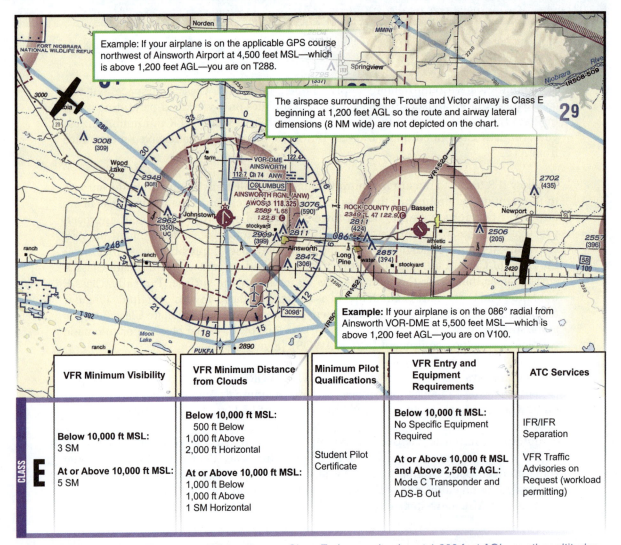

Example: If your airplane is on the applicable GPS course northwest of Ainsworth Airport at 4,500 feet MSL—which is above 1,200 feet AGL—you are on T288.

The airspace surrounding the T-route and Victor airway is Class E beginning at 1,200 feet AGL so the route and airway lateral dimensions (8 NM wide) are not depicted on the chart.

Example: If your airplane is on the 086° radial from Ainsworth VOR-DME at 5,500 feet MSL—which is above 1,200 feet AGL—you are on V100.

	VFR Minimum Visibility	VFR Minimum Distance from Clouds	Minimum Pilot Qualifications	VFR Entry and Equipment Requirements	ATC Services
CLASS E	Below 10,000 ft MSL: 3 SM At or Above 10,000 ft MSL: 5 SM	Below 10,000 ft MSL: 500 ft Below 1,000 ft Above 2,000 ft Horizontal At or Above 10,000 ft MSL: 1,000 ft Below 1,000 ft Above 1 SM Horizontal	Student Pilot Certificate	Below 10,000 ft MSL: No Specific Equipment Required At or Above 10,000 ft MSL and Above 2,500 ft AGL: Mode C Transponder and ADS-B Out	IFR/IFR Separation VFR Traffic Advisories on Request (workload permitting)

Figure 4-62. Surrounding T-routes and Victor airways, Class E airspace begins at 1,200 feet AGL, or other altitudes specified on the chart.

Due to ATC requirements in a few remote and coastal areas, you might encounter Class E airspace that begins at 14,500 feet MSL or that has a floor indicated in feet MSL noted on the chart. As with other Class E airspace, the ceiling of these areas is the floor of Class A airspace at 18,000 feet MSL. [Figure 4-63]

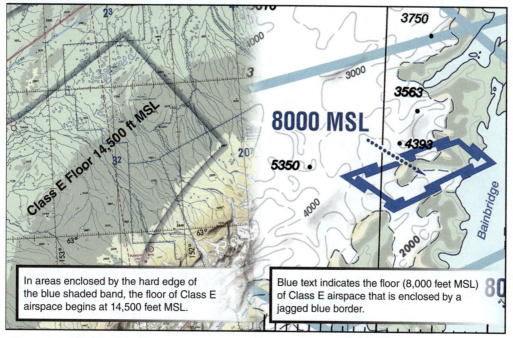

In areas enclosed by the hard edge of the blue shaded band, the floor of Class E airspace begins at 14,500 feet MSL.

Blue text indicates the floor (8,000 feet MSL) of Class E airspace that is enclosed by a jagged blue border.

Figure 4-63. Class E airspace might be depicted on the chart in unique ways in remote and coastal areas.

CLASS D AIRSPACE

An airport that has an operating control tower, but does not provide Class B or C airspace ATC services, is surrounded by **Class D airspace**. The control tower provides sequencing and traffic advisories to VFR aircraft operating into and out of the airport, and IFR traffic separation. You must establish two-way radio communication with the tower prior to entering Class D airspace and maintain radio contact during all operations to, from, or on that airport.

You have established two-way radio communication only if the controller responds to your radio call with your call sign. For example, if, after transmitting to a control tower, you hear the response *"Cessna 123 Whiskey Tango, standby"* radio communication has been established and you may enter the airspace. In contrast, if you hear *"Aircraft calling Janesville tower, standby"* you may not enter the airspace because the controller did not state your call sign. As a general rule, you should avoid Class D airspace except to take off or land at an airport within the area.

The airspace at an airport with a part-time control tower is designated as Class D only when the tower is in operation. At airports where the tower operates part time, the airspace changes to Class E, or a combination of Class E and Class G when the tower is closed. For these airports, check the Airport/Facility Directory in the Chart Supplement for the tower's hours of operation and the airspace designation.

Class D airspace can be various sizes and shapes, depending on the instrument approach procedures established for that airport. Most Class D airspace is a circle with a radius of approximately 4 NM extending up to 2,500 feet AGL. [Figure 4-64]

In some Class D airspace areas, a satellite airport may be located within the airspace designated for the primary airport. If a temporary control tower is in operation at the satellite airport, you should contact it for arrival and departure. When the satellite airport is a nontowered field, you must establish contact with the primary airport's control tower.

SECTION D ■ Airspace

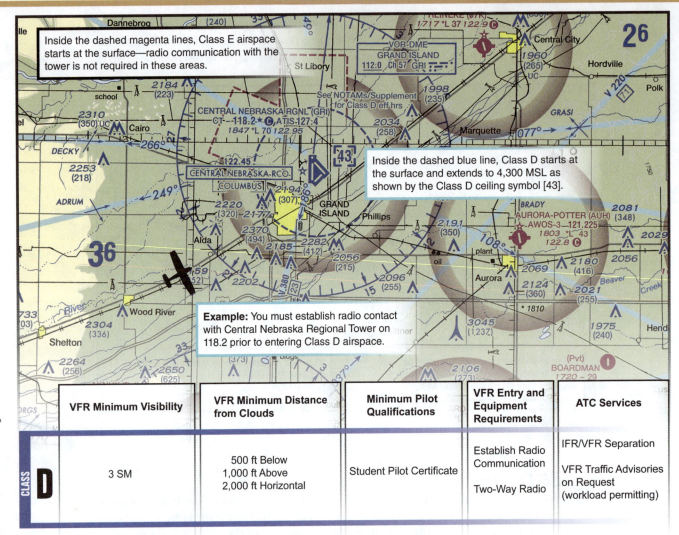

Inside the dashed magenta lines, Class E airspace starts at the surface—radio communication with the tower is not required in these areas.

Inside the dashed blue line, Class D starts at the surface and extends to 4,300 MSL as shown by the Class D ceiling symbol [43].

Example: You must establish radio contact with Central Nebraska Regional Tower on 118.2 prior to entering Class D airspace.

	VFR Minimum Visibility	VFR Minimum Distance from Clouds	Minimum Pilot Qualifications	VFR Entry and Equipment Requirements	ATC Services
CLASS D	3 SM	500 ft Below 1,000 ft Above 2,000 ft Horizontal	Student Pilot Certificate	Establish Radio Communication Two-Way Radio	IFR/VFR Separation VFR Traffic Advisories on Request (workload permitting)

Figure 4-64. Class D airspace is around an airport where you must be in radio contact with the control tower.

When departing a nontowered satellite airport in Class D airspace, contact the controlling tower on the ground if possible, or as soon as practicable after takeoff. To the maximum extent practical, and consistent with safety, satellite airports are excluded from Class D airspace. For example, airspace might be carved out of a Class D area to allow traffic to arrive and depart from a nontowered satellite airport.

CLASS C AIRSPACE

Factors considered in designating controlled airspace include safety, users' needs, and the volume of air traffic. Because of these considerations, many busy airports are surrounded by **Class C airspace**. Within a Class C area, ATC provides radar service to all IFR and VFR aircraft, and participation in this service is mandatory. An outer area normally extends out to 20 nautical miles from the primary airport, where Class C services are available but not mandatory.

Class C areas usually have similar dimensions from one location to another, although some may be modified to fit unique aspects of a particular airport's location. A 5 nautical mile radius core area extends from the surface to 4,000 feet above the elevation of the primary airport. A 10 nautical mile radius shelf area usually extends from 1,200 feet to 4,000 feet above the airport elevation. Aeronautical charts depict the MSL altitudes of the floor and ceiling of each segment of Class C airspace.

Prior to entering Class C airspace, you must establish two-way communication with the ATC facility having jurisdiction and maintain it while you are operating within the airspace. When you are departing the primary airport, you must maintain radio contact with ATC until you are clear of the area. In addition to the two-way radio requirement, all aircraft operating in a Class C area, and in all airspace above it, must be equipped with ADS-B Out and a transponder with Mode C capability. Aircraft operating in the airspace beneath a Class C shelf area are not required to have ADS-B Out or a Mode C transponder. [Figure 4-65]

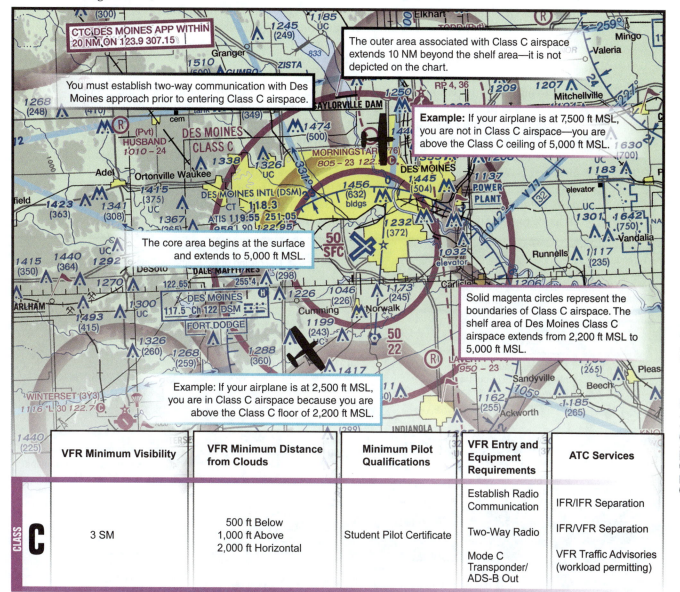

	VFR Minimum Visibility	VFR Minimum Distance from Clouds	Minimum Pilot Qualifications	VFR Entry and Equipment Requirements	ATC Services
CLASS **C**	3 SM	500 ft Below 1,000 ft Above 2,000 ft Horizontal	Student Pilot Certificate	Establish Radio Communication Two-Way Radio Mode C Transponder/ ADS-B Out	IFR/IFR Separation IFR/VFR Separation VFR Traffic Advisories (workload permitting)

Figure 4-65. You are required to communicate with ATC in the core and shelf areas of Class C that typically extend to 10 NM from the airport.

A few Class C ATC facilities are part-time, so some services are not always available. If the ATC facility is closed, the operating rules for the Class C area not in effect. Hours of operation for ATC facilities are listed in the Airport/Facility Directory in the Chart Supplement.

CLASS B AIRSPACE

At some of the country's major airports, **Class B airspace** has been established to separate all arriving and departing traffic. Although each Class B area usually is designated for a major terminal, it typically serves several airports in the area. Each Class B area is individually designed to serve the needs of the particular airport that it surrounds. Terrain, the amount and flow of air traffic, and the location of other airports all influence each design. Generally, you will find that Class B airspace surrounds the busiest airports in the country.

A Mode C transponder and ADS-B Out are required within 30 nautical miles of the Class B area's primary airport, from the surface to 10,000 feet MSL. In addition, a VOR or TACAN is required for IFR operations. In order to fly within Class B airspace, or to take off or land at an airport within that airspace, you must possess at least a private pilot certificate. In certain Class B areas, student pilots may be permitted to conduct flight operations by obtaining specified training and a logbook endorsement from a certificated flight instructor. However, student pilot operations are prohibited at designated major airports within the nation's busiest Class B areas. Refer to FAR Part 91 for specific rules pertaining to student pilot operations within Class B airspace.

Prior to entering any part of Class B airspace, you are required to obtain a clearance from ATC on the appropriate frequency. You must advise ATC of your intended altitude and route of flight before departing an airport in a Class B area. ATC permission is required before you can fly through Class B airspace, even after a departure from an airport that is other than the primary airport. Whenever you are flying VFR in or around Class B airspace, a VFR terminal area charts (TAC) helps significantly with orientation and navigation. TACs show the lateral limits of the various sections of the Class B area with more detail than sectional charts. Sectional charts display a white border around Class B airspace to indicate the area covered by a TAC. [Figure 4-66]

SECTION D ■ Airspace

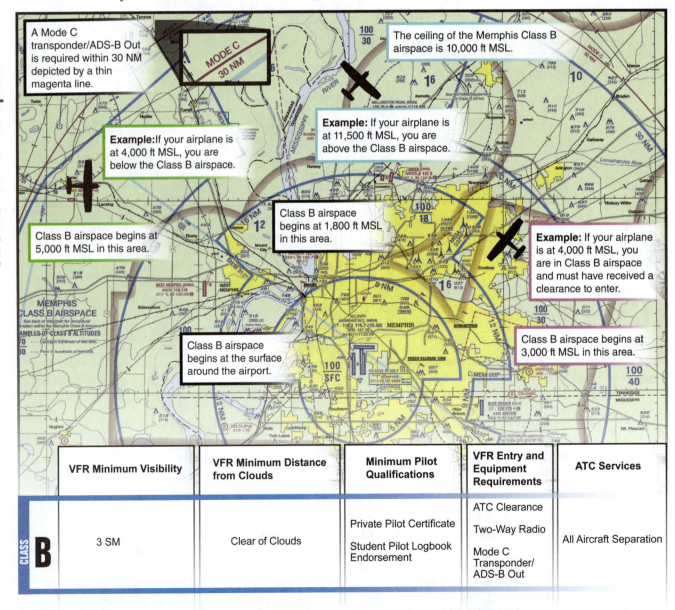

A Mode C transponder/ADS-B Out is required within 30 NM depicted by a thin magenta line.

The ceiling of the Memphis Class B airspace is 10,000 ft MSL.

Example: If your airplane is at 11,500 ft MSL, you are above the Class B airspace.

Example: If your airplane is at 4,000 ft MSL, you are below the Class B airspace.

Class B airspace begins at 1,800 ft MSL in this area.

Class B airspace begins at 5,000 ft MSL in this area.

Example: If your airplane is at 4,000 ft MSL, you are in Class B airspace and must have received a clearance to enter.

Class B airspace begins at 3,000 ft MSL in this area.

Class B airspace begins at the surface around the airport.

	VFR Minimum Visibility	VFR Minimum Distance from Clouds	Minimum Pilot Qualifications	VFR Entry and Equipment Requirements	ATC Services
CLASS B	3 SM	Clear of Clouds	Private Pilot Certificate Student Pilot Logbook Endorsement	ATC Clearance Two-Way Radio Mode C Transponder/ADS-B Out	All Aircraft Separation

Figure 4-66. The TAC shows ceiling and floor in hundreds of feet MSL is show for each Class B airspace sector is labeled with its MSL altitude on this TAC.

VFR flyway planning charts are published on the reverse side of some VFR terminal area charts or can be downloaded with these charts. The flyway planning charts show VFR routes for transitioning around, under, and through Class B airspace. These routes are not intended to discourage requests for VFR operations, but are designed to help you avoid heavily congested areas, such as IFR arrival and departure routes.

Flyway charts omit most of the terrain features and geographic information found on terminal area charts because they are for planning, not navigating. However, major landmarks are shown as visual aids to orientation. Routes shown on flyway charts include **VFR transition routes**, **VFR flyways**, and **special flight rules areas (SFRAs)**. Some SFRAs were originally designed as less restrictive VFR corridors. [Figure 4-67]

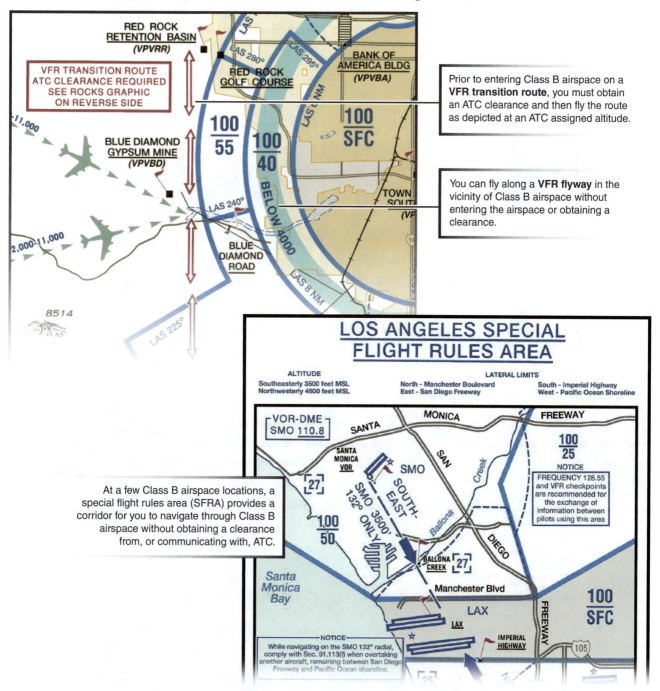

Prior to entering Class B airspace on a **VFR transition route**, you must obtain an ATC clearance and then fly the route as depicted at an ATC assigned altitude.

You can fly along a **VFR flyway** in the vicinity of Class B airspace without entering the airspace or obtaining a clearance.

At a few Class B airspace locations, a special flight rules area (SFRA) provides a corridor for you to navigate through Class B airspace without obtaining a clearance from, or communicating with, ATC.

Figure 4-67. Flyway planning charts help you navigate around and through Class B airspace.

SECTION D ■ Airspace

CLASS A AIRSPACE

The airspace that extends from 18,000 feet MSL up to and including FL600 is defined as **Class A airspace**. It covers the majority of the contiguous states and Alaska, as well as the area extending 12 nautical miles out from the U.S. coast. To operate within Class A airspace, you must be instrument rated and your aircraft must be transponder equipped, operate on an IFR flight plan, and be under positive ATC control. Because of the high speeds of airplanes operating in Class A airspace and the corresponding increase of the closure rates between these aircraft, VFR flight is not allowed. Jet routes are designed to serve aircraft operations from the floor of Class A airspace up to and including FL450.

Within Class A airspace, you are required to set your altimeter to the standard setting of 29.92 inches Hg so that all pilots maintain their assigned altitudes using the same altimeter reference. Altitudes within Class A airspace are expressed to ATC by using the term flight level (FL). [Figure 4-68]

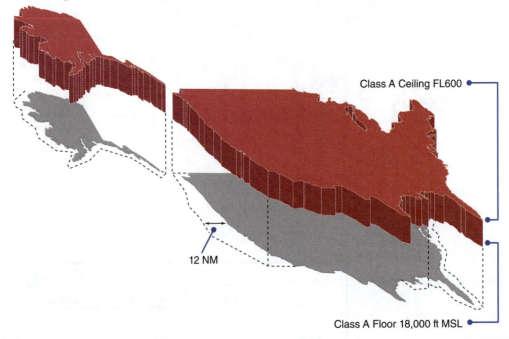

Class A Ceiling FL600

12 NM

Class A Floor 18,000 ft MSL

CLASS	VFR Minimum Visibility	VFR Minimum Distance from Clouds	Minimum Pilot Qualifications	VFR Entry and Equipment Requirements	ATC Services
A	N/A	N/A	Private Pilot Certificate Instrument Rating	IFR Flight Plan IFR Clearance	All Aircraft Separation

Figure 4-68. Class A airspace extends above 18,000 feet MSL over the majority of the U.S. so it is not shown on aeronautical charts.

SPECIAL VFR

In addition to maintaining the VFR minimums already discussed, you may only operate within the areas of Class B, C, D, or E airspace that extend to the surface around an airport, when the ground visibility is at least 3 statute miles and the cloud ceiling is at least 1,000 feet AGL. If ground visibility is not reported, you can use flight visibility. When the weather is below these VFR minimums, and there is no conflicting IFR traffic, you may obtain a **special VFR clearance** from the ATC facility having jurisdiction over the affected airspace. A special VFR clearance allows you to enter, leave, or operate within most Class D and Class E surface areas and in some Class B and Class C surface areas if the flight visibility is at least 1 statute mile and you can remain clear of clouds. At least 1 statute mile of ground visibility is required for takeoff and landing. However, if ground visibility is not reported, you must have at least 1 statute mile flight visibility.

As a private pilot, you may obtain a special VFR clearance only during the daytime. Because of the difficulty in seeing clouds at night, special VFR is not permitted between sunset and sunrise unless you have a current instrument rating and the aircraft is equipped for instrument flight. At some major airports, special VFR clearances are not available. [Figure 4-69]

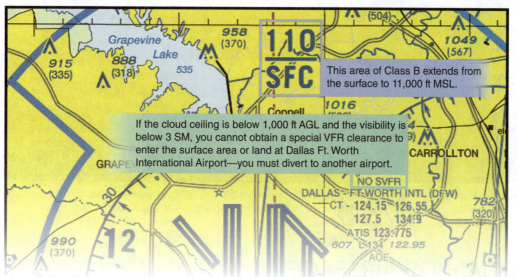

This area of Class B extends from the surface to 11,000 ft MSL.

If the cloud ceiling is below 1,000 ft AGL and the visibility is below 3 SM, you cannot obtain a special VFR clearance to enter the surface area or land at Dallas Ft. Worth International Airport—you must divert to another airport.

Figure 4-69. The phrase NO SVFR indicates that you cannot obtain a special VFR clearance to operate at the airport.

Scud Running

Imagine that you are on a cross-country flight. Everything has been progressing smoothly but as you scan the horizon, you notice some clouds up ahead. *"It must just be a high scattered layer,"* you say to yourself. *"My briefing didn't mention anything about low clouds."* As you proceed on course, you realize that the clouds are at your altitude. *"No problem,"* you think, *"I'll just drop down a little lower, stay underneath this weather until it clears up. The forecast for the airport was clear."* As you continue to descend lower and lower to avoid the clouds, visual references begin to disappear and the sky turns white. Panic sets in…

You have just attempted what many pilots refer to as "scud running"—trying to stay below the clouds while continuing into deteriorating weather conditions. When inexperienced, noninstrument-rated pilots find themselves in this situation, the outcome often is fatal. NTSB statistics indicate that approximately 25% of all general aviation accidents are weather related, as well as nearly 40% of all fatal accidents. The NTSB cites "continued VFR flight into adverse weather/IMC" as the primary cause in many of these accidents. Inadvertent entry into instrument weather conditions can result in either flying into terrain or experiencing spatial disorientation.

How can you avoid an unplanned flight into instrument conditions? The first step is to define safe personal weather minimums for yourself and stick to them. Set conservative ceiling and visibility values, which are higher than those required by the FARs. Start with a thorough weather briefing prior to your flight and if there is any question about the conditions, don't go. Learn how to obtain weather information enroute so you can keep updated on changing conditions. If you do inadvertently enter IFR weather, maintain control of the airplane and make a 180° turn back to VFR conditions. If you cannot maintain VFR, do not let your pride keep you from contacting ATC or Flight Service for assistance.

AIRSPEED LIMITATIONS

Because the airspace at lower altitudes, and especially in the vicinity of airports, tends to be congested, the FAA has established aircraft speed restrictions. In general, flights below 10,000 feet MSL are limited to a maximum indicated airspeed of 250 knots. When operating in Class C or D airspace, at or below 2,500 feet above the surface, and within 4 nautical miles of the primary airport, you must not exceed 200 knots indicated airspeed. This 200-knot restriction also applies in airspace underlying a Class B area and in VFR corridors through Class B airspace. [Figure 4-70]

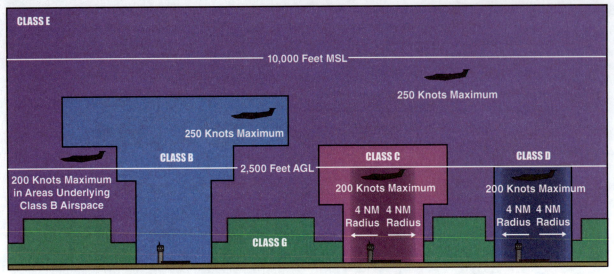

Figure 4-70. Although you might not approach these maximum airspeeds in your training airplane, these limitations will apply when you transition to larger airplanes, especially in the pursuit of a professional pilot career.

SPECIAL USE AIRSPACE

Special use airspace is used to confine certain flight activities and to place limitations on aircraft operations that are not part of these activities. Special use airspace is divided into alert areas, military operations areas (MOAs), warning areas, restricted areas, prohibited areas, controlled firing areas, and national security areas (NSAs). [Figure 4-71]

SPECIAL USE AIRSPACE ON CHEYENNE SECTIONAL CHART

Unless otherwise noted altitudes are
MSL and in feet. Time is local.
"TO" an altitude means "To and including."
FL – Flight Level
NO A/G – No air to ground communications.
Contact Flight Service for information.

† Other times by NOTAM.
NOTAM – Use of this term in Restricted
Areas indicates FAA and DoD NOTAM
systems. Use of this term in all
other Special Use areas indicates the
DoD NOTAM system.

U.S. P–PROHIBITED, R–RESTRICTED, W–WARNING, A–ALERT, MOA–MILITARY OPERATIONS AREA

NUMBER	ALTITUDE	TIME OF USE	CONTROLLING AGENCY/ CONTACT FACILITY	FREQUENCIES
R-7001 A	TO BUT NOT INCL 8000	INTERMITTENT BY NOTAM 24 HRS IN ADVANCE	DENVER CNTR	135.6 385.6
R-7001 B	8000 TO 23,500	INTERMITTENT BY NOTAM 24 HRS IN ADVANCE	DENVER CNTR	135.6 385.6
R-7002 A, B, C	TO 23,500	BY NOTAM 24 HRS IN ADVANCE	DENVER CNTR	

MOA NAME	ALTITUDE*	TIME OF USE†	CONTROLLING AGENCY/ CONTACT FACILITY	FREQUENCIES
POWDER RIVER 2 HIGH	12,000 MSL	0730-1200 & 1800-2330 MON-THU & 0730-1200 FRI BY NOTAM 2 HRS IN ADVANCE; O/T BY NOTAM 4 HRS IN ADVANCE	DENVER CNTR	127.95 338.2
POWDER RIVER 2 LOW	500 AGL TO BUT NOT INCLUDING 12,000 MSL	0730-1200 & 1800-2330 MON-THU & 0730-1200 FRI BY NOTAM 2 HRS IN ADVANCE; O/T BY NOTAM 4 HRS IN ADVANCE	DENVER CNTR	127.95 338.2

*Altitudes indicate floor of MOA. All MOAs extend to but do not include FL 180 unless otherwise indicated in tabulation or on chart.
†Other times by DoD NOTAM.

Figure 4-71. By referring to tables on each sectional chart, you can determine the altitudes, times of use, and controlling agencies for the special use airspace depicted on that specific chart.

ALERT AREAS

Areas shown on aeronautical charts to inform you of unusual types of aerial activities, such as parachute jumping, glider towing, or high concentrations of student pilot training are designated as **alert areas**. Pilots of participating aircraft and pilots transiting the area are equally responsible for collision avoidance, so you should be especially cautious when flying through alert areas. [Figure 4-72]

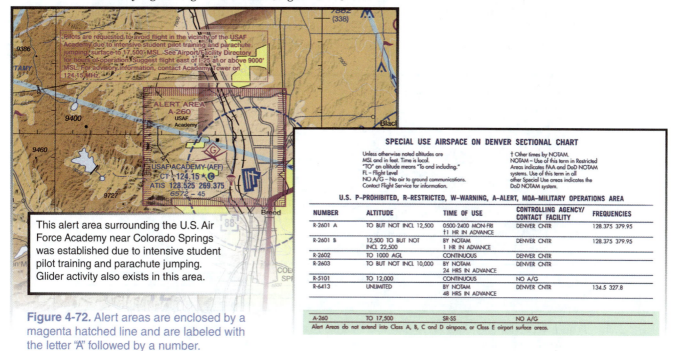

This alert area surrounding the U.S. Air Force Academy near Colorado Springs was established due to intensive student pilot training and parachute jumping. Glider activity also exists in this area.

Figure 4-72. Alert areas are enclosed by a magenta hatched line and are labeled with the letter "A" followed by a number.

SPECIAL USE AIRSPACE ON DENVER SECTIONAL CHART

Unless otherwise noted altitudes are MSL and in feet. Time is local. "TO" an altitude means "To and including." FL – Flight Level NO A/G – No air to ground communications. Contact Flight Service for information.

† Other times by NOTAM. NOTAM – Use of this term in Restricted Areas indicates FAA and DoD NOTAM systems. Use of this term in all other Special Use areas indicates the DoD NOTAM system.

U.S. P–PROHIBITED, R–RESTRICTED, W–WARNING, A–ALERT, MOA–MILITARY OPERATIONS AREA

NUMBER	ALTITUDE	TIME OF USE	CONTROLLING AGENCY/ CONTACT FACILITY	FREQUENCIES
R-2601 A	TO BUT NOT INCL 12,500	0500-2400 MON-FRI †1 HR IN ADVANCE	DENVER CNTR	128.375 379.95
R-2601 B	12,500 TO BUT NOT INCL 22,500	BY NOTAM 1 HR IN ADVANCE	DENVER CNTR	128.375 379.95
R-2602	TO 1000 AGL	CONTINUOUS	DENVER CNTR	
R-2603	TO BUT NOT INCL 10,000	BY NOTAM 24 HRS IN ADVANCE	DENVER CNTR	
R-5101	TO 12,000	CONTINUOUS	NO A/G	
R-6413	UNLIMITED	BY NOTAM 48 HRS IN ADVANCE	DENVER CNTR	134.5 327.8
A-260	TO 17,500	SR-SS	NO A/G	

Alert Areas do not extend into Class A, B, C and D airspace, or Class E airport surface areas.

MILITARY OPERATIONS AREAS

A **military operations area (MOA)** is a block of airspace in which military training and other military maneuvers are conducted. MOAs usually have specified floors and ceilings for containing military activities. VFR aircraft are not prevented from flying through active MOAs, but it is wise to avoid them when possible. If you do choose to fly in an MOA, you should exercise extreme caution when military activity is being conducted. Most military training activities require acrobatic or abrupt flight maneuvers at high speeds. Flight Service can advise you of the hours of operation of an MOA along your route, or you can check the special use airspace panel on the edge of your chart. [Figure 4-73]

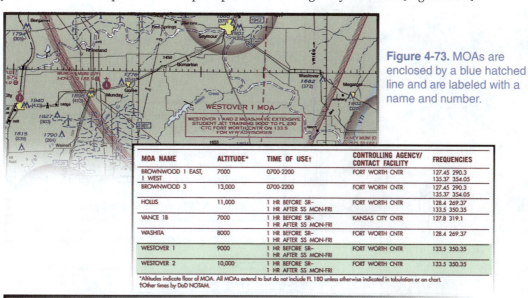

Figure 4-73. MOAs are enclosed by a blue hatched line and are labeled with a name and number.

MOA NAME	ALTITUDE*	TIME OF USE†	CONTROLLING AGENCY/ CONTACT FACILITY	FREQUENCIES
BROWNWOOD 1 EAST, 1 WEST	7000	0700-2200	FORT WORTH CNTR	127.45 290.3 135.37 354.05
BROWNWOOD 3	13,000	0700-2200	FORT WORTH CNTR	127.45 290.3 135.37 354.05
HOLLIS	11,000	1 HR BEFORE SR– 1 HR AFTER SS MON-FRI	FORT WORTH CNTR	128.4 269.37 133.5 350.35
VANCE 1B	7000	1 HR BEFORE SR– 1 HR AFTER SS MON-FRI	KANSAS CITY CNTR	127.8 319.1
WASHITA	8000	1 HR BEFORE SR– 1 HR AFTER SS MON-FRI	FORT WORTH CNTR	128.4 269.37
WESTOVER 1	9000	1 HR BEFORE SR– 1 HR AFTER SS MON-FRI	FORT WORTH CNTR	133.5 350.35
WESTOVER 2	10,000	1 HR BEFORE SR– 1 HR AFTER SS MON-FRI	FORT WORTH CNTR	133.5 350.35

*Altitudes indicate floor of MOA. All MOAs extend to but do not include FL 180 unless otherwise indicated in tabulation or on chart.
†Other times by DoD NOTAM.

As shown on the margin of the Dallas Ft. Worth sectional chart, operations in Westover 1 and 2 MOAs begin at 9,000 ft MSL and 10,000 ft MSL, respectively, and extend up to but not including FL180.

Military activities can occur from 1 hour before sunrise to 1 hour after sunset Monday through Friday.

Prior to entering Westover 1 or 2 MOAs if they are active, contact Fort Worth Center for advisories.

WARNING AREAS

A **warning area** is airspace of defined dimensions, extending from three nautical miles outward from the coast of the United States, that contains activity that may be hazardous to nonparticipating aircraft. Warning areas are depicted on aeronautical charts to caution nonparticipating pilots of the potential hazards, such as aerial gunnery and guided missiles. A warning area may be located over domestic or international waters or both. [Figure 4-74]

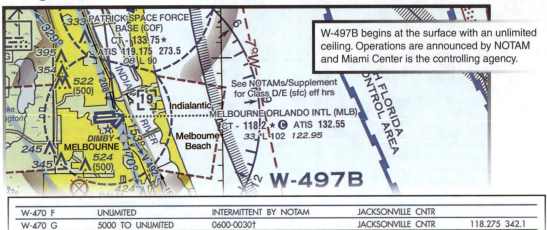

W-497B begins at the surface with an unlimited ceiling. Operations are announced by NOTAM and Miami Center is the controlling agency.

W-470 F	UNLIMITED	INTERMITTENT BY NOTAM	JACKSONVILLE CNTR	
W-470 G	5000 TO UNLIMITED	0600-0030†	JACKSONVILLE CNTR	118.275 342.1
W-497 A, B	UNLIMITED	BY NOTAM	MIAMI CNTR	

Figure 4-74. Warning areas are enclosed with a blue hatched line and labeled with the letter "W" followed by a number.

RESTRICTED AREAS

Restricted areas often have invisible hazards to aircraft, such as artillery firing, aerial gunnery, or guided missiles. Permission to fly through restricted areas must be granted by the controlling agency. The dimensions of most restricted areas are typically easy to plan flights around. However, extra caution is appropriate even when you are flying near the area. [Figure 4-75]

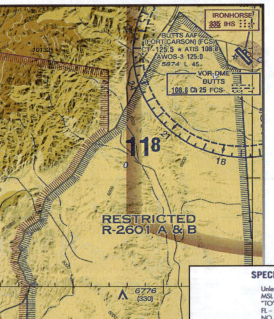

Figure 4-75 Restricted areas are enclosed by a blue hatched line and are labeled with the letter "R" followed by a number.

As shown on the margin of the Colorado Springs terminal area chart:
- R-2601 A extends from the surface up to but not including 12,500 ft MSL and is in use from 0500 to 2400 local time Monday through Friday and by NOTAM.
- R-2601 B extends from 12,500 ft MSL to 22,500 ft MSL and is in use by NOTAM.

SPECIAL USE AIRSPACE ON COLORADO SPRINGS TERMINAL AREA CHART

Unless otherwise noted altitudes are MSL and in feet. Time is local.
"TO" an altitude means "To and including."
FL – Flight Level
NO A/G – No air to ground communications.
Contact Flight Service for information.

† Other times by NOTAM.
NOTAM – Use of this term in Restricted Areas indicates FAA and DoD NOTAM systems. Use of this term in all other Special Use areas indicates the DoD NOTAM system.

U.S. P–PROHIBITED, R–RESTRICTED, W–WARNING, A–ALERT, MOA–MILITARY OPERATIONS AREA

NUMBER	ALTITUDE	TIME OF USE	CONTROLLING AGENCY/ CONTACT FACILITY	FREQUENCIES
R-2601 A	TO BUT NOT INCL 12,500	0500-2400 MON-FRI †1 HR IN ADVANCE	DENVER CNTR	128.375 379.95
R-2601 B	12,500 TO BUT NOT INCL 22,500	BY NOTAM 1 HR IN ADVANCE	DENVER CNTR	128.375 379.95
R-2602	TO 1000 AGL	CONTINUOUS	DENVER CNTR	

PROHIBITED AREAS

Prohibited areas are established for security or other reasons associated with national welfare and they contain airspace within which the flight of aircraft is prohibited. They are shown with the letter "P" followed by a number on charts. You may not operate within a prohibited area without permission from the controlling agency. [Figure 4-76]

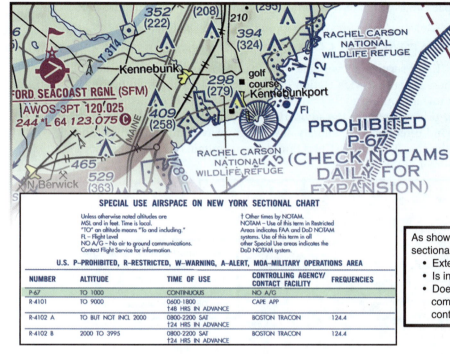

Figure 4-76. Prohibited areas are enclosed with a blue hatched line and labeled with the letter "P" followed by a number.

As shown on the margin of the New York sectional chart, P-67:
- Extends from the surface to 1,000 ft MSL.
- Is in use continuously.
- Does not provide air-to-ground communications. You must contact Flight Service for information.

SPECIAL USE AIRSPACE ON NEW YORK SECTIONAL CHART

Unless otherwise noted altitudes are MSL and in feet. Time is local.
"TO" an altitude means "To and including."
FL – Flight Level
NO A/G – No air to ground communications. Contact Flight Service for information.

† Other times by NOTAM.
NOTAM – Use of this term in Restricted Areas indicates FAA and DoD NOTAM systems. Use of this term in all other Special Use areas indicates the DoD NOTAM system.

U.S. P–PROHIBITED, R–RESTRICTED, W–WARNING, A–ALERT, MOA–MILITARY OPERATIONS AREA

NUMBER	ALTITUDE	TIME OF USE	CONTROLLING AGENCY/ CONTACT FACILITY	FREQUENCIES
P-67	TO 1000	CONTINUOUS	NO A/G	
R-4101	TO 9000	0600-1800 †48 HRS IN ADVANCE	CAPE APP	
R-4102 A	TO BUT NOT INCL 2000	0800-2200 SAT †24 HRS IN ADVANCE	BOSTON TRACON	124.4
R-4102 B	2000 TO 3995	0800-2200 SAT †24 HRS IN ADVANCE	BOSTON TRACON	124.4

NATIONAL SECURITY AREAS

National security areas (NSAs) are established at locations where there is a requirement for increased security and safety of ground facilities. You are requested to voluntarily avoid flying through an NSA. At times, flight through an NSA may be prohibited to provide a greater level of security and safety. A NOTAM is issued to advise you of any changes in an NSA's status. You should comply with all restrictions related to an NSA, although the rules are normally voluntary. [Figure 4-77]

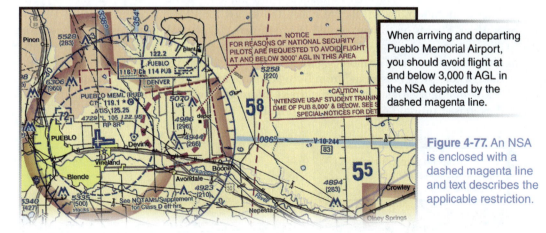

When arriving and departing Pueblo Memorial Airport, you should avoid flight at and below 3,000 ft AGL in the NSA depicted by the dashed magenta line.

Figure 4-77. An NSA is enclosed with a dashed magenta line and text describes the applicable restriction.

CONTROLLED FIRING AREAS

The distinguishing feature of a **controlled firing area**, compared to other special use airspace, is that its activities are discontinued immediately when a spotter aircraft, radar, or ground lookout personnel determines an aircraft might be approaching the area. Because nonparticipating aircraft are not required to change their flight path, controlled firing areas are not depicted on aeronautical charts.

SECTION D ■ Airspace

OTHER AIRSPACE AREAS

Most **other airspace areas** typically do not have the same types of restrictions or hazardous activities that apply to special use airspace areas. Other airspace areas include local airport advisory areas, military training routes, parachute jump aircraft operations areas, and terminal radar service areas. Temporary flight restrictions (TFRs) also are also classified as other airspace areas and have restrictions that are unique to the specific TFR.

LOCAL AIRPORT ADVISORY SERVICE

Local airport advisory (LAA) service is available only in Alaska and extends 10 statute miles from airports where a flight service station (FSS) is located on the field but there is no operating control tower. You normally will contact the FSS on the published CTAF frequency of 123.6 MHz prior to entering the advisory area. The FSS provides local airport information, such as wind direction and velocity, favored runway, altimeter setting, and reported traffic within the area. The Alaska Chart Supplement contains a listing of airports with LAA service. [Figure 4-78]

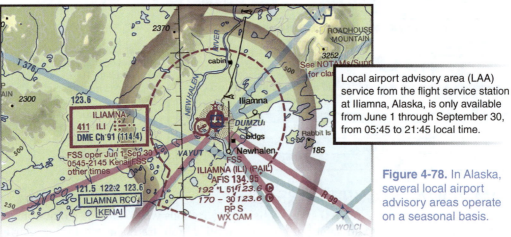

Local airport advisory area (LAA) service from the flight service station at Iliamna, Alaska, is only available from June 1 through September 30, from 05:45 to 21:45 local time.

Figure 4-78. In Alaska, several local airport advisory areas operate on a seasonal basis.

MILITARY TRAINING ROUTES

Low-level, high-speed military training flights are conducted on **military training routes (MTRs)**. Generally, MTRs are established below 10,000 feet MSL for operations at speeds in excess of 250 knots. MTRs are classified as VR for VFR operations and IR for IFR operations. Flights on routes marked IR are under ATC control regardless of the weather. MTRs that are entirely at or below 1,500 feet AGL are identified by four-digit numbers, and those that have one or more segments above 1,500 feet AGL use three-digit numbers. Although you are not restricted from flying through MTRs, it is good operating practice to check with Flight Service to obtain current information about MTR activity in your area, and exercise caution if operating there. [Figure 4-79]

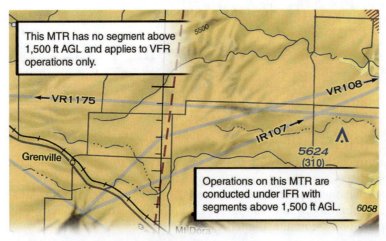

This MTR has no segment above 1,500 ft AGL and applies to VFR operations only.

Operations on this MTR are conducted under IFR with segments above 1,500 ft AGL.

Figure 4-79. High-speed military activity—faster than 250 knots—may occur on military training routes.

PARACHUTE JUMP AIRCRAFT OPERATIONS

Parachute jump aircraft operations areas are tabulated in the Chart Supplement. The busiest periods of activity are normally on weekends and holidays. Times of operation are local, and MSL altitudes are listed unless otherwise specified. Parachute jumping sites that have been used on a frequent basis and that have been in use for at least one year are depicted on sectional charts. [Figure 4-80]

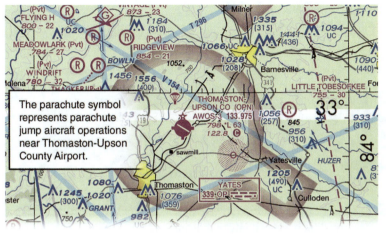

The parachute symbol represents parachute jump aircraft operations near Thomaston-Upson County Airport.

Figure 4-80. You should be alert for pilots announcing parachute activities at the airport when you see the parachute symbol.

TERMINAL RADAR SERVICE AREAS

Terminal radar service areas (TRSAs) do not fit into any of the U.S. airspace classes. Originally part of the terminal radar program at selected airports, TRSAs have never been established as controlled airspace and, therefore, FAR Part 91 does not contain any rules for TRSA operations. By contacting approach control, you can receive radar services within a TRSA, but participation is not mandatory. The primary airport within the TRSA is surrounded by Class D airspace and the outer portion of a TRSA normally overlies Class E airspace beginning at 700 or 1,200 feet AGL. [Figure 4-81]

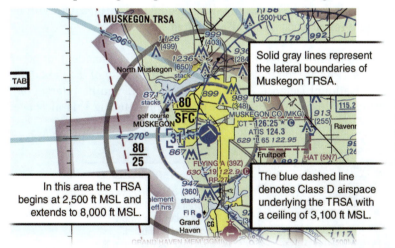

Solid gray lines represent the lateral boundaries of Muskegon TRSA.

In this area the TRSA begins at 2,500 ft MSL and extends to 8,000 ft MSL.

The blue dashed line denotes Class D airspace underlying the TRSA with a ceiling of 3,100 ft MSL.

Figure 4-81. A TRSA is not a specific class of airspace, but an area that is collocated with Class D and E airspace where terminal radar service is available.

TEMPORARY FLIGHT RESTRICTIONS

Temporary flight restrictions (TFRs) are regulatory actions that temporarily restrict certain aircraft from operating within a defined area in order to protect persons or property in the air or on the ground. Several types of TFRs are defined by FARs. [Figure 4-82]

Disaster/hazard TFRs are established near disaster or hazard areas to protect persons or property on the surface or in the air from a hazard associated with an incident on the surface, provide a safe environment for disaster relief aircraft operations, and prevent unsafe congestion of sightseeing or other aircraft above the incident. A separate regulation covers natural disaster areas in Hawaii.

SECTION D ■ Airspace

Figure 4-82. TFRs are implemented for a wide variety of reasons.

Space flight operations TFRs provide a safe environment for space launch operations, typically in Florida, New Mexico, and California. The NOTAMs that create these TFRs usually activate existing special use airspace or airspace adjacent to these areas.

TFRs are issued when high barometric pressure exists. Very cold, dry air masses can produce barometric pressures in excess of 31.00 inches of mercury (inHg), and many aircraft altimeters cannot be adjusted above that setting. When an altimeter cannot be set to pressure settings above 31.00 inHg, the aircraft's true altitude will be higher than the indicated altitude on the altimeter. When any information indicates that the barometric pressure on a route of flight currently exceeds or will exceed 31.00 inHg, a TFR will be issued that establishes flight restrictions in the affected area.

VIP TFRs are flight restrictions in the proximity of the President, Vice President, or other government officials. These TFRs are the most restrictive in the vicinity of the President. VIP TFRs are considered National Defense Airspace and violators can be intercepted, face criminal prosecution, or lose their pilot certificates for entering this airspace. [Figure 4-83]

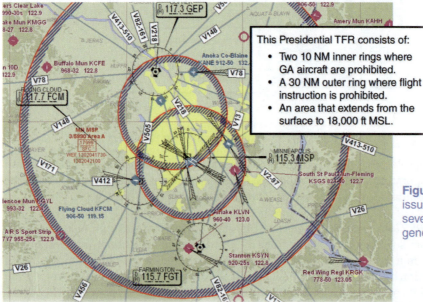

This Presidential TFR consists of:
- Two 10 NM inner rings where GA aircraft are prohibited.
- A 30 NM outer ring where flight instruction is prohibited.
- An area that extends from the surface to 18,000 ft MSL.

Figure 4-83. TFRs that are issued for VIP movement severely restrict or ban general aviation activity.

In response to the 9/11 terrorist attacks, the FAA closed all U.S. airspace for the first time in history. This is an extreme example of the type of situation that would require the FAA to issue emergency air traffic rules. These rules allow the FAA to implement TFRs if an emergency condition exists, or will exist, that prevents safe and efficient operation of the air traffic control system.

On a case-by-case basis, the FAA establishes TFRs for air shows and sporting events. Generally these restrictions encompass the minimum airspace needed for the management of aircraft operations near the event. The FAA also issues TFRs under FAR 99.7 Special Security Instructions. These TFRs address situations determined to be detrimental to the interests of national defense. An example of this type of TFR is described by a standing NOTAM that applies to major sporting events. [Figure 4-84]

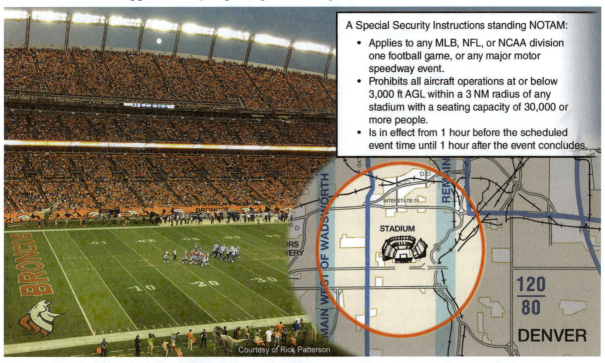

A Special Security Instructions standing NOTAM:

- Applies to any MLB, NFL, or NCAA division one football game, or any major motor speedway event.
- Prohibits all aircraft operations at or below 3,000 ft AGL within a 3 NM radius of any stadium with a seating capacity of 30,000 or more people.
- Is in effect from 1 hour before the scheduled event time until 1 hour after the event concludes.

Courtesy of Rick Patterson

Figure 4-84. You must be aware of this standing NOTAM if you are flying in an area that hosts major sporting events.

TFRs are issued in NOTAMs that specify the dimensions, restrictions, and effective times. To determine if a TFR affects your flight, obtain NOTAMs from Flight Service during your online or phone briefing. You can also obtain a list of TFR NOTAMs with graphic depictions at tfr.faa.gov. The FAA cautions that the depicted TFR data might not be a complete listing, so always follow up with Flight Service during flight planning. [Figure 4-85]

DISCOVERY

Spaceports

The FAA's Office of Commercial Space transportation, referred to as AST, plays a vital role in commercial space launch and reentry operations in the U.S. AST issues licenses for commercial space transportation activities and for spaceport operations. As the commercial space industry continues to grow and evolve, more and more spaceports are established. Facilities, such as the Mojave Air & Space Port and Spaceport America are expanding the horizons of commercial spaceflight. (The Wallops Flight Facility, a spaceport in Virginia is shown in the photo.) During launch and reentry of commercial space vehicles, AST coordinates with ATC to ensure that conventional air traffic can safely operate at the same time. With the help of AST, spaceports around the country will continue to make commercial space endeavors possible, including missions such as public suborbital flight and the establishment of commercial space stations.

Courtesy of NASA

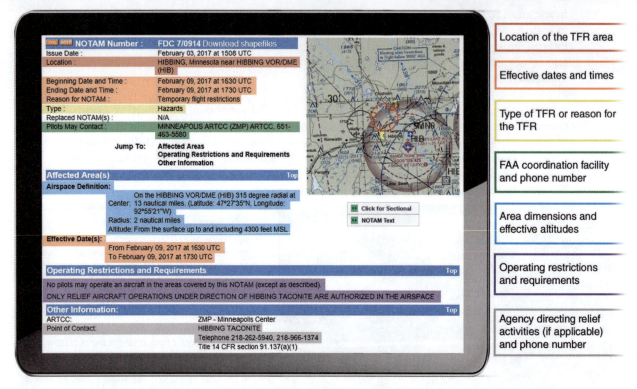

Figure 4-85. At tfr.faa.gov, select each TFR that applies to your flight to view detailed information.

ADIZ

An **Air Defense Identification Zone (ADIZ)** facilitates early identification of all aircraft in the vicinity of a nation's airspace boundaries. The AIM and FAR Part 99 specify requirements to enter a United States ADIZ. The Alaskan ADIZ, which lies along the coastal waters of Alaska has different operating rules than the contiguous U.S. ADIZ. To operate within the Contiguous U.S. ADIZ, you must file an IFR or defense VFR (DVFR) flight plan containing the time and point at which you plan to enter the ADIZ. Set your Mode C transponder to the assigned code prior to entering the ADIZ and maintain two-way communication with the appropriate ATC facility. You must depart the ADIZ within 5 minutes of the estimated departure time in the flight plan. If you are planning to fly in an ADIZ, you should refer to the AIM or the *International Flight Information Manual* for detailed procedural information. [Figure 4-86]

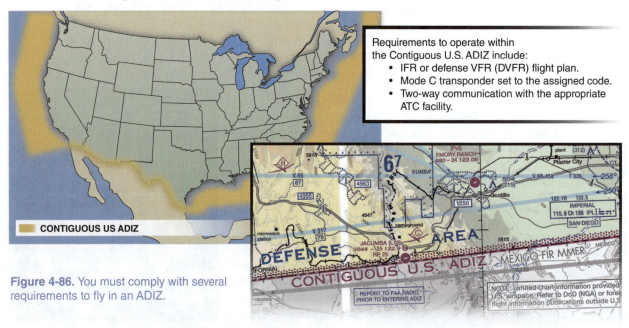

Requirements to operate within the Contiguous U.S. ADIZ include:
- IFR or defense VFR (DVFR) flight plan.
- Mode C transponder set to the assigned code.
- Two-way communication with the appropriate ATC facility.

Figure 4-86. You must comply with several requirements to fly in an ADIZ.

WASHINGTON DC
SPECIAL FLIGHT RULES AREA

The **Washington DC Special Flight Rules Area (SFRA)** is airspace where the ready identification, location, and control of aircraft is required in the interests of national security. Depicted on charts, the SFRA includes all airspace within a 30 nautical mile radius of the Washington DC VOR (DCA) from the surface up to but not including flight level 180 (FL180). This includes areas with additional requirements: the Leesburg Maneuvering Area with its own special procedures, and the **flight restricted zone (FRZ)**—a highly-restricted ring of airspace within 13 to 15 nautical miles of the Washington DC VOR, which is directly over the nation's capital. Only specially authorized aircraft may fly in the FRZ under IFR flight plans. Flight under VFR and general aviation aircraft operations are prohibited. [Figure 4-87]

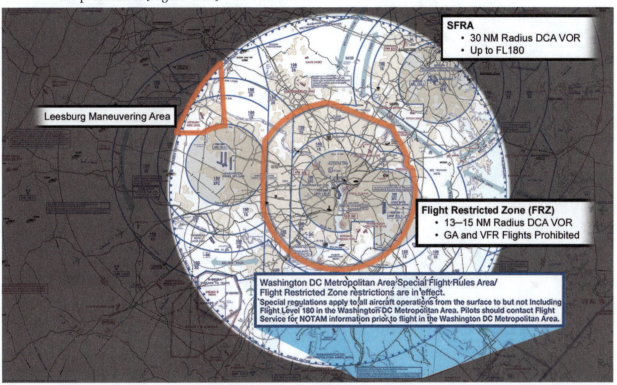

Figure 4-87. The dimensions of the Washington DC SFRA are shown here on the VFR Baltimore/Washington Flyway Planning Chart.

SECTION D ■ Airspace

Visual Warning System

If you were to stray into the Washington DC SFRA without meeting the requirements, the government might point a highly focused laser at your aircraft, warning you to turn away and contact ATC on the appropriate frequency or 121.5. This Visual Warning System (VWS) consists of an alternating red and green signal pattern visible only from an encroaching aircraft.

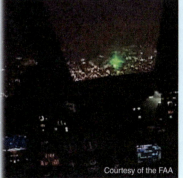

Courtesy of the FAA

DISCOVERY

If you are planning to fly under VFR within 60 nautical miles of the Washington DC VOR you must complete the FAA Special Awareness Training course at faasafety.gov. After you finish the training, print the completion certificate and carry it with you. You must present it at the request of an FAA, NTSB, law enforcement, or TSA authority. The Special Awareness course covers all you need to know to operate in or near the

SFRA. However, you should be aware of some of the basic requirements, including filing a special flight plan with specifies entry and exit points, communication requirements, and speed restrictions. [Figure 4-88]

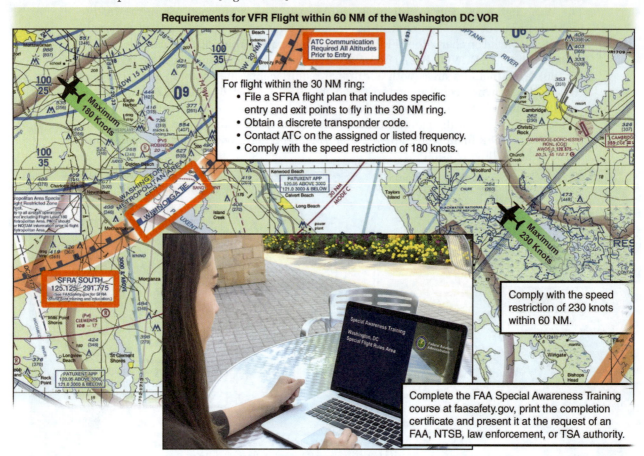

Requirements for VFR Flight within 60 NM of the Washington DC VOR

ATC Communication Required All Altitudes Prior to Entry

Maximum 180 Knots

For flight within the 30 NM ring:
- File a SFRA flight plan that includes specific entry and exit points to fly in the 30 NM ring.
- Obtain a discrete transponder code.
- Contact ATC on the assigned or listed frequency.
- Comply with the speed restriction of 180 knots.

Maximum 230 Knots

Comply with the speed restriction of 230 knots within 60 NM.

Complete the FAA Special Awareness Training course at faasafety.gov, print the completion certificate and present it at the request of an FAA, NTSB, law enforcement, or TSA authority.

Figure 4-88. If you are planning to fly under VFR within 60 NM of the Washington DC VOR, you must meet specific requirements.

INTERCEPT PROCEDURES

If you penetrate an area with security-related flight restrictions, you risk being intercepted by U.S. military or law enforcement aircraft. Review the **intercept procedures** in the AIM regularly. [Figure 4-89]

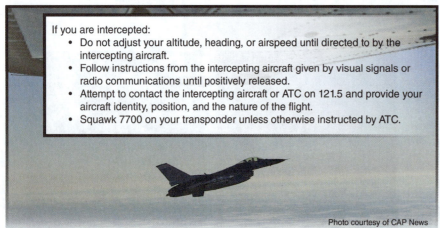

If you are intercepted:
- Do not adjust your altitude, heading, or airspeed until directed to by the intercepting aircraft.
- Follow instructions from the intercepting aircraft given by visual signals or radio communications until positively released.
- Attempt to contact the intercepting aircraft or ATC on 121.5 and provide your aircraft identity, position, and the nature of the flight.
- Squawk 7700 on your transponder unless otherwise instructed by ATC.

Photo courtesy of CAP News

Figure 4-89. Follow the procedures outline in the AIM if your airplane is intercepted.

During an interception, two aircraft will approach from behind to identify your airplane. One aircraft flies to the left of your airplane, matches your speed and heading, rocks its wings or flashes its navigation lights if it is night. You should acknowledge by rocking your wings or flashing your navigation lights at night. If they understand your intentions, the intercepting aircraft might break away by performing a

90° turn without crossing your flight path, which means you are free to go. If this does not occur, you must comply with the intercepting aircraft's instructions. [Figure 4-90]

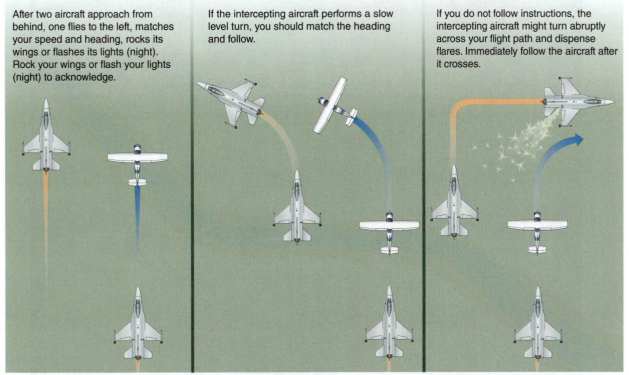

After two aircraft approach from behind, one flies to the left, matches your speed and heading, rocks its wings or flashes its lights (night). Rock your wings or flash your lights (night) to acknowledge.

If the intercepting aircraft performs a slow level turn, you should match the heading and follow.

If you do not follow instructions, the intercepting aircraft might turn abruptly across your flight path and dispense flares. Immediately follow the aircraft after it crosses.

Figure 4-90. If you are not in radio contact with the intercepting aircraft, you must be able to interpret visual signals, and take action to comply.

If the intercepting aircraft circles an airport, lowers its landing gear, and overflies a runway, land on the runway. If you cannot land safely, overfly the runway with your gear up (if applicable), flash your landing light, circle the airport between 1,000 to 2,000 feet AGL, and wait for further instructions. [Figure 4-91]

Intercepting Aircraft Signal	Meaning	Intercepted Aircraft Response
Approaches from behind, normally positions on the left, and rocks its wings. At night, same and flashes its navigation lights at irregular intervals.	You have been intercepted.	Rock your wings. At night, rock your wings and flash your navigation lights at irregular intervals.
After acknowledgement, performs a slow level turn, normally to the left, onto the desired heading.	Follow me. Fly this way.	Match the heading of the intercepting aircraft, and follow.
Turns abruptly across your flight path, and dispenses flares.	Warning! Turn in the direction of the intercepting aircraft immediately.	Immediately follow the intercepting aircraft after it crosses your flight path.
Circles an airport, lowers landing gear, and overflies the runway. At night, landing lights are on.	Land at this airport.	Land on the runway. At night, landing lights on. If you cannot land safely: • Overfly the runway with your gear up. • Flash your landing light. • Circle the airport between 1,000 and 2,000 feet AGL. • Wait for further instructions.
Breaks away by performing a 90° turn without crossing your flight path.	Intercepting aircraft understands your intentions. You are free to go.	Rock your wings to acknowledge, and proceed on course.

Figure 4-91. This table summarizes the intercept signals, their meanings, and proper responses.

Now that you are familiar with the various classes of airspace, Chapter 5 will provide you with additional information regarding specific ATC services available and the radio procedures used to communicate with ATC in the airspace system. You can review the primary classes of airspace, their relationship to one another, and airspace operating requirements by referring to Figures 4-92 and 4-93.

SECTION D ■ Airspace

SECTION D ■ Airspace

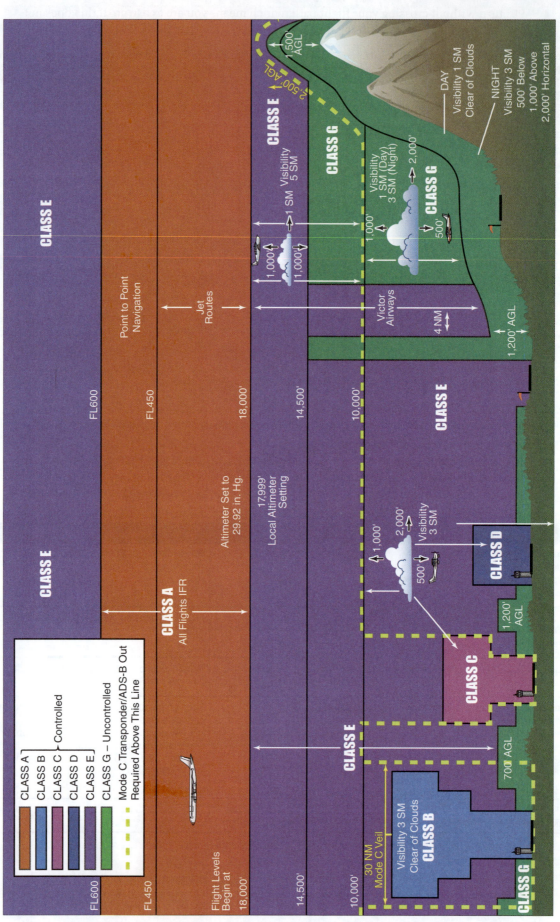

Figure 4-92 This diagram helps you to visualize how the various airspace classes interrelate.

	VFR Minimum Visibility	VFR Minimum Distance from Clouds	Minimum Pilot Qualifications	VFR Entry and Equipment Requirements	ATC Services
CLASS A	N/A	N/A	Private Pilot Certificate Instrument Rating	IFR Flight Plan IFR Clearance	All Aircraft Separation
CLASS B	3 SM	Clear of Clouds	Private Pilot Certificate Student Pilot Logbook Endorsement	ATC Clearance Two-Way Radio Mode C Transponder/ ADS-B Out	All Aircraft Separation
CLASS C	3 SM	500 ft Below 1,000 ft Above 2,000 ft Horizontal	Student Pilot Certificate	Establish Radio Communication Two-Way Radio Mode C Transponder/ ADS-B Out	IFR/IFR Separation IFR/VFR Separation VFR Traffic Advisories (workload permitting)
CLASS D	3 SM	500 ft Below 1,000 ft Above 2,000 ft Horizontal	Student Pilot Certificate	Establish Radio Communication Two-Way Radio	IFR/IFR Separation VFR Traffic Advisories (workload permitting)
CLASS E	**Below 10,000 ft MSL:** 3 SM **At or Above 10,000 ft MSL:** 5 SM	**Below 10,000 ft MSL:** 500 ft Below 1,000 ft Above 2,000 ft Horizontal **At or Above 10,000 ft MSL:** 1,000 ft Below 1,000 ft Above 1 SM Horizontal	Student Pilot Certificate	**Below 10,000 ft MSL:** No Specific Equipment Required **At or Above 10,000 ft MSL and Above 2,500 ft AGL:** Mode C Transponder and ADB-B Out	IFR/IFR Separation VFR Traffic Advisories on Request (workload permitting)
CLASS G	**1,200 ft AGL and Below:** Day 1 SM; Night 3 SM **Below 10,000 ft MSL:** Day 1 SM; Night 3 SM **At or Above 10,000 MSL :** 5 SM (above1,200 ft AGL)	**1,200 ft AGL and Below:** Day; Clear of Clouds Night; 500 ft Below, 1,000 ft Above, 2,000 ft Horizontal **Below 10,000 ft MSL:** 500 ft Below 1,000 ft Above 2,000 ft Horizontal (above 1,200 ft AGL) **At or Above 10,000 ft MSL:** 1,000 ft Below 1,000 ft Above 1 SM Horizontal (above 1,200 ft AGL)	Student Pilot Certificate	**Below 10,000 ft MSL:** No Specific Equipment Required **At or Above 10,000 ft MSL and Above 2,500 ft AGL:** Mode C Transponder	VFR Traffic Advisories on Request (workload permitting)

Figure 4-93. You can use this table as a quick reference for the VFR weather minimums and operating requirements of each airspace class. When trying to remember the weather minimums, it is easier to remember the minimums that are the same (highlighted) and know the few exceptions.

SECTION D ■ Airspace

SUMMARY CHECKLIST

✓ In each class of airspace, you must maintain specific VFR weather minimums (minimum flight visibilities and distances from clouds).

✓ While operating in controlled airspace (Class A, Class B, Class C, Class D, Class E) you are subject to certain operating rules, as well as pilot qualification and aircraft equipment requirements.

✓ Class G airspace typically extends from the surface to 700 or 1,200 feet AGL. In some areas, Class G may extend from the surface to 14,500 feet MSL.

✓ When the weather conditions deteriorate below VFR minimums, all aircraft in controlled airspace must only be flown by instrument-rated pilots in accordance with IFR clearances issued by ATC.

✓ A transponder is an electronic device aboard the airplane that enhances your aircraft's identity on an ATC display and aid in the separation of traffic.

✓ The FARs require an operating transponder with Mode C capability (ADS-B after 2020) when flying at or above 10,000 feet MSL over the 48 contiguous states (excluding the airspace at and below 2,500 feet AGL), in Class A airspace, in and above Class B and Class C airspace, within 30 nautical miles of Class B primary airports.

✓ ATC does not exercise control of air traffic in uncontrolled, or Class G, airspace.

✓ There are no communication requirements to operate within Class E airspace, but you can request traffic advisory services that ATC provides on a workload-permitting basis.

✓ Federal airways or Victor airways are usually 8 nautical miles wide, begin at 1,200 feet AGL, and extend up to but not including 18,000 feet MSL.

✓ You must establish two-way radio communication with the control tower prior to entering Class D airspace and maintain radio contact during all operations to, from, or on that airport.

✓ Prior to entering Class C airspace, you must establish two-way communication with the ATC facility having jurisdiction and maintain it while you are operating within the airspace. Within a Class C area, ATC provides services to all IFR and VFR aircraft.

✓ Class B airspace is established at some of the country's major airports to separate all arriving and departing traffic. Generally, Class B airspace surrounds the busiest airports in the country.

✓ Prior to entering any part of Class B airspace, you are required to obtain a clearance from ATC.

✓ To operate in Class B airspace, you must be at least a private pilot or a student pilot with the appropriate logbook endorsement.

✓ Whenever you are flying VFR in or around Class B airspace, VFR terminal area charts (TACs) will help significantly with orientation and navigation. Sectional charts display a white border around the area covered by a TAC.

✓ VFR flyway planning charts, published on the reverse side of some VFR terminal area charts, show VFR routes for transitioning around, under, and through Class B airspace.

✓ To operate within Class A airspace, you must be instrument rated and your aircraft must be transponder equipped, operate on an IFR flight plan, and be under positive ATC control.

✓ A special VFR clearance must be obtained from ATC to operate within the surface areas of Class B, C, D, or E airspace when the ground visibility is less than 3 statute miles and the cloud ceiling is less than 1,000 feet AGL.

✓ Aircraft speed restrictions include a maximum indicated airspeed of 250 knots below 10,000 feet MSL and a maximum of 200 knots when operating in Class C or D airspace, at or below 2,500 feet above the surface and within 4 nautical miles of the primary airport.

✓ This 200-knot speed restriction applies in airspace underlying a Class B area and in VFR corridors through Class B airspace.

✓ Alert areas are shown on aeronautical charts to inform you of unusual types of aerial activities, such as parachute jumping and glider towing, or high concentrations of student pilot training.

✓ A military operations area (MOA) is a block of airspace in which military operations are conducted. VFR aircraft are not prevented from flying through active MOAs, but it is wise to avoid them when possible.

✓ Warning areas extend from three nautical miles outward from the coast of the United States and contain activity that may be hazardous to nonparticipating aircraft.

✓ Restricted areas often have invisible hazards to aircraft, such as artillery firing, aerial gunnery, or guided missiles. Permission to fly through restricted areas must be granted by the controlling agency.

✓ Prohibited areas are established for security or other reasons associated with the national welfare and contain airspace within which the flight of aircraft is prohibited.

✓ National security areas (NSAs) are established at locations where there is a requirement for increased security and safety of ground facilities. While you are always requested to avoid flying through an NSA, a NOTAM will be issued when flight is prohibited.

✓ Activities within a controlled firing area are discontinued immediately when a spotter aircraft, radar, or ground lookout personnel determines an aircraft might be approaching the area.

✓ Local airport advisory (LAA) areas extend 10 statute miles from airports in Alaska where there is an FSS located on the field and no operating control tower.

✓ Generally, military training routes (MTRs) are established below 10,000 feet MSL for operations at speeds in excess of 250 knots.

✓ Frequently used parachute jump aircraft operations areas that have been in use for at least one year are depicted on sectional charts.

✓ Temporary flight restrictions (TFRs) are regulatory actions that temporarily restrict certain aircraft from operating within a defined area in order to protect persons or property in the air or on the ground.

✓ TFRs are issued in NOTAMs that specify the dimensions, restrictions, and effective times.

✓ Types of TFRs include those that address disaster and hazards, space flight operations, VIP movement, emergency air traffic rules, air shows and sporting events on a case-by-case basis, and major sporting events.

✓ A Special Security Instructions standing NOTAM applies to any MLB, NFL, or NCAA division one football game, or any major motor speedway event.

SECTION D ■ Airspace

✓ Air defense identification zones (ADIZs) are established to facilitate early identification of all aircraft in the vicinity of a nation's airspace boundaries.

✓ Depicted on charts, the Washington DC Special Flight Rules Area (SFRA) includes all airspace within a 30 nautical mile radius of the Washington DC VOR (DCA) from the surface up to but not including FL180.

✓ Flight under VFR and general aviation aircraft operations are prohibited in the flight restricted zone (FRZ) of the SFRA—a highly-restricted ring of airspace within 13 to 15 nautical miles of the Washington DC VOR.

✓ If you are planning to fly under VFR within 60 nautical miles of the Washington DC VOR you must complete the FAA Special Awareness Training course at faasafety.gov.

✓ If you penetrate an area with security-related flight restrictions, you risk being intercepted by U.S. military or law enforcement aircraft.

✓ Attempt contact on 121.5 and follow instructions from the intercepting aircraft given by visual signals or radio communications until positively released.

KEY TERMS

Airspace	Special VFR Clearance
VFR Weather Minimums	Special Use Airspace
Controlled Airspace	Alert Area
Uncontrolled Airspace	Military Operations Area (MOA)
Transponder	Warning Area
Class G Airspace	Restricted Area
Class E Airspace	Prohibited Area
Class E Transition Areas	National Security Area (NSA)
Class E Surface Areas	Controlled Firing Area
Federal Airways	Other Airspace Areas
Victor Airways	Military Training Route (MTR)
T-Routes	Parachute Jump Aircraft Operations
Class D Airspace	Terminal Radar Service Area (TRSA)
Class C Airspace	Temporary Flight Restriction (TFR)
Class B Airspace	Air Defense Identification Zone (ADIZ)
VFR Flyway Planning Chart	
VFR Transition Routes	Washington DC Special Flight Rules Area (SFRA)
VFR Flyways	Flight Restricted Zone (FRZ)
Special Flight Rules Areas (SFRAs)	Intercept Procedures
Class A Airspace	

QUESTIONS

Use the chart excerpt to answer Questions 1 through 9.

1. You are flying south of the city of Timmonsville, maneuvering at 1,000 feet AGL. What airspace class are you in? What minimum cloud clearance and visibility must you maintain?

2. You are on the downwind leg of the traffic pattern at Marion Country Airport (MAO) at 1,000 feet AGL. What minimum cloud clearance and visibility must you maintain?

3. You are flying at 11,500 feet MSL on V437. What minimum visibility and cloud clearance must you maintain?

4. What are the lateral dimensions of V3-157?

5. The ceiling is 900 feet and the visibility is 2 statute miles at Florence Regional Airport and the control tower is in operation. Select the true statement regarding operating at this airport.
   A. You cannot enter the Class D airspace because no special VFR is allowed at this airport.
   B. You must request a special VFR clearance to enter, leave, or operate within the Class D area
   C. Because you will be in Class G airspace in the traffic pattern, you can take off and land as long at you maintain 1 statute mile visibility and remain clear of clouds.

6. What is VR1043?
   A. A Victor airway based on VOR or VORTAC navaids that begins at 1,200 feet MSL and extends to 17,999 feet MSL.
   B. A military training route on which VFR military training flights operate at speeds in excess of 250 knots below 1,500 feet AGL.
   C. A military training route on which VFR military training flights operate at speeds in excess of 250 knots above 1,500 feet AGL.

You are flying over Florence Regional Airport at 5,500 feet MSL. Answer Questions 7, 8, and 9 regarding this scenario.

7. What airspace class are you in?

8. Are you required to be in contact with Florence Regional Tower? Why or why not?

9. What minimum cloud clearance and visibility must you maintain?

SECTION D ■ Airspace

Use the chart excerpt. You are in the position shown by the airplane symbol at 9,500 feet MSL. Answer Questions 10, 11, and 12 regarding this scenario.

10.   What airspace class are you in?

11.   What minimum visibility and cloud clearance must you maintain?

12.   What are the VFR entry and equipment requirements to be in the airspace?

Use the chart excerpt. You are in the position shown by the airplane symbol at 5,000 feet MSL. Answer Questions 13 through 16 regarding this scenario.

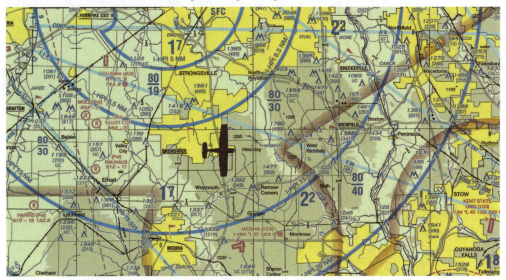

13.   What airspace class are you in?

14.   What minimum visibility and cloud clearance must you maintain?

15.   What are the VFR entry and equipment requirements to be in the airspace?

16.   If you descend to 2,500 feet MSL, what airspace class are you in?

17.   Select the true statement regarding transponder operation.
      A.   To enter Class D airspace, you are required to have a Mode C transponder.
      B.   You are required to have a Mode C transponder from 2,500 feet AGL up to and including 10,000 feet MSL.
      C.   The FARs require that you have an operating Mode C transponder in Class B airspace and within 30 nautical miles of the Class B primary airport.

18. At what altitude should you set the altimeter to 29.92, when climbing to cruising flight level?
    A. 14,500 feet MSL
    B. 18,000 feet MSL
    C. 24,000 feet MSL

19. What is the maximum airspeed that you can fly below 10,000 feet MSL?

20. What is an MOA?

21. What special use airspace often has invisible hazards to aircraft, such as artillery firing, aerial gunnery, or guided missiles and permission is required to enter?
    A. Restricted area
    B. Controlled firing area
    C. Military operations area

22. List at least three types of TFRs.

23. How can you learn the time and location of TFRs?

24. You are flying VFR and not talking to ATC. You observe a military jet fighter off your left wingtip, and it matches your speed and heading. It then rocks its wings, and makes a slow left turn away from you. What must you do?

SECTION D ■ Airspace

CHAPTER 5

Communication and Flight Information

SECTION A
ATC Services

When you are on a flight as a pilot in the United States, you will be operating one of the approximately 7,000 aircraft that are flying at any given time in 5,000,000 square miles of airspace. You will take off and land at one of over 19,000 airports and you will take advantage of the state-of-the art technologies and services that are provided by air traffic control (ATC) as part of the FAA NextGen system.

ADS-B SYSTEM

The **automatic dependent surveillance-broadcast (ADS-B)** system incorporates GPS, aircraft transmitters and receivers, and ground stations to provide pilots and ATC with specific data about the position and speed of aircraft. Two forms of ADS-B equipment apply to aircraft—ADS-B Out and ADS-B In. ADS-B Out signals travel line of sight from transmitting aircraft to ATC ground receivers and aircraft receivers. In order to receive the signal and display traffic information in your aircraft, you must also have ADS-B In capability. [Figure 5-1]

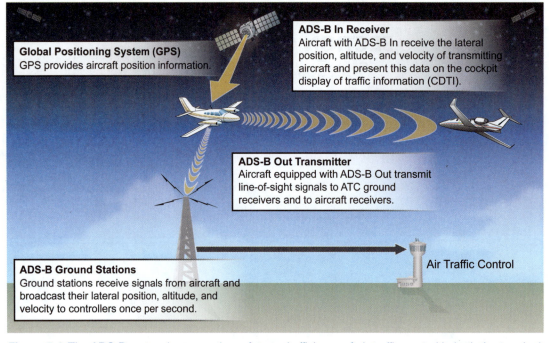

ADS-B In Receiver
Aircraft with ADS-B In receive the lateral position, altitude, and velocity of transmitting aircraft and present this data on the cockpit display of traffic information (CDTI).

Global Positioning System (GPS)
GPS provides aircraft position information.

ADS-B Out Transmitter
Aircraft equipped with ADS-B Out transmit line-of-sight signals to ATC ground receivers and to aircraft receivers.

ADS-B Ground Stations
Ground stations receive signals from aircraft and broadcast their lateral position, altitude, and velocity to controllers once per second.

Air Traffic Control

Figure 5-1. The ADS-B system improves the safety and efficiency of air traffic control in both the terminal and enroute environments.

ADS-B SERVICES

The ADS-B system provides precise real-time data that immediately indicates to controllers when an aircraft deviates from its assigned flight path. With an effective range of 100 nautical miles, ADS-B gives ATC a large area in which to implement traffic conflict detection and resolution. Aircraft equipped with ADS-B In receive traffic data from other ADS-B aircraft on a cockpit display of traffic information (CDTI), which can be a dedicated display or integrated into an existing display, such as a GPS moving map or multi-function display (MFD). [Figure 5-2]

Figure 5-2. ADS-B provides detailed traffic information to both controllers and pilots.

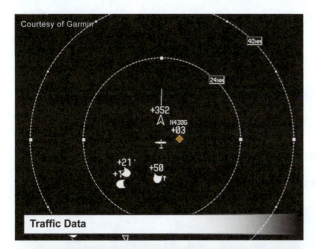

Traffic Data

The ADS-B system displays aircraft with transponders that are not ADS-B equipped using the traffic information service-broadcast (TIS-B). ATC and pilots also can use ADS-B to monitor aircraft surface movement. You can use this feature to increase situational awareness while taxiing and reduce the risk of runway incursion. In addition to traffic data, ADS-B can provide the flight information service-broadcast (FIS-B) to suitably-equipped aircraft. FIS-B delivers a broad range of textual and graphical weather products, as well as other flight information, such as the location of temporary flight restrictions (TFRs) and special use airspace (SUA) status. [Figure 5-3]

Aircraft Surface Movement

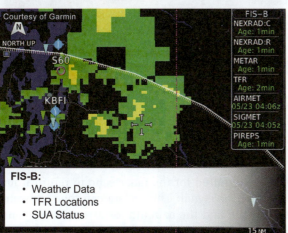

FIS-B:
• Weather Data
• TFR Locations
• SUA Status

Figure 5-3. ADS-B provides you with a variety of flight environment information to increase situational awareness.

ADS-B DATA LINKS

Aircraft transmit and receive information using one or both of two ADS-B data links. Your aircraft equipment can use the 1090 MHz extended squitter (1090ES) link for operations at any altitude and you must use this link at or above 18,000 feet MSL. This link enables TIS-B data but does not enable FIS-B. The 978 MHz universal access transceiver (978 UAT) is limited to altitudes below 18,000 feet MSL. You can receive both TIS-B and FIS-B data using this data link. To fly at any altitude and receive FIS-B data, you can transmit (ADS-B Out) on 1090ES and receive (ADS-B In) on both 1090ES and 978 UAT. [Figure 5-4]

1090ES
- For operations at any altitude.
- Required for operations at or above 18,000 feet MSL.
- Enables TIS-B.
- Does not enable FIS-B.

Both
- For operations at any altitude.
- ADS-B Out on 1090ES.
- ADS-B In on both 1090ES and 978 UAT.

18,000 Feet MSL

978 UAT
- Limited to altitudes below 18,000 feet MSL.
- Enables TIS-B and FIS-B.

Figure 5-4. The altitude at which your airplane operates is a factor in the data link that you must use.

DISCOVERY

B-2

The Northrup Grumman B-2 Bomber is designed to have little or no radar return.

- The exterior design and classified radar-absorbing surface coating produce a radar cross-section of about 0.1 square meters.
- The airframe's geometry reflects radar waves away from the radar transmitter.
- The engine air intakes are positioned to mask the radar-reflective fan blades.

RADAR

In addition to ADS-B, ATC uses **radar** to monitor aircraft—a ground-based synchronized radio transmitter and receiver emits and detects radio waves using a rotating antenna. When the radio waves strike your airplane, some of the waves reflect back to the antenna.

RADAR OPERATION

The primary radar system measures the time required for the radar echo to return from your airplane and determines the direction of the signal. The system can then indicate your airplane's distance from the radar antenna and the azimuth, or direction, of your airplane in relation to the antenna. [Figure 5-5]

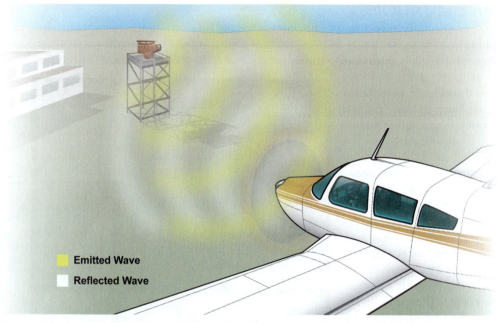

Figure 5-5. The radar antenna is designed to both transmit radio waves and receive the reflected signals.

Secondary surveillance radar, or the **air traffic control radar beacon system (ATCRBS)**, enables ATC to assign a discrete code that you enter on the transponder in your airplane. An interrogator transmits a signal that causes the transponder to reply automatically with a specific coded signal, independent of, and much stronger than the primary radar return. The transponder signal is used to display an identification code, as well as your airplane's altitude and ground speed. The secondary surveillance radar antenna is typically attached to the top of the primary radar antenna and primary and secondary returns are shown together on a single display. [Figure 5-6]

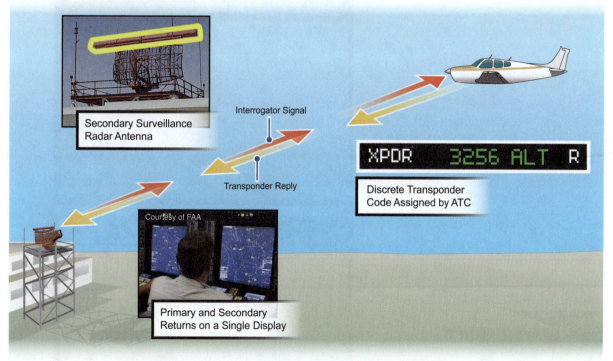

Secondary Surveillance Radar Antenna

Interrogator Signal

Transponder Reply

XPDR 3256 ALT R

Discrete Transponder Code Assigned by ATC

Courtesy of FAA

Primary and Secondary Returns on a Single Display

Figure 5-6. Secondary surveillance radar consists of ground and airborne equipment.

SECTION A ■ **ATC Services**

RADAR LIMITATIONS

Radio waves normally travel in a continuous straight line unless they are "bent" by atmospheric phenomena, such as temperature inversions. This bending is called anomalous propagation. If the radio waves bend toward the ground, extraneous returns known as ground clutter can appear on the radar screen. Radio waves that bend upward reduce the detection range. Radio waves also can be reflected by dense objects, such as heavy clouds, precipitation, ground obstacles, or mountains. The reflected waves can block out aircraft at the same range and greatly weaken or completely eliminate the display of targets at a greater range. Finally, radio waves can be screened by high terrain features. Relatively low altitude aircraft are not seen if they are obstructed by mountains or are below the radio waves due to earth curvature. [Figure 5-7]

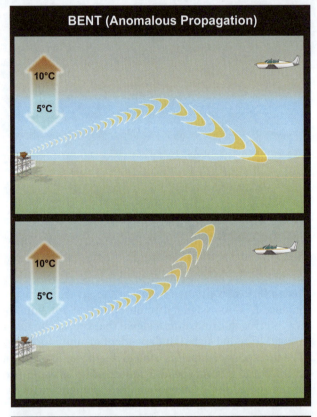

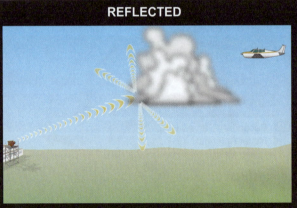

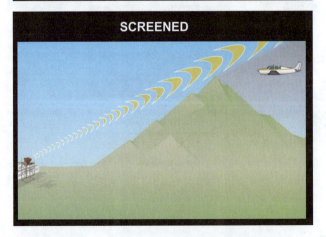

Figure 5-7. Radar has some significant limitations.

TRANSPONDERS

Both ADS-B and radar systems use your airplane's **transponder** to provide precise information, such as your airplane's altitude and speed. Although the appearance of controls and displays vary, most transponders have the same general features. [Figure 5-8]

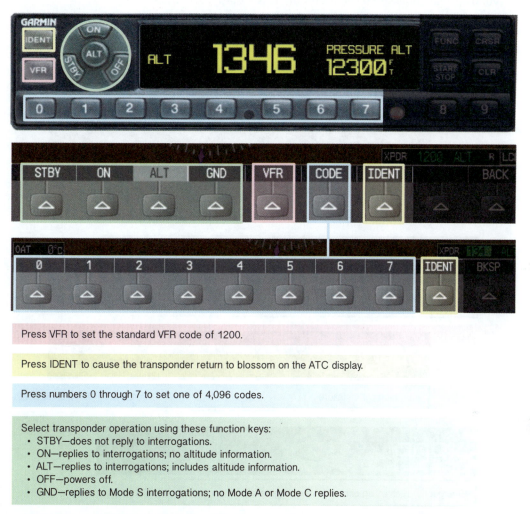

Press VFR to set the standard VFR code of 1200.

Press IDENT to cause the transponder return to blossom on the ATC display.

Press numbers 0 through 7 to set one of 4,096 codes.

Select transponder operation using these function keys:
- STBY—does not reply to interrogations.
- ON—replies to interrogations; no altitude information.
- ALT—replies to interrogations; includes altitude information.
- OFF—powers off.
- GND—replies to Mode S interrogations; no Mode A or Mode C replies.

Figure 5-8. The transponder display and controls might be a separate piece of equipment (top) or incorporated into an digital display, such as the Garmin G1000 (bottom).

TRANSPONDER MODES

The information that appears to the controller depends on the transponder mode. Mode A transmits a four-digit code to ATC for aircraft identification. Mode C transmits your airplane's altitude to ATC. Mode C capability is required in Class A, B, and C airspace, within 30 miles of the primary airport in Class B airspace, above Class B and C airspace, and above 10,000 feet MSL. Starting in 2020, ADS-B Out is required in this same airspace.

A Mode S two-way data link enables the transponder to exchange information with ATC and with other Mode S-equipped aircraft. Mode S is required for ADS-B. A 1090ES Mode S transponder transmits GPS data, such as aircraft position, direction, velocity, and vertical trend at least once per second. This enables ATC and other aircraft to precisely track your airplane's flight path. Mode S is also required for traffic alert and collision avoidance systems (TCAS) to coordinate resolution advisories with other aircraft that have TCAS.

TRANSPONDER CODES

For most VFR operations, set the transponder code to 1200. Many transponders default to this code or you can press a VFR button to set 1200. ATC will assign you a specific, or discrete, code in Class A, B or C airspace, or when you request traffic advisory services. Several special codes are reserved for distress situations. [Figure 5-9]

XPDR   1200   ALT   R

Most VFR operations; might be set by a VFR button or as default

XPDR   3256   ALT   R

Discrete code—assigned by ATC; used in Class B or C airspace or when you request VFR radar services

XPDR   7500   ALT   R

Hijacking

XPDR   7600   ALT   R

Communication failure

XPDR   7700   ALT   R

Emergency, such as an onboard fire or engine failure

Figure 5-9. Use these standard codes when operating your transponder.

To prevent activating an alarm at the ATC facility, avoid accidently selecting 7500, 7600, or 7700. You run a greater risk of this mistake if your airplane has an older model transponder with knobs. Avoid entering seven into the first window as your first step in changing the code. For example, if changing the code from 2700 to 7200, first change the code to 2200 and then change it to 7200; do not switch to 7700 and then to 7200.

TRANSPONDER OPERATION

Controllers use distinctive phraseology when referring to transponder operation. The term **squawk** is used to assign a code and to indicate which transponder function you should select. Many transponders automatically engage the ALT and ADS-B functions when you turn on the equipment. ATC might ask you to *"Ident"* to cause the transponder return to blossom on the ATC display for a few seconds, enabling the controller to easily identify you. If you need to turn the transponder or the ADS-B function to STANDBY or OFF, wait until arriving at parking. [Figure 5-10 and Figure 5-11]

Transponder Phraseology	
Squawk (Number)	Set the designated code.
Ident	Press the IDENT button/key.
Squawk VFR	Set the code to 1200.
Squawk standby	Switch to STBY.
Stop altitude squawk	Turn off the automatic altitude-reporting feature of your transponder. This typically occurs if ATC sees that your altitude is off by more than 300 feet of your reported altitude caused by equipment errors or an incorrect altimeter setting.
Stop squawk	Switch to OFF.
Squawk MAYDAY	Set the code to 7700.

Figure 5-10. This table lists the actions to take when you receive certain transponder instructions from ATC.

Before you taxi from parking:

Set your transponder to ALT, which also typically enables ADS-B transmissions.

When you are airborne:

Operate the transponder on ALT and use the appropriate code or the code assigned by ATC. If the controller requests it, press IDENT.

Figure 5-11. Follow these procedures for operating your transponder depending on your airplane's specific equipment. The Garmin G1000 is shown here.

After landing and when clear of the active runway:

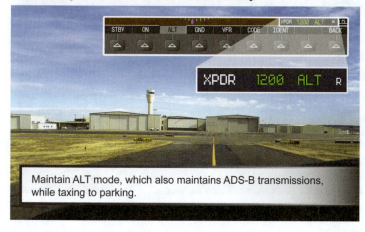

Maintain ALT mode, which also maintains ADS-B transmissions, while taxiing to parking.

SECTION A ■ ATC Services

TRANSPONDER INSPECTION

The FARs require that transponders be tested and inspected every 24 calendar months for operations in controlled airspace. The inspection must be done by a certificated repair station and the results entered into the aircraft logbook. [Figure 5-12]

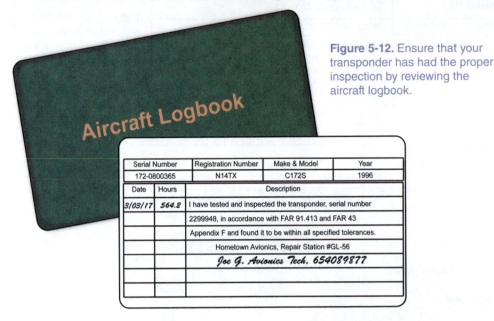

Figure 5-12. Ensure that your transponder has had the proper inspection by reviewing the aircraft logbook.

Serial Number	Registration Number	Make & Model	Year
172-0800365	N14TX	C172S	1996

Date	Hours	Description
3/03/17	564.2	I have tested and inspected the transponder, serial number
		2299948, in accordance with FAR 91.413 and FAR 43
		Appendix F and found it to be within all specified tolerances.
		Hometown Avionics, Repair Station #GL-56
		Joe G. Avionics Tech. 654089877

FLIGHT SERVICE

You use **Flight Service** during each phase of flight from preflight planning to landing at your destination. To plan your flight, obtain a weather briefing and Notices to Air Missions (NOTAMs) on the phone at 1-800-WX-BRIEF or online at 1800wxbrief.com. On this site, you can also use flight plan aids to determine an optimal departure time, altitude, and route.

Next, file a flight plan online or by phone. A flight plan provides Flight Service with essential information, such as your destination, route of flight, time enroute, and the number of people on board. Although not required for flight under VFR, filing a flight plan is strongly recommended so that you receive search and rescue protection. Flight Service also provides Adverse Condition Alerting Service (ACAS), which notifies you by text or email if an adverse condition arises after you file a flight plan. [Figure 5-13]

Figure 5-13. By registering for ACAS, you receive weather or NOTAM alerts specific to your flight plan.

SECTION A ■ ATC Services

Controlled Flight Into Terrain

At 1:45 a.m. on March 16, 1991, a Hawker Jet (HS-125) crashed into the side of Otay Mountain located 8 nautical miles northeast of San Diego's Brown Field Municipal Airport. The crew of 2 and all 8 passengers perished. It was a clear moonless night when the jet departed Brown Field under VFR. Although they were unfamiliar with the area's terrain, the crew opted to stay below the San Diego Class B airspace until they could obtain an IFR clearance. Two minutes after departure and prior to obtaining the instrument clearance from San Diego approach control, the Hawker erupted into a fireball as it flew into the terrain at 3,300 feet MSL, only 72 feet from clearing the top of Otay Mountain.

Learning to effectively use your resources can help you maintain situational awareness and prevent a controlled flight into terrain (CFIT) accident. You should use ATC services to the maximum extent possible, especially in an unfamiliar area. However, do not become complacent while being monitored by ATC. You should plan your flight with a thorough knowledge of the terrain along your route and use in-flight resources, such as aeronautical charts, a moving map with topographic information, and terrain awareness displays to monitor the height of terrain and obstacles.

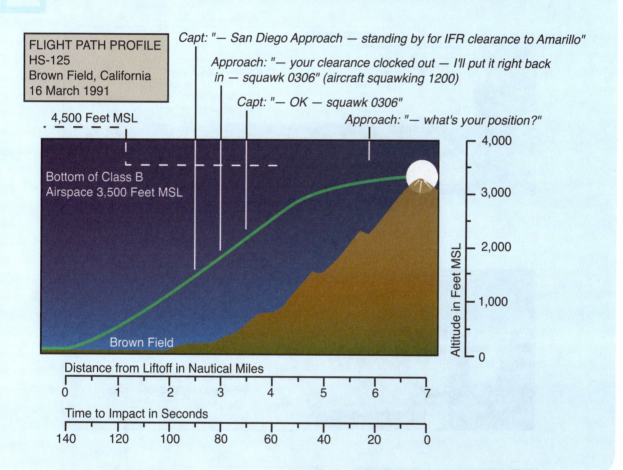

FLIGHT PATH PROFILE
HS-125
Brown Field, California
16 March 1991

Capt: "— San Diego Approach — standing by for IFR clearance to Amarillo"

Approach: "— your clearance clocked out — I'll put it right back in — squawk 0306" (aircraft squawking 1200)

Capt: "— OK — squawk 0306"

Approach: "— what's your position?"

4,500 Feet MSL

Bottom of Class B Airspace 3,500 Feet MSL

Brown Field

Altitude in Feet MSL — 4,000 / 3,000 / 2,000 / 1,000 / 0

Distance from Liftoff in Nautical Miles — 0 1 2 3 4 5 6 7

Time to Impact in Seconds — 140 120 100 80 60 40 20 0

During departure, activate your flight plan on 122.2 MHz or the local Flight Service frequency. You can also use the EasyActivate™ service—you receive an email 30 minutes prior to the estimated time of departure for your filed VFR flight plan. Click on a link in the email to activate the flight plan. While enroute, a Flight Service briefer can provide you updated weather or assistance if you are lost or in an emergency situation. You are encouraged to provide Flight Service with pilot reports (PIREPs) and reports of ADS-B malfunctions, including TIS-B, and FIS-B performance problems.

After arriving at your destination, close your flight plan by phone or use the EasyClose™ service. Click the link in the EasyClose™ email, which is sent 30 minutes prior to your estimated time of arrival (based on your actual departure time). [Figure 5-14]

SECTION A ■ ATC Services

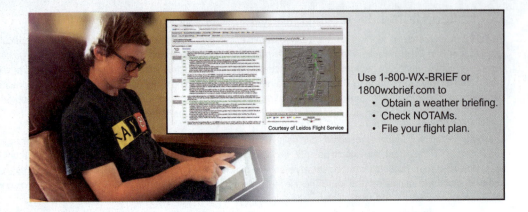

Use 1-800-WX-BRIEF or 1800wxbrief.com to
- Obtain a weather briefing.
- Check NOTAMs.
- File your flight plan.

Courtesy of Leidos Flight Service

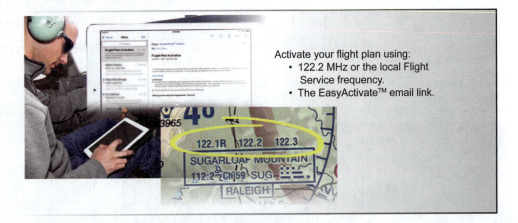

Activate your flight plan using:
- 122.2 MHz or the local Flight Service frequency.
- The EasyActivate™ email link.

Flight Service provides:
- Updated weather.
- Assistance in lost or emergency situations.

You provide:
- Pilot reports (PIREPs).
- Reports of ADS-B malfunctions.

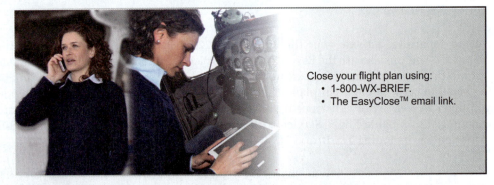

Close your flight plan using:
- 1-800-WX-BRIEF.
- The EasyClose™ email link.

Figure 5-14. Flight Service provides services that are essential to planning and implementing a safe and effective flight.

SEARCH AND RESCUE

Inform Flight Service if, after activating your flight plan, your destination changes or if you will be at least 15 minutes later than planned. If you have not closed your flight plan within 30 minutes after your estimated time of arrival, Flight Service will try to find your airplane—calling your destination airport, other airports on your route, your home base, and the contacts in your flight plan. If the phone search is unsuccessful, Flight Service alerts the **search and rescue (SAR)** system. SAR personnel from organizations, such as the U.S. Coast Guard, the Civil Air Patrol, and law enforcement, are dispatched to search for your airplane and provide survival and medical aid if you have been in an accident. [Figure 5-15]

Courtesy of FAA

Copyright Corel

Courtesy of U.S. Air Force

Figure 5-15. If a phone search to ATC and your contacts is unsuccessful, Flight Service will alert search and rescue personnel.

Flight Service is also teamed with multiple partners to provide surveillance-enhanced search and rescue. If your airplane is properly equipped, position reports are sent to Flight Service during your flight. The system alerts Flight Service if your airplane stops moving or stops sending reports. This enables search procedures that focus on the correct area to be initiated immediately. Visit 1800wxbrief.com for more information about this service.

CONTROL TOWER SERVICES

A **control tower** provides services in the terminal area. Prior to flight from an airport with an operating control tower, you typically listen to the broadcast of **automatic terminal information service (ATIS)** to obtain airport advisory information. ATIS is updated when airport conditions change, when any official weather is received, and normally at the top of each hour.

Each new ATIS broadcast is labeled with successive letters from the phonetic alphabet; for example, Information Alpha, Information Bravo, and so on. ATIS is precorded and broadcast continuously on its own frequency. At larger airports, an ATIS frequency might be designated for departing aircraft and another for arriving aircraft. Anticipating the type of information and its order in the ATIS broadcast helps you to quickly understand the content. For example, if the sky condition and visibility are absent from the broadcast, it means that the visibility is 5 miles or more and the cloud ceiling is 5,000 feet or higher. [Figure 5-16]

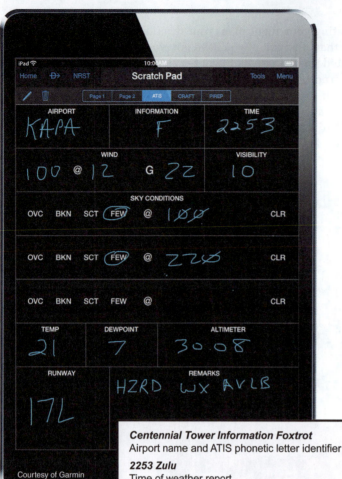

Figure 5-16. Although you can listen to the broadcast as many times as necessary, writing down the ATIS information is a good habit.

Centennial Tower Information Foxtrot
Airport name and ATIS phonetic letter identifier

2253 Zulu
Time of weather report

Wind 100 at 12, gust 22
Magnetic wind direction and velocity

Visibility 10, few clouds 10,000, few clouds 22,000
Visibility, obstructions to visibility, and ceiling or sky condition
Note—Sky condition and visibility absent:
  • Visibility 5 miles or more
  • Ceiling 5,000 feet or higher

Temperature 21—check density altitude, dewpoint 7
Temperature and dewpoint (if available)

Altimeter 30.08
Altimeter setting

Visual approach in use Runway 17 Left, landing and departing Runway 17 Left.
Instrument approach and runways in use

Notice to Air Missions—hazardous weather for Colorado available from Flight Service.
VFR departures, advise ground control direction of flight. All departures advise when runup is complete. Remain on ground control frequency until advised to switch to tower.
Any other pertinent remarks relating to operations on or near the airport

Advise on initial contact you have Information Foxtrot.
Phonetic letter identifier, restated at the end of the broadcast

At busy airports, **clearance delivery** issues IFR clearances and also might provide VFR aircraft with departure instructions and transponder codes to improve traffic coordination. When you are ready to taxi, contact **ground control** to obtain your clearance to taxi to the runway. You must have a clearance to operate in movement areas—runways and taxiways that are not part of loading ramps or parking areas. After reaching the runway, the tower controller clears you for takeoff.

Whether you are departing, arriving or transitioning through Class D airspace enroute, the tower controller provides traffic advisories (workload permitting) and safety alerts if the controller believes that your airplane is in unsafe proximity to terrain, obstructions, or another aircraft. As you near your destination, listen to ATIS and then contact the tower controller for instructions to enter the traffic pattern and for clearance to land. After you land and clear the runway, ground control clears you to taxi to parking. If you are unfamiliar with the airport, you can ask ground control to provide you with progressive taxi instructions—step-by-step routing directions. [Figure 5-17]

Listen to ATIS for airport advisory information.

Contact clearance delivery at busy airports for departure instructions and a transponder code.

Contact ground control for clearance to taxi in movement areas.

Contact the tower controller for takeoff clearance.

Figure 5-17. When operating in Class D airspace, you use control tower services during each phase of flight.

Courtesy of FAA

The tower controller provides traffic advisories and safety alerts.

Listen to ATIS for airport advisory information.

The tower controller directs you to enter the traffic pattern and clears you to land.

Ground control clears you to taxi to parking.

TRACON AND ARTCC SERVICES

A network of facilities across the United States uses the ADS-B system and radar equipment to provide ATC services to both IFR and VFR aircraft. A **terminal radar approach control (TRACON)** provides services in the terminal area of airports typically in Class B, Class C, or TRSA airspace. A TRACON can be combined with a control tower or it can be a separate facility. To receive TRACON services at busy airports and airports in Class B and Class C airspace, obtain a transponder code and a departure frequency from clearance delivery prior to taxi. You must utilize Class B and C services. At airports that are not in Class B or C airspace, you can request optional **flight following**—a service to alert you to relevant traffic—by obtaining the appropriate frequency from the control tower, Flight Service, or airport information sources.

At many airports, TRACON controllers monitor surface movement on traffic displays. For example, airport surface detection equipment–model X collects data from sources such as radar, ADS-B sensors, and transponders to provide detailed coverage of movement on runways and taxiways. The surface visualization tool provides a picture of airport surface traffic from the perspective of the control tower.

After takeoff, the tower controller directs you to contact the departure controller who provides traffic advisories and safety alerts, as well as vectoring and separation from other aircraft in specific locations. For example, controllers provide separation between IFR and VFR aircraft in Class C airspace and between all aircraft in Class B airspace.

While enroute, **air route traffic control centers (ARTCCs)** are the facilities that provide flight following. The primary purpose of ARTCCs is to manage IFR traffic. However, depending on workload, these centers may also provide services to VFR pilots. If you requested flight following on the ground, the departure controller will direct you to contact the controlling center on a specific frequency. Because of their extensive area coverage, centers can also provide assistance if you become lost or disoriented.

As you near a destination with a TRACON, you will contact approach control. In addition to the same services as departure, the approach controller provides sequencing—instructions or vectors that sequence your airplane with others for landing. The approach controller then hands you off to the tower controller, who clears you for landing. [Figure 5-18]

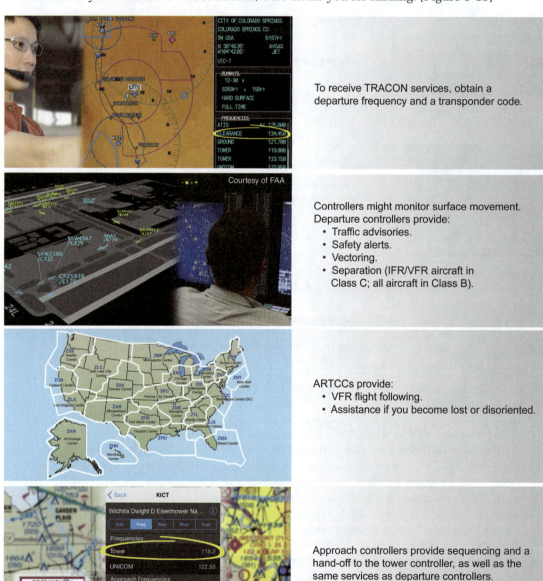

To receive TRACON services, obtain a departure frequency and a transponder code.

Controllers might monitor surface movement. Departure controllers provide:
- Traffic advisories.
- Safety alerts.
- Vectoring.
- Separation (IFR/VFR aircraft in Class C; all aircraft in Class B).

ARTCCs provide:
- VFR flight following.
- Assistance if you become lost or disoriented.

Approach controllers provide sequencing and a hand-off to the tower controller, as well as the same services as departure controllers.

Figure 5-18. TRACONs and ARTCCs provide a variety of VFR services in the terminal and enroute environments.

SECTION A ■ ATC Services

ATC SERVICES — PILOT RESPONSIBILITIES

The ability for ATC to provide services, such as traffic advisories, safety alerts, vectors, sequencing, and separation depends on the type of facility and controller workload. The controller's first responsibility is to aircraft flying on IFR flight plans. Factors such as the volume of traffic or frequency congestion might prevent controllers from providing VFR services.

"78 Juliet Romeo, traffic at 12 o'clock, 4 miles, eastbound, Bonanza at 8,000." **Traffic advisories** alert you to air traffic relevant to your flight, but you are still responsible for collision avoidance. Inform the controller that you are *"looking for traffic," "negative contact"* or *"traffic in sight"* as the case may be. Maintain visual contact until the traffic is no longer a factor.

"78 Juliet Romeo, low altitude alert, climb to 8,000 immediately." **Safety alerts** are mandatory services provided to all aircraft. Controllers warn you if, in their judgment, your airplane is in unsafe proximity to terrain, obstructions, or other aircraft. Keep in mind that safety alerts are contingent upon the capability of the controller to recognize unsafe situations.

"78 Juliet Romeo, turn left heading 270, advise airport in sight." **Vectors** provide navigational guidance on an advisory basis only. The controller might vector you for safety reasons or you might request a vector if you are unfamiliar with the area. When being vectored, flight safety remains your responsibility. For example, advise the controller if the vector will cause you to enter IFR conditions.

"78 Juliet Romeo, follow the Cessna turning final, do you have that traffic in sight?" **Sequencing** provides a safe and orderly flow of arriving aircraft into the traffic pattern. A controller might instruct you to follow other traffic, but that does not authorize you to comply with instructions issued directly to the preceding aircraft, and you are still responsible for flight safety. For example, you must adjust your flight path to avoid wake turbulence.

"78 Juliet Romeo, cleared to enter Class B airspace, maintain 7,500." ATC provides you with **separation** from both IFR and VFR aircraft in Class B airspace, and from IFR aircraft in Class C airspace. You must obtain a clearance to enter and operate within Class B airspace. However, this clearance does not relieve you of your responsibility to see and avoid traffic, avoid wake turbulence, maintain terrain and obstruction clearance, and to maintain VFR weather conditions. Advise ATC and obtain a revised clearance if an assigned route, heading, or altitude will compromise flight safety.

INTERPRETING TRAFFIC ADVISORIES

When giving traffic advisories and safety alerts, ATC references traffic from your airplane as if it were a clock with 12 o'clock at the nose. *"Cessna 78 Juliet Romeo, traffic at 11 o'clock, 2 miles, southbound, Archer at 6,500."* This means that traffic appears to the controller to be 30 degrees left of the airplane nose. You should look for the traffic anywhere between the nose and the left wing of your airplane or between 12 and 9 o'clock. Keep in mind that wind correction angles do not show up on the ATC display. [Figure 5-19]

SECTION A ■ ATC Services

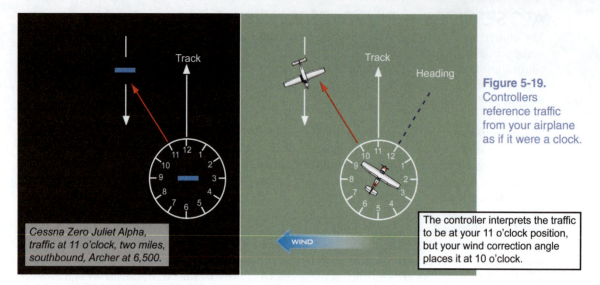

Cessna Zero Juliet Alpha, traffic at 11 o'clock, two miles, southbound, Archer at 6,500.

WIND

The controller interprets the traffic to be at your 11 o'clock position, but your wind correction angle places it at 10 o'clock.

Figure 5-19. Controllers reference traffic from your airplane as if it were a clock.

SUMMARY CHECKLIST

✓ The automatic dependent surveillance-broadcast (ADS-B) system incorporates GPS, aircraft transmitters and receivers, and ground stations to provide pilots and ATC with specific data about the position and speed of aircraft.

✓ ADS-B Out signals travel line-of-sight from transmitting aircraft to ATC ground receivers or aircraft receivers.

✓ Aircraft equipped with ADS-B In receive traffic data from other ADS-B aircraft on a cockpit display of traffic information (CDTI).

✓ The ADS-B system displays aircraft with transponders that are not ADS-B equipped using the traffic information service-broadcast (TIS-B).

✓ The flight information service-broadcast (FIS-B) delivers weather products and other flight information, such as TFR locations.

✓ The 1090 MHz extended squitter (1090ES) link can be used at any altitude and is required above 18,000 feet MSL. The 978 MHz universal access transceiver (978 UAT) is limited to altitudes below 18,000 feet MSL, but is required to receive FIS-B data.

✓ The primary radar system measures the time required for the radar echo to return from your airplane and determines the direction of the signal.

✓ The air traffic control radar beacon system (ATCRBS), enables ATC to assign a discrete transponder code to display your airplane's identification, altitude, and ground speed.

✓ Radar limitations include the bending, reflecting, and screening of radio waves.

✓ You can set up to 4,096 codes on your transponder, which is used with both ADS-B and radar systems to provide precise information about your airplane's position and movement.

✓ Pressing IDENT causes the transponder return to blossom on the ATC display.

✓ Mode C transmits your airplane's altitude to ATC.

✓ A Mode S two-way data link is required for ADS-B and enables the transponder to exchange information with ATC and with other Mode S-equipped aircraft.

✓ Standard transponder codes include: VFR—1200; hijacking—7500; communication failure—7600; and emergency—7700.

✓ Controllers use the term squawk to assign a code and to indicate which transponder function you should select.

✓ ATC assigns a discrete transponder code in Class B or C airspace or when you request traffic advisory services.

✓ The FARs require that transponders be tested and inspected every 24 calendar months for operations in controlled airspace.

✓ To plan your flight, obtain a weather briefing and NOTAMs from Flight Service at 1-800-WX-BRIEF or online at 1800wxbrief.com.

✓ File a VFR flight plan with Flight Service online or by phone to receive search and rescue protection.

✓ If you have not closed your flight plan within 30 minutes after your estimated time of arrival, Flight Service will initiate a phone search to locate your airplane. If a phone search is unsuccessful, Flight Service alerts the search and rescue (SAR) system.

✓ Automatic terminal information service (ATIS) provides airport advisory information at airports with an operating control tower. ATIS is updated when airport conditions change, when any official weather is received, and normally at the top of each hour.

✓ If the sky condition and visibility are absent from the ATIS broadcast, it means that the visibility is 5 miles or more and the cloud ceiling is 5,000 feet or higher.

✓ At busy airports, clearance delivery issues IFR clearances and also might provide VFR aircraft with departure instructions and a transponder code.

✓ Contact ground control to obtain clearance to taxi in movement areas.

✓ A terminal radar approach control (TRACON) provides services in the terminal area of airports, typically Class B, Class C, or TRSA airspace.

✓ While enroute, air route traffic control centers (ARTCCs) are the facilities that provide flight following.

✓ Traffic advisories alert you to air traffic relevant to your flight, but you are still responsible for collision avoidance.

✓ Controllers issue a safety alert if, in their judgment, your airplane is in unsafe proximity to terrain, obstructions, or other aircraft.

✓ Vectors provide navigational guidance on an advisory basis only.

✓ Sequencing provides a safe and orderly flow of arriving aircraft into the traffic pattern.

✓ ATC provides you with separation from both IFR and VFR aircraft in Class B airspace and from IFR aircraft in Class C airspace.

✓ When giving traffic alerts, ATC references traffic from your airplane as if it were a clock with 12 o'clock at the nose. Wind correction angles do not show up on the ATC display.

SECTION A ■ ATC Services

KEY TERMS

Automatic Dependent Surveillance-Broadcast (ADS-B)

Radar

Secondary Surveillance Radar

Air Traffic Control Radar Beacon System (ATCRBS)

Transponder

Squawk

Flight Service

Search and Rescue (SAR)

Control Tower

Automatic Terminal Information Service (ATIS)

Clearance Delivery

Ground Control

Terminal Radar Approach Control (TRACON)

Flight Following

Air Route Traffic Control Center (ARTCC)

Traffic Advisory

Safety Alert

Vectors

Sequencing

Separation

QUESTIONS

1. Select the true statement regarding ADS-B system components.
   A. ADS-B In transmits line-of-sight signals from aircraft to ATC ground receivers and to aircraft receivers.
   B. ADS-B Out transmits line-of-sight signals from aircraft to ATC ground receivers and to aircraft receivers.
   C. ADS-B In receives signals from aircraft and displays their lateral position, altitude, and velocity to controllers.

2. Select the true statement regarding ADS-B services.
   A. Regulations require the CDTI to be a dedicated display.
   B. An effective range of 250 nautical miles enhances traffic conflict detection and resolution.
   C. ADS-B provides data to ATC that immediately indicates when an aircraft deviates from its assigned flight path.

3. Describe three limitations of radar.

4. Which is true about transponder modes?
   A. Mode S provides two-way data link capability and enables ADS-B.
   B. Mode A provides 4,096 available codes and transmits your airplane's altitude.
   C. Mode C transmits your airplane's altitude and provides two-way data link capability.

5. List the standard transponder codes to use for VFR flight, hijacking, communication failure, and emergencies.

6. True/False. After landing and when clear of the active runway, you should set the transponder to STBY or OFF.

7. When should you press IDENT on the transponder and what is the purpose of this function?

8. Which services does Flight Service provide?
   A. Weather briefings, NOTAMs, and flight plan filing at 1-800-FLT-SERV or 1800fltserv.com
   B. Weather briefings, NOTAMs, and flight plan filing at 1-800-WX-BRIEF or 1800wxbrief.com
   C. Updated weather, flight plan filing, and traffic advisories on 122.2 MHz or the local FSS frequency

9. Which is true regarding search and rescue services (SAR)?
   A. Flight Service alerts the SAR system if a phone search to locate your airplane is unsuccessful.
   B. If you have not closed your flight plan within 15 minutes after your ETA, Flight Service begins a phone search.
   C. If you have not closed your flight plan within 30 minutes after your ETA, Flight Service immediately alerts the SAR system.

10. Which is a true regarding the services that a control tower provides?
    A. Clearance delivery provides taxi clearances with step-by-step routing to the runway.
    B. In Class D airspace, tower controllers provide you with separation from other VFR aircraft.
    C. The ATIS broadcasts of airport advisory information are updated when airport conditions change, when any official weather is received, and at the top of each hour.

11. What does it mean if the sky condition and visibility are absent from the ATIS broadcast?

12. Which is a true regarding the services TRACONs and ARTCCs provide?
    A. TRACON and ARTCC controllers provide separation between all aircraft.
    B. TRACON controllers provide traffic advisories, safety alerts, vectoring, and separation between all aircraft in Class B airspace.
    C. You must obtain a transponder code and departure frequency for flight following when operating at any tower-controlled airport.

13. Which is true regarding ATC services?
    A. ATC assumes responsibility for collision avoidance when issuing a traffic advisory.
    B. ATC issues a safety alert if your airplane is in unsafe proximity to terrain, obstructions, or other aircraft.
    C. A clearance to operate in Class B airspace relieves you of the responsibility to maintain VFR weather conditions.

14. You are on a heading of 360°. ATC issues this traffic advisory, *"Traffic 10, o'clock, 2 miles, southbound."* Where should you look?
    A. Northwest—between directly ahead and 90° to the left
    B. Northwest—between directly ahead and 90° to the right
    C. Northeast—between directly ahead and 90° to the right

SECTION A ■ ATC Services

SECTION B
Radio Procedures

Twenty-seven hours after Charles Lindbergh took off in *The Spirit of St. Louis* from New York's Curtiss Field, he spotted several fishing boats, circled low, and shouted, *"Which way is Ireland?"* to one of the fishermen. If today's extensive ATC network had been in operation during the golden age of aviation, Lindbergh would have had a much better chance of pinpointing his position. Nearly any ATC service you obtain during flight requires radio contact, so you need to learn this unique way of communicating.

VHF COMMUNICATION EQUIPMENT

General aviation communication radios use a portion of the **very high frequency (VHF)** range, which includes the frequencies between 118.0 megahertz (MHz) and 135.975 MHz. These radios are classified according to the number of channels they accommodate. A 360-channel radio uses 50 kHz (.05 MHz) spacing between channels, such as 118.05, 118.10, 118.15, 118.20. A 720-channel radio uses 25 kHz (.025 MHz) spacing, such as 118.025, 118.050, 118.075, 118.100. To receive full ATC services, a 720-channel transceiver is a necessity, particularly in busy terminal areas. Because communication radios usually combine a transmitter and receiver, they are called **transceivers**. [Figure 5-20]

Frequency Selector
Turn the outer and inner knobs to change the frequency. On these transceivers, pulling the knob enables you to select .025 MHz fractions.

Standby/Active Button
Switch between active and standby frequencies. Set your next frequency in advance and retain the last frequency until you are sure it is no longer needed.

Volume and Squelch
Turn the volume up or down by rotating the knob. Typically pulling or pushing on the volume knob turns on the squelch, which increases the reception range, enabling you to receive weaker signals. You can also turn on the squelch to adjust the volume when no one is talking.

Figure 5-20. Communication transceivers typically are combined with navigation receivers and might be integrated with GPS equipment as shown here by the lower display.

DISCOVERY

Speaking Tubes

During the 1940's the Stearman, a tandem seat, open cockpit biplane was one of the primary military trainers. A rigid speaking tube, called a gosport tube, enabled the instructor in the forward cockpit to communicate with the student in the rear cockpit. The instructor wore a rubber mask fitted with a flexible hose connected to the gosport tube. When the instructor spoke into the mask, the sound traveled through the tube to ear pieces connected to the student's cloth helmet. Although the communication was only one way, the instructor could see the student by using a mirror mounted to the underside of the Stearman's upper wing. For civilian aircraft, many manufacturers offered portable speaking tubes that provided two-way communication as shown in the illustration.

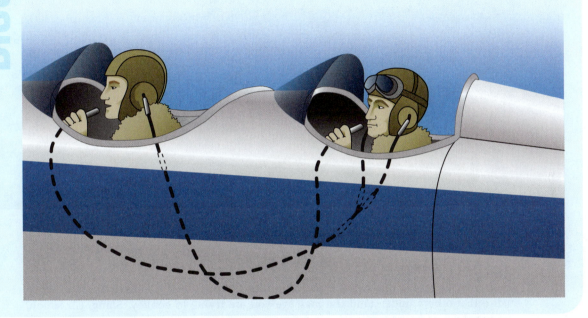

VHF antennas usually are the bent whip rods or plastic-encapsulated blade types that are mounted on top of the cabin. [Figure 5-21]

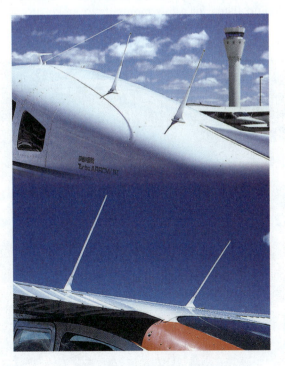

Figure 5-21. Normally, a VHF antenna is installed for each transceiver in the airplane.

The range of VHF transmissions is limited to **line of sight**, which means that obstructions such as buildings, terrain, or the curvature of the earth block the radio waves. [Figure 5-22]

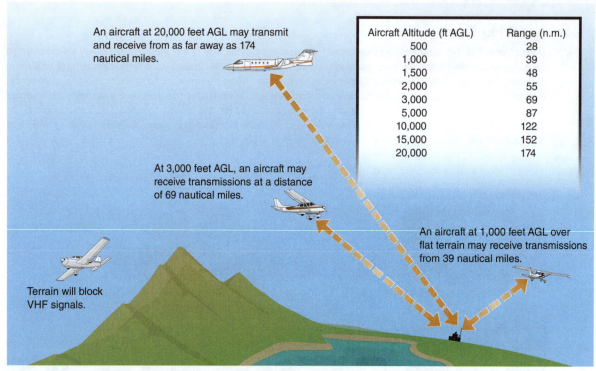

Figure 5-22. Because VHF radio signals are limited to line of sight, aircraft flying at higher altitudes are able to transmit and receive at greater distances.

USING THE RADIO

When you are using the radio, it is important to speak in a professional manner to ensure that others understand the message you are trying to convey. Slang, CB jargon, and incorrect radio procedures can compromise your safety and the safety of others. Radio transmissions should be as brief as possible to help avoid frequency congestion. Before you depress the microphone button (key the mic), think of what you will say and listen for a few moments to make sure that someone else is not already talking or waiting for a response. When making an initial callup to ATC or another facility, make your transmissions as brief as possible. [Figure 5-23]

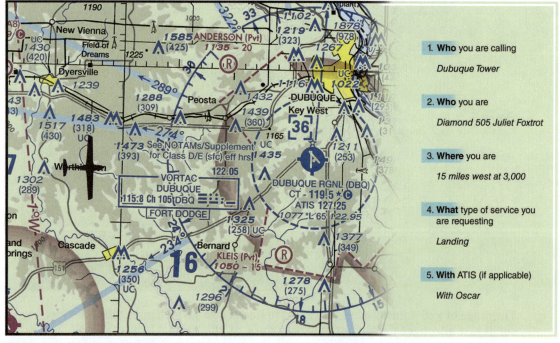

Figure 5-23. This example of an initial callup to a control tower illustrates five key "W" words that help you remember the elements to include.

Squelch Noise

If you have been flying, you are probably familiar with the hissing noise heard on your radio receiver with the squelch on and the volume up. You might have theorized that the hiss is atmospheric noise, picked up by the antenna and processed by the receiver. However, on VHF frequencies practically no atmospheric noise is present for the receiver to pick up. In fact, if the antenna were removed from the receiver, the noise would still be there.

The noise is generated in the receiver by the molecular motion of the materials that make up the electrical components. Although the noise produced by molecular motion is extremely weak, the sensitivity of the VHF receiver is so great that it amplifies the noise enough to make it audible. If the amplification was reduced to remove the hiss, the receiver would be unable to pick up the very weak stations, which are heard as a mixture of station and noise.

When you are ready to talk, hold the mic very close to your lips. Then, key the mic and speak into it in a normal, conversational tone. It might take a few moments for the facility you have called to respond. If you do not receive any response, try again. If you still do not hear anything, determine if the radio is working properly. Make sure the mic is not stuck in the transmitting position because this can block other transmissions and disrupt communication for an extended period of time.

PHONETIC ALPHABET

It would be difficult for pilots to speak all of the languages that could be involved in international flight. Therefore, the English language is recommended by the International Civil Aviation Organization (ICAO) for international air/ground communication. In countries where English is not the official language, ICAO member states have agreed to make English available upon request. ICAO also has adopted a **phonetic alphabet** to be used in radio transmissions. Because letters like B, C, D, and E are sometimes difficult to distinguish over the radio, the phonetic alphabet was developed to avoid misunderstandings. When you identify your aircraft during initial contact with ATC or other facilities, you will routinely use the phonetic alphabet. You should also use the phonetic equivalents for single letters to spell out difficult words or groups of letters. [Figure 5-24]

		ICAO Phonetic Alphabet						
A	Alfa	(**Al**-fah)	· —		**N**	November	(No-**vem**-ber)	— ·
B	Bravo	(**Brah**-voh)	— · · ·		**O**	Oscar	(**Oss**-cah)	— — —
C	Charlie	(**Char**-lee) or (**Shar**-lee)	— · — ·		**P**	Papa	(Pah-**pah**)	· — — ·
					Q	Quebec	(Keh-**beck**)	— — · —
D	Delta	(**Dell**-tah)	— · ·		**R**	Romeo	(**Row**-me-oh)	· — ·
E	Echo	(**Eck**-oh)	·		**S**	Sierra	(See-**air**-rah)	· · ·
F	Foxtrot	(**Foks**-trot)	· · — ·		**T**	Tango	(**Tang**-oh)	—
G	Golf	(Golf)	— — ·		**U**	Uniform	(**You**-nee-form) or (**Oo**-nee-form)	· · —
H	Hotel	(Hoh-**tell**)	· · · ·					
I	India	(**In**-dee-ah)	· ·		**V**	Victor	(**Vik**-tah)	· · · —
J	Juliett	(**Jew**-lee-**ett**)	· — — —		**W**	Whiskey	(**Wiss**-key)	· — —
K	Kilo	(**Key**-loh)	— · —		**X**	X-ray	(**Ecks**-ray)	— · · —
L	Lima	(**Lee**-mah)	· — · ·		**Y**	Yankee	(**Yang**-key)	— · — —
M	Mike	(Mike)	— —		**Z**	Zulu	(**Zoo**-loo)	— — · ·

Figure 5-24. The phonetic alphabet helps you understand letters of the alphabet spoken over the radio or by phone. You also will listen for the Morse code identifiers shown in the table to identify navigation facilities, which typically transmit three-letter codes.

USING NUMBERS ON THE RADIO

When you transmit or receive numbers over the radio, each number is spoken the same way you are used to saying it, with the exception of the number nine. It is spoken as *"niner"* to distinguish it from the German word *"nein,"* which means no. To reduce confusion, certain sets of numbers are spoken as individual digits. When you state radio frequencies, the decimal is pronounced as *"point,"* but the decimal is dropped when you state an altimeter setting.

In the U.S., each aircraft is identified by a registration number, which is painted on the outside of the airplane. Registration numbers usually are a combination of five letters and numbers. They are sometimes referred to as the tail number, or **N-number**, because all U.S.-registered aircraft have an N preceding the number. On initial callups to ATC or other facilities, you should state the name of the facility you are calling and then give your aircraft type, model, or manufacturer and registration number. If you state the manufacturer's name or model, you may drop the N prefix of the registration. [Figure 5-25]

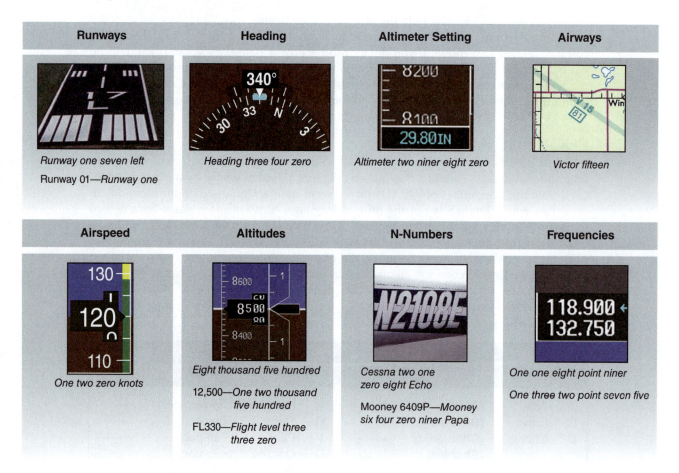

Figure 5-25. By studying these examples, you will become more familiar with the correct pronunciation of numbers.

COORDINATED UNIVERSAL TIME

Because a flight might cross several time zones, estimating the arrival time at your destination using only the local time at the departure airport would be confusing. To overcome this problem, aviation uses the 24-hour clock system, along with an international standard called **coordinated universal time (UTC)**. The 24-hour clock eliminates the need for a.m. and p.m. designations because the 24 hours of the day are numbered consecutively. For instance, 9 a.m. becomes 0900 hours; 1 p.m. becomes 1300 hours, and so on.

Coordinated universal time, which is referred to as **Zulu time** in aviation, places the entire world on one time standard. When a time is expressed in UTC, or Zulu, it is the time at the 0° line of longitude, which passes through Greenwich, England. All of the 24 time zones around the world are based on this reference. In the United States, you add hours to convert local time to Zulu time and, to convert Zulu time to local time, you subtract hours. [Figure 5-26]

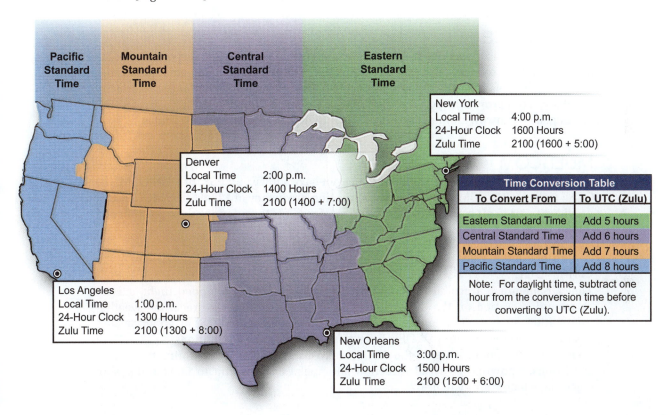

Pacific Standard Time | Mountain Standard Time | Central Standard Time | Eastern Standard Time

New York
Local Time   4:00 p.m.
24-Hour Clock   1600 Hours
Zulu Time   2100 (1600 + 5:00)

Denver
Local Time   2:00 p.m.
24-Hour Clock   1400 Hours
Zulu Time   2100 (1400 + 7:00)

Los Angeles
Local Time   1:00 p.m.
24-Hour Clock   1300 Hours
Zulu Time   2100 (1300 + 8:00)

New Orleans
Local Time   3:00 p.m.
24-Hour Clock   1500 Hours
Zulu Time   2100 (1500 + 6:00)

Time Conversion Table	
To Convert From	**To UTC (Zulu)**
Eastern Standard Time	Add 5 hours
Central Standard Time	Add 6 hours
Mountain Standard Time	Add 7 hours
Pacific Standard Time	Add 8 hours

Note: For daylight time, subtract one hour from the conversion time before converting to UTC (Zulu).

Figure 5-26. ATC operates on UTC, or Zulu time, regardless of the time zone in which the facility is located. When you refer to a specific time, first convert it to the 24-hour clock and then to Zulu.

COMMON TRAFFIC ADVISORY FREQUENCY

To increase safety at airports without operating control towers or when the tower is closed, all radio-equipped aircraft should transmit and receive traffic information on a common frequency. You can broadcast your position and intentions to other aircraft in the area on the **common traffic advisory frequency (CTAF)**. At many airports you can receive airport advisories and activate pilot-controlled lighting on the designated CTAF. You can obtain the CTAF for a specific airport by referring to the Airport/Facility Directory listing of the Chart Supplement, aeronautical charts, and other pilot information sources, such as an electronic flight bag (EFB).

SECTION B ■ Radio Procedures

UNICOM

An aeronautical advisory station, or **UNICOM** (universal communication) is a privately owned air/ground communication station that transmits on a limited number of frequencies—122.7, 122.725, 122.8, 122.975, or 123.0. Announcing your position and intentions is standard procedure at airports where the designated CTAF is a UNICOM. You can also request an airport advisory from the UNICOM operator. Advisories usually include wind direction and speed, favored runway, and known traffic. Because UNICOMs are privately operated, you also can request other information or services, such as refueling.

Some airports are equipped with an automated UNICOM system that provides you with automated weather, airport advisories, and radio checks. Upon initial contact, you receive a general greeting for the airport and instructions on how to access additional information. Automated UNICOM availability is published as AUNICOM in the applicable Airport/Facility Directory listing in the Chart Supplement.

LAA SERVICE

Local airport advisory service (LAA) is provided at the uncontrolled airports in Alaska that have a flight service station (FSS) on the field. The airport advisory area extends out to 10 statute miles from the primary airport. An FSS with a designated CTAF (normally 123.6 MHz) can provide airport information, such as wind direction and velocity, favored or designated runway, altimeter setting, known traffic, NOTAMs, and taxi routes.

MULTICOM

If your flight takes you to an airport that does not have a tower, an FSS, or a UNICOM, the CTAF is the **MULTICOM** (multiple communication) frequency of 122.9 MHz. MULTICOM provides an air-to-air communication frequency for pilots to announce their position and intentions to other aircraft in the area.

CTAF SELF-ANNOUNCE PROCEDURE

The **self-announce procedure** is broadcasting your position or intended flight activity or ground operation on the designated CTAF. Many airports have recorded weather information, such as an automated surface observing system (ASOS) or an automated weather observing system (AWOS). Prior to announcing your intentions on the CTAF frequency, you should listen to the ASOS or AWOS broadcast to determine the airport conditions. Knowing the wind direction and speed enables you to make a decision regarding which runway to use.

Although some procedures might be unique to a specific airport or facility, you should be familiar with the general recommended CTAF procedures. When using a CTAF, make your initial call when you are 10 miles from the airport. You should also report entering the downwind, base, and final legs of the traffic pattern, and when exiting the runway.

During departure, you should monitor and communicate on the CTAF from the time you start the engine, during taxi, and until 10 miles from the airport unless the FARs or local procedures require otherwise. In addition, if you are performing other operations at altitudes used by arriving or departing aircraft, such as practicing maneuvers, or if you are enroute over the area, you should monitor the CTAF and communicate your intentions within 10 miles of the airport. Broadcasting your position at specific, easily identifiable locations, helps other pilots find and maintain visual contact with your aircraft. Because other uncontrolled airports might be within reception range on the same CTAF frequency, it is helpful to repeat the name of the airport at the end of your transmission. [Figure 5-27]

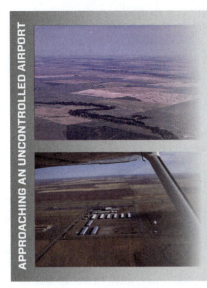

Listen to ASOS/AWOS if applicable. Make your initial call when you are 10 miles from the airport.

Longmont UNICOM, Cessna 50826, 10 miles south, descending through 7,500, landing Longmont, request wind and runway information, Longmont.

Report when you enter the downwind, base, and final legs of the traffic pattern and when exiting the runway.

Longmont traffic, Cessna 50826 entering downwind for Runway 29, full stop, Longmont.

Listen to ASOS/AWOS if applicable. Contact the UNICOM prior to taxi.

Longmont UNICOM, Cessna 50826, at Corporate Air, taxiing to Runway 26, request wind and traffic information, Longmont.

Monitor the CTAF during taxi, departure, and until you are 10 miles from the airport. Announce when you are taxiing on to the runway for departure.

Longmont traffic, Cessna 50826, departing Runway 29, departing the pattern to the south, Longmont.

Figure 5-27. The communication in this example takes place at an airport with a UNICOM.

SECTION B ■ **Radio Procedures**

CONTROLLED AIRPORTS

As you approach a controlled airport you establish communication with a control tower, or approach control at a radar facility. Your initial callup to ATC should include the name of the facility you are trying to contact, your full aircraft identification, the type of message to follow or request if it is short. An example of an initial callup is, *"Great Falls Tower, Cessna 8458 Romeo."* If your message is short, you also may include your request, as well as your position and altitude with the callup. At times, ATC might ask you to *"stand by,"* which means that the controller will get back to you as soon as possible. [Figure 5-28]

An **air traffic control clearance** is an authorization by ATC for you to proceed under specified traffic conditions within controlled airspace. The clearance's purpose is to prevent collisions between known aircraft. If you receive an ATC clearance and do not hear all of it or do not understand it, ask the controller to *"say again"* and the controller will repeat the last message. You also can ask the controller to *"speak slower"* and the controller will repeat the previous transmission more slowly. At times, you or the controller might repeat the transmission to verify that it was heard correctly. The response in these situations usually is, *"that is correct."* The use of the word *"over"* indicates that your transmission

Pilot:
Great Falls Tower, Piper 3064 Papa, 15 south, 5,500, landing, Information Charlie.

Figure 5-28. Shortening your aircraft call sign reduces the amount of air time needed for further transmissions. You should not shorten your call sign until the controller does so first.

ATC:
Piper 3064 Papa, Great Falls Tower, report entering left downwind Runway 21.

After communication is established, ATC may abbreviate your call sign by using the prefix and the last three numbers or letters. You may use the abbreviated call sign in subsequent contacts with the controller.

ATC: *Piper 64 Papa, traffic at 2 o'clock and 3 miles, southbound.*
Pilot: *Piper 64 Papa, traffic in sight.*

is complete and that you expect a response. On subsequent contacts, the ground station name and the word *"over"* may be omitted if the message requires an obvious reply and there is no danger of misunderstanding.

If a controller contacts you with a request, you should acknowledge it and quickly restate any instructions given to reduce the possibility of a misunderstanding. For example, if asked to turn right to a heading of 210°, you should respond with, *"Cessna 58 Romeo, roger, turn right heading 210."*

If you are asked to contact the same controller on a different frequency, the controller will say, *"Cessna 58 Romeo, change to my frequency, 123.4."* In this situation, you can abbreviate your callup by saying, *"Cessna 58 Romeo on 123.4."* At times, such as when you are flying into Class B or C airspace, you might talk to a succession of controllers on different frequencies. Each controller will hand you off to the next controller by telling you when to change to the next frequency and what frequency to use. Before changing frequencies, you should verify that you heard the new frequency correctly with a readback such as, *"Cessna 58 Romeo, roger, contact tower 118.3."* Because the two controllers have already coordinated the hand-off, the tower will be expecting your call.

As a student pilot, you can request additional assistance from ATC simply by identifying yourself as a student pilot. For example, assume you are approaching a controlled airport with heavy traffic and you are unfamiliar with the airport. In this situation, you should make your initial callup as follows: *"Centennial Tower, Cessna 8458 Romeo, student pilot."* This procedure is not mandatory, but it does alert controllers so they can give you extra assistance and consideration, if needed. In addition, identifying yourself as a student might be advantageous when you self-announce your position over the CTAF at uncontrolled airports.

ATC FACILITIES

You learned about the services provided by control tower, TRACON, and ARTCC facilities in Section A of this chapter. You must be familiar with the radio procedures to effectively operate in airspace where you receive these services. The following scenario covers radio procedures for a flight departing Colorado Springs Airport (KCOS) in Class C airspace.

DEPARTURE PROCEDURES

After listening to ATIS, contact clearance delivery prior to taxiing. Procedures for contacting clearance delivery for VFR flights can vary at airports that have this service. However, typically in Class C and Class B airspace you will obtain a departure clearance that includes a transponder code. Include your N-number and aircraft type and specify that you are VFR and have listened to the current ATIS. Finally, identify your destination or direction of flight. This enables departure controllers to improve traffic coordination.

Clearance delivery assigns you a transponder code, provides departure instructions, and then instructs you to contact ground control. At an airport in Class B airspace, you are typically assigned an altitude and initial heading to fly after takeoff. You might also be assigned a heading and altitude at other airports for traffic coordination. You receive a clearance from ground control to taxi to the active runway. You must have a clearance to operate in movement areas—runways and taxiways that are not part of loading ramps or parking areas. [Figure 5-29]

ATIS

Colorado Springs Airport, Information Golf, 1454 Zulu. Wind calm, visibility 10, 15,000 scattered, temperature 14, dewpoint 4, altimeter 30.28. Visual approach is in use, landing and departing Runways 35 Left, 35 Right, and 31. Land and hold short operations in effect. Notice to Airman, Taxiway November closed. Advise on initial contact you have Information Golf.

CLEARANCE DELIVERY

Contact Springs Clearance.

Springs Clearance, Piper 8252 Sierra with Foxtrot, VFR to Centennial.

A departure clearance normally contains a transponder code and departure control frequency and might include a heading, altitude, and additional instructions.

Piper 8252 Sierra, after departure, fly heading 350, climb and maintain 8,500, squawk 3504, departure frequency 124.0. Contact ground control 121.7 when ready to taxi.

Write down the clearance and read it back to the controller.

GROUND CONTROL

Contact Springs Ground.

Springs Ground, Piper 8252 Sierra, at general aviation parking, ready for taxi.

Ground control states the departure runway followed by the specific taxi route. The controller issues hold-short instructions if necessary. You are not authorized to cross or taxi on any runway without a specific clearance to do so.

Piper 8252 Sierra, Runway 35 Left, taxi via Alpha.

Figure 5-29. Ground operations at an airport in Class C airspace include receiving a clearance from clearance delivery and taxi instructions from ground control.

TAXI CLEARANCES

At busy airports, you might be instructed to wait in a holding area near the runway. If you are asked to *"hold short"* of a runway, you must read back the hold short clearance to the controller, stop at the hold lines preceding the runway, check for traffic, and continue only after cleared to do so. At unfamiliar airports, you can request a **progressive taxi** for a controller to provide you with precise taxi instructions or direct you in stages as you proceed. ATC issues an explicit clearance for all runway crossings—do not cross a runway without a clearance. A taxi clearance does not authorize you to enter or cross the assigned departure runway at any point. [Figure 5-30]

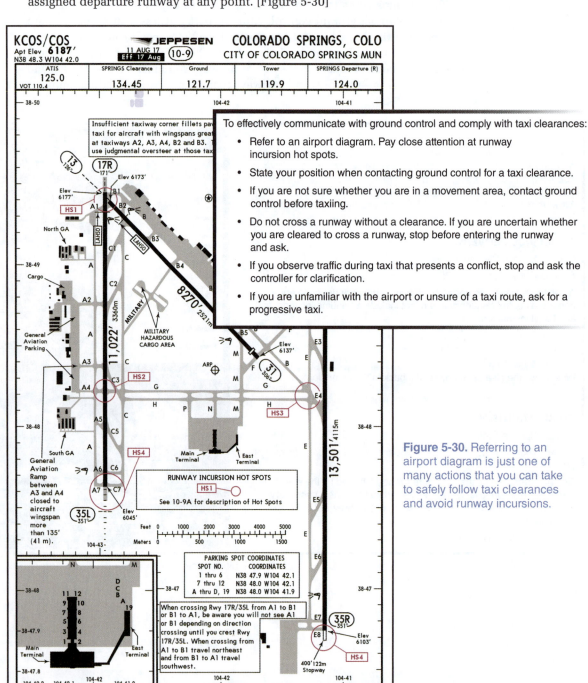

To effectively communicate with ground control and comply with taxi clearances:

- Refer to an airport diagram. Pay close attention at runway incursion hot spots.

- State your position when contacting ground control for a taxi clearance.

- If you are not sure whether you are in a movement area, contact ground control before taxiing.

- Do not cross a runway without a clearance. If you are uncertain whether you are cleared to cross a runway, stop before entering the runway and ask.

- If you observe traffic during taxi that presents a conflict, stop and ask the controller for clarification.

- If you are unfamiliar with the airport or unsure of a taxi route, ask for a progressive taxi.

Figure 5-30. Referring to an airport diagram is just one of many actions that you can take to safely follow taxi clearances and avoid runway incursions.

CONTROL TOWER

After completing the Before-Takeoff checklist at the departure runway, you normally switch to the control tower frequency for takeoff clearance. When you receive takeoff clearance, make a final check for traffic before you taxi onto the runway. You might be cleared to *"…line up and wait."* This means that you position the airplane on the runway for takeoff while waiting for another aircraft to clear the runway. Usually, you can expect to be cleared for takeoff after the other aircraft is clear.

DEPARTURE CONTROL

After takeoff, contact **departure control** and a controller will advise if you are in radar contact. The term **radar contact** is used by ATC to inform you that your airplane has been identified using an approved ATC surveillance source on a controller's display and that flight following will be provided until radar service is terminated. If you receive flight following, the controller is observing the progress of your airplane while you provide your own navigation. The departure controller will hand you off to the ARTCC if you have requested enroute flight following. [Figure 5-31]

CONTROL TOWER

Contact the tower after completing the Before-Takeoff checklist.

Springs Tower, Piper 8252 Sierra, ready for takeoff, Runway 35 Left.

The final approach path is clear and the controller clears you for takeoff.

Piper 8252 Sierra, cleared for takeoff Runway 35 Left.

Shortly after you take off, the tower controller hands you off to Springs Departure.

Piper 52 Sierra, contact Departure 123.7.

DEPARTURE CONTROL

Contact Springs Departure. Be brief—the controller is familiar with your departure clearance.

Springs Departure, Piper 8252 Sierra, climbing through 7,500 for 8,500.

The controller acknowledges that your airplane has been identified.

Piper 8252 Sierra, Springs Departure, radar contact, report reaching 8,500.

When you are clear of the Class C airspace, Springs Departure terminates your radar service.

Piper 52 Sierra, radar service terminated, frequency change approved, squawk VFR.

Figure 5-31. During departure, terrain and obstruction clearance remains your responsibility until the controller begins to provide navigational guidance in the form of radar vectors.

ARRIVAL PROCEDURES

Listen to ATIS and then contact **approach control** while still outside of Class B or Class C airspace. It is good practice at busy Class B airports to contact approach and state your call sign and wait for a response before providing your position, altitude, and request. The controller will provide you with a transponder code and advise that you are in radar contact. To enter Class B airspace, the controller must state that you are *"cleared to enter Class B."* Approach control frequencies are published on sectional charts and broadcast over ATIS. At large terminals, expect different frequencies for approach control, depending on your arrival sector. ATC routinely provides you with wind, runway, and altimeter information unless you indicate that you have received the ATIS. You might hear the phrase *"have numbers."* This phrase means that the pilot has received only wind, runway, and altimeter information, but has not listened to ATIS. [Figure 5-32]

You should respond immediately to time-critical requests by ATC. If, at any time, you are given an instruction that is beyond the capabilities of your airplane, is not safe to

APPROACH CONTROL

When you return to Colorado Springs, listen to ATIS. While you are outside the Class C airspace, contact Springs Approach and state your position. In this case, you are near a designated visual checkpoint (indicated by the flag).

Springs Approach, Piper 8252 Sierra, 15 miles north at the tower, with Hotel, 9,000, landing.

Springs Approach assigns you a transponder code and asks you to ident.

Piper 8252 Sierra, Springs Approach, squawk 3550 and ident.

The controller acknowledges contact and gives you a clearance.

Piper 52 Sierra, radar contact, 15 miles north of Colorado Springs Airport, fly heading 170°, expect left downwind 35 Left.

At the appropriate time, Springs Approach hands you off to the tower controller.

Piper 52 Sierra, contact tower on 119.9.

CONTROL TOWER

After contacting the tower, continue your approach and follow any additional instructions. The tower clears you to land.

Piper 52 Sierra, follow Cessna traffic on base, cleared to land Runway 35 Left.

During your roll-out, the tower gives you instructions for clearing the runway.

Piper 52 Sierra, turn left at Alpha 2 and contact ground, 121.7.

Springs Ground clears you to taxi to the parking area.

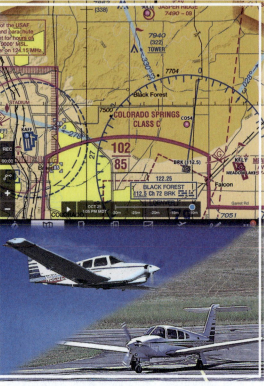

Figure 5-32. When landing at a busy airport, expect to talk to approach control, the tower, and then ground control.

follow, or would cause you to violate an FAR, you must inform the controller that you are *"unable"* to comply with the directions. The controller should then give you an amended clearance with instructions that you can safely follow.

There might be times when you want to stay in the traffic pattern to practice landings. You should advise approach control or the tower on initial contact that you will be *"remaining in the pattern."* The tower controller might ask you to *"make closed traffic."* This means you should remain in the traffic pattern unless you are otherwise instructed. During your last time around the pattern, request a *"full-stop"* landing.

After landing, do not switch to ground control until the tower instructs you to do so. The tower controller might need to issue subsequent instructions to hold short of another runway or taxiway. You must be extra vigilant when on a taxiway between parallel runways. At some airports, these taxiways are under the jurisdiction of the tower controller who will provide taxi instructions to hold short of or cross the parallel runway. Query the tower controller if the controller has not instructed you to taxi or contact ground after a reasonable amount of time.

Formation Flight

A formation flight is comprised of a lead aircraft and one or more wingmen flying in close proximity with coordinated movements. In a flight of two, the wingman uses the lead as a fixed frame of reference to maintain a specific position. Formation pilots must have extensive training, possess tremendous flying ability, and be skilled communicators. Prior to each flight, the pilots discuss and define objectives for the mission. A typical preflight briefing covers subjects from engine start and taxi procedures to radio failure and emergencies.

Copyright Corel

Effective communication during a formation flight is essential. Pilots use the air-to-air frequency of 122.75, and to avoid frequency congestion, a strict format is used. The lead is responsible for traffic avoidance, navigation, and ATC communication. An example of an initial callup made by the lead to a control tower is, *"Centennial Tower, Cessna 6319 Lima, flight of two."*

READING BACK CLEARANCES

If you are operating from an airport with clearance delivery, be prepared to write down and read back your clearance. To help you read back and remember your taxi clearance, refer to a printed or digital airport diagram. Read back the runway assignment, significant parts of the route, hold short instructions, and clearances to enter or cross a runway. If you forget to read back hold short instructions, the controller is required to solicit the readback from you. During flight, repeat any altitude and heading assignments and set bugs on the heading indicator and altimeter, if applicable. As you approach the airport, read back the tower instruction for pattern entry and the runway assignment. Then, use your chart and the compass card on your heading indicator to visualize the runway direction and your traffic pattern entry.

LOST COMMUNICATION PROCEDURES

As you are aware, establishing two-way radio communication with the control tower is required before you enter Class D airspace. If your communication radios become inoperative, it is still possible to land at an airport with an operating control tower by following **lost communication procedures**. If you believe that your radio has failed, set your transponder to code 7600. If you are in an area of radar coverage, the code 7600 will alert ATC of your radio failure. [Figure 5-33]

Figure 5-33. If you are unable to make contact with ATC, you can follow several steps to determine if your radio is inoperative.

If you are unable to contact ATC:

- Ensure that you are using the correct frequency. Try a different frequency for the ATC facility, if available.

- Check the volume and squelch on your transceiver.

- Check the switch position on your audio control panel.

- Ensure that the headset mic and speaker plugs (or the handheld mic plug, if applicable) are properly inserted into the jacks.

- Try the handheld mic if you are using headsets.

- If your aircraft is equipped with more than one radio, try the alternate transceiver.

- If it is within range, try requesting assistance from the last ATC facility with which you had contact.

If after taking these steps, you still are unable to contact ATC, follow the lost communication procedures.

DISCOVERY

Avianca Flight 052

In 1990, Avianca Airlines Flight 052 crashed into a residential area near Long Island, New York after running out of fuel. There were 73 fatalities including the 8 crew members. The NTSB determined that in addition to mismanagement of the airplane's fuel load, the accident was caused by the crew's failure to communicate an emergency fuel state to ATC before fuel exhaustion occurred.

At JFK International Airport, Flight 052 was placed into a holding pattern three times by ATC. Although the crew stated that they needed priority, they also indicated that they could remain in the holding pattern for at least five minutes. During a poorly executed ILS approach, the Captain elected to go around and at this time finally stated to the First Officer to *"tell them we are in an emergency."* As the airplane was being vectored around for a second approach, the First Officer radioed, *"We just lost two engines and we need priority please."*

Although a long chain of events led to this accident, this tragedy could have been prevented if the crew of Flight 052 would have properly alerted ATC to their situation by declaring a fuel emergency prior to being instructed to enter the holding pattern. Although you may be hesitant to admit an error, you should never let pride or fear of repercussions get in the way of notifying ATC in the event of an emergency situation.

If neither your transmitter nor your receiver is working, you need to remain outside or above the Class D airspace until you have determined the direction and flow of traffic. Then, join the airport traffic pattern and maintain visual contact with the tower to receive light signals. During the daytime, you should acknowledge tower transmissions or light signals by rocking your wings and at night by blinking your landing light or navigation lights. [Figure 5-34]

Figure 5-34. In the event of a radio failure, a tower controller can provide light signals to your aircraft. Each color or color combination has a specific meaning for an aircraft in flight, or on the airport surface.

COLOR AND TYPE OF SIGNAL	MEANING	
	On the Ground	**In Flight**
Steady Green	Cleared for takeoff	Cleared to land
Flashing Green	Cleared to taxi	Return for landing (to be followed by steady green at proper time)
Steady Red	Stop	Give way to other aircraft and continue circling
Flashing Red	Taxi clear of landing area (runway) in use	Airport unsafe — do not land
Flashing White	Return to starting point on airport	(No assigned meaning)
Alternating Red and Green	Exercise extreme caution	Exercise extreme caution

It is possible for only your radio transmitter or receiver to fail. For example, if you are fairly certain that only the receiver is inoperative, you should remain outside or above the Class D airspace until you have determined the direction and flow of traffic. Then, advise the tower of your aircraft type, position, altitude, intention to land, and request to be controlled by light signals. When you are approximately three to five miles away from the airport, advise the tower of your position and join the traffic pattern. Watch the tower for light signals, and if you fly a complete pattern, self-announce your position when you are on downwind and/or turning base.

If only your transmitter is inoperative, follow the same procedure that you would when the receiver is not working, but do not self-announce your intentions. Monitor the airport frequency for landing or traffic information, and look for a light signal that might be addressed to your airplane.

EMERGENCY PROCEDURES

An emergency can be either a distress or an urgency condition. The *Aeronautical Information Manual* defines **distress** as a condition of being threatened by serious or imminent danger and requiring immediate assistance, such as fire, mechanical failure, or structural failure. You are experiencing an **urgency** situation the moment you become doubtful about your position, fuel endurance, weather, or any other condition that could adversely affect flight safety. If you become apprehensive about your safety for any reason, you should request assistance immediately. Do not wait until the situation has developed into a distress condition.

The frequency of 121.5 MHz is used across the United States for transmitting emergency messages. Although range is limited to line-of-sight, 121.5 MHz is guarded (listened to) by military towers, most civil towers, Flight Service, and radar facilities. In a distress situation, using the word MAYDAY commands radio silence on the frequency in use. When you hear the words PAN-PAN, the urgency situation in progress has priority over all other communication and warns other stations not to interfere with these transmissions. [Figure 5-35]

Emergency Message	
Distress or Urgency	"MAYDAY, MAYDAY, MAYDAY (or PAN-PAN, PAN-PAN, PAN-PAN),
Name of station addressed	Denver Radio,
Identification and type of aircraft	5674R Cessna 172,
Nature of distress or urgency	trapped above overcast,
Weather	marginal VFR,
Your intentions and request	request radar vectors to nearest VFR airport,
Present position and heading	Newberg VOR, heading 253°,
Altitude	6,500,
Fuel remaining in hours and minutes	estimate 30 minutes fuel remaining,
Number of people aboard	three people aboard,
Any other useful information	squawking 7700."

Figure 5-35. Emergency communications begin with the word MAYDAY or PAN-PAN repeated three times, followed by information about your situation and the assistance that you require.

Radar-equipped ATC facilities can provide radar assistance and navigation service to you within an area of radar coverage. Changing your transponder code to 7700 triggers an alarm, or special indication, at all radar facility control positions. Although you may not be sure if your aircraft is within radar coverage, it is a good idea to squawk 7700 to alert any ATC facility that might be in the area. If you are under radar control and in contact with ATC, continue squawking the code assigned, unless instructed otherwise.

If you are lost, keep in mind, the **five Cs—climb, communicate, confess, comply, and conserve**. Climb for better radio and navaid reception, as well as increased radar coverage. Communicate with any nearby facility using frequencies shown on your sectional chart or another flight information source. Confess that you are lost when contacting ATC or another ground facility, and if your situation is threatening, clearly explain your problem, using the emergency frequency 121.5 MHz, if necessary. Then, comply with assistance instructions, and consider reducing your power setting to conserve fuel. In addition, ensure that the mixture is leaned properly to extend your range and endurance.

Emergency locator transmitters (ELTs) are emergency signaling devices to help locate downed aircraft. Required for most general aviation airplanes, these electronic, battery-operated transmitters emit a distinctive audio tone on designated emergency frequencies, typically 243 MHz and 406 MHz. (Older ELTs transmit on 121.5 MHz.) If armed and subjected to crash-generated forces, ELTs are designed to activate automatically. The transmitters should operate continuously for at least 48 hours over a wide range of temperatures. A properly-installed and maintained ELT can expedite search and rescue operations and save your life. Most aircraft have a remote ELT switch in the cockpit or are designed to provide pilot access to the ELT. If necessary after a crash landing, you can manually activate the ELT. [Figure 5-36]

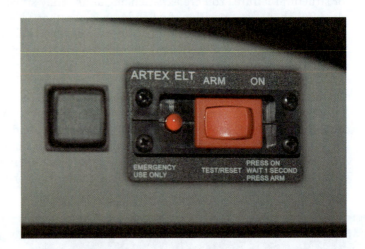

Figure 5-36. This remote ELT switch in the cockpit enables you to activate the ELT in an emergency.

ELTs must be tested and maintained according to the manufacturer's instructions. The FARs require that the ELT battery be replaced, or recharged if the battery is rechargeable, after one-half of the battery's useful life, or when the transmitter has been in use for more than one cumulative hour. ELTs should be tested in a screened room to prevent broadcast of signals, but when this cannot be done, you can conduct ELT testing in your aircraft only during the first five minutes after the hour, and for no longer than three audible sweeps. Airborne tests of ELTs are not allowed.

SUMMARY CHECKLIST

✓ Communication radios in general aviation aircraft use a portion of the very high frequency (VHF) range, which includes the frequencies between 118.0 MHz and 135.975 MHz.

✓ The range of VHF transmissions is limited to line of sight, which means that obstructions such as buildings, terrain, or the curvature of the earth block radio waves.

✓ Your initial callup to ATC or another facility should include who you are calling; who you are; where you are; what type of service you are requesting; and with ATIS (if applicable).

✓ ICAO has adopted a phonetic alphabet to be used in radio transmissions.

✓ Aviation uses the 24-hour clock system and coordinated universal time (UTC), or Zulu time, which places the entire world on one time standard.

✓ Broadcast your position and intentions to other aircraft on the common traffic advisory frequency (CTAF) at airports without operating control towers or when the tower is closed.

✓ A UNICOM is a privately owned air/ground communication station that can provide airport advisory information.

✓ At an airport that does not have a tower, an FSS, or a UNICOM, the CTAF is the MULTICOM frequency, 122.9 MHz.

✓ The self-announce procedure is broadcasting your position or intended flight activity or ground operation on the designated CTAF.

✓ An air traffic control clearance is an authorization by ATC for you to proceed under specified traffic conditions within controlled airspace.

✓ Clearance delivery provides a clearance that includes a transponder code and departure frequency.

✓ When you contact clearance delivery, state your N-number, type of aircraft, that you are VFR, have listened to ATIS, and your destination or direction of flight.

✓ You receive a clearance from ground control to taxi to the active runway in movement areas—runways and taxiways that are not part of loading ramps or parking areas.

✓ If you are asked to *"hold short"* of a runway, you must read back the hold short clearance to the controller, stop at the hold lines preceding the runway, check for traffic, and continue only after cleared to do so.

✓ ATC issues an explicit clearance for all runway crossings—do not cross a runway without a clearance.

✓ The term radar contact is used by ATC to inform you that your aircraft has been identified using an approved ATC surveillance source on a controller's display and that flight following will be provided.

✓ After landing, do not switch to ground control until the tower instructs you to do so.

✓ Listen to ATIS and contact approach control while still outside Class B or Class C airspace. The controller will provide you with a transponder code and advise that you are in radar contact.

✓ To enter Class B airspace, the controller must state *"cleared to enter Class B."*

✓ If you are unable to contact ATC, troubleshoot to determine the cause of radio failure by taking actions such as checking the volume and squelch, verifying that the mic is properly plugged in, and trying an alternate transceiver.

✓ To land at a tower-controlled airport if your communication radios become inoperative, set your transponder code to 7600, and follow the lost communication procedures.

✓ In the event of a radio failure, a tower controller can provide light signals to direct your aircraft.

✓ If armed and subjected to crash-generated forces, ELTs are designed to automatically emit a distinctive audio tone on designated emergency frequencies.

✓ The FARs require that the ELT battery must be replaced, or recharged if the battery is rechargeable, after one-half of the battery's useful life, or if the transmitter has been used for more than one cumulative hour.

SECTION B ■ **Radio Procedures**

KEY TERMS

Very High Frequency (VHF)

Transceiver

Line of Sight

Phonetic Alphabet

N-Number

Coordinated Universal Time (UTC)

Zulu Time

Common Traffic Advisory Frequency (CTAF)

UNICOM

MULTICOM

Self-Announce Procedure

Air Traffic Control Clearance

Progressive Taxi

Departure Control

Radar Contact

Approach Control

Lost Communication Procedures

Distress

Urgency

Five Cs—Climb, Communicate, Confess, Comply, and Conserve

Emergency Locator Transmitter (ELT)

QUESTIONS

1. Select the true statement regarding general aviation aircraft communication radios.
    A. The antennas are usually fin shaped and mounted below the cabin.
    B. The radios use a portion of the very high frequency (VHF) range from 118.0 MHz to 135.975 MHz.
    C. The range of transmissions is not limited by obstructions, such as buildings, terrain, or the curvature of the earth.

2. Using the phonetic alphabet, write out the pronunciation of the following aircraft call sign: Cessna 649SP.

3. What is the correct method of stating 10,500 feet MSL to ATC?
    A. *Ten thousand, five hundred*
    B. *Ten point five*
    C. *One zero thousand, five hundred*

Refer to the time conversion table below to answer questions 4 and 5.

Time Conversion Table	
To Convert From:	**To UTC (Zulu)**
Eastern Standard Time	Add 5 Hours
Central Standard Time	Add 6 Hours
Mountain Standard Time	Add 7 Hours
Pacific Standard Time	Add 8 Hours
NOTE: For daylight time, subtract one hour from the conversion time before converting to UTC (Zulu).	

4. If you depart an airport in the Central Standard Time zone at 1:00 p.m. and arrive in a city on the west coast two hours later, what is your arrival time in Zulu?

5. If you fly from a city in the Pacific Daylight Time zone to a city in the Central Daylight Time zone, and you leave at 0700 PDT with an estimated time enroute of three hours, what is your estimated time of arrival in Zulu?

SECTION B ■ Radio Procedures

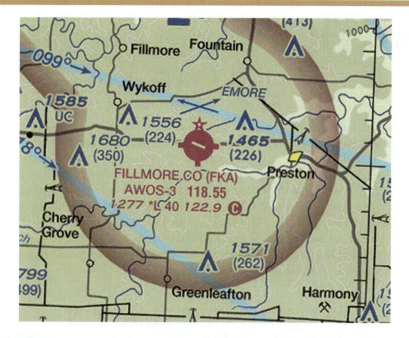

6. What is the recommended communications procedure for landing at Fillmore County Airport?
   A. Listen to AWOS and transmit intentions on 122.9 MHz at 10 miles out and give position reports in the traffic pattern.
   B. Contact the tower on 118.55, state your call sign, position, and intention to land. Follow the tower's instructions for entering the traffic pattern.
   C. Contact the UNICOM on 122.9 at 10 miles out, receive wind and runway information, and make position reports when in the traffic pattern.

7. Select the actions that you would take to comply with the following ATC clearance: *"Cessna 52 Sierra, Runway 17, taxi via Bravo, hold short Runway 26 Right."*
   A. Taxi onto Runway 17.
   B. Taxi along taxiway Bravo toward Runway 17, stopping at the hold line for Runway 26 Right and waiting for ATC clearance to cross Runway 26 Right.
   C. Taxi on Runway 17, crossing taxiway Bravo, to Runway 26 Right, stopping at Runway 26 Right until cleared to proceed by ground control.

8. After landing at a tower-controlled airport, when should you contact ground control?
   A. Prior to turning off the runway.
   B. When advised by the tower to do so.
   C. After reaching a taxiway that leads directly to the parking area.

9. Describe the lost communication procedure used for landing at a Class D airspace primary airport.

10. Which is a correct at is the meaning of these light signals in flight and on the ground?
    • Steady green
    • Flashing green
    • Steady red
    • Flashing red
    • Flashing white
    • Alternating red and green

11. How often must the ELT battery be replaced or recharged if the battery is rechargeable?

SECTION B ■ Radio Procedures

SECTION C
Sources of Flight Information

Your destination is Greater Green River Intergalactic Spaceport, located not in the outer reaches of the galaxy, but in the southwest corner of Wyoming. Although the airport is depicted on your sectional chart, without any additional information, you feel as if you are voyaging into uncharted territory. You are wondering if any factors affect your plans, such as navaids shut down for maintenance, parachute operations near your route, or runway closures? You have not flown into an uncontrolled airport recently. What are the procedures for entering the traffic pattern and announcing your intentions? Can you get fuel at the airport? You can answer these and many other questions by using sources of flight information.

LOCATING FLIGHT INFORMATION

At **faa.gov**, you can search for most of the flight information that you need to fly safely in the National Airspace System, download many publications for free, or order printed versions from approved providers. **Electronic flight bags (EFBs)**, such as ForeFlight, provide a wide variety of flight information similar to that found on the FAA website, such as aeronautical charts, airport data, and essential FAA publications. Additional EFB features include a moving map, weather information, aircraft performance data, and a logbook. You also can obtain flight environment, weather and airport information on GPS units in your airplane and find printed flight information products in pilot supply stores. [Figure 5-37]

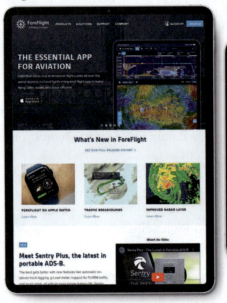
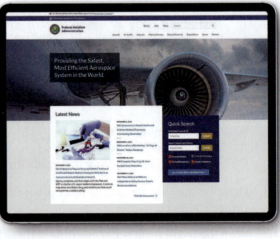

Figure 5-37. Search faa.gov or use an EFB for a wide variety of flight planning information.

AERONAUTICAL CHARTS

Aeronautical charts—sectional charts and terminal area charts (TACs)—are your primary VFR navigation references and depict topographic information, visual landmarks, and airport data. Updated every 56 days, you can download these charts from faa.gov or purchase printed versions. In addition, these same charts are also available in EFB form. A ForeFlight subscription enables you to regularly download the latest changes to the charts. One of the many great advantages to ForeFlight is the capability to overlay features like radar and traffic on aeronautical charts to help improve your situational awareness in flight. [Figure 5-38]

Figure 5-38. Updated every 56 days, sectional charts and terminal area charts are available in print and digital formats.

CHART SUPPLEMENTS

Chart Supplements include data that cannot be readily depicted in graphic form on charts. This data applies to public and joint-use airports, seaplane bases, and heliports, as well as navaids and airspace. Each of the seven volumes of Chart Supplements covers a specific region of the contiguous United States, Puerto Rico, and the U.S. Virgin Islands. Two additional volumes cover Alaska and the Pacific region, including Hawaii. Similar to the sectional charts, the Chart Supplements are updated every 56 days. [Figure 5-39]

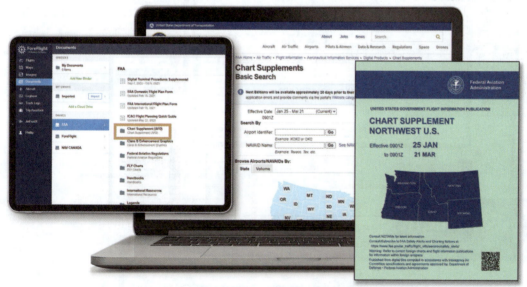

Figure 5-39. Available in both print and digital formats, each of the nine volumes of Chart Supplements include flight information for a specific region.

Each Chart Supplement contains five primary sections:

- The Airport/Facility Directory Legend
- The Airport/Facility Directory
- Notices
- Associated Data
- Airport Diagrams

At faa.gov, you can access the information in a Chart Supplement in several ways. You can download an entire volume or you can search for a specific city or airport. When using the search feature, you can download a specific Airport/Facility Directory page that contains the desired airport listing and download other sections of the Chart Supplement separately. [Figure 5-40]

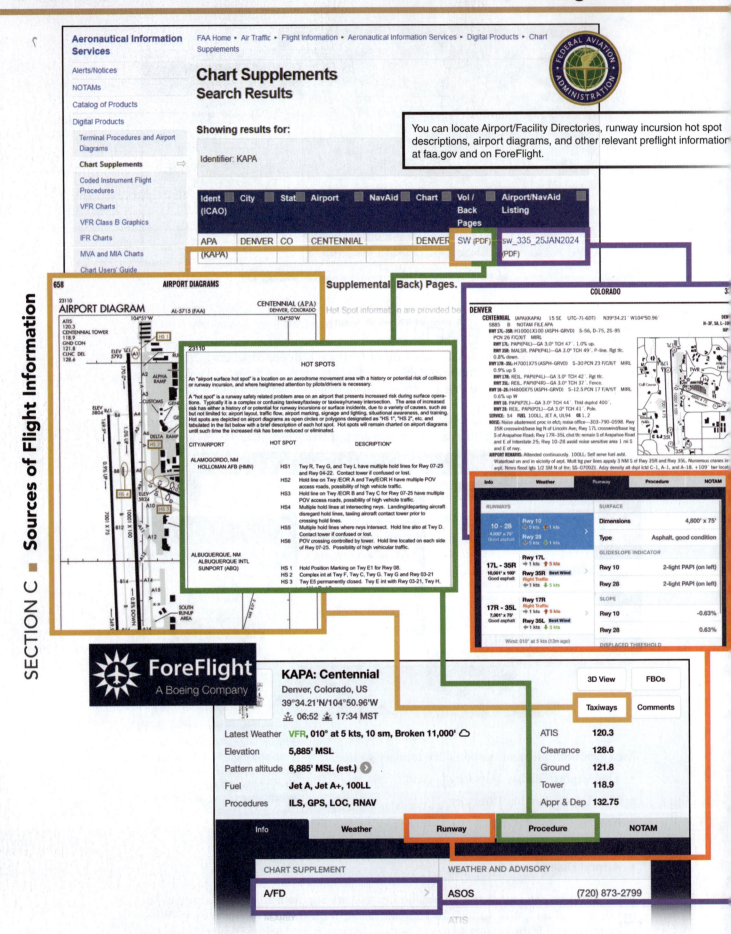

Figure 5-40. You can use menus and search tools on faa.gov and ForeFlight to obtain Chart Supplement information.

AIRPORT/FACILITY DIRECTORY

The **Airport/Facility Directory (A/FD)** in the Chart Supplement contains a descriptive listing of all airports, heliports, and seaplane bases that are open to the public. To effectively interpret the information in the A/FD, you should have a general understanding of the types of data included in an airport listing and refer to the A/FD Legend. Each listing begins with airport identification and location information [Figure 5-41]

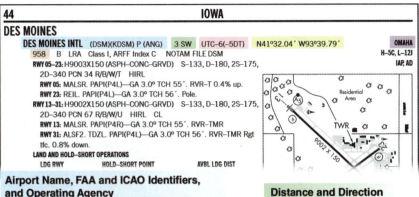

Hours Difference Between Local and Zulu Time
To convert local time to UTC, add 6 hours (5 hours in daylight saving time).

Airport Position/ Airport Reference Point (ARP)
The latitude and longitude of the approximate geometric center of all usable runway surfaces is shown.

Sectional Chart
This airport appears on the Omaha sectional chart.

Airport Elevation
The airport elevation is 958 feet MSL.

Airport Name, FAA and ICAO Identifiers, and Operating Agency
The letter P indicates the airport is a public civil airport with use by transient military aircraft. A military tenant on the field is the Air National Guard (ANG)

Distance and Direction from the Associated City
This airport is three miles southwest of Des Moines.

Figure 5-41. Identification and location information includes the conversion from UTC to local time, airport elevation, and sectional chart name.

In addition to basic runway data, such as length, width, and composition, the A/FD lists the types of runway lighting available, including the type of visual glideslope indicator, if the runway has one. Land-and-hold short operations (LAHSO) information includes available landing distances to the hold-short point. Declared distance information is provided for airlines and other operators with specific regulatory requirements for determining takeoff and landing distances. [Figure 5-42]

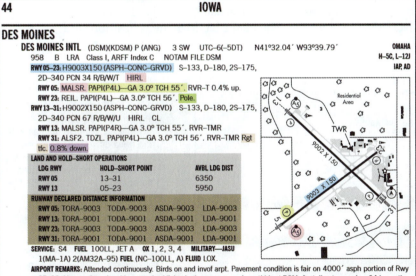

Length, Width, and Composition
Runway 05-23 is 9,003 feet long and 150 feet wide. It is composed of asphalt and concrete and is grooved.

Types of Lighting
Runway 05 has high intensity runway lights (HIRL) and a medium intensity approach lighting system with runway alignment indicator lights (MALSR).

Visual Glideslope Indicator
Runway 05 has a precision approach path indicator (PAPI) with 4 lights on the runway's left side (P4L). The PAPI approach slope/glide angle (GA) is 3.0° with a threshold crossing height (TCH) of 55 feet.

Traffic Pattern
Runway 31 has a right-hand traffic pattern. The other runways have standard left-hand patterns.

Runway Slope
Runway 31 slopes 0.8% downward.

Other Information
A pole is located at the approach end of Runway 23.

LAHSO Information
The available landing distance on Runway 05 to the hold-short point at the intersection of Runway 13-31 is 6,350 feet.

Runway Declared Distance Information
This data includes:
- Take–off run available (TORA).
- Take–off distance available (TODA).
- Accelerate–stop distance available (ASDA).
- Landing distance available (LDA).

Figure 5-42. Runway information helps you make effective decisions about landing and taking off at the airport.

Airport Services information includes the availability of maintenance, fuel, and oxygen. Refer to the A/FD Legend for the meaning of code numbers used to specify the type of maintenance and oxygen services. The Airport Remarks section indicates the days and hours that the airport is attended, which means an operator provides at least minimum services, such as fuel. This section also alerts you to any essential safety information about conditions that are permanent or will last for more than 30 days concerning facilities, services, maintenance, procedures, or hazards. {Figure 5-43]

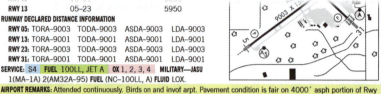

Oxygen Codes

The oxygen codes 1,2, 3, and 4 indicate that you can purchase both high-pressure and low-pressure oxygen refills and bottles at this airport.

Airport Remarks

Airport Remarks include:

- The days and hours that the airport provides minimum services. This airport is attended continuously.
- Any essential safety information. For example, birds are on and in the vicinity of the airport and Runway 05-23 has fair pavement condition.

Maintenance Codes

Maintenance codes from S-1 to S-8 indicate the type of repairs available. S4 means that this airport offers both major airframe and major powerplant repair services.

Fuel Codes

Fuel codes indicate the type of fuel available. You can purchase 100LL and Jet A fuel at this airport.

Figure 5-43. Determining the available services, hours the airport is attended, and safety information helps you effectively assess risk during flight planning.

You can obtain airport weather information by using the phone numbers and frequencies listed as Weather Data Sources. The Communications section shows applicable frequencies for the airport, including ATIS, UNICOM, Flight Service, tower, ground control, approach, and departure. Airspace information indicates the type of airspace surrounding the airport. If that information includes part-time Class C, D, or E airspace, this listing includes the effective times. Navaid information includes the VOR test facility (VOT) frequency, if applicable, and the navaid NOTAM file identifier. The identifier, frequency, position, and any restrictions are listed for each navaid associated with the airport that appears on VFR and IFR charts or that applies to an instrument approach procedure. [Figure 5-44]

time. Fuel will be purchased from FBOs, contract fuel unavbl at KDSM. 132 WG personnel to support aircrews are ltd.
Mil crews can ctc HAWKI CP on 252.9 on apch Tue–Fri 1300–2330Z‡. PPR req for official business only are rqr, DSN 261–8573 or 261–8574.
AIRPORT MANAGER: 515-256-5100
WEATHER DATA SOURCES: ASOS (515) 287–1012 HIWAS 117.5 DSM. WSP.
COMMUNICATIONS: ATIS 119.55 515–974–8046 **UNICOM** 122.95
    RCO 122.65 (FORT DODGE RADIO)
Ⓡ **APP/DEP CON** 123.9
    TOWER 118.3 **GND CON** 121.9 **CLNC DEL** 134.15
AIRSPACE: CLASS C svc ctc **APP CON**
VOR TEST FACILITY (VOT) 109.2
RADIO AIDS TO NAVIGATION: NOTAM FILE DSM.
  (H) VORTACW 117.5 DSM Chan 122 N41º26.26´ W93º38.91´ 347º 5.8 NM to fld. 951/7E. **HIWAS.**
  VOR unusable:
    095º–150º
  FOREM NDB (LOM) 344 DS N41º28.93´ W93º34.84´ 308º 4.8 NM to fld.
  ILS/DME 111.5 I–DWW Chan 52 Rwy 05. Class IE. Glideslope unusable for coupled apchs blo 1,256´ agl.
  ILS 111.9 I–VGU Rwy 13. Class IE.
  ILS/DME 110.3 I–DSM Chan 40 Rwy 31. Class IIIE. LOM FOREM NDB.
COMM/NAV/WEATHER REMARKS: Contact Gnd Control on 121.9 prior to pushback on the terminal apron.

Navaids

- Use the VOT frequency of 109.2 to test your VOR equipment.
- The DSM NOTAM file applies to the listed navaids.
- The DSM VORTAC on frequency 117.5 is located on the 347º bearing 5.8 nm from the airport at 951 feet MSL. The magnetic variation is 7ºE.
- DSM VORTAC is unusable from radial 095º to 150º.

Airspace

- Des Moines International Airport is in Class C airspace.
- Because no effective times are listed, Class C is in effect continuously.
- To receive Class C ATC services, contact approach control when arriving.

Weather Data

- Listen to the automated surface observing system (ASOS) by calling (515) 287-1012.
- Hazardous inflight weather advisory service (HIWAS) is transmitted over the DSM VORTAC frequency of 117.5.
- A weather systems processor (WSP) provides airport wind shear detection.

Communications

- A phone number for ATIS is listed in addition to the frequency.
- This airport has a remote communications outlet (RCO) for Flight Service on 122.65.
- The Ⓡ symbol indicates radar approach and departure control. 05-23 has fair pavement condition.

Figure 5-44. Frequencies that you use to obtain weather, contact ATC, and navigate are included in the A/FD listing.

Airport Qualification Charts

The FARs require air carrier pilots to meet special qualifications to take off and land at certain airports that have unique flight considerations due to surrounding terrain, obstacles, or complex approach and departure procedures. Within 12 calendar months prior to operating at one of these airports, the pilot in command must have completed a takeoff and landing at that airport or have used approved visual training aids, such as Airport Qualification Charts. These charts provide photos of the airport, descriptions of the surrounding terrain and obstacles, typical weather conditions, and other unique airport information.

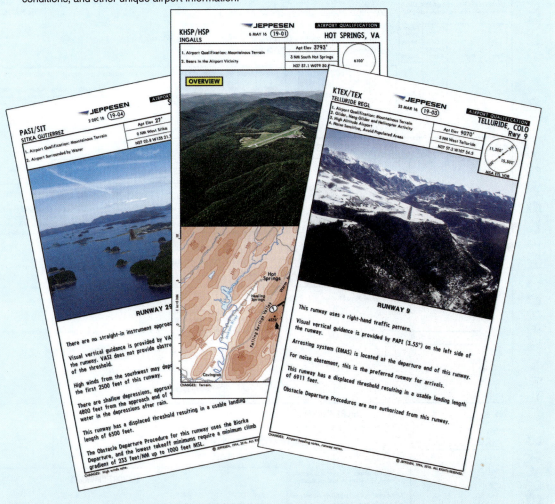

NOTICES, ASSOCIATED DATA, AND AIRPORT DIAGRAMS

In addition to the A/FD listings, refer to the information in the other sections of the Chart Supplement as you prepare for a flight. The Notices section contains two types of notices: Special Notices, and Regulatory Notices. Special Notices describe unique restrictions or operating procedures that apply to specific airports or airspace. Regulatory Notices summarize special air traffic rules and airport traffic patterns established in FAR Part 93 that apply to the airports and airspace in the region.

The Associated Data section contains information such as telephone numbers and frequencies for ATC facilities and Flight Service, VOR receiver checkpoints and VOR test facilities, and parachute jumping areas. Refer to the Airport Diagrams section for diagrams of most towered airports and other selected nontowered airports, as well as information about runway incursion hot spots. These full-page diagrams provide more detail than the smaller airport sketches in the A/FD listings. [Figure 5-45]

SECTION C ■ Sources of Flight Information

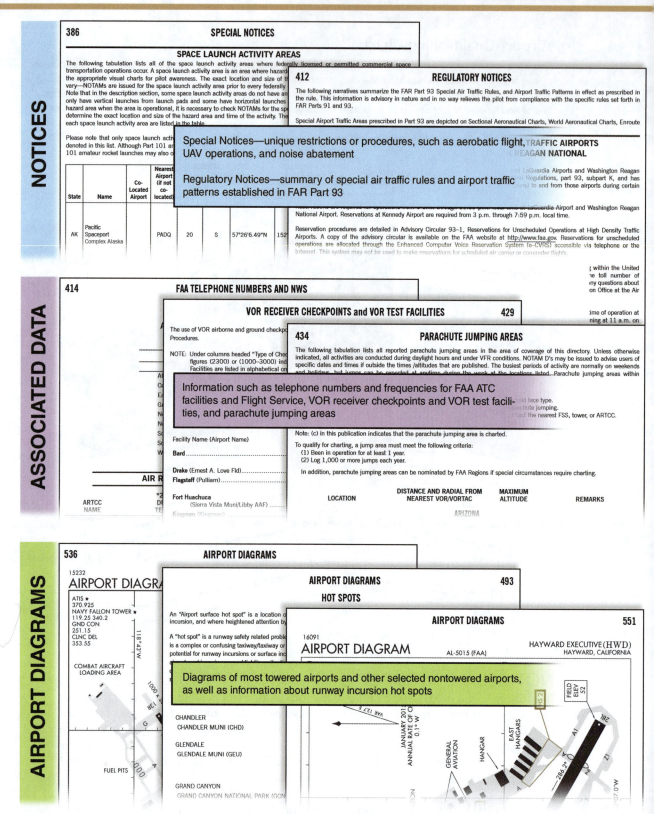

Figure 5-45. Reviewing each section of the Chart Supplement for information about your destination is part of effective preflight planning.

NOTICES TO AIR MISSIONS

Often, changes in flight information are not known far enough in advance to be included in the most recent aeronautical charts or Chart Supplements. **Notices to Air Missions (NOTAMs)** provide time-critical flight planning information regarding a facility, service, procedure, or hazard. Examples include temporary flight restrictions (TFRs), primary runway closures, new obstructions, communication frequency updates, changes in the status of navaids or airspace, radar service availability, and other information essential to pilot decision making. The FAA specifies several types of NOTAMs based on how they are disseminated and whether they are intended for international, domestic, military, or civil operations. The NOTAM is cancelled when the temporary information in the NOTAM is returned to normal status or is published on the next publication of the applicable chart. NOTAMs may be disseminated up to 7 days before the start of the activity.

Four classifications of NOTAMs are: Domestic, Flight Data Center (FDC), International, and Military. Two primary NOTAM types that are important for you to understand to effectively prepare for a flight are the Domestic NOTAM, referred to as a NOTAM (D) and the FDC NOTAM. **NOTAM (D)** information is disseminated for all navigational facilities that are part of the U.S. airspace system, and all public use airports, seaplane bases, and heliports listed in the Chart Supplements. **FDC NOTAMs** contain regulatory information, such as TFRs or amendments to aeronautical charts. A Center Area NOTAM is an FDC NOTAM issued for a condition that is not limited to one airport, therefore this type of NOTAM is filed under the Air Route Traffic Control Center (ARTCC) that controls the airspace involved. TFRs, airway changes, and laser light activity are examples of Center Area NOTAMs.

NOTAMs specific to your proposed flight are shown when you obtain Flight Service online briefing or a briefing using an EFB, such as ForeFlight. You also can search for the latest NOTAMs via the FAA Federal NOTAM System (FNL) search tool. [Figure 5-46]

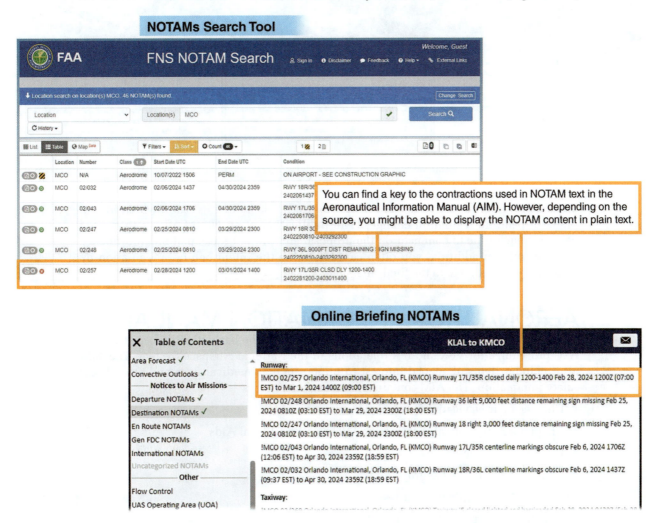

Figure 5-46. As part of your preflight planning, you should obtain all the NOTAMs that affect your flight.

SECTION C ■ Sources of Flight Information

FEDERAL AVIATION REGULATIONS

The Code of Federal Regulations (CFR) contains the official text of public regulations issued by agencies of the federal government. The Code is divided into titles by subject matter. The **Federal Aviation Regulations (FARs)** are under subject title 14, Aeronautics and Space. The FARs are broken down into numbered parts and then divided into sections that refer to a specific regulation. In government documents, a regulation is referred to by the CFR subject title number (14 CFR section 91.3). You normally will see the same regulation referred to as FAR 91.3. You can access individual regulations through a link on faa.gov or purchase the FARs from a commercial publisher.

FAR Parts cover subjects ranging from aircraft certification and maintenance to pilot medical standards and flight rules. As a private pilot, you should pay particular attention to FAR Part 61, FAR Part 91, and NTSB 830. [Figure 5-47]

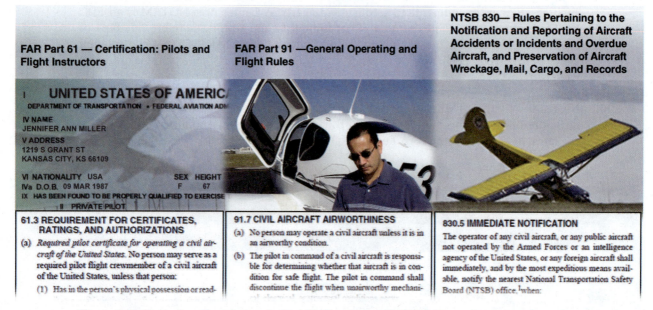

FAR Part 61 — Certification: Pilots and Flight Instructors

FAR Part 91 —General Operating and Flight Rules

NTSB 830— Rules Pertaining to the Notification and Reporting of Aircraft Accidents or Incidents and Overdue Aircraft, and Preservation of Aircraft Wreckage, Mail, Cargo, and Records

61.3 REQUIREMENT FOR CERTIFICATES, RATINGS, AND AUTHORIZATIONS

(a) *Required pilot certificate for operating a civil aircraft of the United States.* No person may serve as a required pilot flight crewmember of a civil aircraft of the United States, unless that person:

  (1) Has in the person's physical possession or read-

91.7 CIVIL AIRCRAFT AIRWORTHINESS

(a) No person may operate a civil aircraft unless it is in an airworthy condition.

(b) The pilot in command of a civil aircraft is responsible for determining whether that aircraft is in condition for safe flight. The pilot in command shall discontinue the flight when unairworthy mechani-

830.5 IMMEDIATE NOTIFICATION

The operator of any civil aircraft, or any public aircraft not operated by the Armed Forces or an intelligence agency of the United States, or any foreign aircraft shall immediately, and by the most expeditious means available, notify the nearest National Transportation Safety Board (NTSB) office,[1] when:

Figure 5-47. Examples of regulations in Part 61, Part 91, and NTSB 830 are shown here.

The **Federal Register** contains **Notices of Proposed Rulemaking (NPRM),** which inform pilots of pending regulation changes. The publication of an NPRM allows interested parties enough time to comment on the proposal and present ideas that might influence the final rule. In addition, you often can determine the intent behind a regulation by reviewing the preamble to the final rule.

AERONAUTICAL INFORMATION MANUAL

The *Aeronautical Information Manual* **(AIM)** contains the basic flight information and ATC procedures for VFR and IFR operations in the National Airspace System. The AIM also includes items of special interest to pilots, such as medical facts and other flight safety information. The AIM contains these chapters:

- Chapter 1 Air Navigation
- Chapter 2 Aeronautical Lighting and Other Airport Visual Aids
- Chapter 3 Airspace
- Chapter 4 Air Traffic Control
- Chapter 5 Air Traffic Procedures
- Chapter 6 Emergency Procedures
- Chapter 7 Safety of Flight
- Chapter 8 Medical Facts for Pilots

- Chapter 9 Aeronautical Charts and Related Publications
- Chapter 10 Helicopter Operations

DISCOVERY

FAA Safety Briefing

FAA Safety Briefing is a bi-monthly magazine for non-commercial general aviation that focuses on safety and regulatory issues, explains FAA resources, and encourages continued training. The magazine covers topics such as preflight inspections, ATC services, special use airspace, flight deck technology, and medical certification. In addition, GA Safety Enhancement Topic Fact Sheets explore subjects such as mountain flying, personal minimums, and maneuvering flight. You can download the magazine and fact sheets at faa.gov.

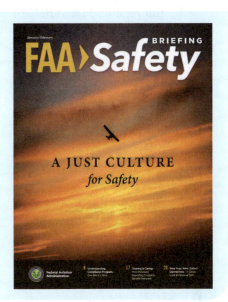

The AIM is revised several times a year and you can download the current version from faa.gov. An explanation of changes is included in the front of the AIM and changes are also marked in the content. The FARs and the AIM are often published together as the FAR/AIM. [Figure 5-48]

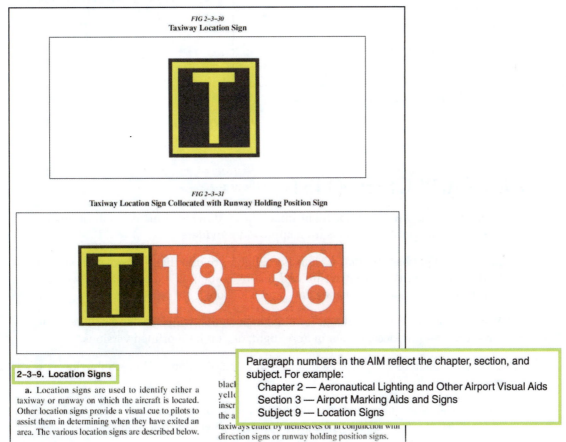

Figure 5-48. Use the AIM as a reference for questions, review the content often to refresh your knowledge, and look for changes to stay current on a wide variety of subjects.

SECTION C ■ Sources of Flight Information

ADVISORY CIRCULARS

The FAA issues **advisory circulars (ACs)** to provide nonregulatory guidance and information in a variety of subject areas. ACs also explain methods for complying with the FARs. Unless incorporated into a regulation by reference, the contents of an AC are not binding. You can obtain ACs from faa.gov. [Figure 5-49]

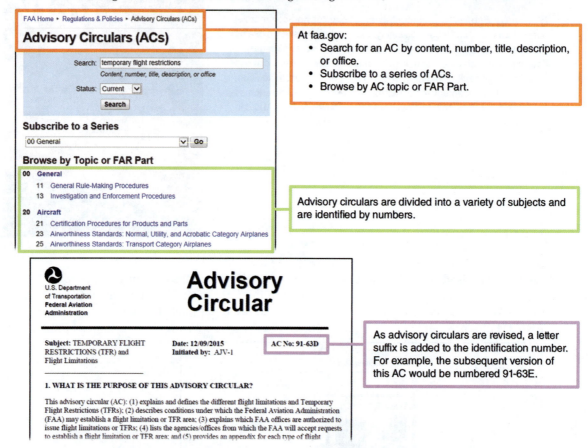

Figure 5-49. ACs are an effective way to learn how to enhance flight safety and interpret regulations.

SUMMARY CHECKLIST

✓ At faa.gov, you can search for flight information, download many publications for free, and order printed versions from approved providers.

✓ Electronic flight bags (EFBs), such as ForeFlight, provide flight information similar to that found on the FAA website and features, such as a moving map, weather information, aircraft performance data, and a logbook.

✓ You can download sectional and terminal area charts, which are updated every 56 days, from faa.gov, access them in ForeFlight, or purchase printed versions.

✓ Chart Supplements include data that cannot be readily depicted in graphic form on charts.

✓ Each of the nine volumes of Chart Supplements are available in both print and digital formats, include flight information for a specific region, and are updated every 56 days.

✓ At faa.gov, you can access the information in a Chart Supplement in several ways, including searching for a specific city or airport.

✓ The Airport/Facility Directory (A/FD) within the Chart Supplement contains a descriptive listing of all airports, heliports, and seaplane bases that are open to the public.

✓ A/FD listings include airport identification and location information, runway data, and airport services—the availability of maintenance, fuel, and oxygen.

✓ A/FD listings include weather data sources, communication and navaid frequencies, and airspace information.

✓ The Notices section of the Chart Supplement includes Special Notices and Regulatory Notices.

✓ Special Notices list unique restrictions or procedures, such as aerobatic flight, UAV operations, and noise abatement.

✓ Regulatory Notices include a summary of special air traffic rules and airport traffic patterns established in FAR Part 93.

✓ The Associated Data section contains information such as, telephone numbers and frequencies for FAA ATC facilities and Flight Service, VOR receiver checkpoints and VOR test facilities, and parachute jumping areas.

✓ Refer to the Airport Diagrams section for diagrams of most towered airports and other selected nontowered airports and runway incursion hot spot information.

✓ Notices to Air Missions (NOTAMs) provide time-critical flight planning information regarding a facility, service, procedure, or hazard.

✓ NOTAM (D) information is disseminated for all navigational facilities that are part of the U.S. airspace system and all public use airports, seaplane bases, and heliports listed in the Chart Supplements.

✓ FDC NOTAMs, issued by the National Flight Data Center, contain regulatory information, such as TFRs or amendments to aeronautical charts.

✓ Review the NOTAMs applicable to your flight when you obtain an online weather briefing, request NOTAMs during a phone briefing, or use an FAA tool to search for NOTAMs online.

✓ The Federal Aviation Regulations (FARs) are under subject title 14, Aeronautics and Space of the Code of Federal Regulations (CFR).

✓ The FARs are broken down into numbered parts and then divided into sections that refer to a specific regulation, such as FAR 91.3.

✓ As a private pilot, you should pay particular attention to FAR Part 61, FAR Part 91, and NTSB 830.

✓ The Federal Register contains Notices of Proposed Rulemaking (NPRM), which inform pilots of pending regulation changes.

✓ The Aeronautical Information Manual (AIM) contains the basic flight information and ATC procedures for VFR and IFR operations in the National Airspace System.

✓ Advisory circulars (ACs) provide nonregulatory guidance and information in a variety of subject areas and explain methods for complying with FARs.

SECTION C ■ **Sources of Flight Information**

KEY TERMS

Faa.gov	FDC NOTAM
Electronic Flight Bag (EFB)	Federal Aviation Regulations (FARs)
Aeronautical Charts	Federal Register
Chart Supplements	Notice of Proposed Rulemaking (NPRM)
Airport/Facility Directory (A/FD)	Aeronautical Information Manual (AIM)
Notice to Air Missions (NOTAM)	Advisory Circular (AC)
NOTAM (D)	

QUESTIONS

1. Name at least two sources of flight information.

2. Select the true statement regarding Chart Supplements.
   A. The seven volumes of Chart Supplements are updated every year.
   B. Chart Supplements include data that cannot be readily depicted in graphic form on charts.
   C. Each Chart Supplement contains an Airport/Facility Directory, airport diagrams, and sectional charts that cover a specific region.

Refer to the Airport/Facility Directory listing to answer questions 2 through 6.

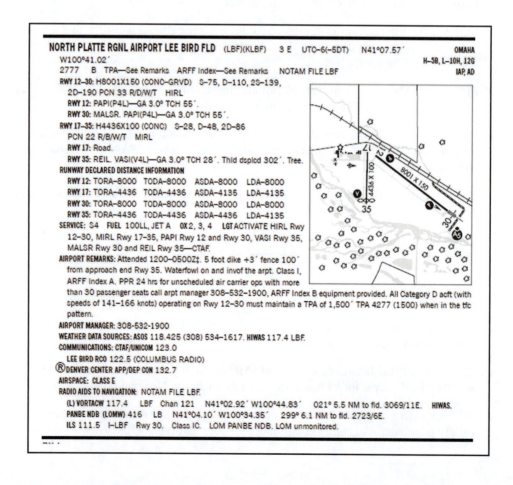

3. What is the distance and direction of Lee Bird Field from the city? What is the airport elevation?

4. How many feet long and wide is the longest runway at Lee Bird Field?

5. You are landing at Lee Bird Field at 10:00 p.m. local time. Will you be able to obtain fuel? What information in this listing did you use to determine this?

6. What are the frequencies for ASOS, UNICOM, and the navaid, LBF VORTAC?

7. In what section of a Chart Supplement can you find information about runway incursion hot spots?
   A. Special Notices
   B. Associated Data
   C. Airport Diagrams

8. Explain the differences between a NOTAM (D) and an FDC NOTAM.

9. Select the true statement about the Federal Aviation Regulations (FARs).
   A. FAR 2-3-9 is a standard way to list a specific regulation.
   B. FAR Part 91 covers the certification of pilots and flight instructors.
   C. The FARs are under subject title 14, Aeronautics and Space in the Code of Federal Regulations (CFR).

10. What do FAA issued advisory circulars (ACs) provide?

SECTION C ■ Sources of Flight Information

SECTION C ■ **Sources of Flight Information**

PART III

Aviation Weather

Know the signs of the sky and you will far the happier be.

— Benjamin Franklin

PART III

It is one of the largest variables affecting any flight, yet its fickle nature is one of the things that makes flying so challenging and exciting. It, of course, is the atmosphere and the weather that occurs within it. The information in Part III will provide you with the tools necessary to ensure that your flights in the ever-changing atmosphere are safe and enjoyable. In *Meteorology for Pilots*, you will discover how weather forms and how its hazards can affect aircraft operations. By exploring *Interpreting Weather Data*, you will learn how to obtain and decipher the wide variety of weather products designed for pilots to determine the current and forecast conditions that impact your flying.

CHAPTER 6

Meteorology For Pilots

SECTION A
Basic Weather Theory

Weather and flying are inextricably linked. You cannot take to the air without being affected by the environment through which you are flying. In order to determine the impact of weather on your flight, you will need to evaluate day-to-day elements like clouds, wind, and rain. To do so, a fundamental understanding of the atmosphere and its dynamic nature is essential.

THE ATMOSPHERE

The **atmosphere** is a remarkable mixture of life-giving gases surrounding our planet. Without the atmosphere, there would be no clouds, oceans, or protection from the sun's intense rays. Although this protective blanket is essential to life on earth, it is extraordinarily thin—nearly 99% of the atmosphere exists within 30 kilometers (about 100,000 feet) of the surface. That's roughly equivalent to a piece of paper wrapped around a beach ball. [Figure 6-1]

Courtesy of NASA

Figure 6-1. When viewed from space, the most dense part of the earth's atmosphere is seen as a thin blue area near the surface.

DISCOVERY

Why is the Sky Blue?

As sunlight passes through air, the wavelengths of light are scattered by the atmospheric gases. Colors with short wavelengths such as blue, green, and violet are scattered more than others and, consequently, enter the eye from a variety of angles. Because the human eye is more sensitive to the wavelength associated with blue, it seems as if blue light is coming from all directions, and so that appears to be predominant color of the sky.

Of course, the shade of blue can vary widely from one day to the next. In general, the lighter the shade of blue, the more likely that contaminants are present in the air. When relatively large particles, such as dust, become suspended in the atmosphere, all the wavelengths of light are scattered, which makes the sky appear as milky white, or hazy.

ATMOSPHERIC LEVELS

There is no specific upper limit to the atmosphere; it simply thins until it fades away into empty space. As you progress outward from the earth's surface, the atmosphere displays different properties, including a fluctuating distribution of temperatures. These temperature variances are the most common basis for classifying the atmosphere into layers, or spheres.

The **troposphere** (from the Greek word, trope, meaning "turn" or "change") is the layer extending from the surface to an average altitude of about 36,000 feet. Above the troposphere is the **tropopause**—a level, not a layer—which acts as a lid to confine most of the water vapor, and the associated weather, to the troposphere. The height of the tropopause varies with the season and location over the globe.

The tropopause is lower near the poles and in the winter; it is higher near the equator and in the summer. Above the tropopause are three more atmospheric layers. The first is the **stratosphere**, which has much the same composition as the troposphere and extends to a height of approximately 160,000 feet. Above the stratosphere are the **mesosphere** and **thermosphere**, which have little practical influence over weather. [Figure 6-2]

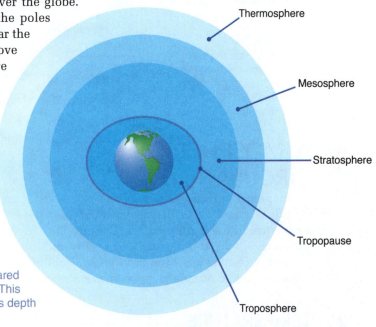

Thermosphere
Mesosphere
Stratosphere
Tropopause
Troposphere

Figure 6-2. The atmosphere, compared to the size of the earth, is very thin. This diagram dramatically exaggerates its depth to show the various layers.

SECTION A ■ **Basic Weather Theory**

DISCOVERY

Flying with Sunglasses

In the presence of bright glare, a good pair of sunglasses can usually improve visual acuity, reduce fatigue, and help your dark adaptation later in the day. There are, however, some things you should consider when selecting and wearing sunglasses. For example, sunglasses should be a neutral gray color in order to allow the widest spectrum of light to reach the eye. A green lens, for example, will tend to wash out any object colored green. This of course can be critical if a particular cockpit display uses green as a primary color. A neutral gray lens, on the other hand, filters out equal amounts of all colors, and provides you with the truest representation of the entire light spectrum. [Figure A]

In addition, you should understand that your depth of field is reduced when wearing sunglasses. This is analogous to a camera lens. As the aperture of the lens is opened to allow more light in (similar to the eye's iris opening as sunglasses are donned) objects will remain focused to a lesser depth. [Figure B] As you can imagine, this can be of particular importance in the landing pattern.

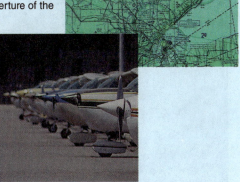

COMPOSITION OF THE ATMOSPHERE

If you could capture a cubic foot of atmosphere and analyze it, you would find it is composed of about 78% nitrogen and 21% oxygen. The remaining 1% is made up of several other gases, primarily argon and carbon dioxide. This cubic foot of atmosphere would also contain anywhere from almost zero to about 4% water vapor by volume. This relatively small amount of water vapor is responsible for major changes in the weather. [Figure 6-3]

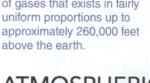

Atmosphere

78% Nitrogen
1% Other Gases
21% Oxygen

260,000 ft

Figure 6-3. The atmosphere is a mixture of gases that exists in fairly uniform proportions up to approximately 260,000 feet above the earth.

ATMOSPHERIC CIRCULATION

Atmospheric **circulation** may be considered simply to be the movement of air relative to the earth's surface. Because the atmosphere is fixed to the earth by gravity and rotates with the earth, there would be no circulation without forces that upset the atmosphere's equilibrium. The dynamic nature of the atmosphere is due, in a large part, to unequal temperatures at the earth's surface.

TEMPERATURE

As the earth rotates about the sun, the length of time and the angle at which sunlight strikes a particular portion of the earth's surface changes. Over the course of a year, this variance in solar energy is the reason we experience seasons. In general, however, the most direct rays of the sun strike the earth in the vicinity of the equator while the poles receive the least direct light and energy from the sun. [Figure 6-4]

The sun's energy is more concentrated near the equator. At higher latitudes the solar radiation spreads over a much greater surface area.

Courtesy of NASA

Figure 6-4. More of the sun's heat is absorbed at the Equator because of the angle at which the sun's rays strike the earth's surface.

Temperature in the Upper Atmosphere

In the upper reaches of the atmosphere, the air temperature can approach 1500°F, yet, if a thermometer were shielded from the sun, it would display a temperature close to absolute zero (–459.67°F). How is this possible? The apparent paradox stems from the meaning of air temperature and how it is measured. Normally, a thermometer measures temperature by displaying the average kinetic energy of the air molecules striking the thermometer's bulb. Although the few molecules in the thin upper atmosphere are moving at speeds corresponding to a very high temperature, the traditional method of measuring temperature doesn't work well because there simply aren't enough molecules to heat an object (such as a thermometer's bulb). Temperatures at extremely high altitudes are more closely related to radiation gain and loss between the portion of an object facing the sun and the opposite side. The harsh environment requires that space suits be designed to keep astronauts comfortable even when the difference between sun and shade could be the equivalent of nearly 2000°!

Courtesy of NASA Ames Home Page (image no. AS11-44-6557).

CONVECTION

To compensate for heating inequities, heat is transported, or circulated, from one latitude to another. When air is heated, its molecules spread apart. As the air expands, it becomes less dense and lighter than the surrounding air. As air cools, the molecules become packed more closely together, making it denser and heavier than warm air. As a result, the cool, heavy air tends to sink and replace warmer, rising air. This circulation process is known as **convection**.

To understand how convection contributes to atmospheric circulation, it is easiest to start with a simplified example. In this convection-only model, a stationary earth uniformly covered with water is warmed by energy radiating from the sun positioned directly over the equator. This solar energy strikes equatorial regions in much greater concentrations, resulting in much higher temperatures than at the poles. As a result, cold, dense air from the poles sinks and flows toward the equator, where it displaces rising air that is warmer and less dense. [Figure 6-5]

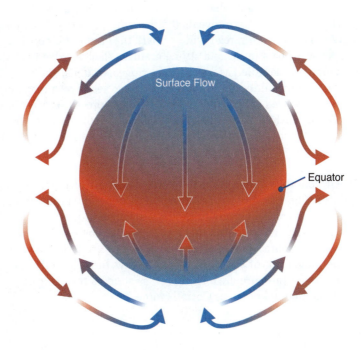

Figure 6-5. If the earth did not rotate, a huge convective circulation pattern would develop as air flowed from the poles to the equator and back again.

Surface Flow

Equator

THREE-CELL CIRCULATION PATTERN

Of course, the earth doesn't stand still in space. As the earth rotates, the single-cell circulation breaks up into three cells per hemisphere that distribute heat energy. In the Hadley cell, warm air rises and moves toward the poles until reaching approximately 30° latitude where the air cools and sinks back to the surface. Some of this air moves poleward until, at about 60° latitude, it meets cooler air migrating from the poles. The two air masses of different temperatures are forced upward creating the circulation associated with the Ferrel and Polar cells. This cell pattern is mirrored in the southern hemisphere. [Figure 6-6]

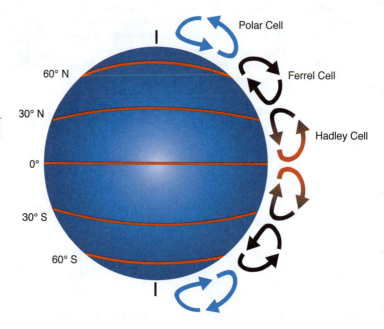

Figure 6-6. Three cells per hemisphere distribute heat energy. The Hadley and Ferrel cells are named for scientists who proposed and researched this circulation model.

ATMOSPHERIC PRESSURE

The unequal heating of the surface not only modifies air density and creates circulation patterns, it also causes changes in pressure. This is one of the main reasons for differences in altimeter settings between weather reporting stations. Meteorologists plot these pressure readings on weather maps and connect points of equal pressure with lines called **isobars**. Isobars are labeled in millibars and are usually drawn at four-millibar intervals. The resulting pattern reveals the **pressure gradient**, or change in pressure over distance. When isobars are spread widely apart, the gradient is considered to be weak, while closely spaced isobars indicate a strong gradient. Isobars also help to identify pressure systems, which are classified as highs, lows, ridges, troughs, and cols. A **high (H)** is a high pressure center surrounded on all sides by lower sea-level pressure. Conversely, a **low (L)**, is a low pressure center surrounded by higher pressure. A **ridge** is an elongated area of high pressure, while a **trough** is an elongated area of low pressure. A **col** can designate either a neutral area between two highs and two lows, or the intersection of a ridge and a trough. [Figure 6-7]

Air generally flows from the cool, dense air of highs into the warm, less dense air of lows. The speed of the resulting wind depends on the strength of the pressure gradient. A strong gradient tends to produce strong wind, while a weak gradient results in lighter winds. The force behind this movement is caused by the pressure gradient and is referred to as pressure gradient force. If **pressure gradient force** were the only force affecting the movement of air, wind would always blow directly from the higher pressure area to the lower pressure area. However, as soon as the air begins to move, it is deflected by a phenomenon known as Coriolis force.

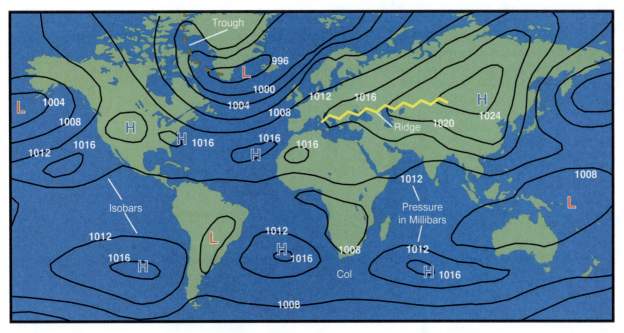

Figure 6-7. Pressure patterns provide key information about weather patterns and are plotted on many charts. Isobars are analogous to contour lines on a topographical chart.

Balloon Flight World Records

Maxie Anderson, Ben Abruzzo, and Larry Newman made the first successful Atlantic crossing in a manned balloon, *Double Eagle II*, in August 1978. They landed near Paris some 200 years after the first manned balloon flight, which also occurred in France. From 1859 until the successful crossing, balloon flight across the Atlantic had been attempted 17 times—with at least 7 fatalities—and with Anderson and Abruzzo themselves failing on their first attempt in *Double Eagle I*.

Three years later, in 1981, Abruzzo and Newman with two new crew members, Ron Clark and Rocky Aoki, flew the *Double Eagle V* across the Pacific Ocean.

The next step—circumnavigating the globe in a balloon—was another difficult challenge. Creating a balloon that could make the trip would not be easy—it had to be thinner than a piece of paper, yet strong enough to support a crew capsule weighing several tons while withstanding hurricane-force winds.

Unlike powered around-the-world flight, an attempt in a balloon is at the mercy of the winds and politics. Several crews attempted it between 1996 and 1999, including in the *Solo Challenger*, *Solo Spirit*, *Global Hilton*, *ICO Global Challenger*, and the *Cable and Wireless* balloon. *Breitling Orbiter 1* and *2* also tried. The first had a fuel leak and ditched in the Mediterranean Sea. The second was forced down because China would not allow overflight.

Crew Capsule

Breitling Orbiter 3 finally succeeded on March 21, 1999. Having taken off from Chateaux d'Oeux, Switzerland with crewmembers Bertrand Piccard and Brian Jones, the cabin become cold and miserable as the balloon stalled over Central America in stagnant winds prior the last part of the trip. Eventually, the winds improved and carried Orbiter 3 across the Atlantic at 200 km/hr to a landing in Egypt after 19 days, 21 hours, and 55 minutes in the air.

In June 2002, Steve Fossett followed with a *solo* flight around the world in a manned balloon, the *Bud Light Spirit of Freedom*.

SECTION A ■ Basic Weather Theory

CORIOLIS FORCE

Coriolis force, named after the French scientist who first described it in 1835, affects all objects moving freely across the face of the earth. Essentially, things such as ocean currents and airplane flight paths, which would otherwise follow a straight line, end up tracing a curved path due to the earth's rotation. [Figure 6-8]

A man stands on the edge of a spinning merry-go-round.

A ball is thrown to the man by a person standing on the opposite side of the merry-go-round.

Figure 6-8. Coriolis force explains why a ball thrown across a spinning merry-go-round appears to curve away from the intended target.

To an outside observer, the ball travels in a straight line after it is thrown.

To the people on the merry-go-round, the ball seems to curve away from the intended path, missing the target.

The outside observer realizes that the ball is not actually deflected; rather, the thrower and catcher change positions during the time it takes the ball to travel across the merry-go-round.

Because the earth rotates only once every 24 hours, the effect of Coriolis force is much weaker than what was displayed in the merry-go-round example in Figure 6-8. In fact, Coriolis is only significant when an object such as a parcel of air moves over a large distance (several hundred miles or more). The amount of deflection produced by Coriolis force also varies with latitude. It is zero at the equator and increases toward the poles. [Figure 6-9] In the northern hemisphere, any deviation will be to the right of its intended path while the opposite will occur in the southern hemisphere. In addition, the magnitude of the curve caused by Coriolis force varies with the speed of the moving object—the greater the speed, the greater the deviation.

Actual Flight Path

Destination

Figure 6-9. An airplane traveling either east or west directly above the equator will not experience any deviation from its intended flight path due to the effects of Coriolis force. Closer to the poles, the effects of Coriolis force become increasingly apparent.

SECTION A ■ **Basic Weather Theory**

FRICTIONAL FORCE

Pressure gradient and Coriolis forces work in combination to create wind. Pressure gradient force causes air to move from high pressure areas to low pressure areas. As the air begins to move, Coriolis force deflects it to the right in the northern hemisphere. This results in a clockwise flow around a high pressure area. The deflection continues until pressure gradient force and Coriolis force are in balance, and the wind flows roughly parallel to the isobars. As the air flows into a low pressure area, it moves counterclockwise around the low. This generally holds true for upper air winds. However, within about 2,000 feet of the ground, friction caused by the earth's surface slows the moving air. This frictional force reduces the Coriolis force. Because the pressure gradient force is now greater than Coriolis force, the wind is diverted from its path along the isobars toward the lower pressure. [Figure 6-10]

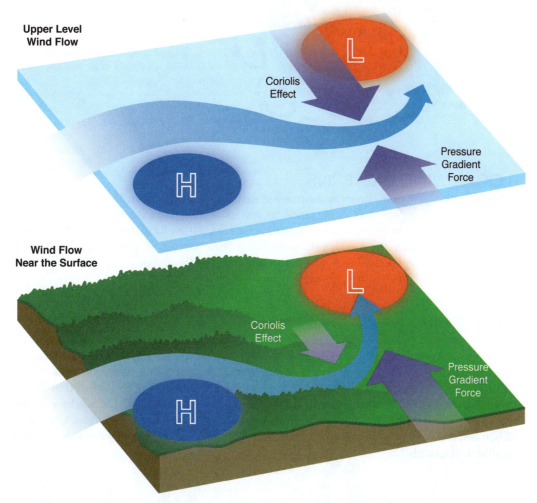

Figure 6-10. As surface friction retards the airflow, Coriolis force is weakened. This allows pressure gradient force to shift the airflow toward areas of lower pressure, causing the wind to blow at an angle across the isobars.

GLOBAL WIND PATTERNS

The three-cell circulation pattern produces semi-permanent low and high pressure areas. As tropical air rises and flows northward, it is deflected to the right by Coriolis force. Around 30° north, the flow is eastward, causing air to pile up (converge) at this latitude, resulting in a high pressure area. At the surface, air flows back toward the equator where low pressure exists. This low-level southerly flow is deflected to the west creating the northeast trade winds, named for the steady winds provided to ships sailing to the New World. The low-level air flowing northward from the high pressure at 30° also is deflected to the right by Coriolis. This creates the prevailing westerlies common to the middle latitudes. The cold, high-pressure polar air that flows southward is deflected to the right to create the polar easterlies. As this cold air meets the relatively warmer air from the prevailing westerlies, a low pressure area is formed around 60° north latitude. The line separating the air masses in this region is called the polar front. A mirror image of these pressure and wind patterns exists in the southern hemisphere. [Figure 6-11]

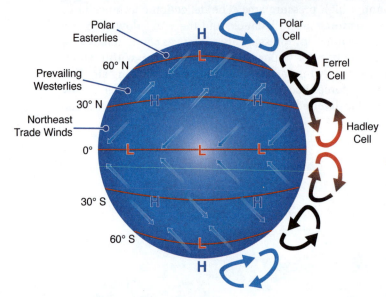

Figure 6-11. The three-cell circulation pattern produces semipermanent areas of high and low pressure as well as wind patterns.

LOCAL WIND PATTERNS

Although global wind patterns influence the earth's overall weather, local wind patterns may be of greater practical importance to you as a pilot because they usually cause significant changes in the weather of a particular area. These localized wind patterns are caused by terrain variations such as mountains, valleys, and water. The force behind these winds—cool air replacing warm air—is the same as it is for global wind patterns, but on a much smaller scale.

SEA BREEZE

During the day, a low-level sea breeze flows from sea to land due to the pressure gradient from the temperature differential between the land and the water. The temperature differential exists because land surfaces warm more rapidly than water surfaces so the land is usually warmer than water during the day. The **sea breeze** is a wind that blows from the cool water to warmer land. As afternoon heating increases, the sea breeze can reach speeds of 10 to 20 knots. A well-developed sea breeze exists between 1,500 and 3,000 feet AGL. A return flow forms above the sea breeze due to the reversal of the pressure gradient. [Figure 6-12]

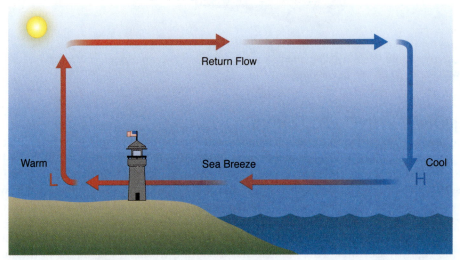

Figure 6-12. Coastal sea breezes occur when the low pressure over the warm land draws the cool air off the ocean.

LAND BREEZE

At night, land cools faster than water, and a **land breeze** blows from the cooler land to the warmer water. The pressure gradients are a reversal of what occurs during the day, however, because the temperature contrasts are smaller at night, the land breeze is generally weaker than the sea breeze. The land breeze can reach an altitude of 1,000 to 2,000 feet AGL and extend between 5 and 100 nautical miles inland, depending on the conditions and the location. [Figure 6-13]

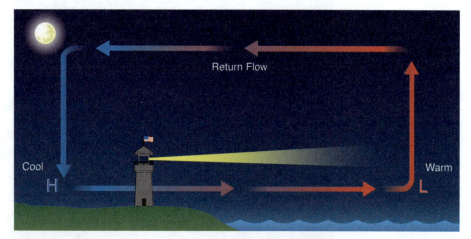

Figure 6-13. The land breeze is capped by a weaker onshore return flow.

VALLEY BREEZE

As mountain slopes are warmed by the sun during the day, the adjacent air also is heated. Because the heated air is less dense than the air at the same altitude over the valley, an upslope flow known as a **valley breeze** is created. Typical valley breezes reach speeds of between 5 and 20 knots with the maximum winds occurring a few hundred feet above the surface. [Figure 6-14]

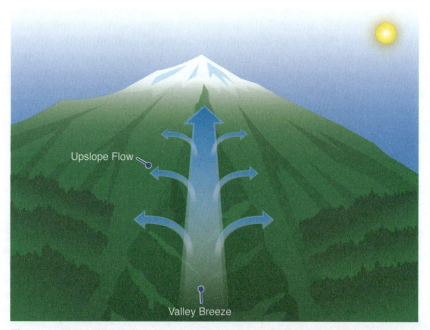

Figure 6-14. At low levels, air typically flows up the valley and up the warm slopes during the day.

MOUNTAIN BREEZE

At night, the high terrain cools off and eventually becomes cooler than the air over the valley. The pressure gradient reverses and a downslope flow, or **mountain breeze** develops. Prior to sunrise, speeds of 5 to 15 knots are common with greater speeds at the mouth of the valley, sometimes exceeding 25 knots. [Figure 6-15]

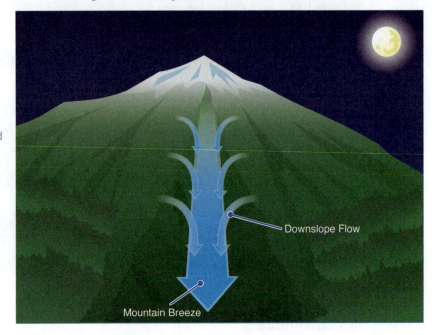

Figure 6-15. As the ground cools at night, air flows down the slope and away from the higher terrain.

Downslope Flow

Mountain Breeze

KATABATIC WINDS

Technically, any downslope wind can be classified as a **katabatic wind**. However, in many cases, the term katabatic is used to refer to downwind flows that are stronger than mountain breezes. Katabatic winds can be either warm or cold and some are even given special names in areas where they are particularly severe.

COLD DOWNSLOPE WINDS

When large ice and snow fields accumulate in mountainous terrain, the overlying air becomes extremely cold and a shallow dome of high pressure forms. This pressure gradient force pushes the cold air through gaps in the mountains. Although the air may be warmed during its descent, it's still colder than the air it displaces. If the wind is confined to a narrow canyon it can dramatically increase in velocity. Combined with the force of gravity, some winds can reach speeds in excess of 100 knots. [Figure 6-16] Cold downslope winds that occasionally occur around the world include the *bora* in Croatia, the *mistral* in the Rhone Valley of France, and the *Columbia Gorge wind* in the northwestern United States.

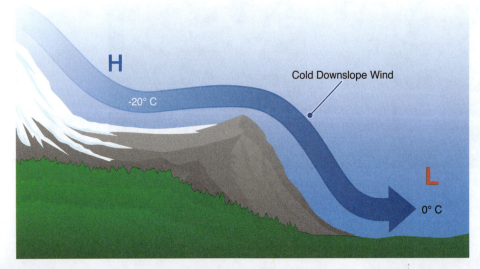

H

-20° C

Cold Downslope Wind

L

0° C

Figure 6-16. A cold downslope wind can become quite strong as it flows downhill from snow-covered plateaus or steep mountain slopes.

WARM DOWNSLOPE WINDS

When a warm air mass moves across a mountain range at high levels, it often forms a trough of low pressure on the downwind, or lee, side which causes a downslope wind to develop. As the air descends the lee side, it is compressed, which results in an increase in temperature. The warmer wind can raise temperatures over 20° in an hour. Wind speed is typically 20 to 50 knots although, in extreme cases, speeds can reach nearly 100 knots. [Figure 6-17] Well-known winds of this type include the *Chinook*, which occurs along the eastern slopes of the Rocky Mountains, the *foehn* in the Alps, and the *Santa Ana* in southern California.

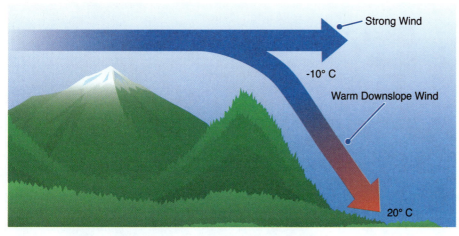

Strong Wind

-10° C

Warm Downslope Wind

20° C

Figure 6-17. A fast-moving wind down a mountain slope will warm as it descends, sometimes dramatically raising the temperature at the base of the mountain.

SUMMARY CHECKLIST

✓ The troposphere is the atmospheric layer extending from the surface to an average altitude of about 36,000 feet. Above the troposphere is the stratosphere, mesosphere, and the thermosphere.

✓ Because of heating inequities, heat is transported, or circulated, from one latitude to another by a process known as convection.

✓ In the three-cell circulation model, the Hadley, Ferrel, and Polar cells generate predictable wind patterns and distribute heat energy.

✓ Pressure readings on weather maps connect points of equal pressure with lines called isobars. When isobars are spread widely apart, the pressure gradient is considered to be weak, while closely spaced isobars indicate a strong gradient.

✓ A high is a center of high pressure surrounded on all sides by lower pressure. Conversely, a low is an area of low pressure surrounded by higher pressure.

✓ A ridge is an elongated area of high pressure, while a trough is an elongated area of low pressure. A col can designate either a neutral area between two highs and two lows, or the intersection of a ridge and a trough.

✓ Coriolis force causes all free-moving objects to trace a curved path due to the earth's rotation. In the northern hemisphere, the deviation will be to the right of its intended path while the opposite will occur in the southern hemisphere.

✓ Frictional force causes a wind to shift directions when near the earth's surface.

✓ A sea breeze blows from the cool water to the warmer land during the day. At night, a land breeze blows from the cooler land to the warmer water.

✓ A cold downslope wind flows downhill from snow-covered plateaus or steep mountain slopes.

✓ Warm, downslope winds sometimes dramatically raise the temperature at the base of the mountain.

SECTION A ■ **Basic Weather Theory**

KEY TERMS

Atmosphere

Troposphere

Tropopause

Stratosphere

Mesosphere

Thermosphere

Circulation

Convection

Isobars

Pressure Gradient

High (H)

Low (L)

Ridge

Trough

Col

Pressure Gradient Force

Coriolis Force

Frictional Force

Sea Breeze

Land Breeze

Valley Breeze

Mountain Breeze

Katabatic Wind

QUESTIONS

1. Identify the four major layers of the atmosphere depicted in the accompanying illustration.

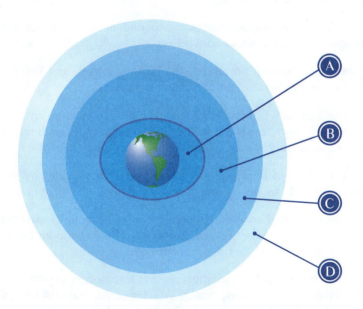

2. Which layer of the atmosphere contains most of the earth's weather?

3. Describe the atmospheric convection process.

4. What is the primary cause of changing altimeter settings between weather reporting points?

5. True/False. Closely spaced isobars on a weather map is an indicator of light surface winds.

6. What three factors affect the amount of deflection caused by Coriolis force?

7. Describe how frictional force causes wind to shift near the earth's surface.

Match the following types of breezes with the most appropriate descriptor.

8. Sea Breeze   A. Occurs at night as land cools faster than water

9. Land Breeze   B. Occurs during the day as the sun warms mountain slopes

10. Valley Breeze   C. Occurs at night as the high terrain cools
                              (relative to air over the valley)

11. Mountain Breeze   D. Occurs during the day as the sun warms the land

SECTION B
Weather Patterns

It was a stormy night in late October 1743 when Benjamin Franklin ventured outside hoping to catch a glimpse of a lunar eclipse. Although he never saw it, he learned later that clear skies in Boston the same night had given a friend an unobstructed view of the rare occurrence. His friend also related that the day following the eclipse, the weather took a turn for the worse. This prompted Franklin to study weather reports in an effort to document the storm's movement. His subsequent findings led to a more complete understanding of weather patterns. Since then, much more has been learned about the nature of the atmosphere and the weather it produces. A good grasp of these atmospheric principles can help you evaluate weather conditions prior to and during your flights.

ATMOSPHERIC STABILITY

Stability is the atmosphere's resistance to vertical motion. A stable atmosphere does not necessarily prevent air from moving vertically, but it does make that movement more difficult. In most cases, the vertical motions present in a stable environment are very small, resulting in a generally smooth airflow. In an unstable atmosphere, convection is the rule. The air rises because it is warmer than its surroundings. In comparison with vertical motions in a stable environment, unstable vertical movements are large, and the airflow is turbulent. This instability can lead to significant cloud development, turbulence, and hazardous weather.

Air that moves upward expands due to lower atmospheric pressure. When air moves downward, it is compressed by the increased pressure at lower altitudes. As the pressure of a given portion of air changes, so does its temperature. The temperature change is caused by a process known as **adiabatic heating** or **adiabatic cooling**, which is a change in the temperature of dry air during expansion or compression.

The adiabatic process takes place in all upward and downward moving air. When air rises into an area of lower pressure, it expands to a larger volume. As the molecules of air expand, the temperature of the air lowers. As a result, when a parcel of air rises, pressure decreases, volume increases, and temperature decreases. When air descends, the opposite is true. The rate at which temperature decreases with an increase in altitude is referred to as its **lapse rate**. As you ascend through the atmosphere, the average rate of temperature change is 2°C (3.5°F) per 1,000 feet.

Because water vapor is lighter than air, moisture decreases air density, causing it to rise. Conversely, as moisture decreases, air becomes denser and tends to sink. Because moist air cools at a slower rate, it's generally less stable than dry air because the moist air must rise higher before its temperature cools to that of the surrounding air. The dry adiabatic lapse rate (unsaturated air) is 3°C (5.4°F) per 1,000 feet. [Figure 6-18] The moist adiabatic lapse rate varies from 1.1°C to 2.8°C (2°F to 5°F) per 1,000 feet.

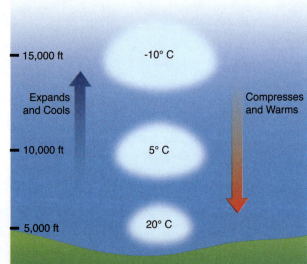

Figure 6-18. As long as a parcel of air remains unsaturated, it will expand and cool at a rate of 3°C per 1,000 feet. As the air descends, it will compress and warm at the same rate.

Overall, the combined effects of temperature and moisture determine the stability of the air and, to a large extent, the type of weather produced. The greatest instability occurs when the air is both warm and moist. Tropical weather, with its almost daily thunderstorm activity, is a perfect example of weather that occurs in very unstable air. Air that is both cool and dry resists vertical movement and is very stable. A good example of this can be found in arctic regions in winter, where stable conditions often result in very cold, generally clear weather.

TEMPERATURE INVERSIONS

Although temperature usually decreases with an increase in altitude, the reverse is sometimes true. When temperature increases with altitude, a **temperature inversion** exists. Inversions are usually confined to fairly shallow layers and might occur near the surface or at higher altitudes. Temperature inversions act as a lid for weather and pollutants. Below the inversion, visibility is often restricted by fog, haze, smoke, and low clouds. Temperature inversions occur in stable air with little or no wind and turbulence. [Figure 6-19]

Figure 6-19. Visibility is often very poor in a temperature inversion.

One of the most familiar types of inversions is the one that forms near the ground on cool, clear nights when the wind is calm. As the ground cools, it lowers the temperature of the adjacent air. If this process of terrestrial radiation continues, the air within a few hundred feet of the surface may become cooler than the air above it. An inversion can also occur when cool air is forced under warm air, or when warm air spreads over cold. Both of these are called frontal inversions.

SECTION B ■ Weather Patterns

Smoke Plumes and Atmospheric Stability

What can you tell about the changing stability of the atmosphere from smoke patterns? From the typical temperature inversion in the early morning, through the reversal of the vertical temperature profile around mid-day, to the reappearance of the temperature inversion in the evening, the stability of the atmosphere changes. Coupled with a steady breeze, these changes produce characteristic patterns as shown in the accompanying illustration.

DISCOVERY

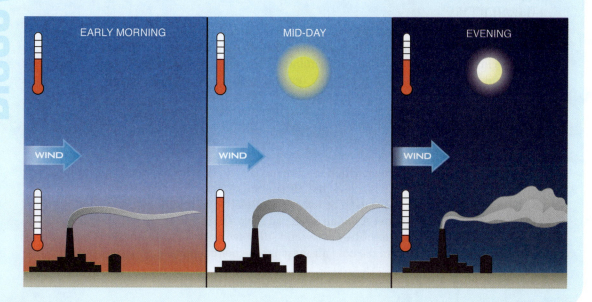

MOISTURE

Even over tropical rain forests, moisture only accounts for a small percentage of the total volume of the atmosphere. Despite this small amount of moisture, water vapor is still responsible for many of the flight hazards encountered in aviation operations. Generally speaking, if the air is very moist, poor, or even severe weather can occur; if the air is dry, the weather usually will be good.

CHANGE OF STATE

Water is present in the atmosphere in three states: solid, liquid, and gas. All three states are found within the temperature ranges normally encountered in the atmosphere, and the change from one to another happens readily. Changes in state occur through the processes of evaporation, condensation, sublimation, deposition, melting, and freezing. As water changes from one physical state to another, an exchange of heat takes place.

Evaporation is the changing of liquid water to invisible water vapor. As water vapor forms, heat is absorbed from the nearest available source. For example, as perspiration evaporates from your body, you feel cooler because some of your body heat has been absorbed by the water vapor. This heat exchange is known as the latent heat of evaporation. The reverse of evaporation is **condensation**. It occurs when water vapor changes to a liquid, as when water drops form on a cool glass on a warm day. When condensation takes place, the heat absorbed by water vapor during evaporation is released. The heat released is referred to as the latent heat of condensation, and is an important factor in cloud development.

Sublimation is the changing of ice directly to water vapor, while the transformation of water vapor to ice is known as **deposition**. In both cases, the liquid state is bypassed. The changes from ice to water and water to ice are well known to most everyone as **melting** and **freezing**, respectively. The heat exchange that occurs during melting and freezing is small—about 15% of the liquid-to-vapor change—and has relatively little effect on weather. [Figure 6-20]

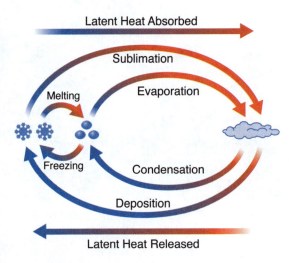

Figure 6-20. Heat is absorbed when water melts or evaporates and is released when water condenses or freezes.

HUMIDITY

Humidity simply refers to moisture in the air. For example, on warm, muggy days when you perspire freely, the air is said to be humid. **Relative humidity** is the actual amount of moisture in the air compared to the total amount that could be present at that temperature. It's important to remember that relative humidity tells you nothing about the actual amount of water vapor in the air. For example, at the same 100% relative humidity levels, air at -4°F in Alaska only has about one twentieth of the water vapor as 68°F air in Florida.

DEWPOINT

Dewpoint is the temperature at which air reaches a state in which it can hold no more water. When the dewpoint is reached, the air contains 100% of the moisture it can hold at that temperature, and it is said to be **saturated**. Relative humidity and dewpoint both relate moisture to temperature. A hot air mass can have a higher humidity, that is absolute amount of water vapor, than a cold air mass, and still have lower relative humidity because that amount of water vapor in the cold air mass is a higher percentage of what it can hold. [Figure 6-21]

32°C Temperature
10°C Dewpoint
26% Relative Humidity

Copyright Corel

-5°C Temperature
-5°C Dewpoint
100% Relative Humidity

Copyright Corel

Figure 6-21. The higher temperature and dewpoint in the desert shown on the left means that the air can and does hold more moisture. The area on the right has less total moisture even though the air is saturated.

When warm, moist air begins to rise in a convective current, clouds often form at the altitude where its temperature and dewpoint reach the same value. When lifted, unsaturated air cools at about 5.4°F per 1,000 feet, and the dewpoint temperature decreases at about 1°F per 1,000 feet. Therefore, the temperature and dewpoint converge at 4.4°F per 1,000 feet. You can use these values to estimate cloud bases. For example, if the surface temperature is 80°F and the surface dewpoint is 62°F, the spread is 18°F. This difference, divided by the rate that the temperature approaches the dewpoint (4.4°F), helps you judge the approximate height of the base of the clouds in thousands of feet (18 ÷ 4.4 = 4 or 4,000 feet AGL).

DEW AND FROST

On cool, still nights, surface features and objects may cool to a temperature below the dewpoint of the surrounding air. Water vapor then condenses out of the air in the form of **dew**, which explains why grass is often moist in the early morning. **Frost** forms when water vapor changes directly to ice on a surface that is below freezing. From late fall through early spring, you may frequently encounter frost on your airplane in the early morning.

Frost can pose a serious hazard during takeoffs. It interferes with the smooth airflow over the wings and can cause early airflow separation, resulting in a loss of lift. It also increases drag and, when combined with the loss of lift, may prevent the aircraft from becoming airborne. Always remove all frost from the aircraft surfaces before flight.

CLOUDS

As air cools to its saturation point, condensation changes invisible water vapor into a visible state. Most commonly, this visible moisture takes the form of clouds or fog. Clouds are composed of very small droplets of water or, if the temperature is low enough, ice crystals. The droplets condense on very small particles of solid matter in the air. These particles, called **condensation nuclei**, can be dust, salt from evaporating sea spray, or products of combustion. When clouds form near the surface, they are referred to as fog. [Figure 6-22]

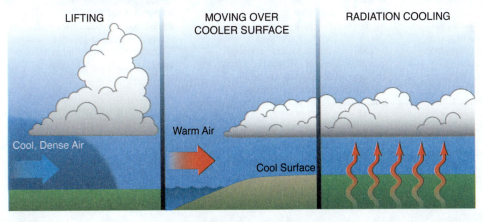

LIFTING

MOVING OVER COOLER SURFACE

RADIATION COOLING

Cool, Dense Air

Warm Air

Cool Surface

Figure 6-22. Air can be cooled by lifting, by moving over a cooler surface, or by cooling from the underlying surface. Depending upon the temperature of the air, water vapor may condense into visible water droplets or form ice crystals.

DISCOVERY

Frost Formation

How is it possible for frost to form on the ground when the thermometer reads 40°F (4°C)? Because ice doesn't form at temperatures above freezing, the object on which frost forms must be below 32°F (0°C). At night, variations in temperature can occur over very short distances. As night falls, heat is lost from the ground at a faster rate than the air several feet above it. This rapid cooling decreases the temperature immediately above the ground, possibly below freezing, even though the temperature at the level of a thermometer located in a standard instrument shelter 6 feet above the surface is greater than 32°F. [Figure A] This can result in frost forming on the ground even though the minimum reported overnight temperature never dropped below freezing. [Figure B]

Clouds and fog usually form as soon as the air becomes saturated. You can anticipate the formation of fog or very low clouds by monitoring the difference between surface temperature and dewpoint, usually referred to as the **temperature/dewpoint spread**. When the spread reaches 4°F (2°C) and continues to decrease, the air is nearing the saturation point, increasing the probability of fog and low clouds.

TYPES OF CLOUDS

Clouds are your weather signposts in the sky. They provide a visible indication of the processes occurring in the atmosphere. To the astute observer, they give valuable information about current and future conditions. Clouds are divided into four basic groups, or families, depending upon their characteristics and the altitudes where they occur. Clouds are classified as low, middle, high, and clouds with vertical development.

In general, clouds are named using Latin words. Sheet-like clouds are referred to as *stratus* (meaning "layer"); *cumulus* translates as "heap" and refers to puffy clouds; *cirrus* ("ringlet") is used to designate wispy clouds; and rain clouds contain the prefix or suffix *nimbus,* which means "violent rain."

LOW CLOUDS

Low clouds extend from near the surface to about 6,500 feet AGL. Low clouds usually consist almost entirely of water but sometimes may contain supercooled water, which can create an icing hazard for aircraft. Types of low clouds include **stratus**, **stratocumulus**, and **nimbostratus**. [Figure 6-23]

LOW CLOUDS

Stratus Clouds
Stratus clouds are layered clouds that form in stable air near the surface due to cooling from below. Stratus clouds have a gray, uniform appearance and generally cover a wide area. Although turbulence in these clouds is low, they usually restrict visual flying due to low ceilings and visibility. Icing conditions are possible if temperatures are at or near freezing. Stratus clouds may form when moist stable air is lifted up sloping terrain, or when warm rain evaporates as it falls through cool air.

Nimbostratus Clouds
Nimbostratus clouds are gray or black clouds that can be more than several thousand of feet thick, contain large quantities of moisture, and produce widespread areas of rain or snow. If temperatures are near or below freezing, they may create heavy aircraft icing.

Stratocumulus Clouds
Stratocumulus clouds are white, puffy clouds that form as stable air is lifted. They often form as a stratus layer breaks up or as cumulus clouds spread out.

Figure 6-23. Low clouds are found at altitudes extending from the surface to about 6,500 feet AGL.

FOG

Technically, **fog** is a low cloud that has its base within 50 feet of the ground. If the fog is less than 20 feet deep, it is called **ground fog**. Fog is classified according to the way it forms. **Radiation fog** forms over low-lying, fairly flat surfaces on clear, calm, humid nights. As the surface cools by radiation, the adjacent air also is cooled to its dewpoint. Radiation fog usually occurs in stable air associated with a high pressure system. As early morning temperatures increase, the fog begins to lift and usually "burns off" by mid-morning. If higher cloud layers form over the fog, visibility will improve more slowly. [*Figure 6-24*]

Figure 6-24. Radiation fog is often found in river valleys where cool air pools and moisture is abundant.

Advection fog is caused when a low layer of warm, moist air moves over a cooler surface. It is most common under cloudy skies along coastlines where wind transports air from the warm water to the cooler land. Winds up to about 15 knots will intensify the fog. Above 15 knots, turbulence creates a mixing of the air, and it usually lifts sufficiently to form low stratus clouds. **Upslope fog** forms when moist, stable air is forced up a sloping land mass. Like advection fog, upslope fog can form in moderate to strong winds and under cloudy skies.

Steam fog, which is often called sea smoke, occurs as cold, dry air moves over comparatively warmer water. The warm water evaporates and rises upward resembling rising smoke. It is composed entirely of water droplets that often freeze quickly and fall back into the water as ice particles. This can produce an icing hazard to aircraft. In addition, aircraft may experience low-level turbulence in steam fog because it forms in relatively unstable air. [Figure 6-25]

Figure 6-25. Cold, dry air moving over warmer water in this lake creates steam fog.

MIDDLE CLOUDS

Middle clouds have bases that range from about 6,500 to 20,000 feet AGL. They are composed of water, ice crystals, or supercooled water, and may contain moderate turbulence and potentially severe icing. **Altostratus** and **altocumulus** are classified as middle clouds. [Figure 6-26]

SECTION B ■ **Weather Patterns**

MIDDLE CLOUDS

Altostratus Clouds
Altostratus clouds are flat, dense clouds that cover a wide area. They are a uniform gray or gray-white in color. Although they produce minimal turbulence, they may produce moderate aircraft icing.

Altocumulus Clouds
Altocumulus clouds are gray or white, patchy clouds of uniform appearance that often form when altostratus clouds start to break up. They usually extend over a wide area, produce light turbulence, and may contain supercooled water droplets.

Figure 6-26. Middle clouds are found at altitudes extending from 6,500 feet to 20,000 feet AGL.

HIGH CLOUDS

High clouds have bases beginning above 20,000 feet AGL. They are generally white to light gray in color and form in stable air. They are composed mainly of ice crystals and seldom pose a serious turbulence or icing hazard. The three basic types of high clouds are called **cirrus**, **cirrostratus**, and **cirrocumulus**. [Figure 6-27]

HIGH CLOUDS

Cirrus Clouds
Cirrus clouds form in stable air at high altitudes. They are thin and wispy and usually form above 30,000 feet. White or light gray in color, they often exist in patches or narrow bands that cross the sky. Since cirrus clouds are sometimes blown from the tops of thunderstorms, they can be an advance warning of approaching bad weather.

Cirrostratus Clouds
Cirrostratus clouds also are thin, white clouds that often form in long bands or sheets against a deep blue background. Although they may be several thousands of feet thick, moisture content is low and they pose no icing hazard.

Cirrocumulus Clouds
Cirrocumulus clouds are white patchy clouds that look like cotton. They form as a result of shallow convective currents at high altitude and may produce light turbulence.

Figure 6-27. High clouds are found at altitudes extending above 20,000 feet AGL.

CLOUDS WITH VERTICAL DEVELOPMENT

Cumulus clouds are puffy white clouds with flat bases that can start off as harmless fair weather clouds and build vertically into **towering cumulus** or even giant **cumulonimbus** clouds. The bases of clouds with vertical development are found at altitudes associated with low to middle clouds, and their tops extend into the altitudes associated with high clouds. Frequently, these cloud types are obscured by other cloud formations. When this happens, they are said to be **embedded**. [Figure 6-28]

CLOUDS WITH VERTICAL DEVELOPMENT

Cumulus Clouds
Cumulus clouds form in convective currents resulting from the heating of the earth's surface. They usually have flat bottoms and dome-shaped tops. Widley spaced cumulus clouds that form in fairly clear skies are called fair weather cumulus and indicate a shallow layer of instability. You can expect turbulence, but little icing and precipitation.

Towering Cumulus
Towering cumulus clouds look like large mounds of cotton with billowing cauliflower tops. Their color may vary from brilliant white at the top to gray near the bottom. Towering cumulus clouds indicate a fairly deep area of unstable air. They contain moderate to heavy turbulence with icing and often develop into thunderstorms.

Cumulonimbus Clouds
Cumulonimbus clouds, which are more commonly called thunderstorms, are large, vertically developed clouds that form in very unstable air. They are gray-white to black in color and contain large amounts of moisture. Many flying hazards are linked with cumulonimbus clouds.

Figure 6-28. Clouds with vertical development, also known as cumuliform clouds, indicate some instability.

PRECIPITATION

Precipitation can be defined as any form of particles, whether liquid or solid, that fall from the atmosphere. Whether it reaches the ground or evaporates before it reaches the surface, precipitation contributes to many aviation weather problems. It can reduce visibility, affect engine performance, increase braking distance, and cause dramatic shifts in wind direction and velocity. Under the right conditions, precipitation can freeze on contact, affecting airflow over aircraft wings and control surfaces.

PRECIPITATION CAUSES

Although a cloud usually forms when the atmosphere is saturated, it doesn't necessarily mean that the cloud will produce precipitation. For precipitation to occur, water or ice particles must grow in size until they can no longer be supported by the atmosphere. There are three ways by which precipitation-size particles can be produced. In the condensation/deposition method, water droplets or ice crystals simply continue to grow by the same processes as they were initially formed until they are large enough to fall out of the cloud. The second process, known as coalescence, generally occurs when the initial cloud water droplets are of different sizes. The larger drops fall faster, growing as they collide and capture the smaller ones.

The first two processes, while important, are usually too slow to allow much precipitation to fall within the normal lifetime of a rain cloud. A more efficient ice-crystal process occurs primarily in the middle to high latitudes where the clouds can extend upward into areas well below freezing. These clouds not only contain ice crystals, but also water droplets. Although it may seem contradictory, water droplets can exist in temperatures below freezing. In fact, studies show that, depending on the conditions, these **supercooled water droplets** can stay in a liquid form in temperatures as low as –40°C. During the ice-crystal process, water vapor given up by the evaporating water droplets causes the ice crystals to grow to precipitation-sized particles in a relatively short time period.

SECTION B ▪ **Weather Patterns**

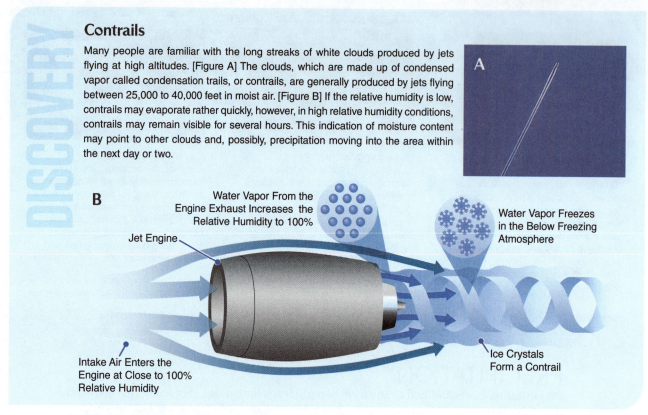

Contrails

Many people are familiar with the long streaks of white clouds produced by jets flying at high altitudes. [Figure A] The clouds, which are made up of condensed vapor called condensation trails, or contrails, are generally produced by jets flying between 25,000 to 40,000 feet in moist air. [Figure B] If the relative humidity is low, contrails may evaporate rather quickly, however, in high relative humidity conditions, contrails may remain visible for several hours. This indication of moisture content may point to other clouds and, possibly, precipitation moving into the area within the next day or two.

A

B

Water Vapor From the Engine Exhaust Increases the Relative Humidity to 100%

Water Vapor Freezes in the Below Freezing Atmosphere

Jet Engine

Intake Air Enters the Engine at Close to 100% Relative Humidity

Ice Crystals Form a Contrail

DISCOVERY

TYPES OF PRECIPITATION

As they fall, snowflakes and raindrops may change into other types of precipitation depending on the atmospheric conditions beneath the cloud. In addition to snow and rain, falling moisture also can take the form of drizzle, ice pellets, or hail.

DRIZZLE AND RAIN

Drizzle is distinguished by very small droplets (less than 0.02 inches in diameter). It is commonly associated with fog or low stratus clouds. Falling drops of liquid precipitation is considered to be rain when it is 0.02 inches in diameter or greater. Rain generally falls at a relatively steady rate and stops gradually. Rain showers refer to liquid precipitation that starts, changes intensity, and stops suddenly. The largest drops and greatest short-term precipitation amounts typically occur with rain showers associated with cumulus clouds and thunderstorms. At times, falling rain may never reach the ground due to rapid evaporation. The smaller the drops become, the slower their rate of fall, hence the appearance of rain streaks hanging in the air. These streams of evaporating precipitation are called **virga**. [Figure 6-29]

When warm rain or drizzle falls through a layer of cooler air near the surface, evaporation from the falling precipitation may saturate the cool air, causing **precipitation-induced fog** to form. This fog can be very dense, and usually does not clear until the rain moves out of the area.

Freezing drizzle and freezing rain maintain the same general characteristics as described above except that they freeze upon contact with the ground or other objects such as power lines, trees, or aircraft. Freezing rain can produce black ice (clear ice on black pavement). Because it's difficult to distinguish, black ice can be a serious hazard to aircraft ground operations.

Figure 6-29. Streaks of rain that evaporate before reaching the ground are known as virga.

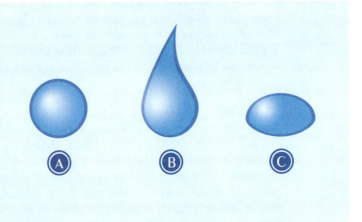

What Shape are Raindrops?

Of the three illustrations to the right, which do you think most closely resembles an actual raindrop? Most people pick drop B. Did you? Well, contrary to popular conception, raindrops don't look like tears. Their real shape depends on the size of the drop. A raindrop will take the shape shown as drop A when it is less than about 0.08 inches (2 mm) in diameter. Larger drops tend to be shaped like drop C.

ICE PELLETS AND HAIL

If rain falls through a temperature inversion, it may freeze as it passes through the underlying colder air, striking the ground as **ice pellets**. In some cases, water droplets that freeze in clouds with strong upward currents may grow in size as they collide with other freezing water droplets. Eventually they become too large for air currents to support, and they fall as **hail**. Hailstones can grow to more than five inches in diameter with weights of more than one and one-half pounds.

SNOW

Snow is precipitation composed of ice crystals. Snow and snow showers are distinguished in the same way as rain showers. Snow grains are the solid equivalent of drizzle. They are very small, white, opaque particles of ice. They are different from ice pellets in that they are flatter and they neither shatter nor bounce when they strike the ground. Ice crystals that descend from cirrus clouds are called cirrus **fallstreaks**, or mare's tails. [Figure 6-30]

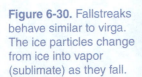

Figure 6-30. Fallstreaks behave similar to virga. The ice particles change from ice into vapor (sublimate) as they fall.

AIR MASSES

An **air mass** is a large body of air with fairly uniform temperature and moisture content. It may be several hundred miles across and usually forms where air remains stationary, or nearly so, for at least several days. During this time, the air mass takes on the temperature and moisture properties of the underlying surface.

SOURCE REGIONS

The area where an air mass acquires the properties of temperature and moisture that determine its stability is called its **source region**. An ideal source region is a large area with fairly uniform geography and temperature. A source region is usually located where

SECTION B ■ **Weather Patterns**

air tends to stagnate. The best areas for air mass development are in the regions where atmospheric circulation has caused the buildup of semipermanent areas of high pressure. This often occurs in snow and ice covered polar regions, over tropical oceans, and in the vicinity of large deserts. The middle latitudes are poor source regions because of the strong westerly winds and the continual mixing of tropical and polar air masses.

CLASSIFICATIONS

Air masses are classified according to the regions where they originate. They are generally divided into polar or tropical to identify their temperature characteristics, and continental or maritime to identify their moisture content. A continental polar air mass, for example, originates over a polar land mass and contains cold, dry, and stable air. A stable air mass generally exhibits widespread stratiform clouds, restricted visibility, smooth air, and steady rain or drizzle. A maritime tropical air mass originates over water and contains warm, moist, and unstable air. The instability associated with a warm air mass tends to result in the formation of cumuliform clouds with showers, turbulence, and good surface visibility. [Figure 6-31]

SECTION B ■ **Weather Patterns**

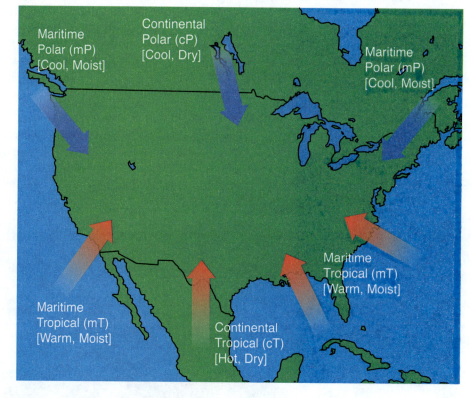

Figure 6-31. Air mass source regions surround North America. As air masses move out of their source regions, they often converge and give birth to the continent's major weather systems.

MODIFICATION

As an air mass moves out of its source region, it is modified by the temperature and moisture of the area over which it moves. The degree to which an air mass is changed depends on several factors including its speed, the nature of the region it moves over, the depth of the air mass, and the temperature difference between the air mass and the new surface.

WARMING FROM BELOW

As an air mass moves over a warmer surface, its lower layers are heated and vertical movement of the air develops. Depending on temperature and moisture levels, this can result in extreme instability. As shown in Figure 6-32, the Great Lakes modify the continental polar air masses moving out of Canada. In early winter, the lakes heat and moisten air near the surface, destabilizing the air and producing heavy lake-effect snow over the lakes and on the lee shores.

Figure 6-32. Air mass modification by the Great Lakes produces heavy lake-effect snow.

COOLING FROM BELOW

When an air mass flows over a cooler surface, its lower layers are cooled and vertical movement is inhibited. As a result, the stability of the air is increased. If the air is cooled to its dewpoint, low clouds or fog may form. This cooling from below creates a temperature inversion and can result in low ceilings and visibility for long periods of time. Figure 6-33 shows how this phenomena creates dense and persistent winter fog in California's Sacramento Valley.

Figure 6-33. In winter months, warm, moist Pacific air flowing into the Sacramento Valley is cooled to its dewpoint, resulting in widespread stratus clouds or fog that can persist for weeks.

FRONTS

When an air mass moves out of its source region, it comes in contact with other air masses that have different moisture and temperature characteristics. The boundary between air masses is called a **front**. Because the weather along a front often presents a serious hazard to flying, you need to have a thorough understanding of the associated weather.

SECTION B ■ **Weather Patterns**

TYPES OF FRONTS

Fronts are named according to the temperature of the advancing air relative to the temperature of the air it is replacing. In a **cold front**, cold air is moving to displace warmer air. In a **warm front**, warm air is replacing cold air. A **stationary front** has no movement. Cold fronts are usually fast moving and often catch up to and merge with a slower moving warm front. When cold and warm fronts merge, they create an **occluded front**.

FRONTAL DISCONTINUITIES

In weather terminology, frontal discontinuities refer to the comparatively rapid changes in the meteorological characteristics of an air mass. When you cross a front, you move from one air mass into another air mass with different properties. The changes between the two may be very abrupt, indicating a narrow frontal zone. On the other hand, the changes may occur gradually, indicating a wide and, perhaps diffused frontal zone. These changes can give you important cues to the location and intensity of the front.

TEMPERATURE

Temperature is one of the most easily recognized discontinuities across a front. At the surface, the temperature change is usually very noticeable and may be quite abrupt in a fast-moving front. With a slow-moving front, it usually is less pronounced. When you are flying through a front, you can observe the temperature change on the outside air temperature gauge. However, the change may be less abrupt at middle and high altitudes than it is at the surface.

WIND

The most reliable indications that you are crossing a front are a change in wind direction and, less frequently, wind speed. Although the exact new direction of the wind is difficult to predict, the wind always shifts to the right in the northern hemisphere. When you are flying through a front at low to middle altitudes, you will always need to correct to the right in order to maintain your original ground track.

PRESSURE

As a front approaches, atmospheric pressure usually decreases, with the area of lowest pressure lying directly over the front. Pressure changes on the warm side of the front generally occur more slowly than on the cold side. When you approach a front toward the cool air, pressure drops slowly until you cross the front, then rises quickly. When you are crossing toward the warm air, pressure drops abruptly over the front and then rises slowly. The important thing to remember is that you should update your altimeter setting as soon as possible after crossing a front.

FRONTAL WEATHER

The type and intensity of frontal weather depend on several factors. Some of these factors are the availability of moisture, the stability of the air being lifted, and the speed of the frontal movement. Other factors include the slope of the front and the moisture and temperature variations between the two fronts. Although some frontal weather can be very severe and hazardous, other fronts produce relatively calm weather.

COLD FRONTS

A cold front separates an advancing mass of cold, dense, and stable air from an area of warm, lighter, and unstable air. Because of its greater density, the cold air moves along the surface and forces the less dense, warm air upward. In the northern hemisphere, cold fronts are usually oriented in a northeast to southwest line and may be several hundred miles long. Movement is usually in an easterly direction. A depiction of the typical cold front and a summary of its associated weather is shown in Figure 6-34.

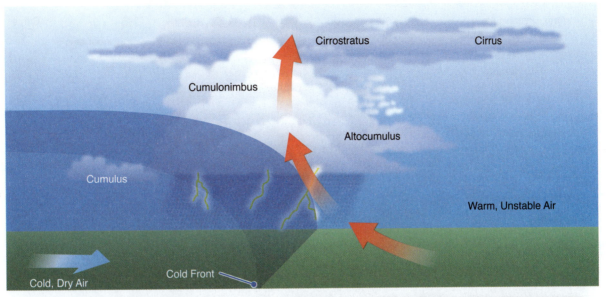

TYPICAL COLD FRONT WEATHER			
	Prior to Passage	**During Passage**	**After Passage**
Clouds	• Cirriform • Towering cumulus and/or cumulonimbus	• Towering cumulus and/or cumulonimbus	• Cumulus
Precipitation	• Showers	• Heavy showers • Possible hail, lightning, and thunder	• Slowly decreasing showers
Visibility	• Fair in haze	• Poor	• Good
Wind	• SSW	• Variable and gusty	• WNW
Temperature	• Warm	• Suddenly cooler	• Continued cooler
Dewpoint	• High	• Rapidly dropping	• Continued drop
Pressure	• Falling	• Bottoms out, then rises rapidly	• Rising

Figure 6-34. Cumuliform clouds and showers are common in the vicinity of cold fronts.

FAST-MOVING COLD FRONTS

Fast-moving cold fronts are pushed along by intense high pressure systems located well behind the front. Surface friction acts to slow the movement of the front, causing the leading edge of the front to bulge out and to steepen the front's slope. These fronts are particularly hazardous because of the steep slope and wide differences in moisture and temperature between the two air masses.

SLOW-MOVING COLD FRONTS

The leading edge of a slow-moving cold front is much shallower than that of a fast-moving front. This produces clouds that extend far behind the surface front. A slow-moving cold front meeting stable air usually causes a broad area of stratus clouds to form behind the front. When a slow-moving cold front meets unstable air, large numbers of vertical clouds often form at and just behind the front. Fair weather cumulus clouds are often present in the cold air, well behind the surface front.

WARM FRONTS

Warm fronts occur when warm air overtakes and replaces cooler air. They usually move at much slower speeds than cold fronts. The slope of a warm front is very gradual, and the warm air may extend up over the cool air for several hundred miles ahead of the front. A depiction of the typical warm front and a summary of its associated weather is shown in Figure 6-35.

SECTION B ■ **Weather Patterns**

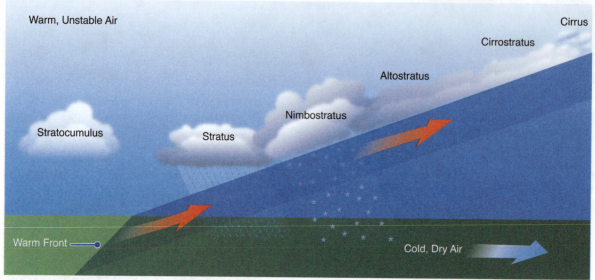

Figure 6-35. Although stratus clouds usually extend out ahead of a slow-moving warm front, cumulus clouds sometimes develop along and ahead of the surface front if the air is unstable.

TYPICAL WARM FRONT WEATHER			
	Prior to Passage	**During Passage**	**After Passage**
Clouds	• Cirriform • Stratiform • Fog • Possible cumulonimbus in the summer	• Stratiform	• Stratocumulus • Possible cumulonimbus in the summer
Precipitation	• Light-to-moderate rain, drizzle, sleet, or snow	• Drizzle, if any	• Rain or showers, if any
Visibility	• Poor	• Poor, but improving	• Fair in haze
Wind	• SSE	• Variable	• SSW
Temperature	• Cold to cool	• Rising steadily	• Warming, then steady
Dewpoint	• Rising steadily	• Steady	• Rising, then steady
Pressure	• Falling	• Becoming steady	• Slight rise, then falling

STATIONARY FRONTS

When the opposing forces of two air masses are relatively balanced, the front that separates them may remain stationary and influence local flying conditions for several days. The weather in a stationary front is usually a mixture of that found in both warm and cold fronts.

OCCLUDED FRONTS

A frontal occlusion occurs when a fast-moving cold front catches up to a slow-moving warm front. The difference in temperature within each frontal system is a major factor that influences which type of front and weather are created. A **cold front occlusion** develops when the fast-moving cold front is colder than the air ahead of the slow-moving warm front. In this case, the cold air replaces the cool air at the surface and forces the warm front aloft. A **warm front occlusion** takes place when the air ahead of the slow-moving warm front is colder than the air within the fast-moving cold front. In this case, the cold front rides up over the warm front, forcing the cold front aloft. When the air being lifted by a cold front occlusion is moist and stable, the weather will be a mixture of that found in both a warm and a cold front. When the air being lifted by a warm front occlusion is moist and unstable, the weather will be more severe than that found in a cold front occlusion. [Figure 6-36]

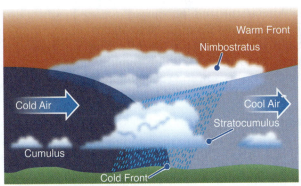

COLD FRONT OCCLUSION

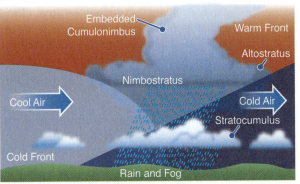

WARM FRONT OCCLUSION

Figure 6-36. Both cold and warm front occlusions can produce severe weather.

TYPICAL OCCLUDED FRONT WEATHER			
	Prior to Passage	**During Passage**	**After Passage**
Clouds	• Cirriform • Stratiform	• Nimbostratus • Possible towering cumulus and/or cumulonimbus	• Nimbostratus • Altostratus • Possible cumulus
Precipitation	• Light-to-heavy precipitation	• Light-to-heavy precipitation	• Light-to-moderate precipitation, then clearing
Visibility	• Poor	• Poor	• Improving
Wind	• SE to S	• Variable	• W to NW
Temperature	• Cold Occlusion: Cold to Cool • Warm Occlusion: Cold	• Cold Occlusion: Falling • Warm Occlusion: Rising	• Cold Occlusion: Colder • Warm Occlusion: Milder
Dewpoint	• Steady	• Slight drop	• Rising, then steady
Pressure	• Falling	• Becoming steady	• Slight drop; however, may rise after passage of warm occlusion

SUMMARY CHECKLIST

✓ Stability is the atmosphere's resistance to vertical motion.

✓ The rate at which temperature decreases with an increase in altitude is referred to as its lapse rate. As you ascend in the atmosphere, temperature decreases at an average rate of 2°C (3.5°F) per 1,000 feet.

✓ When temperature increases with altitude, a temperature inversion exists.

✓ Evaporation is the changing of liquid water to invisible water vapor. Condensation occurs when water vapor changes to a liquid. Sublimation is the changing of ice directly to water vapor, while the transformation of water vapor to ice is known as deposition. In both cases, the liquid state is bypassed.

✓ Relative humidity is the actual amount of moisture in the air compared to the total amount that could be present at that temperature.

✓ The temperature at which air reaches a state where it can hold no more water is called the dewpoint.

✓ Frost forms on aircraft when the temperature of the collecting surface is at or below the dewpoint of the surrounding air and the dewpoint is below freezing. If frost is not removed from the wings before flight, it may decrease lift and increase drag to a point that seriously compromises safety.

✓ When the temperature/dewpoint spread reaches 4°F (2°C) and continues to decrease, the air is nearing the saturation point and the probability of fog and low clouds forming increases.

SECTION B ■ **Weather Patterns**

✓ Because they normally form below 6,500 feet AGL, stratus, stratocumulus, and nimbostratus are classified as low clouds. Altostratus and altocumulus are classified as middle clouds and have bases that range from about 6,500 to 20,000 feet AGL. High clouds have bases beginning at altitudes above 20,000 feet AGL. The three basic types of high clouds are called cirrus, cirrostratus, and cirrocumulus. Extensive vertical development is characteristic of cumulus, towering cumulus, and cumulonimbus clouds.

✓ Fog is a low cloud that has its base within 50 feet of the ground. If the fog is less than 20 feet deep, it is called ground fog.

✓ Although a cloud usually forms when the atmosphere is saturated, it doesn't necessarily mean that the cloud will produce precipitation. For precipitation to occur, water or ice particles must grow in size until they can no longer be supported by the atmosphere.

✓ As they fall, snowflakes and raindrops may change into other types of precipitation depending on the atmospheric conditions beneath the cloud. In addition to snow and rain, falling moisture also can take the form of drizzle, ice pellets, or hail.

✓ An air mass is a large body of air with fairly uniform temperature and moisture content. As an air mass moves, it is modified by the temperature and moisture of the area over which it moves.

✓ Stable air is generally smooth with layered or stratiform clouds. Visibility is usually restricted, with widespread areas of clouds and steady rain or drizzle. Moist unstable air causes the formation of cumuliform clouds, showers, turbulence, and good surface visibility.

✓ A cold front is one where cold air is moving to displace warmer air. In a warm front, warm air is replacing cold air. A stationary front has no movement. When cold and warm fronts merge, they create an occluded front.

✓ Frontal discontinuities are the comparatively rapid changes in the characteristics of an air mass. When you cross a front, you move from one air mass into another and will normally experience changes in temperature, pressure, and wind.

KEY TERMS

Stability	Relative Humidity
Adiabatic Heating	Dewpoint
Adiabatic Cooling	Saturated
Lapse Rate	Dew
Temperature Inversion	Frost
Evaporation	Condensation Nuclei
Condensation	Temperature/Dewpoint Spread
Sublimation	Stratus
Deposition	Stratocumulus
Melting	Nimbostratus
Freezing	Fog
Humidity	Ground Fog

Radiation Fog

Advection Fog

Upslope Fog

Steam Fog

Altostratus

Altocumulus

Cirrus

Cirrostratus

Cirrocumulus

Cumulus

Towering Cumulus

Cumulonimbus

Embedded

Precipitation

Supercooled Water Droplets

Virga

Precipitation-Induced Fog

Ice Pellets

Hail

Fallstreaks

Air Mass

Source Region

Front

Cold Front

Warm Front

Stationary Front

Occluded Front

Cold Front Occlusion

Warm Front Occlusion

QUESTIONS

1. What is the average rate of temperature change associated with a change in altitude?

2. Describe the pressure and temperature changes that take place in ascending and descending air.

3. What two processes add water vapor to the atmosphere?

4. At what height above the ground would you expect to find the base of the clouds if the surface temperature is 65°F and the surface dewpoint is 56°F?

5. Identify the clouds in the following photos.

SECTION B ■ **Weather Patterns**

6. True/False. Advection fog normally occurs when the wind is calm.

7. What must happen for a cloud to precipitate?

8. What is the difference between rain and rain showers?

9. Describe the weather characteristics of a stable air mass.

Match the following fronts with the weather characteristics you would expect to see as the front approaches.

10. Warm Front

    A. Nimbostratus clouds, light-to-heavy precipitation, poor visibility in precipitation, steady dewpoint

11. Cold Front

    B. Stratus clouds, fog, light-to-moderate rain, poor visibility, steadily rising dewpoint

12. Occluded Front

    C. Towering cumulus clouds, short periods of showers, fair visibility in haze, high dew-point

SECTION C
Weather Hazards

After brewing for several days, severe weather exploded over the central plains of the United States on April 3 and 4, 1974. The churning atmosphere created huge areas of thunderstorms and spawned 127 tornadoes, some containing winds in excess of 260 miles per hour. When it was over, 315 people in 11 states were killed and 6,142 were injured. Of course, the magnitude of destruction resulting from this single weather occurrence is rare. Nonetheless, the atmosphere continues to produce hazardous weather on a near daily basis. You can avoid the worst of nature's fury by combining a thorough understanding of its characteristics with a healthy respect for its power.

THUNDERSTORMS

Thunderstorms are arguably the single greatest threat to aircraft operations. They may contain strong wind gusts, icing, hail, driving rain, lightning, and sometimes tornadoes. Before a thunderstorm capable of exhibiting these hazards can develop, three conditions must be present — air that has a tendency toward instability, some type of lifting action, and relatively high moisture content. The lifting action may be provided by several factors, such as rising terrain (orographic lifting), fronts, or heating of the earth's surface (convection).

TYPES OF THUNDERSTORMS

While all thunderstorms usually have similar physical features, they are generally classified as one of two types depending on the intensity of the conditions occurring within them. The scattered thunderstorms that are common during summer afternoons, or in coastal areas at night, are frequently described as **air mass thunderstorms**. They are relatively short-lived storms that rarely produce large hail or strong winds. On the other hand, violent thunderstorms with wind gusts of 50 knots or more, hail 3/4 inches in diameter or larger, and/or tornadoes are referred to as **severe thunderstorms**.

A thunderstorm may exist as a single cell, supercell, or, if combined with others, in a multicell form. A **single-cell** thunderstorm lasts less than one hour. In contrast, a **supercell** severe thunderstorm may last two hours. A **multicell** storm is a compact cluster of thunderstorms. It is usually composed of air mass thunderstorms in different stages of development. These cells interact to cause the duration of the cluster to be much longer than any individual cell. In some cases, thunderstorms may form in a line, called a **squall line**. While it often forms 50 to 300 miles ahead of a fast-moving cold front, the existence of a front is not necessary for a squall line to form. This continuous line of non-frontal thunderstorms can range in distance from about one hundred to several hundred miles in length. [Figure 6-37] Depending on the degree of instability, thunderstorms along a squall line may be ordinary multicell, supercell, or a mixture of storms. The most severe weather conditions, such as destructive winds, heavy hail, and tornadoes are generally associated with squall lines.

Courtesy of NASA

Figure 6-37. A squall line is clearly visible in this photograph from space.

The terminology, **frontal thunderstorms** is sometimes used to refer to storms that are associated with frontal activity. Those that occur with a warm front are often obscured by stratiform clouds. When there is showery precipitation near a warm front, thunderstorms should be expected. In a cold front, the cumulonimbus clouds are often visible in a continuous line parallel to the frontal surface. Depending on the conditions, thunderstorms also may be present in an occluded front.

LIFE CYCLE

About fifty years ago it was discovered that thunderstorms progress through three definite stages—cumulus, mature, and dissipating. Certain characteristics, such as cloud shape, air current direction, and precipitation intensity, are associated with each stage. [Figure 6-38]

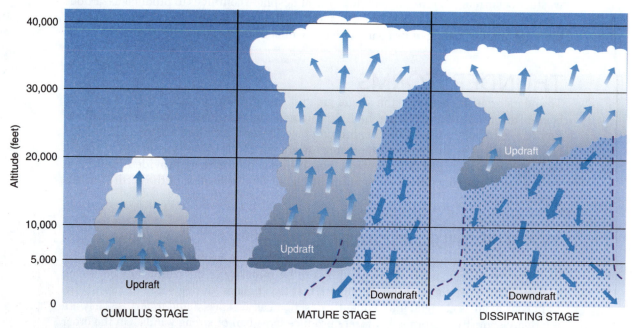

Figure 6-38. These distinctive cloud shapes signal the stages of a thunderstorm. In some cases, other weather phenomena may prevent you from seeing these characteristic shapes.

CUMULUS STAGE

In the **cumulus stage**, a lifting action initiates the vertical movement of air. Towering cumulus clouds are characteristic of the cumulus stage of an air mass thunderstorm. Although not every cumulus cloud develops into a thunderstorm, all thunderstorms begin at the cumulus stage. As the air rises and cools to its dewpoint, water vapor condenses into small water droplets or ice crystals. If sufficient moisture is present, heat released by the condensing vapor provides energy for continued vertical growth of the cloud. Because of strong updrafts, precipitation usually does not fall. Instead, the water drops or ice crystals rise and fall within the cloud, growing larger with each cycle. Updrafts as great as 3,000 feet per minute can begin near the surface and extend well above the cloud top. During the cumulus stage, the convective circulation grows rapidly into a towering cumulus (TCU) cloud that typically grows to 20,000 feet in height and 3 to 5 miles in diameter. The cloud reaches the mature stage in about 15 minutes. [Figure 6-39]

Figure 6-39. A towering cumulus cloud like this one could be the start of an air mass thunderstorm.

MATURE STAGE

At the **mature stage**, the water drops in the cloud grow too large to be supported by the updrafts and rain falls from the cloud. The TCU becomes a cumulonimbus cloud with precipitation creating a downward motion in the surrounding air, creating the storm's most violent stage. The relatively warm updraft and the cool, precipitation-induced downdraft exist side by side, causing severe turbulence. At the surface, the down-rushing

air spreads outward, producing a sharp drop in temperature, a rise in pressure, and strong, gusty surface winds. The leading edge of this wind is referred to as a **gust front**. As the thunderstorm advances, a rolling, turbulent, circular-shaped cloud may form at the lower leading edge of the cloud. This is called the **roll cloud**. The gust front and roll cloud are generally associated with large multicell and supercell thunderstorms. [Figure 6-40]

Figure 6-40. A gust front is the sharp boundary found on the edge of the pool of cold air that is fed by the downdrafts and spreads out below a thunderstorm.

The top of the mature cell can reach as high as 40,000 feet. [Figure 6-41] The highest portion of the cloud may develop a cirriform appearance because of the very cold temperatures and the strong stability of atmosphere above it. As the vertical motions slow near the top of the storm, the cloud spreads out horizontally, forming the well-known anvil shape. The anvil top is an indicator of upper-level winds and points in the approximate direction of the storm's movement.

Figure 6-41. A cumulonimbus cloud indicates that the thunderstorm has reached at least the mature stage.

An exception to the model of the mature thunderstorm sometimes occurs in arid regions. In these areas, lightning and thunder may occur, but the precipitation often evaporates before reaching the ground, creating virga. Below the virga, an invisible downdraft will often continue to the ground below. This combination of gusty winds, lack of precipitation reaching the ground, and lightning is often the cause of forest fires.

SECTION C ■ **Weather Hazards**

DISSIPATING STAGE

Fifteen to thirty minutes after it reaches the mature stage, the single-cell air mass thunderstorm reaches the **dissipating stage**. As the storm develops, more and more air aloft is disturbed by the falling drops. Eventually, the downdrafts begin to spread out within the cell, taking the place of the weakening updrafts. Because upward movement is necessary for condensation and the release of the latent energy, the entire thunderstorm begins to weaken. When the cell becomes an area of predominant downdrafts, it takes on a stratiform appearance, gradually dissipating. Because the anvil top is an ice cloud, it often lasts longer than the rest of the cell.

Occasionally, a severe thunderstorm does not dissipate in the typical manner. If winds become markedly stronger with altitude, the upper portion of the cloud may be "tilted," or blown downwind. In this case, precipitation falls through only a small portion of the rising air, or it may fall completely outside the cloud. As a result, the updrafts may continue to maintain their strength, prolonging the mature stage. [Figure 6-42]

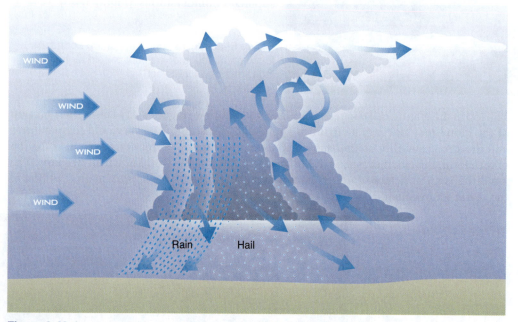

Figure 6-42. In a severe thunderstorm, upper-level winds can tilt the up- and downdrafts, causing the precipitation to fall outside of the updrafts and prolonging the storm sometimes by several hours.

THUNDERSTORM HAZARDS

The weather hazards associated with thunderstorms are not confined to the cloud itself. For example, you can encounter turbulence in clear conditions as far as 20 miles from the storm. You can think of a cumulonimbus cloud as the visible part of a widespread system of turbulence and other weather hazards.

Inside a Thunderstorm

On July 27, 1959, U.S. Marine Corps Lieutenant Colonel William H. Rankin was forced to eject from his jet at 47,000 feet. Generally, most people would be incapacitated by the low atmospheric pressure at that altitude. However, Lt. Col. Rankin was not so lucky. He remained conscious as he descended into a well-developed severe thunderstorm. Although it should have taken him only 10 minutes to pass through the storm, the updrafts kept him inside the cloud for 40 minutes. During that time he battled for survival in temperatures as low as –57°C (–71°F), all the while being pelted with hailstones, blinded by lightning, deafened by thunder, and soaked by driving rain. At one point, the rain was so torrential that Lt. Col. Rankin thought he might survive everything else only to be drowned. To this day, he might be the only person ever to have survived such an ordeal.

DISCOVERY

TURBULENCE

Thunderstorm turbulence develops when air currents change direction or velocity rapidly over a short distance. The magnitude of the turbulence depends on the differences between the two air currents. Within the thunderstorm's cumulonimbus cloud, the strongest turbulence occurs in the zone between the updrafts and downdrafts. Near the surface, there is a low-level area of turbulence that develops as the downdrafts spread out at the surface. These create a **shear zone** between the surrounding air and the cooler air of the downdraft. The resulting area of gusty winds and turbulence can extend outward for many miles from the center of the storm.

LIGHTNING

Lightning is always associated with thunderstorms and can occur in several forms including in-cloud, cloud-to-cloud, cloud-to-ground, and occasionally, between the cloud and clear air. Regardless of the type, a lightning discharge involves a voltage difference between both ends of the lightning stroke of about 300,000 volts per foot. Air along the discharge channel is heated to more than 50,000°F causing a rapid expansion of air and the production of a shock wave that you eventually hear as thunder. For all its power, lightning rarely causes crew injury or substantial damage to aircraft structures. However, lightning can cause temporary loss of vision, puncture the aircraft skin, or damage electronic navigation and communications equipment. [Figure 6-43]

Figure 6-43. The typical thunderstorm generates three to four lightning discharges per minute.

HAIL

Hail can occur at all altitudes within or outside a thunderstorm. You can encounter it in flight, even when no hail is reaching the surface. In addition, large hailstones have been encountered in clear air several miles downwind from a thunderstorm. While any hail can be dangerous, large hail with diameters in excess of 3/4 inches can inflict enormous damage to aircraft. [Figure 6-44]

Figure 6-44. Hail can cause significant damage to aircraft.

SECTION C ■ **Weather Hazards**

TORNADOES

Funnel clouds are violent, spinning columns of air that descend from the base of a cloud. Wind speeds within them may exceed 200 knots. If a funnel cloud reaches the earth's surface, it is referred to as a **tornado**. [Figure 6-45] If it touches down over water, it is called a **waterspout**.

Courtesy of NOAA

Figure 6-45. A severe thunderstorm's rotational circulation pattern can produce extremely powerful tornadoes.

TURBULENCE

In addition to turbulence in and near thunderstorms, three other categories of turbulence affect aviation operations: Low-level turbulence, clear air turbulence, and mountain wave turbulence. The effects of turbulence can vary from occasional light bumps to severe jolts that can cause personal injury to occupants and/or structural damage to the airplane. If you enter turbulence or expect that you will encounter it during flight, slow the airplane to maneuvering speed or less, attempt to maintain a level flight attitude, and accept variations in airspeed and altitude. If you encounter turbulent or gusty conditions during an approach to a landing, you should consider flying a power-on approach at an airspeed slightly above the normal approach speed.

LOW-LEVEL TURBULENCE

While **low-level turbulence (LLT)** is often defined as turbulence below 15,000 feet MSL, most low-level turbulence originates due to surface heating or friction within a few thousand feet of the ground. LLT includes mechanical turbulence, convective turbulence, frontal turbulence, and wake turbulence.

MECHANICAL TURBULENCE

When obstacles such as buildings or rough terrain interfere with the normal wind flow, turbulence occurs. This phenomenon, referred to as **mechanical turbulence**, is often experienced in the traffic pattern when the wind forms eddies as it blows around hangars, stands of trees, or other obstructions. [Figure 6-46] As the winds grow stronger, mechanical turbulence extends to greater heights. For example, when the surface winds are 50 knots or greater, significant turbulence due to surface effects can reach altitudes in excess of 3,000 feet AGL.

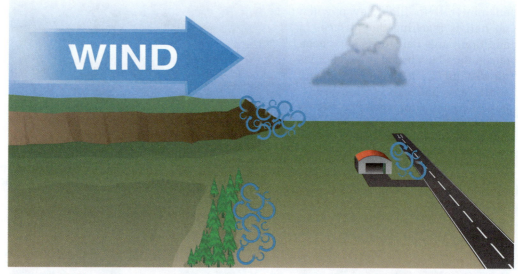

Figure 6-46. Mechanical turbulence is produced downwind of obstructions such as a line of trees, buildings, and hills.

Mechanical turbulence also occurs when strong winds flow nearly perpendicular to steep hills or mountain ridges. In comparison to flat ground, the relatively larger size of the hills produces greater turbulence. In addition, steep hillsides can produce stronger turbulence because the sharp slope encourages the wind flow to separate from the surface. Steep slopes on either side of a valley can produce particularly dangerous turbulence. [Figure 6-47]

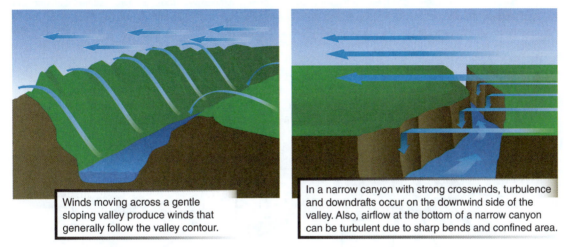

Winds moving across a gentle sloping valley produce winds that generally follow the valley contour.

In a narrow canyon with strong crosswinds, turbulence and downdrafts occur on the downwind side of the valley. Also, airflow at the bottom of a narrow canyon can be turbulent due to sharp bends and confined area.

Figure 6-47. Strong winds across steep canyons can cause potentially hazardous turbulence.

CONVECTIVE TURBULENCE

Convective turbulence, which is also referred to as thermal turbulence, is typically a daytime phenomena that occurs over land in fair weather. It is caused by currents, or thermals, which develop in air heated by contact with the warm surface below. This heating can occur when cold air is moved horizontally over a warmer surface or when the ground is heated by the sun. When the air is moist, the currents may be marked by build-ups of cumulus cloud formations.

With typical upward gusts ranging from 200 to 2,000 ft/min, thermals are great for glider pilots to gain altitude. However, in a powered aircraft, your passengers might object to the continuous bumpiness produced by these thermals. You might find relief by climbing into the **capping stable layer** that begins at the top of the convective layer. You can sometimes observe a layer of cumulus clouds, haze, or dust marking the altitude where smooth air begins. The height of the capping layer is typically a few thousand feet above the ground, although it can exceed to 10,000 feet AGL over the desert in the summer. [Figure 6-48]

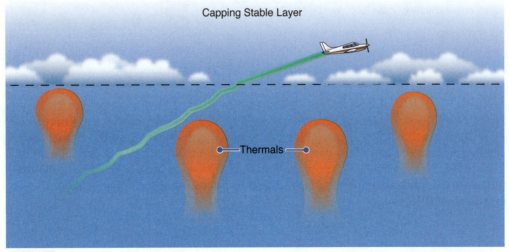

Capping Stable Layer

Thermals

Figure 6-48. By climbing into the capping stable layer, you might find relief from convective turbulence.

DISCOVERY

Hurricane Hunters

When most city inhabitants flee from a hurricane, others are headed in the opposite direction, flying right into the eye of the storm! Some call them crazy, but to the people who make the journey, it's not only exciting and fascinating, but it's also necessary.

Whenever a hurricane threatens the country, Air Force Reserve Lockheed WC-130s and two National Oceanic and Atmospheric Administration (NOAA) Lockheed WP-3D aircraft weave their way into the heart of the storm. While both types of airplanes are used to gather data for forecasters, the NOAA WP-3Ds also carry scientists who conduct hurricane research.

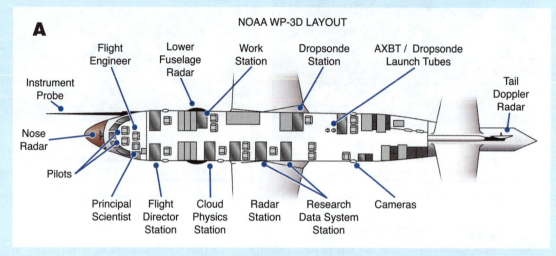

A NOAA WP-3D LAYOUT

NOAA has flown the heavily instrumented WP-3Ds since 1975. [Figure A] While inside the hurricane, scientists gather information such as temperature, pressure, wind speed and direction, and humidity. For information on other portions of the storm, scientists launch weather instruments, called dropsondes, through the airplane's floor. As the devices parachute through the clouds, they radio back meteorological data. Scientists also can launch another instrument package, called an AXBT, which floats on the ocean's surface to collect more data. Information gathered by scientists is sent to the Hurricane Center in Miami, Florida via satellite links. In one instance, observations made by hurricane hunters resulted in an upgraded forecast credited with saving 10,000 lives.

Certainly, flying into a hurricane is not routine; however, thorough planning, weather radar, and constant crew coordination combine to make most flights relatively uneventful. In heavy turbulence, the pilot flying the airplane only tries to maintain wings level. Few, if any, corrections are made for altitude deviations. The flight engineer, sitting behind the two pilots, monitors the instruments and manipulates the four throttles. The pilot-not-flying backs up both the pilot and flight engineer. The worst of the turbulence is in the wall of the eye, but once established inside the eye, it's smooth enough for crewmembers to sip coffee while gazing down on the churning ocean 5,000 feet below. [Figure B]

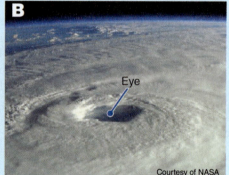

Courtesy of NASA

FRONTAL TURBULENCE

Frontal turbulence occurs in the narrow zone just ahead of a fast-moving cold front where updrafts can reach 1,000 ft/min. When combined with convection and strong winds across the front, these updrafts can produce significant turbulence. Over flat ground, any front moving at a speed of 30 knots or more will generate at least a moderate amount of turbulence. A front moving over rough terrain will produce moderate or greater turbulence, regardless of its speed.

WAKE TURBULENCE

Whenever an airplane generates lift, air spills over the wingtips from the high pressure areas below the wings to the low pressure areas above them. This flow causes rapidly rotating whirlpools of air called wingtip vortices, or **wake turbulence**. The intensity of the turbulence depends on aircraft weight, speed, and configuration.

The greatest wake turbulence danger is produced by large, heavy aircraft operating at low speeds, high angles of attack, and in a clean configuration. Because these conditions are most closely duplicated on takeoff and landing, you should be alert for wake turbulence near airports used by large airplanes. In fact, wingtip vortices from large commercial jets can induce uncontrollable roll rates in smaller aircraft. Although wake turbulence settles, it persists in the air for several minutes, depending on wind conditions. In light winds of three to seven knots, the vortices may stay in the touchdown area, sink into your takeoff or landing path, or drift over a parallel runway. The most dangerous condition for landing is a light, quartering tailwind. It can move the upwind vortex of a landing aircraft over the runway and forward into the touchdown zone.

If you are in a small aircraft approaching to land behind a large aircraft, controllers must ensure adequate separation. However, if you accept a clearance to follow an aircraft you have in sight, the responsibility for wake turbulence avoidance is transferred from the controller to you. On takeoff, controllers will sequence you to provide an interval behind departing heavy aircraft. Although you may waive these time intervals, this is not a wise decision. [Figure 6-49]

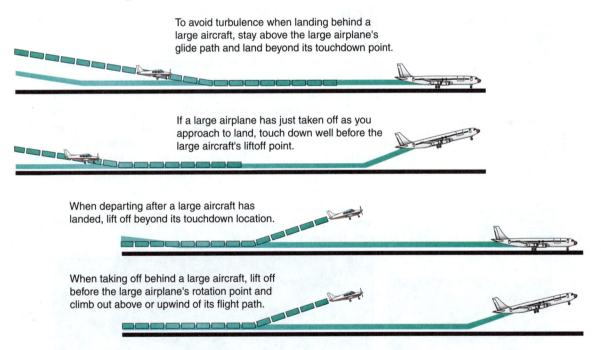

To avoid turbulence when landing behind a large aircraft, stay above the large airplane's glide path and land beyond its touchdown point.

If a large airplane has just taken off as you approach to land, touch down well before the large aircraft's liftoff point.

When departing after a large aircraft has landed, lift off beyond its touchdown location.

When taking off behind a large aircraft, lift off before the large airplane's rotation point and climb out above or upwind of its flight path.

Figure 6-49. Maintaining a safe distance from large aircraft can be critical. Research has shown that a vortex has the potential to "bounce" twice as high as the wingspan of the generating aircraft.

SECTION C ■ **Weather Hazards**

TAKEOFF THRUST

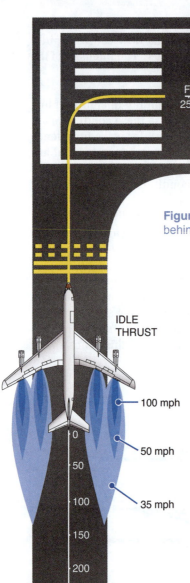

Figure 6-50. Hurricane force winds can be encountered over 200 feet behind a jet using takeoff power.

Jet engine blast is a related hazard. It can damage or even overturn a small airplane if it is encountered at close range. To avoid excessive jet blast, you must stay several hundred feet behind a jet with its engines operating, even when it is at idle thrust. [Figure 6-50]

CLEAR AIR TURBULENCE

Clear air turbulence (CAT) is normally a high altitude phenomenon, occurring above 15,000 feet. However, it can take place at any altitude and normally with no visual warning. Although it occurs in clear skies by definition, CAT can also be present in nonconvective clouds. Clear air turbulence can be caused by the interaction of layers of air with differing wind speeds, convective currents, or obstructions to normal wind flow. It often develops in or near the **jet stream**, which is a narrow band of high altitude winds near the tropopause. CAT tends to be found in thin layers, typically less than 2,000 feet deep, less than 20 miles wide and more than 50 miles long. CAT often occurs in sudden bursts as aircraft intersect thin, sloping turbulent layers. [Figure 6-51]

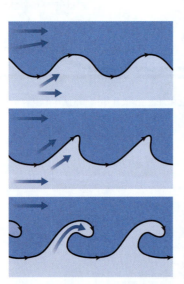

Figure 6-51. Although you normally cannot see clear air turbulence, it is sometimes marked by clouds with distinctive waves.

DISCOVERY

Dust Devils

Although it might look like a twister, the whirling column of dust you see on warm summer days doesn't come anywhere near producing the violence associated with a tornado. Widely known as dust devils, the whirlwinds form as heated air rises in thermals due to convention. Occasionally, strong winds become partially blocked by an obstruction, such as a stand of trees or a small hill. Air sweeping around the sides of the obstruction causes eddies to form downwind. This causes the thermal to rotate, stretch vertically, and shrink horizontally. You see the same effect when a figure skater pulls his or her arms toward the body, increasing the rate of rotation.

Once formed, dust devils move in the direction of the average wind in the layer that they occupy. Normally, they don't inflict much damage but, with wind speeds of up 50 knots, dust devils can still be a threat to aircraft operations.

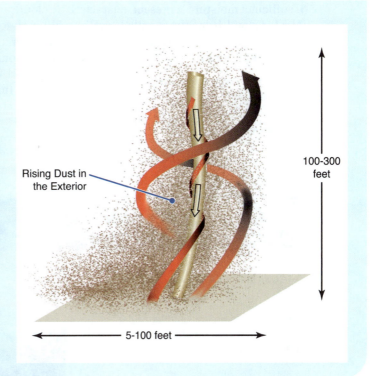

Rising Dust in the Exterior

100-300 feet

5-100 feet

MOUNTAIN WAVE TURBULENCE

When stable air crosses a mountain barrier, the airflow is smooth on the windward side. Wind flow across the barrier is laminar—that is, it tends to flow in layers. The barrier can set up waves, called **mountain waves**. Mountain wave turbulence is possible as the stable air moves across a ridge and the wind is 40 knots or greater. The wave pattern can extend 100 miles or more downwind, with crests as high as 100,000 feet. An area of rotary circulation, or **rotor**, often exists below the crest of each wave. Both the rotor and the waves can create violent turbulence along the lee sides of mountains. Mountain wave turbulence has caused serious injuries and, in rare cases, total loss of *air carrier* aircraft—and it is much more dangerous to general aviation airplanes operating at low altitudes. [Figure 6-52]

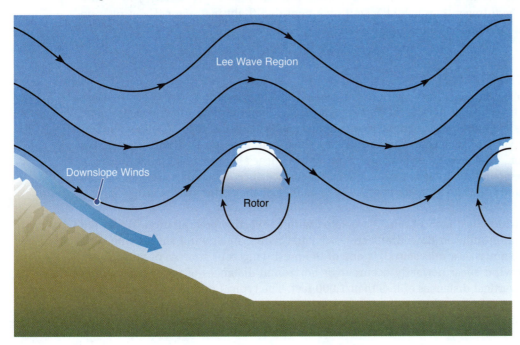

Lee Wave Region

Downslope Winds

Rotor

Figure 6-52. Mountain wave turbulence normally is most significant within a few thousand feet AGL, as shown here; however some violent wave patterns can extend well into the flight levels.

If sufficient moisture is present, characteristic clouds will warn you of the mountain wave. A rotor cloud (sometimes called a roll cloud) may form in the rotors. The crests of the waves may be marked by lens-shaped, or lenticular, clouds. Although they may contain winds of 50 knots or more, they may appear stationary because they form in updrafts and dissipate in downdrafts. Because of this, they are sometimes referred to as standing lenticulars. Another cloud that can signal the presence of mountain wave turbulence is called a cap cloud. In some instances, cap clouds may obscure the mountain peaks. [Figure 6-53]

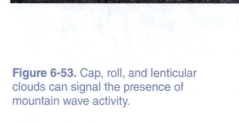

Figure 6-53. Cap, roll, and lenticular clouds can signal the presence of mountain wave activity.

You should anticipate some mountain wave activity whenever the wind is in excess of 25 knots and is blowing roughly perpendicular to mountain ridges. As wind speed increases, so does the associated turbulence. When conditions indicate a possible mountain wave, recommended cruising altitudes are at least 3,000 to 5,000 feet above the peaks. You should climb to the selected altitude while approximately 100 miles from the range, depending on wind and aircraft performance. Approach the ridge from a 45° angle to permit a safer retreat if turbulence becomes too severe. If winds at the planned flight altitude exceed 30 knots, the FAA recommends against flight over mountainous areas in small aircraft. Because local conditions and aircraft performance vary widely, you should consider scheduling a thorough checkout by a qualified flight instructor if you plan on flying in mountainous terrain.

WIND SHEAR

Wind shear is a sudden, drastic shift in wind speed and/or direction that may occur at any altitude in a vertical or horizontal plane. It can subject your aircraft to sudden updrafts, downdrafts, or extreme horizontal wind components, causing loss of lift or violent changes in vertical speeds or altitudes. Wind shear can be associated with convective precipitation, a jet stream, or a frontal zone. Wind shear also can materialize during a low-level temperature inversion when cold, still surface air is covered by warmer air that contains winds of 25 knots or more at 2,000 to 4,000 feet above the surface.

Generally, wind shear is most often associated with convective precipitation. While not all precipitation-induced downdrafts are associated with critical wind shears, one such downdraft, known as a **microburst**, is one of the most dangerous sources of wind shear. Microburst wind shear normally occurs over horizontal distances of one nautical mile or less and vertical distances of less than 1,000 feet. The typical microburst seldom lasts longer than 15 minutes with an average peak wind speed of about 25 knots. While winds in excess of 100 knots are possible, the average microburst will produce a headwind change of approximately 45 knots. The downdrafts within a microburst can be as strong as 6,000 ft/min. The intense downdrafts and wind shifts make the microburst particularly dangerous, especially when encountered close to the ground. [Figure 6-54]

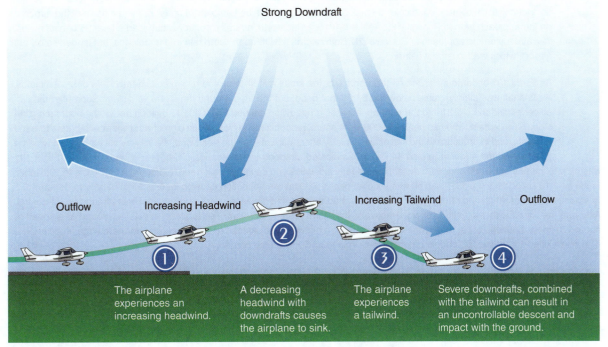

Strong Downdraft

Outflow   Increasing Headwind   Increasing Tailwind   Outflow

① The airplane experiences an increasing headwind.

② A decreasing headwind with downdrafts causes the airplane to sink.

③ The airplane experiences a tailwind.

④ Severe downdrafts, combined with the tailwind can result in an uncontrollable descent and impact with the ground.

Figure 6-54. The unfavorable wind shear that occurs across a microburst, together with intense downdrafts, can drive an airplane into the ground.

LOW-LEVEL WIND SHEAR ALERT SYSTEM

To help detect hazardous wind shear associated with microbursts, **low-level wind shear alert systems (LLWAS)** have been installed at many airports. The LLWAS uses a system of anemometers placed at strategic locations around the airport to detect variances in the wind readings. Many systems operate by sending individual anemometer readings every 10 seconds to a central computer that evaluates the wind differences across the airport. A wind shear alert is usually issued if one reading differs from the mean by at least 15 knots. If you are arriving or departing from an airport equipped with LLWAS, you will be advised by air traffic controllers if an alert is posted. You also will be provided with wind velocities at two or more of the sensors.

TERMINAL DOPPLER WEATHER RADAR

In addition to LLWAS, **terminal Doppler weather radar (TDWR)** systems are being installed at airports with high wind shear potential. These radar systems use a more powerful and narrower radar beam than conventional radar. The TDWR can provide a clearer, more detailed picture of thunderstorms, which allows for better probability of predicting the presence of wind shear.

IN-FLIGHT VISUAL INDICATIONS

In areas not covered by LLWAS or TDWR, you may only be able to predict the presence of wind shear using visual indications. In humid climates where the bases of convective clouds tend to be low, wet microbursts are associated with a visible rain shaft. In the

SECTION C ■ **Weather Hazards**

Flight 191 Wind Shear Accident

It was a typical August day in Texas. The forecast called for the standard isolated or widely scattered rain showers and thunderstorms when the Lockheed L-1011 departed Fort Lauderdale, Florida enroute to Dallas/Fort Worth. By the time Flight 191 approached its destination, thunderstorms had built up east of the airport. While the flight crew concentrated on avoiding the well-developed cells, a small area of rain showers began to show up on radar just north of the airport and on the approach path to Runway 17L. As the L-1011 skirted north, air traffic control sequenced a Lear 25 behind a Boeing 727 already on the approach to Runway 17L. Both aircraft flew through some rain and turbulence, but landed safely at the airport, which was in good weather.

As Flight 191 became stabilized on the approach, the small rain shower had grown considerably—in only 12 minutes. A rainshaft was plainly visible below the 6,000-foot base, but the area around the storm was clear. Soon after the first officer piloted the L-1011 into the rain, an increasing headwind caused Flight 191 to balloon above the approach glide slope. The captain, who had nearly 30,000 flight hours, recognized the classic first sign of wind shear and began to issue warnings and instructions to the first officer. Quickly after encountering the headwind, the jet was buffeted by a downdraft that caused the airplane to lose 44 knots of indicated airspeed in 10 seconds. Then, in one short second, Flight 191 was hit with intense gusts, which turned the airplane nearly sideways and caused its airspeed to drop from 140 knots to 120 knots. In the next 13 seconds, the L-1011 was hit with six changes in vertical wind direction causing load factor to vary from negative 0.3 to a positive 2.0 G's. At this point, the captain called for a go-around, but it was too late—the huge jet was already out-of-control. Although the crew fought valiantly to effect a recovery, the storm refused to relent. The airplane struck the ground at a speed of about 170 knots, bounced once, then came back to earth on a highway and skidded to a stop near the northern edge of the airport. The crash killed 134 of the 165 people on board the airplane and one on the ground.

In its report, the National Transportation Safety Board (NTSB) stated the probable causes of the accident as: "the flight crew's decision to initiate and continue the approach into a cumulonimbus cloud they observed to contain visible lightning, the lack of specific guidelines, procedures, and training for avoiding and escape from low-altitude wind shear, and the lack of definitive, real-time wind shear information. This resulted in the aircraft's encounter at low altitude with a microburst-induced, severe wind shear from a rapidly developing thunderstorm located on the final approach course."

What can be learned from this tragedy? The NTSB may have answered that question best when it said, "The circumstances of this accident indicate that there is an apparent lack of appreciation on the part of some, and perhaps many, flight crews of the need to avoid thunderstorms and to appraise the position and severity of the storms pessimistically and cautiously."

drier climates of the deserts and mountains of the western United States, the higher thunderstorm cloud bases result in the evaporation of the rain shaft producing a dry microburst. The only visible indications under these conditions may be virga at the cloud base and a dust ring on the ground. It's important to note that because downdrafts can spread horizontally across the ground, low-level wind shear may be found beyond the boundaries of the visible rain shaft. You should avoid any area you suspect could contain a wind shear hazard.

ICING

Ice can build up on any exposed surface of an aircraft during flights in areas of visible moisture, when the temperature of the aircraft surface is 0°C or colder. Aircraft are affected by structural ice in a number of ways: thrust is reduced, drag and weight is increased, and lift is decreased. These effects combine to increase stall speed and reduce overall aircraft performance. In extreme cases, it can take as little as 5 minutes for 2 to 3 inches of ice to accumulate on the leading edge of the airfoil. Some aircraft may experience as much as a 50 percent decrease in lift after the build-up of only 1/2 inch of ice. There are three general types of ice—rime, clear, and a mixture of the two.

Rime ice normally is encountered in stratus clouds and results from instantaneous freezing of tiny supercooled water droplets striking the aircraft surface. It has an opaque appearance caused by air being trapped in the water droplets as they freeze. Rime ice is particularly hazardous due to its ability to change the shape of an airfoil and destroy its lift. Because rime ice freezes instantly, it builds up on the leading edge of airfoils, but it does not flow back following the basic curvature of the wing and tail surfaces. Rime ice normally forms in temperatures between −15°C and −20°C.

Clear ice may develop in areas of large supercooled water droplets that are in cumulus clouds or in freezing rain beneath a warm front inversion. The highest accumulation rate generally occurs in freezing rain. When the droplets flow over the aircraft structure and slowly freeze, they can glaze the aircraft's surfaces. Clear ice is the most serious form of ice because it adheres tenaciously to the aircraft and is difficult to remove. The formation of clear ice is likely when temperatures are between 0°C to –10°C. **Mixed ice** is possible in visible moisture between –10°C and –15°C.

RESTRICTIONS TO VISIBILITY

Particles that can absorb, scatter, and reflect light are always present in the atmosphere. The fact that the amount of particles in the air varies considerably explains why visibility is better on some days than others. Restrictions to visibility can take many forms, the most common of which are haze, smoke, smog, and dust.

HAZE

Haze is caused by a concentration of very fine dry particles. Individually, they are invisible to the naked eye, but in sufficient numbers, can restrict your visibility. Haze particles may be composed of a variety of substances, such as salt or dust particles. It occurs in stable atmospheric conditions with relatively light winds. Haze is usually no more than a few thousand feet thick, but it may occasionally extend to 15,000 feet. Visibility above the haze layer is usually good; however, visibility through the haze can be very poor. Dark objects tend to be bluish, while bright objects, like the sun or distant lights, have a dirty yellow or reddish hue.

DISCOVERY

Icing Research Aircraft

Artificial icing facilities such as wind tunnels or airborne spray tankers provide a unique capability to test and evaluate aircraft systems in an authentic, yet controlled icing environment. One such facility in use is the U.S. Army Helicopter Icing Spray System (HISS), which is installed in a modified JCH-47C Chinook helicopter. Flying ahead of another aircraft, the helicopter produces a spray cloud in flight. In figure A, a DH-6 Twin Otter airplane is shown flying behind a Chinook equipped with a HISS.

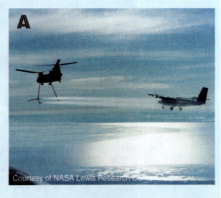

A

Courtesy of NASA Lewis Research Center

B

Courtesy of NASA Lewis Research Center

The DH-6 is used for icing research, including the characterization of icing clouds and the aerodynamic effects of icing. To make the ice accumulations more readily visible, a dye is mixed with the water. This can be seen as the yellow-green ice on the DH-6 shown in figure B.

SMOKE

Smoke is the suspension of combustion particles in air. The impact of smoke on visibility is determined by the amount of smoke produced, wind velocity, diffusion by turbulence, and distance from the source. You can often identify smoke by a reddish sky as the sun rises or sets, and an orange-colored sky when the sun is well above the horizon. When smoke travels distances of 25 miles or more, large particles fall out and the smoke tends to become more evenly distributed, giving the sky a grayish or bluish appearance, similar to haze.

SMOG

Smog, which is a combination of fog and smoke, can spread very poor visibility over a large area. In some geographical areas, topographical barriers, such as mountains, may combine with stable air to trap pollutants. This results in a build-up of smog, further reducing visibility.

DUST

Dust refers to fine particles of soil suspended in the air. When the soil is loose, the winds are strong, and the atmosphere is unstable, dust may be blown for hundreds of miles. Dust gives a tan or gray tinge to distant objects while the sun may appear as colorless, or with a yellow hue. Blowing dust is common in areas where dry land farming is extensive, such as the Texas panhandle.

VOLCANIC ASH

While lava from volcanoes generally threatens areas only in the immediate vicinity of the volcano, the ash cloud can affect a much more widespread area. **Volcanic ash**, which consists of gases, dust, and ash from a volcanic eruption, can spread around the world and remain the stratosphere for months or longer. [Figure 6-55] Due to its highly abrasive characteristics, volcanic ash can pit the aircraft windscreens and landing lights to the point they are rendered useless. Under severe conditions, the ash can clog pitot-static and ventilation systems as well as damage aircraft control surfaces. Piston aircraft are less likely than jet aircraft to lose power due to ingestion of volcanic ash, but severe damage is possible, especially if the volcanic cloud is only a few hours old.

Courtesy of NASA

Figure 6-55. These two images show the same eruption of the Kliuchevskoi volcano in Kamchatka, Russia. The left photo from the space shuttle Endeavor shows what the astronauts saw from space. The right image shows how radar can see through the ash and smoke to reveal the land underneath.

An ash cloud may not be easily distinguishable from ordinary clouds when approached from a distance. However, if you suspect that you are in the vicinity of an ash cloud you should attempt to stay upwind. If you inadvertently enter a volcanic ash cloud you should not attempt to fly straight through or climb out of the cloud because ash cloud may be hundreds of miles wide and extend to great heights. You should reduce power to a minimum, altitude permitting, and reverse course to escape the cloud.

SUMMARY CHECKLIST

✓ Air mass thunderstorms are relatively short-lived storms, which rarely produce large hail or strong winds. Severe thunderstorms contain wind gusts of 50 knots or more, hail 3/4 inch in diameter or larger, and/or tornadoes.

✓ The life of a thunderstorm passes through three distinct stages. The cumulus stage is characterized by continuous updrafts. When precipitation begins to fall, the thunderstorm has reached the mature stage. As the storm dies during the dissipating stage, updrafts weaken and downdrafts become predominant.

✓ Some weather hazards associated with thunderstorms, such as turbulence, lightning, and hail are not confined to the cloud itself.

✓ If you encounter turbulence during flight, you should establish maneuvering speed and try to maintain a level flight attitude.

✓ Mechanical turbulence is often experienced in the traffic pattern when wind forms eddies as it blows over hangars, stands of trees, or other obstructions.

✓ When sufficient moisture is present, cumulus cloud build-ups indicate the presence of convective turbulence.

✓ Wingtip vortices are created when an airplane generates lift. The greatest vortex strength occurs when the generating aircraft is heavy, slow, and in a clean configuration.

✓ Mountain wave turbulence can be anticipated when the winds across a ridge are 40 knots or more, and the air is stable. The crests of mountain waves may be marked by lens-shaped, or lenticular, clouds.

✓ Wind shear can exist at any altitude and may occur in a vertical or horizontal direction. A microburst is one of the most dangerous sources of wind shear.

✓ The three types of structural ice are rime, clear, and mixed.

✓ Volcanic ash clouds may be hundreds of miles wide and thousands of feet thick.

KEY TERMS

Air Mass Thunderstorm

Severe Thunderstorm

Single-Cell

Supercell

Multicell

Squall Line

Frontal Thunderstorm

Cumulus Stage

Mature Stage

Gust Front

Roll Cloud

Dissipating Stage

Shear Zone

Tornado

Waterspout

Low-Level Turbulence (LLT)

Mechanical Turbulence

Convective Turbulence

Capping Stable Layer

Frontal Turbulence

Wake Turbulence

Jet Engine Blast

Clear Air Turbulence (CAT)

Jet Stream

SECTION C ■ **Weather Hazards**

Mountain Waves

Clear Ice

Rotor

Mixed Ice

Wind Shear

Haze

Microburst

Smoke

Low-Level Wind Shear Alert Systems (LLWAS)

Smog

Dust

Terminal Doppler Weather Radar (TDWR)

Volcanic Ash

Rime Ice

QUESTIONS

1. What are the three basic ingredients needed for the formation of a thunderstorm?

2. What causes the thunderstorm's anvil to form? What can it tell you about the thunderstorm's movement?

3. Recall the general procedures you should use when encountering turbulence in flight.

4. True/False. Cumulus clouds indicate the presence of mechanical turbulence.

5. What is the most dangerous condition for landing with respect to wingtip vortices?

6. What technique should be used to avoid wake turbulence during takeoff behind a large aircraft?
   A. Climb on the flight path of the preceding aircraft.
   B. Climb below the flight path of the preceding aircraft.
   C. Climb above the flight path of the preceding aircraft.

7. What kind of turbulence is indicated by the presence of rotor, cap, and lenticular clouds?

8. Discuss the in-flight visual indications of possible wind shear.

Match the following types of structural icing with its characteristic.

9. Clear

A. Occurs in temperatures between -10°C and -15°C

10. Rime

B. Develops in an area of large supercooled water droplets

11. Mixed

C. Normally is encountered in stratus clouds

12. What is the recommended course of action if you inadvertently enter a volcanic ash cloud?
    A. Reverse course
    B. Attempt to climb up and out of the cloud
    C. Continue straight ahead to exit on the opposite side

CHAPTER 7

Interpreting Weather Data

SECTION A
The Forecasting Process

Can you accurately predict the future? Most people would answer, "No," but ask a meteorologist and the response might be different. It's true that weather forecasts are not always totally precise, however scientists' ability to decipher atmospheric clues and produce a reasonably accurate forecast continues to improve. Weather forecasts, on which thousands of decisions are based every day, are generated through an elaborate process involving individual observers and complex computer programs.

FORECASTING METHODS

Predicting the weather can be accomplished through a multitude of methods with wide variances in accuracy. Some methods are best suited for short-term forecasts while others are more adequate for making long-range predictions. The methods can be used alone, or in combination, to create a picture of future weather conditions.

PERSISTENCE FORECAST

With very little meteorological information or knowledge of weather theory you can still make a weather forecast by simply looking out the window. The **persistence method** of forecasting involves simply predicting that the weather you are experiencing at the moment will continue to prevail. For example, if you wake up in the morning to brilliant sunshine, it would be reasonable for you to predict that it will remain sunny for the remainder of the morning. As you might expect, these types of forecasts are usually only accurate over a short period of time, such as a few hours.

TREND FORECAST

The **trend forecast** employs the assumption that weather systems that are moving in one direction and speed will continue to do so in the absence of any other intervening circumstances. For example, suppose a cold front is moving toward an airport at 30 miles per hour. If the front is 60 miles away, you could predict that the front will begin to pass over the airport in 2 hours. This type of forecasting is typically used to predict events that will occur over the next few hours.

CLIMATOLOGICAL FORECAST

Weather predictions also can be made by basing the outlook on the average weather (climatology) in a region. For example, historical records indicate there is, on average, less than one day of rain a month during the summer in San Francisco, California. Using the **climatological forecast** method, a meteorologist could confidently predict a remote chance of rain during next year's July 4th weekend in San Francisco. This type of forecasting is usually only reliable during specific times in areas that experience little change in day-to-day weather conditions.

ANALOGUE FORECAST

Combining historical information with other prevailing weather elements can help refine a forecast. The **analogue forecast** compares the features of a current weather chart with those of a similar chart from the past. The weather conditions produced as a result of the historical chart can be used as a guide to predicting what will occur in the future. Unfortunately, regardless of how close two situations seem, there is rarely, if ever, an instance where everything is exactly the same. Despite these limitations, the analogue method can be used to predict things such as minimum and maximum temperatures. For example, a meteorologist can statistically correlate the minimum temperature on a particular date with other local weather factors such as humidity, wind, and cloud cover. Then, the forecaster can use the relationships in combination with current weather data to predict the minimum temperature for the day.

METEOROLOGICAL FORECAST

A **meteorological forecast** is generally more accurate than many other methods because it uses the forecaster's scientific knowledge of atmospheric processes to generate a weather prediction. In many cases, an experienced meteorologist can produce an accurate 6- to 12-hour forecast simply by analyzing a variety of weather charts and other data.

NUMERICAL WEATHER PREDICTION

Increased forecast accuracy can be accomplished using powerful computers and complex software. This process, known as **numerical weather prediction**, uses mathematical equations that relate atmospheric conditions with other variables. The system of equations is referred to as a numerical prediction model and represents the physical laws that govern the behavior of the atmosphere. The computer applies the model to the current atmospheric conditions to develop a forecast for the very near future (5 to 10 minutes). The resulting forecast in fed back into the computer to generate a prediction for the following 5 to 10 minutes. This process is repeated over and over again until a prediction for the next day or two has been developed. These forecasts are an integral part of today's overall weather data compilation and processing system.

COMPILING AND PROCESSING WEATHER DATA

Predicting weather conditions begins with an analysis of present and past conditions. From the compilation of the weather data until a final forecast is disseminated, the information is passed among various agencies for processing, analysis, and creation of weather charts, graphics, and text. [Figure 7-1]

FORECASTING ACCURACY AND LIMITATIONS

Despite advances in forecasting methods, especially with the aid of computers, there are limits to the accuracy of predictions. To some degree, this is due to the large distances between weather reporting stations and the length of time between weather reports. Also, the atmosphere occasionally does not behave exactly like computer models predicts it should. These flawed computer predictions are one of the primary reasons a forecasted day of sunshine turns out to be rainy.

SECTION A ■ **The Forecasting Process**

GENERATING WEATHER DATA

OBSERVATIONS

The World Meteorological Organization (WMO) standardizes observation procedures. Observations are collected from satellites, buoys, ships, aircraft, and land-based observation stations. Airports generally take an observation on an hourly basis. The GOES satellite network provides advanced imagery and atmospheric measurements of the Earth's weather. Observations are transmitted by phone, computer, or satellite to a communication substation for subsequent relay to three World Meteorological Centers (WMCs).

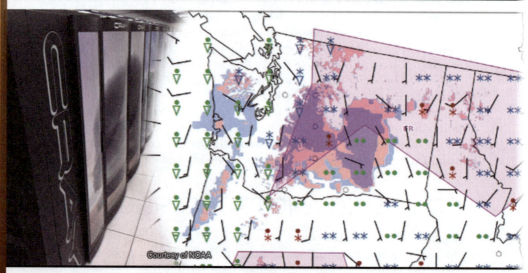

PROCESSING

In the U.S., weather observation information is transmitted to the National Oceanic and Atmospheric Administration (NOAA). A supercomputer enters the observations into a series of algorithms that represent the atmosphere's physical properties. The resulting equations help predict the formation, intensity and track of complex weather systems—taking into account how these systems influence each other and the underlying atmospheric patterns driving their behavior. Using this data NOAA's National Weather Service (NWS) generates thousands of textual and graphic weather products.

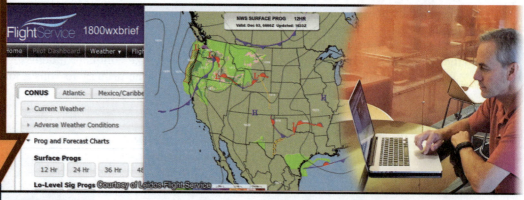

DISSEMINATION

Information is sent to Weather Forecast Offices (WFOs) across the country, facilities such as Flight Service, and other public and private agencies worldwide. Forecasters use a variety of numerical, statistical, and conceptual models and local experience to determine how the current conditions will change with time.

Figure 7-1. The forecasts you use to plan your flights are the result of the coordinated interaction of several agencies.

Usually, forecasters can accurately forecast the occurrence of large-scale weather events such as cold waves and significant storms several days in advance. On the other hand, smaller, more short-term weather phenomena such as tornadoes and some types of wind shear are much more difficult to predict. Regardless of the type of weather, the further ahead in time the forecast is, the less accurate it becomes. [Figure 7-2]

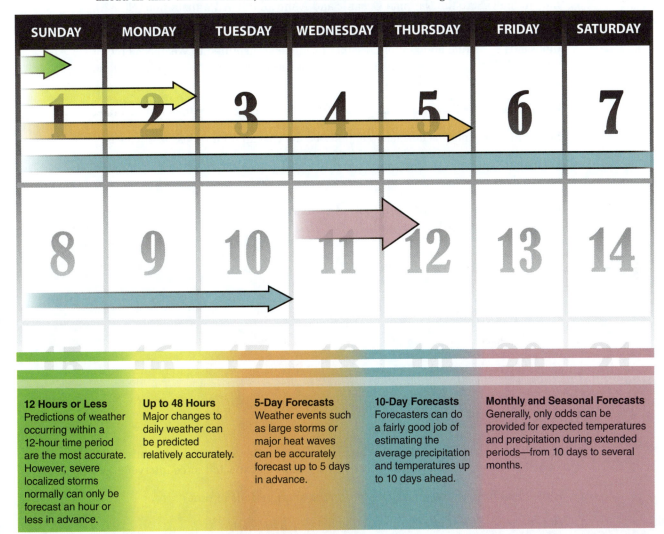

12 Hours or Less
Predictions of weather occurring within a 12-hour time period are the most accurate. However, severe localized storms normally can only be forecast an hour or less in advance.

Up to 48 Hours
Major changes to daily weather can be predicted relatively accurately.

5-Day Forecasts
Weather events such as large storms or major heat waves can be accurately forecast up to 5 days in advance.

10-Day Forecasts
Forecasters can do a fairly good job of estimating the average precipitation and temperatures up to 10 days ahead.

Monthly and Seasonal Forecasts
Generally, only odds can be provided for expected temperatures and precipitation during extended periods—from 10 days to several months.

Figure 7-2. Although the accuracy of weather forecasts continues to improve, long-term forecasts are still relatively unreliable.

SECTION A ■ **The Forecasting Process**

SUMMARY CHECKLIST

✓ Predicting that the weather you are experiencing at the moment will continue to prevail is referred to as the persistence method of forecasting.

✓ The trend forecast assumes that weather systems that are moving in one direction and speed will continue to do so in the absence of any other intervening circumstances.

✓ Climatological forecasts are based on the average weather in a region.

✓ The analogue forecast uses past weather patterns as a guide to predict what will occur in the future.

✓ A meteorological forecast uses the forecaster's scientific knowledge of the atmosphere and its processes to generate a weather prediction.

✓ Numerical weather prediction develops a forecast using mathematical equations that relate atmospheric conditions with other variables.

✓ Weather observations are collected from satellites, buoys, ships, aircraft, and land-based observation stations. Airports generally take an observation on an hourly basis.

✓ Observations are transmitted to three World Meteorological Centers (WMCs) and then in the U.S. to the National Oceanic and Atmospheric Administration (NOAA).

✓ After processing by a supercomputer, the National Weather Service (NWS) generates thousands of textual and graphic weather products transmitted to Weather Forecast Offices (WFOs) across the country, facilities such as Flight Service, and other public and private agencies worldwide.

✓ Predictions of weather occurring within a 12-hour time period are the most accurate.

✓ Severe localized storms normally can only be forecast an hour or less in advance.

✓ Weather events such as large storms or major heat waves can be accurately forecast up to 5 days in advance.

✓ Forecasters can do a fairly good job of estimating the average precipitation and temperatures up to 10 days ahead.

KEY TERMS

Persistence Method   Analogue Forecast

Trend Forecast   Meteorological Forecast

Climatological Forecast   Numerical Weather Prediction

QUESTIONS

1. True/False. A persistence forecast assumes that weather systems will continue to move in the same direction and speed, unless some unexpected force intervenes.

2. Is the trend forecast more accurate for long or short periods of time?

3. How does numerical weather prediction develop a forecast?

4. Normally, how often do airports record a weather observation?
   A. Every hour
   B. Every 4 hours
   C. Every 12 hours

5. What type of weather events can be accurately predicted four days in advance?

SECTION B
Aviation Weather Reports and Forecasts

Aviation weather reports and forecasts are disseminated in coded formats and in plain text. This section introduces you to the elements in each of the textual products you commonly use to prepare for a flight—the reports and forecasts are presented in the coded format followed by an interpretation of the data. [Figure 7-3]

> ▸ MVFR   New Ulm Municipal, New Ulm, MN (KULM). Jul 17, 1635Z, Automated. Light Rain, Wind from 130° at 15 knots, 4 statute miles visibility, Clear Skies, Temperature 21°C, Dewpoint 21°C, Altimeter is 29.76. Remarks: automated station with precipitation discriminator 0.01 inches precipitation (water equivalent) in past hour lightning distant S 0.01 inches precipitation (water equivalent) in past hour
>
> ▸ VFR   Redwood Falls Municipal, Red[wood Falls, MN...] Light Rain, Wind from 170° a[t...] 5,500 feet, Ceiling is Broken [at...] Temperature 22°C, Dewpoint [...] station with precipitation disc[...] 0.01 inches precipitation (wa[ter...] dewpoint 21.7°C
>
> ▸ VFR   Olivia Regional, Olivia, MN (K[...] at 8 knots with gusts to 15 k[nots...] Temperature 22°C, Dewpoint [...] station with precipitation disc[...]
>
> ▸ LIFR   Glencoe Municipal, Glencoe, [...] Wind from 140° at 7 knots, 5 [...] feet, Temperature 23°C, Dew[...] automated station with precip[...] 22.8°C
>
> ▸ VFR   Litchfield Municipal, Litchfield[...] 130° at 9 knots, 10 statute m[...] Temperature 24°C, Dewpoint 23°C, Altimeter is 29.81. Remarks: automated station with precipitation discriminator
>
> **Destination:**
>
> ▸ LIFR   Hutchinson Muni/Butler Field, Hutchinson, MN (KHCD). Jul 17, 1635Z, Automated. Mist, Wind from 120° at 13 knots, 5 statute miles visibility, Ceiling is Overcast at 300 feet, Temperature 23°C, Dewpoint 23°C, Altimeter is 29.81. Remarks: automated station with precipitation discriminator hourly temp 22.8°[...] dewpoint 22.5°C

▸ MVFR	KULM	171635Z AUTO	13015KT 4SM -RA CLR 21/21 A2976 RMK AO2 P0001 LTG DSNT S P0001
▸ VFR	KRWF	171553Z AUTO	17005KT 10SM -RA FEW055 BKN100 OVC110 22/22 A2981 RMK AO2 SLP091 P0001 T02220217
▸ VFR	KOVL	171635Z AUTO	13008G[...]
▸ LIFR	KGYL	171636Z AUTO	14007KT[...] T022702[...]
▸ VFR	KLJF	171636Z AUTO	13009KT[...]

Destination:

▸ LIFR	KHCD	171635Z AUTO	12013KT[...] T022802[...]

Alternate 1:

▸ IFR	SP KMKT	171646Z AUTO	18003KT[...]

Figure 7-3. These plain text and coded METARs are from a Flight Service briefing at 1800wxbrief.com.

> Aviation weather reports and forecasts are presented in both coded and plain-text formats. METARs and TAFs often include color-coded notations so you can quickly picture the overall weather conditions for your flight.
>
> **VFR** — ceiling greater than 3,000 feet and visibility greater than 5 miles
>
> **MVFR** — (marginal VFR) ceiling 1,000 to 3,000 feet and/or visibility 3 to 5 miles inclusive
>
> **IFR** — ceiling 500 to less than 1,000 feet and/or visibility 1 to less than 3 miles
>
> **LIFR** — (low IFR) ceiling less than 500 feet and/or visibility less than 1 mile

OBSERVATIONS

In general, a weather report records measurements and observations of existing conditions at a particular time and place. An observation differs from a forecast, which is a prediction of conditions expected sometime in the future. A variety of reports disseminate information gathered by trained observers, automated systems, radar, and pilots. Some of the reports you might encounter include aviation routine weather reports and pilot weather reports.

AVIATION ROUTINE WEATHER REPORT

An **aviation routine weather report (METAR)** is an observation of surface weather that is reported in a standard format. Although the METAR code has been adopted worldwide, each country is allowed to make modifications or exceptions to the code. Such modifications usually accommodate local procedures or particular units of measure. The following discussion covers the elements in a METAR originating in the United States. [Figure 7-4]

Type of Report	Station Identifier	Time of Report	Modifier (Not Shown)	Wind Information	Visibility
METAR	KTPA	122150Z		08020G38KT	1/2SM R36L/2400FT
+TSRA	SCT008 OVC012CB	20/18	A2995	RMK TSB24RAB24 SLP134	
Weather	Sky Condition	Temperature / Dewpoint	Altimeter	Remarks	

Figure 7-4 Although the content might vary somewhat, a typical METAR contains several distinct elements. Elements that cannot be observed at the time of the report are omitted.

TYPE OF REPORT
METAR KTPA...

The two types of reports are the METAR, which is taken every hour, and the non-routine (special) aviation weather report (SPECI). The SPECI weather observation is an unscheduled report indicating a significant change in one or more elements.

STATION IDENTIFIER
METAR **KTPA** 122150Z...

Each reporting station has a four-letter International Civil Aviation Organization (ICAO) identifier. In the contiguous 48 states, station identifiers begin with the letter K, which precedes the three-letter domestic location identifier. For example, the domestic identifier for Tampa International Airport is TPA, and the ICAO identifier is KTPA. In other areas of the world, the first two letters indicate the region, country, or state. Alaska identifiers begin with PA, Hawaii identifiers begin with PH, and the Canadian prefixes are CU, CW, CY, and CZ. You can find the identifiers for specific reporting stations or decode unfamiliar identifiers using the Chart Supplement or the NOAA/NWS website.

TIME OF REPORT
...KTPA **122150Z** 08020G38KT...

The date (day of the month) and time of the observation follows the station identifier. The time is given in UTC, or Zulu, as indicated by a Z following the time. The report in the example was issued on the 12th of the month at 2150Z.

MODIFIER
When a METAR is created by a totally automated weather observation station, the modifier AUTO follows the date/time element (e.g., ...251955Z AUTO 30008KT...). Automated stations are classified by the type of sensor equipment they use, and AO1 or AO2 will be noted in the remarks section of the report. AO2 in the remarks indicates that the station has a precipitation discriminator (which can determine the difference between liquid and frozen/freezing precipitation). AO1 indicates that the automated station did not use a precipitation discriminator.

The modifier COR is used to indicate a corrected METAR, which replaces a previously disseminated report. There is no station type designator in the remarks section of a corrected report from an automated station. A METAR with no modifier (as in this example) means that the observation was taken at a manual station or that an automated station had manual input.

WIND INFORMATION
...122150Z **08020G38KT** 1/2SM R36L/2400FT...

The wind direction and speed are reported in a five digit group, or six digits if the speed is over 99 knots. The first three digits represent the direction from which the wind is blowing, in reference to true north.

The next two (or three) digits show the speed in knots (KT). Calm winds are reported as 00000KT. Gusty winds are reported with a G, followed by the highest gust. In the example, wind was reported to be from 080° true at 20 knots with gusts to 38 knots.

If the wind direction varies 60 degrees or more and the speed is more than six knots, a variable group follows the wind group. The extremes of wind direction are shown separated by a V. For example, if the wind is blowing from 020°, varying to 090°, it is reported as 020V090. A wind of 6 knots or less that is varying in direction is indicated with the contraction VRB. [Figure 7-5]

CODED METAR DATA	EXPLANATIONS
00000KT	Wind calm
20014KT	Wind from 200° at 14 knots
15010G25KT	Wind from 150° at 10 knots, gusts to 25 knots
VRB04KT	Wind variable in direction at 4 knots
210103G130KT	Wind from 210° at 103 knots with gusts to 130 knots

Figure 7-5. These examples help you to become familiar with the METAR formats for wind direction and speed.

VISIBILITY

...08020G38KT **1/2SM R36L/2400FT** +TSRA...

Prevailing visibility is the greatest distance an observer can see and identify objects through at least half of the horizon. To determine prevailing visibility, an observer looks at distinctive objects, such as towers or smokestacks, that are a known distance from the observation site. At night, observers use lighted objects to determine visibility. When the prevailing visibility is greater in one direction than in another, the visibility in the majority of the sky is reported. If visibility varies significantly, the observer can report the visibility in individual sectors in the remarks section of the METAR. [Figure 7-6]

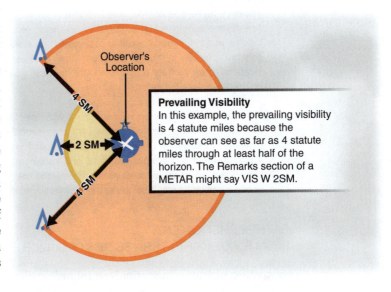

Prevailing Visibility
In this example, the prevailing visibility is 4 statute miles because the observer can see as far as 4 statute miles through at least half of the horizon. The Remarks section of a METAR might say VIS W 2SM.

Figure 7-6. This example depicts how prevailing visibility is determined.

Visibility is reported in statute miles (SM) in the United States and in meters elsewhere in the world. For example, 1/2SM indicates one-half statute mile and 4SM indicates 4 statute miles. Sometimes **runway visual range (RVR)** is reported following prevailing visibility. RVR is based on what a pilot in a moving aircraft should see when looking down the runway. RVR is designated with an R, followed by the runway number, a slant (/), and the visual range in feet (FT). In the example, R36L/2400FT means Runway 36 Left visual range is 2,400 feet. Variable RVR is shown as the lowest and highest visual range values separated by a V. Outside the United States, RVR is normally reported in meters.

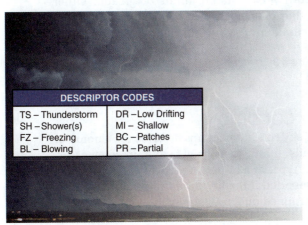

DESCRIPTOR CODES	
TS – Thunderstorm	DR – Low Drifting
SH – Shower(s)	MI – Shallow
FZ – Freezing	BC – Patches
BL – Blowing	PR – Partial

WEATHER

Weather or obstructions to vision that are present at the time of the observation are reported immediately after the visibility in the following order: intensity or proximity, descriptor, precipitation, obstruction to visibility, and any other weather phenomena. The intensity or proximity and/or descriptor are used to qualify the precipitation, obscuration, or other weather phenomena. [Figure 7-7]

Figure 7-7. The type of precipitation or obscuration is described by a two-letter code.

QUALIFIER AND DESCRIPTOR
...1/2SM R36L/2400FT **+TS**RA SCT008 OVC012CB 20/18...

In the example, a thunderstorm (TS) was reported. Intensity or proximity of precipitation is shown just before the precipitation codes. The indicated intensity applies only to the first type of precipitation reported. Intensity levels are shown as light (–), moderate (no sign), or heavy (+). In the example, the precipitation was reported as heavy.

Weather occurring near, but not at, the airport is prefixed with VC, for "in the vicinity" of the airport. The vicinity is typically between five and 10 statute miles of the observation point. VC is not used when an intensity qualifier is reported.

Determining Visibility

Without a frame of reference, determining how far you can see down a runway is just a guess. To determine ground visibility at some airports, weather observers have several selected objects at known distances from a specific location. When visibility is restricted, the farthest visible object determines the reported visibility..

A much better way of determining visibility is with an optical instrument called a transmissometer. A transmissometer measures runway visual range (RVR) down a specific runway. It is made up of two components. The first is a projector that sends a beam of light, at a known intensity, down the length of the runway. The second component, the receiver, contains a photoelectric cell that measures the amount of light penetrating the obscuring phenomena. A computer then converts the amount of light received into runway visual range.

WEATHER PHENOMENA
...1/2SM R36L/2400FT +TS**RA** SCT008 OVC012CB...

The weather phenomena section can include codes for nine types of precipitation, eight kinds of obscurations, and six rather uncommon weather events. Up to three types of precipitation can be coded in a single grouping of present weather conditions. When more than one is reported, they are shown in order of predominance. In the example, rain (RA) was observed in connection with the thunderstorm.

WEATHER PHENOMENA CODES	
Precipitation	
RA — Rain	GR — Hail (> 1/4")
DZ — Drizzle	GS — Small Hail/Snow Pellets
SN — Snow	SG — Snow Grains
IC — Ice Crystals	PL — Ice Pellets
UP — Unknown Precipitation	
Obstructions to Visibility	
FG — Fog	PY — Spray
BR — Mist	SA — Sand
FU — Smoke	DU — Dust
HZ — Haze	VA — Volcanic Ash
Other Weather Phenomena	
SQ — Squall	SS — Sandstorm
DS — Duststorm	PO — Dust/Sand Whirls
FC — Funnel Cloud	
+FC — Tornado or Waterspout	

Obscurations are factors that limit visibility, and are shown after any reported precipitation. Fog (FG) is listed when the visibility is less than 5/8 statute mile. If the visibility is between 5/8 and 6 statute miles, the code for mist (BR) is used. Shallow fog (MIFG), patches of fog (BCFG), or partial fog (PRFG) might be coded if the prevailing visibility is 7 statute miles or greater. Other weather phenomena can be listed following the obscurations. [Figure 7-8]

Figure 7-8. Precipitation, obscurations, and other weather phenomena all have unique two-letter codes.

SKY CONDITION

The sky condition groups describe the amount of clouds, if any, their heights and, in some cases, their type. In addition, a vertical visibility might be reported if the height of the clouds cannot be determined due to an obscuration.

AMOUNT
...+TSRA **SCT**008 **OVC**012CB 20/18...

The amount of clouds covering the sky is reported in eighths, or octas, of sky cover. Each layer of clouds is described using a code that corresponds to the octas of sky coverage. If the sky is clear, it is designated by SKC in a manual report and CLR in an automated report. Because an automated station cannot detect clouds above 12,000 feet, a report of clear indicates there were no clouds detected below 12,000 feet. FEW is used when cloud coverage is greater than zero to 2/8 of the sky. Scattered clouds, which cover 3/8 to 4/8 of the sky, are shown by SCT. Broken clouds, covering 5/8 to 7/8 of the sky, are designated by BKN, and an overcast sky is reported as OVC. In the example, there is a layer of scattered clouds (SCT) and an overcast layer (OVC).

The sky cover condition for a cloud layer represents total sky coverage, which includes any lower layers of clouds. Whether human or automated, the observer can only report conditions that are visible from a point on the ground. The observer cannot determine the extent of an upper layer that is partly hidden by a lower layer, so the observer must add the amount of sky covered by the lower layers to the amount of sky covered the upper layer. [Figure 7-9]

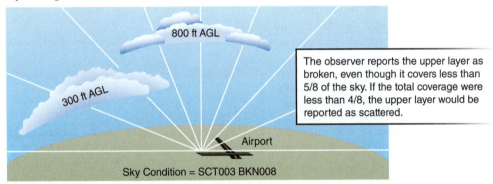

800 ft AGL

300 ft AGL

The observer reports the upper layer as broken, even though it covers less than 5/8 of the sky. If the total coverage were less than 4/8, the upper layer would be reported as scattered.

Airport

Sky Condition = SCT003 BKN008

Figure 7-9. The sky coverage reported for a higher layer of cloud includes the area of the sky obscured by lower layers.

HEIGHT, TYPE, AND VERTICAL VISIBILITY
...+TSRA SCT**008** OVC**012CB** 20/18...

The height of clouds or the vertical visibility into obscuring phenomena is reported with three digits in hundreds of feet above ground level (AGL). To determine the cloud height, add two zeros to the number given in the report. When more than one layer is present, the layers are reported in ascending order. Automated stations can only report a maximum of three layers at 12,000 feet AGL and below. Human observers can report up to six layers of clouds at any altitude. In the example, the scattered layer was at 800 feet AGL and the overcast layer was at 1,200 feet AGL.

In a manual report, a cloud type can be included if towering cumulus clouds (TCU) or cumulonimbus clouds (CB) are present. The code follows the height of their reported base. In the example, the base of the reported cumulonimbus clouds was at 1,200 feet AGL.

When more than half of the sky is covered by clouds, a **ceiling** exists. By definition, a ceiling is the AGL height of the lowest layer of clouds that is reported as broken or overcast, or the vertical visibility into an obscuration, such as fog or haze. Human observers may rely simply on their experience and knowledge of cloud formations to determine ceiling heights, or they can combine their experience with the help of reports from pilots, balloons, or other instruments. The ceiling and visibility determine if flight conditions are VFR. To depart under VFR from a controlled airport (without a special VFR clearance), the ceiling must be at least 1,000 feet and the visibility at least 3 statute miles.

Unlike clouds, an obscuration does not have a definite base. Obscuration is caused by phenomena such as fog, haze, or smoke that extend from the surface to an indeterminable height. In METAR reports, a total obscuration is shown with a VV followed by three digits indicating the vertical visibility in hundreds of feet. For example, VV006 describes an indefinite ceiling at 600 feet AGL. Obscurations that do not cover the entire sky might be reported in the remarks section of the METAR.

CLOUD LAYER THICKNESS

If you know the MSL height of the top of a cloud layer, you can easily determine the layer's thickness by adding the airport elevation (MSL) to the height of the cloud base (AGL), and then subtracting this number from the height of the cloud tops. For example, if an overcast begins at 700 feet AGL, the field elevation is 620 feet MSL, and the tops of the overcast layer are reported at 6,500 feet MSL, then the cloud layer is 5,180 feet thick:

$$
\begin{array}{r}
620 \text{ feet (MSL field elevation)} \\
+ 700 \text{ feet (AGL height above cloud base)} \\
\hline
1,320 \text{ feet (MSL height of cloud base)}
\end{array}
$$

$$
\begin{array}{r}
6,500 \text{ feet (MSL top of overcast)} \\
- 1,320 \text{ feet (MSL height of cloud base)} \\
\hline
5,180 \text{ feet (thickness of cloud layer)}
\end{array}
$$

TEMPERATURE AND DEWPOINT
...SCT008 OVC012CB **20/18** A2995...

METARs list the observed air temperature and dewpoint in degrees Celsius immediately following the sky condition. In the example, the temperature and dewpoint are reported as 20°C and 18°C, respectively. Temperatures below 0° Celsius are prefixed with an M to indicate minus. For instance, M10 indicates a temperature of 10°C below zero. Temperature and dewpoint readings to the nearest 1/10°C can be included in the remarks section of the METAR.

ALTIMETER
...20/18 **A2995** RMK TSB24RAB24 SLP134

The letter A followed by a four-digit group reports the altimeter setting in inches of mercury with the decimal omitted. In the example, the altimeter setting was 29.95 in. Hg.

REMARKS
...A2995 **RMK TSB24RAB24 SLP134**

The beginning of the remarks section is indicated by the code RMK. The remarks section reports weather conditions, which are not covered in the previous sections of the METAR, that are considered significant to aircraft operations. The remarks section can include wind data, variable visibility, beginning and ending times of a particular weather phenomena, pressure information, and precise temperature/dewpoint readings.

The beginning of an event is shown by a B, followed by the time in minutes after the hour. If the event ended before the observation, the ending time in minutes past the hour is noted by an E. In the example, a thunderstorm began at 24 minutes past the hour (TSB24). Rain also began at 24 minutes past the hour (RAB24). Additionally, the sea level pressure (SLP) was 1013.4 millibars, or hectoPascals (hPa). A hectopascal is the metric equivalent of a millibar (1 mb = 1 hPa). Examples of other coded remarks are shown in Figure 7-10.

CODED DATA	EXPLANATIONS
A02	Automated station with precipitation discriminator
PK WND 20032/25	Peak wind from 200° at 32 knots, 25 minutes past the hour
VIS 3/4V1 1/2	Prevailing visibility variable 3/4 to 1 and 1/2 miles
FRQ LTG NE	Frequent lightning to the northeast
FZDZB45	Freezing drizzle began at 45 minutes past the hour
RAE42SNB42	Rain ended and snow began at 42 minutes past the hour
PRESFR	Pressure falling rapidly
SLP045	Sea level pressure in millibars (hPa), 1004.5 mb (hPa)
WSHFT 30 FROPA	Wind shift occurred at 30 minutes past the hour due to frontal passage
T00081016	Temperature/dewpoint in tenths °C, .8 °C/–1.6 °C (Since the first digit after the T is a 0, it indicates that the temperature is positive; the dewpoint in this example is negative since the fifth digit is a 1.)

Figure 7-10. Examples of coded remarks are shown in the left column, with the corresponding explanations on the right.

PILOT WEATHER REPORTS

Pilot weather reports (PIREPs) are often your best source to confirm such information as the bases and tops of cloud layers, in-flight visibility, icing conditions, wind shear, and turbulence. When significant conditions are reported or forecast, ATC facilities are required to solicit PIREPs. Anytime you encounter unexpected weather conditions, you are encouraged to make a pilot report. When you make a PIREP, the ATC facility or Flight Service adds your report to the distribution system so it can be used to brief other pilots or provide in-flight advisories. [Figure 7-11]

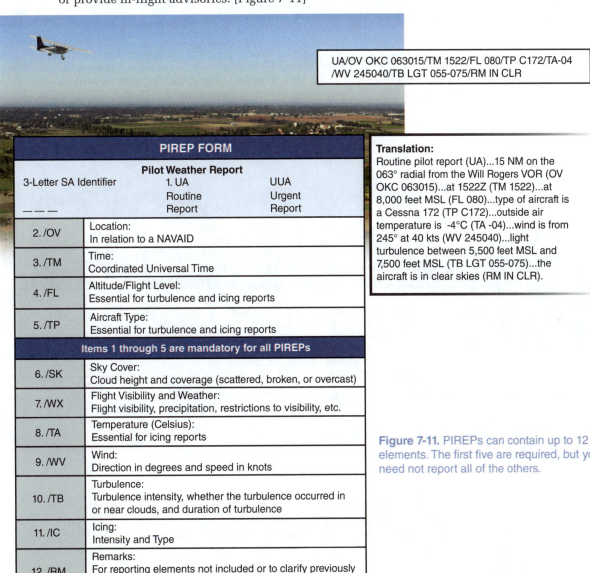

UA/OV OKC 063015/TM 1522/FL 080/TP C172/TA-04
/WV 245040/TB LGT 055-075/RM IN CLR

PIREP FORM

	Pilot Weather Report		
3-Letter SA Identifier	1. UA Routine Report	UUA Urgent Report	
— — —			
2. /OV	Location: In relation to a NAVAID		
3. /TM	Time: Coordinated Universal Time		
4. /FL	Altitude/Flight Level: Essential for turbulence and icing reports		
5. /TP	Aircraft Type: Essential for turbulence and icing reports		
Items 1 through 5 are mandatory for all PIREPs			
6. /SK	Sky Cover: Cloud height and coverage (scattered, broken, or overcast)		
7. /WX	Flight Visibility and Weather: Flight visibility, precipitation, restrictions to visibility, etc.		
8. /TA	Temperature (Celsius): Essential for icing reports		
9. /WV	Wind: Direction in degrees and speed in knots		
10. /TB	Turbulence: Turbulence intensity, whether the turbulence occurred in or near clouds, and duration of turbulence		
11. /IC	Icing: Intensity and Type		
12. /RM	Remarks: For reporting elements not included or to clarify previously reported items		

Translation:
Routine pilot report (UA)...15 NM on the 063° radial from the Will Rogers VOR (OV OKC 063015)...at 1522Z (TM 1522)...at 8,000 feet MSL (FL 080)...type of aircraft is a Cessna 172 (TP C172)...outside air temperature is -4°C (TA -04)...wind is from 245° at 40 kts (WV 245040)...light turbulence between 5,500 feet MSL and 7,500 feet MSL (TB LGT 055-075)...the aircraft is in clear skies (RM IN CLR).

Figure 7-11. PIREPs can contain up to 12 elements. The first five are required, but you need not report all of the others.

SECTION B ■ **Aviation Weather Reports and Forecasts**

Make your PIREP concise and as complete as possible, but do not be overly concerned with strict format or terminology. The important thing is to make the report so that other pilots can benefit from your observation. The Aeronautical Information Manual (AIM) contains detailed information on the interpretation of PIREPs.

If you cannot make a PIREP before landing, you should submit the report of the conditions that you experienced during the flight, after you land. You also can use tablet and smartphone-friendly PIREP submission tools. In addition, the FAA has an electronic PIREP submission tool on the Aviation Weather Center (AWC) website.

Another type of PIREP is called an AIREP or air report. Disseminated electronically, these reports are used almost exclusively by commercial airlines. However, when accessing the internet for current weather you might see the abbreviation ARP followed by a weather report that was captured from an Aeronautical Radio Incorporated (ARINC) communications addressing and reporting system (ACARS) transmission.

FORECASTS

Many of the reports of observed weather conditions are used to develop forecasts of future conditions. Every day, Weather Forecast Offices (WFOs) prepare over 2,000 forecasts for specific airports, over 900 route forecasts, and a variety of other forecasts for flight planning purposes. The forecasts that pilots use most often include the terminal aerodrome forecast, and the wind and temperature aloft forecast.

TERMINAL AERODROME FORECAST

The **terminal aerodrome forecast (TAF)** is a concise statement of the expected meteorological conditions within a 5 statute mile radius from the center of an airport's runway complex. TAFs normally are valid for a 24-hour period and scheduled four times a day at 0000Z, 0600Z, 1200Z, and 1800Z. Each TAF contains these elements: type, ICAO station identifier, issuance date and time, valid period, and the forecast. [Figure 7-12]

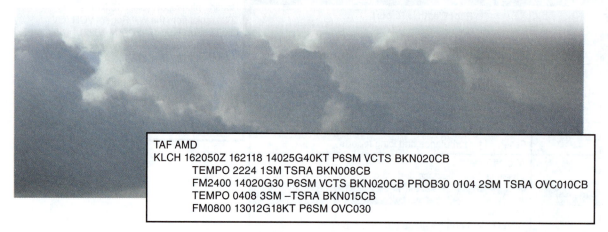

```
TAF AMD
KLCH 162050Z 162118 14025G40KT P6SM VCTS BKN020CB
     TEMPO 2224 1SM TSRA BKN008CB
     FM2400 14020G30 P6SM VCTS BKN020CB PROB30 0104 2SM TSRA OVC010CB
     TEMPO 0408 3SM −TSRA BKN015CB
     FM0800 13012G18KT P6SM OVC030
```

Figure 7-12. With a few exceptions, the codes used in the TAF are similar to those used in the METAR.

TYPE OF FORECAST
TAF AMD

KLCH...

The TAF normally is a routine forecast. However, an amended TAF (TAF AMD) might be issued when the current TAF no longer represents the expected weather. TAF or TAF AMD appears in a header line prior to the text of the forecast. The abbreviations COR and RTD indicate a corrected or a delayed TAF, respectively. The example depicts an amended TAF.

STATION IDENTIFIER AND ISSUANCE DATE/TIME
TAF AMD

KLCH 162050Z 162118...

The four letter ICAO location identifier code is the same as that used for the METAR/SPECI. The first two numbers of the date/time group represent the day of the month, and the next four digits are the Zulu time that the forecast was issued. The example TAF is a forecast for Lake Charles Regional Airport (KLCH) that was issued on the 16th day of the month at 2050Z.

VALID PERIOD
...162050Z **162118** 14025G40KT...

Normally, the forecast is valid for 24 hours. The first two digits represent the valid date. Next is the beginning hour of the valid time in Zulu, and the last two digits are the ending hour. The forecast for Lake Charles Regional Airport is valid from 2100Z on the 16th of the month to 1800Z the next day. Because the TAF in this example was amended, the valid period is less than 24 hours. At an airport that is open part time, the TAFs issued for that location will have the abbreviated statement AMD NOT SKED AFT (closing time) Z added to the end of the forecast text. For TAFs issued when these airports are closed, the word NIL appears in place of the forecast text.

FORECAST
...162050Z 162118 **14025G40KT P6SM VCTS BKN020CB...**

The body of the TAF contains codes for forecast wind, visibility, weather, and sky condition. Weather, including obstructions to visibility, is added to the forecast when it is significant to aviation.

The forecast wind is depicted by a 5-digit group—the first 3 digits represent the wind direction and the last 2 digits or 3 digits (for winds 100 knots or more) represent the wind speed in knots (KT). The prevailing wind direction is forecast for any speed greater than or equal to 7 knots. When the prevailing wind direction is variable (variations in wind direction of 30 degrees or more), the wind direction is encoded as VRB. Two conditions where this can occur are very light winds (1 to 6 knots inclusive) and convective activity. A calm wind is forecast as 00000KT. Wind gusts are noted by the letter G appended to the wind speed followed by the highest expected gust. In this example, the winds are predicted to be from 140°at 25 knots with gusts to 40 knots.

When the expected prevailing visibility is 6 statute miles (SM) or less it is included in the forecast, followed by the letters, SM. Visibilities greater than (plus) 6 statute miles are indicated in the forecast by the letter P (P6SM).

The weather phenomena and sky conditions are given next using the same format, qualifiers, and contractions as the METAR reports. If no significant weather is expected to occur during a specific time period in the forecast, the weather group is omitted for that time period. CLR is never used in the TAF. In this example, thunderstorms are forecast to be in the vicinity (VC) and a broken cloud layer, made up of cumulonimbus clouds, is forecast at 2,000 feet AGL.

Low-level wind shear that is not associated with convective activity might be included using the code WS followed by a three digit height (up to and including 2,000 feet AGL), a forward slash (/) and the winds at the height indicated. For example, WS010/18040KT indicates low-level wind shear at 1,000 feet AGL, wind 180° at 40 knots.

SECTION B ■ **Aviation Weather Reports and Forecasts**

FORECAST CHANGE GROUPS

When a significant change to the weather conditions is expected during the valid time, a change group is used. The changes can be temporary, rapid, or gradual. Each change indicator marks a time group within the TAF.

TEMPORARY FORECAST

...TEMPO 2224 1SM TSRA BKN008CB

FM2400 14020G30 P6SM VCTS BKN020CB PROB30 0104 2SM TSRA OVC010CB

TEMPO 0408 3SM -TSRA BKN015CB

FM0800 13012G18KT P6SM OVC030

Wind, visibility, weather, or sky conditions that are expected to last less than an hour are described in a temporary (TEMPO) group, followed by beginning and ending times. The first temporary group in the example predicts that between 2200Z and 2400Z, visibility is expected to be reduced to 1 statute mile in moderate rain associated with a thunderstorm. A broken layer of cumulonimbus clouds with bases at 800 feet is also predicted to occur. It's important to remember that these conditions only modify the previous forecast and are expected to last for periods of less than one hour. The second temporary group predicts that between 0400Z and 0800Z, the visibility is likely to be 3 statute miles in light rain associated with a thunderstorm. The ceiling is expected to be a broken layer of cumulonimbus clouds with the bases at 1,500 feet.

FROM OR BECOMING FORECAST

...TEMPO 2224 1SM TSRA BKN008CB

FM2400 14020G30 P6SM VCTS BKN020CB PROB30 0104 2SM TSRA OVC010CB

TEMPO 0408 3SM -TSRA BKN015CB

FM0800 13012G18KT P6SM OVC030

If a rapid change, usually within one hour, is expected, the code for from (FM) is used with the time of change. The conditions listed following the FM will continue until the next change group or the end of the valid time of the TAF. In the example, the first forecast change group indicates from 2400Z, the wind will be from 140° at 20 knots with gusts to 30 knots, the visibility will be greater than 6 statute miles, with thunderstorms in the vicinity and a cloud layer, made up of cumulonimbus clouds, is expected to be broken at 2,000 feet AGL. From 0800Z, the wind is forecast to be from 130° at 12 knots gusting to 18 knots, visibility will improve to greater than 6 statute miles and the sky will be overcast at 3,000 feet AGL.

A more gradual change in the weather, taking about two hours, is coded as BECMG, followed by beginning and ending times of the change period. The gradual change is expected to occur at an unspecified time within this time period. All items, except for the changing conditions shown in the BECMG group, are carried over from the previous time group. For instance, BECMG 2310 24007KT P6SM NSW indicates a gradual change in conditions is expected to occur between 2300Z and 1000Z. Sometime during this period, the wind will be from 240° at 7 knots, visibility is predicted to improve to greater than 6 miles with no significant weather (NSW). The NSW code is used only after a time period in which significant weather was forecast. NSW only appears in BECMG or TEMPO groups.

PROBABILITY FORECAST

...FM2400 14020G30 P6SM VCTS BKN020CB **PROB30 0104 2SM TSRA OVC010CB**

TEMPO 0408 3SM -TSRA BKN0015CB...

TAFs can include the probability of thunderstorms or precipitation events with the associated wind, visibility, and sky conditions. A PROB group is used when the probability of occurrence is between 30 and 49%. The percentage is followed by the beginning and ending time of the period during which the thunderstorm or precipitation is expected. In the example, there is a 30% chance of a thunderstorm with moderate rain and 2 statute miles visibility between 0100Z and 0400Z. In addition, there is a 30% probability of an overcast layer of cumulonimbus clouds with bases at 1,000 feet AGL during the four hour time period.

WIND AND TEMPERATURE ALOFT FORECAST

A **wind and temperature aloft forecast** provides an estimate of wind direction in relation to true north, wind speed in knots, and the temperature in degrees Celsius for selected altitudes. Depending on the station elevation, winds and temperatures are usually forecast for levels between 3,000 and 53,000 feet. [Figure 7-13]

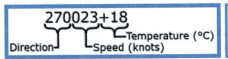

270023+18
Direction — Speed (knots) — Temperature (°C)

Key
Direction is in true degrees. A plus or minus sign precedes the temperature.

Filed Altitude
When obtaining a briefing, your proposed or filed enroute altitude is highlighted.

PART 1 - Flight Levels 030-180 (PART 2 - Flight Levels 240-530 available below)

Station ID	030	055 Filed-4k	060	075 Filed-2k	090	095 Filed	115 Filed+2k	120	135 Filed+4k	180
051800Z 1400Z-2100Z										
DEN					230011-04	236013-05	263023-09	270026-10	270030-13	270042-23
PUB					260013-01	263014-02	276020-06	280022-08	275025-11	260037-21
GLD			030010	320013	250016-03	251016-04	258018-08	260019-09	257023-12	250037-22
GCK			040008+01		LGTVRB-01			260014-06	255019-09	240035-20
SLN	010012	018012	020013+03	355009+01	330006-01	318007-02	271011-06	260013-07	255019-10	240039-20
ICT	020015	036015	040016+05	325010+02	250005+00	250008+00	250020-04	250023-05	250025-08	250031-19

PART 2 - Flight Levels 240-530 (PART 1 - Flight Levels 030-180 available above)

Station ID	240	300	340	390	450	530
051800Z 1400Z-2100Z						
DEN	260052-			56	260072-52	260063-57
PUB	250044-			57		
GLD	250047-			55	250061-52	260055-59
GCK	250042-32	260048-47	260056-56	260066-55	260060-52	260058-61
SLN	240043-33	250047-48	260055-56	260059-54		
ICT	250032-32	260040-47	260050-55	260055-53	260058-53	250054-62

Observation and Forecast Time
The forecast valid time is on the 5th at 1800Z and the forecast is intended for use between 1400Z to 2100Z.

Omissions
Temperatures are not forecast for the 3,000-foot level or for any level within 2,500 feet of the station elevation. Wind groups are omitted when the level is within 1,500 feet of the station elevation or if forecast winds are not available at a specific level for that station.

Figure 7-13. This wind and temperature aloft forecast is shown in a typical format when obtaining an online weather briefing from Flight Service.

The presentation of wind information is similar to other reports and forecasts. The first two numbers indicate the true direction from which the wind is blowing. For example, 2635-08 indicates the wind is from 260° at 35 knots and the temperature is –8°C. Quite often you must interpolate between two levels.

In some formats, wind speeds between 100 and 199 knots are encoded so direction and speed can be represented by four digits. This is done by adding 50 to the two-digit wind direction and subtracting 100 from the velocity. For example, a wind of 270° at 101 knots is encoded as 7701 (27 + 50 = 77 for wind direction, and 101 – 100 = 01 for wind speed). A code of 9900 indicates light and variable winds (less than five knots). However, wind speeds of 200 knots or more are encoded as 199.

AIRMETS AND SIGMETS

AIRMETs, SIGMETs, and convective SIGMETs are textual forecasts that advise enroute aircraft of the development of potentially hazardous weather. You can get them from FIS-B during flight and, in some cases, ATC will broadcast the alert. You also can get these

SECTION B ■ Aviation Weather Reports and Forecasts

advisories during your preflight weather briefing to learn of the latest adverse conditions affecting your flight. These advisories use the same location identifiers (VORs, airports, and well-known geographic areas) to describe the location of the hazardous weather.

AIRMET

AIRMET is an acronym for airman's meteorological information. **AIRMETs (WAs)** are issued every six hours, with amendments issued as necessary, for weather phenomena that are of operational interest to all aircraft, but hazardous mainly to light aircraft. The three types of AIRMETs are labeled:

- Sierra—IFR conditions (ceilings less than 1,000 feet or visibility less than 3 miles) and/or extensive mountain obscuration).

- Tango—Moderate turbulence, sustained surface winds of 30 knots or greater, or nonconvective low-level wind shear.

- Zulu—moderate icing and freezing level heights. AIRMETs are numbered sequentially for easier identification.

If obtaining an online briefing, you can view the affected areas of an AIRMET on an interactive map, either from Flight Service at 1800wxbrief.com or from the Aviation Weather Center at aviationweather.gov. The depictions of the affected areas are easier to visualize on these graphs than from textual or verbal boundary descriptions. [Figure 7-14]

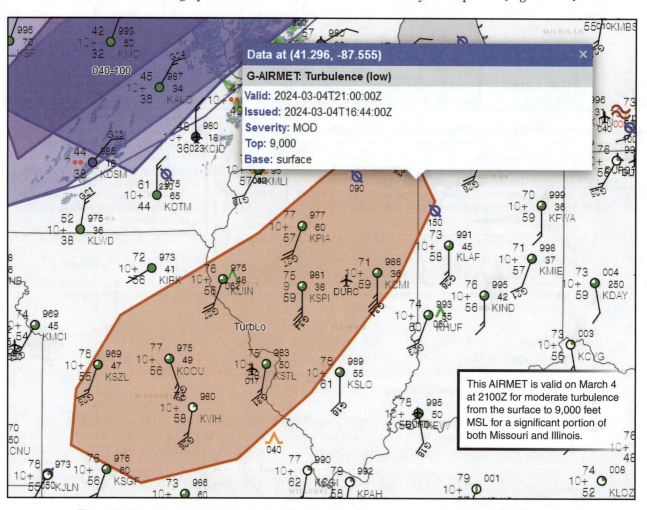

Data at (41.296, -87.555)

G-AIRMET: Turbulence (low)

Valid: 2024-03-04T21:00:00Z
Issued: 2024-03-04T16:44:00Z
Severity: MOD
Top: 9,000
Base: surface

This AIRMET is valid on March 4 at 2100Z for moderate turbulence from the surface to 9,000 feet MSL for a significant portion of both Missouri and Illinois.

Figure 7-14. In this example from the Aviation Weather Center, you can read the AIRMET text, view the lateral boundaries on the map, and easily determine what airports are affected.

SIGMET

SIGMETs (WSs) are issued for hazardous weather (other than convective activity) that is considered significant to all aircraft. SIGMET stands for significant meteorological information, and these advisories include:

Courtesy of NASA

- Severe icing.
- Severe and extreme turbulence.
- Clear air turbulence (CAT).
- Dust storms and sandstorms lowering visibility to less than three miles.
- Volcanic ash. [Figure 7-15]

Figure 7-15. SIGMETs are issued for severe weather phenomena, such as dust storms with visibility less than 3 miles.

SIGMETs are unscheduled forecasts that are valid for four hours, but if the SIGMET relates to hurricanes, it is valid for six hours. SIGMETs use alphanumeric designators November through Yankee, excluding Sierra and Tango. [Figure 7-16]

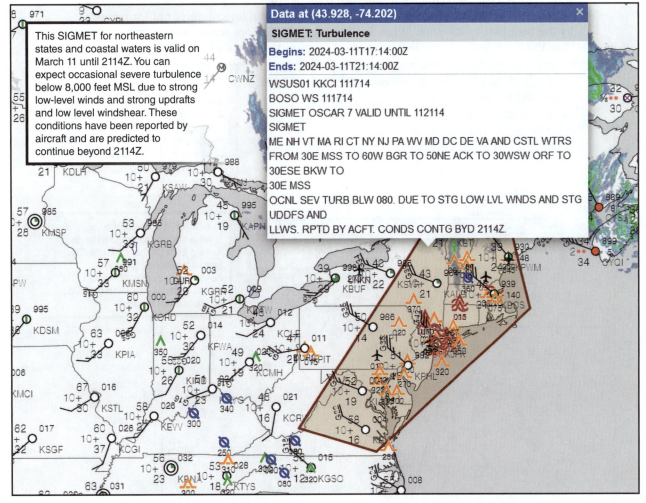

This SIGMET for northeastern states and coastal waters is valid on March 11 until 2114Z. You can expect occasional severe turbulence below 8,000 feet MSL due to strong low-level winds and strong updrafts and low level windshear. These conditions have been reported by aircraft and are predicted to continue beyond 2114Z.

Data at (43.928, -74.202)

SIGMET: Turbulence

Begins: 2024-03-11T17:14:00Z
Ends: 2024-03-11T21:14:00Z

WSUS01 KKCI 111714
BOSO WS 111714
SIGMET OSCAR 7 VALID UNTIL 112114
SIGMET
ME NH VT MA RI CT NY NJ PA WV MD DC DE VA AND CSTL WTRS
FROM 30E MSS TO 60W BGR TO 50NE ACK TO 30WSW ORF TO 30ESE BKW TO
30E MSS
OCNL SEV TURB BLW 080. DUE TO STG LOW LVL WNDS AND STG UDDFS AND
LLWS. RPTD BY ACFT. CONDS CONTG BYD 2114Z.

Figure 7-16. In this example from the Aviation Weather Center, you can read the SIGMET text, view the lateral boundaries on the map, and easily determine what airports are affected.

CONVECTIVE SIGMET

Convective SIGMETs (WSTs) are issued for hazardous weather related to thunderstorms that is significant to the safety of all aircraft. WSTs always imply severe or greater turbulence, severe icing, and low-level wind shear, so these items are not specified in the advisory. A WST consists of either an observation and a forecast or simply a forecast. A WST is issued for any of the following phenomena:

- Tornadoes.
- Lines of thunderstorms
- Thunderstorms over a wide area
- Embedded thunderstorms.
- Hail ¾ inch in diameter or more.
- Wind gusts to 50 knots or greater. [Figure 7-17]

Courtesy of NASA

Figure 7-17. Convective SIGMETs are issued for clusters of thunderstorms. This photograph, taken from the Space Shuttle Orbiter, displays a variety of weather elements including overshooting thunderstorm tops, squall lines, and areas of probable high-speed downdrafts or microbursts.

Convective SIGMETs are issued for the Eastern, Central, or Western United States. Convective SIGMETs are not issued for Alaska or Hawaii, where convective SIGMET conditions are included in (non-convective) SIGMETs. Convective SIGMETs are issued at 55 minutes past each hour, and numbered sequentially for each area each day. Special bulletins are issued as required. SIGMET forecasts are valid for 2 hours or until superseded by the next hourly issuance. When convective SIGMETs are not necessary, the message CONVECTIVE SIGMET...NONE is issued at 55 minutes after each hour. [Figure 7-18]

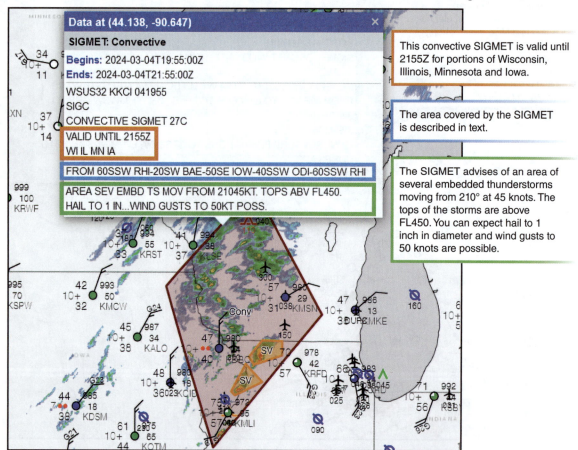

Data at (44.138, -90.647)

SIGMET: Convective

Begins: 2024-03-04T19:55:00Z
Ends: 2024-03-04T21:55:00Z

WSUS32 KKCI 041955
SIGC
CONVECTIVE SIGMET 27C
VALID UNTIL 2155Z
WI IL MN IA

FROM 60SSW RHI-20SW BAE-50SE IOW-40SSW ODI-60SSW RHI

AREA SEV EMBD TS MOV FROM 21045KT. TOPS ABV FL450. HAIL TO 1 IN...WIND GUSTS TO 50KT POSS.

This convective SIGMET is valid until 2155Z for portions of Wisconsin, Illinois, Minnesota and Iowa.

The area covered by the SIGMET is described in text.

The SIGMET advises of an area of several embedded thunderstorms moving from 210° at 45 knots. The tops of the storms are above FL450. You can expect hail to 1 inch in diameter and wind gusts to 50 knots are possible.

Figure 7-18. You can interpret the location of the conditions described by the convective weather SIGMET in text. However, viewing the lateral dimensions on a map helps you to better visualize the location of the weather phenomenon.

SUMMARY CHECKLIST

✓ An aviation routine weather report (METAR) is an observation of surface weather written in a standard format that typically contains 10 or more separate elements.

✓ A non-routine aviation weather report (SPECI) is issued when a significant change in one or more of the elements of a METAR has occurred.

✓ Prevailing visibility is the greatest distance an observer can see and identify objects through at least half of the horizon.

✓ Runway visual range (RVR) is based on what a pilot in a moving aircraft should see when looking down the runway. If included in a METAR, RVR is reported following prevailing visibility.

✓ A ceiling is the height above ground level of the lowest layer of clouds aloft that is reported as broken (BKN) or overcast (OVC), or the vertical visibility (VV) into an obscuration.

✓ The bases and tops of cloud layers, in-flight visibility, icing conditions, wind shear, and turbulence can be included in a pilot weather report (PIREP).

✓ Terminal aerodrome forecasts (TAFs) predict the weather at a specific airport for a 24-hour period of time.

✓ You can find an estimate of wind direction in relation to true north, wind speed in knots, and the temperature in degrees Celsius for selected altitudes in the wind and temperature aloft forecast (FB).

✓ AIRMETs, SIGMETs, and convective SIGMETs are textual forecasts that predict the development of potentially hazardous weather. Although these are text products, you can view the affected areas on an interactive map when online.

✓ AIRMETs are issued every six hours for weather phenomena that are of operational interest to all aircraft, but hazardous mainly to light aircraft.

✓ SIGMETs are issued for hazardous weather that affects all aircraft, including severe icing, severe and extreme turbulence, clear air turbulence (CAT), dust storms, sandstorms and volcanic ash.

✓ Convective SIGMETs are issued for hazardous thunderstorm-related weather that affects the safety of all aircraft. They include tornadoes, thunderstorms in lines or over a wide area, embedded thunderstorms, hail ¾ inch in diameter or more, and 50-knot wind gusts.

KEY TERMS

Aviation Routine Weather Report (METAR)

Non-Routine (Special) Aviation Weather Report (SPECI)

Prevailing Visibility

Runway Visual Range (RVR)

Ceiling

Pilot Weather Report (PIREP)

Terminal Aerodrome Forecast (TAF)

Winds And Temperatures Aloft Forecast (FB)

AIRMET (WA)

SIGMET (WS)

Convective SIGMET (WST)

SECTION B ■ Aviation Weather Reports and Forecasts

QUESTIONS

Use the following METAR for Ponca City Municipal Airport (KPNC), to answer questions 1 through 5.

METAR KPNC 161954Z 03015G27KT 2SM -RA BR BKN007 BKN017 OVC030 07/06 A2978 RMK PK WND 06030/51 SLP086 T00670061

1. What are the reported winds?
   A. 03° at 015 knots with gusts from 220° at 7 knots
   B. 300° at 15 knots with gusts at 27 knots
   C. 030° at 15 knots with gusts to 27 knots

2. What is the reported intensity of the rain in Ponca City, Oklahoma?
   A. Moderate
   B. Light
   C. Heavy

3. What is the height of the lowest ceiling at Ponca City Municipal?
   A. 700 feet
   B. 1,700 feet
   C. 3,000 feet

4. What is the actual temperature/dewpoint at Ponca City Municipal Airport?

5. What is the sea level pressure in millibars?

6. True/False. Altitudes given in PIREPs are in hundreds of feet MSL.

7. In a TAF what does the code TEMPO indicate?

Use the following TAF for Dallas Fort Worth International Airport (KDFW) to answer questions 8 through 11.

TAF

KDFW AMD 161849Z161918 14015G20KT P6SM VCTS BKN015CB OVC040

       FM2100 17015G20KT P6SM VCTS BKN020CB OVC050

       FM0200 20008KT P6SM BKN012 OVC030

       TEMPO 0206 2SM -RA OVC008

       FM0600 24012KT P6SM BKN012 OVC030 PROB30 0612 -RA

       FM1400 24010KT P6SM SCT020 BKN060

8. What is the valid period for the KDFW TAF?
   A. 1900Z to 1800Z the following day
   B. 1600Z to 1900Z the following day
   C. 1600Z to 1800Z the following day

9. What weather conditions are forecast to exist between 0200Z and 0600Z?

10. What does the statement **PROB30 0612 -RA** indicate?
    A. There is a probability of light rain over 30% of the vicinity
    B. There is a 30% probability of light rain during the forecast
    C. There is a 30% probability of light rain occurring between 0600Z and 1200Z

11. True/False. After 1400Z the visibility at KDFW will be less than 6 statute miles.

12. In the wind and temperature aloft forecast at 30,000 feet, how is 751015 decoded?

13. True/False. Weather phenomena that are of operational interest to all aircraft are reported in an AIRMET.

14. Select the weather phenomena that can initiate the issuance of a SIGMET.
     A. Severe icing
     B. Embedded thunderstorms
     C. Hazardous convective weather

15. Convective SIGMETs include information on which weather conditions?
     A. Thunderstorms, super cells, and tornadoes
     B. Tornadoes, embedded thunderstorms, and lines of thunderstorms
     C. Embedded thunderstorms, severe thunderstorms with hail greater than or equal to 1/2 inch in diameter, and/or wind gusts 50 knots or greater

SECTION B ■ Aviation Weather Reports and Forecasts

SECTION C
Graphic Weather Products

You can develop a more complete picture of the weather that will affect your flights by combining the information contained in aviation weather reports and forecasts with graphic portrayals of the current and predicted weather. These graphic weather products use information gathered from ground observations, weather radar, satellites, and other sources to give you a pictorial view of large scale weather patterns and trends.

The National Weather Service (NWS) and the National Oceanic and Atmospheric Administration (NOAA) produce text, digital, and graphical observations, analyses, and forecasts of aviation-related weather variables. You can obtain graphic weather products online from Aviation Weather Center (AWC) at aviationweather.gov and from Flight Service at 1800wxbrief.com. Flight planning apps and electronic flight bags (EFBs), such as ForeFlight, also provide graphic weather products.

As new weather products are disseminated, they might be labeled with restrictions until proven and accepted. FAA policy regarding which weather products are acceptable for pilot briefings is evolving. As pilot in command, you must decide which weather graphics meet your preflight requirements and comply with any restrictions displayed on government-sourced charts. Many digital graphic weather products are interactive and enable you to select multiple layers of weather data, display specific information for an airport, route, or region, and zoom to increase the level of detail. [Figure 7-19]

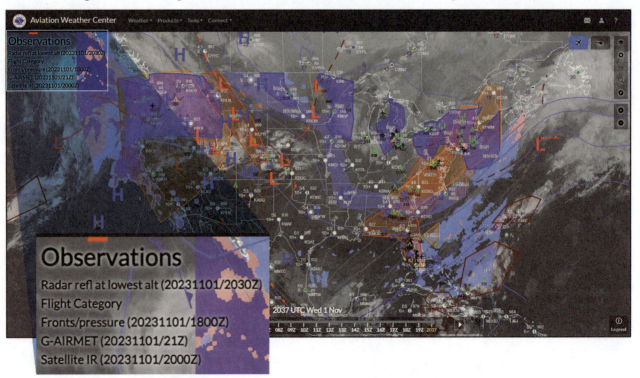

Figure 7-19. This interactive map on the Aviation Weather Center website shows selected layers of observations, which include weather radar, flight categories, fronts and pressure systems, G-AIRMETs, and satellite imagery.

This section covers the primary graphic weather products that you use when obtaining a preflight weather briefing. You can expect the appearance and functionality of these weather products to evolve and interactive weather maps can include information found on multiple charts covered in this section.

OBSERVATIONS

Observations are weather data collected by one or more sensors, and are the basic information upon which forecasts and advisories are made. The most common types of weather observations include radar observations, satellite imagery, and surface observations.

RADAR OBSERVATIONS

Weather **radar observations** are graphical displays of precipitation and non-precipitation targets detected by weather radar. A network of weather surveillance radar—1988 Doppler (WSR-88D), also known as next generation weather radar (NEXRAD)—is continuously generating radar observations. The WSR-88D uses radar antennas that automatically raise to higher and higher preset angles, or elevation scans, as they rotate. Each radar observation, called a volume scan, consists of 5 to 14 separate elevation tilts, and takes between 4 and 11 minutes to generate, depending on the radar's mode of operation. Radar observation times are not standard—the valid time of the observation is the end of the last radar scan.

In general, the amount of reflective power (reflectivity) received is proportional to the intensity of the precipitation. The colors on radar images represent the reflectivity of the precipitation target measured in decibels (dBZ). [Figure 7-20]

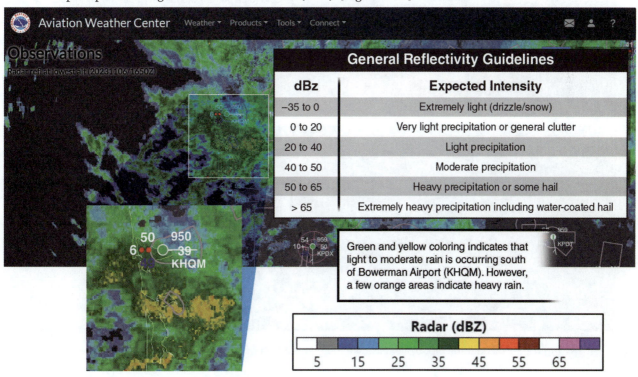

Figure 7-20. Use the color legend on radar images to determine the precipitation intensity.

SATELLITE IMAGERY

Some of the most recognizable weather products come from satellites. Specialized weather satellites not only generate photos, but also record temperatures, humidities, wind speeds, and water vapor locations. Three types of **satellite imagery** are available from weather satellites—visible, infrared (IR), and water vapor.

SECTION C ■ **Graphic Weather Products**

You primarily use visible images to determine the presence of clouds and the cloud shape and texture. IR photos depict the heat radiation emitted by the various cloud tops and the earth's surface. You can use water vapor imagery, which displays the quantity of water vapor generally located in the middle and upper troposphere, to determine the locations and movements of weather systems, jet streams, and thunderstorms. Because visible imagery is produced by reflected sunlight (radiation), the images are only available during daylight hours. IR and water vapor imagery are available day or night. [Figure 7-21]

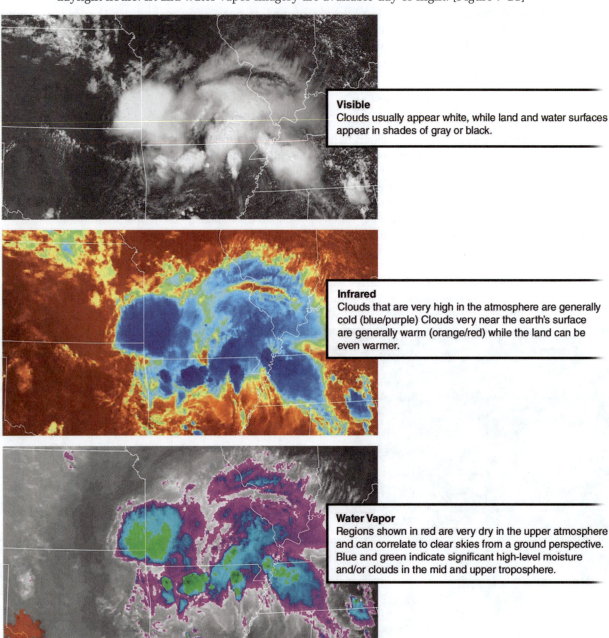

Visible
Clouds usually appear white, while land and water surfaces appear in shades of gray or black.

Infrared
Clouds that are very high in the atmosphere are generally cold (blue/purple) Clouds very near the earth's surface are generally warm (orange/red) while the land can be even warmer.

Water Vapor
Regions shown in red are very dry in the upper atmosphere and can correlate to clear skies from a ground perspective. Blue and green indicate significant high-level moisture and/or clouds in the mid and upper troposphere.

Figure 7-21. The three types of satellite imagery are shown here.

GRAPHICAL DEPICTIONS OF SURFACE OBSERVATIONS

Graphical depictions of surface observations that use METAR data are available from the AWC and Flight Service in several formats. Surface observations for the valid time shown on the chart are displayed as color-coded **station models** that depict weather flying categories (LIFR, IFR, MVFR, and VFR) and provide a summary of surface conditions using text and symbols. You might be able to obtain graphical depictions of both METARs and TAFs for a location on the same chart. [Figures 7-22 and 7-23]

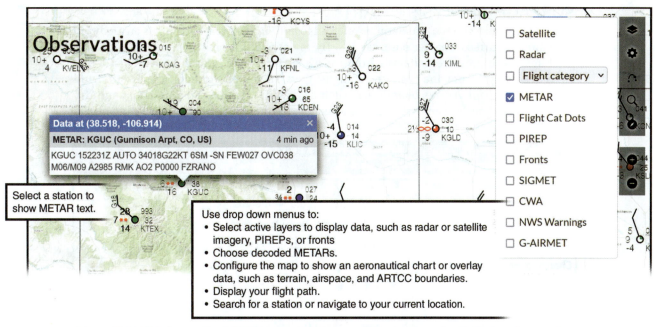

Figure 7-22. This interactive METAR map provides a variety of options for displaying observed weather data.

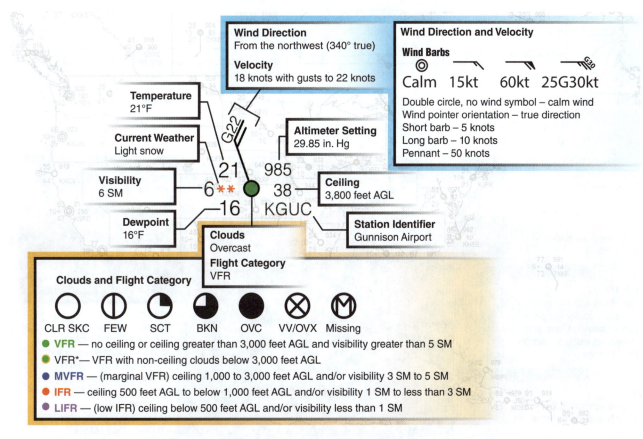

Figure 7-23. This sample station model shows data that is depicted on graphical METARs and other weather charts. Note that the temperature and dewpoint are shown in °F while textual METARs use °C. Always refer to the legend for the specific product that you are using.

ANALYSIS

An analysis is an enhanced depiction of observed data—a map or chart—that can also include interpretation by a weather specialist. Analysis charts show weather conditions that are observed or measured and then plotted graphically for easy interpretation. The most commonly used graphic analysis is the surface analysis chart.

SECTION C ■ **Graphic Weather Products**

SURFACE ANALYSIS CHART

The **surface analysis chart**, sometimes referred to as a surface weather map, is an analyzed depiction of surface weather observations. By reviewing this chart, you obtain a picture of atmospheric pressure patterns at the earth's surface. Standard chart symbols depict areas of equal pressure, the positions of highs, lows, ridges, and troughs, the location and type of fronts, and various boundaries, such as drylines, which separate moist and dry air masses. The surface analysis chart is issued at various time periods depending on the provider. [Figures 7-24 and 7-25]

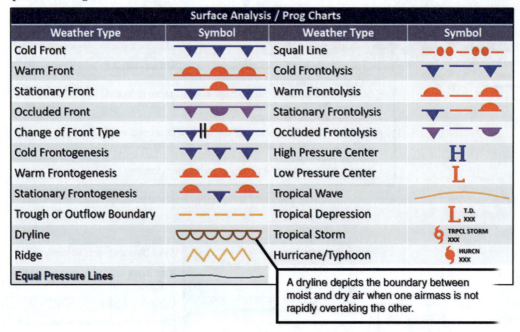

Surface Analysis / Prog Charts			
Weather Type	**Symbol**	**Weather Type**	**Symbol**
Cold Front		Squall Line	
Warm Front		Cold Frontolysis	
Stationary Front		Warm Frontolysis	
Occluded Front		Stationary Frontolysis	
Change of Front Type		Occluded Frontolysis	
Cold Frontogenesis		High Pressure Center	H
Warm Frontogenesis		Low Pressure Center	L
Stationary Frontogenesis		Tropical Wave	
Trough or Outflow Boundary		Tropical Depression	L T.D. XXX
Dryline		Tropical Storm	TRPCL STORM XXX
Ridge		Hurricane/Typhoon	HURCN XXX
Equal Pressure Lines			

A dryline depicts the boundary between moist and dry air when one airmass is not rapidly overtaking the other.

Figure 7-24. The symbols and colors depicted on surface analysis charts are standardized and are used on other graphic products, such as surface prognostic charts.

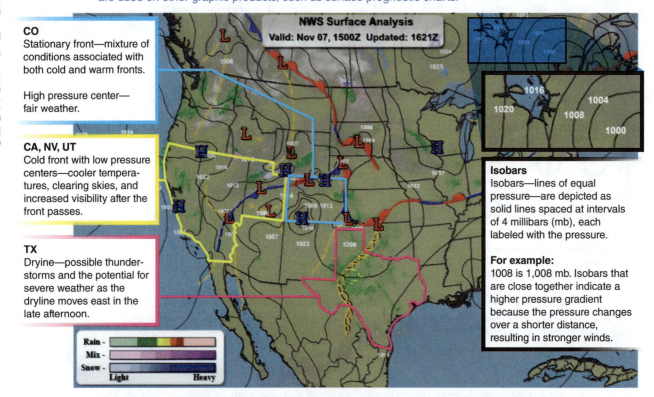

CO
Stationary front—mixture of conditions associated with both cold and warm fronts.

High pressure center—fair weather.

CA, NV, UT
Cold front with low pressure centers—cooler temperatures, clearing skies, and increased visibility after the front passes.

TX
Dryline—possible thunderstorms and the potential for severe weather as the dryline moves east in the late afternoon.

Isobars
Isobars—lines of equal pressure—are depicted as solid lines spaced at intervals of 4 millibars (mb), each labeled with the pressure.

For example:
1008 is 1,008 mb. Isobars that are close together indicate a higher pressure gradient because the pressure changes over a shorter distance, resulting in stronger winds.

Figure 7-25. The surface analysis chart depicts isobars, pressure systems, and fronts.

FORECASTS

A forecast is a prediction of the development and movement of weather phenomena based on meteorological observations and mathematical models. Graphic weather products help you visualize the forecast VFR and IFR ceilings and visibilities, surface winds, turbulence, winds and temperatures aloft, hazardous weather, and icing conditions.

The precipitation symbols used on graphic forecasts are standardized to appear on a wide variety of weather graphics. The number of symbols indicates the intensities of drizzle, rain, and snow: two—light, three—moderate, and four—heavy. A mixture of two precipitation types is shown as two symbols separated by a slash. [Figure 7-26]

Symbol	Meaning		Symbol	Meaning		Symbol	Meaning
=	Mist BR		๏𝄌	Freezing Drizzle FZDZ		�	Moderate Icing
≡	Fog FG		●𝄌	Freezing Rain FZRA			Severe Icing
∞	Haze HZ		⬠	Ice Pellets PE PL		‿	Moderate Turbulence
⌒	Smoke FU VA		✳	Snow SSN		ᐱ	Severe Turbulence
❟	Drizzle DZ		▽	Rain Shower SHRA		⅃	Tropical Storm
●	Rain RA		▽	Snow Shower SHSN		⅃	Hurricane (Typhoon)
			⌐	Thunderstorms TS			

Figure 7-26. Some of the symbols used on graphic forecast products are shown here. Always refer to the legend for the specific product that you are using.

CLOUDS FORECAST

The **clouds forecast** provides cloud coverage, bases, layers, and tops and includes depictions of AIRMET boundaries for mountain obscuration and icing up to 18 hours in the future. The clouds forecast uses color to designate the amount of cloud coverage and the AIRMET type. [Figure 7-27]

For 18 hours into the future, the **surface forecast** provides visibility, weather phenomena, precipitation, and winds (including wind gusts) with depictions of the boundaries of AIRMETs for IFR conditions and sustained surface winds of 30 knots or more. The surface forecast uses color to indicate the flight category and AIRMET type with specific symbols representing the type of precipitation. [Figure 7-28]

SECTION C ■ **Graphic Weather Products**

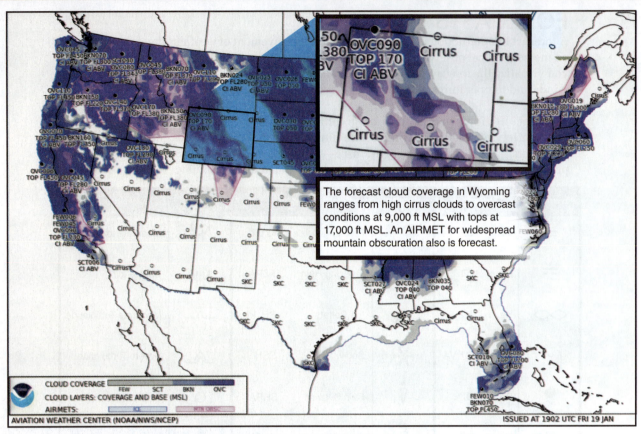

The forecast cloud coverage in Wyoming ranges from high cirrus clouds to overcast conditions at 9,000 ft MSL with tops at 17,000 ft MSL. An AIRMET for widespread mountain obscuration also is forecast.

Figure 7-27. You can see at a glance areas of extensive cloud coverage and the location of AIRMETs. Station models provide more detailed information about forecast clouds.

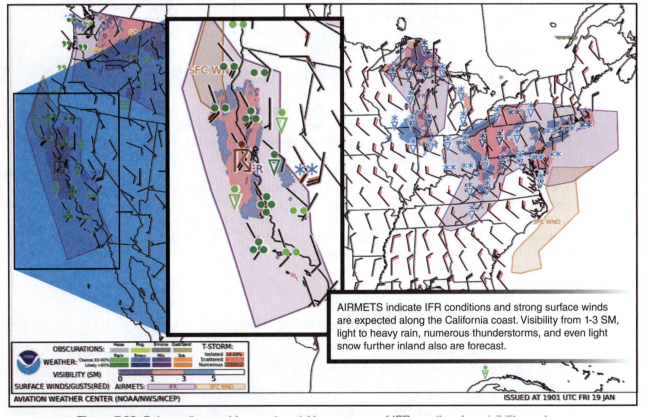

AIRMETS indicate IFR conditions and strong surface winds are expected along the California coast. Visibility from 1-3 SM, light to heavy rain, numerous thunderstorms, and even light snow further inland also are forecast.

Figure 7-28. Color coding enables you to quickly see areas of IFR weather, low visibility, and strong surface winds. The multiple wind barbs show the trend of surface wind direction and speed over large areas.

SURFACE PROGNOSTIC (PROG) CHART

Use a **surface prognostic (prog) chart** to obtain an overview of the progression of surface weather—the change in position, size, and intensity of conditions over time. These charts depict forecast surface pressure systems, fronts, and precipitation over several days. The charts contain standard symbols for fronts, isobars, and low and high pressure centers. [Figure 7-29]

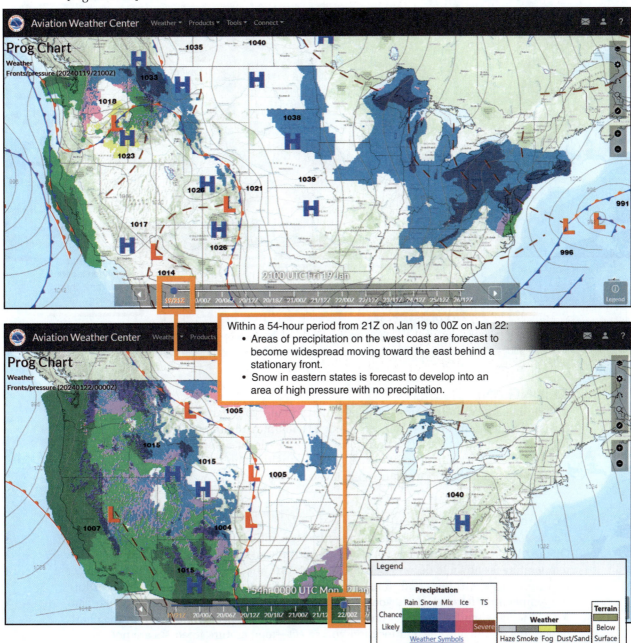

Figure 7-29. On the AWC website, you can select times within an 8 day period to view the trend for surface conditions and the movement of fronts and pressure centers.

AWC FORECAST GRAPHICS

On the AWC website, you can select a wide variety of customizable weather graphics that use color and symbols to depict forecasts for conditions such as ceilings, visibilities, wind, and temperature. You also can view specific forecast maps for potentially hazardous weather, such as turbulence, thunderstorms, and icing, and conditions contained in AIRMETs. Refer to more than one forecast graphic and overlay several data sources on a single forecast chart to gain a comprehensive understanding of the conditions that might affect your proposed flight. For example, if the winds forecast shows that strong winds

SECTION C ■ Graphic Weather Products

are predicted at your planned enroute altitude, you might suspect that turbulence will be a factor during your proposed flight. Refer to the turbulence forecast to confirm this prediction. [Figure 7-30]

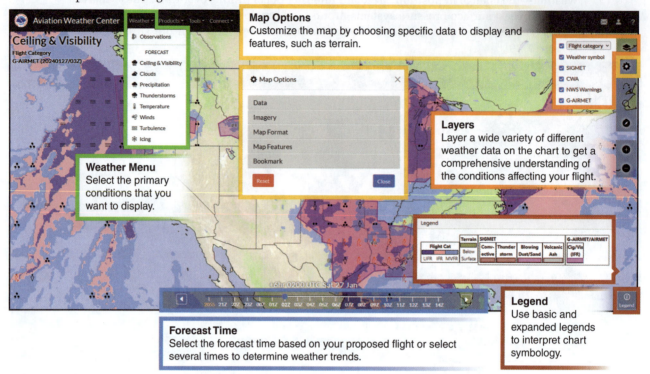

Figure 7-30. During flight planning, take advantage of all the features of AWC forecasts to improve your situational awareness of weather conditions and enhance your flight safety.

CEILING AND VISIBILITY

The ceiling and visibility forecast is designed to help you plan flights to avoid areas of low visibilities and ceilings. You can select a time up to 18 hours in the future to depict forecast IFR and marginal VFR conditions. A U.S. low-level significant weather (SIGWX) chart is another graphic weather product that provides similar information and shows forecast areas of turbulence and freezing levels. [Figure 7-31]

TEMPERATURE

The temperature forecast presents temperature at specified times, altitudes, and locations. On the AWC interactive display, you select the altitude and time frame for the forecast. The temperature is shown in degrees Celsius. Temperature affects your airplane's performance and influences weather conditions. In addition, if you are an instrument-rated pilot, the temperature forecast alerts you to areas of below freezing temperatures, which can lead to icing when flying in IFR conditions. Temperature might not appear on the chart depending on the altitude and location you select—temperature forecasts are not issued for altitudes within 2,500 feet of the surface. [Figure 7-32]

WINDS

The winds forecast presents wind direction, wind speed, and temperature at specified times, altitudes, and locations. You can use this information to visualize the wind at your proposed enroute altitude. Winds aloft affect your heading and groundspeed on a cross-country flight and strong winds can alert you to the potential for turbulence. As on other charts, the forecast wind direction (true) is depicted by a stem (line) pointed in the direction from which the wind is blowing with pennants and barbs to indicate the wind speed. Wind might not appear on the chart depending on the altitude and location you select—wind forecasts are not issued for altitudes within 1,500 feet of a location's elevation. [Figure 7-33]

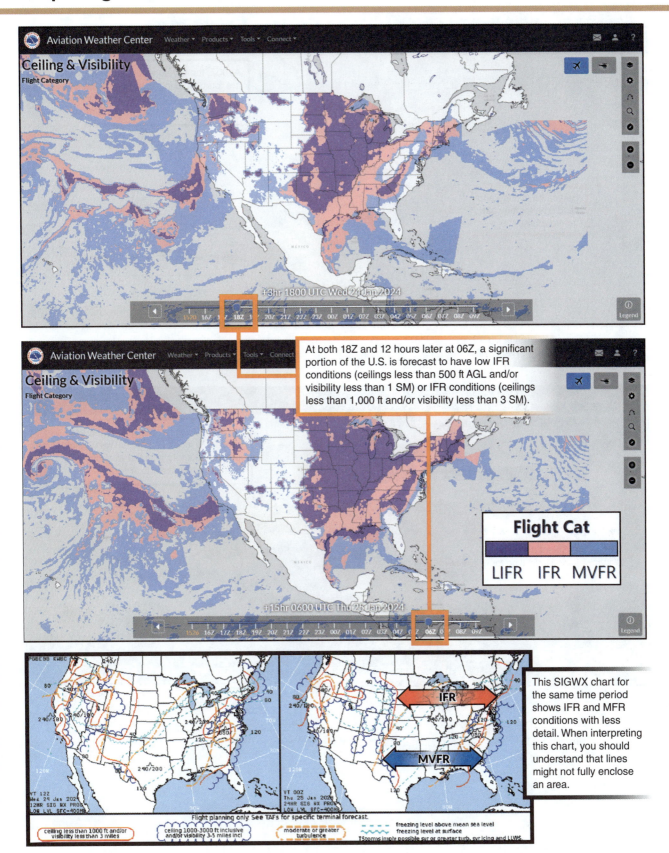

At both 18Z and 12 hours later at 06Z, a significant portion of the U.S. is forecast to have low IFR conditions (ceilings less than 500 ft AGL and/or visibility less than 1 SM) or IFR conditions (ceilings less than 1,000 ft and/or visibility less than 3 SM).

This SIGWX chart for the same time period shows IFR and MFR conditions with less detail. When interpreting this chart, you should understand that lines might not fully enclose an area.

Figure 7-31. Color coding alerts you to areas of low IFR, IFR, and MVFR conditions on the ceiling and visibility forecast. Red and blue outlines indicate IFR and MVFR conditions for 12 and 24 hours in the future on the SIGWX chart.

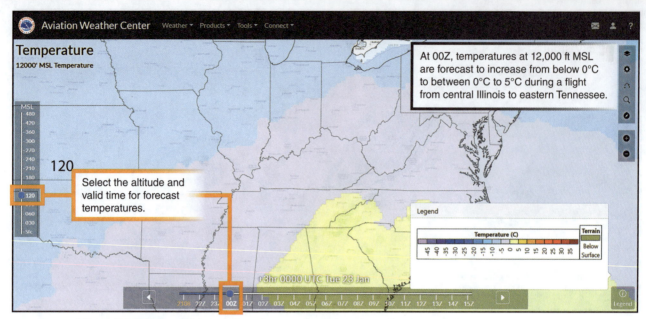

Figure 7-32. Using the temperature forecast in conjunction with other weather forecasts provides greater detail regarding the overall conditions that can affect your flight.

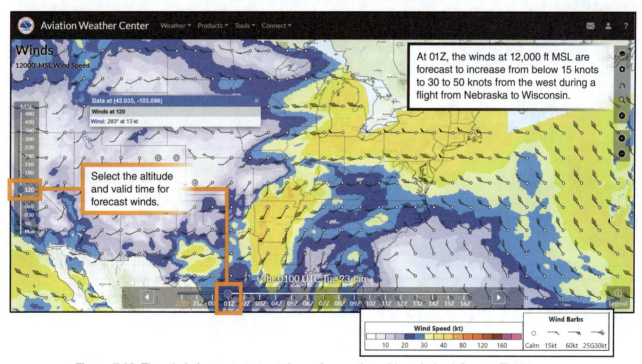

Figure 7-33. The winds forecast starts at the surface and provides winds aloft up to FL480.

SECTION C ■ **Graphic Weather Products**

TURBULENCE

The **turbulence forecast** enables you to select a time up to 18 hours in the future and choose a specific altitude to determine if turbulence will affect your proposed flight. Color-coding shows areas of forecast turbulence and additional layers show PIREPs, SIGMETs, G-AIRMETs, and other turbulence advisories and warnings. [Figure 7-34]

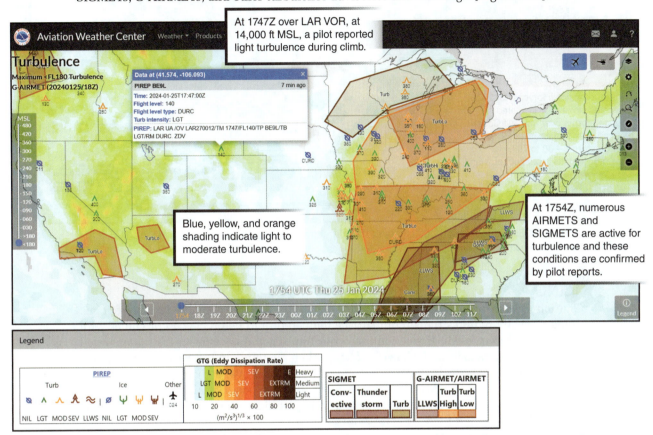

Figure 7-34. You can add symbols for PIREPs as a layer on the chart. Select a symbol to see the PIREP text.

THUNDERSTORMS

The **thunderstorms forecast** uses color-coding to indicate forecasts for isolated, scattered, or numerous thunderstorm over an 18-hour period. You can layer SIGMETs and other warnings and thunderstorm advisories on this chart. The **SPC convective outlook** issued by the NWS Storm Prediction Center can be overlaid on the chart. Convective outlooks are issued in panels over an 8-day period. For example, the Day 1 panel depicts predicted general and severe thunderstorm activity and is issued 5 times daily, with the first issuance valid beginning at 1200Z and all 5 issuances valid until 1200Z the following day.

The NWS **traffic flow management convective forecast (TCF)** is another source of data that you can be view with the thunderstorm forecast. The TCF is a high confidence graphical representation of forecasted convection meeting specific criteria of coverage, intensity, and echo top height. Be sure to refer to the legend when interpreting the thunderstorm forecast because you can overlay several products with different colors and meanings. [Figure 7-35]

SECTION C ■ **Graphic Weather Products**

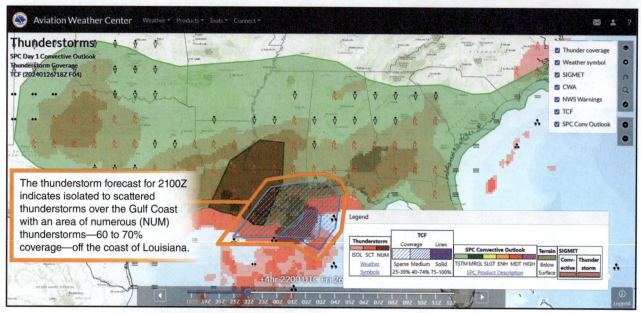

Figure 7-35. In this example, all layers of thunderstorm data are displayed including the TCF and SPC convective outlook.

ICING

The **icing forecast** provides data to help you avoid areas with the potential for aircraft icing. Icing normally requires the aircraft surface temperature to be 0°C (32°F) or colder with visible moisture present. If you continue your training and become an instrument-rated pilot, you must be alert to the hazard of icing when operating in IFR weather conditions. The freezing level is the lowest altitude in the atmosphere over a given location at which the air temperature reaches 0°C. Icing is most likely to occur at altitudes with below-freezing temperatures. You can use freezing-level graphics to help assess the risk of icing when planning an IFR flight. [Figure 7-36]

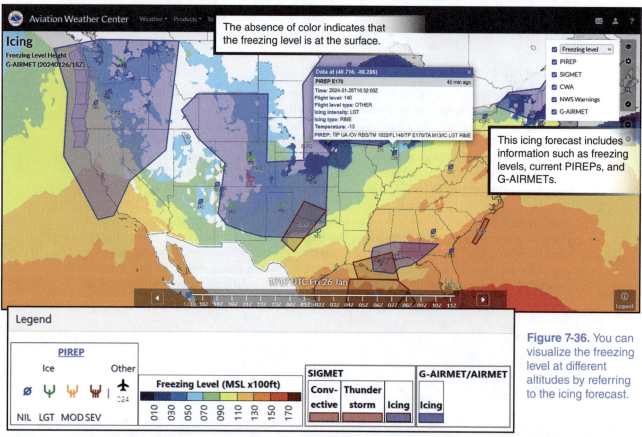

Figure 7-36. You can visualize the freezing level at different altitudes by referring to the icing forecast.

G-AIRMETS

The graphical AIRMET (**G-AIRMET**) is a decision-making tool that depicts various aviation weather hazards, including IFR ceilings and visibility, mountain obscuration, icing, freezing levels, turbulence, low-level wind shear, and strong surface winds. G-AIRMETs that coincide with the issuance of textual AIRMETs identify hazardous weather in space and time more precisely than textual products, enabling you to maintain high safety margins while flying more efficient routes. [Figure 7-37]

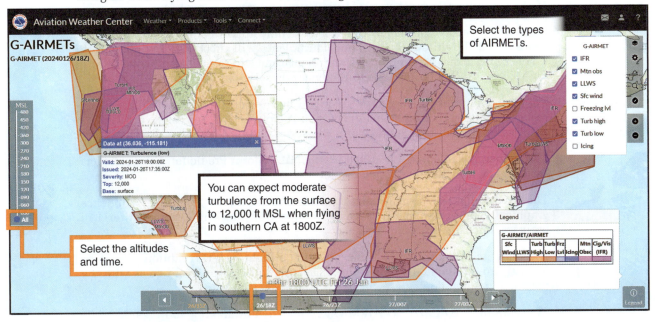

Figure 7-37. G-AIRMETs depict a variety of weather conditions, including areas of IFR conditions, mountain obscuration, turbulence, freezing levels, and icing.

SUMMARY CHECKLIST

✓ Weather surveillance radar—1988 Doppler (WSR-88D), also known as next generation weather radar (NEXRAD) continuously generates weather radar observations.

✓ Radar image colors represent the reflectivity of precipitation targets in decibels (dBZ).

✓ Use visible images from satellites to determine the presence of clouds and the cloud shape and texture. Satellite infrared (IR) photos depict the heat radiation emitted by cloud tops and the earth's surface. Use water vapor satellite imagery to determine the locations and movements of weather systems, jet streams, and thunderstorms.

✓ Graphical depictions of surface observations are displayed as color-coded station models that depict weather flying categories (IFR, LIFR, VFR, and MVFR) and provide a summary of METAR data using text and symbols.

✓ Standard chart symbols on the surface analysis chart depict areas of equal pressure, the positions of highs, lows, ridges, and troughs, the location and type of fronts.

✓ Isobars spaced at intervals of 4 millibars (mb) are depicted on surface analysis charts as solid lines, each labeled with the pressure.

✓ Standardized symbols on weather graphics are used to portray precipitation types, turbulence, and icing.

✓ The cloud forecast provides cloud coverage, bases, layers, and tops and depicts AIRMETs for mountain obscuration and icing.

✓ The surface forecast provides visibility, weather phenomena, and winds (including wind gusts) with depictions of the boundaries of AIRMETs for IFR conditions and sustained surface winds of 30 knots or more.

SECTION C ■ Graphic Weather Products

✓ Surface prognostic (prog) charts depict the forecast for the progression of fronts, and low and high pressure centers over several days.

✓ The ceiling and visibility forecast helps you plan flights to avoid areas of low visibilities and ceilings. You can select a time up to 18 hours in the future to depict forecast LIFR, IFR, and MVFR conditions.

✓ A U.S. low-level significant weather (SIGWX) chart depicts forecast areas of IFR and MVFR conditions, forecast areas of turbulence, and freezing levels.

✓ The turbulence forecast enables you to select a time up to 18 hours in the future and choose a specific altitude to determine if turbulence will affect your proposed flight. Color-coding shows areas of forecast turbulence and additional layers show PIREPs, SIGMETs, G-AIRMETs, and other turbulence advisories and warnings.

✓ The thunderstorms forecast uses color-coding to indicate forecasts for isolated, scattered, or numerous thunderstorm over an 18-hour period. You can layer SIGMETs and other warnings and thunderstorm advisories on this chart.

✓ The SPC convective outlook issued by the NWS Storm Prediction Center depicts predicted general and severe thunderstorm activity.

✓ The NWS traffic flow management convective forecast (TCF) is a high confidence graphical representation of forecasted convection meeting specific criteria of coverage, intensity, and echo top height.

✓ If you are an instrument-rated pilot planning a flight in IFR conditions, the icing forecast provides data, including freezing levels to help you avoid areas with the potential for aircraft icing.

✓ The graphical AIRMET (G-AIRMET) is a decision-making tool that depicts various aviation weather hazards, including IFR ceilings and visibility, mountain obscuration, icing, freezing levels, turbulence, low-level wind shear, and strong surface winds.

KEY TERMS

Radar Observations	Temperature Forecast
Satellite Imagery	Winds Forecast
Graphical Depictions of Surface Observations	Turbulence Forecast
Station Models	Thunderstorms Forecast
Surface Analysis Chart	SPC Convective Outlook
Clouds Forecast	Traffic Flow Management Convective Forecast (TCF)
Surface Forecast	Icing Forecast
Surface Prognostic (Prog) Chart	G-AIRMET
Ceiling and Visibility Forecast	
U.S. Low-Level Significant Weather (SIGWX) Chart	

SECTION C ■ **Graphic Weather Products**

QUESTIONS

Refer to the radar image to answer question 1.

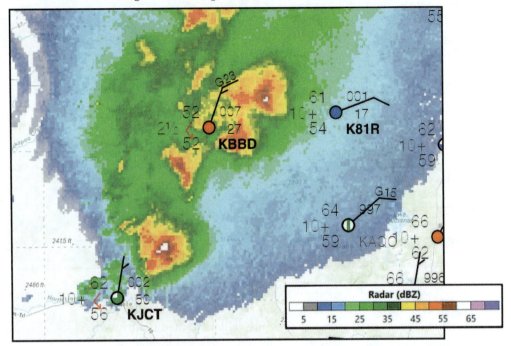

1. Which of these precipitation conditions exists?
   A. Very light precipitation over K81R and KBBD
   B. Moderate to heavy precipitation west of K81R and KBBD
   C. Moderate to heavy precipitation northeast of KBBD and KJCT

2. Select the true statement regarding satellite imagery.
   A. IR and water vapor imagery are only available during daylight hours.
   B. In IR images, clouds appear white and land and water surfaces appear in shades of gray or black.
   C. You can use water vapor imagery to determine the movements of weather systems, jet streams, and thunderstorms.

3. What items are shown on a surface analysis chart?
   A. Areas of VFR, MVFR, IFR, and LIFR
   B. Forecast surface winds and temperatures at selected reported stations
   C. Positions of highs, lows, ridges, and troughs and the location and type of fronts

SECTION C ■ Graphic Weather Products

Refer to the surface analysis chart to answer question 4.

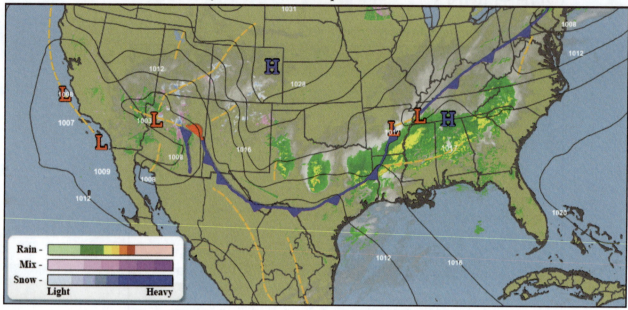

4. Which weather conditions apply to portions of Louisiana, Mississippi, and Alabama?
    A. Heavy precipitation
    B. A stationary front with low pressure centers located to the north
    C. Light to moderate rain extending north of an elongated area of low pressure

Refer to the clouds forecast to answer question 5.

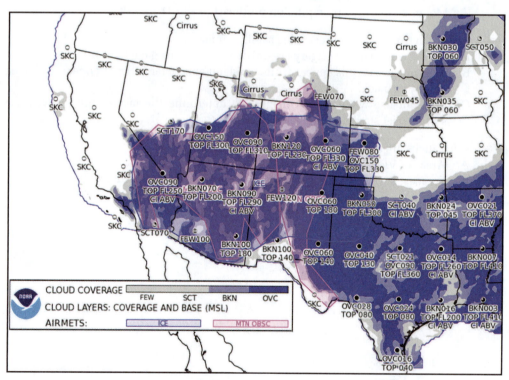

5. Which forecast conditions apply?
    A. AIRMETs for icing and mountain obscuration over Arizona
    B. Overcast skies with cloud bases at 4,000 feet MSL over central Oklahoma
    C. Overcast skies with cloud bases at 9,000 feet AGL over southwest California and Nevada

Refer to the station models to answer questions 6 through 12.

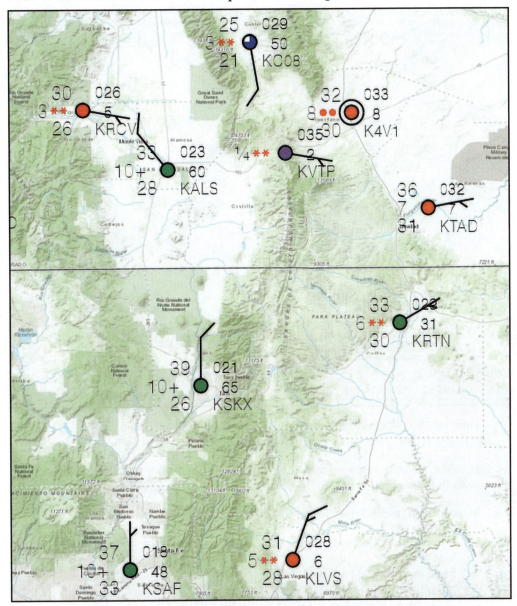

6. Which airport(s) are reporting calm wind conditions?

7. Which airport(s) have ceilings between 1,000 to 3,000 feet and/or visibility 3 to 5 miles inclusive?

8. What is the wind direction and speed at KALS?
   A. 300° at 5 knots
   B. 320° at 10 knots
   C. 040° at 10 knots

9. What weather phenomena is occurring at KRTN and KVTP?

10. What is the altimeter setting for KLVS?

11. What is the ceiling and visibility at KC08?

12. Which airport(s) are reporting VFR conditions?

Refer to the surface forecast to answer question 13 through 16.

13. Describe the forecast weather conditions for the southeast coast of Texas in terms of weather phenomena, visibility, winds, and advisories.

14. In which state are surface winds forecast in excess of 30 knots that are not associated with thunderstorms?

15. What type of precipitation is forecast in Colorado?

16. What is the approximate primary direction and speed of the forecast surface winds in Oklahoma?
    A. From the northeast at 20 knots
    B. From the northeast at 10 to 15 knots with gusts to 20 knots.
    C. From the southwest at 10 to 15 knots with gusts to 20 knots.

Refer to the accompanying low-level significant weather (SIGWX) chart to answer questions 18 through 21.

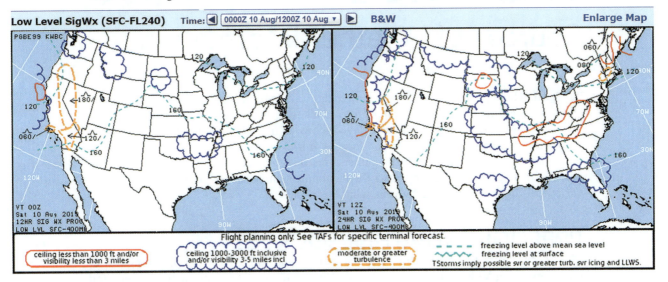

17. The low-level significant weather chart is for use from the surface up to what altitude?

18. What condition is forecast for southern California at 0000Z?
    A. Freezing level at the surface
    B. Moderate or greater turbulence below 12,000 feet MSL
    C. Ceilings less than 1,000 feet and/or visibility less than 3 miles

19. Which forecast is correct for Kansas?
    A. VFR at 0000Z and IFR at 1200Z
    B. VFR at 0000Z and MVFR at 1200Z
    C. MVFR at 0000Z and VFR at 1200Z

20. At what altitude is the forecast freezing level in Colorado at 1200Z?

21. Select the true statement regarding graphic weather products.
    A. G-AIRMETs display the lateral boundaries of weather hazards, such as icing, thunderstorms, and sandstorms.
    B. Surface prog charts enable you to obtain an overview of the progression of pressure systems, fronts, and precipitation over several days.
    C. The SPC convective outlook depicts radar images of current thunderstorm precipitation and a forecast for convective activity for the next 5 hours.

SECTION C ■ **Graphic Weather Products**

SECTION D
Sources of Weather Information

Part of being a safe pilot is keeping up to date on the latest weather developments and maintaining the ability to adjust your plans to mitigate risks. The weather briefing process usually begins several days before your flight, when you look at mass-disseminated weather information and form an initial opinion regarding the feasibility of your flight. As the flight gets closer, gather more detailed information about the weather along your route. Then, during your flight, update your weather information using the various in-flight sources to determine what lies ahead.

PREFLIGHT WEATHER SOURCES

The FARs require that you obtain weather reports and forecasts when operating IFR or when operating VFR away from an airport. However, you should know the current and forecast weather conditions whenever you fly. Preflight weather sources include Flight Service, the NOAA/NWS Aviation Weather Center, and private sources. All of these sources offer a variety of weather observations and forecasts for a large snapshot of an area down to detailed information about a specific airport. [Figure 7-38]

Figure 7-38. Flight Service, National Weather Service, Aviation Weather Center, and even certain EFBs, such as ForeFlight, are valuable sources of weather information.

FLIGHT SERVICE

Flight Service is a well-recognized FAA source of preflight weather information. You can call for a phone briefing for your specific flight at **1-800-WX-BRIEF** or obtain an online weather briefing and other weather information at **1800wxbrief.com**. After you register and create a user account online, you will gain access to a variety of pilot briefing and flight planning tools.

PREFLIGHT WEATHER BRIEFING

You can choose from three types of briefings—standard, abbreviated, or outlook. Obtain a **standard briefing**—the most complete briefing—when you are planning a flight and have not obtained preliminary weather or a previous briefing. Obtaining a standard briefing assumes you have do not already have a detailed comprehensive weather picture for your flight. When you need only one or two specific items or would like to update weather information from a previous briefing or other weather sources, obtain an **abbreviated briefing**. If your proposed departure time is six or more hours in the future, obtain an **outlook briefing.** An outlook briefing provides forecast information that helps you initially determine the feasibility of your flight. Always obtain a standard briefing closer to your departure time.

The online briefing feature of Flight Service is an FAA-approved briefing for which official records are kept. To obtain a preflight weather briefing online, fill out a preliminary flight plan at 1800wxbrief.com. [Figure 7-39] By registering for the Adverse Condition Alerting Service (ACAS), you are notified by text or email when a new adverse condition arises along your planned route of flight after you have received a briefing and filed a flight plan online.

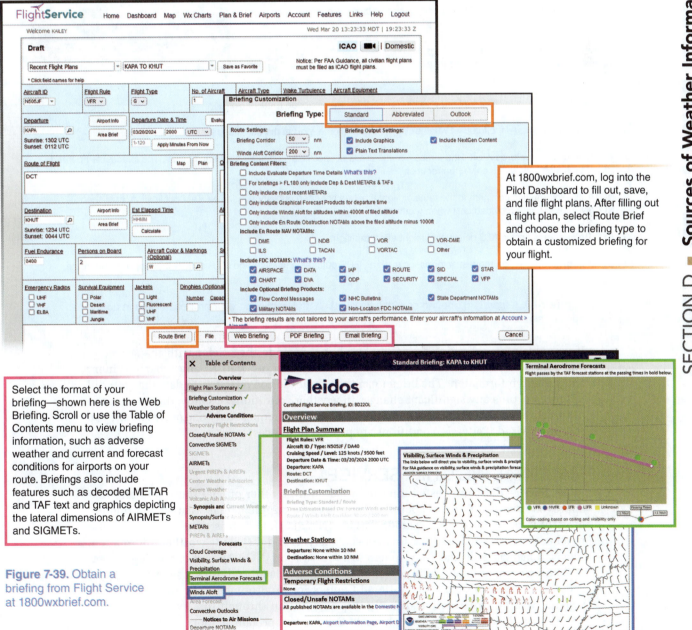

At 1800wxbrief.com, log into the Pilot Dashboard to fill out, save, and file flight plans. After filling out a flight plan, select Route Brief and choose the briefing type to obtain a customized briefing for your flight.

Select the format of your briefing—shown here is the Web Briefing. Scroll or use the Table of Contents menu to view briefing information, such as adverse weather and current and forecast conditions for airports on your route. Briefings also include features such as decoded METAR and TAF text and graphics depicting the lateral dimensions of AIRMETs and SIGMETs.

Figure 7-39. Obtain a briefing from Flight Service at 1800wxbrief.com.

Courtesy of Leidos

If you prefer to call Flight Service, perform a self-brief first to become familiar with the conditions that apply to your route of flight. You will better understand the information provided by the briefer and be able to ask more effective questions if you self-brief. When you call for a briefing, an automated phone system asks which state you are departing from and then routes your call, workload permitting, to a briefer who specializes in the weather for that area. When you reach the briefer, identify yourself as a pilot and supply the following information: type of flight planned (VFR or IFR); aircraft number or your name; aircraft type; departure airport; route of flight; destination; flight altitude(s); estimated time of departure (ETD); and estimated time enroute (ETE). Whether you obtain an online briefing or speak to a briefer by phone, the basic elements of the briefing are the same. [Figure 7-40]

1. **Adverse Conditions**—information that might influence your decision to cancel or alter the route of flight, such as hazardous weather or airport closings.

2. **VFR Flight Not Recommended**—the weather for the route of flight is below VFR minimums or, in the briefer's judgment, it is doubtful the flight can be made under VFR conditions due to the forecast weather. Although the final decision to conduct the flight rests with you, this advisory should be taken seriously.

3. **Synopsis**—a broad overview of fronts and major weather systems that affect the proposed flight.

4. **Current Conditions**—the current ceilings, visibility, winds, and temperatures. If the departure time is more than 2 hours away, current conditions are not included in the briefing.

5. **Enroute Forecast**—a summary of the weather forecast for the proposed route of flight.

6. **Destination Forecast**—the forecast weather for the destination airport at the estimated time of arrival (ETA).

7. **Winds and Temperatures Aloft**—the forecast of the winds at specific altitudes for the route of flight. The forecast temperature information aloft is provided only upon request.

8. **NOTAMs**—information pertinent to the route of flight that has not been published in the NOTAM publication. Published NOTAM information is provided only when requested.

9. **ATC Delays**—any known ATC delays that might affect the flight.

10. **Other Information**—upon request, information such as GPS outages and MOA and MTR activity that apply to your route.

Figure 7-40. A standard briefing provides information that applies to your proposed flight in a specific sequence.

If you obtain an abbreviated briefing by phone, provide the briefer with the source of and time you received the prior weather information, as well as any other pertinent background information. The briefer can then limit the conversation to data that you did not receive, plus any significant changes in weather conditions. Usually, the sequence of information follows that of the standard briefing. If you request only one or two items, you still are advised if adverse conditions are present or forecast.

NATIONAL WEATHER SERVICE

Flight Service disseminates weather information from other official sources in an FAA-approved briefing. You can obtain preliminary and supplemental weather information directly from the National Weather Service (NWS). The NWS is an essential part of a number of interrelated agencies that collect, analyze, and distribute weather information. The National Oceanic and Atmospheric Administration (NOAA), part of the Department of Commerce, coordinates U.S. government weather-related activities. Under NOAA, the National Weather Service and National Centers for Environmental Prediction (NCEP), provide weather information directly to pilots online through the Aviation Weather Center (AWC), or indirectly through Flight Service and private vendors. [Figure 7-41]

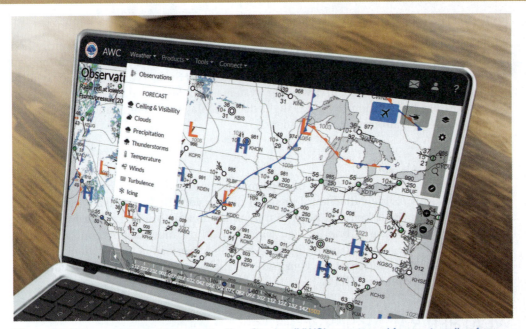

Figure 7-41. You can obtain National Weather Service (NWS) reports and forecasts online from the Aviation Weather Center.

PRIVATE INDUSTRY SOURCES

A wide variety of companies provide weather information to the aviation industry. Some of these services are integrated into flight planning apps or electronic flight bags (EFBs). You can find dozens of websites that provide weather information and many of these sites are directed towards aviation. However, you must understand that weather information received from non-FAA or non-NWS sites might not be current, accurate, or relevant. Therefore, ensure that you are using weather information from reliable and trusted sources. Many private industry sources, such as Jeppesen, meet FAA requirements for weather information and conform to ICAO requirements. By using an approved EFB such as ForeFlight, you can access NWS and NOAA weather graphics, overlay weather information on your route, and generate a weather briefing specific to your flight, similar to Flight Service. [Figure 7-42]

Figure 7-42. ForeFlight enables you to view a general synopsis of weather and specific conditions for airports along your route through a variety of formats.

SECTION D ■ Sources of Weather Information

IN-FLIGHT WEATHER SOURCES

Because forecasting is an inexact science, weather conditions can change rapidly and unexpectedly over a few hours. This uncertainty requires updating weather information while in flight. You can obtain weather updates from Flight Service, air route traffic control centers (ARTCCs), and automated weather systems. In addition, data link services can deliver weather information to digital flight deck displays or a tablet and your airplane might have its own airborne weather radar.

FLIGHT SERVICE

Although some pilots might think of it primarily as a preflight planning tool, Flight Service can also provide valuable information during your flight. Typically, you should contact Flight Service when you need to update a previous briefing. After establishing contact, specify the type of briefing you want as well as information about your flight similar to what you supply to a preflight briefer. Flight Service is an excellent source of current weather information along your route of flight because it is a central collection and distribution point for PIREPs.

WEATHER RADAR SERVICES

The NWS operates a nationwide network of weather radar sites that provides real-time information for textual and graphic weather products, and that also furnishes Flight Service with data for in-flight advisories. Because weather radar can detect coverage, intensity, and movement of precipitation, a Flight Service specialist might be able to provide you with suggested routing around areas of hazardous weather. It is important to remember, however, that simply avoiding areas highlighted by weather radar does not guarantee clear weather conditions.

DISCOVERY

Forecasting Tornadoes

Tornadoes carry some of the most devastating forces available in nature. The amount of energy in an F5 tornado's 250 mph wind is four times that of a 125 mph wind encountered in a hurricane. It is little wonder that tornadoes pick up houses, buildings, and large trucks and scatter them like toothpicks.

How do they measure such wind speeds? Even if a tornado should pass directly over traditional measuring equipment, it is unlikely the equipment could survive such an encounter. Professor T. Fujita of the University of Chicago, who developed the widely-used Fujita scale for measuring tornado intensity, originally determined wind speeds by analyzing film of tornadoes and calculating the velocity of the flying debris.

Now, doppler radar accurately measures wind speed inside tornadoes. Forecasters not only can determine the location, speed and direction of movement of a tornado, but also can predict how devastating it is likely to be, and warn people accordingly. The National Weather Service's WSR-88D, or NEXRAD, radar system, together with new geostationary operational environmental satellites (GOES) and the automated surface observation system (ASOS) network, is expected to save many lives by pinpointing the time and place where severe weather is likely to strike.

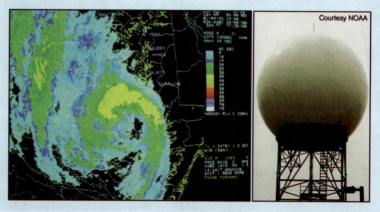

Courtesy NOAA

CENTER WEATHER ADVISORIES

A **center weather advisory (CWA)** is an unscheduled weather advisory issued by an ARTCC to alert pilots of existing or anticipated adverse weather conditions within the next two hours. A CWA can be initiated when a SIGMET has not been issued but, based on current PIREPs, conditions meet those criteria. Additionally, a CWA can be issued to supplement an existing in-flight advisory as well as any conditions that currently or will soon adversely affect the safe flow of traffic within the ARTCC area of responsibility.

ARTCCs broadcast CWAs as well as SIGMETs, convective SIGMETs, and AWWs once on all but emergency frequencies when any part of the area described is within 150 miles of the airspace under the ARTCC jurisdiction. In terminal areas, local control and approach control might limit these broadcasts to weather occurring within 50 miles of the airspace under their jurisdiction. These broadcasts contain the advisory identification and a brief description of the weather activity and general area affected.

AUTOMATED WEATHER REPORTING SYSTEMS

Although human weather observers contribute to weather reports at larger airports, many surface weather observation tasks are automated. The FAA operates the automated weather observing systems and the NWS operates the automated surface observing systems. Although their complexity and capabilities vary, these systems are primarily designed to provide weather information for aviation.

AUTOMATED WEATHER OBSERVING SYSTEM

The **automated weather observing system (AWOS)** uses various sensors, a voice synthesizer, and a radio transmitter to provide real-time weather data. There are four types of AWOS—AWOS-A only reports the altimeter setting; AWOS-1 also measures and reports wind speed, direction and gusts, temperature, and dewpoint; AWOS-2 provides visibility information in addition to everything reported by an AWOS-1; and the most capable system, AWOS-3, also includes cloud and ceiling data.

AUTOMATED SURFACE OBSERVING SYSTEM

The network of **automated surface observing system (ASOS)** monitoring stations located at airports across the country is a joint effort between the FAA and the NWS. ASOS is the primary surface weather observing system in the U.S. [Figure 7-43]

Figure 7-43. ASOS stations like this one are the primary source of U.S. weather observations.

OBTAINING AUTOMATED WEATHER INFORMATION

ASOS/AWOS information is broadcast over discrete VHF frequencies or the voice portions of local navaids. You can normally receive these weather report transmissions within 25 miles of the site and up to 10,000 feet AGL. Locations and frequencies are shown on aeronautical charts and listed in the applicable A/FD listing of the Chart Supplement. You can also listen to ASOS/AWOS by telephone using the telephone numbers listed in the A/FD.

At many airports around the U.S. that have both a part-time tower and an automated weather system, the automated weather information is automatically broadcast over the ATIS frequency when the tower is closed.

Human Weather Observations

Both human observers and automated systems have limitations when it comes to determining the complete weather picture. The measurements from AWOS/ASOS systems are completely objective, and in the strictest sense, error free. However, the AWOS/ASOS systems are limited in their ability to deduce an overall view of the weather at various locations in the airport vicinity. A human observer is able to look around the sky and quickly make a subjective judgment as to sky condition, visibility and present weather. However, depending on the complexity of the weather, and on other duties, a human observer can only make one or two observations per hour. An automated system looks at a single part of the sky over a period of time, and averages this data.

According to the FAA, the fixed time, spatial averaging technique used by human observers, and the automated system's fixed location, time averaging technique, yield remarkably similar results. Keep in mind, it is possible for an automated system to report inaccurately if there is a localized condition near the reporting equipment and little or no cloud movement.

DATA LINK WEATHER

Data link weather services are often included with GPS and EFB flight deck or tablet display systems. The weather information displayed using a data link is near real-time but should not be thought of as instantaneous, up-to-date information. Each type of weather display is stamped with the age information, which is typically the time in Zulu when the information was assembled at the ground station. Flight information service-broadcast (FIS-B) is an example of a data-link weather service you can receive if your airplane has ADS-B capability and the proper transceiver. [Figure 7-44]

Courtesy of Garmin

Figure 7-44. FIS-B provides free in-flight weather information over an ADS-B datalink.

Data link weather services provide a variety of broadcast weather and aeronautical information products that can include:

- METARs
- TAFs
- PIREPs
- Winds and temperatures aloft
- NOTAMs
- AIRMETs
- SIGMETs and convective SIGMETs
- Special use airspace (SUA)
- Temporary flight restrictions (TFRs)
- NEXRAD

The NEXRAD system is comprised of a series of Weather Surveillance Radar–1988 Doppler (WSR-88D) sites installed throughout the United States, as well as selected overseas sites. The NEXRAD radar image that you view on your flight deck display is not real time and can be up to 5 minutes old. Even small-time differences between the age indicator and actual conditions can be critical to flight safety, especially when considering fast-moving weather hazards, quickly developing weather scenarios, and/or fast-moving aircraft. At no time should you use the images as storm penetrating radar to navigate through a line of thunderstorms.

Note: Use caution with regard to relying on data link weather products, especially those related to thunderstorms. The NTSB published a safety alert after investigating fatal accidents in which in-flight deck NEXRAD mosaic imagery was available to pilots operating near quickly-developing and fast-moving convective weather. In one of these accidents, the images were from 6 to 8 minutes old. The NTSB warned that NEXRAD data can, in rare cases, be as much as 20 minutes older than the age indicated on the flight deck display. During that time delay, dramatic changes in the thunderstorm location and intensity can occur.

SECTION D ■ Sources of Weather Information

AIRBORNE WEATHER RADAR

Airborne weather radar operates under the same principle as ground-based radar. The directional antenna, normally behind the airplane's fiberglass nose cone, transmits pulses of energy out ahead of the aircraft. The signal is reflected by water and ice, and is picked up by the antenna from which it was transmitted. The bearing and distance of the weather is plotted, normally on a color display. These displays are useful for indicating different degrees of precipitation severity. A radar unit might use green to indicate light precipitation, yellow for moderate rain, and red to depict heavy storms. [Figure 7-45]

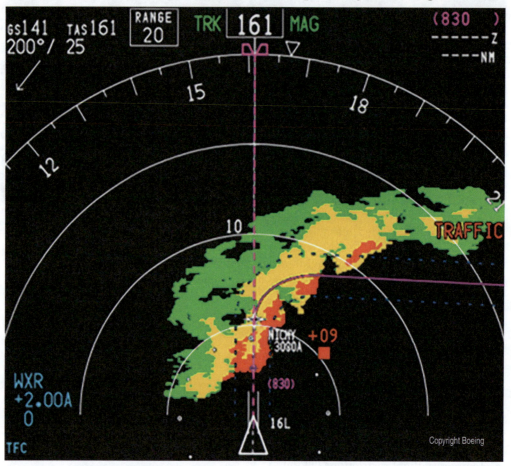

Figure 7-45. Airborne weather radar allows pilots to see and avoid many thunderstorms which do not appear on ground-based radar.

Aircraft radar generally use one of two frequency ranges—X-band and C-band. X-band systems, which are more common in general aviation aircraft, transmit on a frequency of 9,333 gigahertz (GHz), which has a wavelength of only 0.03 mm. This extremely short wave is reflected by very small amounts of precipitation. Due to the high amount of reflected energy, X-band systems provide a higher resolution and "see" farther than C-band radars. A disadvantage is that very little energy can pass through one storm to detect another that might be behind the first. The C-band frequency (5.44 GHz) can penetrate farther into a storm, providing a more complete picture of the storm system. This capability makes C-band weather radar systems better for penetration into known areas of precipitation. Consequently, C-band radars are more likely to be found on large commercial aircraft.

Aircraft radar is prone to many of the same limitations as ground-based systems. It cannot detect water vapor, lightning, or wind shear. Training and experience as well as other on-board equipment and ATC radar are important tools for enhancing your mental picture of the weather ahead.

LIGHTNING DETECTION EQUIPMENT

Because lightning is always associated with severe thunderstorms, systems that detect lightning can reliably indicate the active parts of these storms. [Figure 7-46] Lightning detection equipment is more compact, uses substantially less power than radar, and requires less interpretation by the pilot. The equipment is designed to help a pilot completely avoid storm cells. However, lightning detection equipment does not directly indicate areas of heavy precipitation, hail, and wind shear and might not provide as accurate information as radar regarding these hazards.

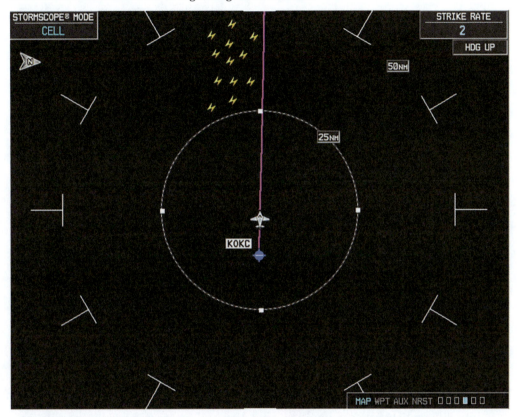

Figure 7-46. Lightning detection equipment provides many of the benefits of airborne radar at much lower cost, smaller size, and lower power consumption.

SUMMARY CHECKLIST

✓ Preflight weather sources include Flight Service, the NOAA/NWS Aviation Weather Center, and private sources, such as the electronic flight bag (EFB) ForeFlight.

✓ You can call Flight Service for a phone briefing for your specific flight at 1-800-WX-BRIEF, obtain an online weather briefing at 1800wxbrief.com, and sign up for updates through the Adverse Condition Altering Service (ACAS).

✓ Obtain a standard briefing when you are planning a flight and have not obtained preliminary weather or a previous briefing.

✓ Obtain an abbreviated briefing when you need only one or two specific items or to update weather information from a previous briefing or other weather sources.

✓ If your proposed departure time is six or more hours in the future, obtain an outlook briefing, which provides forecast information that helps you make an initial judgment about the feasibility of your flight.

SECTION D ■ **Sources of Weather Information**

✓ When you call a Flight Service briefer at 1-800-WX-BRIEF, supply the following information: type of flight planned (VFR or IFR), aircraft number or your name, aircraft type, departure airport, route of flight, destination, flight altitude(s), estimated time of departure (ETD), and estimated time enroute (ETE).

✓ The basic elements of a briefing include: adverse conditions, VFR flight not recommended (if applicable), synopsis, current conditions, enroute forecast, destination forecast, winds and temperatures aloft, NOTAMs, ATC delays, and other information.

✓ Obtain preliminary and supplemental weather information directly from the National Weather Service (NWS) online at the Aviation Weather Center (AWC) site.

✓ By using an approved EFB such as ForeFlight, you can access NWS and NOAA weather graphics, overlay weather information on your route, and generate a weather briefing specific to your flight, similar to Flight Service.

✓ When you need to update a previous briefing during flight, contact Flight Service for current weather information along your route, including PIREPs.

✓ Weather radar sites operated by the National Weather Service provide real-time information for textual and graphic weather products.

✓ A center weather advisory (CWA) is an unscheduled weather advisory issued by an ARTCC to alert pilots of existing or anticipated adverse weather conditions within the next two hours.

✓ Automated weather observation systems currently in use are the automated weather observing system (AWOS) and the automated surface observing system (ASOS).

✓ The information broadcast by AWOS depends on the type of system. AWOS-3 can report the altimeter setting, wind speed, direction and gusts, temperature, dew point, visibility, and cloud and ceiling data.

✓ ASOS measures and reports variable cloud height, variable visibility, and rapid pressure changes, as well as precipitation type, intensity, accumulation, and beginning and ending times.

✓ ASOS is capable of measuring wind shifts and peak winds and some ASOS stations can differentiate between liquid precipitation and frozen or freezing precipitation.

✓ You can obtain weather data from automated weather observation stations by tuning to discrete VHF frequencies or the voice portions of local navaids, or by calling the telephone number for the station.

✓ Data link weather services are often included with GPS and EFB flight deck or tablet display systems. The weather information displayed using a data link is near real-time but should not be thought of as instantaneous, up-to-date information.

✓ The NEXRAD radar image that you view on your flight deck display can be up to 5 minutes old. Even small-time differences between the age indicator and actual conditions can be critical to flight safety. At no time should you use the images as storm penetrating radar to navigate through a line of thunderstorms.

✓ Airborne radar or lightning detection equipment usually can locate areas of hazardous weather ahead of your aircraft with greater reliability than ground-based radar.

KEY TERMS

Flight Service

1-800-WX-BRIEF

1800wxbrief.com

Standard Briefing

Abbreviated Briefing

Outlook Briefing

Aviation Weather Center (AWC)

Center Weather Advisory (CWA)

Automated Weather Observing System (AWOS)

Automated Surface Observing System (ASOS)

Data Link Weather

Airborne Weather Radar

Lightning Detection Equipment

QUESTIONS

1. What information should you provide to a preflight weather briefer?

2. What are the three types of weather briefings? Explain the circumstances under which you would obtain each type of briefing.

3. Select the true statement regarding surface weather observing systems.
   A. ASOS is capable of measuring wind shifts and peak winds.
   B. ASOS uses surface observation trends to forecast conditions for the airport.
   C. AWOS can determine the difference between liquid precipitation and frozen or freezing precipitation.

4. True/False. Data link weather is instantaneous, up-to-date weather information.

5. What is a precaution that you should consider when using NEXRAD information on a flight deck display?

SECTION D ■ Sources of Weather Information

SECTION D ■ **Sources of Weather Information**

PART IV

Performance and Navigation

SUCCESS FOUR FLIGHTS THURSDAY MORNING ALL AGAINST TWENTY ONE MILE WIND STARTED FROM LEVEL WITH ENGINE POWER ALONE AVERAGE SPEED THROUGH AIR THIRTY ONE MILES LONGEST 59 SECONDS INFORM PRESS HOME CHRISTMAS

OREVELLE WRIGHT

— Telegram message received by Orville Wright's father. Note the telegrapher's misspelling of Orville's name.

PART IV

The preflight preparations for a space mission are complex and detailed. Without dozens of technicians and engineers calculating performance and navigation data, successful spaceflight would be virtually impossible. While the preflight preparations of general aviation pilots are less visible, they are no less important. Part IV introduces techniques that not only reduce your workload in the air, but also result in a safer, more enjoyable flight. *Airplane Performance* will show you how to get the most out of your airplane, whether that means the most speed or the most economy, the shortest takeoffs or the longest range. In *Navigation* you will learn to find your way from place to place using the latest technology and explore fundamental navigation techniques that will never go out of date.

CHAPTER 8

Airplane Performance

SECTION A
Predicting Performance

To describe the effectiveness of an aircraft in the jobs it was designed to accomplish, we use the term **performance**. Different designs emphasize speed, maneuverability, load-carrying capability, or the ability to handle short, rough fields. Aircraft designers usually try to accentuate specific performance characteristics at the expense of others. For instance, a competitive aerobatic plane is extremely maneuverable, but has virtually no payload capability other than the pilot. On the other hand, a design optimized for cross-country cargo hauling will sacrifice maneuverability for payload, stability, and economy of operation. [Figure 8-1]

Aerobatic Aircraft – This aerobatic airplane is designed for maneuverability and agility.

Copyright Corel

Fighter Aircraft – The F-16 was designed for vertical penetration, maneuverability, and quick acceleration.

Courtesy USAF, Senior Airman Greg L. Davis

Airliner – The airlines require their aircraft to have the ability to haul a large payload economically and in all weather conditions at a moderate speed.

Agricultural Aircraft – Aircraft designed for agricultural operations must be capable of flying at low altitudes and carrying large loads of chemicals.

General Aviation Trainer – A training aircraft is engineered to be relatively stable with average speed and endurance characteristics.

Figure 8-1. Different kinds of performance are emphasized in the variety of aircraft designs.

Your ability to predict the performance of an airplane is extremely important. It allows you to determine how much runway you need for takeoff, if you can safely clear obstacles in your departure path, how long it will take you to reach a destination, the quantity of fuel required, and how much runway you will need for landing. Aircraft manufacturers provide much of this information in **performance charts**, usually located in the Performance section of the POH.

AIRCRAFT PERFORMANCE AND DESIGN

In developing performance charts, airplane manufacturers make certain assumptions about the condition of the airplane and ability of the pilot. The pilot is expected to follow normal checklist procedures and to perform each of the required tasks correctly and at the appropriate time. Manufacturers also assume the airplane to be in good condition, with a properly tuned engine and all systems operating normally.

With the aid of these assumptions, the manufacturer develops performance data for the airplane based on actual flight tests. Rather than test the airplane under each and every condition shown on the performance charts, manufacturers evaluate specific flight test data and mathematically derive the remaining information. This data is provided for your use in the form of tables and graphs in the POH.

DISCOVERY

Flight Testing and Test Pilots

Flight testing is the process of gathering information that will accurately describe the performance of a particular type of airplane. This information can then be used to predict the capabilities of the aircraft.

Test pilots specialize in many different kinds of flight testing. Experimental test pilots fly research aircraft to gather information that will be used to improve aircraft designs. Engineering test pilots evaluate newly designed and experimental aircraft, determine how well they comply with design standards, and make recommendations for improvements. Production test pilots fly new aircraft as they come off assembly lines to make sure they are airworthy and ready to turn over to customers. Test pilots for the airlines not only check airplanes after major overhauls to be sure they are ready to return to service, but also test new aircraft to make sure they are up to standards before the airline accepts them from the manufacturer.

Experimental test pilot Milt Thompson, shown in the accompanying photo, made 14 flights in the X-15 research airplane. He gathered data on aerodynamics, thermodynamics, rocket propulsion, flight controls, and the physiological aspects of high speed, high altitude flight.

Courtesy of NASA Dryden Flight Research Center

SECTION A ■ **Predicting Performance**

DISCOVERY

Test Pilot Gives Free Lesson!

Here is your chance to learn from veteran experimental test pilot Scott Crossfield as he makes the second-ever dead stick (engine out) landing in the North American F-100.

I called Edwards and declared an emergency. All airborne planes in the vicinity were warned away. I held the ailing F-100 on course, dropping swiftly, lining up for a dead stick landing. I flared out and touched down smoothly. It was in fact one of the best landings I ever made. I then proceeded to violate a cardinal rule of aviation: never try tricks with a compromised airplane. I had already achieved the exceptional, now I would end it with a flourish. I would snake the stricken F-100 right up the ramp and bring it to a stop immediately in front of the NACA hangar…

Courtesy of NASA Dryden Flight Research Center

According to the F-100 handbook, the hydraulic brake system was good for three "cycles" (pumps on the brake) engine out. The F-100 was moving at about 15 mph when I turned up the ramp. I hit the brakes once, twice, three times, the plane slowed but not enough. I hit the brakes a fourth time—and my foot went clear to the floorboards. The hydraulic fluid was exhausted. The F-100 rolled on, straight between the yawning hangar doors!

The NACA hangar was then crowded with expensive research tools the Skyrocket … the X-3, X-4, and X-5. Yet somehow, my plane, refusing to halt, squeezed by them all and bored steadily on toward the side wall of the hangar.

The nose of the F-100 crunched through the corrugated aluminum, punching out an eight-inch steel I-beam. I was lucky.

— Scott Crossfield, *Always Another Dawn*

Even test pilots need to stay within limits—and that includes personal limits as well as those of the aircraft. Although a superb pilot and fully aware of his situation, Scott Crossfield gave in to the urge to show off a little—with nearly catastrophic results. No matter how skilled you might become, your attitude as pilot in command will often determine the safety of the flight.

CHART PRESENTATIONS

To be a well-informed pilot, you need to know how to find and interpret published performance information, as well as how to operate the aircraft within the performance limitations imposed by aircraft design and atmospheric conditions. Keep in mind that all performance charts apply to specific aircraft, and the ones you see in this section are only samples. Because performance data can vary significantly between similar models, or even from one model year to the next, you should only refer to the POH for the particular airplane you intend to fly.

Performance charts generally present their information in either table or graph format. The table format usually contains several notes that require you to make adjustments for various conditions that are not accounted for in the body of the chart. Graph presentations usually incorporate more variables, reducing the required adjustments. To get as close to stated performance as possible, you must follow all of the chart procedures and conditions. [Figure 8-2]

TAKEOFF DISTANCE

CONDITIONS:
Flaps 10°
Full throttle prior to brake release
Paved, dry, level runway
Zero wind

NOTES:
1. Short field technique.
2. Prior to takeoff from fields above 3000 feet elevation, the mixture should be leaned to give maximum RPM in a full throttle, static runup.
3. Decrease distances 10% for each 9 knots headwind. For operation with tailwinds up to 10 knots, increase distances by 10% for each 2 knots.
4. For operation on a dry, grass runway, increase distances by 15% of the "ground roll" figure.

WEIGHT LBS	TAKEOFF SPEED KIAS		PRESSURE ALTITUDE FEET	0° C		10° C		20° C		30° C		40° C	
	LIFT OFF	AT 50 FEET		GROUND ROLL FEET	TOTAL FEET TO CLEAR 50-FOOT OBSTACLE	GROUND ROLL FEET	TOTAL FEET TO CLEAR 50-FOOT OBSTACLE	GROUND ROLL FEET	TOTAL FEET TO CLEAR 50-FOOT OBSTACLE	GROUND ROLL FEET	TOTAL FEET TO CLEAR 50-FOOT OBSTACLE	GROUND ROLL FEET	TOTAL FEET TO CLEAR 50-FOOT OBSTACLE
2400	51	56	S. L.	800	1465	865	1575	925	1690	1000	1815	1070	1950
			1000	880	1610	945	1730	1020	1865	1090	2005	1175	2160
			2000	965	1775	1040	1915	1120	2065	1200	2220	1295	2400
			3000	1060	1965	1145	2125	1235	2300	1330	2485	1430	2690
			4000	1170	2190	1265	2370	1360	2575	1470	2795	1580	3035
			5000	1290	2450	1395	2665	1505	2900	1625	3165	1750	3460
			6000	1430	2760	1545	3020	1665	3300	1805	3625	1945	3995
			7000	1585	3145	1715	3455	1855	3810	2005	4225	—	—
			8000	1760	3620	1910	4020	2065	4485	—	—	—	—

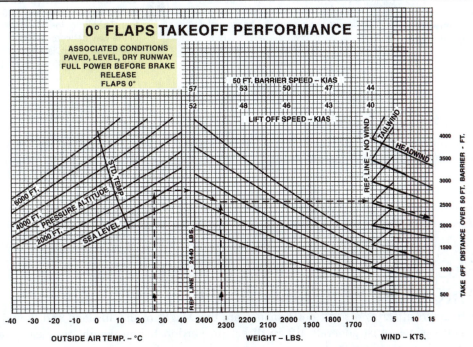

Figure 8-2. Both tables and graphs specify a set of conditions under which the chart is valid.

TABLE FORMAT

Using the table is straightforward. Find the row and column that most closely match the conditions, and read the appropriate values. The table cannot provide every possible altitude and temperature, but you can determine values for conditions in between the values in the table by interpolation—the process of finding an unknown value between two known values. [Figure 8-3]

In practice, pilots often round off values from tables to the more conservative figure. Using values that reflect slightly more adverse circumstances provides a reasonable estimate of performance data, and gives a modest margin of safety.

TAKEOFF DISTANCE

CONDITIONS:
Flaps 10°
Full throttle prior to brake release
Paved, dry, level runway
Zero wind

NOTES:
1. Short field technique.
2. Prior to takeoff from fields above 3000 feet elevation, the mixture should be leaned to give maximum RPM in a full throttle, static runup.
3. Decrease distances 10% for each 9 knots headwind. For operation with tailwinds up to 10 knots, increase distances by 10% for each 2 knots.
4. For operation on a dry, grass runway, increase distances by 15% of the "ground roll" figure.

WEIGHT LBS	TAKEOFF SPEED KIAS		PRESSURE ALTITUDE FEET	0° C		10° C		20° C		30° C		40° C	
	LIFT OFF	AT 50 FEET		GROUND ROLL FEET	TOTAL FEET TO CLEAR 50-FOOT OBSTACLE	GROUND ROLL FEET	TOTAL FEET TO CLEAR 50-FOOT OBSTACLE	GROUND ROLL FEET	TOTAL FEET TO CLEAR 50-FOOT OBSTACLE	GROUND ROLL FEET	TOTAL FEET TO CLEAR 50-FOOT OBSTACLE	GROUND ROLL FEET	TOTAL FEET TO CLEAR 50-FOOT OBSTACLE
2400	51	56	S. L.	800	1465	865	1575	925	1690	1000	1815	1070	1950
			1000	880	1610	945	1730	1020	1865	1090	2005	1175	2160
			2000	965	1775	1040	1915	1120	2065	1200	2220	1295	2400
			3000	1060	1965	1145	2125	1235	2300	1330	2485	1430	2690
			4000	1170	2190	1265	2370	1360	2575	1470	2795	1580	3035
			5000	1290	2450	1395	2665	1505	2900	1625	3165	1750	3460
			6000	1430	2760	1545	3020	1665	3300	1805	3625	1945	3995
			7000	1585	3145	1715	3455	1855	3810	2005	4225	—	—
			8000	1760	3620	1910	4020	2065	4485	—	—	—	—

 Looking at the sample takeoff chart, you can see that the given pressure altitude of 1,500 feet falls between the 1,000- and 2,000-foot pressure altitude values.

 This means if the outside air temperature is 30°C, your ground roll distance will fall between 1,090 and 1,200 feet.

Figure 8-3. Interpolation can find values between rows or columns in a table.

 To solve for ground roll, interpolation is necessary. You must first compute the differences between the known values.

 The 1,500 foot airport pressure altitude is 50% of the way between 1,000 and 2,000 feet. Therefore, the ground roll also is 50% of the way between 1,090 and 1,200 feet. The answer then, is 1,145 (110-foot difference x 0.5 + 1,090 feet = 1,145 feet).

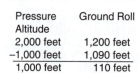

	Pressure Altitude	Ground Roll
	2,000 feet	1,200 feet
	−1,000 feet	1,090 feet
Difference	1,000 feet	110 feet

GRAPH FORMAT

Performance graphs come in many arrangements and configurations, but all are designed to make the process of compensating for several variables fast and accurate. Because a graphic chart has more variables built into it, you must use extra care when determining performance values. You might be tempted to estimate where two lines meet on the chart. This can lead to substantial errors, and with charts that incorporate several sequential steps, a small error at the beginning can lead to a much larger error in the final result.

FACTORS AFFECTING AIRCRAFT PERFORMANCE

Two factors affecting aircraft performance are the weight of the airplane and the wind. Because wings must generate lift in direct proportion to the weight they carry, any increase in weight carries a corresponding penalty in performance, because energy used for lift is unavailable for thrust. Likewise, because it takes more energy to accelerate a heavy airplane to takeoff speed, or to slow it down after landing, runway requirements are greater at heavier weights. The motion of the air itself (wind) can be a help or a hindrance. Airplanes taking off or landing into a strong wind have reduced ground rolls. In cruising flight, the groundspeed and time en route vary depending on the direction and speed of the wind. The wind is an important consideration in planning fuel requirements, because fuel consumption is proportional to flight time.

Atmospheric conditions can decrease air density, increasing the apparent altitude. As pressure decreases, there are fewer air molecules in a given volume, so air density decreases. Because air expands when heated, a cubic foot of air on a hot day will contain fewer air molecules than the same cubic foot of air on a cooler day. Also, air containing water vapor is less dense than dry air.

Because aircraft performance diminishes with altitude, it follows that decreases in air density due to temperature, pressure, or humidity will also reduce performance. For example, when the air is less dense, wings must move through the air faster to develop enough lift for takeoff, resulting in a longer takeoff roll. Lower air density also reduces engine power, because the engine must take in a larger volume of air to get enough air molecules for combustion. Because the propeller works on the same principle as the wings, propeller efficiency also drops. Although lower air density also reduces drag, this results in a relatively minor performance benefit. Decreased air density affects performance in all flight regimes, but the effects are most apparent during takeoff and climb.

When altitude is corrected for nonstandard pressure, the result is pressure altitude. **Density altitude** is the term for pressure altitude that has been corrected for nonstandard temperature. At standard temperatures, pressure altitude and density altitude are the same. On a hot day the density altitude at an airport might be 2,000 or 3,000 feet higher than the field elevation, and as a result, your airplane will perform as though the airport were at the higher elevation. Density altitude differs from field elevation whenever temperature differs from standard conditions, which is most of the time. If you know the field elevation, altimeter setting, and temperature, you can use a chart to find density altitude. [Figure 8-4]

SECTION A ■ Predicting Performance

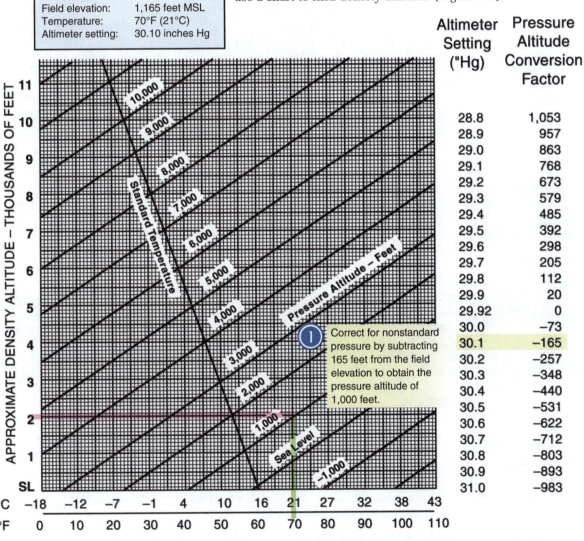

Conditions
Field elevation: 1,165 feet MSL
Temperature: 70°F (21°C)
Altimeter setting: 30.10 inches Hg

Altimeter Setting ("Hg)	Pressure Altitude Conversion Factor
28.8	1,053
28.9	957
29.0	863
29.1	768
29.2	673
29.3	579
29.4	485
29.5	392
29.6	298
29.7	205
29.8	112
29.9	20
29.92	0
30.0	–73
30.1	–165
30.2	–257
30.3	–348
30.4	–440
30.5	–531
30.6	–622
30.7	–712
30.8	–803
30.9	–893
31.0	–983

① Correct for nonstandard pressure by subtracting 165 feet from the field elevation to obtain the pressure altitude of 1,000 feet.

② Enter the chart at the bottom, just above the temperature of 70°F (21°C). Proceed up the chart vertically until you intercept the diagonal 1,000-foot pressure altitude line.

③ Move horizontally to the left and read the density altitude of approximately 2,000 feet. This means your airplane will perform as if it were at 2,000 feet MSL on a standard day.

Figure 8-4. The primary reason for computing density altitude is to help determine aircraft performance.

Humidity usually has a relatively small effect on performance, so it is ordinarily disregarded in density altitude computations. Even so, very high humidity can reduce engine horsepower by as much as 7% and reduce an airplane's takeoff and climb performance by as much as 10%.

Many performance charts do not require you to compute density altitude. Instead, compensation is built into the performance chart itself. All you do is apply the correct pressure altitude and temperature. If the chart you are using does not ask you for these variables, you should compute density altitude before using it.

SECTION A ■ Predicting Performance

DISCOVERY

Lake County Airport

Lake County Airport, located in Leadville, Colorado, has a field elevation of 9,927 feet MSL. The standard temperature at this elevation is 24°F (−4°C), which means that even if the temperature is at freezing, +32°F (0°C), the density altitude is more than 500 feet above the field elevation.

On a typical summer afternoon, with a low barometric pressure and a temperature of 78°F (26°C), the density altitude at Lake County Airport can be as high as 14,000 feet—more than 4,000 feet above field elevation. Because 14,000 feet is near the service ceiling of many single-engine airplanes, liftoff and climb might not be possible.

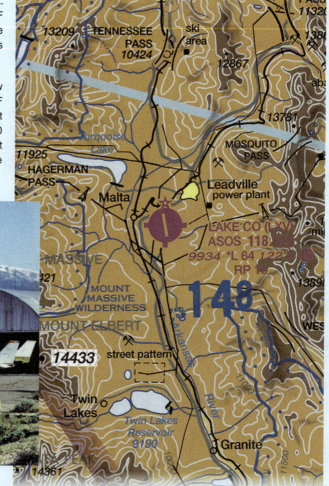

TAKEOFF AND LANDING PERFORMANCE

In addition to density altitude, takeoff performance depends on several factors you can measure or calculate in advance, such as aircraft weight, wind, and runway conditions. Most of these factors also affect landing distances. Under some conditions, takeoff might be impossible within the limits of the available runway.

AIRCRAFT WEIGHT AND CONFIGURATION

To generate sufficient lift for flight, a heavily loaded airplane must accelerate to a higher speed than the same airplane with a light load. Because acceleration will also be slower, the airplane might need significantly more runway for takeoff. You can readily see the effect of weight in the takeoff distance chart. [Figure 8-5]

TAKEOFF DISTANCE

CONDITIONS:
Flaps 10°
Full throttle prior to brake release
Paved, dry, level runway
Zero wind

NOTES:
1. Short field technique.
2. Prior to takeoff from fields above 3000 feet elevation, the mixture should be leaned to give maximum RPM in a full throttle, static runup.
3. Decrease distances 10% for each 9 knots headwind. For operation with tailwinds up to 10 knots, increase distances by 10% for each 2 knots.
4. For operation on a dry, grass runway, increase distances by 15% of the "ground roll" figure.

WEIGHT LBS	TAKEOFF SPEED KIAS LIFT OFF	TAKEOFF SPEED KIAS AT 50 FEET	PRESSURE ALTITUDE FEET	0° C GROUND ROLL FEET	0° C TOTAL FEET TO CLEAR 50-FOOT OBSTACLE	10° C GROUND ROLL FEET	10° C TOTAL FEET TO CLEAR 50-FOOT OBSTACLE	20° C GROUND ROLL FEET	20° C TOTAL FEET TO CLEAR 50-FOOT OBSTACLE	30° C GROUND ROLL FEET	30° C TOTAL FEET TO CLEAR 50-FOOT OBSTACLE	40° C GROUND ROLL FEET	40° C TOTAL FEET TO CLEAR 50-FOOT OBSTACLE
2400	51	56	S. L.	800	1465	865	1575	925	1690	1000	1815	1070	1950
			1000	880	1610	945	1730	1020	1865	1090	2005	1175	2160
			2000	965	1775	1040	1915	1120	2065	1200	2220	1295	2400
			3000	1060	1965	1145	2125	1235	2300	1330	2485	1430	2690
			4000	1170	2190	1265	2370	1360	2575	1470	2795	1580	3035
			5000	1290	2450	1395	2665	1505	2900	1625	3165	1750	3460
			6000	1430	2760	1545	3020	1665	3300	1805	3625	1945	3995
			7000	1585	3145	1715	3455	1855	3810	2005	4225	—	—
			8000	1760	3620	1910	4020	2065	4485	—	—	—	—
2200	49	54	S. L.	655	1200	705	1285	750	1380	810	1475	870	1580
			1000	715	1315	770	1410	830	1515	890	1620	955	1740
			2000	785	1445	845	1550	910	1665	980	1790	1050	1920
			3000	860	1590	930	1710	1000	1840	1075	1980	1155	2135
			4000	950	1755	1025	1895	1105	2045	1185	2210	1275	2380
			5000	1045	1950	1130	2110	1215	2280	1310	2470	1410	2670
			6000	1155	2175	1245	2360	1345	2560	1450	2780	1560	3025
			7000	1275	2445	1380	2660	1490	2895	1610	3160	1735	3455
			8000	1415	2765	1530	3020	1655	3310	1790	3635	1930	4010
2000	46	51	S. L.	530	975	570	1040	605	1115	655	1190	700	1270
			1000	575	1065	620	1140	670	1220	715	1300	770	1390
			2000	630	1165	680	1245	730	1335	785	1430	845	1530
			3000	695	1275	745	1370	805	1470	865	1575	925	1690
			4000	760	1405	820	1505	885	1620	950	1740	1020	1870
			5000	835	1550	905	1665	975	1795	1045	1930	1125	2075
			6000	925	1715	995	1850	1075	1995	1155	2150	1240	2320
			7000	1020	1905	1100	2060	1185	2230	1280	2410	1375	2610
			8000	1130	2130	1220	2310	1315	2505	1415	2720	1525	2955

Figure 8-5. Increased weight results in an increased ground roll.

If you find you will be unable to safely take off at a particular airport at the airplane's proposed weight, you should consider reducing the weight of the airplane, perhaps by carrying less fuel. Consulting the takeoff performance charts in the POH will tell you how much difference a weight reduction would make in the takeoff roll, and you might find that you would be able to safely take off at that airport with a lower total weight. If you are in doubt, it might be best to delay your takeoff for more favorable density altitude or wind conditions.

Because stall speed is also affected by weight, approach and landing speeds will be higher in a heavily loaded airplane. After touchdown, the ground roll will be longer in a heavily loaded airplane due to the additional kinetic energy that must be dissipated by the brakes and wheels.

In most airplanes, the aerodynamic configuration can be changed to enhance takeoff and landing performance. While large aircraft employ a wide variety of devices, most training aircraft simply use wing flaps. Many high performance light airplanes partially extend the flaps on takeoff to provide greater lift at low speeds. Flaps are used during landing approaches to steepen the glide path and to permit lower touchdown speeds. Your use of flaps on landing approach will vary with field conditions and length. Normally, the use of flaps and a lower indicated approach speed are desirable when landing on a short runway or soft runway surface, such as grass.

SECTION A ■ Predicting Performance

SURFACE WINDS

Takeoff and landing distances are influenced by both the speed and direction of surface winds. Because a headwind reduces the amount of speed the airplane must gain to attain flying speed, it reduces ground roll. During landing, a headwind reduces the groundspeed at touchdown, so the landing roll will also be shorter. Because surface winds will not always be exactly aligned with the runway in use, you need a method of determining what portion of the wind is acting along the runway and what portion is acting across it. The **headwind component** refers to that portion of the wind that acts straight down the runway toward the airplane. The **crosswind component** is the portion of the wind that acts perpendicular to the runway. Most airplanes have a maximum demonstrated crosswind component stated in the POH. You can easily compute headwind and crosswind components by using a wind component chart. [Figure 8-6]

SECTION A ■ **Predicting Performance**

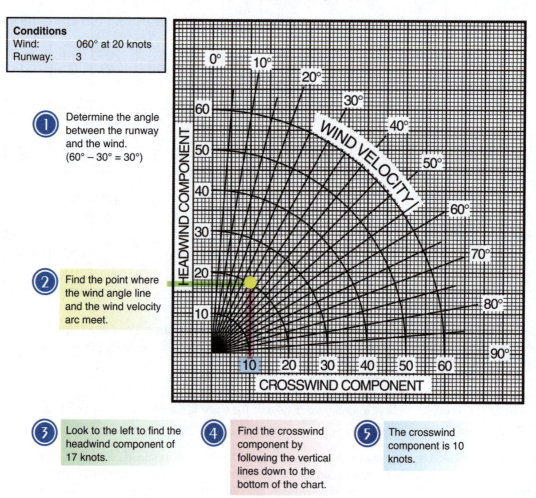

Conditions
Wind: 060° at 20 knots
Runway: 3

① Determine the angle between the runway and the wind. (60° − 30° = 30°)

② Find the point where the wind angle line and the wind velocity arc meet.

③ Look to the left to find the headwind component of 17 knots.

④ Find the crosswind component by following the vertical lines down to the bottom of the chart.

⑤ The crosswind component is 10 knots.

Figure 8-6. When you use a wind component chart, remember that both the runway number and surface winds are given in magnetic direction.

The **tailwind component** is the portion of the wind that acts directly on the tail of the airplane. Attempting to take off with a tailwind component (downwind) adds much more ground roll than the same amount of headwind would reduce it. For example, if you look at Note 3 on the chart in figure 8-5, you will see that taking off in this airplane with a 9 knot headwind reduces ground roll by 10%, but taking off with a 9 knot tailwind increases the ground roll by 45%. Tailwind components have a similar effect on landing distances. Another insidious effect of downwind landings is the pilot's perception of speed over the ground. Because of the greater groundspeed compared to normal approach speeds, some pilots unconsciously slow to a speed that looks right outside the cockpit, but that actually can be dangerously close to the stall speed.

RUNWAY GRADIENT AND SURFACE

Runway conditions relating to aircraft performance data generally specify a paved and level runway with a smooth, dry surface. If any of these conditions are different for the runway you use, you need to adjust the takeoff and landing distances using the methods described in the chart notes.

The **runway gradient**, or **runway slope**, refers to the amount of change in runway height over its length. Gradient is expressed as a percentage—for example, a 2 percent gradient means the runway elevation changes 2 feet for each 100 feet of runway length. A positive gradient means elevation increases along the runway, while a negative gradient means it decreases. A positive gradient is unfavorable for takeoff because the airplane must accelerate uphill, but is favorable for landing because decelerating uphill reduces ground roll. A negative gradient has the opposite effect on both takeoffs and landings. Runway gradient is listed in the Airport/Facility Directory when it is 0.3 percent or more. [Figure 8-7]

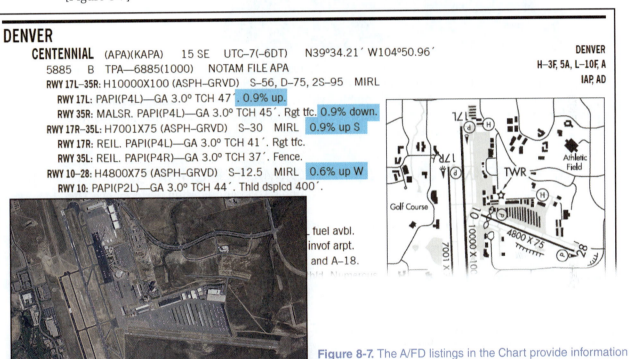

Figure 8-7. The A/FD listings in the Chart provide information regarding runway gradient.

Generally, any runway surface that is not hard, smooth, and dry will increase the takeoff roll. This is due to the inability of the tires to roll smoothly along the runway. For example, on runways that are muddy, or covered with grass or snow, the tires can sink slightly into the ground. This reduces the airplane's acceleration, sometimes to the extent that it might be impossible to accelerate to takeoff speed.

The condition of the runway surface also affects the landing roll and braking. **Braking effectiveness** refers to how much braking power you can apply without skidding the tires. For the most part, it depends on the amount of friction between the tires and the runway. Braking effectiveness is considered normal on a dry runway, but if the runway is wet, less friction is available and your landing roll-out will increase. In some cases, you can lose braking effectiveness because of **hydroplaning**, which happens when a thin layer of water separates the tires from the runway. [Figure 8-8]

Figure 8-8. Exercise caution when operating on a wet runway because of the potential for hydroplaning.

Braking effectiveness also can be completely lost on ice-covered runways. If you must operate in conditions where braking effectiveness is reduced, be sure the runway length is adequate and surface wind is favorable. Although mud, grass, and snow can reduce the friction between tires and the runway, in some cases they might reduce the landing roll. This is because they act as obstructions to the tires.

TAKEOFF AND LANDING PERFORMANCE CHARTS

Most manufacturers supply charts for determining takeoff and landing performance The use of a table for determining takeoff performance is shown in Figure 8-9. Determining landing distance from a graph is explained in Figure 8-10.

Takeoff Conditions
Pressure Altitude: 2,000 feet
Temperature: 30°C
Wind: Calm
Flaps: 10°
Weight: 2,400 lb
Runway Conditions: Paved, level, and dry

TAKEOFF DISTANCE

CONDITIONS:
Flaps 10°
Full throttle prior to brake release
Paved, dry, level runway
Zero wind

 A quick check of the conditions and takeoff weight indicate you are using the correct chart.

WEIGHT LBS	TAKEOFF SPEED KIAS		PRESSURE ALTITUDE FEET	0° C		10° C		20° C		30° C		40° C	
	LIFT OFF	AT 50 FEET		GROUND ROLL FEET	TOTAL FEET TO CLEAR 50-FOOT OBSTACLE	GROUND ROLL FEET	TOTAL FEET TO CLEAR 50-FOOT OBSTACLE	GROUND ROLL FEET	TOTAL FEET TO CLEAR 50-FOOT OBSTACLE	GROUND ROLL FEET	TOTAL FEET TO CLEAR 50-FOOT OBSTACLE	GROUND ROLL FEET	TOTAL FEET TO CLEAR 50-FOOT OBSTACLE
2400	51	56	S. L.	800	1465	865	1575	925	1690	1000	1815	1070	1950
			1000	880	1610	945	1730	1020	1865	1090	2005	1175	2160
			2000	965	1775	1040	1915	1120	2065	1200	2220	1295	2400
			3000	1060	1965	1145	2125	1235	2300	1330	2485	1430	2690
			4000	1170	2190	1265	2370	1360	2575	1470	2795	1580	3035
			5000	1290	2450	1395	2665	1505	2900	1625	3165	1750	3460
			6000	1430	2760	1545	3020	1665	3300	1805	3625	1945	3995
			7000	1585	3145	1715	3455	1855	3810	2005	4225	—	—
			8000	1760	3620	1910	4020	2065	4485	—	—	—	—

 Note the takeoff speed of 51 knots and the speed of 56 knots shortly after takeoff at 50 feet.

 Enter the tabular data of the pressure altitude of 2,000 feet. Proceed horizontally to the column for 30°C. The ground roll distance is 1,200 feet, and the total distance to clear a 50-foot obstacle is 2,220 feet.

Figure 8-9. The takeoff distance table enables you to look up the ground roll distance and the total distance to clear a 50-foot obstacle for a given altitude and temperature.

SECTION A ■ Predicting Performance

Landing Conditions
Pressure Altitude: 8,000 feet
Temperature: 13°C
Weight: 2,800 lb
Wind: 2 knot tailwind

LANDING DISTANCE

ASSOCIATED CONDITIONS:

POWER	RETARDED TO MAINTAIN 900 FT FINAL APPROACH
FLAPS	DOWN
LANDING GEAR	DOWN
RUNWAY	PAVED, LEVEL, DRY SURFACE
APPROACH SPEED	IAS AS TABULATED
BRAKING	MAXIMUM

WEIGHT ~ POUNDS	SPEED AT 50 FT	
	KNOTS	MPH
2950	70	80
2800	68	78
2600	65	75
2400	63	72
2200	60	69

EXAMPLE:

OAT	25°C (77 °F)
PRESSURE ALTITUDE	3965 FT
WEIGHT	2814 LB
WIND COMPONENT	9.0 KNOTS (HEADWIND)
GROUND ROLL	1080 FT
TOTAL OVER 50 FT OBSTACLE	1700FT
APPROACH SPEED	68 KNOTS (78 MPH)

⑥ For the total distance over a 50-foot obstacle, follow the diagonal guide lines up to the distance scale.

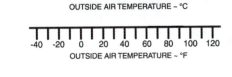

③ Move horizontally to the reference line, then diagonally to the line for the total weight.

② Go up to the diagonal line for the pressure altitude.

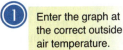

④ Follow the horizontal line to the next reference line. Apply the correction for wind.

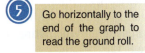

Total Over 50-foot Obstacle = 2,050 feet

Ground Roll = 1,400 feet

OUTSIDE AIR TEMPERATURE ~ °C

OUTSIDE AIR TEMPERATURE ~ °F

WEIGHT ~ POUNDS

WIND COMPONENT ~ KNOTS

OBSTACLE HEIGHT ~ FEET

DISTANCE ~ FEET

① Enter the graph at the correct outside air temperature.

⑤ Go horizontally to the end of the graph to read the ground roll.

Figure 8-10. A graph like this one is another way of determining takeoff or landing distance. To use this landing distance graph, start with the temperature at the lower left and move through the graph as shown.

Results
Total Over 50-ft Obstacle: 2,050 feet
Ground Roll: 1,400 feet

SECTION A ■ **Predicting Performance**

Some pilot's operating handbooks recommend specific **approach airspeeds** for various flap settings and aircraft weights. In general, these recommended airspeeds should be used regardless of temperature and altitude combinations. As you recall, operating at higher density altitudes will result in higher than indicated true airspeeds. This is important to remember when landing at airports at higher elevations than those to which you have become accustomed. If you do not monitor the airspeed indicator, the higher groundspeed could lead you to slow to a dangerously low airspeed. This effect is similar to the downwind landing situation described earlier.

CLIMB PERFORMANCE

Most of the factors affecting takeoff performance also affect the climb capability of an aircraft. The pilot's operating handbook for the airplane lists airspeeds for a variety of climbing flight conditions. Two of the most important are the **best angle-of-climb airspeed (V_X)**, and the **best rate-of-climb airspeed (V_Y)**. [Figure 8-11]

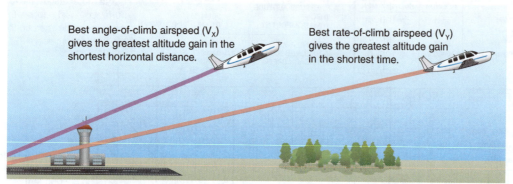

Best angle-of-climb airspeed (V_x) gives the greatest altitude gain in the shortest horizontal distance.

Best rate-of-climb airspeed (V_y) gives the greatest altitude gain in the shortest time.

Figure 8-11. Best angle-of-climb airspeed (V_X) results in a steeper climb path, although the airplane takes longer to reach the same altitude than it would at the best rate-of-climb airspeed (V_Y).

CLIMB SPEEDS

The best angle-of-climb airspeed (V_X) is normally used for obstacle clearance immediately after takeoff. Because of the increased pitch attitude at V_X, your forward visibility is limited. Best angle-of-climb speed should be used anytime you need to gain the maximum amount of altitude in the minimum horizontal distance.

Normally, you use best rate-of-climb (V_Y) after you have cleared all obstacles during departure. Best rate-of-climb speed gives the greatest altitude gain in a given time. After traffic pattern departure, you might use a cruise climb when climbing to cruising altitude. You can also use cruise climb during the enroute portion of a flight to climb to a higher altitude. **Cruise climb speed** is generally higher than V_X or V_Y, and the rate of climb is lower. In addition to better engine cooling and improved forward visibility, faster climb speeds provide better cross-country speeds, cutting down the total time enroute. [Figure 8-12]

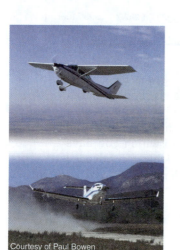

Courtesy of Paul Bowen

Normal Cruise Climb

Best Angle of Climb

Figure 8-12. Climb airspeeds vary between aircraft; for example, the Pilatus PC-12 (bottom) requires higher speeds than the Cessna 172 (top).

Before an airplane can climb, it must have a reserve of power or thrust. At any given speed, more power is required for a sustained climb than for unaccelerated level flight. Because propeller-driven airplanes lose power and thrust with increasing altitude, both the best angle-of-climb and best rate-of-climb speeds change as you climb. When the airplane is unable to climb any further, it has reached its **absolute ceiling**. Another important altitude, known as the **service ceiling**, refers to the altitude where a single-engine airplane is able to maintain a maximum climb of only 100 feet per minute. This altitude is more commonly used than absolute ceiling, since it represents the airplane's practical ceiling. [Figure 8-13]

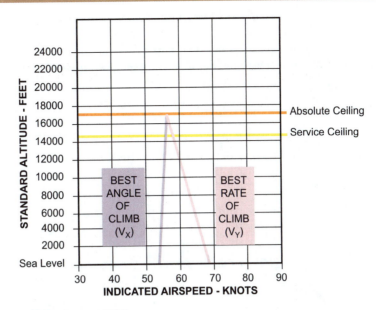

Figure 8-13. As altitude increases, the speed for best angle-of-climb increases, and the speed for best rate-of-climb decreases. The point at which these two speeds meet is the absolute ceiling of the airplane.

CLIMB PERFORMANCE CHARTS

Climb performance data included in the pilot's operating handbook provides the approximate performance that can be expected under various conditions. Many pilot's operating handbooks provide time, fuel, and distance-to-climb data. This table differs from other performance tables because you use it in two stages. First, determine the time, fuel, and distance to climb from sea level to your cruising altitude. Then, subtract the time, fuel, and distance to climb to the elevation of your departure point, because you have essentially already climbed to that altitude. [Figure 8-14]

TIME, FUEL, AND DISTANCE TO CLIMB

Climb Conditions	
Airport Elevation:	2,000 feet MSL
Cruising Altitude:	8,000 feet MSL

CONDITIONS:
Flaps UP
Full throttle
Standard temperature

NOTES:
Add 1.1 gallons of fuel for engine start, taxi and takeoff allowances.
Mixture leaned above 3000 feet for maximum RPM.
Increase time, fuel and distance by 10% for each 10° above standard temperature.
Distances shown are based on zero wind.

 Verify that you have the proper chart and the conditions specified on the chart are met.

 Read the time, fuel, and distance to climb to 8,000 feet.

WEIGHT LB	PRESSURE ALTITUDE FEET	TEMP °C	CLIMB SPEED KIAS	RATE OF CLIMB FT/M	FROM SEA LEVEL		
					TIME MINUTES	FUEL USED GALLONS	DISTANCE NM
2400	S. L.	15	76	700	0	0.0	0
	1000	13	76	650	1	0.3	2
	2000	11	75	605	3	0.6	4
	3000	9	75	555	5	1.0	6
	4000	7	74	510	7	1.4	9
	5000	5	74	465	9	1.7	11
	6000	3	73	420	11	2.2	14
	7000	1	72	370	14	2.6	18
	8000	-1	72	325	17	3.1	22
	9000	-3	71	280	20	3.7	27
	10,000	-5	71	235	24	4.3	33
	11,000	-7	70	185	29	5.0	39
	12,000	-9	70	140	35	5.9	48

 Determine the time, fuel, and distance credits to be applied for departing an airport located at 2,000 feet.

 After you subtract the credits, the net values are 14 minutes, 2.5 gallons, and 18 miles.

A check of the notes, however, indicates that you must add an additional 1.1 gallons of fuel for the engine start, taxi, and takeoff allowances.

The values to climb from 2,000 feet to 8,000 feet are a total of 14 minutes, 3.6 gallons, and 18 miles.

Figure 8-14. When finding the time, fuel, and distance to climb to your cruising altitude, you must subtract the time, fuel, and distance to climb from sea level to your airport elevation.

SECTION A ■ **Predicting Performance**

The type of climb performance information provided in figure 8-14 is helpful in preflight planning. It permits accurate estimates of three important factors affecting the climb segment of a flight: time, fuel, and distance. A climb performance graph allows you to determine the best rate of climb using temperature and pressure altitude. [Figure 8-15]

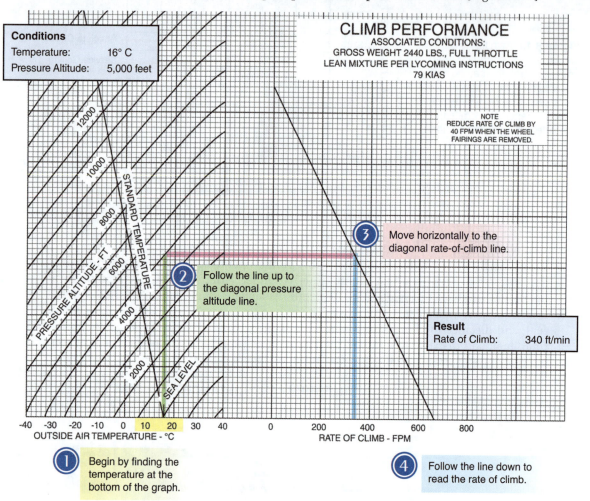

Conditions
Temperature: 16° C
Pressure Altitude: 5,000 feet

CLIMB PERFORMANCE
ASSOCIATED CONDITIONS:
GROSS WEIGHT 2440 LBS., FULL THROTTLE
LEAN MIXTURE PER LYCOMING INSTRUCTIONS
79 KIAS

NOTE
REDUCE RATE OF CLIMB BY
40 FPM WHEN THE WHEEL
FAIRINGS ARE REMOVED.

③ Move horizontally to the diagonal rate-of-climb line.

② Follow the line up to the diagonal pressure altitude line.

Result
Rate of Climb: 340 ft/min

OUTSIDE AIR TEMPERATURE - °C

RATE OF CLIMB - FPM

① Begin by finding the temperature at the bottom of the graph.

④ Follow the line down to read the rate of climb.

Figure 8-15. This graph enables you to determine rate of climb at a given altitude and temperature.

CRUISE PERFORMANCE

The manufacturers of today's light airplanes provide cruise performance charts to indicate rate of fuel consumption, true airspeed, range, and endurance. They will give you a close approximation of the performance you can expect at different altitudes and power settings. Any deviation from the specific information upon which the chart computations are based will affect the accuracy of the results. For example, many cruise performance charts are based on standard atmospheric conditions with zero wind. During flight planning, you must compensate for nonstandard conditions, and use flight times that take into account the effects of predicted winds. You should remember that wind has a significant effect on the distance an aircraft can fly, but no effect on its rate of fuel consumption or the total time it can remain aloft.

In selecting your **cruising speed**, you usually want to cover the distance to be traveled in the shortest period of time, but there are many other factors that might influence your decision. There might be times when you want to use the minimum amount of fuel, or stay aloft for the greatest period of time. If you understand the trade-offs you make between time, power, fuel consumption, range, and speed, you will be able to make choices that maximize the pleasure and utility you get from flying.

DISCOVERY

Max Conrad—Performance Records

Max Conrad knew about cruise. With a total of over 200 ocean crossings, both Atlantic and Pacific, and a solo flight around the world in less than nine days, Conrad set many records that stand to this day.

The distance flights started in 1950, when Conrad decided to fly to visit his wife and children, who were then living in Switzerland. He flew a 135 HP Piper Pacer from New York across the Atlantic to Geneva. The U.S. Air Force intercepted the Pacer midway across the Atlantic and forced Conrad to land in Greenland, where he was detained on suspicion of being a communist spy.

Conrad worked out an arrangement with Piper to promote their aircraft by making more long distance flights. As a pilot for Piper, he delivered many aircraft to customers overseas. To gain attention for the new Piper Apache light twin, he flew one nonstop from New York to Paris in 1954, beating Lindbergh's time by about ten hours.

In an effort to beat an existing distance record, Conrad planned a flight from Casablanca in northern Africa to El Paso, Texas, a distance of 5,500 miles. When he arrived over El Paso, however, he had plenty of fuel remaining, so Conrad continued on to Los Angeles. His total solo flight— a nonstop distance of almost 6,000 miles. He had been aloft for more than 58 hours.

Recognizing that efficiency, not power, is the key to long distance flight, Conrad had the 250 HP engine in his Piper Comanche replaced with a 180 HP engine. Then he flew it for 60 hours straight to set a new closed course record.

Copyright Piper Museum

More long distance flights followed. In 1961, Conrad flew a Piper Aztec around the world at the equator. His final goal, at the age of 65, was to fly around the world via the North and South poles. Despite careful preparations and skilled flying, he had to abort because of bad weather after making it 3/4 of the way. Undaunted, he tried again the next year, but had to abandon his Aztec at the South Pole.

Conrad retired from flying with more than 50,000 hours in the air, after setting dozens of world records, and earning a number of prestigious aviation honors and awards.

PERFORMANCE SPEEDS

In addition to the selection of cruising speeds provided in the charts, there are some specific power settings and speeds that result in maximum performance. Three important speeds are maximum level flight speed, maximum endurance speed, and maximum range speed. These speeds have been determined by the manufacturer, and balance the combination of available power, fuel economy, lift and drag to obtain the best possible performance for the situation described.

In level flight, the maximum speed of the airplane is limited by the amount of power produced by the engine and the total drag generated by the airplane. If thrust exceeds total drag when you apply power, the airplane accelerates. When the force of total drag equals the force of full thrust, the airplane is flying at its **maximum level flight speed**. [Figure 8-16]

Figure 8-16. The curved power-required line shows the amount of power necessary to maintain level flight at various speeds. The power-available line is also curved, because power available in the typical single-engine airplane is a function of airspeed. The point at which these two curves cross is where the forces of thrust and drag are in balance and where maximum level flight speed occurs.

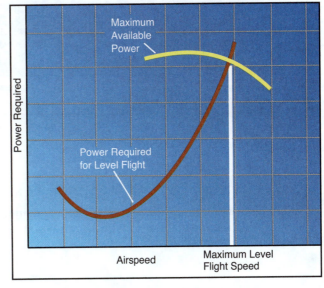

The Power Curve

More correctly called the power-required curve, it shows the amount of power necessary to maintain level flight at airspeeds throughout the aircraft performance envelope. The left end of the curve is the stall speed. Flight at low airspeeds requires a high angle of attack, and a great deal of power is required to overcome the resulting induced drag. As airspeed increases, the wing generates more lift with less induced drag, and less power is needed for level flight until, at the low point of the graph, the highest efficiency is reached. This is the point where the airplane will maintain level flight with the least amount of power. As speed increases past this point, the additional power is being used to overcome increased parasite drag.

You might hear other pilots talk about operating on the back side of the power curve. This refers to the curve to the left of the low point, where any reduction in airspeed requires an increase in power to maintain level flight. Through the majority of the graph (the front side of the curve) a decrease in airspeed means that less power is needed for level flight. Flying on the back side of the curve is discouraged, because a reduction in speed can demand more power than the engine can supply, or an unplanned drop in power could result in an involuntary descent. Because these speeds are relatively close to the stall, even minor engine trouble could leave you with two choices: descend or stall. You should try to avoid situations where you are dependent on engine power to prevent a stall.

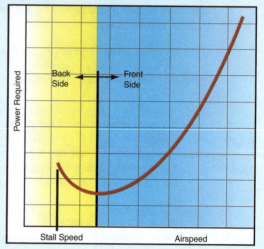

The **maximum range speed** lets you travel the greatest distance for a given amount of fuel. You can think of it as getting the most miles per gallon out of the airplane. This speed is determined by considering the speed and rate of fuel consumption at a given power setting.

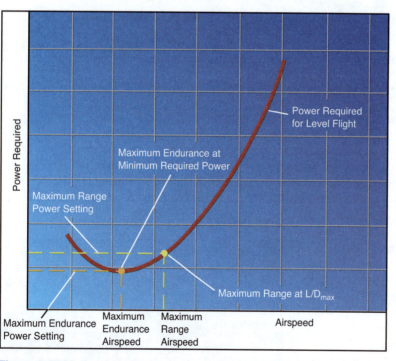

The setting that yields the greatest distance traveled per gallon of fuel burned is the power setting that provides maximum range speed. This speed produces the minimum total drag with enough lift to maintain altitude. It is where the lift-to-drag ratio is greatest, and is referred to as L/D_{max}. The speed and power setting that allows the airplane to remain aloft for the longest possible time is called the **maximum endurance speed**. It uses the minimum amount of power necessary to maintain level flight—think of it as the speed that provides the most hours per gallon. [Figure 8-17]

Figure 8-17. Maximum endurance speed occurs at the lowest point on the power-required curve.

USING CRUISE PERFORMANCE CHARTS

Cruise performance charts vary considerably. A cruise performance table used to determine fuel flow, fuel consumption, true airspeed, and manifold pressure is shown in Figure 8-18.

CRUISE POWER SETTINGS
65% MAXIMUM CONTINUOUS POWER (OR FULL THROTTLE)
2800 POUNDS

Conditions
Pressure Altitude: 8,000 feet
Temperature: Standard
Power Setting: 2,450 RPM

PRESS ALT.	IOAT		ENGINE SPEED	MAN PRESS	FUEL FLOW PER ENGINE		TAS		IOAT		ENGINE SPEED	MAN PRESS	FUEL FLOW PER ENGINE		TAS		IOAT		ENGINE SPEED	MAN PRESS	FUEL FLOW PER ENGINE		TAS	
	ISA –20 °C (–36 °F)								STANDARD DAY (ISA)								ISA +20 °C (+36 °F)							
FEET	°F	°C	RPM	IN HG	PSI	GPH	KTS	MPH	°F	°C	RPM	IN HG	PSI	GPH	KTS	MPH	°F	°C	RPM	IN HG	PSI	GPH	KTS	MPH
SL	27	-3	2450	20.7	6.6	11.5	147	169	63	17	2450	21.2	6.6	11.5	150	173	99	37	2450	21.8	6.6	11.5	153	176
2000	19	-7	2450	20.4	6.6	11.5	149	171	55	13	2450	21.0	6.6	11.5	153	176	91	33	2450	21.5	6.6	11.5	156	180
4000	12	-11	2450	20.1	6.6	11.5	152	175	48	9	2450	20.7	6.6	11.5	156	180	84	29	2450	21.3	6.6	11.5	159	183
6000	5	-15	2450	19.8	6.6	11.5	155	178	41	5	2450	20.4	6.6	11.5	158	182	79	26	2450	21.0	6.6	11.5	161	185
8000	-2	-19	2450	19.5	6.6	11.5	157	181	36	2	2450	20.2	6.6	11.5	161	185	72	22	2450	20.8	6.6	11.5	164	189
10000	-8	-22	2450	19.2	6.6	11.5	160	184	28	-2	2450	19.9	6.6	11.5	163	188	64	18	2450	20.3	6.5	11.4	166	191
12000	-15	-26	2450	18.8	6.4	11.3	162	186	21	-6	2450	18.8	6.1	10.9	163	188	57	14	2450	18.8	5.9	10.6	163	188
14000	-22	-30	2450	17.4	5.8	10.5	159	183	14	-10	2450	17.4	5.6	10.1	160	184	50	10	2450	17.4	5.4	9.8	160	184
16000	-29	-34	2450	16.1	5.3	9.7	156	180	7	-14	2450	16.1	5.1	9.4	156	180	43	6	2450	16.1	4.9	9.1	155	178

NOTES: 1. Full throttle manifold pressure settings are approximate.
2. Shaded area represents operation with full throttle.

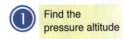 **1** Find the pressure altitude

 2 Use the table for the appropriate temperature. (If the temperature is between the given values, use interpolation.)

 3 Read the true airspeed and fuel consumption for your chosen power setting.

Figure 8-18. This cruise performance table shows the true airspeed and fuel consumption that result from various combinations of altitude, temperature, manifold pressure (MP) and RPM.

Resulting Performance
True Airspeed: 161 knots
Fuel Consumption Rate: 11.5 gal/hr

A range graph is used to determine aircraft range based on specific conditions. In the range graph shown in Figure 8-19, you can determine horsepower combinations and expected range in nautical miles for various temperatures and pressure altitudes.

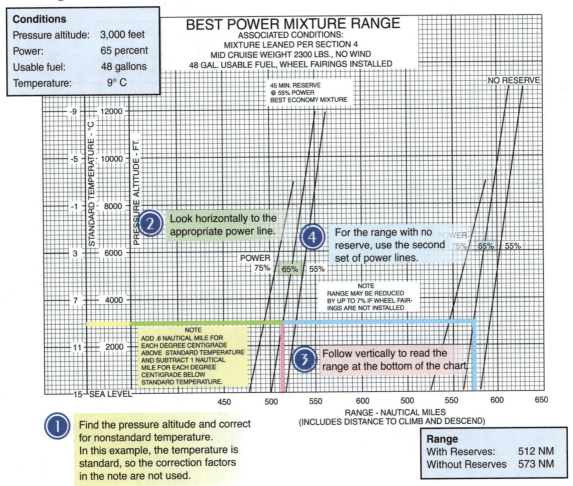

Conditions
Pressure altitude: 3,000 feet
Power: 65 percent
Usable fuel: 48 gallons
Temperature: 9° C

BEST POWER MIXTURE RANGE
ASSOCIATED CONDITIONS:
MIXTURE LEANED PER SECTION 4
MID CRUISE WEIGHT 2300 LBS., NO WIND
48 GAL. USABLE FUEL, WHEEL FAIRINGS INSTALLED

Range
With Reserves: 512 NM
Without Reserves 573 NM

Figure 8-19. This graph shows range with or without fuel reserves.

To determine the amount of power for various altitude/RPM combinations, you should consult either an engine or cruise performance chart. Formats vary, but a typical chart is shown in Figure 8-20.

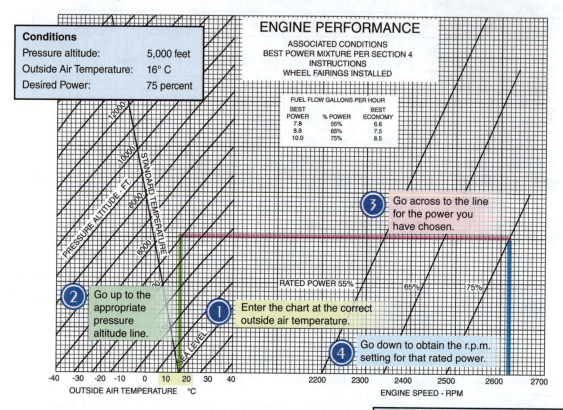

Figure 8-20. Engine performance charts help you determine the engine settings required to achieve a specific cruise power output.

SUMMARY CHECKLIST

✓ The pilot's operating handbook presents numerous charts that allow you to predict the airplane's performance accurately. They pertain to the takeoff, climb, cruise, descent, and landing phases of flight.

✓ Density altitude, wind, and runway conditions can greatly affect airplane performance.

✓ Takeoff performance depends mainly upon factors that can be measured or calculated in advance, such as density altitude, pressure altitude, temperature, wind, aircraft weight, and runway gradient or surface.

✓ You can easily break down wind direction and speed into headwind and crosswind components by using a wind component chart.

✓ Best angle-of-climb airspeed (V_X) is used to gain the most altitude in the shortest horizontal distance.

✓ The best rate-of-climb airspeed (V_Y) gives the maximum altitude gain in the least amount of time.

✓ Typically, a normal or cruise climb airspeed is used when climbing for prolonged periods of time.

✓ Climb performance data is included in the POH to provide you with an idea of the approximate performance that can be expected under various conditions.

✓ When choosing a cruising speed, you should consider fuel consumption, range, and the effects of winds.

KEY TERMS

Performance

Performance Charts

Interpolation

Density Altitude

Headwind Component

Crosswind Component

Tailwind Component

Runway Gradient

Runway Slope

Braking Effectiveness

Hydroplaning

Approach Airspeeds

Best Angle-of-Climb Airspeed (V_X)

Best Rate-of-Climb Airspeed (V_Y)

Cruise Climb Speed

Absolute Ceiling

Service Ceiling

Cruising Speed

Maximum Level Flight Speed

Maximum Range Speed

Maximum Endurance Speed

SECTION A ■ Predicting Performance

QUESTIONS

1. Where can you normally find the performance charts for your airplane?

2. Describe how density altitude affects aircraft performance.

3. True/False. Takeoff performance depends mainly upon factors that can be measured or calculated in advance, such as density altitude, pressure altitude, temperature, wind, aircraft weight, and runway gradient or surface.

4. Refer to the wind component chart shown below to determine the headwind and crosswind component for a departure on Runway 18 with a reported wind of 210° at 20 knots.

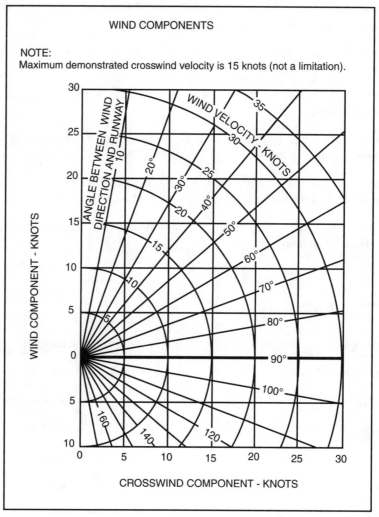

5. What is the runway gradient?
   A. The material used to surface the runway
   B. The amount of change in runway height over its length
   C. The direction of the runway as it relates to magnetic north

6. Use the chart shown below and the following conditions to determine the ground roll and distance necessary to clear a 50-foot obstacle.
   Conditions
Pressure Altitude:	6,000 feet
Temperature:	20°C
Flaps:	10°
Runway:	Paved, level, and dry
Wind:	Calm
Weight:	2,400 pounds

TAKEOFF DISTANCE

CONDITIONS:

Flaps 10°

Full throttle prior to brake release

Paved, dry, level runway

Zero wind

NOTES:

1. Short field technique.
2. Prior to takeoff from fields above 3000 feet elevation, the mixture should be leaned to give maximum RPM in a full throttle, static runup.
3. Decrease distances 10% for each 9 knots headwind. For operation with tailwinds up to 10 knots, increase distances by 10% for each 2 knots.
4. For operation on a dry, grass runway, increase distances by 15% of the "ground roll" figure.

WEIGHT LBS	TAKEOFF SPEED KIAS		PRESSURE ALTITUDE FEET	0° C		10° C		20° C		30° C		40° C	
	LIFT OFF	AT 50 FEET		GROUND ROLL FEET	TOTAL FEET TO CLEAR 50-FOOT OBSTACLE	GROUND ROLL FEET	TOTAL FEET TO CLEAR 50-FOOT OBSTACLE	GROUND ROLL FEET	TOTAL FEET TO CLEAR 50-FOOT OBSTACLE	GROUND ROLL FEET	TOTAL FEET TO CLEAR 50-FOOT OBSTACLE	GROUND ROLL FEET	TOTAL FEET TO CLEAR 50-FOOT OBSTACLE
2400	51	56	S. L.	800	1465	865	1575	925	1690	1000	1815	1070	1950
			1000	880	1610	945	1730	1020	1865	1090	2005	1175	2160
			2000	965	1775	1040	1915	1120	2065	1200	2220	1295	2400
			3000	1060	1965	1145	2125	1235	2300	1330	2485	1430	2690
			4000	1170	2190	1265	2370	1360	2575	1470	2795	1580	3035
			5000	1290	2450	1395	2665	1505	2900	1625	3165	1750	3460
			6000	1430	2760	1545	3020	1665	3300	1805	3625	1945	3995
			7000	1585	3145	1715	3455	1855	3810	2005	4225	—	—
			8000	1760	3620	1910	4020	2065	4485	—	—	—	—

7. True/False. As altitude increases, the best angle-of-climb speed will decrease, and the best rate-of-climb speed will increase.

8. Name the three important factors affecting the climb segment of a flight.

9. Select the items that would be found on a cruise performance chart.
   A. Time, fuel, and distance to climb
   B. Fuel consumption and true airspeed at various power settings
   C. Power required for level flight, including maximum level flight speed and L/D$_{max}$

SECTION A ■ Predicting Performance

8-23

10. From the following range graph, determine the expected range in nautical miles with and without reserve.

      Pressure Altitude: 4,000 feet

      Standard Temperature

      75% Power

      48 Gallons Usable Fuel

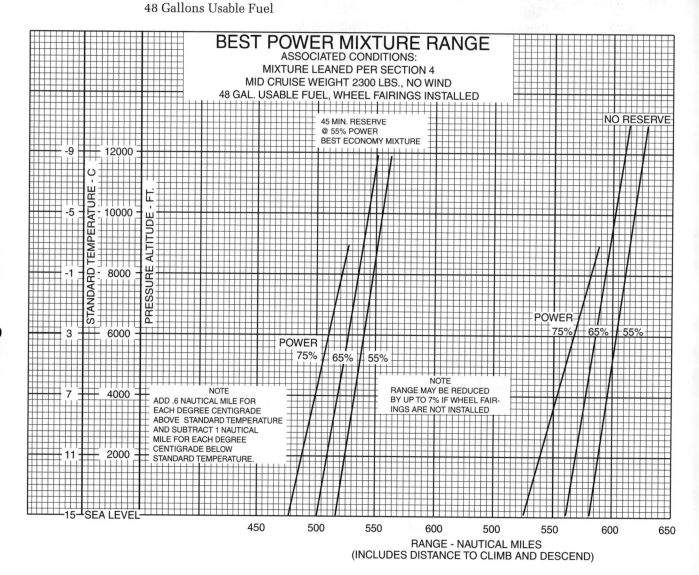

SECTION A ■ Predicting Performance

11. True/False. Landing weight is not a factor that must be considered during an approach.

SECTION B
Weight and Balance

The earliest airplanes could barely lift the pilot and enough fuel for a few minutes of flight. Many could not even manage to get airborne at all on a warm day! Although aircraft performance capabilities continue to improve, pilots still need to keep weight within safe limits, and balance the loads carried to maintain control of the airplane. [Figure 8-21]

Fuel Tank

Figure 8-21. The first Wright Flyer could carry the pilot and a few ounces of fuel — if the headwind was at least 11 miles per hour.

IMPORTANCE OF WEIGHT

Almost every aspect of performance is influenced by the weight of the aircraft and its contents. For example, compared to a properly loaded airplane, an overweight airplane has a longer takeoff run, higher takeoff speed, reduced angle and rate of climb, reduced cruising speed, shorter range, higher stalling speed, and longer landing roll. Loading an aircraft too heavily can dangerously decrease its performance, and increase the risk of structural damage if you encounter turbulence or make a hard landing, or even when maneuvering. As you might expect, a severely overloaded airplane will not fly at all.

Aircraft manufacturers do extensive testing to establish safe limits for aircraft loading. Such limits may include maximum takeoff and landing weights. These weights are approved by the FAA during initial airworthiness certification, and are established as operating limitations after consideration of factors such as performance, structural strength, and type of operation.

IMPORTANCE OF BALANCE

In addition to checking weight, you need to know that the aircraft is balanced within approved limits. It's not just a matter of how much you put into the plane, but where you put it. You can check the balance condition of an airplane by locating its **center of gravity (CG)**, which is the imaginary point where the aircraft would balance if suspended. [Figure 8-22] The location of this point is critical to an airplane's stability and elevator (or stabilator) effectiveness. Improper balance of the airplane's load can result in serious control problems. You can avoid these problems by taking the time to determine the location of the CG before each flight and then making sure it is within the limits provided by the manufacturer. The **CG limits** are the forward and aft center of gravity locations within which the aircraft must be operated at a given weight. Also called the CG envelope or range, the CG limits are established by the manufacturer. Many aircraft are certificated in more than one category, and will have different CG limits depending on the category in which the airplane is operated. If the CG is located within these limits, the airplane can be flown safely. If it is located outside these limits, you will need to rearrange fuel, passengers, or cargo in order to move the CG within acceptable limits. It's also important to maintain this distribution of weight during flight, because any movement of passengers or cargo will change the location of the CG.

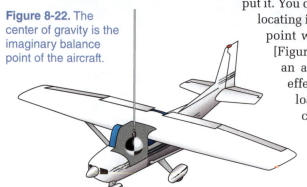

Figure 8-22. The center of gravity is the imaginary balance point of the aircraft.

Figure 8-23 shows an example of an airplane certified in both the normal and utility categories. It is permitted a maximum weight of 2,550 pounds when operated in the green-tinted CG range for normal-category operations. The same airplane may be operated in the utility category if the total weight is kept under 2,200 pounds and the CG is kept within the narrower, yellow range.

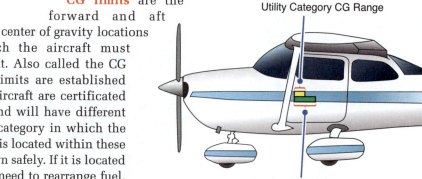

Utility Category CG Range

Normal Category CG Range

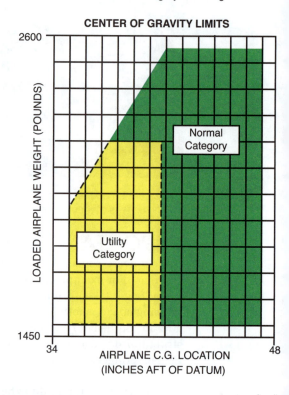

CENTER OF GRAVITY LIMITS

Normal Category

Utility Category

LOADED AIRPLANE WEIGHT (POUNDS)

AIRPLANE C.G. LOCATION (INCHES AFT OF DATUM)

Figure 8-23. To operate this airplane in the utility category, the weight and CG must be in the narrower yellow range; for practical purposes, that means no passengers or cargo may be in the back seat or cargo area.

WEIGHT AND BALANCE TERMS

Fortunately, the concepts and terms for weight and balance are straightforward and sensible. Some of the terms might be familiar to you from science classes, but as a pilot you will need to know their precise meanings in the aviation context.

REFERENCE DATUM

The center of gravity limits usually are specified in inches from a **reference datum**. This is an imaginary vertical plane from which all horizontal distances are measured for balance purposes. Common locations are the nose of the airplane, the engine firewall, the leading edge of the wing, or even somewhere in space ahead of the airplane. The location of the datum is established by the manufacturer and is defined in the POH or in the airplane's weight and balance papers. [Figure 8-24]

The datum is 66.25 inches ahead of the wing's leading edge.

TERMS DESCRIBING THE EMPTY AIRPLANE

Basic empty weight includes the weight of the standard airplane, optional equipment, unusable fuel, and full operating fluids including full engine oil. The **unusable fuel** is the small amount of fuel in the tanks that cannot be safely used in flight or drained on the ground. Older airplanes might use the term **licensed empty weight**, which is similar to basic empty weight except that it does not include full engine oil. Instead, it only counts the weight of undrainable oil. You can obtain basic empty weight by simply adding the weight of the oil to the licensed empty weight.

The weight of the empty airplane might change many times during its lifetime as equipment is installed and removed. This equipment could include new instruments, radios, wheel fairings, engine accessories, or other modifications. Anything that significantly changes the weight or center of gravity must be documented by a mechanic in the aircraft weight and balance papers. This could be something as minor as a new antenna, or as major as removing the landing gear and installing floats. As a pilot, you should be sure you are always using the most recent weight and balance information for your calculations.

The datum is at the front face of the firewall.

The datum is 109.7 inches ahead of the center of the main gear.

Figure 8-24. Although the datum is different in each of these airplanes, it is used in the same way for weight and balance calculations.

TERMS DESCRIBING THE LOADED AIRPLANE

Ramp weight is the term used to describe the airplane loaded for flight prior to engine start. Subtracting the fuel burned during engine start, runup, and taxi, yields the **takeoff weight**. This is the weight of the airplane just before you release the brakes to begin the takeoff

SECTION B ■ **Weight and Balance**

Tail Stands

Some airplanes have a tail stand, a temporary support to hold up the rear fuselage during loading. When the back of the airplane is loaded before the front, it can rock back on its tail before enough weight can be loaded in front to balance things out. Installing the stand keeps the airplane level until the load is balanced. But sometimes people forget to prop up the tail, and...

roll. **Landing weight** is the takeoff weight minus the fuel burned enroute. To determine an airplane's **useful load**, either prior to engine start or at takeoff, you must subtract the basic empty weight from ramp weight or takeoff weight respectively. The useful load includes the weight of the flight crew and usable fuel, as well as any passengers, baggage, and cargo. **Payload** is the term used for the weight of only the passengers, baggage, and cargo. Adding the weight of the flight crew and usable fuel to the payload is another way to determine useful load. Some POHs might refer to total weight or gross weight, which are general terms used to describe the weight of the airplane and everything carried in it.

Aircraft manufacturers try to build some mission flexibility into their designs. This means you might have the choice of filling all the seats with passengers if you carry a reduced fuel load, or of having a long cruising range with full fuel tanks and less payload. Because very few airplanes can handle a full cabin and full fuel tanks, you must balance your needs within the capabilities of your airplane and your flying skills.

Saturn 5 Payload Vs. Useful Load

The Saturn 5 moon rocket was an extreme example of the difference between useful load and payload. The payload was the Apollo Command Module, the Service Module, and the Lunar Excursion Module. These components weighed about 109,000 pounds, depending on the mission. The weight of the fuel (and oxidizer) added about 5,785,500 pounds, for a useful load of 5,894,500 pounds. The empty weight of the boosters added another 528,500 pounds, for a total weight of 6,423,000 pounds at liftoff.

Courtesy of NASA

Many aircraft have different maximum weight limits for different stages of flight. For example, the POH might list a **maximum ramp weight**, which is the maximum allowed for ground operations, such as taxiing. It's usually just a little more than the **maximum takeoff weight**. The difference allows for the weight of fuel used in engine start, taxiing to the runway, and run-up checks. Although it does not amount to much in a relatively fuel-efficient light airplane, it might be several thousand pounds for a large transport. An aircraft's **maximum landing weight** is determined by the manufacturer and is based on the amount of stress the aircraft's structure can withstand during landing. In most small aircraft, the difference between the maximum landing weight and other maximum weight limitations is typically only a few pounds. Anytime an overweight landing occurs, maintenance personnel should be notified because serious damage to the aircraft structure and landing gear is a distinct possibility. It should be noted that maximum gross weight is a general term used in some POHs to describe the airplane's maximum weight limitation set by the manufacturer. [Figure 8-25]

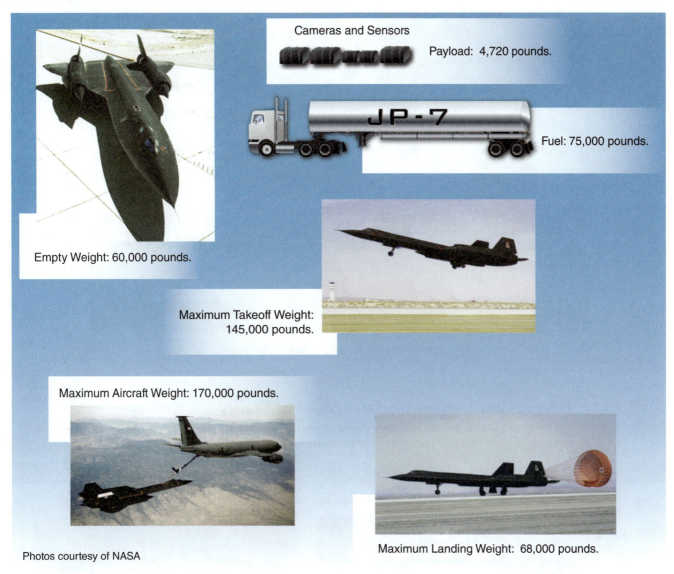

Cameras and Sensors

Payload: 4,720 pounds.

JP-7

Fuel: 75,000 pounds.

Empty Weight: 60,000 pounds.

Maximum Takeoff Weight:
145,000 pounds.

Maximum Aircraft Weight: 170,000 pounds.

Photos courtesy of NASA

Maximum Landing Weight: 68,000 pounds.

Figure 8-25. Inflight refueling makes it possible for the Lockheed SR-71's maximum weight in flight to exceed the maximum takeoff weight by 25,000 pounds. The maximum landing weight indicates that the airplane lands with little fuel remaining in the tanks.

SECTION B ■ **Weight and Balance**

The fuel available during flight is called **usable fuel**, and you will need to account for its weight in your weight and balance calculations. Gasoline weighs 6 pounds per gallon. This makes it very easy to calculate fuel weight by multiplying the number of gallons by 6 to get the weight in pounds. Oil weighs more—7 1/2 pounds per gallon. Because oil is usually measured in quarts, each quart weighs 1 7/8 pounds.

PRINCIPLES OF WEIGHT AND BALANCE

Children on a seesaw illustrate the basic ideas involved in balance theory. The seesaw is balanced when children who weigh the same amount sit an equal distance from the fulcrum. [Figure 8-26] Because children of different sizes often play together, they have to compensate by shifting their weight to make the seesaw balance. Although unaware of it, what they are doing is moving the center of gravity so it is directly over the fulcrum. While the principles of weight and balance are straightforward, applying them is not exactly child's play. With a small amount of work you can apply these principles to loading your airplane safely.

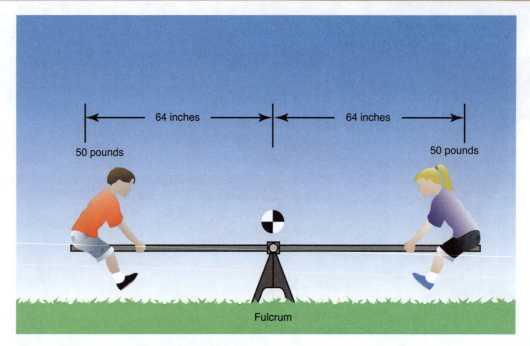

Figure 8-26. To balance the seesaw, the center of gravity must be over the fulcrum.

Landing Gear Configuration and Center of Gravity

The center of gravity is ahead of the main wheels in airplanes with tricycle landing gear. Airplanes with tailwheels (conventional gear) have their CG behind the main wheels.

This Piper design was created with conventional gear as the Pacer in 1950. When tricycle gear was offered as an option in 1952, it became the Tri-Pacer. [Figure A] The owner of this Tri-Pacer has converted his airplane to the tailwheel configuration by installing new landing gear legs that move the main wheels ahead of the CG. [Figure B]

Courtesy of Brian Thomas

ARM AND MOMENT

The name for a distance from the datum is **arm**. On airplanes, the distances are generally measured in inches, and by tradition, distances aft of the datum are positive numbers and forward of the datum are negative. Fuselage station (abbreviated F.S. or sta.) is another term for the arm.

A weight on the end of an arm creates a **moment**. Moment is a measurement of the tendency of the weight to cause rotation at the fulcrum. While an arm is simply a length, a moment is the length multiplied by a weight. When the weight in pounds and the arm in inches are multiplied together, the resulting moment is expressed in pound-inches. [Figure 8-27]

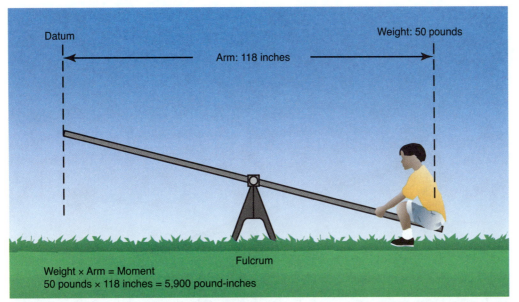

Figure 8-27. Moment is an arm multiplied by the weight. In this example the datum is the left end of the seesaw, and any actual rotation would occur at the fulcrum.

CALCULATING THE POSITION OF THE CG

The position of the CG of the airplane usually is expressed in inches from the datum. Because a measurement from the datum is an arm, the term CG arm also is used to describe the location of the CG. To find the center of gravity of an object or a group of objects, the moments of all the parts are added, and this total is divided by the total weight of the parts. Figure 8-28 shows how to do this using the seesaw example.

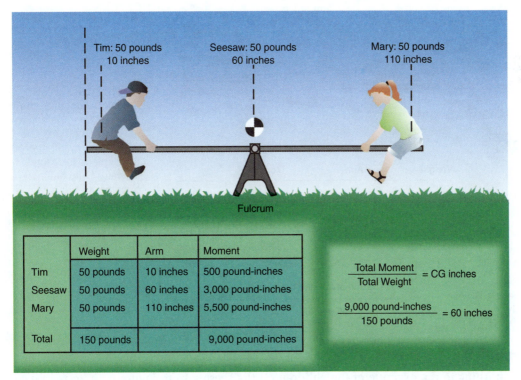

Figure 8-28. Using the left end of the seesaw as the datum, the position of the CG is calculated using the children's weights and the weight of the seesaw itself.

When looking at Figure 8-28 you could probably guess that the CG was over the fulcrum, because the seesaw was in balance. Now, try to calculate the CG location using two new children with different weights. [Figure 8-29]

SECTION B ■ **Weight and Balance**

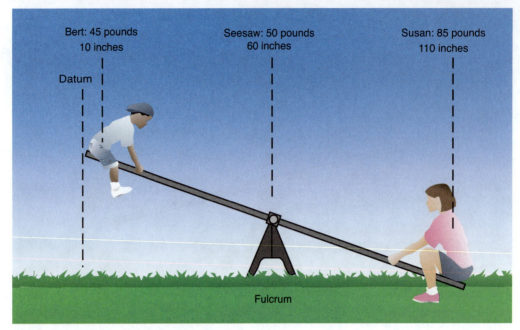

Figure 8-29. Where is the CG? You should get a total moment of 12,800 and a total weight of 180 pounds, for an answer of 71.1 inches.

You can use the same technique to find the CG of an airplane. Substitute the weight and moment of the airplane for the weight and moment of the seesaw. Combine it with the weight and moment of the pilot to obtain the new CG location, as shown in figure 8-30.

Empty Airplane:
Arm : 70 inches
Weight: 1,150 pounds

Pilot:
Arm : 73 inches
Weight: 135 pounds

Figure 8-30. Multiply the pilot's weight by the distance from the datum to get her moment. The weight and moment of the airplane are found in its weight and balance documents.

	Weight		Arm		Moment	
Empty Airplane	1,150 pounds	X	70 inches	=	80,500	pound - inches
Pilot	135 pounds	X	73 inches	=	9,855	pound - inches
Total	1,285 pounds				90,355	pound - inches

$$\frac{\text{Total Moment}}{\text{Total Weight}} = \text{CG Arm} \qquad \frac{90,355 \text{ pound-inches}}{1,285 \text{ pounds}} = 70.3 \text{ inches}$$

SHIFTING WEIGHT TO MOVE THE CG

Returning to the example of the children on the seesaw from Figure 8-29, to rearrange the children so that the seesaw balances, you should move the CG to the fulcrum. To do this, move the larger child, Susan, toward the datum so that her weight acts through a shorter arm, reducing her moment. [Figure 8-31] Of course, all of this can be described mathematically, as you will learn later.

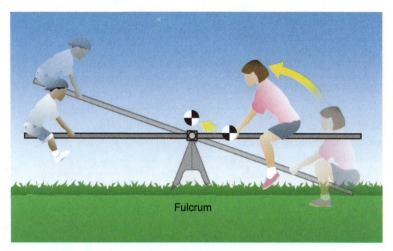

Figure 8-31. Shortening the arm reduces the moment. As Susan moves toward the datum, her moment is reduced, which moves the CG toward the fulcrum until the seesaw is balanced.

Fulcrum

DETERMINING TOTAL WEIGHT AND CENTER OF GRAVITY

In the seesaw examples, you used what is called the computation method. It demonstrates the principles of weight and balance most thoroughly. Using the computation method for airplanes requires multiplying and adding up large numbers, and there are plenty of opportunities to make mistakes, even if you use your calculator. To simplify the process, many manufacturers provide tables and/or graphs in the POH. You should be able to use all three methods (computation, table, or graph), because weight and balance information in your POH may be in any of the different formats.

Fortunately, there is only one way to calculate total weight. All you have to do is add the weight of your passengers, baggage, cargo, fuel, and yourself to the empty weight of the airplane. If you are working from a licensed empty weight, you would also include oil. You should try to be as accurate as possible and use the actual weight of each person and item of baggage, rather than trying to estimate.

COMPUTATION METHOD

Normally, the first step in any weight and balance computation is to see if the weight of what you want to load is within the maximum weight limits. Begin with the weight of the empty airplane and make a list of all the people and items you intend to load, noting the actual weight of each and including the weight of fuel. You might choose to add the weight of usable fuel at first and then subtract fuel to reduce the total weight to the maximum permitted, or you might calculate the total amount of fuel needed for the flight (with reserves) and add its weight to your list of passengers and baggage. You can use a printed weight and balance worksheet, or use the simple format shown here. [Figure 8-32]

If the total weight is greater than the maximum weight limit, you will have to leave something behind; either payload or fuel or some of each. Once you have the total weight within the limits, calculate the location of the center of gravity. You will need to fill out the remaining two columns on your original list. One column is for the arm (in inches) of each item.

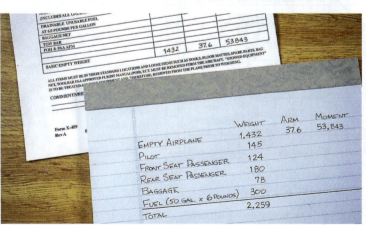

Figure 8-32. You can find the empty weight and CG location on the airplane's weight and balance form. This airplane has a maximum takeoff weight of 2,500 pounds, so the total weight of the listed items is below the maximum.

	WEIGHT	ARM	MOMENT
EMPTY AIRPLANE	1,432	37.6	53,843
PILOT	145		
FRONT SEAT PASSENGER	124		
REAR SEAT PASSENGER	180		
BAGGAGE	78		
FUEL (50 GAL. × 6 POUNDS)	300		
TOTAL	2,259		

SECTION B ■ **Weight and Balance**

8-33

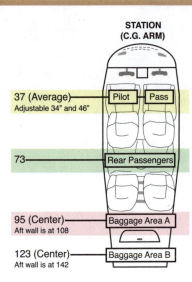

**STATION
(C.G. ARM)**

37 (Average)
Adjustable 34" and 46"

73

95 (Center)
Aft wall is at 108

123 (Center)
Aft wall is at 142

LOADING ARRANGEMENTS

- Usable fuel CG arm is 46 inches.
- Maximum baggage in Area A is 120 pounds.
- Maximum baggage in Area B is 50 pounds.
- Maximum combined baggage
  (Areas A and B) is 120 pounds.

Remember, the arm is just the distance from the datum, and the manufacturer usually provides a diagram to help you to find it. If the chart gives a fore and aft position for adjustable seats, you can adjust the arm dimensions for people whose seats will be further forward or aft than average. For instance, if you always fly with the seat one inch back from the full-forward position, you would use 35 inches as your own arm in the example shown, because it is one inch behind the forward position of 34 inches. If your front seat passenger runs the seat all the way aft for an in-flight nap, you would use 46 inches for her arm.

Arm for pilot and front seat passenger is 37 inches.

Arm for rear passenger is 73 inches.

Arm for baggage is 95 inches.

Arm for fuel is 46 inches.

In the last column, enter the moment for each item, obtained by multiplying each weight by the corresponding arm (weight arm = moment). Adding up the moment column gives the total moment in pound-inches. Divide the total moment by the total weight to get the arm of the overall center of gravity in inches aft of the datum. Then check to see that the center of gravity falls between the forward and aft CG limits. [Figure 8-33]

	WEIGHT	ARM	MOMENT
EMPTY AIRPLANE	1,432	37.6	53,843
PILOT	145	37	5,365
FRONT SEAT PASSENGER	124	37	4,588
REAR SEAT PASSENGER	180	73	13,140
BAGGAGE	78	95	7,410
FUEL (50 GAL. x 6 POUNDS)	300	46	13,800
TOTAL	2,259		98,146

TOTAL MOMENT ÷ TOTAL WEIGHT = CG INCHES
98,146 ÷ 2,259 = 43.5 INCHES

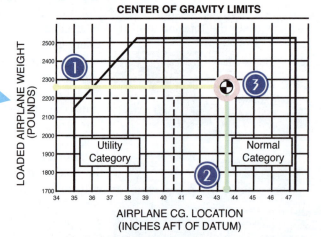

CENTER OF GRAVITY LIMITS

LOADED AIRPLANE WEIGHT (POUNDS)

AIRPLANE CG. LOCATION
(INCHES AFT OF DATUM)

Utility Category

Normal Category

Figure 8-33. Use the diagram from the POH to find the arm for each item. When you have calculated the loaded CG, use the CG limits graph from the POH to see if the loading is acceptable.

1. Enter the graph at your total weight on the left side.

2. Draw a line from the bottom at your calculated CG.

3. The intersection of the lines is within the CG limits, so the loading is acceptable.

SECTION B ■ Weight and Balance

TABLE METHOD

The table method uses a series of tables provided by the manufacturer to eliminate the multiplication and division, but not the addition. A **moment table** is provided for each of the most common payload areas, such as front seats, rear seats, usable fuel, and baggage area. The manufacturer has taken various weights and multiplied them by the arm for that location to obtain the numbers in the table. To find the center of gravity, start with your list of weights as before. Then look up the weight for each item in the appropriate table and read the moment given next to it. Write the moment for each item on your worksheet list and total it. Then, find the total weight on the **moment limits table** and read across to see if your number falls between the limits given for that weight. [Figure 8-34] If not, you will need to rearrange the load. The table method uses rounded weights and approximate arms, so it is not as precise as the computation method, but this method eliminates some of the chances for arithmetic errors. Notice that the table does not state the CG range in inches from the datum. Instead, it gives moment limits for a whole range of total weights. This eliminates the computation method's final step of dividing the moment by the total weight.

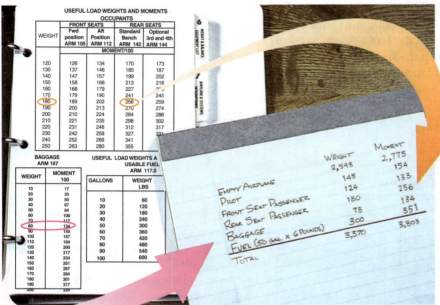

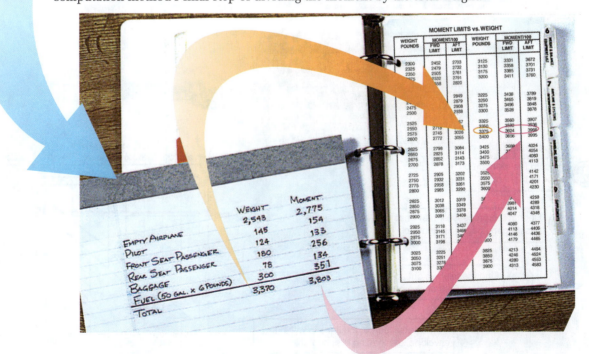

Figure 8-34. The airplane builder supplies tables with the moment already multiplied for you. Read the moment next to the weight for each item in the useful load. If the total moment falls between the moment limits in the table for that weight, the airplane will be loaded safely.

SECTION B ■ **Weight and Balance**

You might have noticed that the moments you were using in examples of the computation method were much larger than the moments provided in the tables in figure 8-34. When the manufacturer multiplies these rounded weights by the average arms, the last digits are inevitably zeros. To make the numbers more manageable, they drop these zeros from the table, in effect dividing the moments by 100 or 1,000. There will always be a note on the table to indicate this reduction factor. The reduction factor can be ignored when using the moment limits table, because the moments listed are reduced by the same factor. Remember to correct for the reduction factor by adding the appropriate number of zeros to the moment if you calculate the CG location by dividing total moment by total weight. It also is important to keep track of reduction factors when using electronic flight computers to solve weight and balance problems.

GRAPH METHOD

This method is similar to the table method, except that the values from the tables have been combined and plotted on a graph for you. This allows you to use values between the increments published in a table, so you can use actual weights instead of rounded values. Only two graphs are necessary for this method. The **loading graph** is used to find the moment for the loads you intend to put into the airplane, and the **center of gravity moment envelope** tells you if your proposed loading is within the weight and balance limits. [Figures 8-35 and 8-36] As with the table method, the graphs usually employ a reduction factor of 100 or 1,000.

Figure 8-35. Use the graph to find moments for each item in the load. After adding them, check the total moment against the total weight on the CG moment envelope graph.

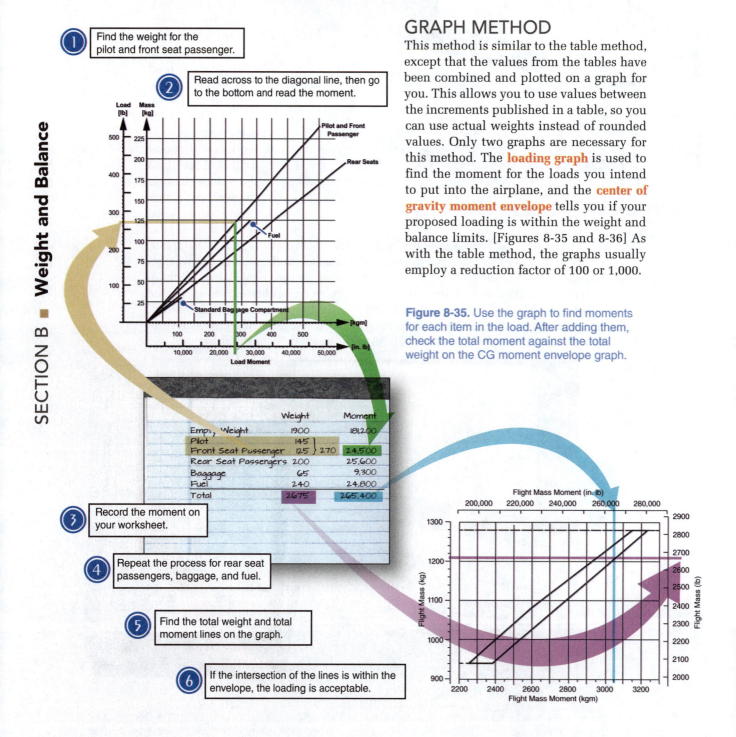

SECTION B ■ Weight and Balance

1. Find the weight for the pilot and front seat passenger.

2. Read across to the diagonal line, then go to the bottom and read the moment.

3. Record the moment on your worksheet.

4. Repeat the process for rear seat passengers, baggage, and fuel.

5. Find the total weight and total moment lines on the graph.

6. If the intersection of the lines is within the envelope, the loading is acceptable.

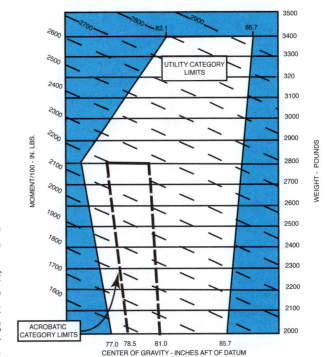

Figure 8-36. Some POHs use a different style for the graph. Gross weight is just another term for total weight.

GROSS WEIGHT MOMENT LIMITS

WEIGHT-SHIFT FORMULA

Sometimes during weight and balance computations, you will find that the CG falls outside acceptable limits. You could rearrange the weights of passengers and baggage and calculate the CG location over and over, but there is an easier way. The following formula helps you to compute exactly what is necessary to bring the CG within limits.

$$\frac{\text{Weight Moved}}{\text{Weight of Airplane}} = \frac{\text{Distance CG Moves}}{\text{Distance Between Arms}}$$

To use the formula, you supply three of the variables and solve for the fourth. [Figure 8-37]

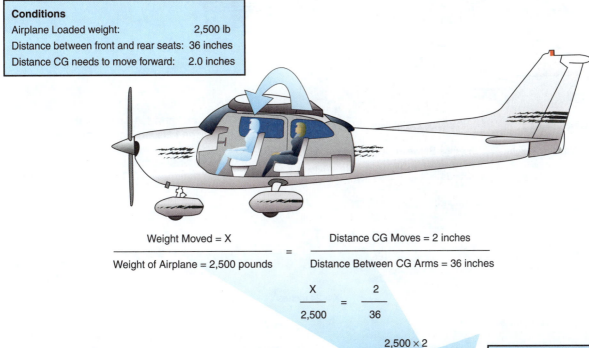

Conditions

Airplane Loaded weight:	2,500 lb
Distance between front and rear seats:	36 inches
Distance CG needs to move forward:	2.0 inches

$$\frac{\text{Weight Moved} = X}{\text{Weight of Airplane} = 2,500 \text{ pounds}} = \frac{\text{Distance CG Moves} = 2 \text{ inches}}{\text{Distance Between CG Arms} = 36 \text{ inches}}$$

$$\frac{X}{2,500} = \frac{2}{36}$$

$$X = \frac{2,500 \times 2}{36}$$

Moving 138.8 pounds to the front seat will change the CG by the required two inches.

Figure 8-37. Solving the weight-shift formula for weight to be moved tells you how many pounds must be moved from one seat to the other to bring the CG within limits.

$$X = 138.8 \text{ pounds}$$

SECTION B ■ **Weight and Balance**

You also can solve for the distance a specific weight would need to move. On the other hand, you can solve for how much the CG will change if you move a given weight a specified distance. [Figures 8-38 and 8-39]

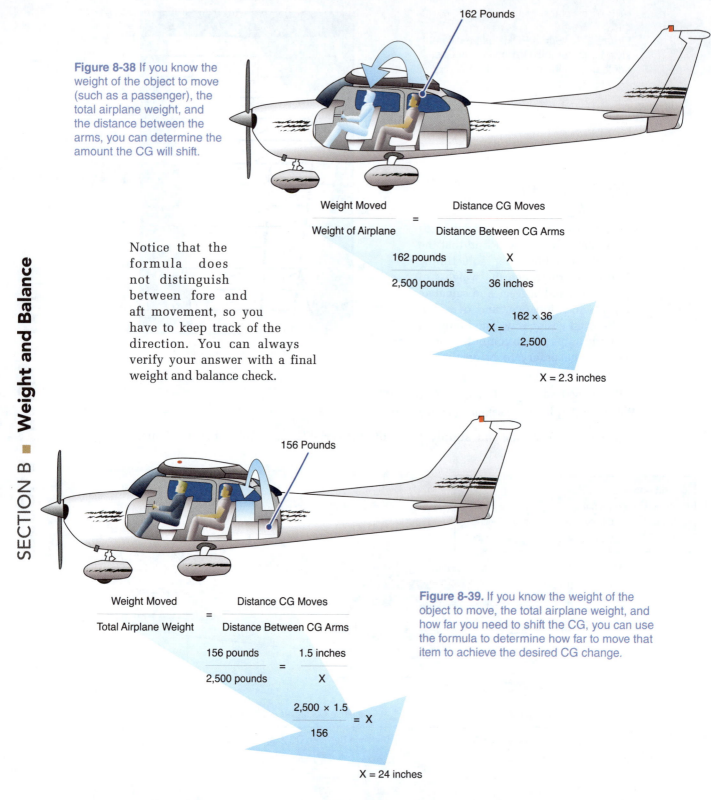

Figure 8-38 If you know the weight of the object to move (such as a passenger), the total airplane weight, and the distance between the arms, you can determine the amount the CG will shift.

162 Pounds

$$\frac{\text{Weight Moved}}{\text{Weight of Airplane}} = \frac{\text{Distance CG Moves}}{\text{Distance Between CG Arms}}$$

$$\frac{162 \text{ pounds}}{2,500 \text{ pounds}} = \frac{X}{36 \text{ inches}}$$

$$X = \frac{162 \times 36}{2,500}$$

$$X = 2.3 \text{ inches}$$

Notice that the formula does not distinguish between fore and aft movement, so you have to keep track of the direction. You can always verify your answer with a final weight and balance check.

156 Pounds

$$\frac{\text{Weight Moved}}{\text{Total Airplane Weight}} = \frac{\text{Distance CG Moves}}{\text{Distance Between CG Arms}}$$

$$\frac{156 \text{ pounds}}{2,500 \text{ pounds}} = \frac{1.5 \text{ inches}}{X}$$

$$\frac{2,500 \times 1.5}{156} = X$$

$$X = 24 \text{ inches}$$

Figure 8-39. If you know the weight of the object to move, the total airplane weight, and how far you need to shift the CG, you can use the formula to determine how far to move that item to achieve the desired CG change.

DISCOVERY

Lilienthal Glider Shifting CG

Before the Wright brothers flew, German aviation pioneer Otto Lilienthal designed and flew a series of successful gliders. The gliders had no flight controls and no movable control surfaces. The pilot controlled them by swinging his legs and body to shift the center of gravity.

Professor R.H. Wood of Johns Hopkins University was a friend of Lilienthal, and had this to say after watching him fly:

What impressed me most was the tremendous amount of athletic work necessary to balance the machine. He was never still a moment, swinging his legs from side to side, and on landing he was all out of breath, though I doubt if he was in the air thirty seconds. It seemed to require as much exercise as a 100 yard dash.

One of your responsibilities as pilot in command is to assess your own physical condition and fitness to fly. You might not have to be as athletic as Herr Lilienthal to control your aircraft, but at the very least you should be rested, alert, and healthy.

DISCOVERY

"I Had All These Goats"

Merle K. Smith, an Alaskan bush pilot (who later became president of Cordova Airlines) recalls some CG problems:

In the old days we used to have what is now called unusual cargo. We called it good paying freight then.

There was a homesteader who decided that he needed fresh milk, so he sent to Seattle and ordered some goats. I loaded these goats into one of the old planes and started up there. Well, I had forgotten that goats like rope, and I had tied these goats to various places in the airplane with ropes so that they wouldn't get loose and annoy me or endanger the flight.

In about twenty minutes they had all the ropes chewed through. So then I had all these goats. They'd all frolic together. There must have been eight hundred pounds of goats; and they'd all get in the back of the airplane and I couldn't hold the nose down. They would all come up and chew on me a little bit, and chew my shirt, and I'd slap them away. But I decided not to do that any more, because the first time I did that, they all went to the back of the airplane. It got to be kind of a serious situation. By the time we got to where I could land these goats they had most of the fabric eaten off the inside of the airplane, and my shirt; and the seat I was sitting on was pretty well stripped of upholstery. But these little things happened. If you came out all right, as in this instance— well, it was funny.

As quoted in *The American Heritage History of Flight*

EFFECTS OF OPERATING AT HIGH TOTAL WEIGHTS

When you learned about the four forces, you found that lift must equal weight in level flight. If more weight is added, the wing needs to generate more lift. Changing the position of the CG also affects the total lift the wing must produce, even when the airplane's weight remains constant, because the wing must also counteract the tail-down force. You know that lift can be increased somewhat with additional speed or a higher angle of attack, but there is an upper limit to how much weight any wing can carry. You also know that the effective weight varies when you maneuver the airplane in flight. Remember how the load factor increases in turns, or in the pullout from a dive? If the wing is near its maximum load-carrying capacity, and the load factor is increased by a turn, a sudden pull-up, or turbulence, the structure of the wing could be damaged or fail. Engineers create an airplane with specific load-factor capabilities in mind. The landing gear is designed to support a certain amount of weight and absorb reasonable landing loads, and wings are designed to lift the airplane and its contents and still allow for load factors imposed by gusts, turbulence and maneuvers. Even though airplanes are built with a considerable margin of safety, the pilot is responsible for seeing that load factors remain within safe limits.

The specific problems of operating at or near the maximum weight limit are generally related to the wing having to fly faster or at a higher angle of attack to generate the additional lift required. When operating near the maximum weight limit, the takeoff roll is longer, because the airplane accelerates more slowly and also needs to reach a higher speed to generate enough lift for takeoff. Both angle and rate of climb are reduced from that achieved at lower weights. In cruise, range is reduced and speed is lower at any given power setting, because more energy is being used to generate lift and overcome the resulting induced drag. Because the wing is already flying at a higher angle of attack just to maintain level flight, it is that much closer to its stalling angle, and stalls at a higher speed than when lightly loaded. As you would expect, the brakes have to work harder to slow down a heavy plane. This, combined with the higher touchdown speeds, results in longer landing distances. Most of the performance numbers in the POH are found by testing the airplane at maximum weight, but the person who flies an overloaded airplane has just decided to be a test pilot, and is flying the airplane outside of the envelope, in areas that might not have been explored in flight testing.

FLIGHT AT VARIOUS CG POSITIONS

As you learned from the discussion of longitudinal stability, most airplanes are designed to fly with the CG slightly forward of the center of pressure. This gives the airplane a small nose-down moment. This tendency to nose down is counteracted by the tail exerting a down force. It acts as a sort of upside-down wing generating a small amount of lift in a downward direction. Arranging the forces this way makes the airplane more stable. If the nose is pitched up, the angle of attack of the tail is reduced, so it generates less down force, allowing the nose to drop. On the other hand, if the nose is pitched down, the tail creates more down force, raising the nose. This stabilizing effect increases as the CG is moved farther forward. A certain amount of stability is helpful, especially in turbulence or when flying by reference to instruments, but too much actually makes the aircraft harder to control, because elevator control input is resisted by the stability of the airplane. With the airplane nose-heavy, the stall speed also is higher. The forward CG requires a greater tail down-force, and this force is equivalent to adding weight. Because the wing must fly at a higher angle of attack to generate the lift to counteract the greater tail-down force, it is closer to its stalling angle of attack for any given speed.

With the CG at or behind the center of pressure, a conventional airplane is unstable in pitch. Because there is no automatic restoring force, when a small bump or control input starts the nose up, the nose continues to pitch up more and more unless the pilot acts. This can happen very quickly, and it is possible the force required to push the nose back down could exceed the aerodynamic capability of the elevators or stabilator. Although the control forces actually become lighter as the CG moves aft, the pilot's workload can be much higher. An unstable airplane requires constant attention and continuous control input even in calm air, and it is all too easy to overcontrol. Overcontrolling can result in dangerously high flight load factors, structural damage, and the breakup of the airframe. An aft CG makes it much easier to enter an accidental stall or spin, and the tail-heavy condition might cause the spin to be flat, characterized by a nearly level pitch attitude, rapid rotation rate, and high sink rate. Recovery might be impossible.

What about lateral stability and CG location? In most light airplanes, you control this by managing the fuel load between wing tanks. Small differences in weight distribution left or right of the CG will cause wing heaviness and reduce flight efficiency, but large imbalances can compromise stability and control. [Figure 8-40]

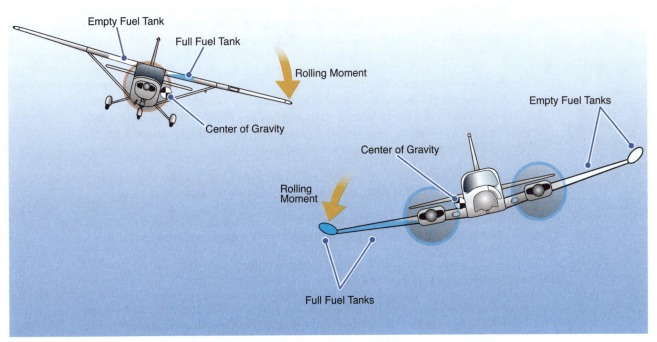

Figure 8-40. The main influence on lateral CG is fuel balance between the wings.

Cross-Country Racers

Flying the airplane with the CG forward enhances stability, because the tail-down force increases the tendency of the airplane to return to level flight if disturbed. However, it reduces fuel economy and speed slightly. On the other hand, cross-country racers try to fly with their CG at its aft limit, very close to the center of pressure of the wing. Because the wing is lifting only the weight of the airplane (and not the additional "weight" of tail down-force), the wing can be flown at a lower angle of attack, with less induced drag, giving a little more speed and better fuel efficiency. The difference is not great, but in a race, every little improvement helps. Monocoupes like these have been raced since the 1930's.

SUMMARY CHECKLIST

✓ Both the amount and the distribution of weight affect aircraft performance.

✓ The reference datum is the location from which all horizontal distances are measured for weight and balance purposes.

✓ An arm is a distance from the datum. Measurements aft of the datum are generally positive numbers, while those forward of the datum are negative numbers. A moment is a weight multiplied by an arm.

✓ To compute the location of the CG, add the moments for each item of useful load to the moment of the empty airplane and divide the total moment by the total weight.

✓ Ramp weight is the term used to describe the airplane loaded for flight prior to engine start. Subtracting the fuel burned during engine start, runup, and taxi, yields the takeoff weight. Landing weight is the takeoff weight minus the fuel burned enroute.

✓ To determine an airplane's useful load, either prior to engine start or at takeoff, you must subtract the basic empty weight from ramp weight or takeoff weight respectively. The useful load includes the weight of the flight crew and usable fuel, as well as any passengers, baggage, and cargo. Payload is the term used for the weight of only the passengers, baggage, and cargo.

✓ The maximum weight may be divided into categories such as maximum ramp weight, maximum takeoff weight, and maximum landing weight.

✓ When performing calculations, the empty weight, moment, and center of gravity information is obtained from the individual aircraft's weight and balance records.

✓ The pilot's operating handbook provides tables and/or graphs to help find the moment of occupants, baggage and fuel.

✓ An overloaded airplane will have diminished performance. It will have a longer takeoff roll, lower angle and rate of climb, higher stall speed, reduced range and cruise speed, and a longer landing roll than a properly loaded airplane.

✓ Moving the CG forward increases stability, due to the increased tail-down force required for trimmed flight. The airplane will also stall at a higher speed, due to the increased wing loading.

✓ If the CG is located ahead of the established CG range, the elevator might not have sufficient force to raise the nose for landing.

✓ If an airplane is flown with the CG aft of the CG range, it will be less stable in pitch. It will be difficult to control, and if a stall or spin is entered, it might be impossible to recover.

✓ Even when an airplane is loaded within CG limits, its handling characteristics will vary with the location of the CG.

KEY TERMS

Center of Gravity (CG)

CG Limits

Reference Datum

Basic Empty Weight

Unusable Fuel

Licensed Empty Weight

Ramp Weight

Takeoff Weight

Landing Weight

Useful Load

Payload

Maximum Ramp Weight

Maximum Takeoff Weight

Maximum Landing Weight

Usable fuel

Arm

Moment

Moment Table

Moment Limits Table

Loading Graph

Center of Gravity Moment Envelope

QUESTIONS

1. What is the term for the reference plane from which all horizontal measurements are made for weight and balance calculations?

2. How is unusable fuel defined?

3. What constitutes the difference between basic empty weight and licensed empty weight?
   A. The weight of usable oil
   B. The weight of unusable oil
   C. The weight of unusable fuel

4. If your weight check shows that the airplane will exceed the maximum ramp weight specified in the POH, which of the following actions would be appropriate?
   A. Take off at that weight, but not taxi.
   B. Rearrange the load to bring it within CG limits.
   C. Remove passengers, cargo, or fuel until the weight limit is reached.

5. What is the weight of 42 gallons of aviation gasoline?
   A. 242 pounds
   B. 248 pounds
   C. 252 pounds

SECTION B ■ **Weight and Balance**

6. The moment for a rear-seat passenger is 14,800. If this passenger sits in a front seat, will her moment be higher or lower?

7. True/False. Moment is usually expressed in inches.

8. You are planning a trip with three friends. You and your front seat passenger weigh a total of 375 pounds. The rear seat passengers weigh 250 pounds. The duffle bags in the standard baggage compartment weigh 65 pounds. Using the accompanying graph, find the total moment of the passengers and bags.

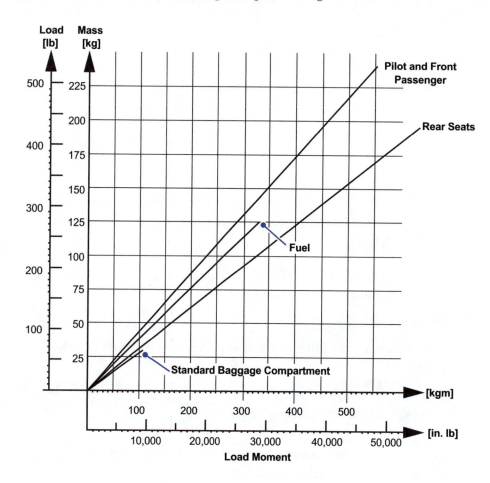

9. The airplane described in Question 8 has an empty weight of 1,900 pounds, and the empty moment is 181,200 inch-pounds. If it has a maximum weight of 2,822 pounds, how much usable fuel can be loaded and still remain within weight and CG limits? (Use the accompanying moment envelope.)

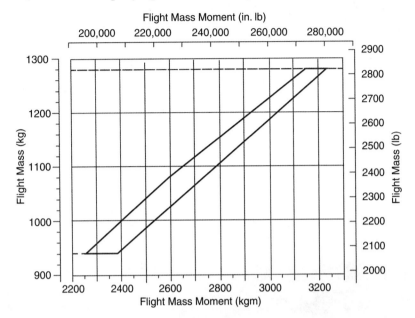

10. The center of gravity of the fully loaded airplane in Question 9 will be how many inches from the datum?

11. While preparing for a flight in another airplane, you discover that the proposed loading places the CG one inch behind the rear CG limit. The total weight is 2,655 pounds. Use the weight-shift formula to determine how much weight needs to be shifted from the baggage area (arm=140) to the back seat (arm=114) to bring the CG within limits.

12. If flight had been attempted without changing the loading described in Question 11, what flight characteristics could be expected?
    A. Tail-heaviness, requiring additional forward trim
    B. Light control forces, longitudinal (pitch) instability, and a higher cruise speed for the power setting
    C. All of the above

13. Compared to a lightly loaded airplane, what flight characteristics could be expected from a heavily loaded airplane?
    A. It will stall at a higher speed, cruise at a lower speed, and have less range.
    B. It will land at a lower speed, cruise at a higher speed, and stall at a higher angle of attack.
    C. It will have a higher cruise speed and range, a reduced rate of climb, and stall at the same speed.

14. If the center of gravity is too far forward, what flight characteristics would you expect?
    A. The airplane would be less stable, and easy to overcontrol.
    B. The airplane would be more stable, and stall at a higher speed.
    C. The airplane would be less stable, and stall at a higher angle of attack.

SECTION B ■ **Weight and Balance**

SECTION C
Flight Computers

When you plan a long car trip, you estimate how far you can drive in a day and where you intend to stop. You might not worry about fuel stops and wind is not a concern unless blowing snow or dust makes driving difficult. By necessity, flight planning is more demanding and requires a certain degree of precision. The consequences of sloppy planning are seen in news reports of airplanes that make forced landings after running out of fuel and airliners that mistakenly land at the wrong airports.

With a flight computer you perform a wide variety of flight planning calculations prior to your flight and you compute in-flight values to ensure your flight is progressing as planned. The term flight computer can be used to describe hand-held mechanical devices, electronic versions of these devices, flight planning apps, electronic flight bags (EFBs), and GPS equipment. [Figure 8-41]

Pay close attention to the details to ensure you are using the correct data throughout your calculations. For example, are temperatures Fahrenheit or Celsius? Are distances statute or nautical miles? Are courses true or magnetic direction?

Figure 8-41. The solutions you get from any computer, electronic or mechanical, are only as accurate as the data that your enter.

FLIGHT PLANNING CONCEPTS

Regardless of the type of flight computer that you use, the problems you must solve and their underlying principles are the same. Some basic concepts apply to the calculations to plan a cross-country flight. For example, you must understand how the wind affects your heading and groundspeed to determine your time enroute and fuel requirements.

SECTION C ■ Flight Computers

EFFECT OF THE WIND

When an airplane moves through air that is also moving, its speed and path over the ground are affected. If you just point the airplane in the direction you want to go, you are likely to miss your destination. Because the air is almost always moving, you must routinely compensate for the effects of wind. The relationship between these different motions can be diagrammed with arrows drawn to scale—vectors—that represent the speeds and directions involved. Adding the vector for the aircraft's airspeed to the wind vector gives a resultant vector that shows the speed and direction of the airplane affected by the wind. [Figures 8-42 and 8-43]

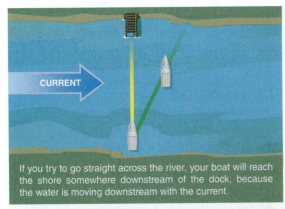

If you try to go straight across the river, your boat will reach the shore somewhere downstream of the dock, because the water is moving downstream with the current.

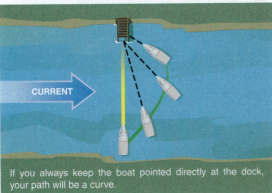

If you always keep the boat pointed directly at the dock, your path will be a curve.

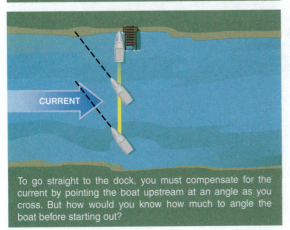

To go straight to the dock, you must compensate for the current by pointing the boat upstream at an angle as you cross. But how would you know how much to angle the boat before starting out?

Figure 8-42. When piloting a boat to a dock straight across the river, the boat moves in relation to the water, but the water is also moving.

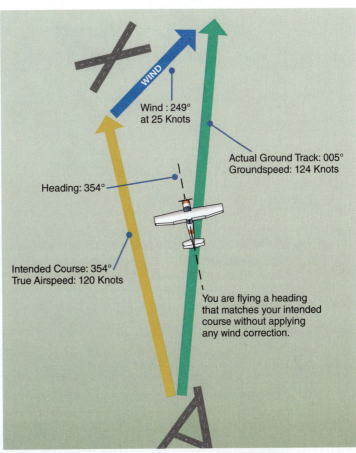

Wind : 249° at 25 Knots

Actual Ground Track: 005°
Groundspeed: 124 Knots

Heading: 354°

Intended Course: 354°
True Airspeed: 120 Knots

You are flying a heading that matches your intended course without applying any wind correction.

Figure 8-43. Like the boat on the river, an airplane in flight moves through a medium (air) that is moving. The relationship between the airplane's flight path and the wind can be represented by vectors.

SECTION C ■ **Flight Computers**

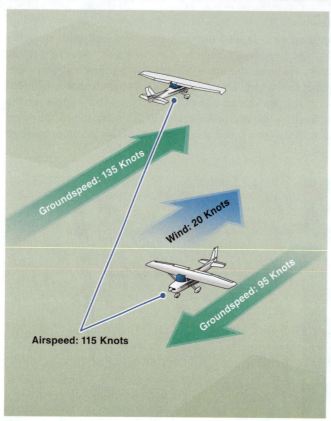

Figure 8-44. Although both airplanes have the same airspeed, their groundspeeds differ by 40 knots.

COMPENSATING FOR WIND

Wind that is aligned with the direction of flight, only acts to increase or decrease the **groundspeed**. [Figure 8-44] Headwinds reduce your groundspeed, and tailwinds increase it. When the wind is not directly aligned with your course, the airplane drifts off course. To prevent drift and stay on course, you must compensate by pointing the nose of the airplane a few degrees into the wind. Although the airplane is in coordinated flight and staying on course, a difference exists between its heading and its course. This difference is the **wind correction angle (WCA)**. Your course, corrected for wind, equals your heading. [Figure 8-45]

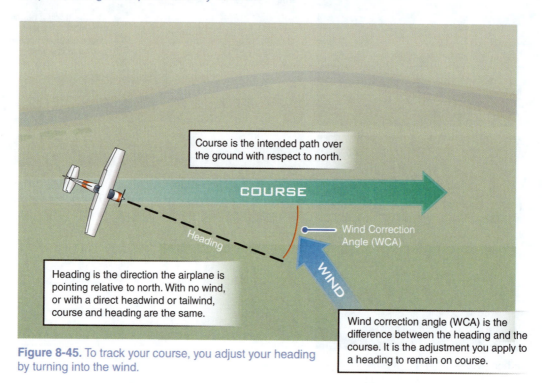

Figure 8-45. To track your course, you adjust your heading by turning into the wind.

Vectors also are used to separate the effect of the wind into two components. One component acts in the direction of flight, either increasing or decreasing the groundspeed. This part of the wind is called the **headwind component** or **tailwind component**. The other component acts perpendicular to the direction of flight. The **crosswind component** causes the airplane to drift off course. [Figure 8-46]

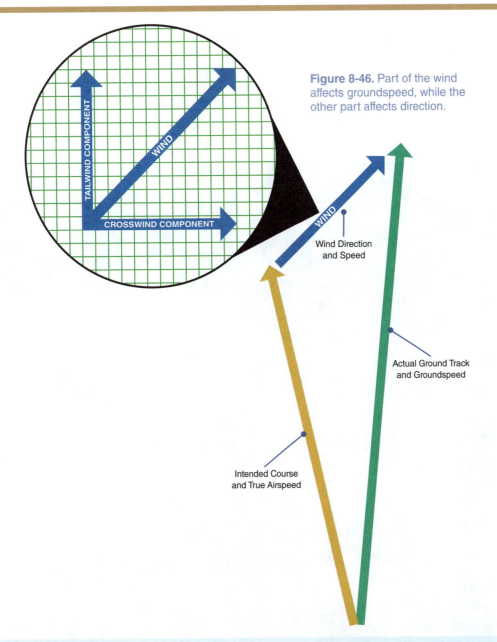

Figure 8-46. Part of the wind affects groundspeed, while the other part affects direction.

TAILWIND COMPONENT

WIND

CROSSWIND COMPONENT

WIND

Wind Direction and Speed

Actual Ground Track and Groundspeed

Intended Course and True Airspeed

Compressibility

When an airplane moves through the air at relatively low speeds, air behaves as an incompressible fluid. The air molecules move out of the way as the airplane passes. As speed increases, the molecules are unable to move aside quickly enough, and air pressure begins to build up in front of the airplane. This effect is called compressibility, because the air begins behaving as a compressible gas.

Compressibility affects the accuracy of the airspeed indicator because the increased pressure at the pitot tube results in a higher airspeed indication. Air friction at high speeds also heats the outside air temperature probe, giving higher readings. These effects are negligible at speeds below 200 knots and pressure altitudes below 20,000 feet, but can increase the indicated airspeed by as much as 19% at 530 knots and 50,000 feet. Many mechanical flight computers have a chart for applying conversion factors for compressibility at various altitudes and speeds.

COMPRESSIBILITY CORRECTION

PRESSURE ALTITUDE	F CORRECTION FACTORS FOR **TAS** CALIBRATED AIRSPEED IN KNOTS							
	200	250	300	350	400	450	500	550
10,000 FT.	1.00	1.00	.99	.99	.98	.98	.97	.97
20,000 FT.	.99	.98	.97	.97	.96	.95	.94	.93
30,000 FT.	.97	.96	.95	.94	.92	.91	.90	.89
40,000 FT.	.96	.94	.92	.90	.88	.87	.87	.86
50,000 FT.	.93	.90	.87	.86	.84	.84	.84	.84

USE CALIBRATED AIRSPEED AND PRESS. ALT. TO OBTAIN **F** FACTOR. MULTIPLY **F** FACTOR BY **TAS** TO OBTAIN THE **TAS** CORRECTED FOR COMPRESSIBILITY

DISCOVERY

ELECTRONIC FLIGHT COMPUTERS

If you are using an electronic flight computer, you must thoroughly read the instructions and understand the features of your specific app or EFB. The following discussion presents scenarios that explore the types of problems that you must complete to plan a flight on the ground and to verify your data in flight. When you use a flight planning app or EFB, many of these individual calculations are performed for you after you create a route. [Figure 8-47]

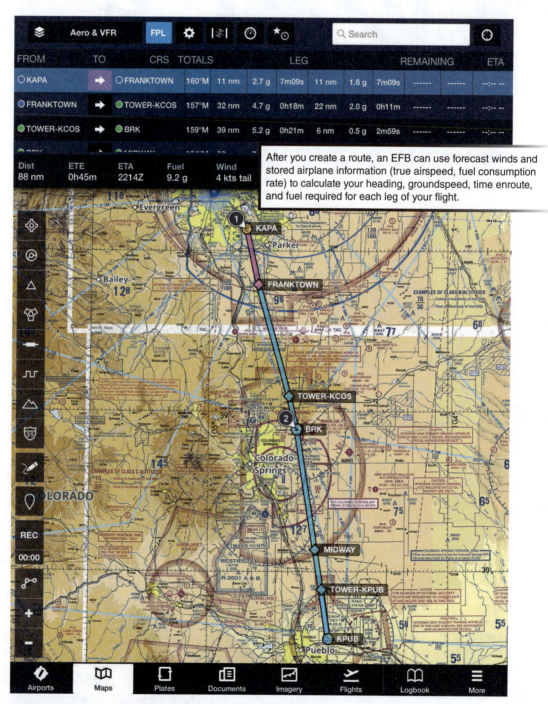

After you create a route, an EFB can use forecast winds and stored airplane information (true airspeed, fuel consumption rate) to calculate your heading, groundspeed, time enroute, and fuel required for each leg of your flight.

Figure 8-47. When you create a route in an EFB, such as ForeFlight, review the data to ensure that each value makes sense based on your airplane's performance and the flight conditions.

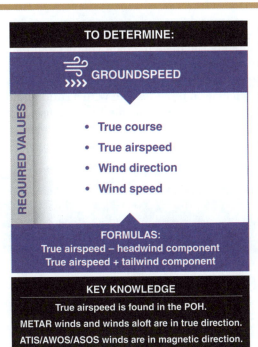

GROUNDSPEED

To determine essential data for a cross-country flight, you must first calculate your airplane's groundspeed based on your true airspeed, true course, and the winds aloft in true direction. You then use your groundspeed to calculate the time and fuel required for each leg of your flight. You obtain your true airspeed—the speed of your airplane through the air—from cruise performance charts in your POH. Your true course (TC) is the desired direction of flight as measured on a chart clockwise from true north. You will learn how to determine your true course in the Pilotage and Dead Reckoning Section of Chapter 9. [Figures 8-48 and 8-49]

Figure 8-48. Before you use any flight computer, you must understand the math and variables involved in each calculation.

Figure 8-49. Determine groundspeed by entering true course, true airspeed, and the winds aloft direction and speed.

SECTION C ■ **Flight Computers**

TIME, SPEED, AND DISTANCE

Flight computers help you quickly and accurately calculate the data you need to plan a flight. However, many calculations involve simple math that you can accomplish with a basic electronic calculator, pencil and paper, or even in your head. Knowing the basic math and the variables to use for a specific calculation provides you with a solid foundation for understanding and verifying the results regardless of the method you use. **Time, speed, and distance (TSD)** relationships are probably the most frequent problems you will solve. If you know two of these variables, you can quickly find the third. [Figure 8-50]

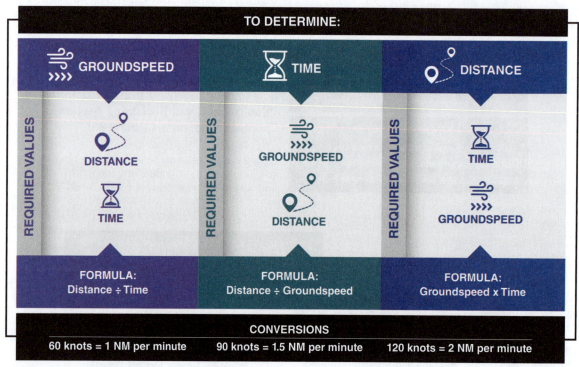

Figure 8-50. The formulas for time, speed, and distance problems involve simple multiplication and division.

RULE OF 60

Most aviation calculations involve time in hours and minutes. Because an hour consists of 60 minutes, typically these calculations require multiplying or dividing by 60, to convert to or from minutes. The first step is to find the relationship to one minute. For example, if you have flown 8 NM in 6 minutes, you would divide 8 by 6 to calculate that you have flown 1.33 NM in one minute. To find the distance you would fly in one hour, you would multiply 1.33 by 60 to get the result of 80 nautical miles—a groundspeed of 80 knots.

FLIGHT TIME

During flight planning, you typically perform time, speed, and distance problems to determine the time required to fly each leg of your flight and the total time of your proposed trip. Dividing the total distance by your speed equals the flight time. For example, a flight of 350 nautical miles will take approximately 3 hours of flight time at a speed of 120 knots (350 ÷ 120 = 2.9 hours). During flight, you monitor the time between checkpoints along your route to confirm your preflight planning. [Figure 8-51]

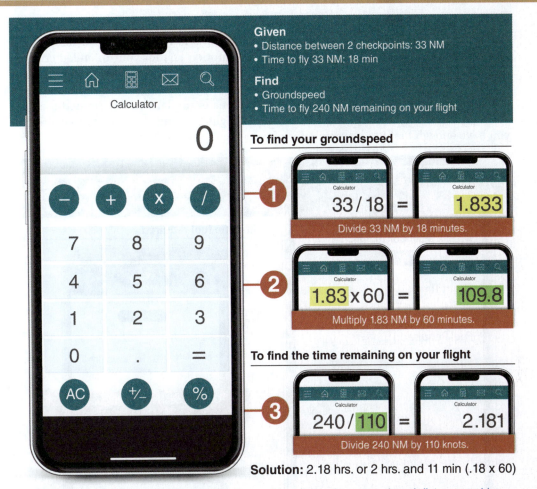

Given
• Distance between 2 checkpoints: 33 NM
• Time to fly 33 NM: 18 min

Find
• Groundspeed
• Time to fly 240 NM remaining on your flight

To find your groundspeed

1 33 / 18 = **1.833**
Divide 33 NM by 18 minutes.

2 **1.83** x 60 = **109.8**
Multiply 1.83 NM by 60 minutes.

To find the time remaining on your flight

3 240 / **110** = 2.181
Divide 240 NM by 110 knots.

Solution: 2.18 hrs. or 2 hrs. and 11 min (.18 x 60)

Figure 8-51. In this example, a calculator is used to solve a time, speed, and distance problem during flight.

FUEL CONSUMPTION

Fuel consumption calculations are similar to time, speed, and distance problems. Generally, your goal is to determine how much fuel will be used over a specific time period or distance or to determine your airplane's endurance. The key value in these computations is the **fuel consumption rate** (fuel flow). [Figure 8-52]

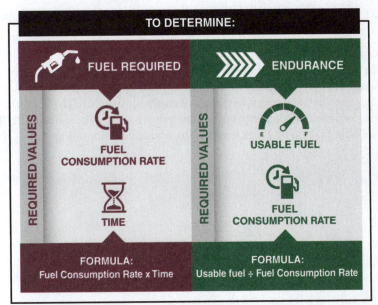

TO DETERMINE:

FUEL REQUIRED	ENDURANCE
FUEL CONSUMPTION RATE	USABLE FUEL
TIME	FUEL CONSUMPTION RATE
FORMULA: Fuel Consumption Rate x Time	FORMULA: Usable fuel ÷ Fuel Consumption Rate

Figure 8-52. Determining the fuel required or your endurance for a flight is based on the fuel consumption rate of your airplane.

SECTION C ■ **Flight Computers**

FUEL REQUIRED

The first step in most fuel consumption calculations is to refer to your airplane's POH to obtain the correct power setting and fuel consumption rate for your intended cruising altitude. With knowledge of your fuel consumption rate (fuel flow), and either time or distance, you can calculate the **fuel required** for each leg of your flight, total fuel required, and the endurance of your airplane. In the previous example, you determined that you have approximately 2 hours and 11 minutes of flight time left. You need to ensure that you have enough fuel for the remainder of your flight. [Figure 8-53]

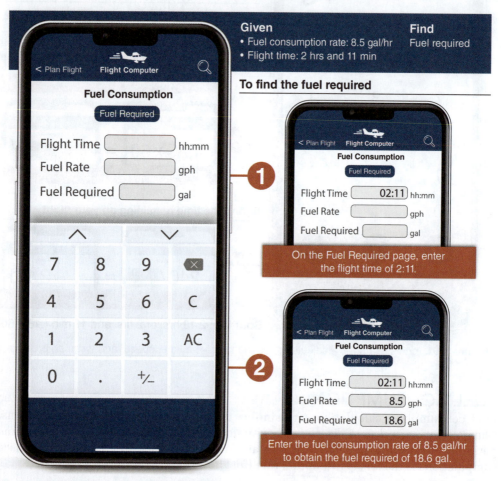

Figure 8-53 In this example, a flight computer app is used to determine the fuel required.

ENDURANCE

Endurance is the amount of time you can remain aloft, based on a known fuel quantity and consumption rate. Because fuel consumption is not constant for the entire flight, think of endurance as an approximation, not an absolute value. This calculation is similar to the previous problem, except that after entering the rate of consumption, the number of gallons available (usable fuel) determines the endurance time. [Figure 8-54]

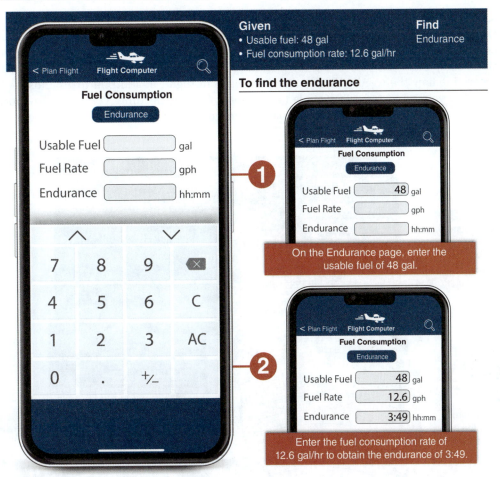

Figure 8-54. In this example, a flight computer app is used to determine endurance.

MECHANICAL FLIGHT COMPUTERS

Using a mechanical flight computer becomes easy with practice and can help you understand the math and variables that apply to each flight planning calculation. A common mechanical flight computer is the **E6B**. The E6B has two sides: the **computer side**, with several scales and small cutout windows, and the **wind side**, with a large transparent window and a sliding grid. Use the computer side for solving ratio-type problems such as time, speed, and distance, fuel consumption, and various conversions. Use the wind side to determine your heading and groundspeed based on the winds aloft.

Birth of the E6B

In the days before pocket calculators, anyone who needed to make fast, accurate computations used a slide rule. As late as the 1970s these were a primary means of making quick calculations, from high school classrooms to the engineering departments of advanced aerospace companies. A circular slide rule (with the ends of the scales joined to read continuously in powers of ten) became a very handy device in the high-performance airplanes of the 1930s. It was easy to use, and could be operated with one hand.

Back in 1936, a Naval Reserve Lieutenant named Phillip Dalton had the idea of adapting the plotting board he had invented for aircraft carriers for use in airplanes. When this device was combined with a circular slide rule, it became a compact and versatile flight computer. Hundreds of thousands were made for the military as the E6B. The mechanical flight computers of today are identical to the E6B in most respects, right down to the big ring that was used to hang the computer on a hook in large airplanes or from a lanyard around the pilot's neck in single-seaters.

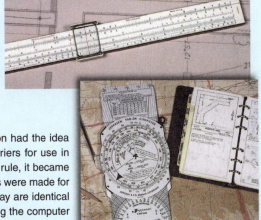

SECTION C ■ Flight Computers

GROUNDSPEED

During flight planning, you must first determine your groundspeed based on your true course, true airspeed, and the winds aloft. The wind side of the computer separates the wind components for you. Using the **azimuth plate**, **true index**, and a **wind dot**, you can determine how the wind will affect your airplane in flight. In addition to groundspeed, you find your wind correction angle to calculate your true heading. You will learn more about using your true heading to calculate the magnetic heading to fly in the Pilotage and Dead Reckoning Section of Chapter 9. [Figure 8-55]

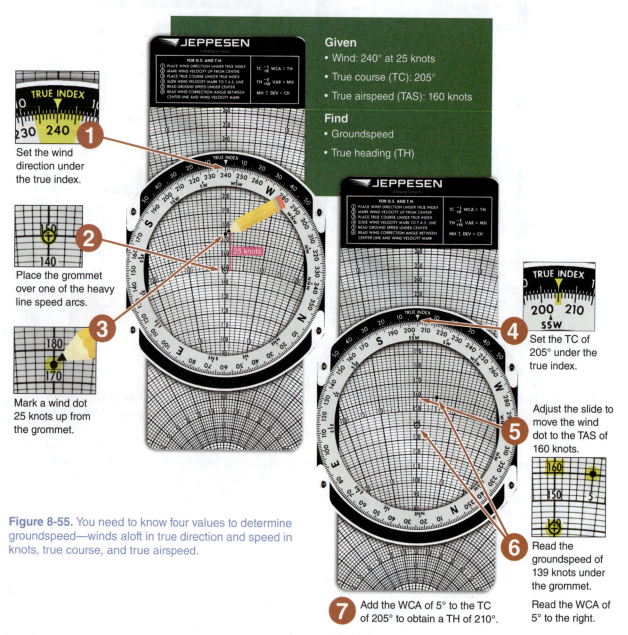

Given
- Wind: 240° at 25 knots
- True course (TC): 205°
- True airspeed (TAS): 160 knots

Find
- Groundspeed
- True heading (TH)

① Set the wind direction under the true index.

② Place the grommet over one of the heavy line speed arcs.

③ Mark a wind dot 25 knots up from the grommet.

④ Set the TC of 205° under the true index.

⑤ Adjust the slide to move the wind dot to the TAS of 160 knots.

⑥ Read the groundspeed of 139 knots under the grommet.

Read the WCA of 5° to the right.

⑦ Add the WCA of 5° to the TC of 205° to obtain a TH of 210°.

Figure 8-55. You need to know four values to determine groundspeed—winds aloft in true direction and speed in knots, true course, and true airspeed.

USING THE COMPUTER SIDE

To get acquainted with the computer side, start by turning the inner disc until the pointer (called the **speed index** or 60 index) points to 60 on the outer scale. Notice that the two adjacent scales match all the way around. The non-rotating outer scale is called the A scale, and the outermost rotating scale is the B scale. The A and B scales are identical. The scales are logarithmic—graduations that are an equal distance apart, represent values which are in an equal ratio. You might notice that the amount of increments between numbers varies. For example, nine marks are between 13 and 14 and only four marks are between 18 and 19, and none at all between 80 and 81. A full set of marks would become too difficult to read as the numbers get closer together. To avoid mistakes, keep these differences in mind when you perform problems. [Figure 8-56]

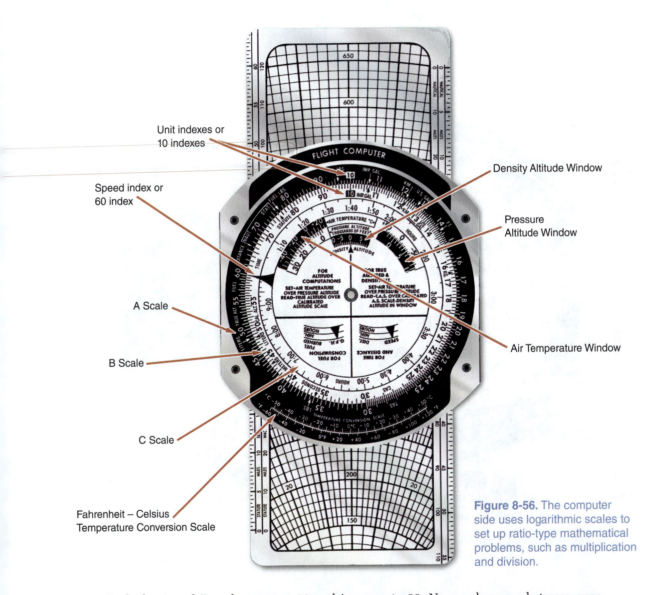

Unit indexes or 10 indexes

Speed index or 60 index

A Scale

B Scale

C Scale

Fahrenheit – Celsius Temperature Conversion Scale

Density Altitude Window

Pressure Altitude Window

Air Temperature Window

Figure 8-56. The computer side uses logarithmic scales to set up ratio-type mathematical problems, such as multiplication and division.

Both the A and B scales start at 10 and increase to 99. No numbers are between zero and 9, and the next mark after 99 is 10. If you read the 10 as 100, you can go around the scale again—then 11 becomes 110, 12 becomes 120, and so on. When you get all the way around and pass 10 again, it represents 1,000. This characteristic makes the scale infinite. This method works in both directions, so that if you start at 10 and go counterclockwise, 90 becomes 9, 80 becomes 8, all the way back to 11 becoming 1.1, 10 becoming 1.0, and 90 becoming 0.9. No limit applies to the range of the numbers your computer can handle, but you have to keep track of the decimal place yourself, which is not as difficult as it sounds. If you are solving for groundspeed and get an answer of 15.6, you know that the answer is not 1,560 knots or 15.6 knots, so 156 knots remains as the only reasonable choice.

A box around the 10 on each scale is called the 10 index or unit index You can use this box for conventional multiplication and division problems. The next scale inward from B is called the C scale, or hours scale. The big arrow at 60, called the speed index, is the unit index for the time scale. Because there are 60 minutes in an hour, you can either read minutes on the B scale, or hours plus minutes on the C scale. Thus, any combination of hours and minutes you find on the C scale accompanies the corresponding total number of minutes on the B scale. A handy rule of thumb for common calculations is that the speed index always points to a rate: knots, gallons per hour, etc.

TIME, SPEED, AND DISTANCE

Time, speed, and distance (TSD) relationships are probably the most frequent problems you will solve with your computer. For TSD problems, the A scale is for distance, the B scale represents time in minutes, and the speed index always points to the rate of speed. [Figure 8-57]

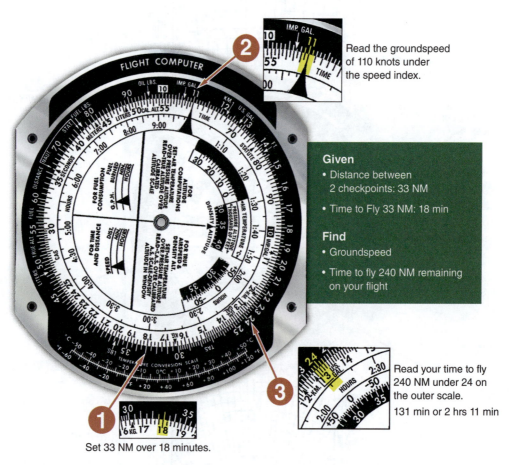

Read the groundspeed of 110 knots under the speed index.

Given
- Distance between 2 checkpoints: 33 NM
- Time to Fly 33 NM: 18 min

Find
- Groundspeed
- Time to fly 240 NM remaining on your flight

Read your time to fly 240 NM under 24 on the outer scale.

131 min or 2 hrs 11 min

Set 33 NM over 18 minutes.

Figure 8-57. Time, fuel and distance problems are as simple as setting up ratios of distance over time.

FUEL CONSUMPTION

Fuel consumption problems are similar to time, speed, and distance problems. The difference is that the speed index indicates gallons per hour, and the A scale is used to represent gallons of fuel consumed instead of distance traveled.

FUEL REQUIRED

The POH provides the rate of fuel consumption (fuel flow) at various cruise power settings. With this information and an accurate flight time, you can use the computer to solve for the amount of fuel required. [Figure 8-58]

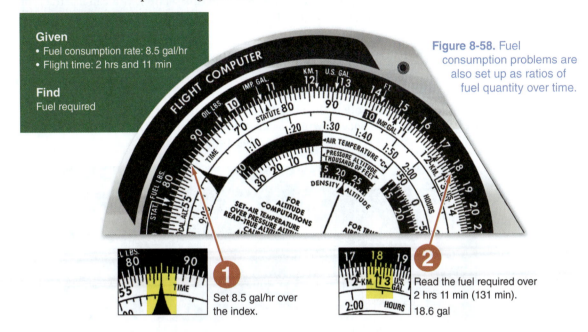

Given
- Fuel consumption rate: 8.5 gal/hr
- Flight time: 2 hrs and 11 min

Find
Fuel required

Figure 8-58. Fuel consumption problems are also set up as ratios of fuel quantity over time.

1 Set 8.5 gal/hr over the index.

2 Read the fuel required over 2 hrs 11 min (131 min). 18.6 gal

ENDURANCE

Endurance is the amount of time you can remain aloft, based on a known fuel quantity and consumption rate. This calculation is similar to the previous example, except that after you set the rate of consumption, the number of gallons available (usable fuel) determines the endurance time. [Figure 8-59]

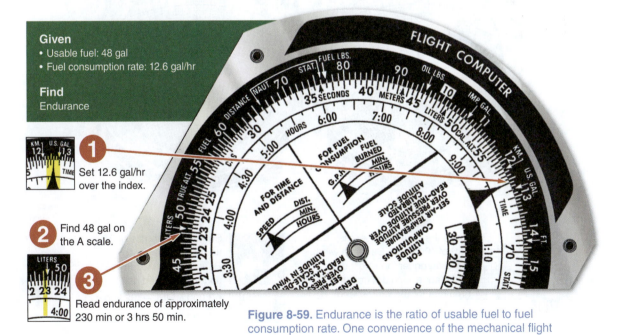

Given
- Usable fuel: 48 gal
- Fuel consumption rate: 12.6 gal/hr

Find
Endurance

1 Set 12.6 gal/hr over the index.

2 Find 48 gal on the A scale.

3 Read endurance of approximately 230 min or 3 hrs 50 min.

Figure 8-59. Endurance is the ratio of usable fuel to fuel consumption rate. One convenience of the mechanical flight computer is that the C scale provides hours and minutes at a glance.

SECTION C ■ **Flight Computers**

OTHER CALCULATIONS

In addition to basic flight planning data, you can use a flight computer for other calculations that provide you with valuable information to enhance your ability to maintain situational awareness. As with the previous problems presented in this section, you can use the type of flight computer that you prefer to solve the problems.

MOST FAVORABLE WINDS

One of the factors you should consider when selecting a cruising altitude is the effect of wind on groundspeed. The mechanical flight computer offers a fairly simple method of determining which altitude offers the most favorable winds for a given course. Of course, you must balance the influence of winds against other factors, such as safe terrain clearance, climb times, and VFR cruising altitude rules. [Figure 8-60]

SECTION C ■ **Flight Computers**

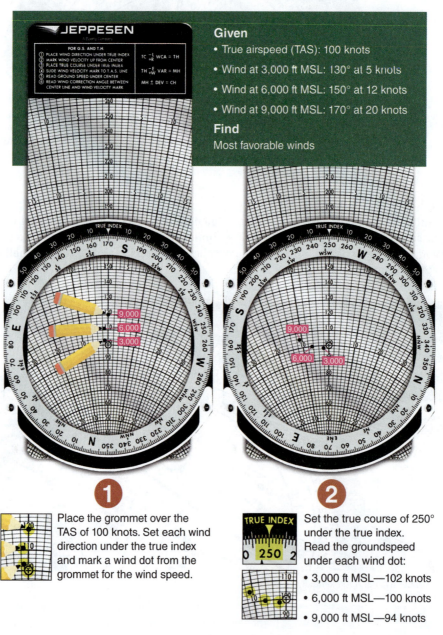

Given
- True airspeed (TAS): 100 knots
- Wind at 3,000 ft MSL: 130° at 5 knots
- Wind at 6,000 ft MSL: 150° at 12 knots
- Wind at 9,000 ft MSL: 170° at 20 knots

Find
Most favorable winds

1 Place the grommet over the TAS of 100 knots. Set each wind direction under the true index and mark a wind dot from the grommet for the wind speed.

2 Set the true course of 250° under the true index. Read the groundspeed under each wind dot:
- 3,000 ft MSL—102 knots
- 6,000 ft MSL—100 knots
- 9,000 ft MSL—94 knots

Figure 8-60. Using forecast wind directions and speeds for various altitudes might help you to choose a cruising altitude that provides the most beneficial (or least detrimental) winds.

ACTUAL WINDS ALOFT

During cross-country flights, you will discover that the actual winds aloft might vary from the forecasted winds. As a result, you can usually expect your actual groundspeed to vary somewhat from your preflight estimate. If your groundspeed is slower than planned, your airplane uses more fuel than predicted, so your reserve is reduced. The additional time needed to complete the flight translates directly into additional fuel consumed.

Solving for actual winds aloft is essentially the reverse of predicting groundspeed from forecast winds aloft. In this case, you will start with your groundspeed and wind correction angle, and use these values to find wind direction and speed. [Figure 8-61]

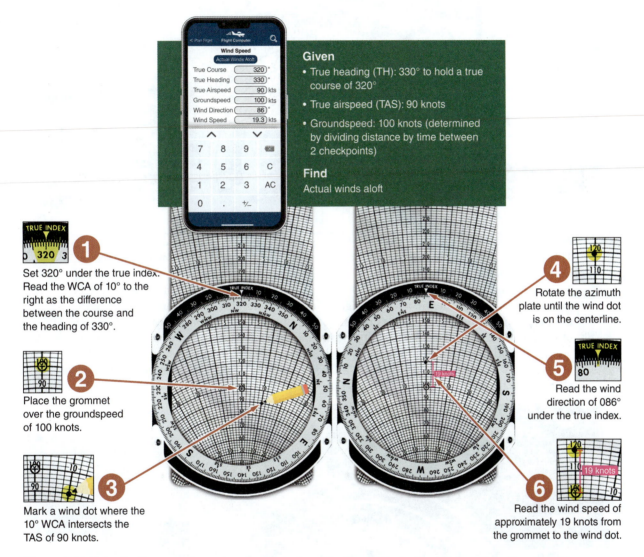

Given
- True heading (TH): 330° to hold a true course of 320°
- True airspeed (TAS): 90 knots
- Groundspeed: 100 knots (determined by dividing distance by time between 2 checkpoints)

Find
Actual winds aloft

1 Set 320° under the true index. Read the WCA of 10° to the right as the difference between the course and the heading of 330°.

2 Place the grommet over the groundspeed of 100 knots.

3 Mark a wind dot where the 10° WCA intersects the TAS of 90 knots.

4 Rotate the azimuth plate until the wind dot is on the centerline.

5 Read the wind direction of 086° under the true index.

6 Read the wind speed of approximately 19 knots from the grommet to the wind dot.

Figure 8-61. Solve for the groundspeed on the computer side, then solve for the winds aloft on the wind side.

SECTION C ■ **Flight Computers**

ACTUAL FUEL CONSUMPTION RATE

After completing a flight, you might want to determine the actual fuel consumption rate based on the total amount of fuel used. Compare the result against the planned fuel consumption rate and consider any variation for future flight planning. Set the number of gallons consumed on the A scale above the time on the B or C scale, then read the actual consumption rate over the speed index. [Figure 8-62]

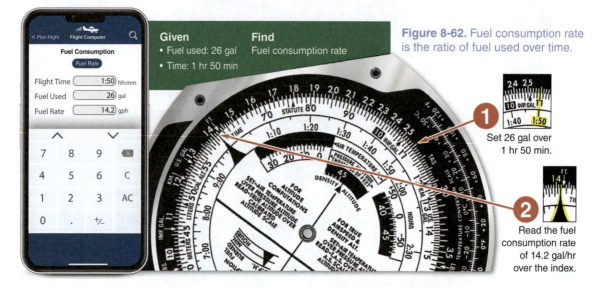

Given
• Fuel used: 26 gal
• Time: 1 hr 50 min

Find
Fuel consumption rate

Figure 8-62. Fuel consumption rate is the ratio of fuel used over time.

1 Set 26 gal over 1 hr 50 min.

2 Read the fuel consumption rate of 14.2 gal/hr over the index.

AIRSPEED AND DENSITY ALTITUDE CALCULATIONS

The airspeed indicator is a sensitive and precise instrument, but as air density decreases, indicated airspeed (IAS) is lower than true airspeed (TAS) by about 2% per 1,000 feet of altitude. Many airspeed indicators include a scale to set outside air temperature (OAT) to determine actual true airspeed in flight. However, you also can determine TAS with the mechanical flight computer using the temperature and altitude scales in the center of the computer. To determine density altitude, use the same scales. For both problems, you must find pressure altitude. In flight, the easiest way to find pressure altitude is to momentarily set your altimeter to the standard pressure of 29.92. Do not forget to reset the altimeter to the correct setting. [Figure 8-63]

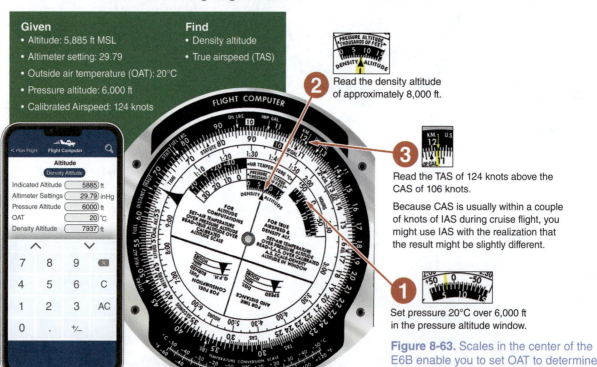

Given
• Altitude: 5,885 ft MSL
• Altimeter setting: 29.79
• Outside air temperature (OAT): 20°C
• Pressure altitude: 6,000 ft
• Calibrated Airspeed: 124 knots

Find
• Density altitude
• True airspeed (TAS)

2 Read the density altitude of approximately 8,000 ft.

3 Read the TAS of 124 knots above the CAS of 106 knots.

Because CAS is usually within a couple of knots of IAS during cruise flight, you might use IAS with the realization that the result might be slightly different.

1 Set pressure 20°C over 6,000 ft in the pressure altitude window.

Figure 8-63. Scales in the center of the E6B enable you to set OAT to determine values, such as density altitude.

CONVERSIONS

Flight computers incorporate many different conversion scales, including Celsius/Fahrenheit temperature and nautical/statute mile conversions. Most flight computers can also convert between aviation gasoline quantities of gallons and pounds, statute miles and kilometers, U.S. gallons and Imperial gallons, U.S. gallons and liters, feet and meters, pounds and kilograms, and various other units. Because the procedures vary, read the computer's instruction manual for detailed instructions. [Figure 8-64]

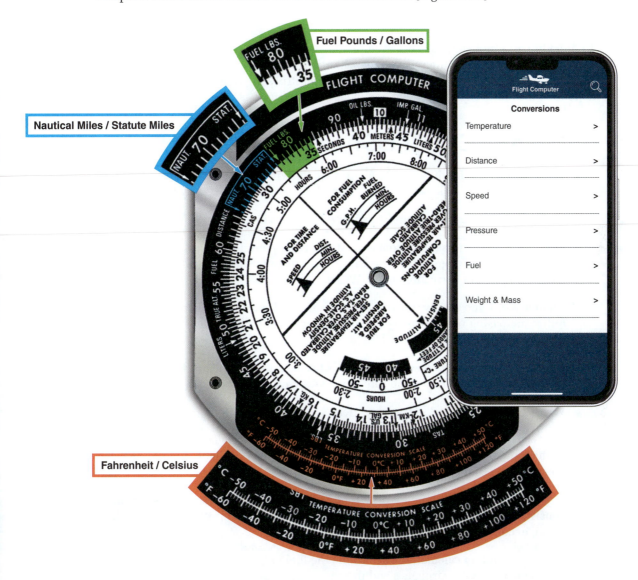

Figure 8-64. On a mechanical flight computer, you can perform many conversions by lining up the arrows marked on the A and B scales.

SECTION C ■ **Flight Computers**

MENTAL MATH

In some cases, certain rules of thumb or simple math can enable you to make a quick approximate mental conversion or calculation. For example, you can use the following rules of thumb for temperature conversions:

$$°C = [(°F - 32) + 0.10 \times (°F - 32)] \div 2$$

$$°F = [(°C \times 2) - 0.10 \times (°C \times 2)] + 32$$

To quickly perform a statute/nautical mile conversion, know that a nautical mile is 15 percent longer than a statute mile. And, when flight planning, know that most Avgas weighs 6 pounds per gallon. Use simple math to multiply the number of gallons by 6 to determine the fuel weight for weight and balance computations.

SUMMARY CHECKLIST

✓ Adding the vector for the aircraft's airspeed to the wind vector speed gives a resultant vector that shows the speed and direction of the airplane affected by the wind.

✓ Headwinds reduce your groundspeed, and tailwinds increase it.

✓ To prevent drift and stay on course, you must compensate by pointing the nose of the airplane a few degrees into the wind. The difference between your heading and course is the wind correction angle (WCA).

✓ The tailwind and headwind components of wind act in the direction of flight, either increasing or decreasing the groundspeed. The crosswind component acts perpendicular to the direction of flight.

✓ During flight planning, the required values to calculate groundspeed are true course (TC), true airspeed (TAS), wind direction, and wind speed.

✓ Winds aloft are reported in true direction.

✓ Time, speed, and distance (TSD) relationships enable you to solve problems, such as the time for your proposed flight, actual groundspeed in flight, and the distance your airplane can travel in a specific amount of time.

✓ Most aviation calculations involve time in hours and minutes. Because an hour consists of 60 minutes, typically these calculations require multiplying or dividing by 60, to convert to or from minutes.

✓ Fuel consumption calculations determine the fuel usage over a specific time period or distance and your airplane's endurance.

✓ The required values to determine the fuel required are fuel consumption rate and time.

✓ The required values to determine endurance are the fuel consumption rate and usable fuel.

✓ The wind side of the mechanical computer uses an azimuth plate, true index, and a wind dot to determine how the wind will affect your airplane in flight.

✓ The computer side of the mechanical flight computer uses a speed index and rotating A, B and C scales to solve time, speed, and distance problems and to determine fuel required and endurance.

✓ The mechanical flight computer offers a fairly simple method of determining which altitude offers the most favorable winds for a given course.

✓ To solve for actual winds aloft, use your groundspeed and wind correction angle to find wind direction and speed.

✓ After completing a flight, determine the actual fuel consumption rate based on the total amount of fuel used you can consider this rate into consideration for future flight planning. Set the number of gallons consumed on the A scale above the time on the B or C scale, then read the actual consumption rate over the speed index.

✓ As air density decreases, indicated airspeed (IAS) is lower than true airspeed (TAS) by about 2% per 1,000 feet of altitude.

✓ In flight, you can determine your actual TAS with the mechanical flight computer by using your indicated airspeed and the outside air temperature (OAT).

✓ The required values to determine density altitude are pressure altitude and OAT.

✓ Flight computers incorporate many different conversion scales, including Celsius/Fahrenheit temperature and nautical/statute mile conversions.

KEY TERMS

Groundspeed	Endurance
Wind Correction Angle (WCA)	E6B
Wind Correction Angle (WCA)	Computer Side
Headwind Component	Wind Side
Tailwind Component	Azimuth Plate
Crosswind Component	True Index
Time, Speed, and Distance (TSD)	Wind Dot
Fuel Consumption Rate	Speed Index
Fuel Required	

SECTION C ■ **Flight Computers**

QUESTIONS

Solve the following problems with either an electronic or a mechanical flight computer.

1. If you cover 61 nautical miles in 41.5 minutes, what is your groundspeed?

2. After landing from a 362 nautical mile flight, you find that it takes 32.5 gallons to refill the fuel tanks. If the tanks were full when you took off, and your groundspeed was 138 knots, what was the average fuel consumption rate for the flight?

3. At a pressure altitude of 6,900 feet and an outside temperature of 7° C, your airspeed indicator shows 96 knots. What is your approximate true airspeed? Why is this number approximate?

4. Your POH provides a fuel consumption figure of 8.7 gallons per hour for your planned power setting and altitude. How much fuel would be used in 2 hours, 30 minutes?

5. How many nautical miles are in 148 statute miles?

6. With a true airspeed of 88 knots and winds from 240° at 16 knots, what heading would be required to maintain a course of 137°? What groundspeed would you expect?

7. What is the estimated time enroute for a flight of 194 nautical miles on a course of 084°. Your true airspeed is 129 knots, and pressure altitude is 5,200 feet. The wind is from 233° at 18 knots, and outside air temperature is 10° C.

8. Suppose you have been holding a heading of 140° to maintain a course of 133° between 2 checkpoints 128 nautical miles apart. It took 37 minutes to cover that distance at a true airspeed of 198 knots. What is the wind direction and speed?

9. What heading should you fly to hold a course of 355° if your true airspeed is 116 knots and the winds are from 050° at 15 knots? What groundspeed would you expect?

CHAPTER 9

Navigation

SECTION A
Pilotage and Dead Reckoning

Your introduction to navigation initially takes you back to the methods used by the earliest cross-country pilots, as they learned to find their way from town to town by following roads and looking for landmarks from the air. This method was adequate for flights over distinctive terrain, but as pilots began to consider more challenging flights, over water or featureless deserts for instance, they borrowed the well-proven techniques used for centuries by ocean navigators. The methods of pilotage and dead reckoning normally are used together, each acting as a cross-check of the other.

Three basic tasks of navigation are to create a course, fly the airplane so as to stay on the course, and make position checks to confirm that you are remaining on course. Finding your way to safety if you get lost also is an important element. Two terms used in connection with finding your position in flight are line of position and fix. A line of position is the simple concept that the airplane is located somewhere along a specific line. A **line of position (LOP)** does not establish the exact position of the airplane, but rather a line of possible positions, one of which is the airplane's actual position.

The intersection of two different lines of position is a **fix,** which establishes your position at a definite location. If you are calling to report a fire in your neighborhood, and you tell the fire department it is on Elm Street, you have established a line of position. It does not identify the position of the fire, but it creates a string of possible locations (all the addresses on Elm Street) while eliminating other possible locations (Maple Street, Locust Street, Spruce Street, etc.). If you report the fire as being at Elm and Main, then you have created a fix, and the firefighters can go to a definite location. These terms are important in radio navigation as well as in pilotage and dead reckoning.

DISCOVERY

Jean Louis Conneau's Maps

French pilot Jean Louis Conneau was a pioneer of aerial navigation, and one of the first to use maps in flight. In 1911 he overcame the problem of the map blowing away by mounting it on a pair of rollers that let him read it like a scroll as the flight progressed. He won air races against faster airplanes by using his map to fly straight lines between cities instead of following roads, an almost revolutionary concept at the time.

Courtesy of Terry Gwynn-Jones

PILOTAGE

Before World War I, aviators seldom carried maps, relying on their familiarity with landmarks such as rivers or railroads. Since then, excellent aeronautical charts have been developed that make visual navigation practical in most areas of the country, whether the pilot is familiar with the local landmarks or not. Navigating by reference to landmarks is called **pilotage**. Most pilots delight in looking at the ground as they fly, so pilotage can be a very interesting and enjoyable way to navigate. Your sectional charts provide a much more accurate and detailed representation of the landscape than the automobile road maps often used by early pilots.

Preflight planning for pilotage begins with obtaining the correct charts. The high level of detail and the relatively large scale make sectional charts the best choice for visual navigation. Make sure that the charts you use are current. Although large permanent features such as mountains do not change much between chart editions, new towers are erected, ponds form or dry up, and airports open or close. You can get a good idea of which charts you will need for a trip by checking the chart coverage diagram on the front panel of any sectional chart.

COURSE CONSIDERATIONS

You might choose to use paper charts or digital charts that are part of a flight planning app. To line up adjacent printed charts precisely, match the latitude and longitude lines near the edges of each chart. If it happens that your departure and destination airports are on opposite sides of the same chart, follow the instructions printed on the margin of the chart to plot your course. [Figure 9-1]

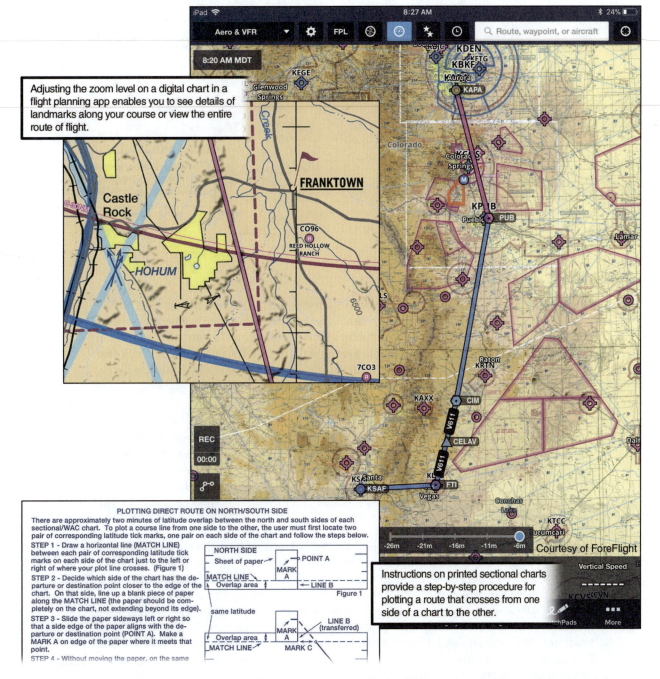

Figure 9-1. You might need more than one printed chart for a flight. Flight planning apps enable you to seamlessly integrate multiple charts.

Review the general route and decide whether to avoid certain areas, such as special-use airspace, mountains, or a large expanse of open water. [Figure 9-2] In areas where prominent landmarks are sparse, pilotage can be difficult, so you might choose to move your course a few miles to the side to include some unmistakable landmarks. In many cases you simply choose the direct route between your departure and destination airports. Draw your course using a straightedge on a paper chart or plot your course using tools on a digital flight planning app. Use a plotter to measure the course distance in nautical miles, taking care to use the correct scale for the chart. A flight planning app automatically displays the course distance after you identify the departure and destination airports.

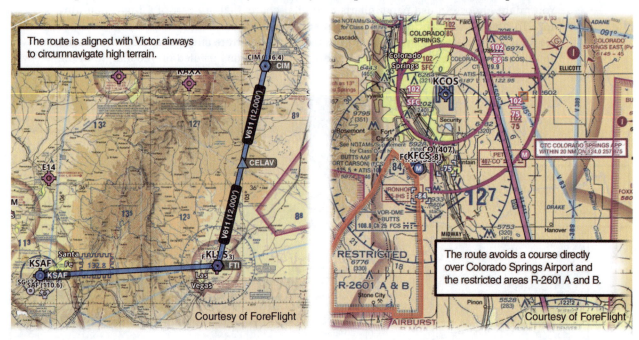

Figure 9-2. You might modify your course to avoid certain areas.

Using information from the POH and winds aloft forecasts, estimate the amount of fuel required for the trip, including takeoff, climb, and reserves. Be generous, because winds might shift direction and speed. If the amount of fuel required makes you think a fuel stop might be necessary, adjust your course to land at an airport with fuel service. Verify that fuel is available by using a source, such as the applicable Airport/Facility Directory listing in the Chart Supplement.

CHECKPOINTS

When you have finalized your course, you can begin selecting **checkpoints**. The best checkpoints are those that cannot be mistaken for any other nearby features, for example, a lake with a distinctive shape, or a major highway crossing a river. Depending on the altitude you choose, you might be able to positively identify landmarks several miles either side of your course. These reference points can provide valuable cues to help you recognize and correct for drift.

Checkpoints are useful only if they can be positively identified from the air, so you should avoid choosing features that could become ambiguous in flight. For example, many small towns look alike from the air, and in many areas of the country they occur at regular intervals along railroads, a legacy of the era when steam locomotives needed periodic stops to take on water. You could be in a situation of knowing you are over one of these railroad towns, but not which one, unless there are other landmarks to distinguish the town from the others up and down the line. On the other hand, small cities have excellent landmark value, because the yellow pattern on the chart attempts to show the actual shape of the city from the air. [Figure 9-3]

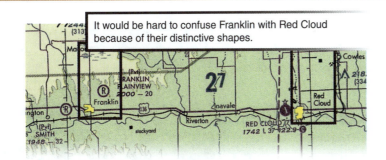

It would be hard to confuse Franklin with Red Cloud because of their distinctive shapes.

Figure 9-3. Small towns can be hard to identify, but small cities can usually be recognized by their distinctive shapes.

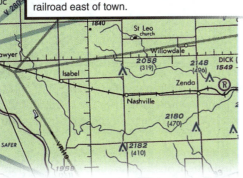

Isabel and Nashville have similar triangles formed by the railroad and the roads north and west of town. Nashville could be identified in flight by the towers north and south of town, as well as the jog in the road where it crosses the railroad east of town.

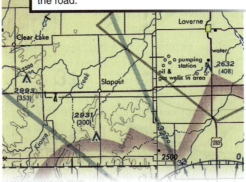

Clear Lake and Slapout could be easily confused from the air, because of the similar road patterns. Both have towers south of the town, but the towers are on opposite sides of the road.

Although major roads are usually good checkpoints, secondary roads can be deceiving. Because only the most prominent secondary roads are portrayed on the chart, you might see many other similar roads in flight. When you consider using a road for a checkpoint, look for features that will be easy to distinguish from the air, such as unique bends or the relative location of railroads, streams, powerlines, etc. Keep in mind that new roads and highways are continuously under construction and might not even appear on a current chart. [Figure 9-4]

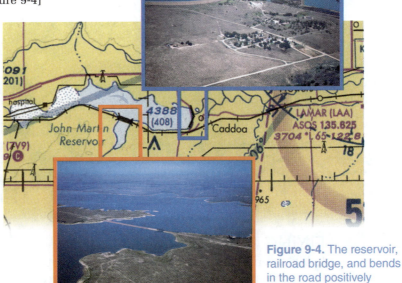

Rivers usually make excellent landmarks, especially when they have distinctive bends or curves. Although the chart depicts most rivers very accurately, be alert for possible seasonal variations. During a flood stage, the appearance from the air might be quite different from the chart pattern. During a drought, on the other hand, the watercourse shown on a sectional might actually be a dried-up river bed. Lakes usually are good references, but in many parts of the country, hundreds of small lakes exist within close proximity, and identifying a particular one can be extremely difficult. In other areas, ponds depicted on the chart might have dried up without a trace. As with small towns and secondary roads, be sure to use other nearby landmarks to confirm identification of small or nondescript features.

Figure 9-4. The reservoir, railroad bridge, and bends in the road positively identify Caddoa.

SECTION A ■ **Pilotage and Dead Reckoning**

DISCOVERY

Water Tower Landmarks

Water towers in small towns often have the town name painted on them. This can sometimes be an aid to pilotage, but you must descend to a low altitude to read the name, if it is there at all.

According to tradition, if a pilot circles the water tower (or the town) before landing at the airport, the local constable comes out and gives the pilot a ride into town. This tradition dates back to the time of the barnstormers.

The number of checkpoints needed for a particular flight is up to you, and depends on your confidence level, the presence or absence of other visual course cues such as fence lines, your altitude, and your use of dead reckoning or radio navigation. Try to plan a couple of prominent checkpoints shortly after takeoff, to help you get established on the correct course. The first checkpoint should be close enough to be easy to locate after takeoff, but far enough that you are clear of the airport traffic pattern before turning your attention to navigation.

When you have decided on your checkpoints, circle them on the chart or create a waypoint at the location in your flight planning app. Pencil lines can be difficult to find on a paper chart, so use a pen or highlighter to make your course and checkpoints easier to see. Choose a color that does not blend in with the colors of chart features. You might also want to place tick marks along your course at 10 or 20 mile intervals, to help estimate distances along the route or to help monitor your progress.

DISCOVERY

Section Lines

Surveyors divide land into townships and sections for legal and real estate purposes. A township consists of 36 sections or subdivisions, each of which is one square mile in area. Farmers often plant their fields in sections, half sections, quarter sections, etc., and roads are usually laid out along the lines that divide sections.

In many parts of the country, these section lines are very conspicuous from the air. By comparing the angle at which your airplane crosses section lines with the course line on your map, section lines can be used as a sort of compass, because the lines run north-south or east-west. If you are flying parallel to the section lines, you can find your groundspeed by timing from one section line to the next. A section is one statute mile on each side, so a time of 30 seconds would indicate a groundspeed of 120 statute miles per hour, or about 104 knots.

FLYING THE COURSE

In flight, your objective is to make your path over the ground match the course line drawn on your chart. The easiest way to remain on course is to stay continuously aware of your position. If you orient the chart so that the course line is aligned with the direction of flight, landmarks on the ground appear in the same relative positions as their respective chart symbols. Mark each checkpoint as you pass it, and if you see that you have drifted to one side of your course, mark your actual position rather than the intended checkpoint. Correct for drift by adjusting your heading. [Figure 9-5]

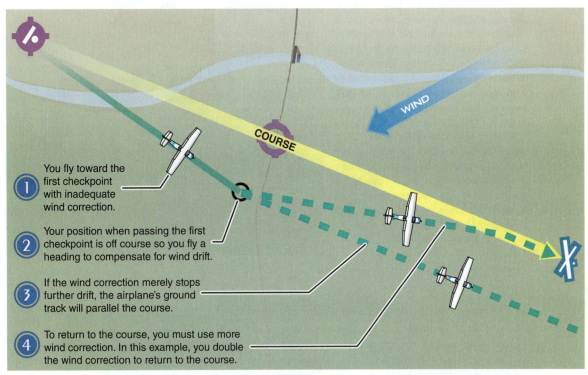

WIND

COURSE

1 You fly toward the first checkpoint with inadequate wind correction.

2 Your position when passing the first checkpoint is off course so you fly a heading to compensate for wind drift.

3 If the wind correction merely stops further drift, the airplane's ground track will parallel the course.

4 To return to the course, you must use more wind correction. In this example, you double the wind correction to return to the course.

Figure 9-5. If you drift off course, you must apply enough wind correction to return to your original route.

Even if you are only off course by a small amount, make a correction to your heading, otherwise you will be off even farther at the next checkpoint, possibly far enough to miss the checkpoint entirely. Between checkpoints you can draw visual lines of position from landmarks that are to one side or the other of your course to find intermediate position fixes. For example, if you find that you are a few miles directly south of a certain town, and you are also passing over a railroad track that runs straight to another town to the southeast, you have established a fix from two lines of position.

When flying by pilotage, match the chart features as much as possible with what you see on the ground, even when you are between checkpoints. Whenever you can, confirm each checkpoint by comparing it against a combination of several ground features. Remember, aeronautical charts do not show every detail of the earth's surface, and even those that are shown might have changed since the chart was updated.

DEAD RECKONING

The methods used by the navigators who explored the oceans in the 15th and 16th centuries sometimes are regarded as primitive from our modern perspective. In fact, these professionals achieved amazing accuracy without many of the tools we consider indispensable, such as detailed maps or reliable timepieces. With little more than a compass and an understanding of mathematics, they were able to navigate in spite of storms, changing winds, and variable ocean currents across thousands of miles of open sea and make reasonably accurate landfalls. Their methods are the basis of the modern navigation technique called **dead reckoning**. Based on calculations of time, speed, distance, and direction, dead reckoning made possible the historic flights of Charles Lindbergh, Amelia Earhart, and other pioneering pilots. You might never fly over an ocean or a trackless desert, but even in perfect visibility over an ordinary landscape, dead reckoning is a valuable complement to pilotage and provides the basis for understanding other forms of navigation.

SECTION A ■ **Pilotage and Dead Reckoning**

SECTION A ■ **Pilotage and Dead Reckoning**

DISCOVERY

Dead Reckoning Term

How did such a grim term come to be used for a method of navigation? Some sources say that the term comes from a contraction of the word "deduced" to "ded," because a navigator uses this technique to deduce position—which seems a little dubious. Others claim that it was coined by early pilots to emphasize the fact that your reckoning had better be accurate, or you would soon be dead. In fact, the term was in use long before the beginning of human flight—in the seafaring world, like so many other aviation terms and practices. It probably comes from the old use of the word "dead" to mean exact or absolute, as in dead center, dead level, dead on, dead right, and so forth. No matter what the derivation, the term has stuck.

COURSE

In pure pilotage, your course line and landmarks point the direction for you to fly. Because the idea of dead reckoning is to find your way without visual landmarks, the compass provides the necessary directional information. The course direction is measured with a **navigation plotter** on a printed chart. The plotter is a transparent plastic instrument that combines a straightedge for drawing a course line, a protractor for measuring the direction of flight, and distance scales for sectional and VFR terminal area charts in both nautical and statute miles.

Draw your course on the chart as you would for a pilotage flight, including measuring the distance, calculating the fuel required, and planning any fuel stops. Then, use the plotter to measure your **true course** at the longitude line nearest the center of the course. [Figure 9-6]

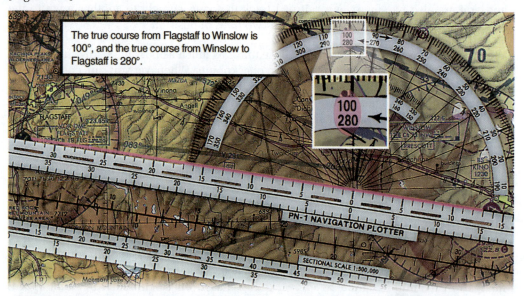

The true course from Flagstaff to Winslow is 100°, and the true course from Winslow to Flagstaff is 280°.

1. Draw a line along your course.
2. Align the protractor center hole over the longitude line closest to the center of the courseline.
3. Read the degrees where the protractor intersects the longitude line.

Figure 9-6. Use a navigation plotter to determine your true course on a paper chart.

Because the longitude lines converge toward the poles in the map projection used for sectional charts, the longitude lines toward the ends of the course would give slightly different true course readings. Measuring near the center of the course will put you on a nearly perfect great circle route between your departure and destination airports. If your course line is more north-south, it might not cross any longitude lines. In this case, place the plotter along your course where it crosses a latitude line, and read the true course from the smaller auxiliary scale on the plotter. [Figure 9-7]

TRUE AND MAGNETIC VALUES

The magnetic compass in an airplane senses magnetic north, which can differ from true north by as much as 20° in the contiguous United States. In dead reckoning you normally fly by magnetic reference, so you will need to correct for the difference between true and magnetic direction, which is called variation. Local magnetic variation is shown on aeronautical charts by a dashed magenta **isogonic line**. [Figure 9-8]

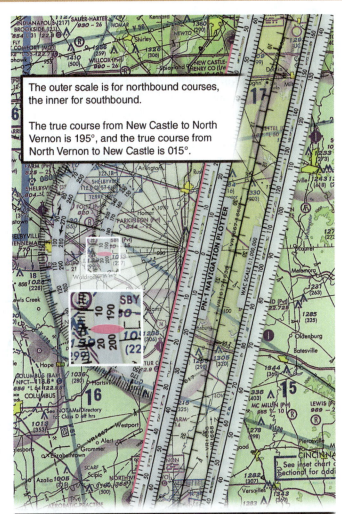

The outer scale is for northbound courses, the inner for southbound.

The true course from New Castle to North Vernon is 195°, and the true course from North Vernon to New Castle is 015°.

1. Draw a line along your course.
2. Align the protractor center hole over the latitude line closest to the center of the courseline.
3. Read the degrees where the protractor intersects the latitude line.

Figure 9-7. Use a latitude line to determine the true course for a north-south route.

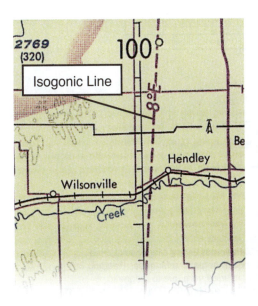

Isogonic Line

Figure 9-8. This isogonic line shows a magnetic variation of 8° east.

Correcting your true course for variation is simply a matter of adding or subtracting the value shown on the isogonic line. If the variation is east of true north, it is subtracted from the true course, and west variation is added to the true course, to obtain magnetic course. East is least and west is best is a memory aid that you can use to remember whether to add or subtract magnetic variation. The magnetic fields of items within the airplane can also affect the accuracy of the magnetic compass. These errors are shown on the compass deviation card near the compass. The deviation is only recorded for certain headings and is typically minor.

SECTION A ■ **Pilotage and Dead Reckoning**

Some pilots choose to correct their true course for wind before correcting for variation, and others correct for variation before wind. The order in which you perform the calculation has no effect on the outcome, but different terms are used to describe some of the intermediate sums, and the wind direction must be changed from true to magnetic if the variation is applied first. A course is always the line drawn on the chart. The true course is measured from true north, and if the correction for variation is applied at this point, the result is expressed as **magnetic course**. A heading is always a direction measured relative to the longitudinal axis of the airplane, or in other words, the direction in which the airplane is pointed. True course corrected for wind gives a **true heading**, and magnetic course corrected for wind gives a magnetic heading. The **compass heading** is always found by correcting the magnetic heading for compass deviation. [Figure 9-9]

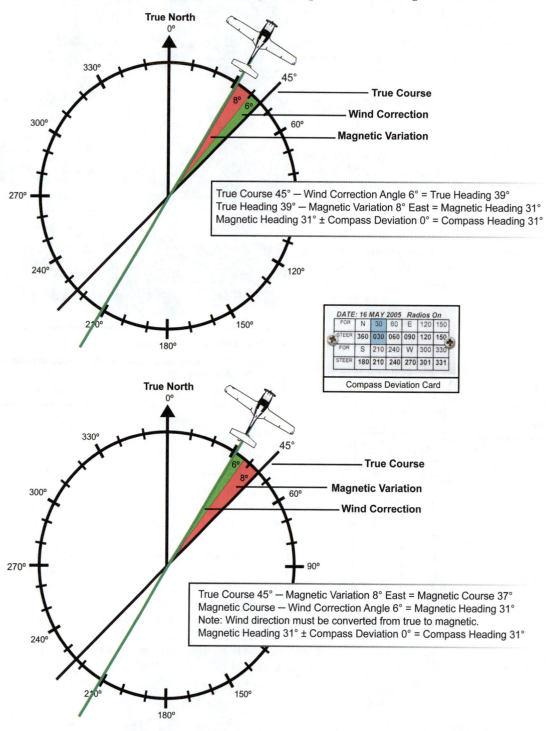

True Course 45° — Wind Correction Angle 6° = True Heading 39°
True Heading 39° — Magnetic Variation 8° East = Magnetic Heading 31°
Magnetic Heading 31° ± Compass Deviation 0° = Compass Heading 31°

DATE: 16 MAY 2005			Radios On		
FOR	N	30	60	E	120	150
STEER	360	030	060	090	120	150
FOR	S	210	240	W	300	330
STEER	180	210	240	270	301	331

Compass Deviation Card

True Course 45° — Magnetic Variation 8° East = Magnetic Course 37°
Magnetic Course — Wind Correction Angle 6° = Magnetic Heading 31°
Note: Wind direction must be converted from true to magnetic.
Magnetic Heading 31° ± Compass Deviation 0° = Compass Heading 31°

Figure 9-9. Whether you compensate for variation or winds first, the same result is obtained. Although this example shows each value as negative, often one or more of the values is positive, in which case it is added.

VFR CRUISING ALTITUDES

There are many factors to consider when selecting your cruising altitude for a cross-country flight. First, consider your height above the terrain and obstructions. Your sectional chart shows terrain heights along your route and tall obstructions, such as towers. In addition to the height of the terrain, evaluate the topography in the area. A flight over mountains with potentially strong downdrafts and nothing but rugged rock below would prompt a much higher cruising altitude than flat terrain with plenty of emergency landing sites within easy gliding distance. In addition, consider the effect of winds on your groundspeed and the performance of your airplane. In general, higher altitudes provide better visibility of checkpoints, better radio range and reception, and more options in the event of an emergency.

Whenever you are in level cruising flight more than 3,000 feet above ground level (AGL), you must comply with the **VFR cruising altitude** rule. VFR aircraft on magnetic courses from 0° to 179° are required to fly at odd thousand-foot altitudes plus 500 feet, such as 3,500 feet MSL, 5,500 feet MSL, 7,500 feet MSL, etc., up to the flight levels. If you fly a magnetic course from 180° to 359°, your choices for VFR cruising altitudes are even thousands plus 500 feet, such as 4,500 feet MSL, 6,500 feet MSL, or 8,500 feet MSL. If you are maneuvering, turning, or changing altitude the rule does not apply. The idea behind this rule is that eastbound and westbound VFR aircraft are theoretically separated by a thousand feet, helping to reduce the possibility of head-on traffic conflicts.

In practice, it is extremely important that you maintain your visual traffic scan at all times, because other aircraft can converge on you from any direction, in full compliance with the rule. For example, suppose you are flying south on a heading of 179° at a proper altitude of 5,500 feet MSL. You can expect VFR aircraft cruising at your altitude to converge on you from any direction to the right of the airplane, from directly behind you to nearly head-on. On courses that are more nearly east or west, be alert for aircraft coming from beside or behind you, where they can be more difficult to see. This is especially important when flying slower aircraft that could be overtaken by faster aircraft. Of course, aircraft changing altitudes can come from any direction, and because the rule only applies to cruising flight, any aircraft that is practicing maneuvers or turning need not comply. Finally, ATC may assign any altitude to IFR aircraft in controlled airspace. [Figure 9-10]

SECTION A ■ **Pilotage and Dead Reckoning**

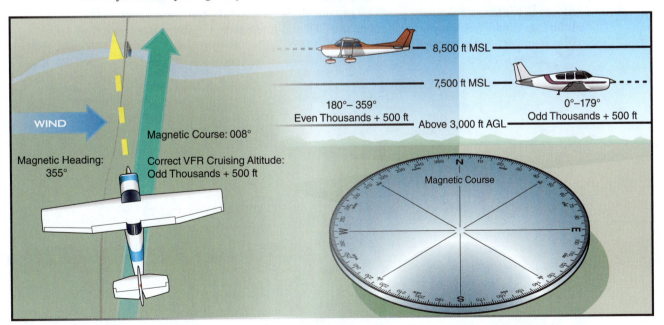

Figure 9-10. You must comply with the VFR cruising altitude rule when you are cruising above 3,000 feet AGL.

FUEL REQUIREMENTS

The FARs require that day VFR flights carry enough fuel to fly to the first point of intended landing at normal cruise speed, and to fly after that for an additional 30 minutes. Night VFR flights must carry a 45-minute reserve. Consider these **fuel reserve** requirements a minimum, and if loading and performance considerations permit, carry larger reserves. An unforecast headwind or a course deviation around an area of poor weather can quickly consume fuel reserves.

FLIGHT PLANNING

The flight planning required for dead reckoning is more involved than for pilotage. Doing as much as possible on the ground before takeoff eases the workload in the air, giving you more time to enjoy the flying and leaving you better able to manage any problems that might arise. When navigating by pilotage, most of the navigation information you need can be found or recorded on your chart. The additional information used in dead reckoning is more easily handled by using a separate **navigation log (nav log)**. This form is a convenient way to organize your preflight navigation planning, as well as to keep track of progress during the flight.

As you recall from pilotage, at times a direct route to your destination is impractical. In these situations, you would plan your flight to include additional legs. You can consider a **leg** on your cross-country flight to be a segment between two checkpoints, to an intermediate stop, or to a location where the course changes. After plotting your true course, list your significant checkpoints on the navigation log, along with the distance for each leg and the distance remaining. Find the true airspeed and fuel consumption in the POH and record it on the log. Next, determine the effects of wind on your airplane using the techniques from Chapter 8. Applying a wind correction angle to your true course gives the true heading. If your trip is not direct, you will need to solve a wind problem for each separate leg. Record your true heading in the appropriate box for each leg of your flight.

Use the information in the POH to determine the time, speed, and distance to climb to your cruising altitude. When you have determined your estimated groundspeed, use it to make a careful estimate of fuel requirements, and decide if you need to plan a fuel stop. Knowing the distance between checkpoints, along with your anticipated groundspeed, you can calculate the estimated time enroute (ETE) for each leg. The first part of your flight should take the lower speed of the climb into consideration as you calculate the ETE and fuel needs. Finally, write down the radio frequencies that you will need. [Figure 9-11]

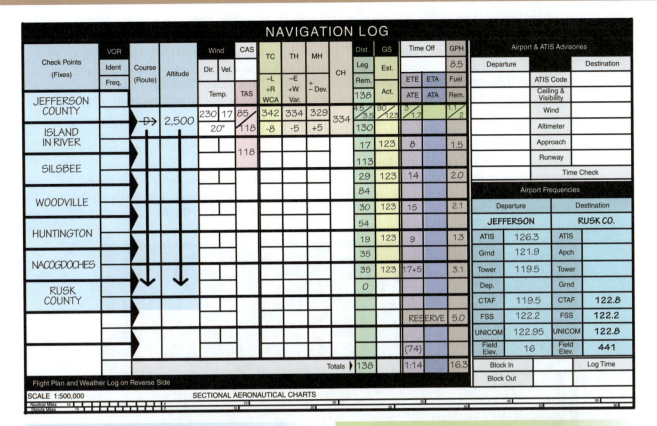

Check Points (Fixes)	VOR Ident / Freq.	Course (Route)	Altitude	Wind Dir. / Vel. / Temp.	CAS / TAS	TC / −L +R WCA	TH / −E +W Var.	MH / + −Dev.	CH	Dist. Leg / Rem.	GS Est. / Act.	Time Off ETE / ATE	ETA / ATA	GPH Fuel / Rem.
JEFFERSON COUNTY		D→	2,500	230 17 / 20°	85 / 118	342 / −8	334 / −5	329 / +5	334	4.5/3.5 138/130	90/123	3/1.7		8.5 1.1/2
ISLAND IN RIVER				/ 118						17 / 113	123	8		1.5
SILSBEE										29 / 84	123	14		2.0
WOODVILLE										30 / 54	123	15		2.1
HUNTINGTON										19 / 35	123	9		1.3
NACOGDOCHES										35 / 0	123	17+5		3.1
RUSK COUNTY		↓ ↓												
												RESERVE		5.0
												(74)		
									Totals ▶	138		1:14		16.3

Airport & ATIS Advisories

Departure	Destination
ATIS Code	
Ceiling & Visibility	
Wind	
Altimeter	
Approach	
Runway	
Time Check	

Airport Frequencies

Departure		Destination	
JEFFERSON		RUSK CO.	
ATIS	126.3	ATIS	
Grnd	121.9	Apch	
Tower	119.5	Tower	
Dep.		Grnd	
CTAF	119.5	CTAF	122.8
FSS	122.2	FSS	122.2
UNICOM	122.95	UNICOM	122.8
Field Elev.	16	Field Elev.	441
Block In		Log Time	
Block Out			

Flight Plan and Weather Log on Reverse Side

SCALE 1:500,000   SECTIONAL AERONAUTICAL CHARTS

Nautical Miles
Statute Miles

1 List your checkpoints, VOR frequencies (if applicable), route, and cruising altitudes in these columns. This example shows direct flight to each checkpoint.

2 List the distance for each leg, along with the distance remaining to the destination.

3 Enter the true airspeed from the POH. (Enter both climb and cruise speed for the first leg.)

4 Use the forecast winds aloft to determine estimated groundspeed and wind correction angle for each leg.

5 Obtain the magnetic variation from the sectional chart and apply it to the true heading to obtain the magnetic heading. Correct for compass deviation to determine the compass heading.

6 Use the groundspeeds that you have calculated to find the estimated time enroute (ETE) between each checkpoint, and total time enroute.

7 Split the first leg into climb and cruise portions. Compute the time, fuel, and distance to climb to your cruising altitude. Subtract the climb portion from the distance to your first checkpoint, then calculate the time to fly the remaining distance at cruise. Add both the climb and cruise portions to find the total time enroute for the first leg. Adjust your calculations if the distance to climb takes you past your first checkpoint.

8 Use the fuel consumption rate from the POH to determine fuel needs. Add at least a 30 minute reserve.

9 List the radio frequencies that you will use.

10 In flight, record your actual time of arrival (ATA) over each checkpoint, find your actual time enroute (ATE) and compute your groundspeed. Use the actual groundspeed to adjust your estimated time of arrival (ETA) as necessary.

Figure 9-11. The navigation log helps you record the known values and compute the unknown values.

SECTION A ■ **Pilotage and Dead Reckoning**

While airborne, mark each checkpoint as you pass it, and make note of the time on your log. Record your actual time enroute (ATE) and compare it to your estimated time to keep track of your flight's progress. To calculate your groundspeed, use the elapsed time between two checkpoints to solve a time-speed-distance problem with your flight computer. If the actual winds aloft are substantially different than forecast, you can use your computer to recalculate your ETE and the fuel needed to reach your destination.

Learning how to determine the information needed for dead reckoning navigation and the ability to fill out a nav log provides you with a solid foundation for flight planning. After you are proficient in this skill, you might decide to use a flight planning app that automatically determines much of the flight planning data and completes a nav log for you as long as you have entered the appropriate airplane performance information. An electronic flight bag (EFB) also keeps track of your flight progress on the nav log in flight. [Figure 9-12]

This NavLog shows information similar to that recorded on a paper nav log. In flight, an EFB with GPS capability will compute your actual groundspeed and update information on your NavLog, such as the ETA to each checkpoint and the ETE for the trip.

Courtesy of ForeFlight

Figure 9-12. You typically can display the nav log and your course on a chart at the same time on an EFB.

FLIGHT PLAN

After completing your nav log, file a **VFR flight plan** with Flight Service. You can file a flight plan online at 1800wxbrief.com or by phone at 1-800-WX-BRIEF. Although not required by the FAA, filing a VFR flight plan provides Flight Service with the details of your flight so they can start a search if you are overdue. You can find specific guidelines for filling out the form at 1800wxbrief.com and in the AIM. For example, enter your initial cruising altitude, even if you plan to change altitude later in the flight. If you will be making a stop on your flight, the FAA recommends that you file a separate flight plan for each leg if the stop is expected to be more than one hour. Enter the usable fuel on board in hours and minutes. The information that you fill out on a flight plan form follows ICAO standards. [Figure 9-13]

When you file a flight plan online or by phone, Flight Service has a record that includes your destination, route of flight, arrival time, and the number of people aboard the airplane. When you are airborne, activate your flight plan on 122.2 MHz or the local Flight Service frequency and inform Flight Service of your actual departure time. You can also use the EasyActivate™ service—you receive an email 30 minutes prior to the estimated time of departure for your filed VFR flight plan. Click on a link in the email to activate the flight plan. After you activate your flight plan, providing position reports (especially in remote areas) keeps Flight Service informed of your aircraft's last known position, which can reduce the size of the search area considerably in the event of an accident.

Flight Plan Form (ICAO)

Aircraft ID	Flight Rule	Flight Type (Optional)	No. of Aircraft	Aircraft Type	Wake Turbulence	Aircraft Equipment
N20JA	VFR	G	1	C172	L	SG

Departure	Airport Info / Area Brief	Departure Date & Time	Evaluate	Cruising Speed	Level	Optimize	Surveillance Equipment
KAPA		06/27/2017 1400 UTC		N0110	A095		EB2

Route of Flight   Map Plan

DCT PUB DCT CIM V611 FTI DCT

Other Information (Optional)

Destination	Airport Info / Area Brief	Total Estimated Elapsed Time
KSAF		0315

Alternate 1 (Optional)   Airport Info / Area Brief

Alternate 2 (Optional)   Airport Info / Area Brief

Fuel Endurance	Persons on Board	Aircraft Color & Markings
0500	2	B/W

Supplemental Remarks (Optional)

Pilot In Command (Optional)
R. ARMSTRONG (303) 555-0182

Emergency Radios
- ☐ UHF
- ☐ VHF
- ☑ ELBA

Survival Equipment
- ☐ Polar
- ☐ Desert
- ☐ Maritime
- ☐ Jungle

Jackets
- ☐ Light
- ☐ Fluorescent
- ☐ UHF
- ☐ VHF

Dinghies (Optional)
Number Capacity Color Covered ☐

Courtesy of Leidos Flight Service

Basic Flight Information
- Aircraft identification number
- VFR when under visual flight rules
- G for general aviation or leave blank
- 1 unless part of formation flight

Aircraft Data
- Approved FAA designator for aircraft type
- Wake turbulence category ("L" for light aircraft)
- Communication/navigation equipment codes ("S" for standard)
- Surveillance equipment codes—transponder mode and ADS-B capability
- Other Information
  - Additional aircraft equipment/capabilities (typically for IFR flights)
  - Special handling requirements
  - Remarks (RMK)

Departure
- Departure ICAO airport identifier
- Planned departure time

Emergency/Survival
- Any emergency/survival equipment
- Supplemental remarks about emergency/survival equipment

Pilot Contact Information
Name and phone number

Enroute and Destination
- True airspeed for cruise flight—"N" for knots and 4-digit number
- Initial cruise altitude—"A" and 3-digit number for hundreds of feet
- Proposed route:
  - Fixes—airports, navaids, GPS waypoints, or lat/long coordinates
  - DCT for direct
  - Victor airway numbers
- Destination ICAO airport identifier
- Estimated time enroute in hours and minutes (including stops of less than an hour)
- Any alternate airports
- Total usable fuel on board in hours and minutes
- Number of persons on board
- Aircraft color(s)

Figure 9-13. This ICAO flight plan form is from 1800wxbrief.com.

After you arrive at your destination, close your flight plan by contacting Flight Service by phone or use the EasyClose™ service. Click the link in the EasyClose™ email, which is sent 30 minutes prior to your estimated time of arrival (based on your actual departure time). If you do not close or extend your flight plan within 30 minutes after your stated ETA, Flight Service begins a preliminary search by phone, then notifies the search and rescue (SAR) system if the phone search is unsuccessful. [Figure 9-14]

SECTION A ■ Pilotage and Dead Reckoning

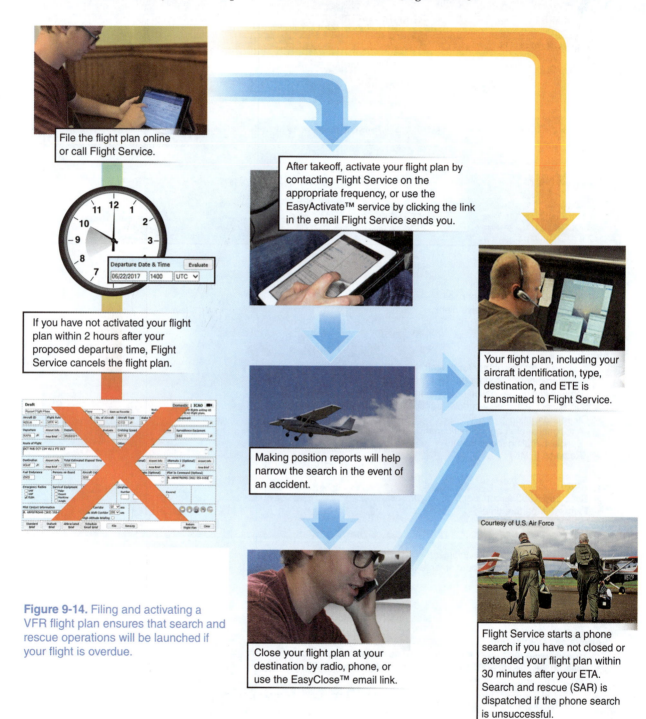

File the flight plan online or call Flight Service.

After takeoff, activate your flight plan by contacting Flight Service on the appropriate frequency, or use the EasyActivate™ service by clicking the link in the email Flight Service sends you.

If you have not activated your flight plan within 2 hours after your proposed departure time, Flight Service cancels the flight plan.

Your flight plan, including your aircraft identification, type, destination, and ETE is transmitted to Flight Service.

Making position reports will help narrow the search in the event of an accident.

Courtesy of U.S. Air Force

Close your flight plan at your destination by radio, phone, or use the EasyClose™ email link.

Flight Service starts a phone search if you have not closed or extended your flight plan within 30 minutes after your ETA. Search and rescue (SAR) is dispatched if the phone search is unsuccessful.

Figure 9-14. Filing and activating a VFR flight plan ensures that search and rescue operations will be launched if your flight is overdue.

LOST PROCEDURES

Recognizing that you have become lost is not as simple as, for example, recognizing that the engine has quit. If a checkpoint does not appear on time, it is easy to wait a little longer to see if it will show up, then wait a little longer still. It is tempting to believe that the two-lane road below is really the superhighway shown on the sectional chart, or that the town on the north side of the road is the same one shown south of the road on the chart, even though it does not have the powerline and radio tower depicted on the chart. If you get lost, do not panic. The five Cs are guidelines to help you take positive action to establish your location: climb, communicate, confess, comply, and conserve.

Climbing to a higher altitude will usually let you see more of the ground, increasing your chances of spotting an identifiable landmark. It improves the reception range of your radios, so you might be able to pick up a VOR station, and it extends the range of your transmitter, which will help with communication.

Communicate with any available facility. If your situation becomes threatening, use the emergency frequency, 121.5 MHz. Confess your situation to any ATC facility, and comply with their suggestions. Conserve fuel by reducing power and airspeed to the values for maximum endurance or range, whichever is most appropriate to your situation.

In addition to the basic five Cs, you could add some others. Check the heading indicator against the magnetic compass and, before you reset it, note the direction of the error. This can help you determine whether you are to the left or right of your intended course. For instance, if the compass indicates 10° more than the heading indicator, you could be to the right of your intended course. Compare the landmarks you see outside to your chart, looking especially for any large, clear landmarks.

SUMMARY CHECKLIST

✓ Pilotage is navigating by reference to landmarks.

✓ The intersection of two different lines of position is a fix, which establishes your position at a definite location.

✓ Paper or digital sectional charts provide a detailed representation of landmarks for cross-country flights.

✓ Select checkpoints that present a number of features to create a unique combination or a distinctive pattern.

✓ Maintaining a constant awareness of your position reduces your chance of becoming lost.

✓ Dead reckoning is navigating by time, speed, distance, and direction calculations.

✓ The navigation plotter combines a straightedge for drawing a course line, a protractor for measuring the direction of flight, and distance scales for sectional and VFR terminal area charts in both nautical and statute miles.

✓ You must correct the true course for magnetic variation, wind drift, and compass deviation to arrive at the compass heading.

✓ Local magnetic variation is shown on aeronautical charts by a dashed magenta isogonic line. East is least and west is best is a memory aid to remember whether to add or subtract magnetic variation.

✓ True course (TC) ± wind correction angle (WCA) = true heading (TH) ± magnetic variation = magnetic heading (MH) ± compass deviation = compass heading (CH).

✓ True course (TC) ± magnetic variation = magnetic course (MC) ± wind correction angle (WCA) = magnetic heading (MH) ± compass deviation = compass heading (CH).(Wind direction must be converted from true to magnetic.)

✓ Factors when choosing a cruising altitude are height above terrain and obstructions, effect of winds on groundspeed and airplane performance, checkpoint visibility, radio reception, emergency options, and the VFR cruising altitude rule.

SECTION A ■ **Pilotage and Dead Reckoning**

✓ The VFR cruising altitude rule dictates that above 3,000 feet AGL, aircraft on magnetic courses from 0° to 179° are required to fly at odd thousand-foot altitudes plus 500 feet and on magnetic courses from 180° to 359°, even thousands plus 500 feet.

✓ Required VFR fuel reserves are 30 minutes for daytime flights and 45 minutes for night flights.

✓ Navigation logs include items such as checkpoints, courses, cruising altitudes, true airspeed, true course, wind correction angle, true heading, magnetic variation, magnetic heading, compass heading, distance between checkpoints, groundspeed, ETE between checkpoints, and fuel consumption.

✓ You can file a VFR flight plan with Flight Service online at 1800wxbrief.com or by phone at 1-800-WX-BRIEF.

✓ Although not required by the FAA, filing a VFR flight plan provides Flight Service with the details of your flight so they can start a search if you are overdue.

✓ When you fill out a flight plan form, enter your initial cruising altitude, even if you plan to change altitude later in the flight.

✓ If you will be making a stop on your flight, the FAA recommends that you file a separate flight plan for each leg if the stop is expected to be more than one hour.

✓ Enter the usable fuel on your flight plan form in hours and minutes.

✓ Activate your flight plan on 122.2 MHz or the local Flight Service frequency or use the EasyActivate™ service and inform Flight Service of your actual departure time.

✓ If you do not close or extend your flight plan within 30 minutes after your stated ETA, Flight Service begins a preliminary search by phone, then notifies the search and rescue (SAR) system.

✓ After you arrive at your destination, close your flight plan by contacting Flight Service by phone or use the EasyClose™ service.

✓ If you become lost, remember the five Cs: climb, communicate, confess, comply, and conserve.

KEY TERMS

Line of Position (LOP)	Magnetic Course
Fix	True Heading
Pilotage	Compass Heading
Checkpoints	VFR Cruising Altitude
Dead Reckoning	Fuel Reserve
Navigation Plotter	Navigation Log (Nav Log)
True Course	Leg
Isogonic Line	VFR Flight Plan

QUESTIONS

1. What is the difference between pilotage and dead reckoning? Explain how these forms of navigation can be used together.

2. True/False. Magnetic variation is the difference between true north and magnetic north.

3. Use the chart excerpt, nav log excerpt, and a navigation plotter to determine the following for a flight from Guntersville-Starnes Airport to Isbell Airport:
   - True Course
   - True Heading
   - Magnetic Variation
   - Magnetic Heading
   - Compass Heading

TC	TH	MH	
			CH
−L	−E		
+R	+W	±Dev.	
WCA	Var.		
+5		+1	

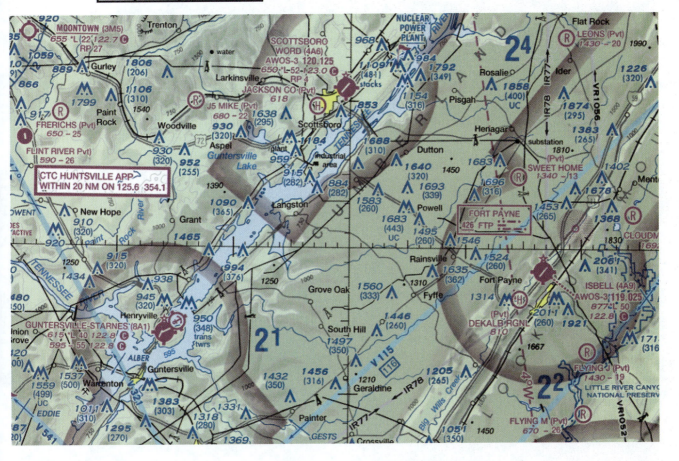

4. Use the information in Question 3. What is the magnetic course?

5. List at least 7 items that you typically include on a nav log during flight planning.

6. Name at least three factors to consider when choosing a cruising altitude.

7. What altitude is required by the VFR cruising altitude rule for an airplane on a magnetic course of 240°?
    A. 3,000 feet AGL
    B. 3,500 feet MSL
    C. 4,500 feet MSL

8. The FARs require that VFR flights carry enough fuel to fly to the first point of intended landing at normal cruise speed, and to fly after that for how many minutes during the day? At night?

9. Your initial cruise altitude is 4,500 feet MSL. You plan to climb to 6,500 feet MSL and then to 8,500 feet MSL for subsequent legs of your flight. What altitude(s) should you enter in the flight plan form?
    A. 4500
    B. 8500
    C. 4500, 6500, 8500

10. Select the true statement regarding filling out the flight plan form.
    A. Enter the usable fuel on your flight plan in gallons.
    B. Include every stop of more than an hour in the route of flight.
    C. File a separate flight plan for a leg if you are planning to stop for more than one hour.

11. Select the true statement regarding closing your flight plan.
    A. If you land at an airport with a control tower, the ground controller automatically closes your flight plan.
    B. After you arrive at your destination, you must close your flight plan by contacting Flight Service by phone or use the EasyClose™ service.
    C. If you do not close or extend your flight plan within 60 minutes after your stated ETA, Flight Service begins a preliminary search by phone.

SECTION A ■ **Pilotage and Dead Reckoning**

SECTION B
VOR Navigation

During aviation's early years, it became evident that a system to assist pilots in navigation would be necessary if the airplane was going to be useful as a transportation tool. Over the years, navigation systems evolved from bonfires to light beacons to radio navigation aids. The **very high frequency omnidirectional range (VOR)** is a common radio navigation system that you will use as a pilot.

GROUND EQUIPMENT

VORs operate in the very high frequency (VHF) range, on frequencies of 108.00 MHz through 117.95 MHz. VHF frequencies offer relatively interference-free navigation, but unlike lower frequency radio waves, which can skip within the atmosphere or travel over the ground for great distances, VOR reception is strictly line of sight. This limits usable signal range at low altitudes or over mountainous terrain. [Figure 9-15] When obstacles reduce VOR reception range below standard values published in the AIM, the affected route and the usable range appears in the applicable Chart Supplement under the individual VOR listings.

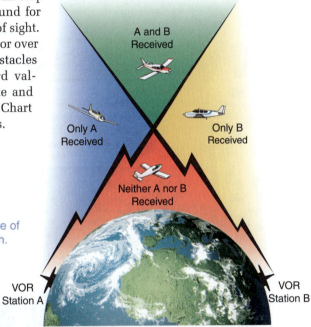

Figure 9-15. Because VOR reception is strictly line of sight, range is limited by the curvature of the earth.

Basic VOR systems only provide course guidance, but **VOR/DME** and **VORTAC** facilities also provide distance information to aircraft equipped with distance measuring equipment (DME). A VORTAC is a collocated VOR and military navigational aid called a TACAN. Civil aviation uses the portion of the TACAN system that provides distance information. [Figure 9-16]

Figure 9-16. Sectional charts depict VORs, VORTACs, and VOR/DMEs with unique symbols.

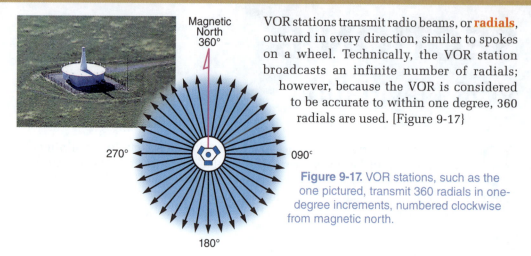

VOR stations transmit radio beams, or **radials**, outward in every direction, similar to spokes on a wheel. Technically, the VOR station broadcasts an infinite number of radials; however, because the VOR is considered to be accurate to within one degree, 360 radials are used. [Figure 9-17]

Figure 9-17. VOR stations, such as the one pictured, transmit 360 radials in one-degree increments, numbered clockwise from magnetic north.

VORs and their associated radials are depicted on sectional charts with circles, graduated in degrees, called **compass roses**. Many VOR stations are connected by specific radials, which form routes called Victor airways. [Figure 9-18]

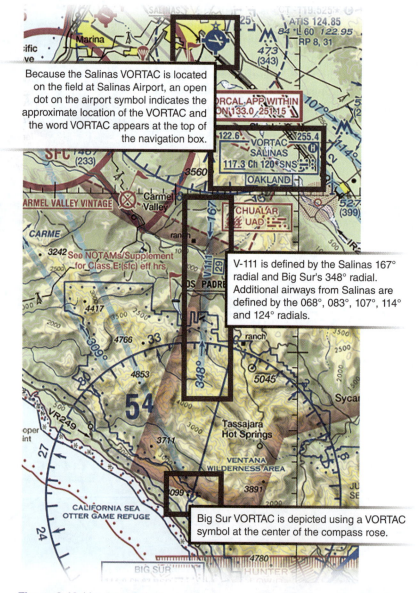

Because the Salinas VORTAC is located on the field at Salinas Airport, an open dot on the airport symbol indicates the approximate location of the VORTAC and the word VORTAC appears at the top of the navigation box.

V-111 is defined by the Salinas 167° radial and Big Sur's 348° radial. Additional airways from Salinas are defined by the 068°, 083°, 107°, 114° and 124° radials.

Big Sur VORTAC is depicted using a VORTAC symbol at the center of the compass rose.

Figure 9-18. You can navigate across the country by flying published magnetic courses from one VOR to the next.

VOR ground stations are divided into different classes according to their normal reception and altitude range. The original service volumes included **terminal VOR (TVOR)**, which were normally located on an airport and were designed to be used within 25 NM and below 12,000 feet AGL. With the new VOR service volumes, you can use a **low altitude VOR (LVOR)** reliably up to 40 NM from the station at altitudes between 1,000 and 5,000 feet AGL and 70 NM from the station from 5,000 to 18,000 feet AGL. A **high altitude VOR (HVOR)** also offers a reception range of 40 NM up to 5,000 feet, 70 NM between 5,000 and 14,500 feet, and 100 NM between 14,500 feet and 18,000 feet. The HVOR's maximum range of 130 NM is available between 18,000 feet and FL450. Between FL450 and FL600, the reception range decreases to 100 NM. You can find the class designation of a VOR facility in the Chart Supplement.

AIRBORNE EQUIPMENT

VOR airborne equipment consists of an antenna, receiver, and indicator. The antenna is easy to recognize—it is shaped like a V. The receiver interprets the signal and sends the resulting course information to the VOR indicator. The VOR indicator consists of the **course deviation indicator (CDI)**, the **TO-FROM indicator**, and the course selector, sometimes referred to as an **omnibearing selector (OBS)**. [Figure 9-19]

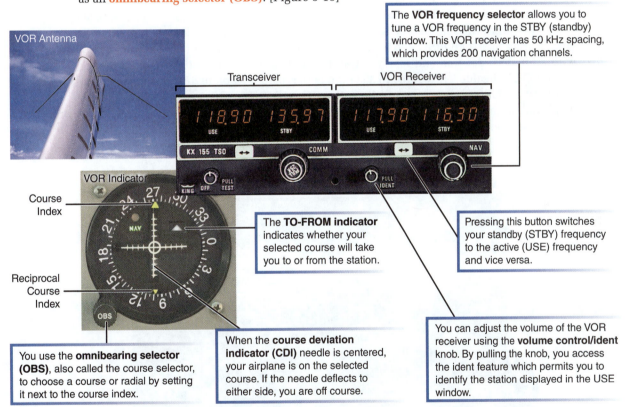

The **VOR frequency selector** allows you to tune a VOR frequency in the STBY (standby) window. This VOR receiver has 50 kHz spacing, which provides 200 navigation channels.

VOR Antenna

Transceiver   VOR Receiver

VOR Indicator

Course Index

Reciprocal Course Index

The **TO-FROM indicator** indicates whether your selected course will take you to or from the station.

Pressing this button switches your standby (STBY) frequency to the active (USE) frequency and vice versa.

You use the **omnibearing selector (OBS)**, also called the course selector, to choose a course or radial by setting it next to the course index.

When the **course deviation indicator (CDI)** needle is centered, your airplane is on the selected course. If the needle deflects to either side, you are off course.

You can adjust the volume of the VOR receiver using the **volume control/ident** knob. By pulling the knob, you access the ident feature which permits you to identify the station displayed in the USE window.

Figure 9-19. Although most VOR equipment is similar in appearance and operation, you should consult your POH or avionics manual to learn the nuances of your aircraft's particular system.

SECTION B ■ **VOR Navigation**

NAVIGATION PROCEDURES

VOR is a relatively easy navigation system to use when you understand the basic navigation procedures. These operations include tuning and identifying a station, interpreting VOR indications, tracking, intercepting a course, and cross checking your position.

IDENTIFYING A STATION

Before using a VOR for navigation, you must tune your VOR receiver and identify the station to ensure you have chosen the right frequency and that the station is working properly. On many receivers you can monitor the station identifier by selecting the ident feature and turning up the volume. If you do not hear the VOR's Morse code identifier or voice identification, you cannot assume a reliable navigation signal. When a station is shut down for maintenance, it might transmit a T-E-S-T signal (— . ... —); at other times there is no identifier at all.

INTERPRETING VOR INDICATIONS

After you tune a VOR frequency, the VOR receiver automatically determines your magnetic direction, or radial, from the ground station. Although some VOR receivers display a digital readout of your radial, many units installed in light aircraft do not directly indicate your radial from the station. To determine your position relative to the station, turn the omnibearing selector (OBS) knob until the course deviation indicator (CDI) needle centers with a FROM indication and read the resulting radial next to the index mark on the top of the VOR indicator. The course to a station is the reciprocal of, or 180° from, the radial. If you set the OBS to center the needle with a TO indication, you can turn to the heading displayed on the VOR indicator and fly on course directly to the station, assuming there is no crosswind. [Figure 9-20]

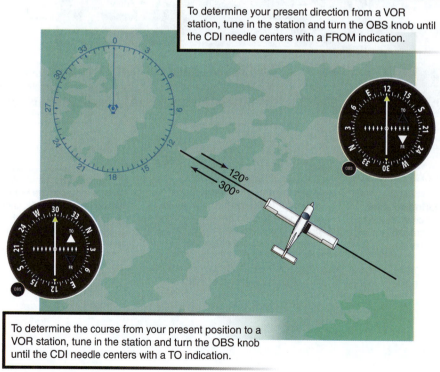

To determine your present direction from a VOR station, tune in the station and turn the OBS knob until the CDI needle centers with a FROM indication.

To determine the course from your present position to a VOR station, tune in the station and turn the OBS knob until the CDI needle centers with a TO indication.

Figure 9-20. You can quickly determine your radial from or course to a VOR station by twisting the OBS knob to center the CDI.

When you are off course, the CDI is designed to point toward your desired course. The scale shows how far you are off course. On a typical analog indicator, each dot on the scale represents a course deviation of two degrees. For example, if your CDI is deflected two dots to the left of center, your desired course is four degrees to your left. [Figure 9-21]

Figure 9-21. To determine your position relative to a VOR, it might help to visualize your airplane at the bottom of the VOR indicator.

This CDI indicates you are on the selected course.

This CDI indicates you are 4 degrees right of your selected course.

REVERSE SENSING

A VOR airborne system does not perceive your aircraft's heading; it only senses your direction from the station and gives the same instrument indications regardless of which way the nose of the aircraft is pointing. If you mistakenly set your VOR indicator to the reciprocal of your desired course, your CDI will be deflected away from the course. In this **reverse sensing** situation, the normal procedure of correcting toward the needle will actually take you farther off course. For correct sensing, you must set the VOR indicator so it generally agrees with your intended course. [Figure 9-22]

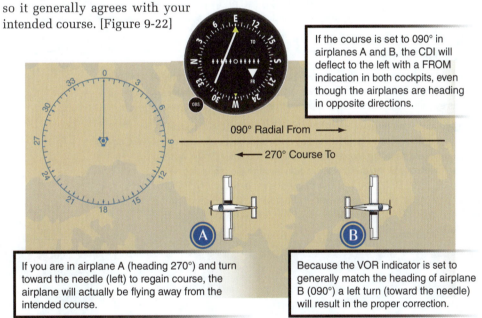

If the course is set to 090° in airplanes A and B, the CDI will deflect to the left with a FROM indication in both cockpits, even though the airplanes are heading in opposite directions.

090° Radial From →

← 270° Course To

If you are in airplane A (heading 270°) and turn toward the needle (left) to regain course, the airplane will actually be flying away from the intended course.

Because the VOR indicator is set to generally match the heading of airplane B (090°) a left turn (toward the needle) will result in the proper correction.

Figure 9-22. To avoid reverse sensing, always set the VOR indicator to approximately agree with your intended course.

OFF INDICATIONS

As you reach a VOR station and fly over it, the TO indication disappears and is replaced briefly by any one of a variety of indications, depending on your equipment. The indications you can encounter include an OFF flag, a NAV indication, a red and white barber pole, or simply the absence of a TO or FROM indication. As you leave the station behind, the OFF flag (or similar indication) is replaced by a FROM indication, signaling that you are traveling away from the station. The area over the station in which the TO-FROM indicator changes is called the **cone of confusion**, or no-signal area. Due to its shape, flight through the cone of confusion can vary from a few seconds at low altitude up to several minutes at high altitude. You might also notice an OFF flag (or similar indication) when the aircraft is abeam the station on the selected course. If an OFF flag appears at any other time, it means you are not receiving a reliable navigation signal. The system might not be tuned or functioning properly. [Figure 9-23]

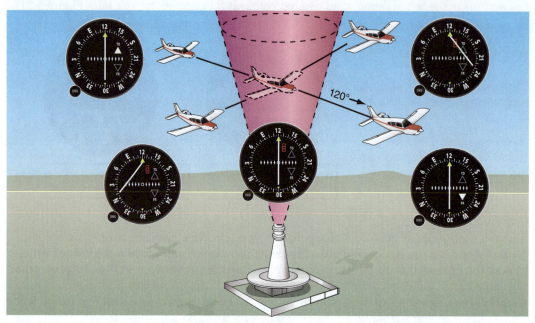

Figure 9-23. This TO-FROM indicator displays a red and white barber pole indicator when flying in the cone of confusion and when the aircraft is directly abeam the station on the selected course of 120°.

Airway Beacons

In the days before radio navigation aids, the airway beacon system consisted of lights installed at approximately 10-mile intervals along routes between major U.S. cities. In 1946, this system consisted of 2,112 beacons that defined 124 airways.

To navigate along the airways, pilots flew from light to light in a sequence defined by the Morse code signal flashed by each beacon. For example, the first beacon would flash the Morse code signal for the letter W; the second would flash the letter U; the third, V; and so on through the remaining seven letters (H, R, K, D, B, G, and M). The sequence repeated every 10 lights until the end of the airway. To remember the code, many pilots used the mnemonic, **W**hen **U**ndertaking **V**ery **H**ard **R**outes **K**eep **D**irections **B**y **G**ood **M**ethods.

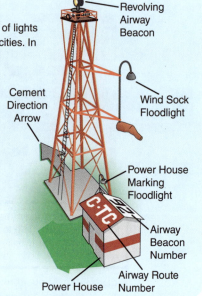

Revolving
Airway
Beacon

Wind Sock
Floodlight

Cement
Direction
Arrow

Power House
Marking
Floodlight

Airway
Beacon
Number

Airway Route
Number

Power House

TRACKING

The most common VOR navigation you will perform is flying from one station to another, using a process called **tracking**. When tracking, you maintain the selected course by keeping the CDI centered. To stay on course in a crosswind, you use a technique called **bracketing**, which involves making a series of corrections to regain and maintain your desired course. [Figure 9-24]

INTERCEPTING A COURSE

In some situations, you might need to track to the station on a different course instead of flying direct from your present position. In these instances, you must fly from the radial

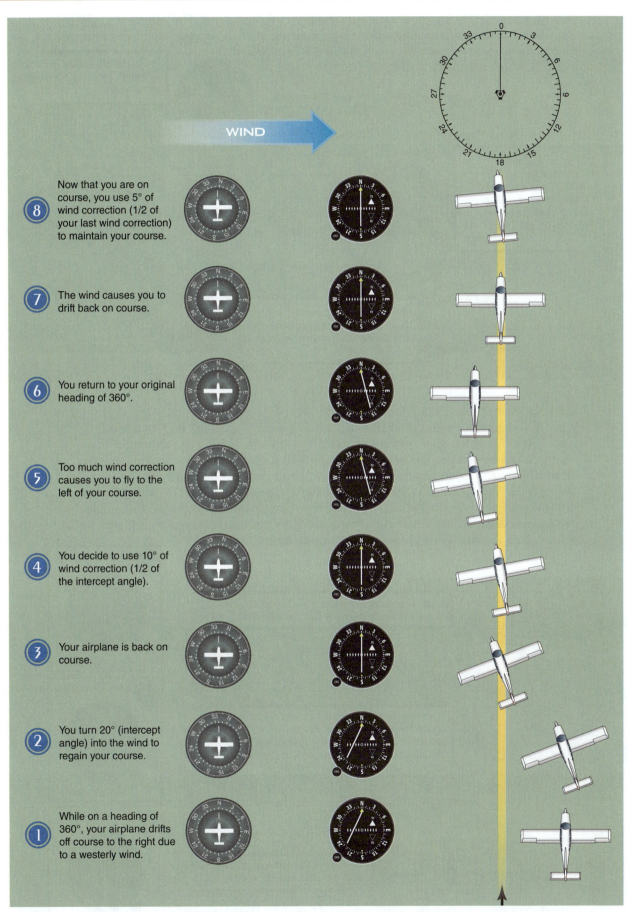

WIND

8 Now that you are on course, you use 5° of wind correction (1/2 of your last wind correction) to maintain your course.

7 The wind causes you to drift back on course.

6 You return to your original heading of 360°.

5 Too much wind correction causes you to fly to the left of your course.

4 You decide to use 10° of wind correction (1/2 of the intercept angle).

3 Your airplane is back on course.

2 You turn 20° (intercept angle) into the wind to regain your course.

1 While on a heading of 360°, your airplane drifts off course to the right due to a westerly wind.

SECTION B ■ VOR Navigation

Figure 9-24. While tracking inbound to the Abbott VORTAC on the 180° radial, you begin to drift right of course due to a crosswind. This example shows the bracketing procedure required to regain the course and determine a heading that will compensate for the crosswind.

you are currently tracking and intercept another radial inbound. The intercept angle you use in a particular situation could range from about 20° when you are close to the station, up to approximately 90° when the station is located a considerable distance away. [Figure 9-25]

When the CDI begins to center, you turn right and track inbound to the station.

You set up an intercept angle by turning left to a heading of 045°. Once established on your intercept course, you turn the OBS to set the new inbound course, 090°, in the VOR indicator.

45° Intercept Angle

090°

270°

070°

250°

While tracking inbound on the 250° radial, you see cumulus clouds ahead and decide to approach the station on the 270° radial.

Figure 9-25. This example uses a 45° angle to intercept a new course due to poor weather along the original route.

CROSS CHECKING YOUR POSITION

When you determine your radial from a station, you only establish that your location is on a line of position (LOP) extending away from the station. You can determine your exact position by cross checking with a second VOR station. To do this, determine your location from the second station and draw a line of position on that radial from the second VOR. Your position is where the two LOPs intersect. Determining your position this way is sometimes called **triangulation** because your position plus the locations of the two navaids make up three points of a triangle. For the most accurate results, you should select radials that are nearly perpendicular to each other. [Figure 9-26]

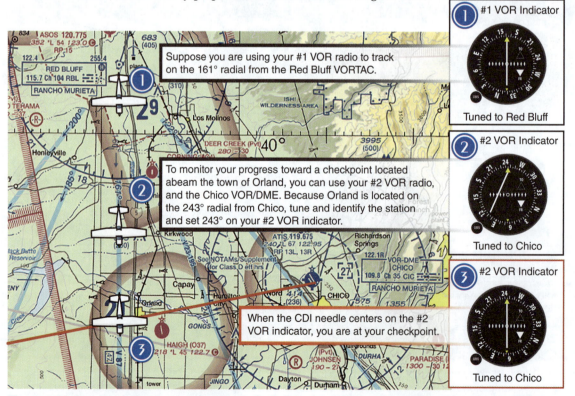

① #1 VOR Indicator

Suppose you are using your #1 VOR radio to track on the 161° radial from the Red Bluff VORTAC.

Tuned to Red Bluff

② #2 VOR Indicator

To monitor your progress toward a checkpoint located abeam the town of Orland, you can use your #2 VOR radio, and the Chico VOR/DME. Because Orland is located on the 243° radial from Chico, tune and identify the station and set 243° on your #2 VOR indicator.

Tuned to Chico

③ #2 VOR Indicator

When the CDI needle centers on the #2 VOR indicator, you are at your checkpoint.

Tuned to Chico

Figure 9-26. You can determine your position by finding the intersection of lines of position from two VORs.

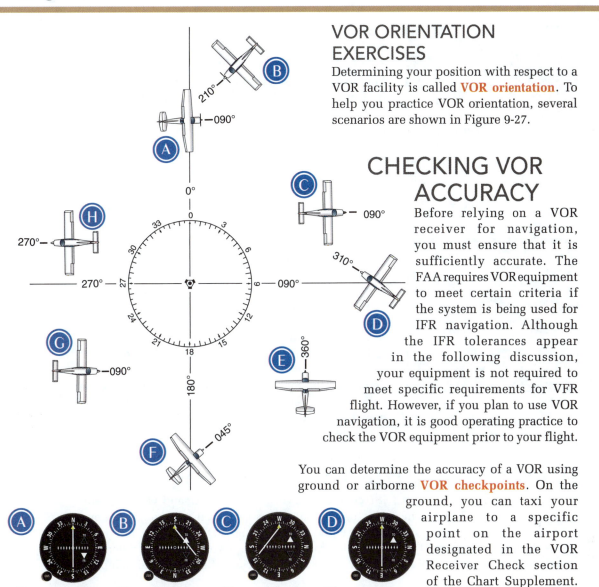

VOR ORIENTATION EXERCISES

Determining your position with respect to a VOR facility is called **VOR orientation**. To help you practice VOR orientation, several scenarios are shown in Figure 9-27.

CHECKING VOR ACCURACY

Before relying on a VOR receiver for navigation, you must ensure that it is sufficiently accurate. The FAA requires VOR equipment to meet certain criteria if the system is being used for IFR navigation. Although the IFR tolerances appear in the following discussion, your equipment is not required to meet specific requirements for VFR flight. However, if you plan to use VOR navigation, it is good operating practice to check the VOR equipment prior to your flight.

You can determine the accuracy of a VOR using ground or airborne **VOR checkpoints**. On the ground, you can taxi your airplane to a specific point on the airport designated in the VOR Receiver Check section of the Chart Supplement. After centering the CDI, compare your VOR course indication to the published radial for that checkpoint. The maximum permissible error for IFR navigation is ± 4°. Airborne checkpoints, also listed in the Chart Supplement, are usually located over easily identifiable terrain or prominent features on the ground. With an airborne checkpoint, the maximum permissible course error is ± 6°. Another way to perform an airborne check is by using a VOR radial that defines the centerline of an airway. To conduct the check, use your sectional chart to locate a prominent terrain feature under the centerline of an airway, preferably 20 miles or more from the facility. Maneuver your aircraft directly over the point, twist the OBS to center the CDI needle and note what course is set on the VOR indicator. The permissible difference between the published radial and the indicated course is ± 6°. [Figure 9-28]

Figure 9-27. Remember as you study these examples that the VOR indications are independent of aircraft heading.

Figure 9-28. In this example, your VOR indicator should display 164° (± 6°) FROM the North Bend VORTAC as you overfly the east edge of the town of Coquille.

You also can conduct a VOR check by comparing the indications of two VOR systems that are independent of each other (except for the antenna). If your aircraft is equipped with two VOR radios, set both to the same VOR facility and note the indicated readings on each. When you check one against the other, the difference should not exceed 4°.

VOR test facilities (VOTs) enable you to make precise VOR accuracy checks regardless of your airplane's position in relation to the facility because VOTs broadcast a signal for only one radial—360°. First, obtain the frequency from the Chart Supplement, tune your VOR receiver, and identify the VOT signal; you should hear a series of dots or a continuous tone. Next, set a course of either 0° or 180° on the VOR indicator. If you set 0°, the CDI should center with a FROM indication. If you set 180°, the CDI should center with a TO indication. If the CDI does not center, you can determine the magnitude of the error by rotating the OBS until the needle moves to the center position. The new course should be within ±4° of your test course (0° or 180°).

HORIZONTAL SITUATION INDICATOR

A **horizontal situation indicator (HSI)** combines the heading indicator and VOR indicator in a single display. The design of this instrument solves nearly all reverse sensing and other visualization problems associated with a conventional VOR indicator, providing you with an easy-to-interpret navigation picture. The HSI uses its own device to determine direction and typically does not need to be adjusted using the magnetic compass. [Figure 9-29]

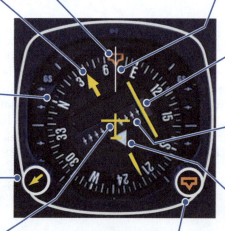

The **course indicating arrow** visually shows the orientation of the selected course relative to your current heading. Because of this, left and right indications on the CDI are always properly oriented.

A rotating **compass card** indicates the aircraft's current magnetic heading. In situations in which a standard VOR indicator gives you reverse sensing, the HSI compass card turns to provide normal sensing.

Use the **course set knob** to position the course indicating arrow.

The **symbolic aircraft** shows your position in relation to the selected course as though you are above the aircraft looking down.

Use the **heading select** bug with the autopilot to automatically turn the aircraft to a newly selected heading.

The **heading set knob** is used to position the heading select bug.

The airplane's heading is displayed under the **heading index**, also called a lubber line.

The **course deviation indicator (CDI)** depicts how far you are off course. When you are on course, the CDI is aligned with the course arrow.

Each dot on the **course deviation scale** represents 2° for VOR navigation.

The **TO-FROM indicator** points to the head of the course arrow when the selected course is inbound to the VOR. When the selected course is outbound from the VOR, the TO-FROM indicator points away from the course arrowhead.

Figure 9-29. This HSI shows the aircraft on a 070° heading, which is a 30° intercept angle for the 040° course from the station.

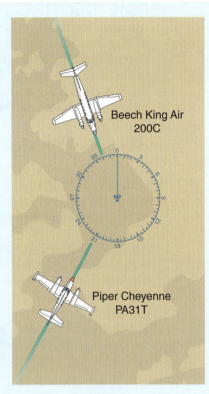

Midair Collision

DISCOVERY

From the files of the NTSB: A Piper PA-31T, N9162Y, and a Beech 200C, N390AC, were involved in a midair collision while both were cruising at 17,500 ft in unlimited visibility. The Piper was tracking inbound on the 210 deg [degree] radial of the Richmond VOR. The Beech was tracking inbound on the 340 deg [degree] radial of the Wilmington VOR. Both acft [aircraft] were substantially damaged, but both aircrew were able to continue flying and land safely.

The plt [pilot] of the Piper said that he saw a tan flash just before the collision. The pilot of the Beech did not see the Piper. The cockpit view of the PA-31 pilot was somewhat restricted by the left, windshield side post. Likewise, the Beech plt's [pilot's] view was somewhat restricted by the windshield center post and the windshield wiper arm. The acft [aircraft] converged on one another with a closure speed of about 420 kts and in the same general directions that the plt's [pilots'] views were obstructed.

If you think of each radial transmitted by a VOR as a highway, every VOR station is analogous to a 360-way road intersection — without a stop sign. Not only can there be a concentration of aircraft in the vicinity of a VOR, but a pilot's workload generally increases with course changes, radio frequency and VOR indicator adjustments, time checks, and other cockpit duties. As you fly near a VOR, concentrate on managing your workload efficiently so you have plenty of time to scan for aircraft on the other 359 highways in your portion of the sky.

Beech King Air 200C

Piper Cheyenne PA31T

DISTANCE MEASURING EQUIPMENT

VOR/DME and VORTAC facilities give you distance information in addition to course guidance. If your aircraft is equipped with **distance measuring equipment (DME)**, you typically can obtain a readout of the distance in nautical miles to the associated VOR/DME or VORTAC site as well as groundspeed and time enroute to the station. To obtain a distance from the station, your aircraft's DME transceiver first transmits an interrogation signal to the station. The ground station then transmits a reply back to the aircraft. The aircraft's equipment measures the round trip time of this signal exchange, computes the distance in nautical miles, and displays it digitally in the cockpit. Depending on altitude and line-of-sight restrictions, you can receive a reliable signal up to 199 miles from the station. [Figure 9-30]

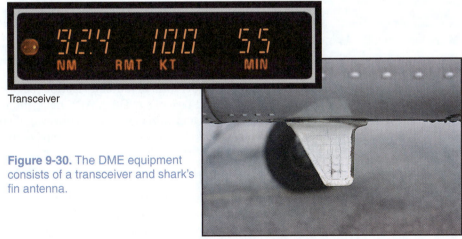

Transceiver

Figure 9-30. The DME equipment consists of a transceiver and shark's fin antenna.

Antenna

SECTION B ■ VOR Navigation

DME IDENTIFICATION

Although you tune the DME using a VOR frequency, remember that the DME is a separate facility, even though it is collocated with a VOR. Each VOR frequency is tied to a specific DME channel under an arrangement called frequency pairing. When you tune and listen to the VOR, you should hear the VOR identifier repeated 3 or 4 times, followed by the DME identifier. A single-coded identification transmitted approximately every 30 seconds signals that the DME is functioning.

DME CAUTIONS

DME can be a very useful navigation aid, however, before using DME, you should understand its limitations. For example, because DME measures groundspeed by comparing the time lapse between a series of pulses, flight in any direction other than directly to or away from the station will result in an unreliable reading.

Although DME normally is accurate to within 1/2 mile or 3% of the actual distance, whichever is greater, you should be aware that DME measures slant range, not horizontal distance to a station. **Slant range distance** is the result of two components — horizontal and vertical distance. The difference between the slant range distance and the horizontal distance is not significant if the aircraft is at least 1 mile from the station for every 1,000 feet of altitude. The error is greatest when the aircraft is directly above the station, where the DME simply indicates the aircraft's altitude in nautical miles. [Figure 9-31]

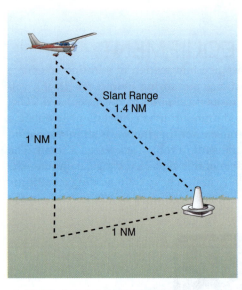

Slant Range
1.4 NM

1 NM

1 NM

Figure 9-31. If you are flying at an altitude of 1 mile at a horizontal distance of 1 mile from the station, your DME will indicate a distance of 1.4 miles.

ADF NAVIGATION

Automatic direction finder (ADF) navigation is an older form of radio navigation that uses signals from ground stations, called non-directional beacons (NDBs). ADF navigation is not widely used today due to the prevalence of VOR and GPS navigation. However, if you have ADF equipment in your airplane, you should know to use this form of navigation.

ADF EQUIPMENT

Nondirectional radio beacons (NDBs) transmit low/medium frequency (L/MF) signals in the range of 190 kHz to 535 kHz. NDBs are shown on VFR aeronautical charts. Your ADF equipment also can receive signals from AM commercial broadcast stations. [Figure 9-32]

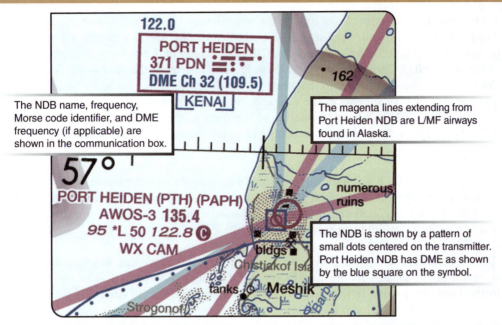

Figure 9-32. This sectional chart excerpt depicts an NDB located in Alaska, where ADF navigation is more common.

ADF equipment in the aircraft permits L/MF signals to be received through the antenna, relayed to the ADF receiver where they are processed, and then sent to the ADF bearing indicator. An ADF requires two antennas—loop and sense—which are either mounted separately or combined into the same unit. Consult the POH or POH supplement to familiarize yourself with the specific equipment installed in your airplane. [Figure 9-33]

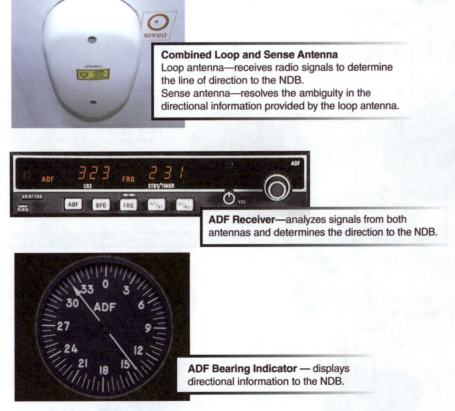

Figure 9-33. Airborne ADF equipment consists of antennas, an ADF receiver, and an ADF bearing indicator.

A bearing—the direction from one point to another—can be measured clockwise in degrees from any reference point. A bearing indicator provides the horizontal direction, or angle, between your airplane and the NDB. The needle of the bearing indicator always points to the NDB. Generally, there are three types of bearing indicators — the fixed-card, the movable-card, and the radio magnetic indicator (RMI). On a fixed-card indicator, the number zero always appears at the top, and the numbers around the 360° azimuth card correspond to an NDB's bearing relative to the nose of the airplane. You can rotate the azimuth card of a movable-card indicator to place the airplane heading under the top index.

An RMI usually has a single-bar and a double-bar needle superimposed over a rotating compass card that is referenced to magnetic north. Although the single-bar needle normally points to a VOR and the double-bar needle points to an NDB, on many RMIs, you can set each needle to either a VOR or NDB.

The fixed-card bearing indicator measures relative bearing—the angular difference between the airplane's longitudinal axis and a straight line drawn from the airplane to the NDB. This value is measured clockwise from the airplane's nose. If you wish to fly directly to the NDB, you must add your magnetic heading to the relative bearing to determine the magnetic bearing to the NDB. Magnetic heading (MH) + relative bearing (RB) = magnetic bearing (MB) to the station. If the total is more than 360°, you will need to subtract 360°. [Figure 9-34]

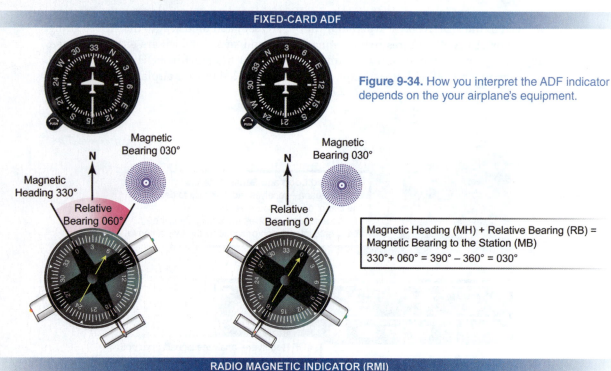

FIXED-CARD ADF

Figure 9-34. How you interpret the ADF indicator depends on the your airplane's equipment.

Magnetic Heading (MH) + Relative Bearing (RB) = Magnetic Bearing to the Station (MB)

330°+ 060° = 390° − 360° = 030°

RADIO MAGNETIC INDICATOR (RMI)

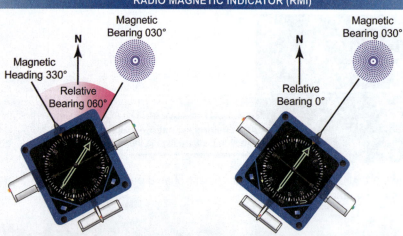

HOMING

A procedure during which you always keep the nose of the airplane pointing directly to the station is called **homing** to the station. To fly to an NDB using the homing procedure simply turn to the magnetic bearing so the head of the ADF needle is on the airplane's nose. In a no-wind situation, a constant magnetic heading will keep the ADF needle positioned at 0° on the fixed-card indicator as you fly inbound to the station. However, in a crosswind situation, the wind will cause your airplane to drift off course, and you must adjust the magnetic heading to keep the nose of the airplane pointing toward the station. [Figure 9-35]

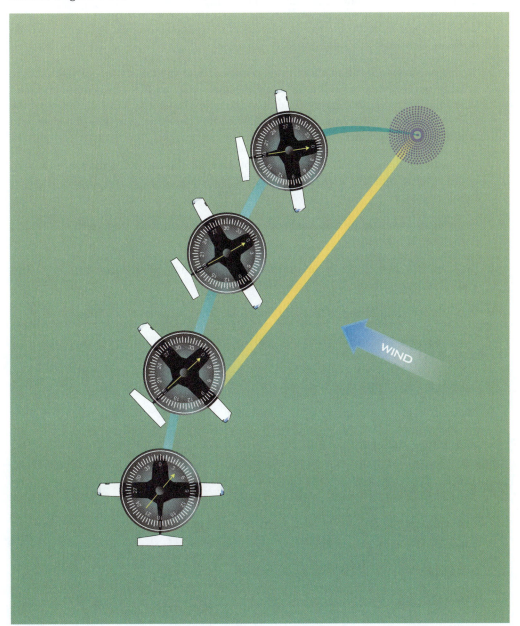

Figure 9-35. The curved line is your flight path when homing to the station in a right crosswind. With no wind, your flight path would follow a straight line.

SUMMARY CHECKLIST

✓ VORs only provide course guidance, but VOR/DMEs and VORTACs also provide distance information.

✓ There are three classes of VORs with different coverage areas — terminal, low altitude, and high altitude.

✓ VOR radials and courses derived from radials are oriented to magnetic north and are depicted on most aeronautical charts using compass roses.

✓ Before using a VOR for navigation, always identify the station using the Morse code or voice identifier.

✓ To determine your location after tuning and identifying a VOR station, turn the course selector or OBS knob until the CDI needle centers with a FROM indication and read the radial next to course index. To determine your course to a VOR station, turn the course selector or OBS knob until the CDI needle centers with a TO indication, and read the magnetic course on the course index.

✓ The indications of a VOR receiver are not directly affected by aircraft heading. To avoid reverse sensing, always set the VOR indicator to generally agree with your intended course.

✓ Tracking involves flying a desired course to or from a station using a sufficient wind correction, if necessary.

✓ Bracketing is the process of determining and applying a wind correction that keeps you on course with the CDI needle centered.

✓ You can determine your position by cross checking between two VORs.

✓ You can check VOR receiver accuracy using ground and airborne checkpoints, or by using a VOT.

✓ An HSI is a VOR indicator combined with a heading indicator.

✓ DME automatically displays your slant range distance to a suitably equipped VOR ground station. Slant range error is greatest when your aircraft is directly over the transmitting station.

KEY TERMS

Very High Frequency Omnidirectional Range (VOR)

VOR/DME

VORTAC

Radial

Compass Rose

Terminal VOR (TVOR)

Low Altitude VOR (LVOR)

High Altitude VOR (HVOR)

Omnibearing Selector (OBS)

Course Deviation Indicator (CDI)

TO-FROM Indicator

Reverse Sensing

Cone of Confusion

Tracking

QUESTIONS

1. What navigation capability does a VORTAC provide?

2. Identify the components of the VOR indicator shown in the accompanying figure.

3. Why is it important to set your VOR indicator to generally agree with your intended course?

4. If the CDI is deflected three dots to the right and your VOR indicator and heading indicator are in general agreement, where is your desired course?
   A. 6° left
   B. 3° right
   C. 6° right

5. True/False. Left and right CDI deflections are always properly oriented to the airplane's heading on an HSI.

6. Approximately, what will a DME display indicate when you are directly over the station at 12,000 feet AGL?

7. What should the OBS and the TO/FROM indicator read when the CDI needle is centered using a VOR test signal (VOT)?
   A. 180° TO, only if the aircraft is directly north of the VOT.
   B. 0° TO or 180° FROM, regardless of the aircraft's position from the VOT
   C. 0° FROM or 180° TO, regardless of the aircraft's position from the VOT

SECTION B ■ **VOR Navigation**

Refer to the following chart excerpt to answer questions 8 through 11.

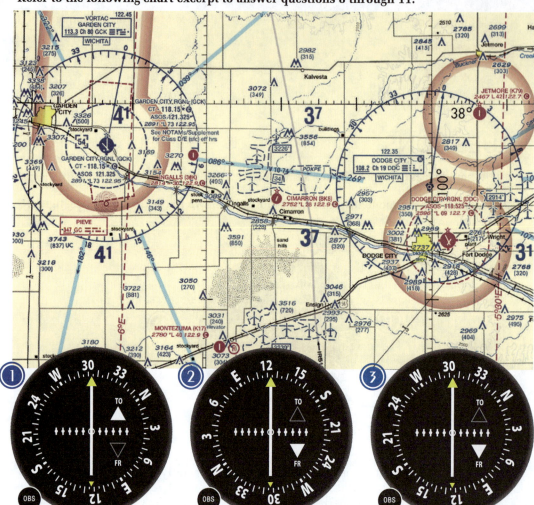

8. The VOR is tuned to Dodge City VORTAC, and the aircraft is positioned over Kalvesta, a small town northeast of the VORTAC next to towers. Which VOR indication is correct?
   A. 1
   B. 2
   C. 3

9. What course should you select on the omnibearing selector (OBS) to make a direct flight from Cimarron Airport to the Dodge City VORTAC with a TO indication?
   A. 077°
   B. 087°
   C. 257°

10. What is the approximate position of your airplane if the VOR receivers indicate the 140° radial of Garden City VORTAC and the 225° radial of Dodge City VORTAC?
    A. Ingalls Airport (30K)
    B. Jetmore Airport (K79)
    C. Montezuma Airport (K17)

11. The VOR is tuned to the Garden City VORTAC. The omnibearing selector (OBS) is set on 150°, with a FROM indication, and a right course deviation indicator (CDI) deflection. What is the aircraft's position from the VORTAC?
    A. North-northeast
    B. South-southeast
    C. South-southwest

SECTION C
Satellite Navigation — GPS

Satellite navigation is based on a network of satellites that transmit radio signals in medium earth orbit. The **global navigation satellite system (GNSS)** is the standard generic term for satellite navigation systems that provide autonomous geo-spatial positioning with global coverage. The United States **global positioning system (GPS)** is a primary satellite navigation system that is globally available.

GPS OPERATION

GPS consists of three segments—space, control, and user. The U.S. Air Force develops, maintains, and operates the space and control segments. The **space segment** contains GPS satellites that fly in medium Earth orbit at an altitude of approximately 12,550 miles (20,200 km). The satellites in the GPS constellation are arranged into 6 equally-spaced orbital planes surrounding the Earth and each plane contains four "slots" occupied by baseline satellites. This 24-slot arrangement ensures users can view at least four satellites from virtually any point on the planet. The Air Force flies more than these 24 satellites to maintain coverage whenever the baseline satellites are serviced or decommissioned. Each satellite circles the Earth twice a day and continuously broadcasts radio signals used by GPS receivers to calculate accurate position information.

The **control segment** is a global network of ground facilities that track the GPS satellites, monitor and analyze their transmissions, and send commands and data to the constellation. The ground facilities include antennas, a master control station, and monitor stations. The **user segment** consists of the GPS receivers that receive the signals from the GPS satellites and use the transmitted information to calculate the user's three-dimensional position. [Figure 9-36]

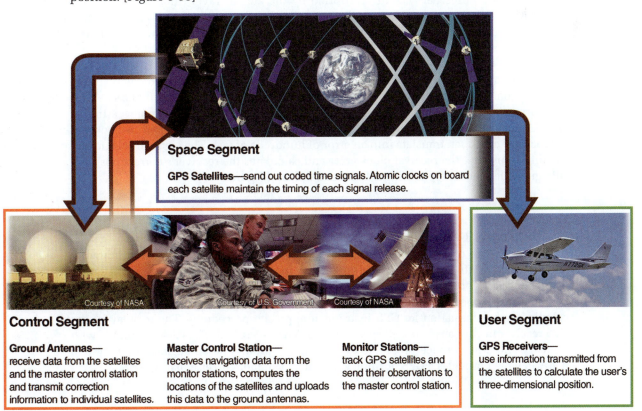

Space Segment

GPS Satellites—send out coded time signals. Atomic clocks on board each satellite maintain the timing of each signal release.

Control Segment

Ground Antennas—
receive data from the satellites and the master control station and transmit correction information to individual satellites.

Master Control Station—
receives navigation data from the monitor stations, computes the locations of the satellites and uploads this data to the ground antennas.

Monitor Stations—
track GPS satellites and send their observations to the master control station.

User Segment

GPS Receivers—
use information transmitted from the satellites to calculate the user's three-dimensional position.

Figure 9-36. GPS operation requires three segments.

DISCOVERY

Atlas V Rocket

United Launch Alliance (ULA) is a joint venture between Lockheed Martin and The Boeing Company to provide reliable, cost-effective access to space for the U.S. government. ULA's Atlas V rocket has delivered GPS satellites to orbit, supplies to the International Space Station, and spacecraft to the moon, the sun, and Mars.

Atlas rockets were an integral part of the space race. In 1962, John Glenn became the first American to orbit the Earth when an Atlas rocket launched his Friendship 7 spacecraft. The main engine of the Atlas V produces nearly 860,000 pounds of thrust to propel the rocket to the speed of sound in approximately 78 seconds. When delivering a GPS satellite, the second stage of the 2-stage rocket rendezvous with the GPS satellite constellation 12,550 miles above the Earth and releases the satellite a little over 3 hours after launch. The 45th Space Wing of the U.S. Air Force provides technical ground support for the GPS satellite missions.

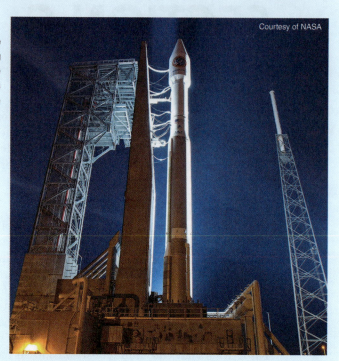

Courtesy of NASA

TRILATERATION

The GPS receiver establishes a position by a process known as **trilateration**. GPS satellites have very precise atomic clocks. Radio wave signals from the satellites travel at the speed of light (186,000 miles per second). The distance of the satellite to the receiver in miles equals 186,000 miles per second multiplied by the signal travel time in seconds. By calculating its distance from three satellites simultaneously, a GPS receiver can determine its general position with respect to latitude, longitude, and altitude. However, a fourth satellite is necessary to determine an accurate position.

Precise radio signal transmission and reception times must be known to calculate the distance from the GPS receiver to each of three satellites. The atomic clocks in the satellites are extremely accurate, but clocks in the GPS receivers are not, which creates a timing error. A mismatch of as little as one millionth of a second between the satellite clock and the receiver clock can translate into an error of hundreds of feet. The signal from a fourth satellite adjusts for the receiver clock error and calculates the receiver's correct position. GPS guarantees accuracy to within 15 meters or less. [Figure 9-37]

WAAS

The accuracy of GPS is enhanced further with the **wide area augmentation system (WAAS)**—a series of ground stations generate a corrective message that is transmitted to aircraft by a geostationary satellite. This corrective message improves navigational accuracy by accounting for positional drift of the satellites and signal delays caused by the ionosphere and other atmospheric factors. In addition, WAAS-certified GPS equipment provides vertical glide path information for GPS instrument approach procedures. The manual for the GPS receiver in your airplane will specify if it is WAAS-certified. The FAA is working with industry and other service providers to develop the ground-based augmentation system (GBAS), which provides a GPS position correction even more precise than WAAS.

DISCOVERY

Atomic Clocks

An atomic clock is an extremely accurate electronic clock regulated by the resonance frequency of atoms of certain elements, such as cesium. When a cesium atom changes from one particular energy level to another, a microwave photon emerges. The wave-like photon oscillates like a pendulum in an old-style clock. When this photon has oscillated precisely 9,192,631,770 times, one atomic second has elapsed. Atomic clocks are so stable that an observer would have to watch the clock for 32,000 years or more to see it gain or lose a single second. Shown here is a NASA mercury-ion atomic clock developed for accurate navigation in deep space.

Courtesy of NASA/JPL

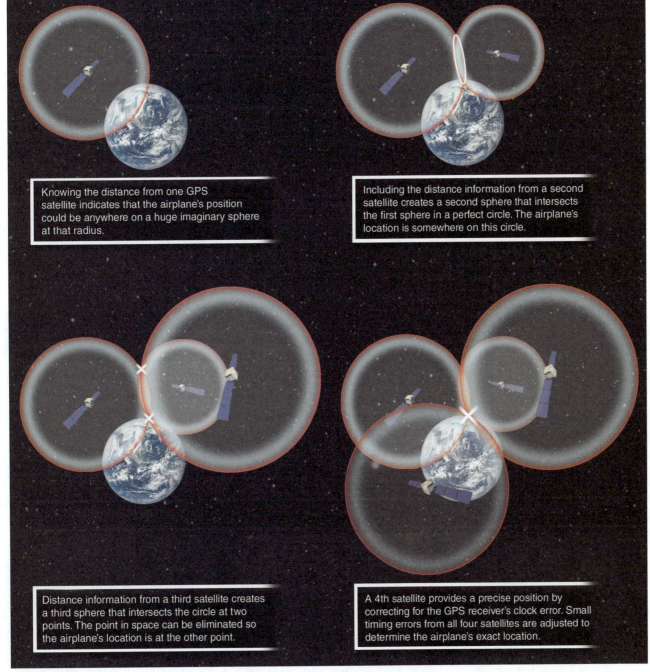

Knowing the distance from one GPS satellite indicates that the airplane's position could be anywhere on a huge imaginary sphere at that radius.

Including the distance information from a second satellite creates a second sphere that intersects the first sphere in a perfect circle. The airplane's location is somewhere on this circle.

Distance information from a third satellite creates a third sphere that intersects the circle at two points. The point in space can be eliminated so the airplane's location is at the other point.

A 4th satellite provides a precise position by correcting for the GPS receiver's clock error. Small timing errors from all four satellites are adjusted to determine the airplane's exact location.

Figure 9-37. A GPS receiver needs at least four satellite signals to calculate a position.

RAIM

GPS receivers used in aircraft continuously verify the integrity (usability) of the signals received from the GPS constellation through **receiver autonomous integrity monitoring (RAIM)**. This means that the equipment monitors and compares signals from multiple satellites to ensure an accurate signal. With RAIM, a fifth satellite monitors the position provided by the other four satellites and alerts you of any discrepancy. You can verify that RAIM will be available by checking NOTAMs, contacting Flight Service, referring to the FAA RAIM prediction website, or by using your GPS receiver's RAIM monitoring and prediction functions. [Figure 9-38]

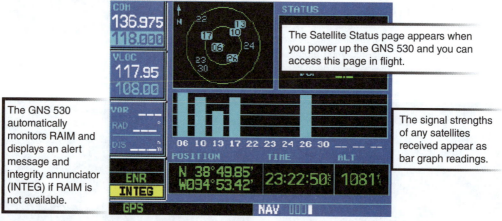

Figure 9-38. The Satellite Status Page from the Garmin GNS 530 provides an example of how a GPS receiver monitors satellite signal integrity.

NAVIGATING WITH GPS

GPS navigation is considered **area navigation (RNAV)**. RNAV equipment computes the aircraft position, actual track, and groundspeed, and then displays distance and time estimates relative to the selected course or waypoint. RNAV enables you to effectively navigate using waypoints—predetermined geographical positions for route definition—without the use of ground facilities.

GPS EQUIPMENT

Throughout this section, a variety of GPS displays are shown to illustrate concepts. To navigate safely and effectively using GPS, you must become thoroughly familiar with the specific GPS equipment that *you* are operating. GPS equipment installed in the airplane includes panel-mounted units that might also include navigation and communication radios and integrated systems that include digital instrument displays. Many of these GPS units are considered **flight management systems (FMS)**. An FMS is a computer system containing a database that enables programming of routes, approaches, and departures that can supply navigation data to the flight director/autopilot from various sources, and can calculate flight data such as fuel consumption and time remaining.

In addition, you might use a hand-held GPS or an electronic flight bag (EFB) with GPS capability for flight planning and in-flight monitoring. Review the GPS operation manual and the AFM or AFM supplement that covers the equipment installation. In addition, practice navigating with a computer-based flight training device or avionics trainer and use the equipment's simulation mode to become familiar with its operation prior to flying. [Figure 9-39]

The information that the GPS equipment displays generally depends on its capabilities and the type of database that is loaded into the unit. Manufacturers divide the information into categories, each of which might contain several pages of information. [Figure 9-40]

Figure 9-39. You must understand the features and limitations of each type of GPS equipment that you operate.

Panel-Mounted GPS with NAV/COM Radios

Electronic Flight Bag (EFB)

Integrated System

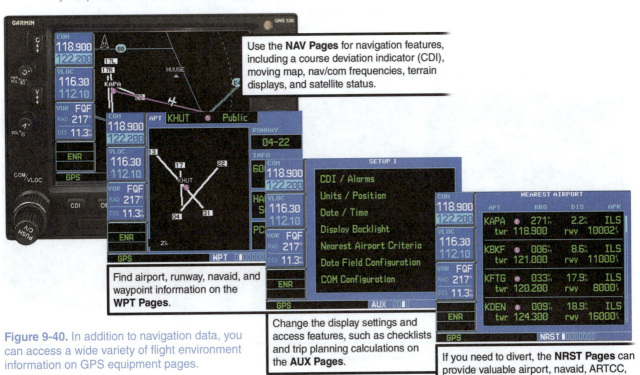

Use the **NAV Pages** for navigation features, including a course deviation indicator (CDI), moving map, nav/com frequencies, terrain displays, and satellite status.

Find airport, runway, navaid, and waypoint information on the **WPT Pages**.

Change the display settings and access features, such as checklists and trip planning calculations on the **AUX Pages**.

If you need to divert, the **NRST Pages** can provide valuable airport, navaid, ARTCC, Flight Service, and airspace information.

Figure 9-40. In addition to navigation data, you can access a wide variety of flight environment information on GPS equipment pages.

NAVIGATION DATABASE

In addition to ensuring that you have an accurate navigational signal, using GPS requires a current **navigation database**. Be sure to understand the limitations of the databases for your specific navigation equipment. A database used for a moving map display is not necessarily identical to a printed chart or procedure. Prior to flight, verify that the navigation database and other databases are current. Database information might include the database type, cycle number, and valid operating dates. [Figure 9-41]

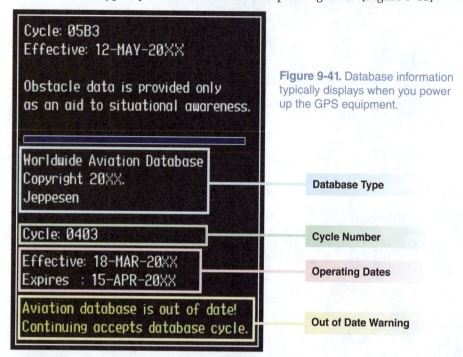

Figure 9-41. Database information typically displays when you power up the GPS equipment.

COURSE DEVIATION INDICATOR

GPS equipment installed in the airplane normally displays a **course deviation indicator (CDI)** on an analog indicator or HSI display, as well as on the GPS unit. You typically have the ability to switch the primary navigation display between a GPS and VOR receiver and an annunciator indicates which equipment is being used as the navigation source. In the case of an integrated display, such as the Garmin G1000, the primary flight display (PFD) incorporates an HSI that provides GPS course deviation information. [Figure 9-42]

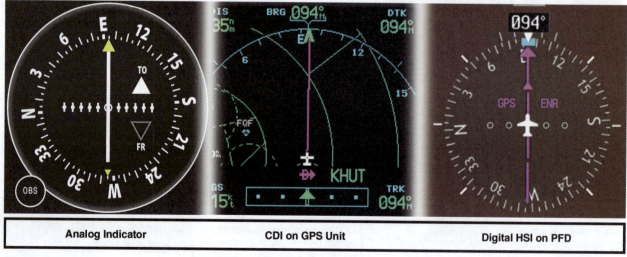

| Analog Indicator | CDI on GPS Unit | Digital HSI on PFD |

Figure 9-42. A CDI for GPS navigation can be displayed in several ways.

When the navigation source is a VOR, the CDI displays the angular deviation from the course. When the navigation source is GPS, the CDI displays the lateral distance from the course. The distance off course is called the cross-track error and is normally displayed in nautical miles. For GPS units, the CDI has three different sensitivities: enroute, terminal, and approach. The enroute sensitivity depends on whether the GPS equipment is WAAS-certified. GPS equipment automatically changes the sensitivity depending on the phase of flight. Approach sensitivity only applies to instrument procedures. [Figure 9-43]

Enroute Mode
- 1 dot = 1.0 NM (WAAS); 2.5 NM (non-WAAS)
- Full-scale deflection at 2 NM (WAAS); 5 NM (non-WAAS)

Terminal Mode
- Within 30 NM of destination
- 1 dot = 0.5 NM
- Full-scale deflection at 1 NM

Approach Mode
- During instrument approach operations—type of approach (i.e. LNAV) or APR shown
- 1 dot = 0.15 NM
- Full-scale deflection at 0.3 NM

Figure 9-43. As you transition from the enroute to the terminal environment, the CDI sensitivity increases.

MOVING MAP

A GPS **moving map** provides a pictorial view of the present position of the aircraft, the programmed route, the surrounding airspace, and topographical features. Moving maps offer options that enable you to specify what information is presented and how it is displayed. In addition, you can select the map orientation, such as north up or track up, the range, and the amount of detail shown on the display (declutter). Do not focus on the moving map to the exclusion of looking outside the aircraft to avoid other aircraft and obstructions. [Figure 9-44]

Use the moving map to:
- Verify the status of the planned route.
- Determine the aircraft position with respect to the course, airspace, and nearby terrain.
- Visualize options in preparation for an emergency.

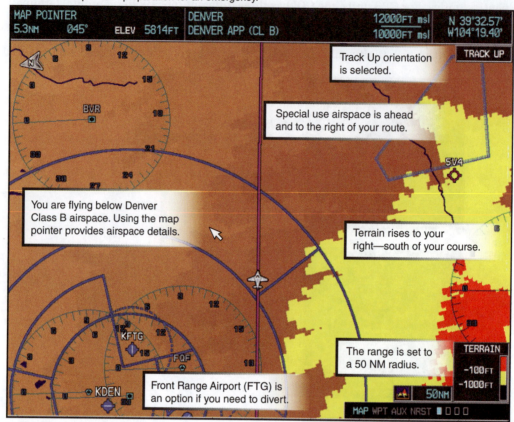

MAP POINTER 5.3NM 045° ELEV 5814FT DENVER DENVER APP (CL B) 12000FT msl 10000FT msl N 39°32.57' W104°19.40'

TRACK UP

Track Up orientation is selected.

Special use airspace is ahead and to the right of your route.

You are flying below Denver Class B airspace. Using the map pointer provides airspace details.

Terrain rises to your right—south of your course.

The range is set to a 50 NM radius.

Front Range Airport (FTG) is an option if you need to divert.

TERRAIN -100FT -1000FT 50NM

MAP WPT AUX NRST ■ ☐ ☐ ☐

Figure 9-44. A moving map helps you maintain situational awareness of the flight environment.

A common error that you should avoid is using the moving map as a primary navigation instrument. The moving map display is designed to provide supplemental navigation data and is not required to meet the certification standards for accuracy or information required for the primary navigation CDI. The apparent accuracy of the moving map display can be affected by factors such as the range setting. For example, an airplane off course by several miles might appear to be centered on course with a wide range setting.

Another common error is overreliance on the moving map leading to complacency. Studies show that pilots monitor navigational information from outside references and primary navigation instruments much less actively when a moving map is available.

DISCOVERY

Moving Map Study

In a NASA study, two groups of pilots were asked to navigate along a circuit of checkpoints during a VFR cross-country flight. One group used pilotage and a sectional chart and the other group had the same sectional chart, an FMS/RNAV computer, and a moving map. After completing the exercise, both groups were asked to navigate the circuit again with no navigational resources. Pilots who had navigated with only the sectional chart found the checkpoints again with reasonable accuracy. However, one-fourth of the pilots who had previously used the FMS/RNAV and moving map made larger errors in identifying the checkpoints and one-fourth were completely unable to find their way back to the departure airport.

This study showed that using FMS/RNAV and a moving map display does not mean that the pilot maintains situational awareness and that the key to the successful use of a moving map is to use the display as a supplement to active involvement in the navigational process. In a second NASA study, pilots who used an FMS/RNAV and moving map display were asked to act as "tour guides," pointing out geographical features to a passenger while navigating the same set of checkpoints. When requested to navigate around the circuit again without the FMS/RNAV and moving map, these pilots performed well—the simple task of pointing out geographical features was enough to avoid losing situational awareness.

WAYPOINTS

One significant benefit of a GPS navigation system is the ability to provide **waypoint** information from its navigation database. The types of waypoints in the database normally include VORs, NDBs, intersections, and airports. You typically create **user waypoints** by entering latitude and longitude coordinates, selecting a position on the map using a cursor or pointer, referencing a bearing and distance from an existing waypoint, or capturing your present position.

VFR waypoints are shown on sectional and terminal area charts to provide a supplementary navigation tool. VFR waypoint names (for computer entry and flight plans) consist of five letters beginning with the letters "VP" and are retrievable from navigation databases. The VFR waypoint names are not intended to be pronounceable, and they are not for use in ATC communication. [Figure 9-45]

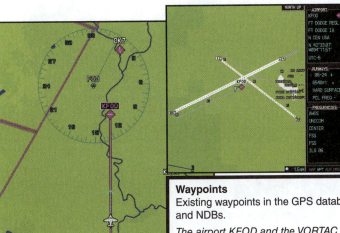

Figure 9-45. For GPS navigation, you might use a variety of different types of waypoints.

Waypoints

Existing waypoints in the GPS database include airports, intersections, VORs, and NDBs.

The airport KFOD and the VORTAC FOD are contained in the GPS database and you can access more detailed information about each waypoint.

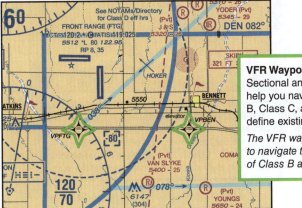

VFR Waypoints

Sectional and terminal area charts depict VFR waypoints that help you navigate in unfamiliar areas and in and around Class B, Class C, and special use airspace. VFR waypoints also define existing visual reporting points.

The VFR waypoints VPFTG and VPBEN enhance your ability to navigate to and from Front Range Airport and in the vicinity of Class B airspace.

User Waypoints

You can use a variety of methods to create user waypoints that are added to your GPS database.

The user waypoint, BYERS corresponds to the visual landmark of the town of Byers. This waypoint enables you to create a course to circumnavigate Class B airspace enroute to your destination.

GPS FLIGHT PLANNING

A fundamental principle of GPS navigation is that you are always flying to a waypoint, never from a waypoint. GPS navigation is TO-TO navigation in contrast to VOR TO-FROM navigation. **Direct-To navigation** enables you to fly from your present position directly to a waypoint. GPS navigation can be as simple as pressing the Direct-To button, selecting your destination, following the CDI, and monitoring the distance and time to the destination. However, entering a flight plan with several waypoints has advantages. You might want to add an airport as a fuel stop or maybe you need to avoid specific airspace or terrain that is enroute to your destination. [Figure 9-46]

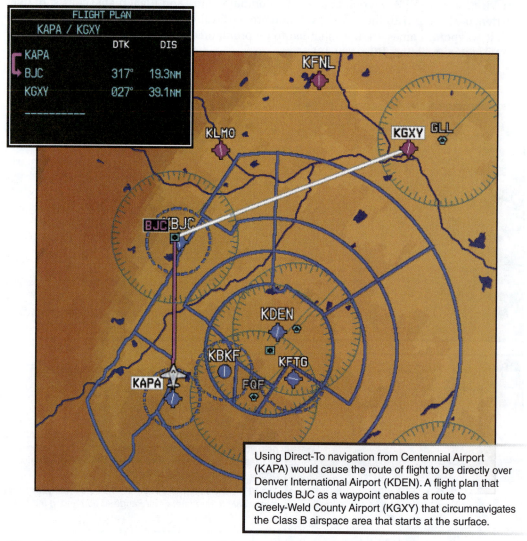

Using Direct-To navigation from Centennial Airport (KAPA) would cause the route of flight to be directly over Denver International Airport (KDEN). A flight plan that includes BJC as a waypoint enables a route to Greely-Weld County Airport (KGXY) that circumnavigates the Class B airspace area that starts at the surface.

Figure 9-46. You can add waypoints to a flight plan to circumnavigate areas in your direct path.

NAVIGATION DATA

When you navigate along your route, the desired track between your previous waypoint and the waypoint you are navigating to is the **active leg**. All the guidance information your GPS provides is based on the active leg. In addition to a CDI and moving map, GPS equipment displays data that helps you navigate on your route. You typically select the data that you want displayed, which ranges from distance and time to the next waypoint to specific information about your airplane's relationship to the desired course. [Figure 9-47]

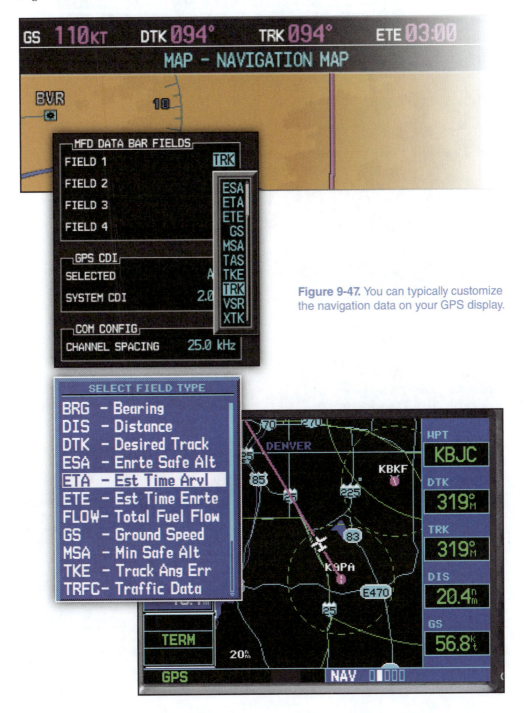

Figure 9-47. You can typically customize the navigation data on your GPS display.

To effectively navigate using GPS, you should be familiar with some basic navigation concepts and how this data is presented on your GPS equipment. The GPS course is often called the desired track and the system calculates it from the latitude and longitude coordinates of each waypoint in your route from one waypoint to the next. Desired track is a number that represents the direction from one waypoint to the next in degrees clockwise from north. It is represented on the map by a line between waypoints. [Figure 9-48]

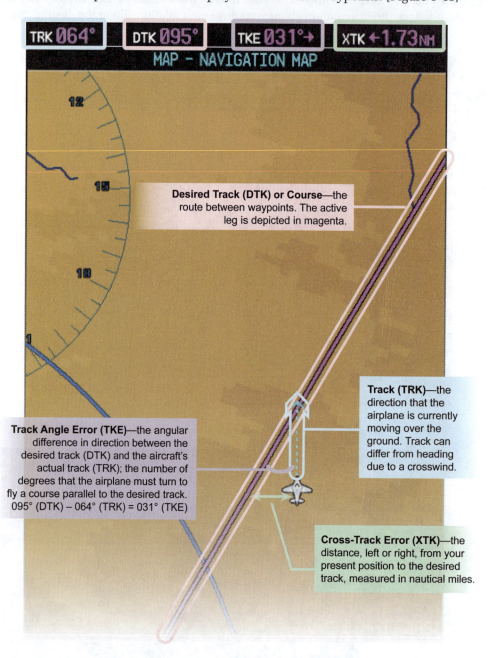

Desired Track (DTK) or Course—the route between waypoints. The active leg is depicted in magenta.

Track (TRK)—the direction that the airplane is currently moving over the ground. Track can differ from heading due to a crosswind.

Track Angle Error (TKE)—the angular difference in direction between the desired track (DTK) and the aircraft's actual track (TRK); the number of degrees that the airplane must turn to fly a course parallel to the desired track. 095° (DTK) − 064° (TRK) = 031° (TKE)

Cross-Track Error (XTK)—the distance, left or right, from your present position to the desired track, measured in nautical miles.

TRK 064°   DTK 095°   TKE 031°→   XTK ←1.73NM
MAP - NAVIGATION MAP

Figure 9-48. Although manufacturer's terms might vary, the terms shown here convey standard navigation concepts.

INTERCEPTING AND TRACKING A COURSE

Many times when you navigate under VFR, you will proceed direct from your position to a waypoint using the Direct-To feature of your GPS equipment. However, if you have programmed and activated a flight plan, you will intercept the course of the first leg of your flight after takeoff. Although you primarily use the CDI for navigation information to intercept a course, include the moving map in your scan (if applicable) to provide increased situational awareness. [Figure 9-49]

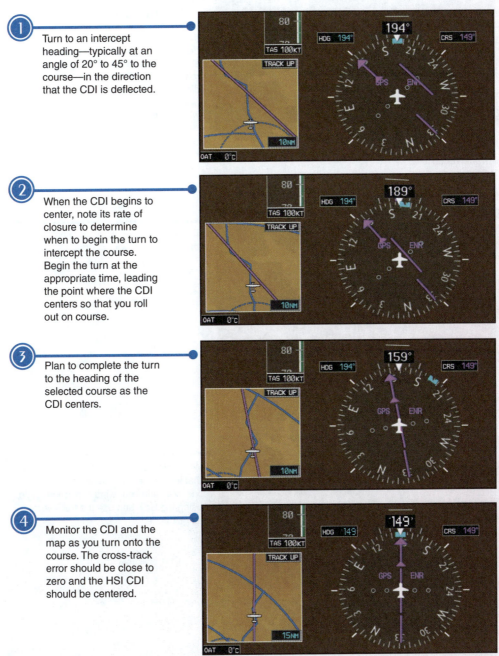

① Turn to an intercept heading—typically at an angle of 20° to 45° to the course—in the direction that the CDI is deflected.

② When the CDI begins to center, note its rate of closure to determine when to begin the turn to intercept the course. Begin the turn at the appropriate time, leading the point where the CDI centers so that you roll out on course.

③ Plan to complete the turn to the heading of the selected course as the CDI centers.

④ Monitor the CDI and the map as you turn onto the course. The cross-track error should be close to zero and the HSI CDI should be centered.

Figure 9-49. Refer to the CDI to intercept a course and monitor the moving map to maintain situational awareness. This example shows a no-wind condition.

SECTION C ■ **Satellite Navigation - GPS**

Although the basic concept of establishing a wind correction angle using the CDI applies to GPS tracking, the equipment typically displays a wide variety of additional information that enables you to maintain courses with increased accuracy. [Figure 9-50]

Wind Direction and Speed

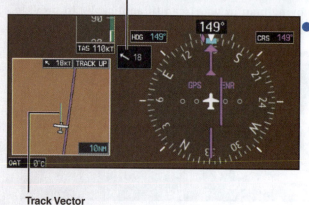

Track Vector

1 You have intercepted your course of 149° but the airplane begins to drift off course due to a strong crosswind from the right. To track the course, you must determine a heading that will compensate for wind so that the airplane remains on the desired track. Use wind direction and speed data to help you anticipate drift and select an appropriate wind correction angle.

2 Turn to a heading (intercept angle) that will re-intercept your course.

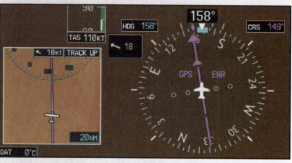

3 When the CDI centers, turn to a heading that keeps the track vector on the course line. Maintain your wind correction angle using the HSI, and use the track vector as guidance.

Figure 9-50. Wind information displayed on your GPS equipment helps you to accurately track a course.

GPS equipment provides **auto-sequencing** of waypoints—when you program a flight plan, the receiver senses when the airplane passes a waypoint and automatically cycles to the next waypoint. Some GPS equipment displays a message advising you to set the next course on the CDI. Integrated systems, such as the G1000, automatically set the course on the HSI. As you approach the next waypoint, the system alerts you to turn onto the new course. The navigation map typically shows a dashed route that cuts the corner to the new course. This is called **turn anticipation**. Waypoint passage is indicated in several ways. [Figure 9-51]

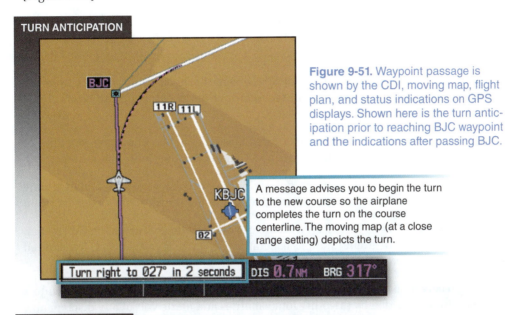

TURN ANTICIPATION

Figure 9-51. Waypoint passage is shown by the CDI, moving map, flight plan, and status indications on GPS displays. Shown here is the turn anticipation prior to reaching BJC waypoint and the indications after passing BJC.

A message advises you to begin the turn to the new course so the airplane completes the turn on the course centerline. The moving map (at a close range setting) depicts the turn.

Turn right to 027° in 2 seconds   DIS 0.7NM   BRG 317°

WAYPOINT PASSAGE

Navigation information changes to the new active leg—BJC to KGXY.

The moving map shows the leg after BJC as the active leg in magenta.

The flight plan window displays the next leg as the active leg.

Navigation indications automatically change to the new course.

SUMMARY CHECKLIST

✓ The global navigation satellite system (GNSS) is the standard generic term for satellite navigation systems. The United States global positioning system (GPS) is a primary satellite navigation system that is globally available.

✓ The space segment contains a minimum of 24 GPS satellites in orbits that ensure users can view at least four satellites from virtually any point on the planet.

✓ The control segment is a global network of ground facilities that track the GPS satellites, monitor and analyze their transmissions, and send commands and data to the constellation.

✓ The user segment consists of the GPS receivers that receive the signals from the GPS satellites and use the transmitted information to calculate the user's three-dimensional position.

✓ By using trilateration, a GPS receiver calculates its distance from three satellites to determine its general position with respect to latitude, longitude, and altitude.

✓ A fourth satellite is necessary to determine an accurate position—small timing errors from all four satellites are adjusted to determine the airplane's exact location.

✓ The accuracy of GPS is enhanced with the use of the wide area augmentation system (WAAS)—a series of ground stations generate a corrective message that is transmitted to aircraft by a geostationary satellite.

✓ Using receiver autonomous integrity monitoring (RAIM), a fifth satellite monitors the position provided by the other four satellites and alerts you of any discrepancy.

✓ RNAV equipment computes the aircraft position, actual track, and groundspeed, and then displays distance and time estimates relative to the selected course or waypoint.

✓ GPS equipment installed in the airplane includes panel-mounted units that might also include navigation and communication radios and integrated systems that include digital instrument displays.

✓ A flight management system (FMS) is a computer system containing a database that enables programming of routes, approaches, and departures that can supply navigation data to the flight director/autopilot from various sources, and can calculate flight data such as fuel consumption and time remaining

✓ Manufacturers divide information into categories, such as NAV, WPT, AUX, and NRST, each of which might contain several pages of information on the GPS display.

✓ Prior to flight, verify that the navigation database and other databases are current. Database information might include the database type, cycle number, and valid operating dates.

✓ GPS equipment installed in the airplane normally displays a course deviation indicator (CDI) on an analog indicator or HSI display, as well as on the GPS unit.

✓ When the navigation source is GPS, the CDI displays cross-track error in nautical miles. The CDI has three different sensitivities: enroute, terminal, and approach.

✓ A GPS moving map provides a pictorial view of the present position of the aircraft, the programmed route, the surrounding airspace, and topographical features.

✓ Two common errors associated with using a moving map are: using the moving map as a primary navigation instrument and overreliance on the moving map leading to complacency.

✓ The types of waypoints in the GPS database normally include VORs, NDBs, intersections, and airports.

✓ You typically create user waypoints by entering latitude and longitude coordinates, selecting a position on the map using a cursor or pointer, referencing a bearing/distance from an existing waypoint, or capturing your present position.

✓ VFR waypoints (five letters beginning with the letters "VP") are shown on sectional and terminal area charts to provide a supplementary navigation tool.

✓ Flying from your present position directly to a waypoint is Direct-To navigation.

✓ All the guidance information your GPS provides is based on the active leg. You typically select the navigation data that you want displayed.

✓ Desired track or course is the route between waypoints.

✓ Track is the direction that the airplane is currently moving over the ground.

✓ Track angle error is the angular difference in direction between the desired track and the aircraft's actual track.

✓ Cross-track error is the distance, left or right, from your present position to the desired track, measured in nautical miles.

✓ Use the CDI for navigation information to intercept a course and include the moving map in your scan to provide increased situational awareness.

✓ Establish a wind correction angle using the CDI to track a course and use additional information, such as a wind vector to maintain a course with increased accuracy.

✓ Auto-sequencing means the receiver senses when the airplane passes a waypoint and automatically cycles to the next waypoint.

✓ GPS equipment provides turn anticipation—a message advises you to begin the turn to the new course so the airplane completes the turn on the course centerline.

✓ Waypoint passage is indicated by changes to the CDI indications, the navigation information, course color on the moving map, and the active leg indication on the flight plan.

KEY TERMS

Global Navigation Satellite System (GNSS)

Global Positioning System (GPS)

Space Segment

Control Segment

User Segment

Trilateration

Wide Area Augmentation System (WAAS)

Receiver Autonomous Integrity Monitoring (RAIM)

Area Navigation (RNAV)

Flight Management System (FMS)

Navigation Database

Course Deviation Indicator (CDI)

Moving Map

Waypoint

User Waypoint

VFR Waypoint

Direct-To Navigation

Active Leg

Auto-Sequencing

Turn Anticipation

SECTION C ■ **Satellite Navigation - GPS**

9-55

QUESTIONS

1. What are the functions of the three segments of GPS?

2. Select the true statement regarding trilateration.
    A. Three satellites determine a precise position by correcting for the GPS receiver's clock error.
    B. Three satellites determine a general three-dimensional position and a fourth satellite is necessary to determine a precise position.
    C. Four satellites are required to determine a general position of latitude, longitude, and altitude and the atomic clock in your GPS is used to determine a more precise position.

3. What is RAIM?
    A. A method by which the GPS receiver computes the aircraft position, track, and groundspeed, and displays distance and time estimates relative to the selected course or waypoint.
    B. A method by which the GPS receiver uses a fifth satellite to continuously verify the integrity of the signals received from the GPS constellation and then alerts you of any discrepancy.
    C. A series of ground stations that generate a corrective message to improve navigational accuracy by accounting for positional drift of the satellites and signal delays caused by the ionosphere and other atmospheric factors.

4. Select the true statement regarding the CDI used for GPS navigation.
    A. The CDI displays the lateral distance from the course.
    B. The CDI displays the angular deviation from the course.
    C. The CDI has three different sensitivities that you must select based on your phase of flight.

5. Name two common errors regarding use of the moving map.

6. Describe three different types of waypoints.

7. Select the true statement regarding GPS navigation.
    A. Direct-To navigation enables you create a route using several waypoints.
    B. The desired track between your previous waypoint and the waypoint to which you are navigating is the active leg normally shown in magenta on a moving map.
    C. To check the currency of the navigation database, you must select the valid operating dates to be displayed with other navigation data, such as track and groundspeed.

8. What is cross-track error?
    A. The distance from your present position to the desired track, measured in nautical miles.
    B. The difference between your heading and the desired course caused by a crosswind.
    C. The difference in direction between the desired track and the aircraft's actual track in degrees.

9. What is auto-sequencing?

10. Name at least two indications of waypoint passage that GPS equipment might display.

PART V

Integrating Pilot Knowledge and Skills

It is possible to fly without motors, but not without knowledge and skill. This I conceive to be fortunate, for man, by reason of his greater intellect, can more reasonably hope to equal birds in knowledge, than to equal nature in the perfection of her machinery.

— Wilbur Wright

PART V

Without knowledge and skill, the art of flying can never truly be mastered. You must apply the knowledge you have gained while exploring this textbook with the skills you have acquired in the cockpit. Part V is designed to help you complete the journey toward your private pilot certificate by integrating the various elements you have already learned. *Applying Human Factors Principles* helps you to improve your judgment as pilot in command by increasing your knowledge of human factors concepts, such as aviation physiology and single-pilot resource management. *Flying Cross-Country* presents a scenario to illustrate how all your knowledge and skills are applied during flight.

CHAPTER 10

Applying Human Factors Principles

SECTION A
Aviation Physiology

The bird is designed for flight. Its skeleton and feathers are strong but extremely light. The bird's system of balance correctly responds to the acceleration forces experienced in flight. An extraordinarily efficient respiratory system includes lightweight air sacs that allow fresh air to continually pass through the bird's lungs. Surface sensors provide the bird with a knowledge of the state of the air flow over its wings so the bird instinctively knows its airspeed, angle of attack and attitude in yaw, pitch, and roll. [Figure 10-1] Humans are terrestrial creatures, designed for earthbound endeavors. To fly, we have had to build efficient aerodynamic structures and invent instruments to tell us visually what the bird knows instinctively. Although we function best when we are on the ground, we have a remarkable ability to adapt to our surroundings. In flight, our bodies must adjust for significant changes in barometric pressure, considerable variation in temperature, and movement at high speed in three dimensions. Aviation would not be possible if we could not compensate for the physiological demands placed upon us by flight. However, there are limitations to the adjustments that the human body can make.

Copyright Corel

Figure 10-1. The wing span of the bald eagle typically reaches seven feet. Although eagles have about 7,000 feathers, together they weigh only approximately one pound and the entire eagle skeleton weighs little more than a half-pound. Eagles can see six to eight times better than humans and can spot a rabbit from as far as a mile away.

VISION IN FLIGHT

The eye works in much the same way as a camera. Both the eye and a camera have an aperture, lens, method of focusing, and a surface for registering images. [Figure 10-2] Vision is the result of light striking the retina after it enters through the cornea and passes through the lens. The **retina** contains many photosensitive cells called cones and rods, which are connected to the optic nerve. The pattern of light that strikes the cones and rods

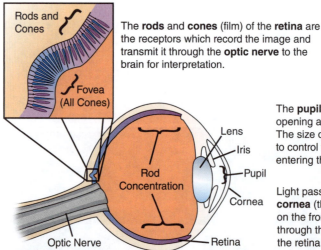

Figure 10-2. Your eyes auto focus so that you can see objects clearly at various distances as the shape of each eye's lens is adjusted by small muscles.

Rods and Cones

Fovea (All Cones)

The **rods** and **cones** (film) of the **retina** are the receptors which record the image and transmit it through the **optic nerve** to the brain for interpretation.

Lens
Iris
Pupil
Cornea

Rod Concentration

Optic Nerve

Retina

The **pupil** (aperture) is the opening at the center of the **iris**. The size of the pupil is adjusted to control the amount of light entering the eye.

Light passes through the **cornea** (the transparent window on the front of the eye) and then through the **lens** to focus on the retina.

is transmitted as electrical impulses by the optic nerve to the brain where these signals are interpreted as an image. The area where the optic nerve is formed on the retina contains no cones or rods, creating a blind spot in vision. Normally, each eye compensates for the other's blind spot. [Figure 10-3]

Figure 10-3. This illustration provides a dramatic example of the eye's blind spot. Cover your right eye and hold this page at arm's length. Focus your left eye on the X in the right side of the windshield and notice what happens to the aircraft as you slowly bring the page closer to your eye.

Cones are concentrated in the center of the retina in a slight depression known as the fovea. The cones gradually diminish and are replaced by rods as the distance from the fovea increases. Cones function well in bright light and are sensitive to colors. Compared to the rod cells, which group together to serve a single neuron, each cone cell has a direct neuron connection, which allows you to detect fine detail. The cones, however, do not function well in darkness, which explains why you cannot see color and detail as vividly at night as you can during the day.

NIGHT VISION

The **rods**, which are concentrated outside the foveal area, react to low light but not to colors. It is estimated that after adapting to darkness, the rods are 10,000 times more sensitive to light than the cones, which make them the primary receptors for night vision. Since the rods are not located directly behind the pupil, they also are responsible for much of your peripheral vision.

The concentration of cones in the fovea causes a night blind spot at the center of your vision. To see an object clearly at night, you must move your eyes to expose the rods to the image. This is accomplished by looking 5° to 10° off center of the object. For example, if you look directly at a dim light in a darkened room, the image can disappear. If you look slightly off center of the light, it becomes clearer and brighter. When scanning for traffic at night, use off-center viewing to focus objects on the rods rather than on the foveal blind spot. [Figure 10-4]

In addition, if you stare at an object at night for more than 2 to 3 seconds, the retina becomes accustomed to the light intensity and the image begins to fade. Continuous scanning and changing the position of your eyes will allow the peripheral vision to keep the object clearly visible.

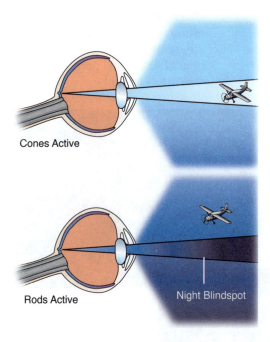

Cones Active

Rods Active

Night Blindspot

Figure 10-4. When you look directly at an object, the image is focused mainly on the cones in the fovea. When it is dark, the cones become ineffective and you depend on the rods outside the fovea for vision.

SECTION A ■ **Aviation Physiology**

DISCOVERY

Visual Field Studies

When you look at an object to your right, both of your eyes are viewing what is called the right visual field, which is processed by your brain's left hemisphere. The information viewed in the left visual field is processed in the right hemisphere. In addition, in approximately 90% of humans, the centers for speech are located in the left hemisphere of the brain. Studies performed by Roger Sperry and coworkers at the California Institute of Technology, provide unique insight into how our brain processes visual information in each hemisphere.

In a series of experiments, observations were made of patients in whom the corpus callosum, the main bundle of neural fibers connecting the left and right hemispheres of the brain, had been severed for the treatment of epilepsy. In a typical experiment, the patient is presented with the word *hatband* flashed on a screen. *Hat* is placed in the left visual field and *band* in the right visual field. The patient reports that he saw the word *band,* but when asked what kind of band he saw, the patient cannot answer correctly. In terms of his ability to communicate verbally, he does not know that the right hemisphere has received a visual impression of the word *hat.* When the patient is asked to write what he saw (with his left hand placed inside a box), he writes the word *hat.* He knows that he has written something, but without seeing the paper, there is no way for the information to reach the left hemisphere, which controls verbal ability.

DARK ADAPTATION

Rods are able to detect images in the dark because they create a chemical called rhodopsin, also referred to as **visual purple**. As visual purple is formed, the rods can take up to 30 minutes to fully adapt to the dark. If you have ever walked from bright sunlight into a dark movie theater, you have experienced this adaptation period. When exposed to bright light, visual purple undergoes a chemical change causing the rods to lose their high sensitivity to light.

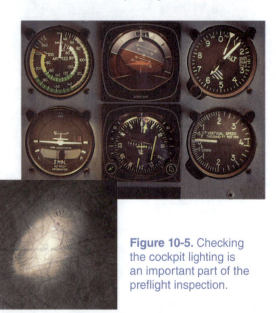

Figure 10-5. Checking the cockpit lighting is an important part of the preflight inspection.

To keep your eyes adapted to the dark, avoid looking directly at any bright light sources such as headlights, landing lights, strobe lights, or flashlights for 30 minutes before a night flight. Red cockpit lighting has been used in the past to help preserve night vision because the rods are least affected by the wavelength of red light. However, because it is difficult to see using red light, especially red instrument markings and map symbols, dim white light is preferred for in-flight use. It is important that you continually reduce the intensity of the light as your eyes adapt, keeping the brightness at the minimum level you need for effective vision. [Figure 10-5]

Your diet and general physical health also affect how well you can see in the dark. Deficiencies in vitamin A affect the eye's ability to produce visual purple. Other factors, such as carbon monoxide poisoning, smoking, alcohol, certain drugs, and a lack of oxygen can greatly decrease your night vision.

AIRCRAFT LIGHTING

According to FAR Part 91, all aircraft operated during the night hours must meet specific lighting and equipment requirements. The regulations also provide a definition of nighttime, describe night currency requirements, and specify minimum fuel reserves for night flights.

The approved aircraft position lights for night operations are a green light on the right wingtip, a red light on the left wingtip, and a white light on the tail. In addition, flashing aviation red or white anticollision lights are required for night flight. These flashing lights can be in a number of locations on the aircraft, but are most commonly found on the wingtips or tail. [Figure 10-6]

If you see a steady green and flashing red light, an aircraft is passing from left to right.

A steady red and flashing red light indicate an aircraft is crossing to your left.

If you see a red position light to the right of a green light, an aircraft is flying toward you.

A steady white light combined with a flashing red light identify an aircraft flying away from you.

Your Airplane

Figure 10-6. By interpreting the position lights of another aircraft, you can determine whether that aircraft is flying away from you or is on a collision course.

SECTION A ■ **Aviation Physiology**

VISUAL ILLUSIONS

Every pilot can experience visual illusions, although they normally go undetected unless an accident or incident occurs. An illusion typically occurs when you do not have the necessary cues for proper interpretation. Understanding the various types of visual illusions and how they occur, as well as preventive measures when appropriate, will help you increase flight safety.

AUTOKINESIS

If you stare at a single point of light against a dark background, such as a ground light or bright star, for more than a few seconds, the light can appear to move. This false perception of movement is called **autokinesis**. To prevent this illusion, you should focus your eyes on objects at varying distances and not fixate on one target, as well as maintain a normal visual scan.

FALSE HORIZONS

Another illusion, **false horizon**, occurs when the natural horizon is obscured or not readily apparent. For example, when flying over a sloping cloud deck, you might try to align the aircraft with the cloud formation. At night, ground lights, stars, and reflections on the windscreen can lead to confusion regarding the position of the horizon. Attempting to align your aircraft with a false horizon can lead to a dangerous flight attitude. [Figure 10-7]

Figure 10-7. Dark terrain, misleading lights on the ground, or a sloping cloud deck can create a false horizon.

LANDING ILLUSIONS

Landing illusions can be caused by a wide variety of factors including runway width, sloping runways and terrain, and weather conditions that reduce visibility. [Figure 10-8] For example, it is not uncommon to find public airports that have runways with a grade, or slope, of 3 percent or more. On a 6,000-foot runway, a 3 percent grade equals a 180-foot elevation difference between the approach and departure ends. Many private airports have even steeper runway grades. When approaching a sloped runway, a pilot tends is to position

Elements that create any type of visual obstruction, such as rain or haze, can cause you to fly a low approach.

Over water, at night, or over featureless terrain, such as snow-covered ground, there is a natural tendency to fly a lower-than-normal approach.

Penetration of fog can create the illusion of pitching up which can cause you to steepen your approach.

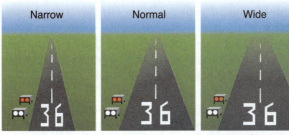
Narrow   Normal   Wide

Figure 10-8. A variety of runway, atmospheric, and terrain conditions can produce visual illusions.

Due to the illusion of greater height, you may fly a lower approach than normal to a narrow runway. A wide runway can have the opposite effect and produce higher-than-normal approaches.

the airplane so that the runway appears as it would for a normal, level runway. If the runway slopes uphill, this results in a dangerously low approach. Conversely, if the runway slopes downhill, the illusion results in a high approach with the possibility of overshooting the runway. [Figure 10-9]

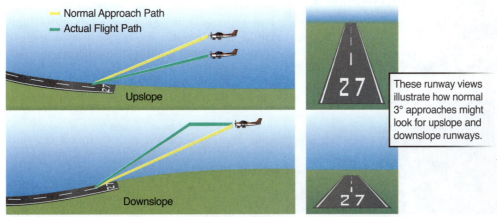
— Normal Approach Path
— Actual Flight Path

Upslope

Downslope

These runway views illustrate how normal 3° approaches might look for upslope and downslope runways.

Figure 10-9. Because pilots rely on the runway sight picture to adjust the height of their glide path, a sloped runway can trick the pilot into flying too low or too high on an approach.

At night, the potential for experiencing a visual illusion during landing increases because you have fewer visual cues to rely on. Your landing approaches at night should be made the same as during the daytime to reduce the effects of landing illusions.

If you are aware of the contributing factors that lead to visual illusions you will be able to identify them long before they become problems. For advance warning of conditions that could cause visual illusions, obtain a thorough weather briefing, examine the applicable aeronautical chart, and consult the Airport/Facility Directory listing for your destination airport in the Chart Supplement. When available, take advantage of a visual glideslope indicator to verify your landing approach angle. In addition, look for other clues, such as steep or featureless surrounding terrain. If you suspect an illusion, fly a normal traffic pattern and avoid long, straight-in approaches.

SECTION A ■ **Aviation Physiology**

FLICKER VERTIGO

A light flickering at a frequency of 4 to 20 flashes per second can produce **flicker vertigo** and although rare, it can lead to convulsions, nausea, or unconsciousness. Flicker vertigo can occur when you are looking through a slow-moving propeller toward the sun or when the sun is behind you, reflecting off the propeller. The best way to prevent flicker vertigo is to avoid looking at a light source through a propeller for any length of time. Making frequent but minor changes in propeller r.p.m. also can decrease your susceptibility to flicker vertigo.

DISORIENTATION

Sensory organs in various parts of your body provide your brain with information about your position in relation to your environment. During flight, you can encounter a variety of conditions that cause the brain to receive conflicting messages from your senses. **Disorientation** is an incorrect mental image of your position, attitude, or movement in relation to what is actually happening to your aircraft. This state of temporary confusion can be caused by misleading information being sent to the brain by your body's various sensory organs. Awareness of your body's position is a result of input from three primary sources: vision, the vestibular system located in your inner ear, and your kinesthetic sense.

Kinesthetic sense is the term used to describe an awareness of position obtained from the nerves in your skin, joints, and muscles. Using this sense is sometimes called "flying by the seat of your pants," and this is literally what you are doing. Kinesthetic sense is unreliable, however, because the brain cannot tell the difference between input caused by gravity and that of maneuvering G-loads.

In good weather and daylight, you obtain your orientation primarily through your vision. At night or in marginal weather conditions there are fewer visual cues, and you rely upon the vestibular and kinesthetic senses to supplement your vision. Because these senses can provide false cues about your orientation, the probability of disorientation occurring at night and especially in IFR weather is quite high. Under these conditions, properly scanning and interpreting your flight instruments is extremely important. [Figure 10-10]

Figure 10-10. Although you may not fly IFR without an instrument rating, you will learn basic flight by reference to instruments so that you can orient yourself at night or in low visibility, as well as maintain aircraft control if you inadvertently enter IFR conditions.

Fatigue, anxiety, heavy pilot workloads, and the intake of alcohol or other drugs increase your susceptibility to disorientation and visual illusions. These factors increase response times, inhibit decision-making abilities, and cause a breakdown in scanning techniques and night vision. Reducing your workload with the use of a simple autopilot and improving your cockpit management skills help prevent pilot overload and the possibility of disorientation.

Lightheadedness, dizziness, the feeling of instability, and the sensation of spinning are often described by pilots and aircraft passengers as vertigo. Although the symptoms can be similar, vertigo usually is caused by a physical disorder, such as a tumor or an infection of the ear or central nervous system. Experiencing these sensations during flight normally is a result of spatial or vestibular disorientation. Although the term spatial disorientation often is used to describe vestibular disorientation, the two terms have different meanings.

SPATIAL DISORIENTATION

When other sensory input is contradictory or confusing, the brain relies primarily upon sight to determine orientation. Your peripheral vision is very strong in relaying body position to your brain. When few outside visual references are available, such as in darkness or areas of limited visibility, you need to rely heavily on your **visual sense** to interpret the flight instruments for accurate information. **Spatial disorientation** can occur when there is a conflict between the information relayed by your central vision scanning the instruments, and your peripheral vision, which has virtually no references with which to establish orientation (as in IFR conditions). The movement of rain or snow seen out the window by your peripheral vision also can lead to a misinterpretation of your own movement and position in space. This is similar to the illusion of motion that you experience when a train next to yours begins pulling away from the terminal. Your peripheral vision can misinterpret this visual cue and lead you to believe that your stationary train car is in motion.

The power of peripheral vision for orientation can be easily demonstrated. Stand on one foot, look straight ahead and focus on a small distant object. Close one eye and wait until your balance stabilizes. Hold your fist a few inches in front of your open eye to block your central vision. You still should feel relatively stable. Now take your fist away and hold a tube, such as the inner tube of a paper towel roll, against your face around your open eye, obstructing your peripheral vision. Because the brain no longer has peripheral vision to orient itself in space, you should feel a balance instability.

VESTIBULAR DISORIENTATION

Located in your inner ear, the **vestibular system** consists of the vestibule and the semicircular canals. The utricle and saccule organs within the vestibule are responsible for the perception of gravity and linear acceleration, which is movement forward and back, side to side, and up and down. A gelatinous substance within the utricle and saccule is coated with a layer of tiny grains of limestone called otoliths. Movement of the vestibule causes the otoliths to shift, which in turn causes hair cells to send out nerve impulses to the brain for interpretation.

The three **semicircular canals**, which are oriented in three planes perpendicular to each other, sense angular acceleration such as roll, pitch, and yaw. Each canal is filled with fluid and contains a gelatinous structure called the cupula. When you maneuver the airplane or move your head, the canal also moves but the fluid lags behind causing the cupula to lean away from the turn. Movement of the cupula results in deflection of hair cells that project into it. This in turn stimulates the vestibular nerve. This nerve transmits impulses to the brain, which interprets the signals as motion around an axis. [Figure 10-11]

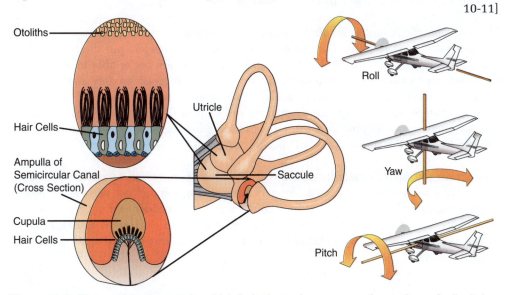

Figure 10-11. The semicircular canals, which lie in three planes, sense the motions of roll, pitch, and yaw. The vestibular nerve transmits impulses from the utricle, saccule, and semicircular canals to the brain to interpret motion.

You can experience a variety of illusions as your brain interprets vestibular signals as specific motions. When subjected to the different forces of flight, the vestibular system can send misleading signals to the brain resulting in **vestibular disorientation**. [Figure 10-12]

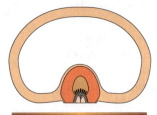 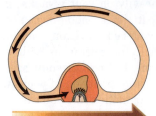 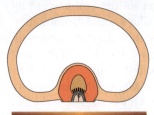 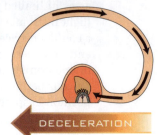

NO ACCELERATION

NO TURN
If no acceleration is taking place, the cupula is stationary and the hair cells are not deflected. No sensation of a turn is felt.

ACCELERATION

INITIATING A CLOCKWISE TURN
A clockwise turn deflects the hair cells in the direction opposite of the acceleration. You experience an accurate sensation of the turn direction.

NO ACCELERATION

PROLONGED CONSTANT-RATE TURN
During a prolonged constant rate turn, you may not sense any motion since the fluid in the canals eventually reaches equilibrium and the hair cells are no longer deflected.

DECELERATION

DECREASE IN RATE OF TURN
If you decrease the rate of turn, the deflection of the hair cells may produce a false sensation of a turn in the opposite direction. In this example, you experience the sensation of a counterclockwise turn.

Figure 10-12. During a prolonged, constant-rate turn, you might not sense any motion since the fluid in the semicircular canals eventually reaches equilibrium and the hair cells are no longer deflected.

ILLUSIONS LEADING TO DISORIENTATION

The majority of the illusions that lead to vestibular disorientation occur when visibility is restricted, either by darkness or by weather. It takes many hours of training and experience before a pilot is competent to fly an aircraft solely by reference to instruments. Each year, many fatalities result from noninstrument-rated pilots continuing flight into deteriorating weather conditions. [Figure 10-13]

MOTION SICKNESS

Because vestibular disorientation can produce severe and intense feelings of instability, it often is the cause of motion sickness. Even experienced pilots can experience motion sickness under flight conditions that disrupt the vestibular system. However, passengers are more susceptible to motion sickness, because they often focus their attention inside the aircraft. Airsickness comes in many forms. Common symptoms of airsickness are general discomfort, paleness, nausea, dizziness, sweating, and vomiting.

Although specific remedies for airsickness can vary among people, there are some actions that generally seem to help. You can suggest that your passengers put their heads back and attempt to relax. With the head reclined, passengers are better able to tolerate the up-and-down motion common to flying in turbulent air. Because anxiety and stress can contribute to motion sickness, keep uneasy or nervous passengers informed on how the flight is progressing, and explain unusual noises such as flap or landing gear retraction and power changes. Opening the fresh-air vents and allowing cool, fresh air into the cabin also can improve the comfort level of your passengers.

Another suggestion that can reduce the possibility of airsickness is that passengers focus on objects outside the airplane. Have passengers follow along and pick out various landmarks on an aeronautical chart or road map. Avoid warm, turbulent air and suggest that your passengers use earplugs. Medications like Dramamine also can prevent airsickness in passengers. In addition, keep in mind that most passengers are not used to steep banks or quick maneuvers.

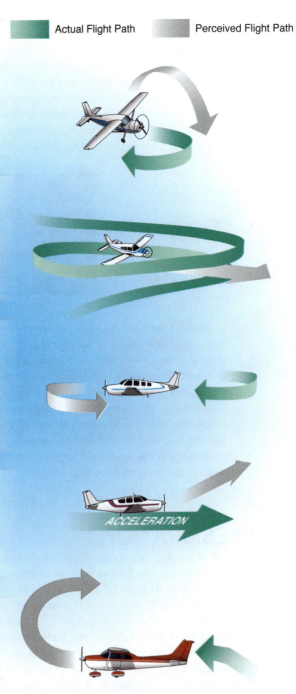

Actual Flight Path   Perceived Flight Path

CORIOLIS ILLUSION

During a constant-rate turn, if you tilt your head down to change a fuel tank or pick up a pencil, the rapid head movement puts the fluid in motion in more than one semicircular canal. This creates an overwhelming sensation of rotating, turning, or accelerating along an entirely different plane. An attempt to stop the sensation by maneuvering the airplane may put it into a dangerous attitude. To avoid this illusion, do not move your head too fast in limited visibility or darkness.

GRAVEYARD SPIRAL

A loss of altitude in a prolonged constant rate turn may be interpreted as a wings-level descent, which can lead you to increase elevator back pressure and tighten the turn, increasing your altitude loss. A recovery to wings-level flight may produce the illusion that the airplane is in a turn in the opposite direction, resulting in a reentry of the spiral. This feeling must be fought until the fluid in your semicircular canals quits moving again.

LEANS

The leans occur when an abrupt recovery or a rapid correction is made to a bank. If you make such a recovery, your semicircular canals sense a roll in the opposite direction. This may cause you to reenter the original attitude. When you return the aircraft to a wings-level condition, you will tend to lean in the direction of the incorrect bank until the semicircular canal fluids return to normal. Maintaining a level attitude for a minute or two generally will stop the leans.

SOMATOGRAVIC ILLUSION

A rapid acceleration can produce the illusion that you are in a nose-high attitude, even though you are still in straight-and-level flight. This may prompt you to lower the nose and enter a dive. A deceleration, such as rapidly retarding the throttle, produces the opposite effect. You may think you are in a dive and raise the nose. If you raise the nose too far, a stall may be produced.

INVERSION ILLUSION

An abrupt change from a climb to straight-and-level flight can create the feeling that you are tumbling backward. The effect may cause you to lower the nose abruptly, which may intensify the illusion.

Figure 10-13. The coriolis illusion is considered to be one of the most deadly.

SECTION A ■ Aviation Physiology

DISCOVERY

Disorienting Experiments

A thread attached to the inside of a glass of water can demonstrate how the semicircular canals sense motion. If you rotate the glass counter-clockwise, the water's inertia prevents the fluid from moving as rapidly as the glass, and the thread leans away from the rotation of the glass. If you continue to turn the glass at a constant rate, the water eventually catches up with the glass, and the thread hangs straight down. When this occurs in a semicircular canal and the hairs no longer lean to one side, the sensory system believes the body is again at rest. If the rotating glass is briefly decelerated, the water's inertia keeps it moving at the original speed and again there is a discrepancy between the speed of the glass and the speed of the fluid. The thread will now lean to the left in the direction of rotation. [Figure A]

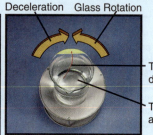

A   Glass Rotation   Deceleration   Glass Rotation

The thread leans away from the direction of rotation.

The fluid lags behind.

The thread leans in the direction of rotation.

The fluid continues to rotate as the glass decelerates.

In another experiment, you can experience the effects of vestibular disorientation. You will need a blindfold, a swivel chair, and some friends. With the blindfold on, tilt your head to one side. Then, have someone spin the chair at a constant rate for 30 to 60 seconds. The chair should then be brought to a gradual stop. When the chair has stopped, you will be told to raise your head. This action should produce the illusion of spinning or rotating. The simulation of disorientation can be so realistic that you should have someone close by to catch you if you fall out of the chair. [Figure B]

B

RESPIRATION

Respiration is the exchange of gases between an organism and its environment. The function of respiration in the human body is to get oxygen into the body and deliver it to the cells and to take carbon dioxide from the cells and remove it from the body. This process is composed of two primary activities—external respiration and internal respiration. External respiration describes the transfer of gases between your lungs and your bloodstream as you inhale and exhale. Internal respiration is the exchange of gases between your blood and your body cells. [Figure 10-14]

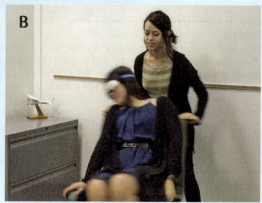

Figure 10-14. Respiration brings oxygen into the body and delivers it to the cells, and then expels carbon dioxide from the body.

Oxygen is inhaled into the lungs and carbon dioxide is exhaled from the lungs.

Oxygen is transferred from the lungs to the bloodstream by diffusion through the thin membranes of small air sacs called alveoli.

The heart pumps blood carrying oxygen through the circulatory system to the body cells.

→ Oxygen (O_2)

→ Carbon Dioxide (CO_2)

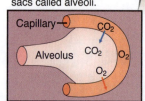

Oxygen diffuses through cell membranes and is exchanged for the waste gas carbon dioxide which is carried by the blood back to the lungs.

Approximately 95% of the oxygen transported in your body is attached to a substance called hemoglobin contained in your red blood cells, while the remaining oxygen is dissolved in the blood plasma. Oxygen and carbon dioxide are transferred through membranes from one part of the body to another by diffusion. This process is described by a physical law that states that a gas of high pressure exerts a force toward a region of lower pressure and, if there is a permeable membrane separating these regions of unequal pressure, the gas of higher pressure will diffuse through the membrane into the region of low pressure.

Each breath you inhale is composed of a mixture of gases. Life-sustaining oxygen makes up only about 21% of each breath, while 78% is nitrogen and 1% is other gases, such as carbon dioxide and argon. Each gas in the atmosphere has its own pressure at any given temperature within a given volume. A principle known as Dalton's Law states that the total pressure of a gas mixture is the sum of the pressure of each gas in the mixture. The pressure exerted by each gas in the mixture is called the partial pressure of that gas.

HYPOXIA

Hypoxia occurs when the tissues in the body do not receive enough oxygen. The symptoms of hypoxia vary with the individual. [Figure 10-15] Hypoxia can be caused by several factors including an insufficient supply of oxygen, inadequate transportation of oxygen, or the inability of the body tissues to use oxygen. The forms of hypoxia are divided into four major groups based on their causes; hypoxic hypoxia, hypemic hypoxia, stagnant hypoxia, and histotoxic hypoxia.

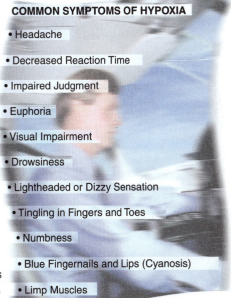

COMMON SYMPTOMS OF HYPOXIA

- Headache
- Decreased Reaction Time
- Impaired Judgment
- Euphoria
- Visual Impairment
- Drowsiness
- Lightheaded or Dizzy Sensation
- Tingling in Fingers and Toes
- Numbness
- Blue Fingernails and Lips (Cyanosis)
- Limp Muscles

Figure 10-15. Hypoxia can cause tunnel vision, slow thinking, and even diminish your sense of pain.

HYPOXIC HYPOXIA

Although the percentage of oxygen in the atmosphere is constant, its partial pressure decreases proportionately as atmospheric pressure decreases. As you ascend during flight, the percentage of each gas in the atmosphere remains the same but there are fewer molecules available at the pressure required for them to pass between the membranes in your respiratory system. This decrease of oxygen molecules at sufficient pressure can lead to **hypoxic hypoxia**.

Hypoxic hypoxia is considered to be the most lethal factor of all physiological causes of accidents. It can occur very suddenly at high altitudes during rapid decompression, or it can occur slowly at lower altitudes when you are exposed to insufficient oxygen over an extended period of time. The **time of useful consciousness** is the maximum time you have to make a rational, life-saving decision and carry it out following a loss of oxygen at a given altitude. You might also hear this referred to as effective performance time. If you go beyond this time, you might not be able to place an oxygen mask over your face. However, recovery from hypoxia usually occurs rapidly after a person has been given oxygen. [Figure 10-16]

Figure 10-16. If a person suffers severe hypoxia, mental and physical performance might be reduced for several hours.

Altitude	Time of Useful Consciousness
45,000 feet MSL	9 to 15 seconds
40,000 feet MSL	15 to 20 seconds
35,000 feet MSL	30 to 60 seconds
30,000 feet MSL	1 to 2 minutes
28,000 feet MSL	2 1/2 to 3 minutes
25,000 feet MSL	3 to 5 minutes
22,000 feet MSL	5 to 10 minutes
20,000 feet MSL	30 minutes or more

SECTION A ■ Aviation Physiology

SECTION A ■ Aviation Physiology

DISCOVERY

Altitude Chambers

A flight in an altitude chamber can give you the chance to experience unpressurized flight, gas expansion, rapid decompression, hypoxia, and the use of oxygen equipment in a controlled and safe environment. An altitude chamber employs a vacuum pump to remove gas/pressure from the chamber to simulate the corresponding pressure of a particular altitude. Figure A shows the interior of the Carter P. Luna Physiology Training Center at Peterson Air Force Base in Colorado Springs, Colorado. [Figure A]

A

Courtesy U.S. Air Force

The first altitude chamber flight for training purposes took place in 1874 when Dr. Paul Bert of France, the first practicing flight surgeon, used a diving bell and a steam-driven vacuum pump to demonstrate the hazards of high altitude balloon flights. The FAA began altitude chamber flights for civilian pilots and crewmembers in 1962 and provides this opportunity through aviation physiology training conducted at the FAA Civil Aeromedical Institute (CAMI) and at many military facilities across the United States. A typical altitude chamber profile used by CAMI is shown in Figure B.

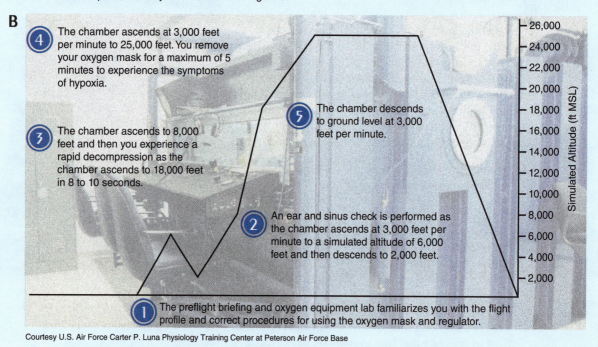

B

4 The chamber ascends at 3,000 feet per minute to 25,000 feet. You remove your oxygen mask for a maximum of 5 minutes to experience the symptoms of hypoxia.

3 The chamber ascends to 8,000 feet and then you experience a rapid decompression as the chamber ascends to 18,000 feet in 8 to 10 seconds.

5 The chamber descends to ground level at 3,000 feet per minute.

2 An ear and sinus check is performed as the chamber ascends at 3,000 feet per minute to a simulated altitude of 6,000 feet and then descends to 2,000 feet.

1 The preflight briefing and oxygen equipment lab familiarizes you with the flight profile and correct procedures for using the oxygen mask and regulator.

Simulated Altitude (ft MSL): 26,000 — 24,000 — 22,000 — 20,000 — 18,000 — 16,000 — 14,000 — 12,000 — 10,000 — 8,000 — 6,000 — 4,000 — 2,000

Courtesy U.S. Air Force Carter P. Luna Physiology Training Center at Peterson Air Force Base

HYPEMIC HYPOXIA

When your blood is not able to carry a sufficient amount of oxygen to the cells in your body, a condition called **hypemic hypoxia** occurs. This type of hypoxia is a result of a deficiency in the blood, rather than a lack of inhaled oxygen and can be caused by a variety of factors. For example, if you have anemia, or a reduced number of healthy functioning blood cells for any reason (disease, blood loss, deformed blood cells, etc.), your blood has a decreased capacity for carrying oxygen. In addition, any factor that interferes or displaces oxygen that is attached to the blood's hemoglobin can cause hypemic hypoxia.

CARBON MONOXIDE

The most common form of hypemic hypoxia is **carbon monoxide poisoning**. Since it attaches itself to the hemoglobin about 200 times more easily than does oxygen, carbon monoxide (CO) prevents the hemoglobin from carrying oxygen to the cells. It can take up to 48 hours for the body to dispose of carbon monoxide. If the poisoning is severe enough, it can result in death.

As it is a form of hypoxia, your susceptibility to carbon monoxide poisoning increases as your altitude increases, requiring you to be especially vigilant for potential carbon monoxide exposure in flight. Carbon monoxide poisoning can result from a faulty aircraft heater. If you begin to experience any of the symptoms of hypoxia, such as a lightheaded sensation or loss of muscular power, and suspect carbon monoxide poisoning, you should turn off the heater immediately, open the fresh air vents or windows, and use supplemental oxygen if it is available.

Approximately 2.5% of the volume of cigarette smoke is carbon monoxide. A blood saturation of 4% carbon monoxide can result from inhaling the smoke of 3 cigarettes at sea level. This causes a reduction in visual acuity and dark adaptation similar to the mild hypoxia encountered at 8,000 feet MSL. Smoking at 10,000 feet MSL produces effects equivalent to those experienced at 14,000 feet MSL without smoking. Heavy smokers can have carbon monoxide blood saturation as high as 8%.

BLOOD DONATION

Hypemic hypoxia also can be caused by the loss of blood that occurs during a blood donation. Your blood can take several weeks to return to normal following a donation. Although the effects of the blood loss are slight at ground level, there are risks when flying during this time.

STAGNANT HYPOXIA

Stagnant hypoxia is an oxygen deficiency in the body due to the poor circulation of the blood. Several different situations can lead to stagnant hypoxia such as shock, the heart failing to pump blood effectively, or a constricted artery. During flight, stagnant hypoxia can be the result of pulling excessive positive Gs. Cold temperatures also can reduce circulation and decrease the blood supplied to extremities.

HISTOTOXIC HYPOXIA

The inability of the cells to effectively use oxygen is defined as **histotoxic hypoxia**. The oxygen can be inhaled and reach the cell in adequate amounts, but the cell is unable to accept the oxygen once it is there. This impairment of cellular respiration can be caused by alcohol and other drugs such as narcotics and poisons. Research has shown that drinking one ounce of alcohol can equate to about an additional 2,000 feet of physiological altitude. [Figure 10-17]

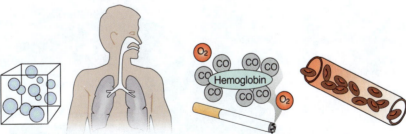

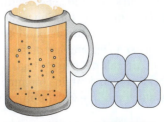

Figure 10-17. A combination of different types of hypoxia affecting your body can cause you to experience symptoms at much lower altitudes than expected.

Hypoxic Hypoxia – Inadequate Supply of Oxygen

Hypemic Hypoxia – Inability of the Blood to Carry Oxygen

Stagnant Hypoxia – Inadequate Circulation of Oxygen

Histotoxic Hypoxia – Inability of the Cells to Effectively Use Oxygen

PREVENTION OF HYPOXIA

You should not assume that if you learn the early symptoms of hypoxia that you will be able to take corrective action whenever they occur. Because judgment and rationality can be impaired when you are suffering from hypoxia, prevention is the best approach. Your susceptibility to hypoxia is related to many factors, many of which you can control. You can increase your tolerance to hypoxia by maintaining good physical condition, eating a nutritious diet, and by avoiding alcohol and smoking. If you live at a high altitude and have become acclimated, you normally have an increased tolerance to the conditions that would lead to hypoxia compared to a person living at a lower altitude.

Your body requires more oxygen during increased physical activity. For example, you can expect a higher risk of becoming hypoxic during a flight when you are flying the aircraft manually in turbulent conditions compared to a smooth flight on autopilot. Temperature extremes in the cockpit can make your body more susceptible to hypoxia. As your body copes with high heat and humidity or shivers when cold, you are using energy, which is comparable to increased activity. The quicker you ascend, the less effective your individual tolerance and you might be less aware of approaching hypoxia. In addition, you can prevent hypoxic hypoxia by flying at low altitudes where hypoxia is not a factor or by using supplemental oxygen.

SUPPLEMENTAL OXYGEN

If you are planning a flight with a cruise altitude over 12,500 feet MSL, you should consult FAR Part 91 for the requirements regarding **supplemental oxygen**. [Figure 10-18] However, consider using supplemental oxygen when you fly above 10,000 feet MSL during the day or above 5,000 feet MSL at night.

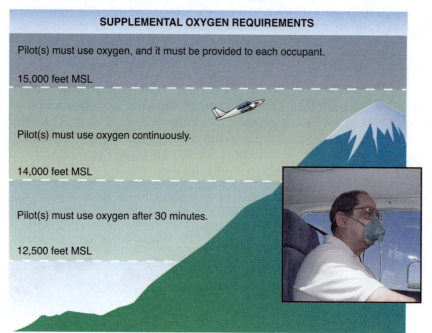

Figure 10-18. You are required by the FARs to follow these supplemental oxygen requirements.

Generally there are three types of oxygen regulators and masks; continuous flow, demand, and pressure demand. The continuous flow regulators provide a flow of 100% oxygen at a rate controlled by turning a valve. A demand regulator provides oxygen only when the user inhales. Pressure demand systems provide a positive pressure application of oxygen to the mask piece and are of great benefit at extreme altitudes such as 40,000 feet MSL or higher. Aircraft oxygen systems should always be filled with aviator's breathing oxygen. Medical oxygen contains too much moisture, which can collect in the valves and lines of the system and freeze, stopping the flow of oxygen.

PRESSURIZATION

Aircraft cabin **pressurization** is the maintenance of a cabin altitude lower than the actual flight altitude by a system that compresses air. Although pressurized aircraft reduce the physiological problems experienced at higher altitudes, the possibility of sudden loss of pressurization exists. **Decompression** occurs when the aircraft's pressurization system is unable to maintain its designed pressure schedule due to a malfunction in the pressurization system or structural damage to the aircraft. The primary danger of decompression is hypoxia. For example, if your aircraft decompresses above 30,000 feet MSL, you will become unconscious in a very short time unless you use supplemental oxygen equipment.

HYPERVENTILATION

The amount of carbon dioxide in your blood stimulates your respiratory system to stabilize your breathing rate at about 12 to 16 breaths per minute in a physically relaxed state. If you become physically active, your body cells use more oxygen and more carbon dioxide is produced. The respiratory system responds to this by increasing the depth and rate of your breathing to remove the excessive carbon dioxide.

Hyperventilation occurs when emotional distress, fear, or pain triggers an increase in the rate and depth of your breathing, despite already low levels of carbon dioxide in your blood. The result is an excessive loss of carbon dioxide from your body, which can lead to unconsciousness due to the respiratory system's overriding mechanism to regain control of your breathing. After becoming unconscious, your breathing rate will be exceedingly low until enough carbon dioxide is produced to stimulate the respiratory center. Because many of the symptoms of hyperventilation are similar to those of hypoxia, it is important to correctly diagnose and treat the proper condition. If you are using supplemental oxygen, check the equipment and flow rate to ensure you are not suffering from hypoxia. [Figure 10-19]

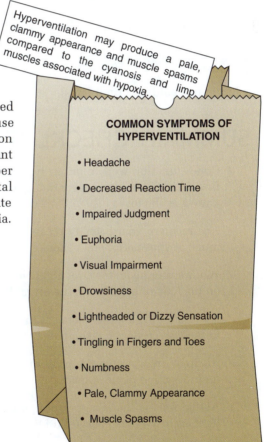

Hyperventilation may produce a pale, clammy appearance and muscle spasms compared to the cyanosis and limp muscles associated with hypoxia.

COMMON SYMPTOMS OF HYPERVENTILATION

- Headache
- Decreased Reaction Time
- Impaired Judgment
- Euphoria
- Visual Impairment
- Drowsiness
- Lightheaded or Dizzy Sensation
- Tingling in Fingers and Toes
- Numbness
- Pale, Clammy Appearance
- Muscle Spasms

Figure 10-19. Hyperventilation has some of the same symptoms as hypoxia, but unlike hypoxia, in which you feel euphoria, hyperventilation is generally associated with anxiety.

SECTION A ■ **Aviation Physiology**

The treatment for hyperventilation involves restoring the proper carbon dioxide level in the body. Breathing normally is both the best prevention and the best cure for hyperventilation. In addition to slowing the breathing rate, you also can breathe into a paper bag or talk aloud to overcome hyperventilation. Recovery is usually rapid once the breathing rate returns to normal.

HYPOTHERMIA

Hypothermia occurs when your body's heat production is outpaced by its heat loss, resulting in a dangerously low body temperature below 95°F (35°C). You and your passengers can be at the risk of hypothermia if you experience a flight deck/cabin heater failure or are exposed to cold temperatures or water after a forced landing or accident. Extended exposure to any surroundings colder than your body, especially without suitable clothing, can induce hypothermia. The onset of hypothermia symptoms is often slow and the associated mental confusion can hinder self-recognition of the condition. Symptoms of hypothermia include:

- Shivering.

- Slurred speech or mumbling.

- Slow, shallow breathing.

- Weak pulse.

- Clumsiness or lack of coordination.

- Drowsiness or very low energy.

- Confusion or memory loss.

- Loss of consciousness.

- Bright red, cold skin (in infants).

SUMMARY CHECKLIST

✓ Cones function well in bright light, are sensitive to colors, and allow you to see fine detail. Cones are concentrated in the center of the retina in a slight depression known as the fovea.

✓ The rods are your primary receptors for night vision and also are responsible for much of your peripheral vision.

✓ While scanning for traffic at night, you should use off-center viewing to focus objects on the rods rather than on the foveal blind spot.

✓ As visual purple—the chemical that enables rods to detect images in the dark—is formed, the rods can take up to 30 minutes to fully adapt to the dark.

✓ At night, interpreting the position lights of other aircraft can help you determine their direction of movement.

✓ Autokinesis is the false perception of movement after staring at a single point of light against a dark background for more than a few seconds.

✓ The false horizon illusion occurs when the natural horizon is obscured or not readily apparent.

✓ Landing illusions can be caused by a wide variety of factors including runway width, sloping runways and terrain, and weather conditions that reduce visibility.

✓ Disorientation is an incorrect mental image of your position, attitude, or movement in relation to what is actually happening to your aircraft.

✓ Kinesthetic sense is the term used to describe an awareness of position obtained from the nerves in your skin, joints, and muscles.

✓ Because you obtain your orientation primarily through your vision, when visual cues are reduced, properly scanning and interpreting the flight instruments is critical.

✓ Spatial disorientation occurs when there is a conflict between the information relayed by your central vision and your peripheral vision.

✓ The utricle and saccule organs within the vestibule are responsible for the perception of gravity and linear acceleration, which is movement forward and back, side to side, and up and down.

✓ The three semicircular canals, which are oriented in three planes perpendicular to each other, sense angular acceleration such as roll, pitch, and yaw.

✓ When subjected to the different forces of flight, the vestibular system can send misleading signals to the brain resulting in vestibular disorientation.

✓ Because vestibular disorientation can produce severe and intense feelings of instability, it often is the cause of motion sickness.

✓ Hypoxic hypoxia is due to a lack of available oxygen molecules at sufficient pressure for the body to use.

✓ Hypemic hypoxia occurs when your blood is not able to carry a sufficient amount of oxygen to the cells in your body.

✓ Because it attaches itself to the hemoglobin about 200 times more easily than does oxygen, carbon monoxide (CO) prevents hemoglobin from carrying oxygen to the body's cells.

✓ Stagnant hypoxia is an oxygen deficiency in the body due to inadequate circulation of the blood.

✓ The inability of the cells to effectively use oxygen is defined as histotoxic hypoxia.

✓ FAR Part 91 lists supplemental oxygen requirements for flights at cabin pressure altitudes above 12,500 feet MSL.

✓ Hyperventilation occurs when emotional distress, fear, or pain triggers an increase in the rate and depth of your breathing, which removes too much carbon dioxide from the blood.

✓ Hypothermia occurs when your body's heat production is outpaced by its heat loss, resulting in a dangerously low body temperature below 95°F (35°C).

SECTION A ■ Aviation Physiology

KEY TERMS

Retina	Vestibular Disorientation
Cones	Respiration
Rods	Hypoxia
Visual Purple	Hypoxic Hypoxia
Autokinesis	Time of Useful Consciousness
False Horizon	Hypemic Hypoxia
Landing Illusions	Carbon Monoxide Poisoning
Flicker Vertigo	Stagnant Hypoxia
Disorientation	Histotoxic Hypoxia
Kinesthetic Sense	Supplemental Oxygen
Visual Sense	Pressurization
Spatial Disorientation	Decompression
Vestibular System	Hyperventilation
Semicircular Canals	Hypothermia

QUESTIONS

1. Explain the difference between the cone and rod cells of the retina.

2. While flying at night, you observe a steady red and flashing red light ahead of your airplane. Based on this light configuration, what is the other aircraft's direction of movement?
   - A. Flying toward you
   - B. Crossing from right to left
   - C. Crossing from left to right

3. What is the term for the visual illusion of movement that occurs when you stare at a fixed light against a dark background for more than a few seconds?

4. What is the tendency when landing on a runway that slopes uphill?

5. Name the three primary sensory sources that provide you with an awareness of your body's position in space.

6. Explain why your vestibular system does not sense any motion during a prolonged constant-rate turn.

7. What is the term used to describe the illusion of rotating in a different plane after moving the head during a constant-rate turn?
   - A. Coriolis illusion
   - B. Graveyard spiral
   - C. Somatogravic illusion

8. What can severe and intense feelings of instability as a result of vestibular disorientation cause?

Match the types of hypoxia to the appropriate descriptions.

9. The inability of the cells to effectively use oxygen

A. Hypemic Hypoxia

10. The inability of the blood to carry sufficient oxygen to the cells due to anemia or carbon monoxide poisoning

B. Histotoxic Hypoxia

11. A decrease of available oxygen molecules at sufficient pressure because of altitude

C. Stagnant Hypoxia

12. Oxygen deficiency due to inadequate circulation of the blood

D. Hypoxic Hypoxia

13. Select the true statement regarding FAR Part 91 supplemental oxygen requirements.
    A. Supplemental oxygen must be used by all aircraft occupants above cabin pressure altitudes of 14,000 feet MSL.
    B. The pilot(s) must use supplemental oxygen for the entire duration of the flight above cabin pressure altitudes of 14,000 feet MSL.
    C. The pilot(s) and aircraft occupants must use supplemental oxygen after 30 minutes of flight duration above cabin pressure altitudes of 12,500 feet MSL.

14. How would you treat a passenger who is suffering from hyperventilation?

15. Shivering, confusion, and clumsiness can all be symptoms of what?

SECTION B
Single-Pilot Resource Management

The person who merely watches the flight of a bird gathers the impression that the bird has nothing to think of but the flapping of its wings. As a matter of fact this is a very small part of its mental labor. To even mention all the things the bird must keep in mind in order to fly securely through air would take a considerable part of the evening.

— Wilbur Wright

The pilot, like the bird, must keep a great many things in mind to safely fly an aircraft. As pilot in command, you are faced with a continuous stream of decisions during each flight. Consider the following examples. What would you do in these situations?

- You are scheduled to fly to an important business meeting in another city. The weather briefing that you obtain an hour before your proposed departure time indicates that marginal VFR weather is forecast along your entire route.

- Shortly after takeoff on a cross-country flight, the low voltage light illuminates on your instrument panel.

- A friend asks you to take him up for a flight in the local area. The only rental aircraft available the day of the proposed flight is an airplane in which you have very little experience.

- You are on a pleasure flight with another pilot who is at the airplane controls. The pilot's reckless attitude and unfamiliarity with the airplane cause you to feel uncomfortable.

- During your aircraft preflight, you check your flight bag only to discover that you do not have your fuel tester with you.

- After receiving an in-flight weather report that indicates clear conditions along your route, you notice building thunderstorms ahead in the direction of your flight.

Although you cannot practice and prepare specifically for every situation that can occur during a flight, you can be prepared to make effective decisions regarding these situations. In Chapter 1, you were introduced to **single-pilot resource management (SRM)** and some of the elements that affect this process. Many of the Discovery Insets throughout this textbook exposed you to examples of decision making and provided additional insight into the many factors which influence pilot judgment. As you read this section, you might want to review the human factors concepts in Chapter 1, Section C.

ACCIDENTS AND INCIDENTS

An aircraft **accident** is an occurrence in which any person on board suffers death or serious injury, or in which the aircraft receives substantial damage. An **incident** is an occurrence other than an accident that affects the safety of operations. The **National Transportation Safety Board (NTSB)** is an independent Federal agency that is responsible for investigating every U.S. civil aviation accident and issuing safety recommendations aimed at preventing future accidents. In addition, the NTSB maintains the government's database on civil aviation accidents and conducts research regarding safety issues of national significance. More than 100,000 aviation accidents have been investigated by the NTSB since its inception. Examining NTSB accident and incident reports can help increase your awareness of how factors such as attitude, workload, situational awareness, fatigue, and stress can affect a pilot's decision-making ability. You also can learn how to recognize the many events that lead to a hazardous situation.

You can view accident and incident reports online at the NTSB Aviation Accident Database. You also can search for individual accident reports dating back to 1962 by using the NTSB Accident Query. [Figure 10-20]

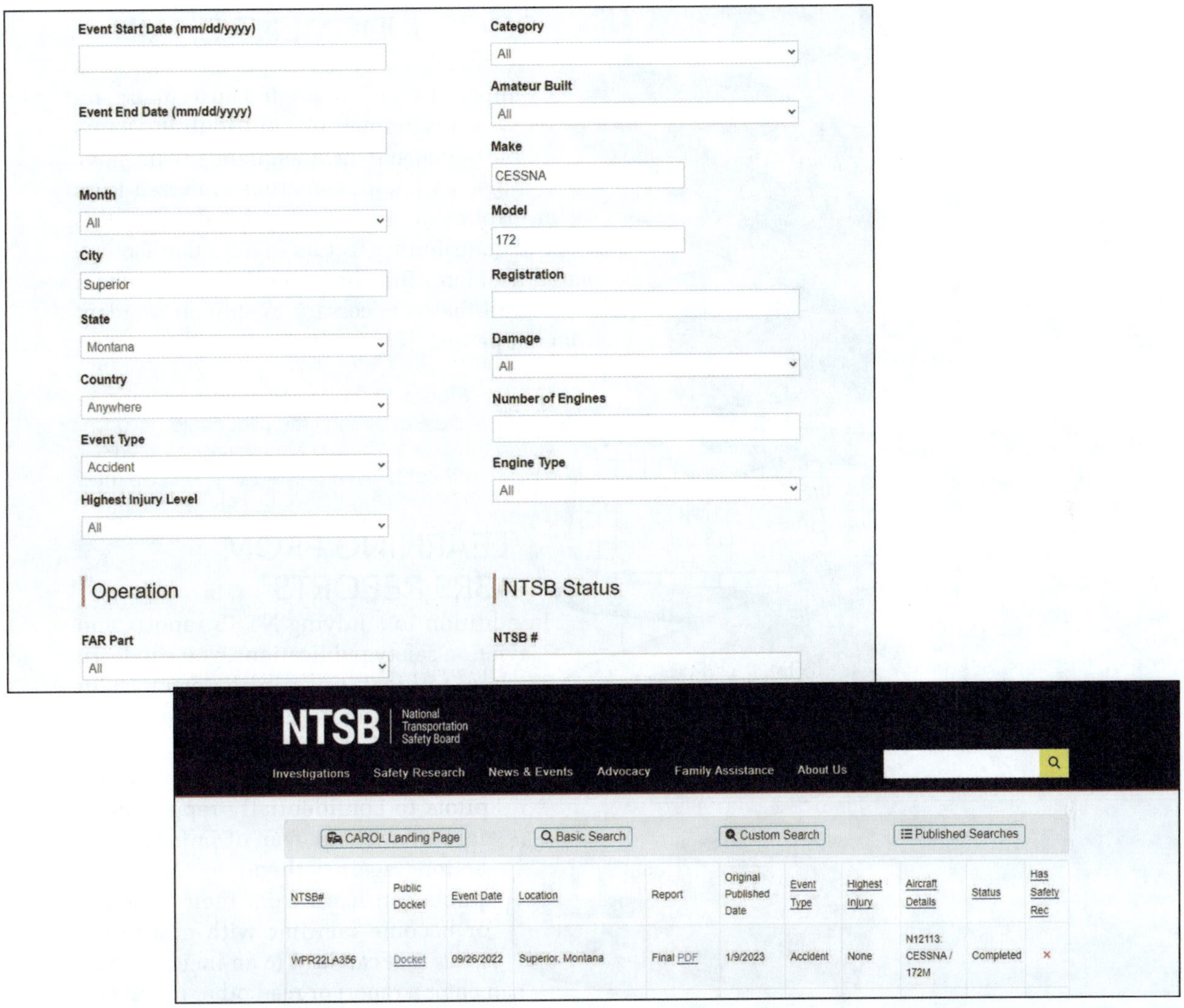

Figure 10-20. The NTSB website contains a query tool that you can use to can find factual accident reports and probable causes from NTSB's extensive database.

Miscommunication Mishap

Many accidents and incidents are the result of a lack of effective communication. For example, a Boeing 727-200 inadvertently landed with its gear retracted after the following error chain of miscommunications.

1. The first officer, who was seated in the captain's seat, gave an order for *"gear down"*. The captain, who was in the right seat and flying the aircraft, assumed the first officer was stating that the gear **was** down.

2. The Before-Landing checklist was interrupted by radio communication and never completed.

3. The ground proximity warning system (GPWS) installed on the aircraft alerted the crew to *"pull up"*, due to the aircraft's proximity to the ground with the gear retracted. However, the flight engineer believed that the GPWS warning was caused by flaps not in the landing position. The flight engineer disengaged the GPWS system by pulling the circuit breaker and the warning was silenced.

4. When the tower observed that the 727 was on final approach with the gear retracted, the controller radioed, *"go around"* but used the wrong aircraft call sign.

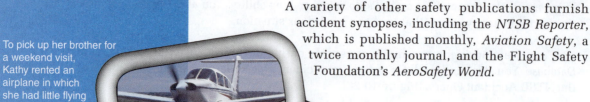

A variety of other safety publications furnish accident synopses, including the *NTSB Reporter*, which is published monthly, *Aviation Safety*, a twice monthly journal, and the Flight Safety Foundation's *AeroSafety World*.

POOR JUDGMENT CHAIN

Although many accidents and incidents appear to have a single cause, there are typically numerous contributing factors that occurred in a sequence. The **poor judgment chain**, sometimes referred to as the error chain, is a term used to describe this concept of contributing factors in a human factors-related accident. Breaking one link in the chain often is all that is necessary to stop an accident from happening. [Figure 10-21]

Figure 10-21. At any point during this chain of events, the pilot could have made a different decision that would probably have prevented this accident.

LEARNING FROM ASRS REPORTS

In addition to studying NTSB reports and aviation safety publications, you can learn how to break links in the error chain by reviewing reports in the NASA **Aviation Safety Reporting System (ASRS)**. The ASRS program allows pilots to confidentially report safety issues, without fear of enforcement actions against them, so that other pilots can learn from their mistakes or become familiar with additional issues that can lead to an incident. You can enter a report or read other reports on the NASA ASRS website. Search the ASRS Database Online for specific reports or report sets that cover certain safety issues. [Figure 10-22]

To pick up her brother for a weekend visit, Kathy rented an airplane in which she had little flying experience.

The night before the trip, Kathy attended a concert and arrived home after midnight.

After a stressful day at the office, Kathy arrived at the airport later than she had anticipated that evening.

Kathy used a performance chart in the airplane's POH to determine fuel burn, but did not note the reduced power setting on which the figure was based.

POWER APPROX. FUEL FLOW
75%   12 GPH
65%   10.8 GPH
55%   9.2 GPH

Already behind schedule, Kathy did not have the aircraft fuel tanks filled. Based on her fuel calculations and the aircraft fuel gauge indications, she concluded that she had plenty of fuel for the flight.

To reach her destination quicker, Kathy used full power and not the reduced power setting on which she had based her fuel calculations.

Upon reaching her destination, Kathy was unable to refuel the airplane since the FBO had closed for the night. Kathy disregarded the low fuel quantity gauge readings and convinced herself that she had enough fuel for the return flight.

Although she felt uneasy during the return flight as the fuel gauge needles bounced on empty, Kathy passed by several airports in an effort to get home quickly.

Not wanting to alert the controller to her fuel situation, Kathy followed ATC instructions to extend her downwind leg in the traffic pattern to follow an aircraft on an instrument approach.

Kathy's airplane ran out of fuel on final approach, two nautical miles short of the runway. While the aircraft sustained substantial damage, fortunately Kathy and her brother suffered only minor injuries.

Figure 10-22. The NASA ASRS site enables pilots and ATC personnel to enter ASRS reports as well as search for reports entered by others.

AERONAUTICAL DECISION-MAKING

To earn a private pilot certificate, you must show sound **aeronautical decision-making (ADM)** ability. You must demonstrate to an FAA examiner or inspector that you can think in the air and make choices that lead to a safe outcome on every flight. And before deciding to fly at all, you must make a competent decision about whether the proposed flight is within your capabilities. If not, do you have the good judgment to make a No-Go decision or fly with an instructor and increase your proficiency before attempting the mission? As you gain experience, the quality of your decisions should improve, enabling you to undertake more challenging trips. That is why you must accumulate a minimum number of hours of flight experience to become eligible for FAA commercial and airline transport pilot certificates.

ADM PROCESS

How do you consistently determine the best course of action in response to a given set of flight circumstances? During each flight, you are required to make decisions that involve interactions between yourself as pilot in command, the aircraft, the environment, and the operation. The **ADM process** involves a systematic evaluation of each of these risk elements to gain an accurate perception of the flight situation.

SECTION B ■ Single-Pilot Resource Management

Although you might reach a decision and implement a course of action, the decision-making process is not complete. It is important to think ahead and determine how your decision could affect the remainder of the flight. As the flight progresses, you need to continue to examine the outcome of your decision to ensure that it is producing the desired result. [Figure 10-23]

1

Recognize a change.

While on a cross-country flight, you discover that your time enroute between two checkpoints is significantly longer than the time you had originally calculated.

ETE	ETA	Fuel
ATE	ATA	Rem
4+5	9:19	.7+.7
11	9:21	38.6
13	9:32	1.8
16	9:37	36.8
15	9:47	2.1
18	9:55	34.7
20	10:07	2.8
		31.9

2

Define the problem.

Based on your insight, your cross-country flying experience, and your knowledge of weather systems, you consider the possibility that you have an increased headwind.

You verify that your original calculations are correct and consider factors that may have lengthened the time between checkpoints, such as a climb or diversion off course. To determine if there is a change in the winds aloft forecast and to check recent pilot reports, you contact Flight Service.

After weighing each information source, you conclude that your headwind has increased. To determine the severity of the problem, you calculate your new groundspeed, and reassess fuel requirements.

3

Choose a course of action.

After considering the expected outcome of each possible action and assessing the risks involved, you decide to refuel at an airport prior to your original destination.

Figure 10-23. The ADM process normally consists of several steps before you choose a course of action.

4

Implement your decision.

You plot the course change and calculate a new estimated time of arrival, as well as contact Flight Service to amend your flight plan and check the weather conditions at your new destination.

5

Evaluate the outcome.

To evaluate your decision and determine if additional steps need to be taken, you monitor your groundspeed, aircraft performance, and the weather conditions as the flight continues.

SELF ASSESSMENT

Just as you ensure that the airplane is airworthy and safe for flight, through **self assessment**, you must ensure that you are current and prepared to act as pilot in command. Having the skills and knowledge to operate an aircraft does not guarantee that you have the good judgment required to be a safe pilot. Judgment is not innate; it is learned, primarily through experience. Your life experiences can positively or negatively affect your ability to exercise good judgment in flight situations. Because judgment often is learned by making mistakes, it is safer for aviators to learn from the experiences of others, such as instructors and other pilots, and by reviewing flight safety publications.

Exercising good judgment must begin prior to taking the controls of your aircraft. You can include **personal minimums and limitations** on a risk management checklist to help determine if you are prepared for a particular flight. For example, your checklist might specify a maximum period of time that may pass between flights without refresher training or a review of the regulations and the POH. Based on your experience and comfort level, you might determine your own weather minimums, which are higher than those listed in the FARs, and set limitations for yourself regarding the maximum crosswind you are comfortable with. Before you leave for the airport on the day of the flight, you can use the FAA's I'M SAFE checklist, covered in Chapter 1, to assess your fitness to fly.

STRESS

Stress is an often-overlooked part of the I'M SAFE checklist. Although you most likely possess good judgment, your ability to make effective decisions during flight can be impaired by **stressors**, which increase your risk of error on the flight deck. [Figure 10-24]

Physical Stress — Stressors from the environment, such as uncomfortable temperature and humidity, noise, vibration, and lack of oxygen

Physiological Stress — Stressors from your physical condition, such as fatigue, lack of physical fitness, sleep loss, missed meals (leading to low blood sugar levels), and illness

Psychological Stress — Stressors from social or emotional factors, such as a death in the family, divorce, sick child, demotion at work, or the mental workload of in-flight situations

Figure 10-24. Factors that increase your stress level can be placed in three categories; physical, physiological, and psychological.

SECTION B ■ **Single-Pilot Resource Management**

DISCOVERY

The Fatigue Factor

The night before Charles Lindbergh's transatlantic flight, the weather forecast changed late in the evening and Lindbergh had to make arrangements late at night for a dawn departure. It was close to midnight before Lindbergh returned to his hotel room. Although he was hoping to get at least 2 1/2 hours of sleep, Lindbergh was disturbed by a friend whom he had posted as a guard outside his door. At 1:40 a.m., without having slept at all, Lindbergh departed for the airport.

Lindbergh was fighting exhaustion only 4 hours into the flight and, after 8 hours, his lack of sleep had become hazardous.

My eyes feel dry and hard as stones. The lids pull down with pounds of weight against their muscles. Keeping them open is like holding arms outstretched without support. I try letting one eyelid close at a time while I prop the other open with my will. But the effort's too much. Sleep is winning. My whole body argues that nothing, nothing life can attain, is quite so desirable as sleep. My mind is losing resolution and control.

After more than 22 hours into the flight, Lindbergh was falling asleep with his eyes open and he began to hallucinate.

These phantoms speak with human voices—friendly, vapor-like shapes, without substance, able to vanish or appear at will, to pass in and out through the walls of the fuselage as though no walls were there...

Although Lindbergh's story has a successful outcome, it graphically illustrates how deeply fatigue can affect a pilot's performance and judgment. If you had been in Lindbergh's place, would you have been so successful? How many pilots have been in a similar situation and have not survived to recount the incident?

Stress is your body's response to physical and psychological demands placed on it. Short-term (acute) stress—the "fight or flight" response—can be beneficial. When faced with a threat, the body releases adrenaline into the blood, increasing metabolism and providing more energy and mental acuity. However, long-term (chronic) stress places an unsustainable burden on a person that severely degrades well-being and performance. Persistent psychological burdens, such as loneliness, financial trouble, and relationship and work problems can produce a cumulative level of stress that exceeds a person's ability to cope, rendering that person incapable of safely piloting an aircraft. Therefore an effective self-assessment must evaluate these stressors—the factors that increase your stress level and degrade your ability to recognize operational pitfalls and hazardous attitudes.

OPERATIONAL PITFALLS

Additional pilot experience is not guaranteed to make you a safer pilot. Experienced pilots can develop dangerous tendencies that the FAA calls **operational pitfalls**. [Figure 10-25]

As you accept new aviation challenges, your focus shifts toward trying to complete a flight as planned, pleasing passengers, and meeting schedules. The desire to meet these goals can cause you to overestimate your piloting skills, which adversely affects safety. It is essential that you identify and counteract the temptations that affect you. [Figure 10-26]

Operational Pitfall	Description
Peer Pressure	Allowing the opinions of coworkers or other pilots to prevent you from evaluating a situation objectively.
Mind-Set	Inability to recognize and cope with changes in a given situation.
Get-There-Itis	Fixating on the original goal or destination, combined with disregarding any alternative course of action.
Duck-Under Syndrome	Succumbing to the temptation to make it into an airport by descending below minimums on an instrument approach. Also applies to VFR pilots who attempt to land under a cloud layer that is lower than VFR minimums.
Scud Running	Trying to maintain visual contact with the terrain at low altitudes when instrument conditions are present.
Continuing VFR Flight into Instrument Conditions	Flying into deteriorating weather by a non-instrument-qualified pilot, entering instrument conditions—resulting in controlled flight into terrain (CFIT) or spatial disorientation and loss of control.
Getting Behind the Aircraft	Becoming reactive instead of proactive, losing the ability to anticipate the next events and being constantly surprised by what happens next; a result of poor workload management.
Loss of Situational Awareness	Losing track of the aircraft's geographical location or becoming unable to recognize other deteriorating circumstances; a consequence of getting behind the aircraft.
Operating Without Adequate Fuel Reserves	Allowing overconfidence, lack of flight planning, or a disregard for regulations to undermine your judgment regarding fuel. Also can occur when encountering unexpected headwinds or delays during flight without recalculating the flight plan.
Descent Below the Minimum Enroute Altitude	Ducking under by a pilot flying IFR.
Flying Outside the Envelope	Overestimating your aircraft's performance or your own flying skills.
Neglect of Flight Planning, Preflight Inspections, and Checklists	Relying on memory, regular flying skills, and familiar routes by an experienced pilot with resulting neglect of established procedures and published checklists.

Figure 10-25. The FAA has identified 12 operational pitfalls that can endanger pilots who become complacent with experience.

OPERATIONAL PITFALLS

REPORT

While flying back from a long cross-country flight and due to poor planning on my side, we landed with minimum fuel in the tanks. Approximately 30 minutes after takeoff I saw that we probably wouldn't make it on time back to our original point of departure, so I decided to proceed direct without a fuel stop. During the entire flight back my student and I kept monitoring our fuel level and decided that if by any chance we couldn't comply with the regulations (because of any unforecasted weather, ATC delays, etc) we would stop and refuel. I made the decision to continue since our fuel level indicated that we have enough to get to [destination] and for an extra 1 hour of flight time. Two days after the flight, while the airplane was being refueled, it came to my attention that the airplane was refueled with 75 gallons of fuel. The total usable fuel [is] 76.4, which means that we landed with only 1.4 gallons although the fuel indication showed us that we had almost 3 gallons in each tank.

ANALYSIS

Because of the *PEER PRESSURE* to return the airplane on time and the decision to skip the planned fuel stop, the instructor ended up *OPERATING WITHOUT ADEQUATE FUEL RESERVES*. The instructor's *GET-THERE-ITIS* led to relying on the overly optimistic indications of the fuel gauges. This event could have been easily prevented if the instructor had simply followed the original flight plan and stopped for refueling.

Figure 10-26. Any pilot can succumb to operational pitfalls, including this flight instructor.

HAZARDOUS ATTITUDES

You were introduced to **hazardous attitudes** in Chapter 1 and might have recognized some of these attitudes as you read the Human Element Insets. Most pilots exhibit one or more hazardous attitudes at some time. Being aware of your own hazardous attitudes is an essential first step in preventing these thoughts from impairing your decision making. To combat a hazardous attitude, first recognize the attitude as hazardous, label the thought as one of the hazardous attitudes, and then recite that attitude's antidote. [Figure 10-27]

Don't tell me.

Flight Service said that VFR is not recommended along my route but the weather doesn't look that bad to me. I don't think the forecast is right.

Follow the rules. They are usually right.

Flight Service is basing its recommendation on valid weather data. They have more information and experience than I have to determine the conditions that will develop.

It won't happen to me.

Those clouds ahead look like they are above me. I shouldn't have any trouble staying below them.

It could happen to me.

I could get caught in IFR conditions if I'm not careful. I need to take action to determine my options before I continue along this course as planned.

I can do it.

I'll be able to get under that ceiling and even if I end up in the clouds for a bit, I'm excellent at attitude instrument flying.

Taking chances is foolish.

Scud running is very dangerous and if I end up in the clouds, I could get disoriented and lose control of the airplane.

Do it quickly.

I'm going to descend right now to get below those clouds so I can see the terrain better.

Not so fast. Think first.

Wait a minute; what other options do I have? There have been a lot of accidents caused by pilots trying to get below a cloud layer and getting into IFR conditions.

What's the use?

There's nothing I can do now but keep descending so I can see something.

I am not helpless. I can make a difference.

I do not have to continue toward my destination. I can turn around and head back to VFR conditions and then get help from ATC to divert.

Figure 10-27. This scenario provides examples of each hazardous attitude and the antidote that can be used to counteract the hazardous thought.

RISK MANAGEMENT

NTSB and other accident research can provide information that supports more effective **risk management**. Accident statistics reveal that certain phases of flight have disproportionately high accident rates and therefore present the highest risk. [Figure 10-28]. And certain types of flight activities are known to have a high correlation with accidents. [Figure 10-29]

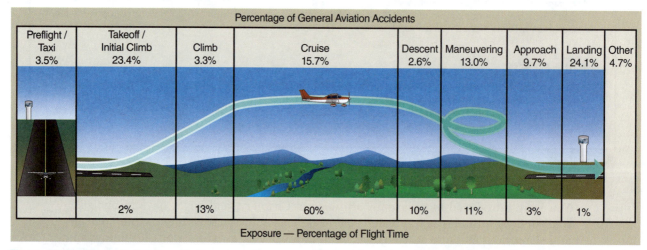

Percentage of General Aviation Accidents

Preflight / Taxi 3.5%	Takeoff / Initial Climb 23.4%	Climb 3.3%	Cruise 15.7%	Descent 2.6%	Maneuvering 13.0%	Approach 9.7%	Landing 24.1%	Other 4.7%
	2%	13%	60%	10%	11%	3%	1%	

Exposure — Percentage of Flight Time

Figure 10-28. The majority of accidents occur when approaching or departing airports. The workload is the greatest at these times, which increases the chance of error.

Section B ■ Single-Pilot Resource Management

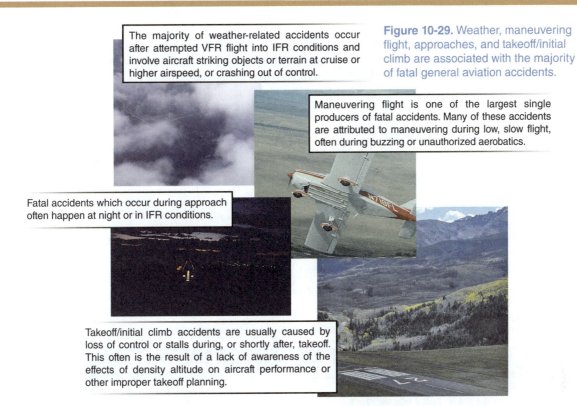

The majority of weather-related accidents occur after attempted VFR flight into IFR conditions and involve aircraft striking objects or terrain at cruise or higher airspeed, or crashing out of control.

Figure 10-29. Weather, maneuvering flight, approaches, and takeoff/initial climb are associated with the majority of fatal general aviation accidents.

Maneuvering flight is one of the largest single producers of fatal accidents. Many of these accidents are attributed to maneuvering during low, slow flight, often during buzzing or unauthorized aerobatics.

Fatal accidents which occur during approach often happen at night or in IFR conditions.

Takeoff/initial climb accidents are usually caused by loss of control or stalls during, or shortly after, takeoff. This often is the result of a lack of awareness of the effects of density altitude on aircraft performance or other improper takeoff planning.

An awareness of the flight operations and phases of flight that are most risky can help you focus risk-mitigation actions on those areas, or simply choose to avoid certain flight operations altogether. And use of the 5Ps, introduced in Chapter 1, is a helpful tool for systematically identifying and mitigating risks.

USING THE 5PS DURING FLIGHT PLANNING

With experience, you will develop your own techniques of managing risk during flight planning. You can consider using **5P checklists**, which provide guidelines on the factors to consider when planning your trip. Create your own checklists or make copies of the checklists at the end of Chapter 1 to use on your flights. Using the flight planning scenario introduced in Figure 10-30, you can explore how to take these steps to use the 5P checklists to manage risk:

1. Identify risk factors by answering the question on each checklist.

2. Mitigate risks by modifying your plans if you answer No to any questions.

3. Make a Go or No-Go decision. If you cannot effectively mitigate all of the risks, make a No-Go decision.

Figure 10-30. How would you use 5P checklists to manage risk for this scenario?

You are a private pilot with 150 hours, and you are planning a VFR flight from Centennial Airport in Denver, Colorado to Spencer, Iowa to attend your high school reunion. You are taking along a passenger who is also a pilot.

SECTION B ■ **Single-Pilot Resource Management**

PILOT

When you first consider flying to Spencer, long before the day of the trip, you look at the 5P Pilot checklist to identify any necessary training and experience you need to make the flight. On the day of your flight, you verify your fitness for flight. The phrase "I'M SAFE" is used to organize this part of the Pilot checklist. Reconsider your plan to fly if you answer No to any one of these questions. [Figure 10-31]

Pilot Risk—Training and Experience
- You have not done any night flying in the last year, so you are not prepared to operate at night or over rugged terrain.
- You have not practiced any crosswind landings in the last year.

Risk Mitigation
- You plan to conduct the flight during daylight hours and land short of your destination if the expected time of arrival at Spencer becomes delayed past sunset.
- You will determine from NOTAMS that the crosswind runway is open at Spencer or that no crosswind exceeding 5 knots is reported.

Pilot Risk—Fitness
You woke up late, grabbed a cup of coffee, and left for the airport without breakfast.

Risk Mitigation
You plan to eat a protein bar at the airport while waiting for your passenger to arrive.

Pilot				
Summary of Training		**Yes**	**No**	**N/A**
Do I have a current flight review?		✓		
Am I current to carry passengers?		✓		
Have I had recent refresher training in this airplane?		✓		
Have I had recent mountain flying training or experience?				✓
Experience	**Personal Minimums**	**Yes**	**No**	**N/A**
Hours in Specific Airplane	10	✓		
Last 90 Days				
• Hours	6	✓		
• Landings	6	✓		
Last 6 Months				
• Night Hours	6		✓	
• Night Landings	6		✓	
• Strong Crosswind/Gusty Landings	2		✓	
• Mountain Flying Hours	1			✓

Fitness — I'M SAFE	Yes	No	N/A
Illness — Am I healthy?	✓		
Medication — Am I free of prescription or over-the-counter drugs?	✓		
Stress — Am I free of pressure (job, financial matters, health problems, or family discord)?	✓		
Alcohol — Have I abstained from alcohol in the previous 24 hours?	✓		
Fatigue — Did I get at least seven hours of sleep?	✓		
Eating — Am I adequately nourished?	✓	~~✓~~	
Emotion — Am I free of upset, distress, or worry?	✓		

Figure 10-31. Consider your recent experience and your fitness for flight to manage pilot risk.

PASSENGERS

The mood and behavior of passengers can affect your flight. Small children, a sick passenger, or someone who is anxious about getting to the destination on time can create significant distractions. It is important to determine your passengers' experience level in small airplanes. [Figure 10-32]

Passenger Risk—Fitness
Your passenger has purchased an advanced reservation for an opening night reunion party and is concerned about missing the event.

Risk Mitigation
You plan to depart for Spencer early in the day and leave enough time to drive if the proposed flight cannot be conducted as planned.

Passengers			
Experience	**Yes**	**No**	**N/A**
Are my passengers comfortable flying? (spent time in small aircraft, certificated pilots, etc.)	✓		
Fitness	**Yes**	**No**	**N/A**
Are my passengers feeling well? (sickness, likely to experience airsickness, etc.)	✓		
Flexibility	**Yes**	**No**	**N/A**
Are my passengers flexible and well-informed about the changeable nature of flying? (arriving late, diverting to an alternate, etc.)	✓	✗	

Figure 10-32. Distractions from passengers can be a risk but a passenger familiar with flying or a copilot can also be a valuable resource.

PLANE

The status of your airplane is one of the most important factors to assess when you are preparing to fly. First, determine whether your airplane is airworthy. During flight planning, determine your airplane performance for the trip. And consider the adequacy of your airplane's equipment for the trip. [Figure 10-33]

Plane Risk—Airworthiness
No significant risks are associated with the airplane to be used for this flight. The airplane is not due for any inspections, and recent squawks regarding the autopilot and a low tire were addressed and the maintenance log states that the airplane has been returned to service. The airplane is easily within weight-and-balance limits and performance for the entire route is ample.

Plane			
Airworthiness	**Yes**	**No**	**N/A**
Are the aircraft inspections current and appropriate to the type of flight? (annual and 100-hour inspections, VOR check, etc.)	✓		
Have all prior maintenance issues been taken care of? (squawks resolved, inoperative equipment placarded, etc.)	✓		
Performance	**Yes**	**No**	**N/A**
Can the aircraft carry the planned load within weight and CG limits?	✓		
Is the aircraft performance (takeoff, climb, enroute, and landing) adequate for the available runways, density altitude, and terrain conditions?	✓		
Is the fuel capacity adequate for the proposed flight legs, including to an alternate airport if required?	✓		
Configuration			
Is the required equipment on board and working for the type of flight? (lights for night flight, on-board oxygen, survival gear, etc.)	✓		

Figure 10-33. Weight and balance, performance, available equipment, and maintenance condition are some of the risk factors that you must consider for the airplane.

PROGRAMMING

The Programming category reminds you to consider all of the avionics systems in your airplane. Ensure that your avionics are airworthy, and verify that your airplane's avionics configuration is appropriate for your navigation requirements. [Figure 10-34]

Programming Risk

No significant risks are associated with programming for this flight. The avionics equipment was operating normally during your last flight and all squawks were addressed. You just updated the databases, you have experience and training operating the GPS, and you practiced autopilot operation during your recent refresher training. You plan to use dead reckoning and pilotage in conjunction with the GPS moving map for navigation, and to use the autopilot to manage workload.

Programming			
Avionics Airworthiness	**Yes**	**No**	**N/A**
Is the avionics equipment working properly? (squawks resolved, autopilot functional)	✓		
Are all databases current? (GPS navdata, terrain, etc.)	✓		
Avionics Operation	**Yes**	**No**	**N/A**
Are you proficient at operating the avionics equipment?	✓		
Avionics Configuration	**Yes**	**No**	**N/A**
Is the avionics configuration appropriate for the navigation required?	✓		

Figure 10-34. To manage risk in the Programming category, verify that you are proficient in operating the avionics, the databases are current, and all the equipment is working.

PLAN

Your plan for the flight to Spencer includes a variety of risk factors. You consider the conditions at both your departure and destination airports. The mission purpose is also a risk factor because of the pressure to get to the destination. A major risk in the Plan category is weather. What you learn in your weather briefing enables you to assess the remaining items on the Plan checklist. [Figure 10-35]

After completing the risk assessment using the 5P checklists during flight planning, you decide to go on your trip.

Plan Risk—Airport Conditions
NOTAMs indicate that Runway 12/30 is closed in Spencer.

Risk Mitigation
Your performance calculations indicate that you will be able to land on Runway 36, even though it is shorter. However, you no longer have a crosswind runway available so you must ensure that the wind is aligned with Runway 36.

Plan Risk—Weather
The forecast wind of 15 gusting to 20 knots at your destination is within your limitations (and is expected to be down the runway), but you have not had recent experience landing in gusty winds.

Risk Mitigation
You plan to check the conditions in flight and to consider landing at an alternate airport if the winds are too high.

Plan			Yes	No	N/A
Airport Conditions			**Yes**	**No**	**N/A**
Do NOTAMs indicate my flight can proceed as planned? (no runway or navaid closures, and so on)			✓		
Are services available at the airport during the appropriate time? (fuel, ATC, Unicom, etc.)			✓		
Terrain/Airspace			**Yes**	**No**	**N/A**
Does the airspace and terrain in the area allow me to fly my route as planned? (Check for mountainous terrain, and areas to avoid, such as TFRs, restricted or prohibited areas).			✓		
Mission			**Yes**	**No**	**N/A**
Do I have alternate plans to manage any commitments that exist at my destination? (reschedule meeting, airline reservations, etc.)			✓		
Did I tell the people whom I'm meeting at my destination that I might be late?			✓		
Do I have an overnight kit containing any necessary prescriptions and toiletries?			✓		
Weather		**Location**	**Yes**	**No**	**N/A**
Are the weather conditions acceptable? (no hazards such as thunderstorms, icing, turbulence, etc.)		departure	✓		
		enroute	✓		
		destination	✓		
Weather Limitations	**Personal Limitations**	**Location**	**Yes**	**No**	**N/A**
Are the weather conditions for my flight within my personal limitations?					
• Minimum Ceiling and Visibility (Day VFR)	2000/10	departure	✓		
		enroute	✓		
		destination	✓		
• Minimum Ceiling and Visibility (Night VFR)	5000/15	departure	✓		
		enroute	✓		
		destination	✓		
• Maximum Surface Wind Speed and Gusts	15G20	departure	✓		
		destination	✓		
• Maximum Direct Crosswind	5	departure	✓		
		destination	✓		

Figure 10-35. The Plan category includes identifying the mission's external pressures, assessing the weather conditions, and evaluating the airport and route selection.

USING THE 5P CHECK IN FLIGHT

After making a Go decision, performing the preflight inspection, and starting the engine, you are ready to begin your flight Spencer. You can continue to manage risk by evaluating your situation using the **5P check** at decision points that correspond to the phases of flight. At each of the five decision points, use the POH checklists and flow patterns to configure the airplane and set up the avionics to help assess your plane, programming, and plan. Then, evaluate your fitness as the pilot and the condition of any passengers. [Figure 10-36]

At each decision point, consider the 5Ps and ask the following questions:

- What is the situation?
- What has changed since my Go/No-Go decision?
- Is the risk associated with a change acceptable?
- What can I do to mitigate risk?

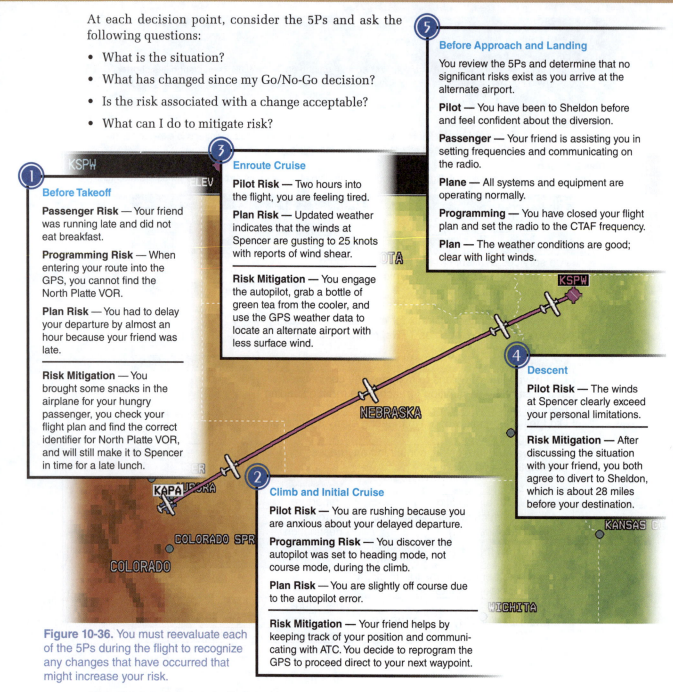

Before Takeoff

Passenger Risk — Your friend was running late and did not eat breakfast.

Programming Risk — When entering your route into the GPS, you cannot find the North Platte VOR.

Plan Risk — You had to delay your departure by almost an hour because your friend was late.

Risk Mitigation — You brought some snacks in the airplane for your hungry passenger, you check your flight plan and find the correct identifier for North Platte VOR, and will still make it to Spencer in time for a late lunch.

Enroute Cruise

Pilot Risk — Two hours into the flight, you are feeling tired.

Plan Risk — Updated weather indicates that the winds at Spencer are gusting to 25 knots with reports of wind shear.

Risk Mitigation — You engage the autopilot, grab a bottle of green tea from the cooler, and use the GPS weather data to locate an alternate airport with less surface wind.

Before Approach and Landing

You review the 5Ps and determine that no significant risks exist as you arrive at the alternate airport.

Pilot — You have been to Sheldon before and feel confident about the diversion.

Passenger — Your friend is assisting you in setting frequencies and communicating on the radio.

Plane — All systems and equipment are operating normally.

Programming — You have closed your flight plan and set the radio to the CTAF frequency.

Plan — The weather conditions are good; clear with light winds.

Descent

Pilot Risk — The winds at Spencer clearly exceed your personal limitations.

Risk Mitigation — After discussing the situation with your friend, you both agree to divert to Sheldon, which is about 28 miles before your destination.

Climb and Initial Cruise

Pilot Risk — You are rushing because you are anxious about your delayed departure.

Programming Risk — You discover the autopilot was set to heading mode, not course mode, during the climb.

Plan Risk — You are slightly off course due to the autopilot error.

Risk Mitigation — Your friend helps by keeping track of your position and communicating with ATC. You decide to reprogram the GPS to proceed direct to your next waypoint.

Figure 10-36. You must reevaluate each of the 5Ps during the flight to recognize any changes that have occurred that might increase your risk.

TASK MANAGEMENT

A pilot forgets to extend the landing gear and lands with the gear up. While searching for information on a chart, a pilot misses several ATC traffic advisories. An airplane's flaps are improperly set for takeoff causing an accident after the aircraft fails to become airborne. A pilot inadvertently enters Class B airspace without a clearance. In each of these cases, steps that were essential to the safety of the flight were missed. Effective **task management** helps ensure that required procedures are accomplished correctly and at the right time. As pilot in command, you must learn how to manage workload and sequence tasks to achieve this goal.

PLANNING AND PRIORITIZING

To effectively manage your tasks on the flight deck, you must anticipate when your workload will be high and prepare for those times during periods of low workload. For example, you know that your workload normally is highest in the vicinity of an airport during departures and approaches. Prior to takeoff, tune each navigation radio and set the heading bug to your departure course. Set the tower and departure frequencies in one communication radio, and the ground and Flight Service frequencies in the second radio.

Before arriving at your destination, review your chart, set radio frequencies, and visualize how you will fly the approach to a particular runway. When within range, monitor the ATIS or ASOS/AWOS and listen to the tower frequency or CTAF to learn what airport and traffic conditions to expect. Perform checklists well in advance so that you have time to focus on traffic and ATC instructions. These procedures are especially important prior to entering a high density traffic area, such as Class B or C airspace. Planning your descent ahead of time to arrive at traffic pattern altitude before reaching the airport also will help alleviate workload as you approach an airport. [Figure 10-37]

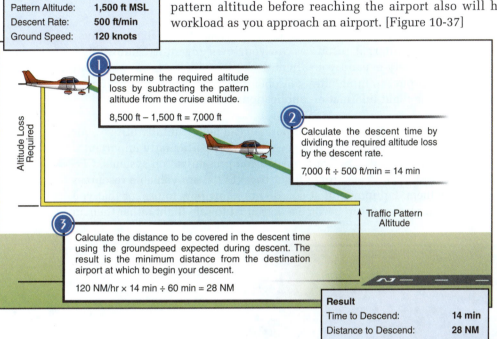

Conditions	
Cruising Altitude:	8,500 ft MSL
Pattern Altitude:	1,500 ft MSL
Descent Rate:	500 ft/min
Ground Speed:	120 knots

1 Determine the required altitude loss by subtracting the pattern altitude from the cruise altitude.

8,500 ft − 1,500 ft = 7,000 ft

2 Calculate the descent time by dividing the required altitude loss by the descent rate.

7,000 ft ÷ 500 ft/min = 14 min

3 Calculate the distance to be covered in the descent time using the groundspeed expected during descent. The result is the minimum distance from the destination airport at which to begin your descent.

120 NM/hr × 14 min ÷ 60 min = 28 NM

Traffic Pattern Altitude

Result	
Time to Descend:	14 min
Distance to Descend:	28 NM

Figure 10-37. Depending on your speed and altitude, your plan to reduce workload can begin 20-30 minutes before your destination when you can begin descending in time to arrive at traffic pattern altitude without an excessive rate of descent.

To decide when to start your descent, you might use a rule of thumb; for example if descending at 120 knots and 500 ft/min, you would start your descent four miles away for every 1,000 feet of altitude to descend. And most GPS systems can help with descent planning by advising you when to begin your descent. [Figure 10-38]

Figure 10-38. GPS vertical navigation features typically calculate the descent based on your current groundspeed, so you might start down a little early if you expect your speed to increase during the descent.

VERTICAL NAVIGATION

TARGET ALTITUDE
1000ft   Above Wpt

TARGET POSITION
2.0ⁿm Before   KFHR

VS PROFILE   VSR
500ft/m   -400ft/m

STATUS
Begin Descent in 00:01:27

MSG   VNAV

OBS   MSG   FPL   VNAV   PROC

Effective task management requires that you effectively prioritize tasks. For example, if you need to perform a go-around, flying the airplane—adding power, gaining airspeed, and properly setting the flaps—is your first priority. Report the go-around to the tower only after these tasks are completed. Priorities can change as your situation changes. If you determine that your fuel quantity is lower than expected on a cross-country flight, you might be looking at the Nearest page on your GPS for an airport at which to refuel. In an emergency situation, like any situation, your first priority is to fly the airplane and maintain a safe airspeed.

SECTION B ■ **Single-Pilot Resource Management**

RESOURCE USE

On a cross-country flight you become disoriented. The landmarks around you do not match what you see on your sectional chart and fuel is running low. What resources do you have to assist you in this situation? In addition to learning how to recognize available resources, you must also be able to evaluate whether you have the time to use a particular resource and the impact that its use will have upon the safety of flight. For example, the assistance of ATC can be very useful if you are lost, however, in an emergency situation when you need to take action quickly, you might not have the time available to contact ATC immediately.

INTERNAL RESOURCES

Internal resources are those that are accessible on the flight deck during flight. Because some of your most valuable internal resources are your own ingenuity, knowledge, and skills, you can dramatically expand your flight deck resources by improving your capabilities—through additional training after you obtain your private pilot certificate, and by frequently reviewing flight information publications, such as the AIM.

Understanding the equipment and systems in your aircraft can help you fully utilize your resources. Knowing the location and function of every switch is particularly important if you fly multiple aircraft. An autopilot, onboard oxygen and pressurization systems, GPS equipment, integrated displays, and electronic flight bags (EFBs) are valuable resources if you know how to use them. Equipment that you do not know how to use, or that you rely on to the point of complacency, can become a detriment to safe flight rather than an asset. [Figure 10-39]

Figure 10-39. Autopilots, GPS equipment, integrated displays, and EFBs are powerful resources for pilots that are trained and skilled in their use.

While it is easy to take checklists for granted, it is important to understand that they are an important flight deck resource to ensure proper execution of normal procedures. The checklists become especially important when a system malfunction or in-flight emergency occurs. Because you do not practice these abnormal procedures frequently, the checklist and the POH are essential for resolving these problems. Other valuable flight deck resources include current aeronautical charts and publications, such as the Chart Supplement or the navigation database of your GPS.

Obviously, another pilot on board the aircraft is an excellent resource, but do not overlook nonpilot passengers as a resource. Passengers can help look for traffic and help you find and fold your charts. They can reach into the back seat for snacks and water bottles. They can also find charts and other information for you on your tablet computer. [Figure 10-40]

Figure 10-40. To help stay focused on your flying, you can delegate potentially distracting tasks to a passenger.

EXTERNAL RESOURCES

Possibly the greatest of your in-flight **external resources** is air traffic control (ATC). You talk to controllers because you are required to in Class B, C, or D airspace, but you can also use ATC at other times for traffic advisories, radar vectors, and assistance in emergency situations. ATC can help you with a wide variety of issues. Did you inadvertently enter IFR conditions? ATC can vector you to VFR conditions, or they might be able to help you operate the autopilot. Are you having difficulty troubleshooting an equipment or system malfunction? ATC might be able to connect you with an AMT or an aircraft manufacturer representative. A controller can summon an ambulance if you have an onboard medical emergency. If you experience any kind of problem onboard the airplane, do not hesitate to ask ATC for help—you might be surprised by what controllers can do for you.

SECTION B ■ Single-Pilot Resource Management

Flight Service specialists can provide updates on weather and answer questions about airport conditions. You also can listen to ASOS/AWOS or ATIS to learn about weather and airport conditions while in flight. FBO personnel that monitor the CTAF (UNICOM) frequency can often provide information to help you operate at an unfamiliar airport. [Figure 10-41]

FREQUENCIES		
ATIS	RX	134.150
PRE-TAXI		118.250
CLEARANCE		118.250
GROUND		121.900
TOWER		120.100
UNICOM		122.950

Figure 10-41. An important part of utilizing external resources is recognizing the vast possibilities of these resources.

FLIGHT DECK MANAGEMENT

Effective use of resources is difficult in a poorly-organized workspace. The flight deck is your workspace for conducting a flight. Be sure that you keep yours organized for safe and efficient operations. Proper **flight deck management** includes the following procedures:

- Ensure that all necessary equipment, documents, checklists, and navigation charts for the flight are on board and accessible during flight.

- If you use portable intercoms, headsets, or hand-held navigation equipment, make sure that the wires and cables do not interfere with the operation of any control.

- Secure everything so that it is not tossed about if you encounter turbulence.

- Make sure that nothing obstructs your vision inside or outside the aircraft.

- Adjust your seat for best visibility and reach of the controls; if necessary, use cushions to provide the best seating position. Make sure that the safety belts and shoulder harnesses do not hamper your ability to reach the controls.

- Ensure that any moveable seat is locked in position. Accidental movement of a seat can cause an accident, especially during takeoff or landing.

- Brief all passengers on how to fasten and unfasten safety belts and shoulder harnesses, how to locate and use safety equipment, and how to exit the airplane in an emergency.

WORK OVERLOAD

Accidents can occur when flying task requirements exceed pilot capabilities. The difference between these two factors is called the **margin of safety**. This margin is normally lowest during the approach and landing. [Figure 10-42]

SECTION B ■ Single-Pilot Resource Management

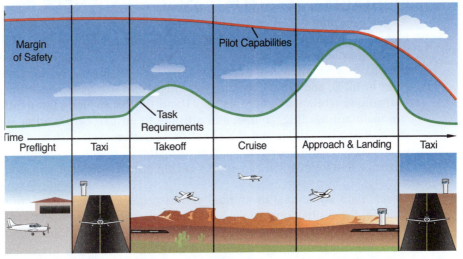

Figure 10-42. During high-workload phases of flight, such as the approach and landing, an emergency or distraction could overtax pilot capabilities, leading to an accident.

When you become task saturated, you stop paying attention to some incoming information and make decisions with incomplete information—increasing the possibility of an error. The first effect of high workload is that you begin to work faster. As the workload increases and you cannot devote your attention to many tasks at one time, you might begin to focus on one item. You need to recognize work overload, stop, think, and slow down. Use the autopilot to decrease your physical workload. If there is another pilot on board, you can delegate specific tasks and you also can enlist the assistance of ATC. For example, if you must divert to an alternate and you are under radar contact, you can request radar vectors rather than trying to navigate to the airport on your own.

SITUATIONAL AWARENESS

To maintain **situational awareness**, you must effectively perform the other SRM tasks, in addition to using your technical skills as a pilot, to maintain an accurate picture of the flight. The risk management process, in which you systematically identify and mitigate risk factors, provides situational awareness regarding risk before and at various points during the flight. Skillful navigation provides situational awareness regarding your position and fuel status. Proficiency operating an integrated flight deck can help you look ahead and gain situational awareness regarding weather and terrain. [Figure 10-43]

Figure 10-43. Situational awareness is using your technical and SRM skills to maintain an accurate perception of the operational and environmental factors that affect the pilot, aircraft, and passengers now and during the remainder of the flight.

OBSTACLES TO SITUATIONAL AWARENESS

As you gain pilot experience, you will encounter a new risk—**complacency**. When your flights become routine and you start using the autopilot to fly most of the time, you can become disconnected from the airplane and lose situational awareness. To counteract complacency, continue looking outside the airplane, identifying landmarks on your VFR chart or MFD, and cross checking your flight and navigation instruments and systems indications while using automation. Hand-fly the airplane frequently to ensure you can maintain heading, course, and altitude to airman certification standards. Even if the airplane is familiar and the situation is routine, establish and adhere to **standard operating procedures (SOPs)**, such as using the 5P Check, performing passenger or self-briefings, and following flow patterns to configure the airplane and avionics.

To maintain situational awareness you also must manage distractions. For example, consider these situations: you extend your downwind at the tower's request and you are unaware of rising terrain; or you are maintaining a higher-than-normal speed on final approach and you end up too fast and high when you reach the end of the runway. To manage distractions, avoid focusing on one task to the exclusion of others. Ensure that controlling the airplane is your priority and maintain proficiency in hand-flying so you can perform other tasks easily. Perform the majority of head-down tasks, such as looking at charts, during low workload periods. Review charts and set frequencies ahead of time so that you can focus on flying a precise traffic pattern and looking for other aircraft.

SITUATIONAL AWARENESS DURING GROUND OPERATIONS

Because a runway incursion can lead to a severe accident, airport situational awareness is an area of special emphasis for the FAA. Plan for the airport surface movement portion of your flight just as you plan for other phases of flight. Prior to taxi, check NOTAMs and ATIS for runway and taxiway closures, construction activity, and other airport-specific risks. Review the current airport diagram before starting the engine—do not wait until you are taxiing, when you need to be looking outside. Understand the taxi route, hold-short positions, crossing runways, and runway incursion hot spots before you start moving. Make sure you know the aircraft's present location and mentally calculate the next location on the route that requires increased attention, such as a turn onto another taxiway, an intersecting runway, or a hot spot. [Figure 10-44]

Take these actions to maintain situational awareness during taxi:

- Prior to entering or crossing any runway, scan the full length of the runway and final approach. If you see a conflicting aircraft, stop taxiing and query ATC.
- Before completing checklists, stop the airplane or ensure you are in a taxiing phase that has no runway incursion risk.
- Be especially vigilant if another aircraft with a similar call sign is on the same frequency.
- Never stop on a runway to communicate with ATC if you become disoriented.
- During landing, do not accept last-minute turnoff instructions unless you are certain that you can safely comply.
- Use caution after landing on a runway that intersects another runway, or on a runway with an exit taxiway in close proximity to another runway's hold short line.
- After landing at a nontowered airport, listen on the CTAF for inbound aircraft. Scan the full length of your landing runway, as well as any runways you intend to cross, including the final approach and departure paths.

Figure 10-44. In addition to knowledge of runway signs and markings, avoiding a runway incursion requires attentiveness during taxi.

STERILE FLIGHT DECK

In commercial flight operations, the **sterile flight deck** rule prohibits crew members from performing nonessential duties or activities during taxi, takeoff, landing, and other noncruise flight operations. Private pilots should also follow this rule to increase safety. Ask passengers to defer questions or conversations during high-workload times when you might miss instructions from ATC, overlook checklist items, or otherwise be distracted from performing essential procedures. You might even need to remind your instructor to postpone critiques or advice if that conversation distracts you from safe flying. [Figure 10-45]

SITUATIONAL AWARENESS

REPORT

I left the FBO ramp at FDK and was directed to several lettered taxiways. There were no markings on any of the taxiways that I was directed to. After going several hundred yards, I stopped and called Ground Control to report that I did not know where I was because there were no markings. I was told by Ground Control to make a right turn and proceed on that taxiway. Part way down that taxiway, I again contacted Ground Control and thought that I was told to proceed across Runway 30 and to make a left turn to Runway 23. In fact, I proceeded across Runway 23 which I only realized after the fact due to a lack of markings, and could only make a left turn which brought me to Runway 30. I called Ground Control and reported that I did not know how to proceed and was advised that I had made a runway incursion. I was told to follow Runway 30 back to the first intersection and make a right on that taxiway. I did so and ended up at the beginning of Runway 23. I did my normal run up and was directed onto Runway 23 for takeoff… The only factors that played a role in this incident were a lack of understanding on my part as to the directions from Ground Control and a complete lack of markings on the taxiways at FDK.

ANALYSIS

Although the pilot did have an airport diagram he could not determine his location on that diagram due to poor airport signage. The pilot asked ground control for assistance but did not continue asking when unable to figure out the instructions—a communication lapse that often occurs when pilots are afraid to appear inept or to overly bother the controller. The problem of poor airport signage exists at many airports, and the FAA requests that pilots report issues when encountering them, as this pilot did. It is important to look for multiple cues to help with taxi situational awareness, such as the heading shown on the instruments and the pavement markings that tell whether the aircraft is on a taxiway or runway.

Figure 10-45. Even though this pilot was using care during taxi, poor signage and lack of followup in communications resulted in a runway incursion.

CONTROLLED FLIGHT INTO TERRAIN AWARENESS

You can help prevent a loss of situational awareness that leads to **controlled flight into terrain (CFIT)** by taking these actions:

- Research the causes of CFIT. Review accident reports and studies to learn about the risk factors that lead to CFIT so you can avoid placing yourself in similar situations.

- Use resources in your airplane effectively. Avoid complacency by actively monitoring navigation equipment and referring to aeronautical charts to know your height above the ground. And, become proficient in the use of advanced avionics features, such as the terrain displays and moving maps, if your airplane has them. [Figure 10-46]

- Avoid behavioral traps (such as the situation described in Figure 10-47) that increase the risk of CFIT:

  ° The pressure to complete a flight as planned, to please passengers, and to meet a schedule.

  ° Flying beneath a lowering ceiling with inadvertent entry into marginal VFR or IFR conditions, sometimes referred to as "scud running."

  ° Flying below FAA-mandated minimum altitudes to show off for friends or family.

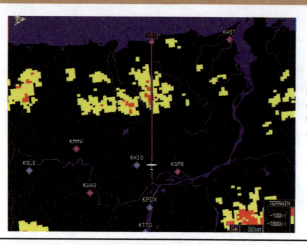

Figure 10-46. This terrain display confirms your flight planning; you know that a climb or horizontal deviation is necessary to avoid terrain ahead.

CONTROLLED FLIGHT INTO TERRAIN

REPORT

The airplane struck a rock outcropping near the peak of a 4,600-foot mountain during a night training flight that included VOR and GPS tracking and intercepting maneuvers in visual meteorological conditions. The airplane had a flight instructor and two pilot-rated students onboard. It could not be determined who was flying the airplane at the time of the accident [because no one survived]. According to the operator, the flight instructor had been counseled not to fly in that particular area at night because it was dark with no ambient surface light to illuminate the area. The flight school had other approved areas designated for night training flights. It is likely that the pilots were practicing a tracking or intercepting maneuver while not adequately monitoring the environment, which lacked illumination, to maintain sufficient altitude as they approached mountainous terrain. The damage to the airplane and associated ground scars were consistent with the airplane flying in a straight-and-level attitude before impacting the rock outcropping. Postaccident examination of the airplane and engine revealed no evidence of mechanical malfunctions or failures that would have precluded normal operation.

ANALYSIS

The flight instructor subjected the flight to CFIT risk by disregarding the flight school's limitations about flying in this dark area. While focusing on the VOR/GPS tracking exercises, neither the instructor nor students paid attention to their location relative to terrain. If they had developed a mental picture of the area as part of their navigation exercise, including awareness of the minimum safe altitude, they could have avoided this accident even if they could not see the terrain.

Figure 10-47. Although CFIT mostly affects IFR operations, lack of situational awareness when operating on a dark night or in reduced visibility does cause fatal accidents under VFR.

AUTOMATION MANAGEMENT

A GPS navigation system with digital displays, a database, moving map, and integrated autopilot adds up to a sophisticated flight management system (FMS) that can decrease workload and increase situational awareness. However, if not used effectively, your FMS could increase workload and decrease situational awareness. Your **automation management** skill determines whether the equipment provides a benefit or a distraction. In addition, if you rely too heavily on your advanced avionics, you might be tempted to operate outside your personal limitations, and you might spend too much time "head down" instead of looking outside the airplane. It is critical that you maintain proficiency in hand-flying the airplane to the airman certification standards.

An autopilot can reduce workload and risk. It can save your life if you inadvertently enter IFR conditions—turning on the autopilot to keep the wings level could help you maintain control of the airplane and it can make the 180-degree turn back to VFR conditions for you. However, the complacency that can develop because of the autopilot's lifesaving capability can increase your risks if it tempts you fly into an area of marginal weather conditions, counting on the autopilot to save you if you get into trouble.

INFORMATION MANAGEMENT

Together with managing the autopilot, you also must manage information. With the vast amount of information available on GPS equipment and multi-function displays, it is sometimes challenging to locate the exact information you need, avoiding the distraction of less-relevant information. Understanding the system and how the information is organized helps you look in the right place for information and solve problems efficiently.

Compared to IFR operations, your VFR automation and information needs are minimal. When operating avionics that are designed for IFR, you could be distracted by some of the system's capabilities—therefore you must stay focused on the information that is relevant to you. Before pushing buttons or turning knobs, decide exactly what you need and then follow the most direct steps to gain access to that information.

Information management is not a unique problem to electronic flight decks. When using paper charts, you must fold them, store them for easy access during flight, and mark your route and other information with a color you can see. However, electronic charts typically require more steps for access, which are more severely disrupted by an error or a hardware/ software anomaly—resulting in distraction and loss of situational awareness. Managing information requires effective task management. Prepare what you need before flight and during low-workload periods, manage and prioritize information flow, understand your limitations, and employ equipment operating levels. [Figure 10-48]

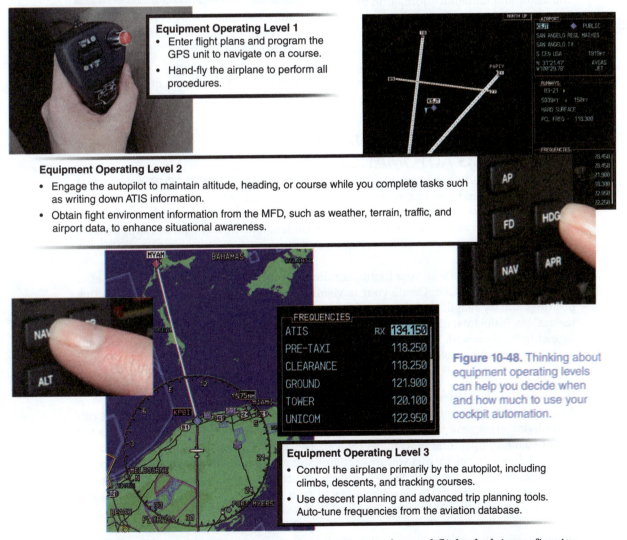

Equipment Operating Level 1
- Enter flight plans and program the GPS unit to navigate on a course.
- Hand-fly the airplane to perform all procedures.

Equipment Operating Level 2
- Engage the autopilot to maintain altitude, heading, or course while you complete tasks such as writing down ATIS information.
- Obtain fight environment information from the MFD, such as weather, terrain, traffic, and airport data, to enhance situational awareness.

FREQUENCIES	
ATIS	RX 134.150
PRE-TAXI	118.250
CLEARANCE	118.250
GROUND	121.900
TOWER	120.100
UNICOM	122.950

Figure 10-48. Thinking about equipment operating levels can help you decide when and how much to use your cockpit automation.

Equipment Operating Level 3
- Control the airplane primarily by the autopilot, including climbs, descents, and tracking courses.
- Use descent planning and advanced trip planning tools. Auto-tune frequencies from the aviation database.

An important part of information management in an advanced flight deck is configuring the screens to provide the data you need. [Figure 10-49] Your customization options on the PFD are limited because the PFD presentation must comply with FAA flight instrument requirements and pilots should not alter how those instruments are displayed. However, you often have a lot of choices for setting up the MFD such as:

- Map orientation—north up or track up.
- Map scale—automatic or manual zoom level.
- Topographic and terrain information.
- Presence of a range ring and track vector.
- Weather information on the primary map page.
- Traffic information display options.
- Content of data fields—ETA, ETE, GS, XTE, VSR—generally you want to set these fields to complement other information shown on the PFD and moving map.
- Content and frequency of in-flight reminders—change fuel tanks, 5P Check, etc.

Figure 10-49. Setting your preferences for the navigation system is an important part of managing information.

You also could tailor the information displayed for a specific flight, for example, to display terrain information on a dark night flight or on a flight through the mountains. With experience, you will learn what works for you and refine your preferences. If you are flying a rental aircraft in which other pilots set their own preferences, try to save your own preferences in a profile with your name.

MAINTAINING SITUATIONAL AWARENESS WHILE USING AUTOMATION

Understanding equipment operating levels can help you use your GPS/FMS in a way that enhances situational awareness. Appropriate training and practice are also essential. If you do not understand the equipment in your airplane, your FMS becomes a distraction that takes away situational awareness. Read and understand the manuals and supplements for any autopilot system and GPS navigation system.

Avoid taking shortcuts with your flight planning, expecting to hastily program a route after you get into the airplane. Create your navigation log with the planned route, including headings and leg length, and then check your programming after entering your route into the system. Is the total distance about what you had planned? Are the waypoints generally logical in location with the correct course and distance between them as shown on your navigation log and on your chart?

STANDARD OPERATING PROCEDURES

Because flight deck automation offers so many options for managing your flight, it can overwhelm an unprepared pilot. An important tool for taking charge of your system is the use of standard operating procedures (SOPs). These are steps you follow, in addition to your checklists, to ensure that you are accomplishing automation management effectively. Consider following a few basic SOPs to prevent errors and maintain situational awareness. In addition you will want to establish a few of your own SOPs that aid you in consistent and efficient operation of your equipment. [Figure 10-50]

If you aspire to become an airline pilot, SOPs that are assigned by your employer will be an essential part of your daily work life. As a private pilot, you create your own SOPs or follow those set by your flight school; but the benefits of following a preestablished routine on every flight will quickly become apparent. [Figure 10-51]

Applying Human Factors Principles

Generally Recommended SOPs

Check the flight routing. Before departure, ensure that the programmed route matches the planned flight route.

Monitor your VOR equipment to back up GPS navigation.

Using verbal callouts to set and verify modes of operation.

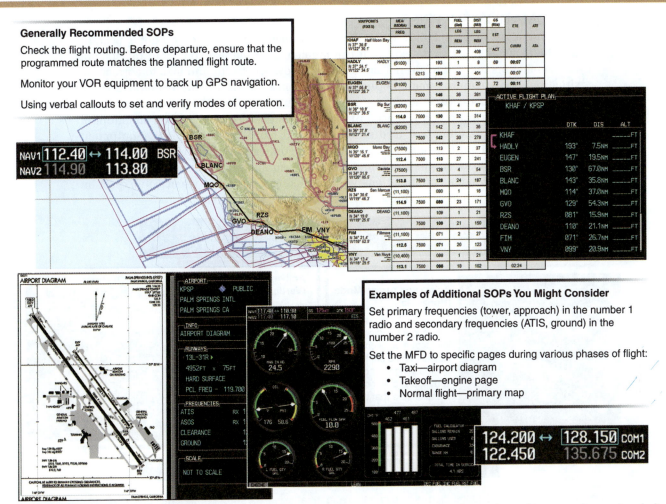

Examples of Additional SOPs You Might Consider

Set primary frequencies (tower, approach) in the number 1 radio and secondary frequencies (ATIS, ground) in the number 2 radio.

Set the MFD to specific pages during various phases of flight:
- Taxi—airport diagram
- Takeoff—engine page
- Normal flight—primary map

Figure 10-50. Standard operating procedures can provide a helpful routine for operating the equipment in your advanced cockpit.

 AUTOMATION MANAGEMENT

REPORT

[I had an] incursion into Class D airspace. Reading the … airport weather page indicates the reporting station for primary and satellite airports. While following the GPS vector from CCC VOR intersection to 87N, I was tracking my route on the weather reporting page. Since 87N utilizes the reporting station from 1N2 the weather reporting station, both airports have the same airport designation [depicted on the moving map]. Consequently, I entered the 1N2 airspace, believing I was nearby 87N, and past the 1N2 airspace.

ANALYSIS

A C172 pilot using a GPS system's weather page to navigate direct to one airport entered the Class D airspace for another airport without ATC authorization, after being confused by the presentation on that map page. Even the best-designed navigation systems can present confusing operational traps. An effective defense against these types of surprises is to study the documentation so that you know exactly what the system shows and under what conditions. Also keep in mind that "specialized" pages, like traffic, terrain, and weather are optimized to handle that specific threat. After using those pages to gain better situational awareness about that threat, consider returning to the "main" map page, where multiple types of information are shown together. This can help avoid fixating on one situation and missing other information that is important.

Figure 10-51. An SOP of monitoring the primary map page might prevent an airspace incursion like this one.

SECTION B ■ Single-Pilot Resource Management

AUTOMATION SURPRISE

An **automation surprise** occurs when the system does something you do not expect, or fails to do something you do expect. It can momentarily confuse you about the state of the automation and you might not immediately know what action to take to correct the situation. For example, some autopilots require you to press a certain combination of buttons to arm a course or altitude capture and if you do not press the buttons at the same time or in the correct order, you get a different mode than expected. When an autopilot surprises you, it is often best to disconnect it and hand fly the airplane (equipment operating level 1) until the airplane is doing what you want. When you have some time, then you can try to set up the autopilot again. You can minimize automation surprise by closely watching mode annunciations while pressing a button or when you expect an event to occur—such as capturing a course or altitude. [Figure 10-52]

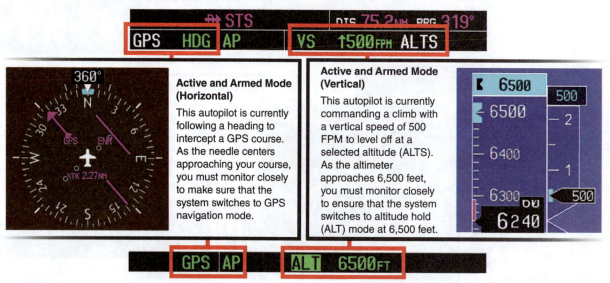

Active and Armed Mode (Horizontal)

This autopilot is currently following a heading to intercept a GPS course. As the needle centers approaching your course, you must monitor closely to make sure that the system switches to GPS navigation mode.

Active and Armed Mode (Vertical)

This autopilot is currently commanding a climb with a vertical speed of 500 FPM to level off at a selected altitude (ALTS). As the altimeter approaches 6,500 feet, you must monitor closely to ensure that the system switches to altitude hold (ALT) mode at 6,500 feet.

Figure 10-52. To maintain situational awareness and avoid an automation surprise, watch your autopilot indications carefully when setting them and when an armed event is scheduled to occur.

APPLYING THREAT AND ERROR MANAGEMENT

Threat and error management (TEM) is a safety management approach that assumes that, as a pilot, you will encounter risk and make mistakes during ground and flight operations. Becoming proficient in SRM, the nine pilot competencies of CBTA (introduced in Chapter 1, Section D), and specific TEM strategies enables you to successfully manage **threats** and **errors** and avoid an **undesired aircraft state (UAS)**—a pilot-induced airplane position/speed deviation, misapplied flight controls, or an incorrect airplane configuration.

A UAS is a transitional state, between a normal operational state (for example, a stabilized approach to a landing), and an outcome. An outcome is an end state, most notably, an incident or an accident. Examples of undesired aircraft states include exceeding speed restrictions, lining up for the wrong runway for takeoff or landing, or unexpectedly approaching an aerodynamic stall.

ROOT CAUSE ANALYSIS AND RISK ANTICIPATION

Root cause analysis (RCA) is the process of examining a problem or why something went wrong to determine how to prevent the issue from reoccurring in the future. After each flight, as part of a debriefing with your instructor or as a self critique, you should consider what you did well during the flight and identify errors that you made. Analyze each error to determine its root cause. Errors can result from insufficient training and experience, inadequate flight planning or preparation, external pressures, and physiological and psychological effects. [Figure 10-53]

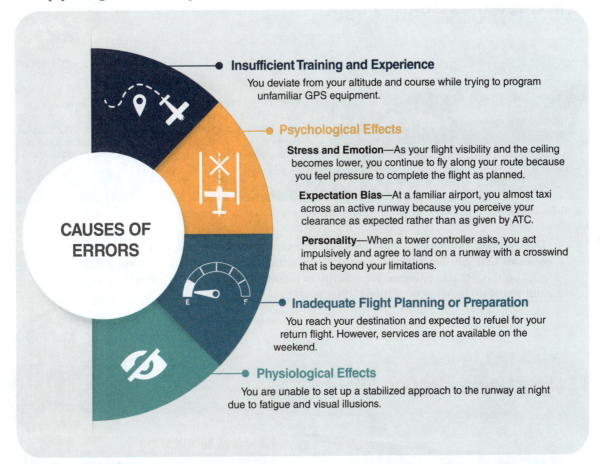

Insufficient Training and Experience
You deviate from your altitude and course while trying to program unfamiliar GPS equipment.

Psychological Effects

Stress and Emotion—As your flight visibility and the ceiling becomes lower, you continue to fly along your route because you feel pressure to complete the flight as planned.

Expectation Bias—At a familiar airport, you almost taxi across an active runway because you perceive your clearance as expected rather than as given by ATC.

Personality—When a tower controller asks, you act impulsively and agree to land on a runway with a crosswind that is beyond your limitations.

Inadequate Flight Planning or Preparation
You reach your destination and expected to refuel for your return flight. However, services are not available on the weekend.

Physiological Effects
You are unable to set up a stabilized approach to the runway at night due to fatigue and visual illusions.

CAUSES OF ERRORS

Figure 10-53. Shown here are examples of pilot errors and causes. Determining the root cause of errors helps you to avoid making similar mistakes in the future.

In addition to examining your own errors, explore accident and incident reports. During your training, your instructor will teach you to consider causal chains in incident and accident scenarios. For example, consider the root causes of a an airplane veering off the runway while landing in a 7-knot crosswind. What was the root cause and chain of events that caused the incident? Lack of pilot experience in crosswind takeoffs and landings? Other poor weather conditions? An airplane malfunction? What could have been done to prevent the incident? This exploration enhances your ability to anticipate risks.

TEM STRATEGIES

Begin employing threat and error management when you are planing a flight. Knowing where to source required information is an observable behavior associated with the CBTA pilot competency of application of knowledge—you should be proficient in locating information relevant to your flight, such as weather and runway conditions. Then, use an SRM tool, such as the 5Ps, to identify and mitigate risks. Finally, employ the TEM strategy of briefing any risks prior to the flight to help you anticipate and be prepared to manage potential threats. [Figure 10-54]

TEM Strategies Before Flight

Anticipate and Mitigate Threats

During flight planning, you identify the risk of a crosswind that exceeds your personal limitations at your destination. You calculate takeoff and landing performance for a shorter runway that you can use to avoid a crosswind. You also determine that you can divert to an alternate airport with forecast calm wind, if necessary.

Figure 10-54. Anticipate and mitigate threats, such as the risks associated with weather conditions, when you are planning a flight.

During flight, being proficient in the CBTA competencies, such as workload management ensures that you have monitoring skills to anticipate and mitigate threats. SRM tools, such as the 5Ps also enhance your ability to maintain situational awareness, manage risk, and adapt to changing conditions. [Figure 10-55]

TEM Strategies During Flight

Predictive Monitoring

- **Anticipate and Mitigate Threats**

  You obtain weather updates during a cross-country flight and determine that the ceiling and visibility are lowering at your destination. You land at an alternate airport that is reporting VFR conditions.

Reactive Monitoring

- **Identify and Mitigate Unexpected Threats**

  An unexpected gust of wind causes your airplane to veer from the runway centerline during final approach. You initiate a go-around.

- **Detect and Correct Errors**

  You inadvertently fly into clouds on a flight under VFR. You perform a 180° turn to return to VFR conditions.

- **Recognize and Recover from a UAS**

  While practicing stalls, you inadvertently enter a spin. You return to straight-and-level flight by following the proper spin recovery procedure for your airplane.

Figure 10-55. Predictive and reactive monitoring are TEM strategies to use in flight.

MANAGING STARTLE EVENTS

A **startle event**, such as a sudden engine failure in flight, and the fight-or-flight response this occurrence might elicit, could cause you to improperly diagnose and manage a problem. When you experience a startle event, your brain initiates a fight-or-flight response, referred to as **amygdala hijack**. The amygdala is part of your brain that is responsible for emotional and behavioral responses. In a startle event, the amygdala hijacks, or shuts down the capabilities of the pre-frontal cortex, which contains working memory, judgment, planning, sequencing of activity, abstract reasoning, and the ability to divide attention—capabilities related to the pilot competencies and crucial for you to remain resilient and successfully respond to an unusual event.

Even with risk anticipation training, startle events still challenge pilots. To develop resilience to startle events, your instructor can provide training scenarios for you on the ground and in the air to practice managing an unexpected or complex situation. Remain open to changing conditions and learn to adapt to a different situations. Exploring scenarios is an effective method for practicing problem solving and decision making, developing TEM strategies, and learning to cope with unexpected events. [Figure 10-56]

1 Root Cause Analysis
Understanding the root cause of accidents and incidents enables you to anticipate risks.

EXAMPLE: You understand that accidents can occur when a loss of power is experienced after takeoff and pilots respond by attempting to return to the runway or pitch up and stall the airplane.

2 Risk Anticipation
If you are able to anticipate a risk, then you increase your situational awareness.

EXAMPLE: You learn the procedures for managing an engine malfunction after takeoff and brief the procedure prior to takeoff.

3 Increased Situational Awareness
Before and during the flight you not only monitor your current situation but also are aware of potential future risks.

EXAMPLE: Because you are aware of the risk, you are vigilant during the takeoff and climb to indications of a loss of power. You are aware of your options for landing the airplane safely considering the terrain near the departure end of the runway.

4 Decreased Startle Effect
If you can anticipate an event, the occurrence can elicit a less intense startle response.

EXAMPLE: Because you are mentally prepared, if a loss of power occurs, you are able to quickly recover from the startle effect and maintain control of the airplane.

5 Ability to Manage the Situation
You quickly recover from the startle effect and are mentally prepared and trained to manage an abnormal or emergency situation.

EXAMPLE: You have prepared for this event and are able to take steps to troubleshoot the power failure, maintain glide speed, and safely land the airplane.

Figure 10-56. As you develop the pilot competencies and learn TEM strategies, you enhance your ability to overcome a startle event and the fight-or-flight response.

Aviate Navigate Communicate is an axiom to remind you of the basic actions to take when you are startled or faced with an emergency or abnormal situation. You should aviate—above all else, maintain control of the airplane; navigate—determine where you are and where you need to go; and communicate—call for help or state your intentions on the radio.

Fly Focus Act (FFA) is a similar axiom created for overcoming anxious moments, physiological responses, overreactions, or hesitancy to take action. The strategy of FFA provides deliberate steps to overcome a startle event or unusual situation to ensure flight safety. [Figure 10-57]

Figure 10-57. Take these actions to manage a startle event.

APPLYING SRM

The following accident accounts provide examples of how human factors can greatly influence aeronautical decision making and the outcome of a flight. Although, these examples are both air carrier accidents, they provide excellent insight into many of the concepts explored in SRM training.

THE CLASSIC HUMAN FACTORS ACCIDENT

As Flight 401, a Lockheed L-1011, was on approach into Miami, the flight crew placed the landing gear handle in the down position. However, the green light that indicates the gear is extended did not illuminate. The crew advised the air traffic controller that they would have to circle because of the light problem.

After climbing to 2,000 feet MSL and following a clearance to proceed west of the airport, the captain instructed the first officer to engage the autopilot. The crew focused their attention on the mechanical problem and the captain instructed the second officer to go below and check the nose gear alignment. During this time, the yoke was inadvertently pushed, which disengaged the autopilot and the aircraft began descending at 250 feet per minute. The crew did not notice the altitude deviation warning and had lost awareness of the aircraft's position. Although the second officer reported that he could not see the nose gear alignment pins, neither crew member nor a maintenance specialist on board, mentioned the special lights available to illuminate the pins.

A Miami controller noticed an altitude readout of 900 feet and indirectly questioned the status of the flight, *"How are things comin' along out there?"* Flight 401 replied that they would *"like to turn around and come back in."* As they started the turn, the first officer said, *"We're still at 2,000 right?"* and the captain exclaimed, *"Hey, what's happening here?"* Seven seconds later, the airplane crashed into the Everglades. Ninety-nine persons were killed.

The NTSB determined the probable cause of this accident was the failure of the flight crew to monitor the flight instruments during the final four minutes of flight and to detect an unexpected descent soon enough to prevent impact with the ground. Preoccupation with a malfunction of the nose gear position indicating system distracted the crew's attention from the instruments and allowed the descent to go unnoticed.

This was one of the primary accidents that motivated airlines to implement human factors training—**crew resource management (CRM)**. Examination of this accident reveals that the crew lost situational awareness because of the distraction of a minor mechanical problem. Poor task management was shown by the captain who delegated duties to crew members without ensuring that someone was flying the airplane. Automation management was an important factor because everyone assumed the autopilot was controlling the airplane and no one was monitoring the instruments. Crew members did not communicate effectively with each other, and pilot-controller communication broke down because the controller did not directly state his concern to the flight crew about Flight 401's dangerously-low altitude. Poor aeronautical decision making was demonstrated, aggravated by lack of airplane knowledge and over involvement of too many crew members, which contributed to improper problem definition.

HUMAN FACTORS TRAINING IN ACTION

On July 19, 1989, United Flight 232, a McDonnell Douglas DC-10, suffered a catastrophic engine failure while cruising at FL370. The destruction of the number two engine's fan rotor caused the loss of all three of the DC-10's redundant hydraulic flight control systems. The aircraft was rendered almost uncontrollable. Because the possibility of a catastrophic failure of all three hydraulic systems was considered by designers to be nearly impossible, there was no procedure in place for this occurrence.

DISCOVERY

Human Factors in Space — Mission to Mars

Managing human factors issues during long stays in space is a unique challenge for scientists. Studies of the crew aboard Skylab, Mir, and most recently, aboard the International Space Station, have shown that in addition to the physiological effects of living in space, there are many psychological factors to consider. [Figure A]

A
Courtesy of NASA

Crew members aboard Mir expressed feelings of isolation and loneliness, and some became lethargic and depressed after as little as two months in space if steps were not taken to boost morale. To prepare for stays aboard Mir, psychologists monitored training sessions and crew members were tested for compatibility. For example, psychologists observed how crew members communicated with one another and how they reacted to stress during days spent in spaceflight simulators. Another test required two crew members to sit at separate control panels and, by turning knobs and dials as a team, they had to bring a single indicator needle to zero. This exercise determined whether crew members could cooperate without succumbing to feelings of frustration or competition.

Astronauts have shown that stays in space of more than 400 days are possible, but what about a longer stay, with more risks, even further away from the home planet? Scientists are preparing for the technological and human challenges posed by a 600 to 1,000-day mission to Mars. It is not fully understood what effect a trip to Mars would have on a crew. Researchers must develop methods to help crew members cope with the severe isolation and boredom, which might result from living millions of miles away from home in a confined area with only electronic methods of contact with earth. In addition, the crew members must have the ability to manage the stress of performing risky and demanding jobs in this unique environment.

B
Courtesy of NASA

In the spirit of the aviation pioneers who overcame daunting obstacles to turn the dream of flight into reality, men and women will boldly face the many challenges that lie ahead to become the first humans to set foot on the surface of another planet. [Figure B]

Against all odds, Captain Alfred C. Haynes and his crew, with the help of a DC-10 instructor pilot who was aboard as a passenger (an internal resource), were able to guide the crippled aircraft to Sioux City, Iowa. Approximately 45 minutes after the hydraulic failure, the aircraft crash-landed at the Sioux City Municipal Airport. Of the 285 passengers and 11 crew members aboard, 175 passengers and 10 crew members survived. The ability of the crew to maintain any control of the airplane and land with survivors when faced with such daunting obstacles was acknowledged by the aviation industry as a feat requiring extraordinary piloting and crew coordination abilities. The degree of success of Flight 232 resulted from the effective application of many human factors-related skills. The NTSB stated that *"the UAL flight crew performance was highly commendable and greatly exceeded reasonable expectations."*

The first priority in an emergency situation is to fly the airplane. The number one and number three engines were still operating so the crew figured out a way to maintain some control of the airplane by adding thrust on one side and reducing thrust on the other to force the airplane in a skid to turn. As pilot in command, Captain Haynes was able to maintain situational awareness and delegate tasks to effectively manage the workload. He was open to input from the crew and assistance from a variety of resources. Excellent communication took place between the crew members so each knew what his responsibility was. The first officer was responsible for communicating with the San Francisco area maintenance facility (an external resource) which brought in a team of experts for assistance. A team of five controllers in the Sioux City Gateway Airport control tower, which was collocated with approach control, worked together to coordinate the aircraft's arrival. Although the heavy workload limited communication between the flight deck and cabin crew, the flight attendants were able to deduce the extent of the emergency just by observing non-verbal cues from the crew as to the intensity of the situation on the flight deck.

I am firmly convinced that CRM played a very important part in our landing at Sioux City with any chance of survival. I also believe that its principles apply no matter how many crew members are in the cockpit. Those who fly single-pilot aircraft sometimes ask, "How does CRM affect me if I fly by myself?" Well, CRM does not just imply the use of other sources only in the cockpit—it is an "everybody resource." To these pilots I say there are all sorts of resources available to them. Ask an astronaut if he thinks he got to the moon by himself. I don't think so—he had a great deal of help. —Captain Alfred C. Haynes

SUMMARY CHECKLIST

✓ Even though you cannot anticipate every situation that will confront you during flight, single-pilot resource management (SRM) can help you make effective decisions regarding challenging situations when they do occur.

✓ The poor judgment chain concept teaches that a human factors-related accident can often be prevented by avoiding one poor decision out of a series of poor decisions that lead to the accident.

✓ Examining NTSB accident and incident reports can help you recognize the many events that can lead to a hazardous situation. NASA Aviation Safety Reporting System (ASRS) reports can also increase your awareness of how factors such as attitude, workload, situational awareness, fatigue, and stress can affect a pilot's decision-making ability.

✓ General steps in the ADM process include to: (1) recognize a change; (2) define the problem; (3) choose a course of action; (4) implement your decision; and (5) evaluate the outcome.

✓ Self assessment is necessary to ensure that you are prepared to act as pilot in command on a particular flight and usually includes evaluating personal minimums and limitations on a risk management checklist.

✓ Experienced pilots can develop dangerous tendencies called operational pitfalls that grow out of the desire to complete a flight as planned, please passengers, and meet schedules. Operational pitfalls can cause you to overestimate your piloting skills, and can adversely affect safety.

✓ Being aware of your hazardous attitudes can help prevent these thoughts from impairing your decision making. To combat a hazardous attitude, recognize the attitude as hazardous, label the thought as one of the hazardous attitudes, and then recite that attitude's antidote.

✓ By using 5P checklists, you can identify risks to consider when planning your trip. Use the 5P check to continue evaluating those risks after making a Go decision during various phases of flight.

✓ Effective task management helps ensure that required procedures are accomplished correctly, at the right time, and in a way that avoids excessive workload.

✓ To maintain situational awareness, you must effectively perform the other SRM tasks—in addition to using your technical skills as a pilot—to maintain an accurate picture of the flight.

✓ Following a sterile flight deck rule can increase safety by controlling distractions that undermine situational awareness during critical phases of flight.

✓ You can help prevent controlled flight into terrain (CFIT) by learning about the risk factors that lead to CFIT so you can avoid placing yourself in similar situations. Actively monitor navigation equipment, including terrain displays, and cross check your position on VFR charts, especially during dark night or low-visibility operations.

✓ Your automation management skill determines whether the equipment provides a benefit or a distraction. Avoid the temptation to operate outside your personal limitations and to spend too much time "head down" instead of looking outside the airplane.

✓ To avoid complacency, maintain proficiency in hand-flying the airplane and use equipment operating levels.

✓ The use of standard operating procedures (SOPs) can help you accomplish automation management effectively.

✓ An automation surprise is confusion that can occur when the system does something you do not expect, or fails to do something you do expect. You can minimize automation surprise by closely watching mode annunciations while pressing a button or when you expect an event to occur.

✓ Becoming proficient in SRM, the nine pilot competencies of CBTA, and specific TEM strategies enables you to successfully manage threats and errors.

✓ Root cause analysis (RCA) is the process of examining a problem or why something went wrong to determine how to prevent the issue from reoccurring in the future.

✓ A startle event, such as a sudden engine failure in flight, and the fight-or-flight response this occurrence might elicit, could cause you to improperly diagnose and manage a problem.

✓ In a startle event, the amygdala hijacks, or shuts down, the capabilities of the pre-frontal cortex, which contains working memory, judgment, planning, sequencing of activity, abstract reasoning, and the ability to divide attention.

SECTION B ■ Single-Pilot Resource Management

✓ When you are startled or faced with an emergency or abnormal situation: aviate— above all else, maintain control of the airplane; navigate—determine where you are and where you need to go; and communicate—call for help or state your intentions on the radio.

✓ The strategy of Fly Focus Act (FFA) provides deliberate steps to overcome a startle event or unusual situation to ensure flight safety.

KEY TERMS

Single-Pilot Resource Management (SRM)

Accident

Incident

National Transportation Safety Board (NTSB)

Poor Judgment Chain

Aviation Safety Reporting System (ASRS)

Aeronautical Decision-Making (ADM)

ADM Process

Self Assessment

Personal Minimums and Limitations

Stressors

Operational Pitfalls

Hazardous Attitudes

Risk Management

5P Checklists

5P Check

Task Management

Internal Resources

External Resources

Flight Deck Management

Margin of Safety

Situational Awareness

Complacency

Standard Operating Procedures (SOPs)

Sterile Flight Deck

Controlled Flight into Terrain (CFIT)

Automation Management

Information Management

Automation Surprise

Threat and Error Management (TEM)

Threats

Errors

Undesired Aircraft State (UAS)

Root Cause Analysis (RCA)

Startle Event

Amygdala Hijack

Aviate Navigate Communicate

Fly Focus Act (FFA)

Crew Resource Management (CRM)

QUESTIONS

1. What is the role of the NTSB regarding aviation accidents?

2. You can learn from other pilots' mistakes or become familiar with issues that can lead to an accident by searching confidential reports from pilot by referring to what online database?

3. What are five general steps in the ADM process?

4. During which phases of flight do most general aviation accidents occur?
   A. Climb and cruise
   B. Maneuvering and descent
   C. Takeoff/initial climb and landing

5. What are some tools you can use to evaluate your fitness to fly?

6. Name at least three operational pitfalls.

Match the thought about the following situation to the appropriate hazardous attitude.

You land at an unfamiliar airport and ask the receptionist at the FBO counter to *"top it off"* and then continue to the pilot's lounge to make a phone call. Returning, you pay the bill and take off without checking the aircraft, the fuel caps, or the fuel.

7. You feel that it is a silly requirement to preflight an aircraft which you have just flown.

   A. Invulnerability

8. You just want to get underway quickly.

   B. Impulsivity

9. You have skipped preflight inspections before and you have never had a problem.

   C. Anti-Authority

10. You feel that a pilot with your skill level can handle anything during the flight that might have been overlooked on the ground.

   D. Resignation

11. Because you pay for services, you feel that the FBO personnel are responsible for ensuring the airplane was fueled correctly.

   E. Macho

12. Give specific examples of at least one risk factor in each of the 5P categories that you should consider during flight planning.

13. Select the true statement about using the 5P check in flight.
    A. Manage risk by using the 5P Check in emergency situations only.
    B. Using the 5P Check is not necessary in flight if you mitigated all risks during flight planning.
    C. Manage risk by using the 5P Check at decision points that correspond to the phases of flight.

14. If you have an in-flight aircraft equipment malfunction and need to divert to an alternate airport, what resources could assist you?

15. During what phases of flight is a sterile flight deck most critical?

16. Which is an action that you can take to maintain situational awareness during ground operations?
    A. Perform your departure briefing during taxi to be prepared when you arrive at the takeoff runway.
    B. No matter what your position is on the airport, stop immediately to communicate with ATC if you become disoriented.
    C. Always know your airplane's present location and mentally calculate the next location in the route that requires increased attention.

17. Which action can help you maintain situational awareness during flight?
    A. Apply SRM skills and follow standard operating procedures.
    B. Hand-fly the airplane at all times to maintain stick-and-rudder skills.
    C. Eliminate all communication with a copilot or passengers during every operation except cruise flight.

18. Scud running and lack of situational awareness when operating in areas of reduced visibility can both lead to what?

19. What is an automation surprise and how can you prevent it?

20. What is root cause analysis?

21. Name four causes of pilot errors.

22. What actions can you initially take to overcome a startle event and manage a challenging situation?

CHAPTER 11

Flying Cross-Country

SECTION A
The Flight Planning Process

Whether you are traveling across the continent, or across the county, the route to an enjoyable and safe flight begins with the flight planning process. This process requires you to apply all the knowledge and skills you have acquired throughout your training. The flight planning process has at least six primary steps. As you complete each step, use an SRM tool, such as the 5P checklists, to assess and mitigate risks. Based on your assessment, you might decide to change, delay, or cancel a portion, or all, of your flight at any point during flight planning. As you gain experience, you might refine and customize your planning procedures, however, the general process should remain similar to that shown in Figure 11-1.

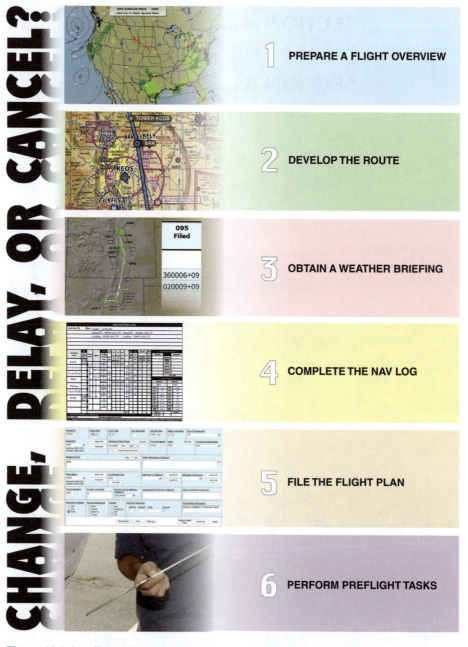

CHANGE, DELAY, OR CANCEL?

1 **PREPARE A FLIGHT OVERVIEW**

2 **DEVELOP THE ROUTE**

3 **OBTAIN A WEATHER BRIEFING**

4 **COMPLETE THE NAV LOG**

5 **FILE THE FLIGHT PLAN**

6 **PERFORM PREFLIGHT TASKS**

Figure 11-1. An efficient flight planning process involves several steps.

In the following scenario, you will plan a solo cross-country in Section A of this chapter from Centennial Airport in Denver Colorado to Pueblo Memorial Airport. [Figure 11-2] A variety of different flight planning tools and flight information sources are shown in this section to illustrate the steps in the flight planning process. You might use one or more of these or similar tools and sources when you plan your own flights. The procedures you use to perform the cross-country are explored in Section B.

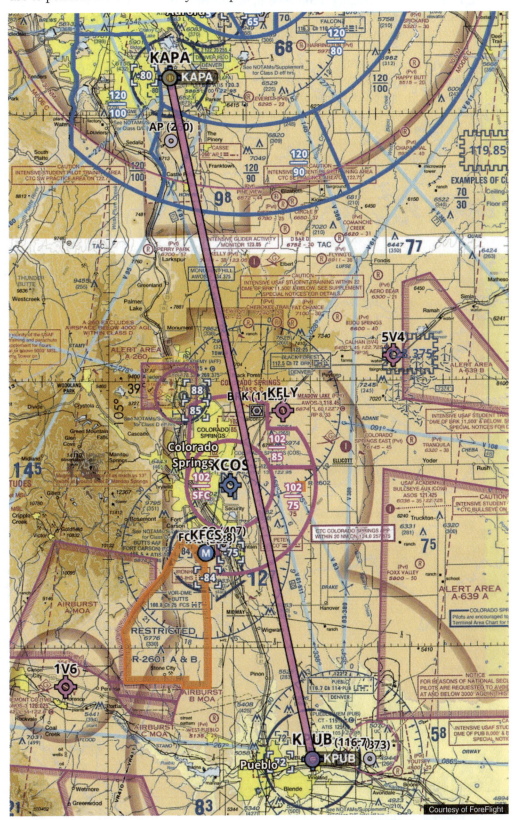

Figure 11-2. Your flight is from Centennial Airport (KAPA) to Pueblo Memorial Airport (KPUB).

SECTION A ■ The Flight Planning Process

PREPARE A FLIGHT OVERVIEW

The flight overview is the first step in the flight planning process and occurs well before the proposed flight. First you must determine if the flight is feasible. Review aeronautical charts to assess the length of the flight, the terrain, and any airspace that might affect your flight planning to ensure that you can accomplish the trip within your capabilities and the limitations of the airplane. Propose a date and time for the flight. Review weather information from online sources, such as Flight Service or the National Weather Service. Check general forecast weather by obtaining an outlook briefing.

After you have determined that the flight is feasible, gather your flight planning materials. Obtain the appropriate aeronautical charts, update your flight planning app (if applicable), and ensure you have current airport information from the Chart Supplement or other appropriate flight information sources. Gather your plotter, a flight computer, a navigation log, and a flight plan form, as applicable. In addition, ensure you have the necessary training and experience. Do you meet the regulatory requirements and your own personal experience minimums that apply to the flight? [Figure 11-3]

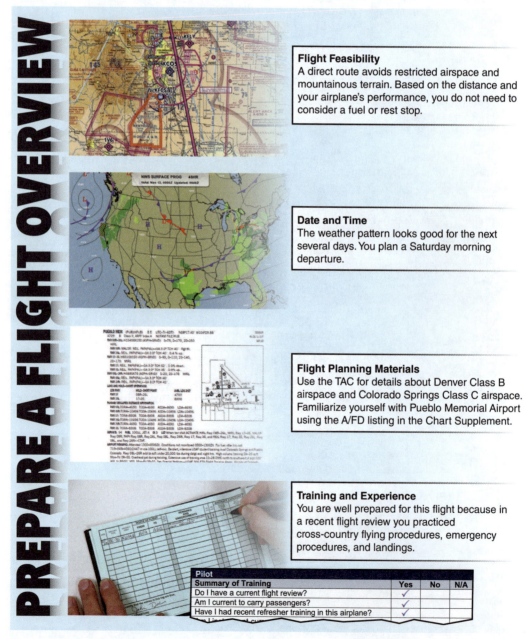

Flight Feasibility
A direct route avoids restricted airspace and mountainous terrain. Based on the distance and your airplane's performance, you do not need to consider a fuel or rest stop.

Date and Time
The weather pattern looks good for the next several days. You plan a Saturday morning departure.

Flight Planning Materials
Use the TAC for details about Denver Class B airspace and Colorado Springs Class C airspace. Familiarize yourself with Pueblo Memorial Airport using the A/FD listing in the Chart Supplement.

Training and Experience
You are well prepared for this flight because in a recent flight review you practiced cross-country flying procedures, emergency procedures, and landings.

Pilot			
Summary of Training	Yes	No	N/A
Do I have a current flight review?	✓		
Am I current to carry passengers?	✓		
Have I had recent refresher training in this airplane?	✓		

Figure 11-3. A flight overview includes ensuring the flight is feasible, selecting the date and time, gathering your flight planning materials, and reviewing your training and experience.

DEVELOP THE ROUTE

As you study the chart, keep in mind that some decisions about your route might change after you receive your weather briefing. Begin with a straight line route and refine it as necessary. As you examine the route features to develop a proposed route, consider significant terrain and obstacles along the route and determine airspace requirements. Decide if you would like to circumnavigate any busy airspace and locate any special use airspace that might affect your route. Modify your course to include appropriate navaids and plan a route that gives you options if you have to divert. [Figure 11-4]

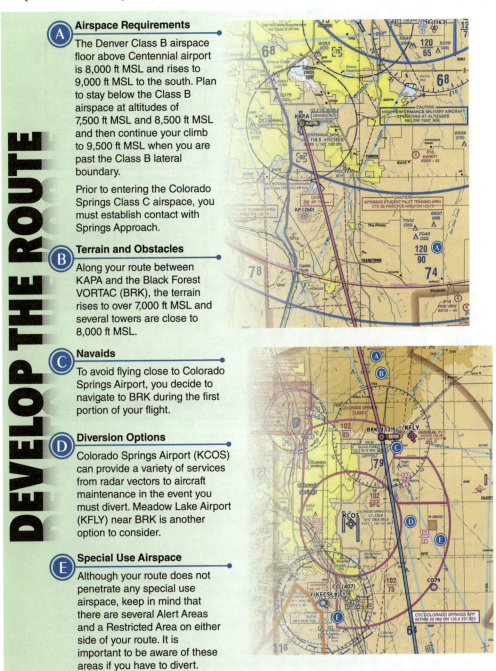

A Airspace Requirements

The Denver Class B airspace floor above Centennial airport is 8,000 ft MSL and rises to 9,000 ft MSL to the south. Plan to stay below the Class B airspace at altitudes of 7,500 ft MSL and 8,500 ft MSL and then continue your climb to 9,500 ft MSL when you are past the Class B lateral boundary.

Prior to entering the Colorado Springs Class C airspace, you must establish contact with Springs Approach.

B Terrain and Obstacles

Along your route between KAPA and the Black Forest VORTAC (BRK), the terrain rises to over 7,000 ft MSL and several towers are close to 8,000 ft MSL.

C Navaids

To avoid flying close to Colorado Springs Airport, you decide to navigate to BRK during the first portion of your flight.

D Diversion Options

Colorado Springs Airport (KCOS) can provide a variety of services from radar vectors to aircraft maintenance in the event you must divert. Meadow Lake Airport (KFLY) near BRK is another option to consider.

E Special Use Airspace

Although your route does not penetrate any special use airspace, keep in mind that there are several Alert Areas and a Restricted Area on either side of your route. It is important to be aware of these areas if you have to divert.

Figure 11-4. When you develop the route, your must consider terrain and obstacles, airspace requirements, special use airspace, navaids, and diversion options.

After examining the route features, continue to develop your route by determining route data. Select easily identifiable checkpoints, such as prominent landmarks and navaids, to keep track of your position, monitor your progress, and provide references for contacting ATC. Determine your magnetic course and total route distance using a plotter or flight planning app. Select a proposed cruise altitude based on VFR cruising altitude

SECTION A ■ The Flight Planning Process

requirements and considering terrain, obstacles, and airspace. Determine an approximate time enroute based on your experience with the airplane's cruise speed. You will need this information to obtain a standard weather briefing before your flight. Use the 5P Plan checklist to ensure that you have mitigated any risks and developed a safe route. [Figure 11-5]

SECTION A ■ The Flight Planning Process

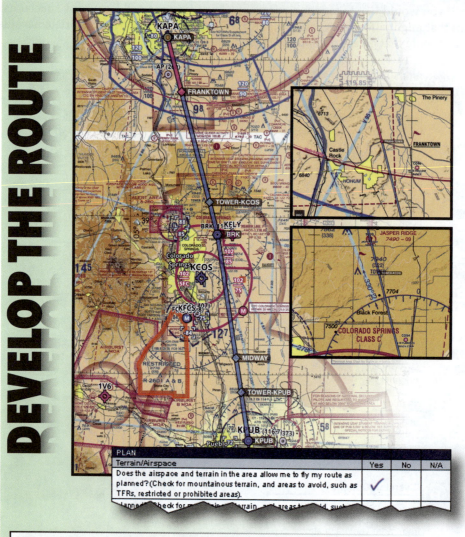

Checkpoints
The town of Franktown, the tower north of KCOS, BRK, abeam the town of Midway, and the tower north of Pueblo are good checkpoints. Use the towers as references for contacting Springs Approach and Denver approach control at Pueblo.

Cruise Altitude
Use intermediate altitudes to stay below Denver Class B airspace as you climb. Then, based on the terrain and obstacles above 7,000 ft MSL along your route, and your magnetic course of 152°, you select a preliminary cruise altitude of 9,500 ft MSL.

Distance and Time Enroute
The route distance is 78 NM, and you typically cruise at about 120 knots. Based on this information, you determine an approximate time enroute of 40 minutes.

Figure 11-5. Select your checkpoints and determine a cruise altitude and time enroute.

OBTAIN A WEATHER BRIEFING

Now you are ready to obtain your FAA-approved weather briefing either by phone or online. You can fill out a preliminary flight plan and select Standard Briefing at 1800wxbrief.com or call a Flight Service briefer at 1-800-WX-BRIEF. Specify whether the flight is VFR or IFR. Provide the aircraft registration number and type, the departure airport, the intended flight route, and the destination airport. Include the intended cruise altitude, the estimated time of departure, and estimated time enroute. The standard briefing contains:

- Any adverse conditions on your route.
- A synopsis of the weather systems or air masses that affect the flight.
- Current weather conditions, including PIREPs.
- Departure, enroute and destination forecasts.
- Winds aloft forecast for the time period you are flying.
- NOTAMs.

To determine if a TFR affects your flight, obtain NOTAMs from Flight Service during your online or phone briefing and obtain TFRs with graphic depictions at tfr.faa.gov. The depicted TFR data might not be a complete listing, so always follow up with Flight Service during flight planning. [Figure 11-6]

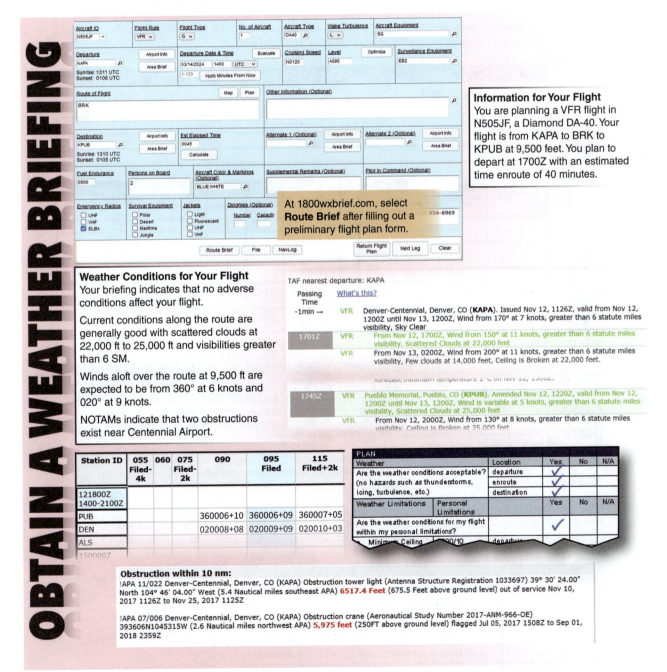

Figure 11-6. Obtain a standard weather briefing by phone or online.

COMPLETE THE NAV LOG

Include a variety of information on your navigation (nav) log for flight planning and to monitor your progress during the flight. First, complete the performance calculations for the flight. Record calculations such as weight-and-balance figures and takeoff and landing distances. Compute cruise performance, including power settings, true airspeed, and fuel consumption so you can complete nav log data. You must use the power settings that you have selected for calculating the airplane performance for accurate cruise performance, including fuel consumption, during flight. Keep in mind that lower power settings increase endurance and decrease engine wear at the expense of true airspeed.

As you determine that airplane performance is sufficient for the flight, check the applicable items on the 5P checklist. To fill out the nav log, add checkpoint information, navaid frequencies, the cruise altitude, and winds aloft for each segment.

Next, complete the navigation and performance data for each leg between checkpoints from left to right. This includes true airspeed, headings, leg distances, groundspeed and estimated time enroute, and fuel consumption. Finally, add airport information, including communication frequencies and airport elevations. [Figure 11-7]

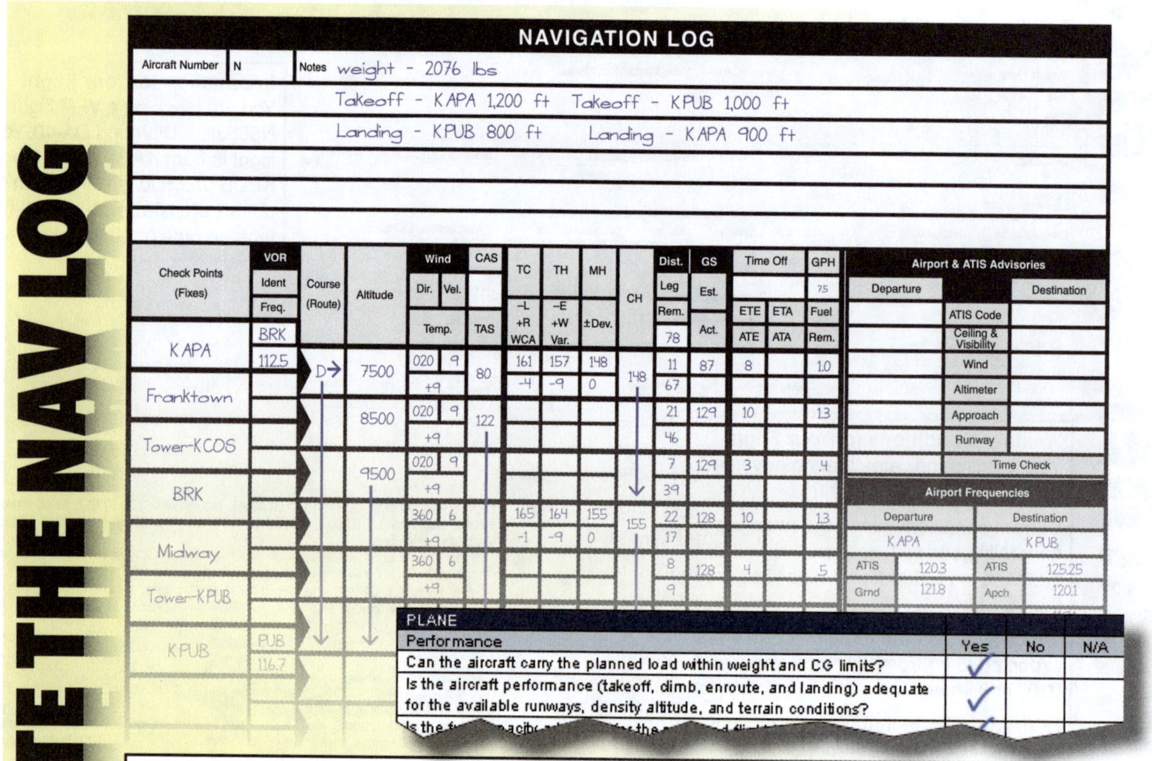

SECTION A ■ The Flight Planning Process

COMPLETE THE NAV LOG

NAVIGATION LOG

Aircraft Number: N
Notes: weight – 2076 lbs
Takeoff – KAPA 1,200 ft   Takeoff – KPUB 1,000 ft
Landing – KPUB 800 ft   Landing – KAPA 900 ft

Check Points (Fixes)	VOR Ident Freq.	Course (Route)	Altitude	Wind Dir. Vel. / Temp.	CAS / TAS	TC / +R WCA	TH / +W Var.	MH / ±Dev.	CH	Dist. Leg Rem.	GS Est. Act.	Time Off ETE ATE	ETA ATA	GPH 7.5 Fuel Rem.	Airport & ATIS Advisories
KAPA	BRK 112.5	D→	7500	020 9 / +9	80	161 / -4	157 / -9	148 / 0	148	78 / 11	87 / 8			1.0	Departure / Destination; ATIS Code; Ceiling & Visibility
Franktown			8500	020 9 / +9	122					67 / 21	129 / 10			1.3	Wind; Altimeter
Tower-KCOS			9500	020 9 / +9						46 / 7	129 / 3			.4	Approach; Runway
BRK		↓		360 6 / +9		165 / -1	164 / -9	155 / 0	155	39 / 22	128 / 10			1.3	Time Check
Midway				360 6 / +9						17 / 8	128 / 4			.5	Airport Frequencies; Departure KAPA / Destination KPUB
Tower-KPUB										9					ATIS 120.3 / ATIS 125.25
KPUB	PUB 116.7	↓	↓												Grnd 121.8 / Apch 120.1

PLANE

Performance	Yes	No	N/A
Can the aircraft carry the planned load within weight and CG limits?	✓		
Is the aircraft performance (takeoff, climb, enroute, and landing) adequate for the available runways, density altitude, and terrain conditions?	✓		

Performance Calculations
Verify that the weight and CG are within limits. Based on takeoff weight and temperature, calculate takeoff and landing distances for both the departure and destination airports.

Checkpoints
Although you will use GPS as your primary navigation method, include your checkpoints and applicable VOR frequencies along your route, such as BRK–112.5 and PUB—116.7 as references.

Leg Navigation and Performance Data
With a flight computer or flight planning app, determine the navigation and performance data for each leg. Use the winds aloft from your weather briefing and the true airspeed and fuel consumption figures from your performance calculations.

Airport Information
Record all of the frequencies available for both airports, even if you do not plan to use them. For example, record the departure control frequency for KAPA just in case you need to communicate with Denver controllers regarding Class B airspace.

Figure 11-7. Use the navigation log to record the flight data and to monitor your progress during the cross-country flight.

After you are familiar with the process of completing a nav log, you might consider using an app that automatically determines much of the flight data. A flight planning app or EFB can use the current and forecast weather information to create a nav log for you after you have entered your airplane's performance and the route of flight. [Figure 11-8]

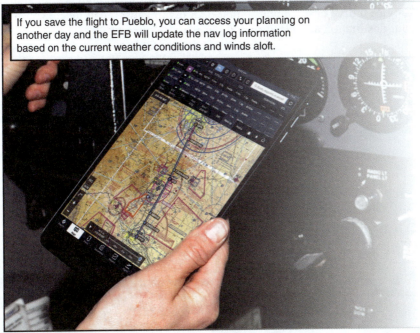

If you save the flight to Pueblo, you can access your planning on another day and the EFB will update the nav log information based on the current weather conditions and winds aloft.

Figure 11-8. The nav log can be displayed with the route on the map and will update as you fly your cross-country.

Vin Fiz

With less than 60 hours of flight experience, Calbraith Perry Rodgers climbed into a canvas-covered biplane on September 11, 1911, and departed from Sheepshead Bay, Brooklyn on what would become the first transcontinental flight. Cal Rodgers' trip was sponsored by the Armour Company, which paid Cal $5.00 for every mile he flew with the company's soft drink name painted on the wing of his Wright EX, the Vin Fiz. On November

5, he landed in Pasadena, California after 82 hours of flight time and 69 stops, many of them unplanned. The journey's hardships, which included 19 crashes, took its toll on both plane and pilot. So many airplane parts were replaced during the excursion that only the rudder and two wing struts remained from the original Vin Fiz. Cal himself did not fare much better—he completed his adventure with a broken leg and a nasty gash across his forehead. While technological advancements can be given a large portion of the credit, thorough preflight planning and sound pilot decision making has played a major, if somewhat uncelebrated role, in making cross-country travel significantly more routine than in Cal Rodgers' day.

FILE THE FLIGHT PLAN

After completing the nav log, modify your preliminary flight plan form as necessary. For example, you might need to change the altitude or time enroute based on the effect of the winds aloft on your groundspeed. You might decide to modify your route or departure time after considering forecast weather or NOTAMs. After your flight plan is finalized, file it with Flight Service online or by phone so that SAR organizations can locate you if you do not reach your destination and close your flight plan. [Figure 11-9]

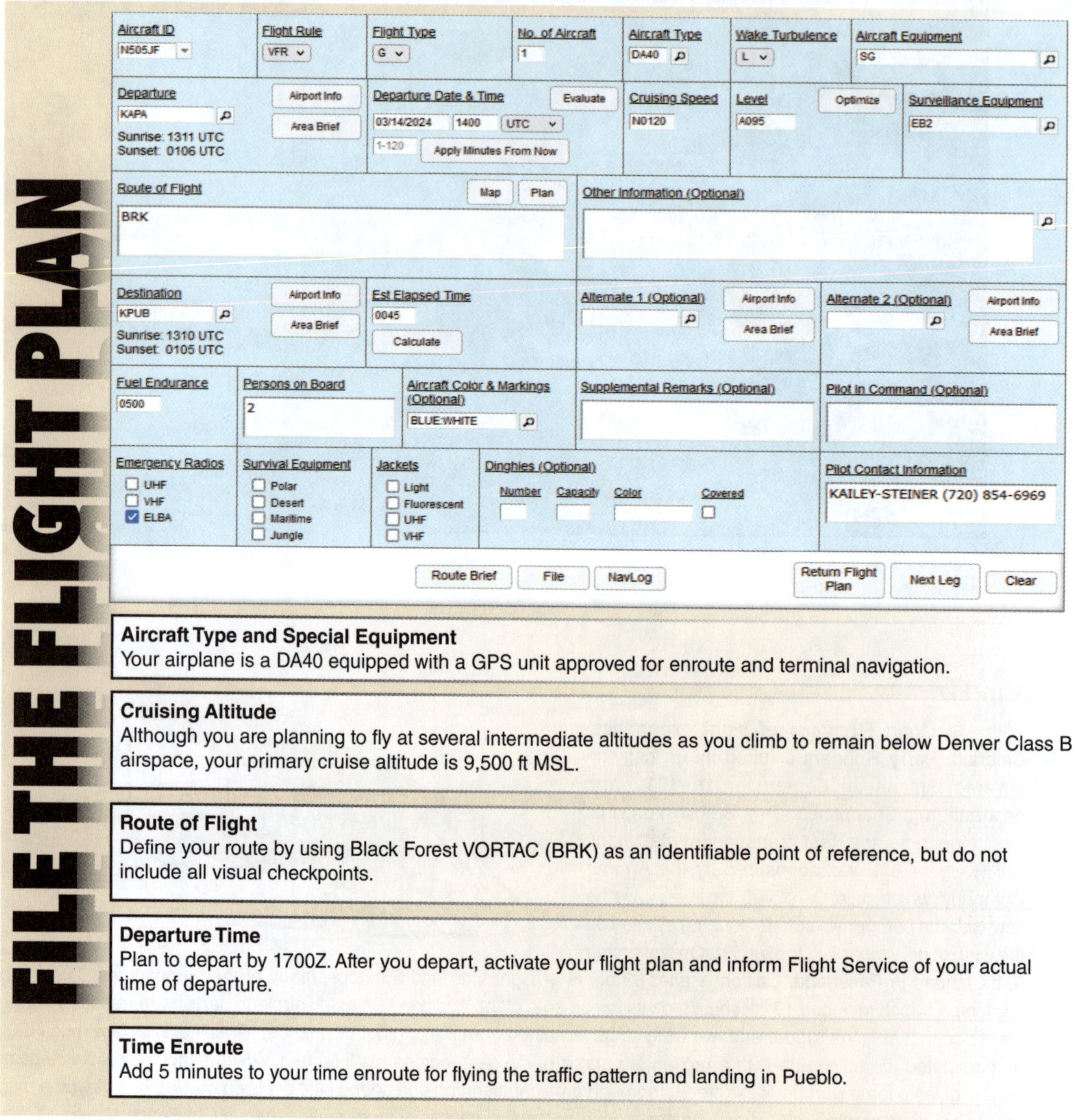

Aircraft Type and Special Equipment
Your airplane is a DA40 equipped with a GPS unit approved for enroute and terminal navigation.

Cruising Altitude
Although you are planning to fly at several intermediate altitudes as you climb to remain below Denver Class B airspace, your primary cruise altitude is 9,500 ft MSL.

Route of Flight
Define your route by using Black Forest VORTAC (BRK) as an identifiable point of reference, but do not include all visual checkpoints.

Departure Time
Plan to depart by 1700Z. After you depart, activate your flight plan and inform Flight Service of your actual time of departure.

Time Enroute
Add 5 minutes to your time enroute for flying the traffic pattern and landing in Pueblo.

Figure 11-9. You can file either a Domestic or ICAO flight plan for your VFR flight at 1800wxbrief.com.

PERFORM PREFLIGHT TASKS

The last step in the flight planning process is to perform preflight tasks—assess your condition and the airplane's airworthiness. First, perform a self-assessment. Evaluate your fitness for flight using the I'M SAFE checklist. Check the aircraft maintenance logbooks to verify that the airplane has been maintained in accordance with FAA requirements. Then perform the preflight inspection. Use a checklist to inspect the aircraft in a logical order without omissions. Be sure that the engine oil level is adequate, and visually check the fuel level. Check off the last items on the 5P Pilot, Plane, and Programming checklists to make the final go/no-go decision. [Figure 11-10]

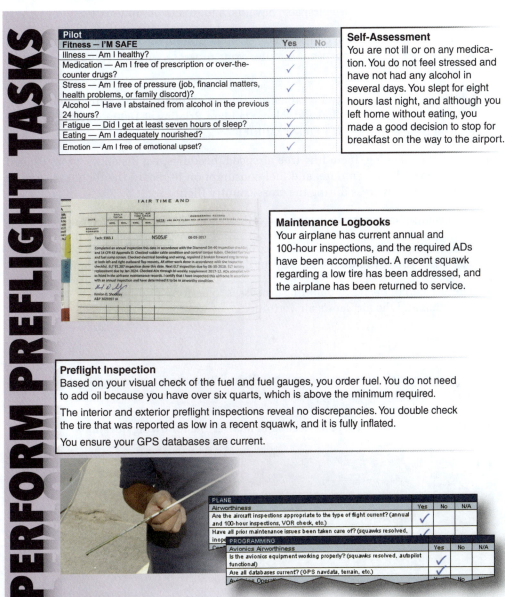

PERFORM PREFLIGHT TASKS

Pilot		
Fitness — I'M SAFE	**Yes**	**No**
Illness — Am I healthy?	✓	
Medication — Am I free of prescription or over-the-counter drugs?	✓	
Stress — Am I free of pressure (job, financial matters, health problems, or family discord)?	✓	
Alcohol — Have I abstained from alcohol in the previous 24 hours?	✓	
Fatigue — Did I get at least seven hours of sleep?	✓	
Eating — Am I adequately nourished?	✓	
Emotion — Am I free of emotional upset?	✓	

Self-Assessment
You are not ill or on any medication. You do not feel stressed and have not had any alcohol in several days. You slept for eight hours last night, and although you left home without eating, you made a good decision to stop for breakfast on the way to the airport.

Maintenance Logbooks
Your airplane has current annual and 100-hour inspections, and the required ADs have been accomplished. A recent squawk regarding a low tire has been addressed, and the airplane has been returned to service.

Preflight Inspection
Based on your visual check of the fuel and fuel gauges, you order fuel. You do not need to add oil because you have over six quarts, which is above the minimum required.

The interior and exterior preflight inspections reveal no discrepancies. You double check the tire that was reported as low in a recent squawk, and it is fully inflated.

You ensure your GPS databases are current.

PLANE			
Airworthiness	Yes	No	N/A
Are the aircraft inspections appropriate to the type of flight current? (annual and 100-hour inspections, VOR check, etc.)	✓		
Have all prior maintenance issues been taken care of? (squawks resolved, inop...	✓		

PROGRAMMING			
Avionics Airworthiness	Yes	No	N/A
Is the avionics equipment working properly? (squawks resolved, autopilot functional)	✓		
Are all databases current? (GPS navdata, terrain, etc.)	✓		

Figure 11-10. Preflight tasks include a self-assessment, checking maintenance logs, and performing a preflight inspection.

SECTION A ■ **The Flight Planning Process**

SUMMARY CHECKLIST

✓ You can decide to change, delay, or cancel a portion or all of your flight at any point during the flight planning process.

✓ A flight overview includes ensuring the flight is feasible, selecting the date and time, gathering your flight planning materials, and reviewing your training and experience.

✓ When you develop the route, you must consider terrain and obstacles, airspace requirements, special use airspace, navaids, and diversion options.

✓ Select easily identifiable checkpoints, such as prominent landmarks and navaids, to keep track of your position, monitor your progress, and provide references for contacting ATC.

✓ Obtain your FAA-approved weather briefing either by phone or online. You can fill out a preliminary flight plan and select Standard Briefing at 1800wxbrief.com or call a Flight Service briefer at 1-800-WX-BRIEF.

✓ Complete a nav log, including true airspeed, headings, leg distances, groundspeed and estimated time enroute, and fuel consumption to record the flight data and to monitor your progress during the cross-country flight.

✓ After completing your nav log modify your preliminary flight plan form as necessary.

✓ File your flight plan with Flight Service online or by phone so that SAR organizations can locate you if you do not reach your destination and close your flight plan.

✓ The last step in the flight planning process is to perform preflight tasks—assess your condition and the airplane's airworthiness.

QUESTIONS

1. You are preparing for a cross-country flight. What are six primary steps you use to perform the flight planning process?

Use the sectional chart excerpt as needed to answer the following questions.

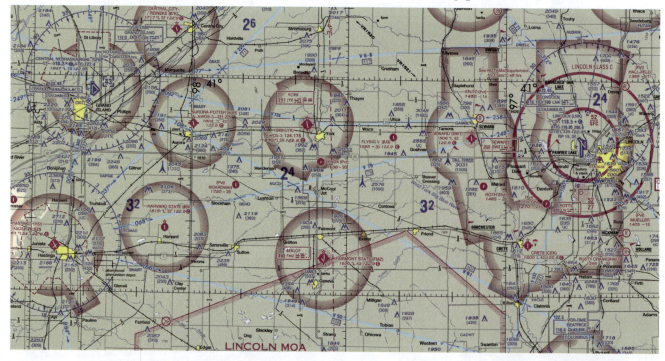

2. It is Monday, and you are planning a flight from Hastings Municipal Airport (KHSI) to Lincoln Airport (KLNK) in Nebraska for the coming weekend. What is an action that you should you take to prepare a flight overview?
   A. Complete the nav log after checking the forecast weather during an outlook briefing.
   B. Check NOTAMs to determine if any TFRs will be active on the proposed day of the flight.
   C. Review an aeronautical chart and determine that the flight is feasible because there is no restricted airspace or terrain that would change your plans.

3. What factors apply to examining the route features as you develop a route from KHSI to KLNK?
   A. A direct course penetrates the lateral boundaries of Lincoln MOA.
   B. The terrain rises from approximately 2,000 feet MSL to over 3,000 feet MSL from Hastings to Lincoln.
   C. You must establish contact with Lincoln approach control prior to entering Lincoln Class C airspace.

4. What factors apply to determining route data as you develop a route from Hastings Municipal (KHSI) to Lincoln Airport (KLNK)?
   A. A cruise altitude of 4,500 feet MSL would allow you to clear any obstacles along the route and meets VFR cruising altitude requirements.
   B. Because the route distance is 77 nautical miles and you typically cruise at about 120 knots, your approximate time enroute is 1 hour 40 minutes.
   C. The tall tower at 3,031 feet MSL about 20 miles from Lincoln airport just north of a direct course would make an excellent visual checkpoint for contacting Lincoln approach.

5. List the information that you should provide to Flight Service to obtain a standard briefing.

6. Which are examples of information that would be provided in a standard briefing for this flight?
   A. AIRMET for moderate turbulence; current conditions at KHSI and KLNK
   B. High-pressure system prevailing over the route; a 5-day forecast KHSI and KLNK
   C. Forecast for winds aloft at 6,000 feet MSL; NOTAM for a runway closure at Omaha

SECTION A ■ The Flight Planning Process

Refer to the nav log for the flight from KHSI to KLNK in a DA40 with 40 gallons of fuel on board as needed to answer the following questions.

NAVIGATION LOG

Aircraft Number	N		Notes												

Takeoff KHSI - 800 ft   Landing KLNK - 700 ft

Check Points (Fixes)	VOR Ident / Freq.	Course (Route)	Altitude	Wind Dir. / Vel. / Temp.	CAS / TAS	TC / -L +R WCA	TH / -E +W Var.	MH / ±Dev.	CH	Dist. Leg / Rem.	GS Est. / Act.	Time Off ETE / ATE	ETA / ATA	GPH / Fuel / Rem.	Airport & ATIS Advisories
												7.5			Departure / Destination
	HSI									77					ATIS Code / Ceiling & Visibility
KHSI	108.8	D→	5,500	320 / 10	120	080 / -4	076			5 / 72	75	4		1.5	Wind
Railroad				20°				0							Altimeter
				320 / 10						11 / 61	125	5		.7	Approach
Harvard State				20°											Runway
				320 / 10						17 / 44		8		1.0	Time Check
Lushton				20°											Airport Frequencies
				320 / 10						12 / 32		6		.8	Departure / Destination
Tower				20°											KHSI / KLNK
				320 / 10						13 / 19		6		.8	ATIS / ATIS 118.05
Tall Tower	LNK / 116.1		↓	20°						19 / 19	9			1.1	Grnd / Apch 128.75
KLNK										0					Tower / Tower 118.5

		Dep.	Grnd 121.9
		CTAF 122.8	CTAF 118.5
		FSS 122.1R	FSS 122.65
		UNICOM 122.8	UNICOM 122.95
Totals ▶	42	64	Field Elev. 1961 / Field Elev. 1219
Flight Plan and Weather Log on Reverse Side		Block In	Log Time
		Block Out	

SCALE 1:500,000   SECTIONAL AERONAUTICAL CHARTS

7. What is the calculated landing distance at KLNK?

8. What is the compass heading?

9. Select the true statement regarding the navigation and performance data.
   A. The groundspeed indicates a 5-knot tailwind.
   B. The groundspeed indicates a 5-knot headwind.
   C. The ETE between each checkpoint is based on the TAS.

10. What is the estimated fuel consumption between Harvard State and Lushton?

11. You have 40 gallons of fuel on board the DA40 prior to takeoff on the flight from KHSI to KLNK. List the following information to include on the flight plan for this cross-country.
   • Type of flight
   • True airspeed
   • Altitude
   • Time enroute
   • Fuel on board

SECTION B

The Flight

This section uses the flight from KAPA to KPUB that was planned in Section A to explore actions that you take during each phase of a cross-country flight, including use of the 5P Check to manage risk. The flight scenario takes place in a Diamond DA40 with a Garmin G1000 integrated display. For the purposes of this lesson, assume that the appropriate airplane checklists are accomplished for each phase of flight and that the preflight inspection, the engine start procedure, and taxiing to the departure runway are complete—you are at the holding area at the end of Runway 10 at KAPA.

PREDEPARTURE

Prior to takeoff, complete as many tasks as possible to decrease your workload during departure. Create a checklist or use a flow pattern designed for your airplane's configuration to ensure you do not miss any tasks. You must tune communication (comm) and navigation (nav) frequencies, program a flight plan if you are using GPS navigation, set course selectors on the navigation indicators, and set heading and altitude bugs as reminders, if applicable. If you are using GPS or VOR navigation in addition to dead reckoning, the initial course on your navigation display might not match the compass heading on your nav log because the forecast winds aloft are not applied. If the actual winds aloft are the same as those forecast, you can anticipate flying a heading close to your planned compass heading to maintain the course shown on your navigation equipment. You might have an EFB that retrieves the winds aloft forecast and provides an initial course for GPS navigation that applies wind correction.

Tune the emergency frequency of 121.5 on COM 2 to monitor any time that you are not in communication with ATC. Controllers attempt contact on this frequency in cases of airspace violations and you can use 121.5 immediately in an emergency. Setting up the avionics might be part of the Before-Takeoff checklist for your airplane. If not, consider adding this essential task to your checklist. [Figure 11-11]

Tune the Nav Frequencies
Although you are using GPS navigation, set nearby Falcon VORTAC (FQF 116.3) and Black Forest VOR/DME (BRK 112.5) as back-up navigation sources.

Tune the Comm Frequencies
Set frequencies in the order that you are going to use them after departure: Centennial Tower (118.9) and Flight Service (122.2) in COM1. In COM2, set 121.5 for monitoring and Colorado Springs ATIS (125.0).

Program the GPS Flight Plan
Program your flight plan from KAPA to BRK to KPUB.

Set the Course Selector
In this case, the HSI on the G1000 automatically sets the initial course to BRK after you enter your flight plan.

Set the Heading and Altitude Bugs
Set the heading bug to 185°—an intercept heading for the course. Enter the initial altitude of 7,500 to stay below the Class B airspace as you fly southeast bound.

Figure 11-11. Although you might have a different avionics configuration, the items shown here should be part of your pre-departure equipment set-up.

Another predeparture action is to perform the takeoff briefing. Include any items specific to your cross-country flight. For example, your initial heading typically is an intercept heading for your course. However, this might not be the case if ATC assigns you a different heading or you have to avoid high terrain and obstacles. In addition, noise abatement procedures might dictate your initial heading. Unless you plan to level off at intermediate altitudes as you climb, your initial altitude typically is the cruise altitude that you selected for your flight planning.

Departure procedures to include in the takeoff briefing are actions to intercept your course, any ATC clearances, and communication frequency changes that might occur during departure, such a switch to departure control or Flight Service. [Figure 11-12]

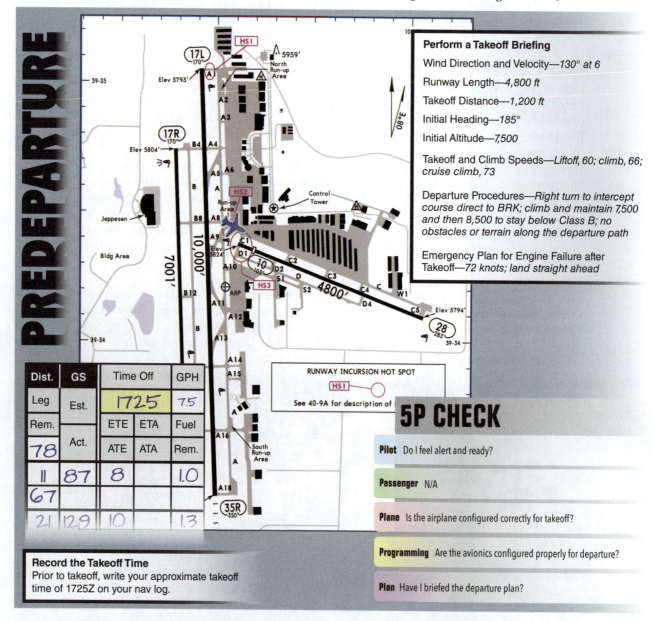

SECTION B ■ The Flight

PREDEPARTURE

Perform a Takeoff Briefing

Wind Direction and Velocity—*130° at 6*

Runway Length—*4,800 ft*

Takeoff Distance—*1,200 ft*

Initial Heading—*185°*

Initial Altitude—*7,500*

Takeoff and Climb Speeds—*Liftoff, 60; climb, 66; cruise climb, 73*

Departure Procedures—*Right turn to intercept course direct to BRK; climb and maintain 7,500 and then 8,500 to stay below Class B; no obstacles or terrain along the departure path*

Emergency Plan for Engine Failure after Takeoff—*72 knots; land straight ahead*

RUNWAY INCURSION HOT SPOT

HS1 ◯——◯

See 40-9A for description of

Dist.	GS	Time Off		GPH
Leg	Est.	**1725**		7.5
Rem.		ETE	ETA	Fuel
78	Act.	ATE	ATA	Rem.
11	87	8		1.0
67				
21	129	10		1.3

Record the Takeoff Time
Prior to takeoff, write your approximate takeoff time of 1725Z on your nav log.

5P CHECK

Pilot Do I feel alert and ready?

Passenger N/A

Plane Is the airplane configured correctly for takeoff?

Programming Are the avionics configured properly for departure?

Plan Have I briefed the departure plan?

Figure 11-12. Perform a takeoff briefing and the 5P Check to ensure you are prepared for takeoff.

CLIMB AND INITIAL CRUISE

Depending on the duration of your climb, you might perform several of the tasks in this phase of flight during the climb and other tasks after you level off at your initial cruise altitude. As you climb, intercept the course to your first checkpoint. The actions that you take to accomplish this task depend on the takeoff runway, the airspace surrounding the airport, significant terrain and obstacles, and any clearance that you might receive from the control tower or departure control, if applicable. For example, instead of immediately turning to an intercept heading for a course using GPS or VOR navigation or flying direct to your first checkpoint using pilotage, you might need to fly a heading to avoid rising terrain or special use airspace. In addition, ATC might provide you with a clearance to a visual checkpoint before you are authorized to navigate on your course or the controller might issue vectors to avoid traffic.

As you navigate toward the first checkpoint, keep in mind that your forward visibility is reduced in the climb. Lower the airplane's nose periodically for just a moment while you clear the area, and then return to the climb attitude. Transition to cruise climb after you are above the traffic pattern altitude to improve forward visibility and engine cooling, and to increase your groundspeed. Contact Flight Service to activate your VFR flight plan when you are outside Class D airspace or the tower approves a frequency change. [Figure 11-13]

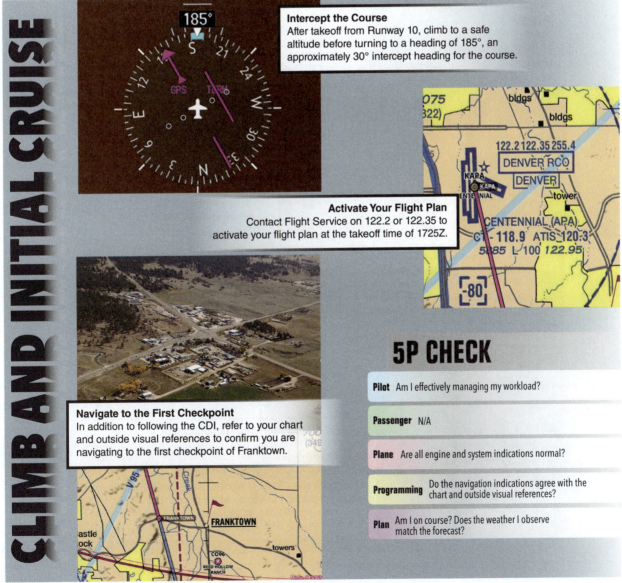

Intercept the Course
After takeoff from Runway 10, climb to a safe altitude before turning to a heading of 185°, an approximately 30° intercept heading for the course.

Activate Your Flight Plan
Contact Flight Service on 122.2 or 122.35 to activate your flight plan at the takeoff time of 1725Z.

Navigate to the First Checkpoint
In addition to following the CDI, refer to your chart and outside visual references to confirm you are navigating to the first checkpoint of Franktown.

5P CHECK

Pilot Am I effectively managing my workload?

Passenger N/A

Plane Are all engine and system indications normal?

Programming Do the navigation indications agree with the chart and outside visual references?

Plan Am I on course? Does the weather I observe match the forecast?

Figure 11-13. The actions that you take after departure might take place as you climb or after you level off at an initial cruise altitude.

ENROUTE

During the enroute phase, monitor the flight progress. As you pass over each checkpoint, note the time and compare it to the nav log estimate. Keep track of your time enroute and your estimated time of arrival using the nav log or GPS equipment. If your flight time is significantly longer than you originally planned, you must determine the reason that the values are different. Are the actual winds aloft stronger or from a different direction than those forecast? If so, you might need to revise your flight plan. Contact Flight Service for updated weather and extend the time enroute, if necessary. On longer flights, you might need to plan for an additional fuel stop.

Your primary enroute activity is to monitor your progress and making adjustments as necessary to stay on course. However, in certain situations, you might need to contact ATC to transition through Class B, Class C, or Class D airspace enroute to your destination. [Figure 11-14]

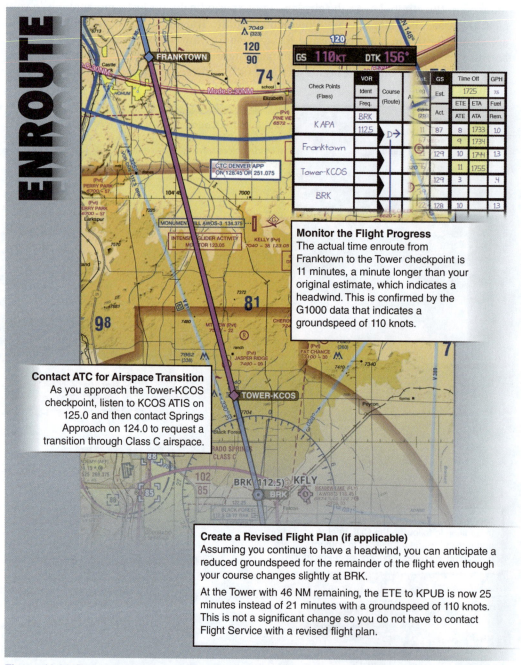

Monitor the Flight Progress
The actual time enroute from Franktown to the Tower checkpoint is 11 minutes, a minute longer than your original estimate, which indicates a headwind. This is confirmed by the G1000 data that indicates a groundspeed of 110 knots.

Contact ATC for Airspace Transition
As you approach the Tower-KCOS checkpoint, listen to KCOS ATIS on 125.0 and then contact Springs Approach on 124.0 to request a transition through Class C airspace.

Create a Revised Flight Plan (if applicable)
Assuming you continue to have a headwind, you can anticipate a reduced groundspeed for the remainder of the flight even though your course changes slightly at BRK.

At the Tower with 46 NM remaining, the ETE to KPUB is now 25 minutes instead of 21 minutes with a groundspeed of 110 knots. This is not a significant change so you do not have to contact Flight Service with a revised flight plan.

Figure 11-14. Enroute activities include monitoring the flight progress, revising your plan if necessary, and contacting ATC for transition through airspace.

If you are not already in contact with approach or departure control during a VFR cross-country, you can contact the local ARTCC for flight following. Flight Service can provide you with the appropriate frequency or can use another flight information source, such as the Chart Supplement or an EFB. Provide ATC with your call sign, position, altitude, and destination; and inform the controller that you are VFR and that you are requesting flight following. ATC will assign you a transponder code and advise when you are in radar contact. Workload permitting, the controller then provides traffic advisories and other services on request, such as weather information, center weather advisories, and the status of special use airspace. ARTCC can provide vectors to airports in an emergency. In addition, if you have a mechanical malfunction that requires an emergency landing, the controller can notify SAR personnel, if necessary.

DIVERSION CONDITIONS

Sometimes circumstances in flight require you to change your flight plan completely and divert to an alternate airport. You must know the conditions that require a diversion and be ready with a contingency plan in case you decide that the safest course of action is to land at the nearest suitable airport. [Figure 11-15]

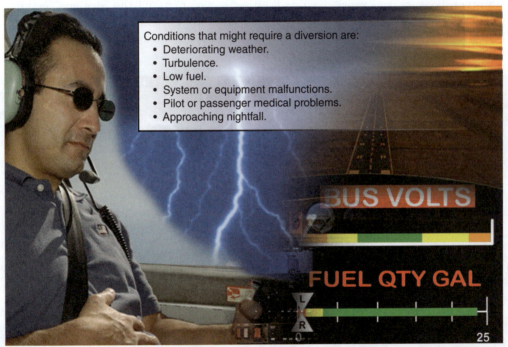

Conditions that might require a diversion are:
- Deteriorating weather.
- Turbulence.
- Low fuel.
- System or equipment malfunctions.
- Pilot or passenger medical problems.
- Approaching nightfall.

Figure 11-15. Conditions that require a diversion range from deteriorating weather to pilot medical problems.

As a VFR pilot, you might be operating in an area where unforecast changes to weather occur or when a forecasted change to the weather occurs earlier than anticipated. In either case, you have to manage a weather situation that prevents the flight from continuing under VFR or avoid a hazard, such as a thunderstorm. You must be prepared for these circumstances and have options at all times, even if the best choice is to turn around.

Aside from convective turbulence, if you happen to be flying over uneven or mountainous terrain, strong winds aloft can create turbulence severe enough that you become uncomfortable with continuing the flight. The potential for the loss of control of the aircraft or structural damage are certainly legitimate reasons to discontinue the flight.

If you effectively monitor your flight using a nav log, GPS equipment, or an EFB, you can determine if you are will burn more fuel than planned and then make the decision to divert early before the fuel on board is critically low. In the event of an equipment failure, you must decide whether you should land the aircraft prior to your planned destination. The checklists in the POH provide guidance in this situation.

By nature, pilots are task oriented and want to complete the flight. However, if you or a passenger is under medical duress, you must decide to divert as soon as you recognize your problem. Finally, if you are planning to land near sunset and are behind your planned ETA, you might consider a diversion. Unfamiliarity with the airport local terrain or obstacles, lack of night flying experience and its associated visual illusions, lack of flying experience by reference to instruments, or the lack of required backup lighting are all reasons to divert if nightfall is approaching.

DIVERSION ACTIONS

If you decide to divert to an airport other than your destination, first, fly the airplane. Stay calm and avoid letting anxiety impair your flying skills. If you are diverting for a fuel stop or to avoid weather well ahead of a critical situation, the diversion is typically not an overly stressful situation.

Turn the airplane toward the airport of intended landing and then complete flight planning details to navigate to the new destination. Computing course, time, speed, and distance information in flight requires the same computations used during preflight planning. However, because you must divide attention between flying the airplane, making calculations, and scanning for traffic, take advantage of all possible shortcuts, such as rule-of-thumb computations, and the Nearest and Direct-To navigation functions of your GPS equipment. [Figure 11-16]

In the event you needed to divert to an alternate airport during your cross-country, use the Nearest feature (NRST) to obtain airport information and the course to the nearest airports from your position.

Keep in mind that the best option might be an airport that is behind you. After you determine the alternate airport, use the Direct-To function to navigate to the alternate airport.

Figure 11-16. Using GPS equipment can decrease your workload during a diversion.

Prioritize tasks and use all available resources to reduce your workload as you proceed to the alternate airport. Engage the autopilot if you have one. Use the POH for checklists and to troubleshoot any system malfunctions. Contact ATC to inform controllers of your situation and to request assistance. Do not hesitate to declare an emergency if the situation warrants it so you receive priority from ATC. Use the 5P Check to maintain situational awareness and be prepared to adapt to changes required by the conditions.

DESCENT

The descent activities depend on the time and distance required for the descent. For example, if you only have 2,000 feet to lose when descending you will be fairly close to the airport when you begin. Therefore, prior to the descent, you will have already accomplished tasks such as performing the Before-Landing checklist and communicating with approach control, the control tower, or CTAF.

In a situation where you have more altitude to lose, you might begin before these tasks are accomplished. For example, a descent from a cruise altitude of 8,500 feet MSL to a pattern altitude of 1,500 feet MSL at a rate of 500 feet per minute at 110 knots requires 14 minutes and should begin approximately 28 nautical miles from the airport. In this case, you most likely will perform many of tasks required before approach and landing as you are descending. Where you begin your descent also depends on the surrounding airspace and terrain. Planning the descent is essential to effectively manage your workload. Determine where to initiate the descent using simple calculations or a vertical navigation (VNAV) feature on your GPS or EFB. [Figure 11-17]

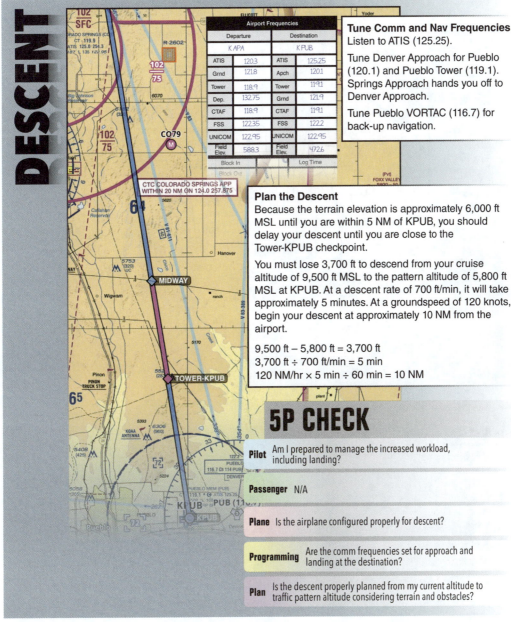

Tune Comm and Nav Frequencies
Listen to ATIS (125.25).

Tune Denver Approach for Pueblo (120.1) and Pueblo Tower (119.1). Springs Approach hands you off to Denver Approach.

Tune Pueblo VORTAC (116.7) for back-up navigation.

Plan the Descent
Because the terrain elevation is approximately 6,000 ft MSL until you are within 5 NM of KPUB, you should delay your descent until you are close to the Tower-KPUB checkpoint.

You must lose 3,700 ft to descend from your cruise altitude of 9,500 ft MSL to the pattern altitude of 5,800 ft MSL at KPUB. At a descent rate of 700 ft/min, it will take approximately 5 minutes. At a groundspeed of 120 knots, begin your descent at approximately 10 NM from the airport.

9,500 ft − 5,800 ft = 3,700 ft
3,700 ft ÷ 700 ft/min = 5 min
120 NM/hr × 5 min ÷ 60 min = 10 NM

5P CHECK

Pilot Am I prepared to manage the increased workload, including landing?

Passenger N/A

Plane Is the airplane configured properly for descent?

Programming Are the comm frequencies set for approach and landing at the destination?

Plan Is the descent properly planned from my current altitude to traffic pattern altitude considering terrain and obstacles?

Figure 11-17. Your descent planning and the tasks performed before or during the descent depend on a variety of factors.

SECTION B ■ The Flight

BEFORE APPROACH AND LANDING

As you near the airport, tasks include scanning for traffic, communicating with ATC or self-announcing on the CTAF, and configuring the airplane for landing. Because of this increased workload, accomplish as many tasks to prepare for approach and landing at your destination as you can before arriving in the terminal area. Listen to ATIS or ASOS/AWSS/AWOS for airport and weather information before you communicate with approach or tower control or announce your intentions at an uncontrolled airport. Visualize your entry into the traffic pattern and perform the before-landing briefing when you have enough information to prepare for traffic pattern entry. [Figure 11-18]

BEFORE APPROACH AND LANDING

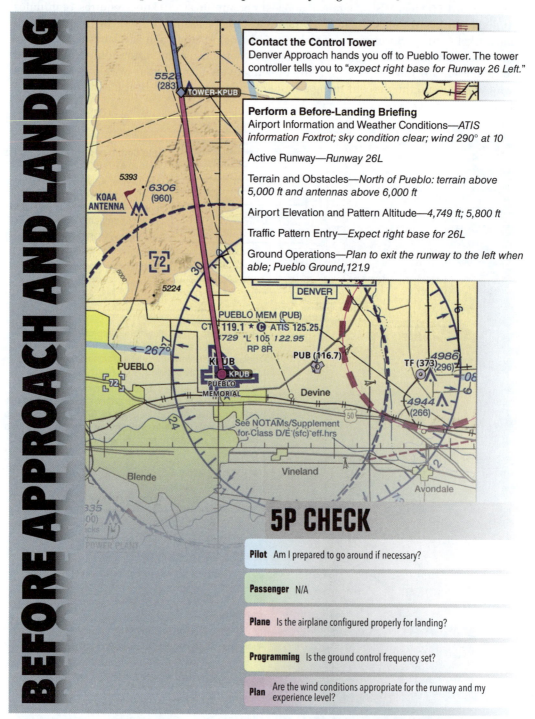

Contact the Control Tower
Denver Approach hands you off to Pueblo Tower. The tower controller tells you to "*expect right base for Runway 26 Left.*"

Perform a Before-Landing Briefing
Airport Information and Weather Conditions—*ATIS information Foxtrot; sky condition clear; wind 290° at 10*

Active Runway—*Runway 26L*

Terrain and Obstacles—*North of Pueblo: terrain above 5,000 ft and antennas above 6,000 ft*

Airport Elevation and Pattern Altitude—*4,749 ft; 5,800 ft*

Traffic Pattern Entry—*Expect right base for 26L*

Ground Operations—*Plan to exit the runway to the left when able; Pueblo Ground, 121.9*

5P CHECK

Pilot Am I prepared to go around if necessary?

Passenger N/A

Plane Is the airplane configured properly for landing?

Programming Is the ground control frequency set?

Plan Are the wind conditions appropriate for the runway and my experience level?

Figure 11-18. Take action to prepare for the approach and landing as soon as possible as you near the airport.

POSTFLIGHT

After landing and clearing the runway, comply with the taxi instructions provided by the control tower or ground control, if applicable. Use an airport diagram for situational awareness and request a progressive taxi if you are unsure of your clearance. Be sure to close your flight plan and service the airplane by adding fuel or oil if necessary. Finally, take time to evaluate the flight and critique your performance. Pilots with years of experience use this technique to increase their professionalism and mitigate any errors on future flights. [Figure 11-19]

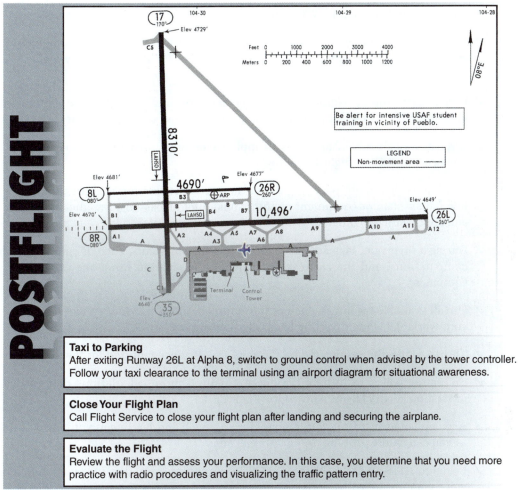

Taxi to Parking
After exiting Runway 26L at Alpha 8, switch to ground control when advised by the tower controller. Follow your taxi clearance to the terminal using an airport diagram for situational awareness.

Close Your Flight Plan
Call Flight Service to close your flight plan after landing and securing the airplane.

Evaluate the Flight
Review the flight and assess your performance. In this case, you determine that you need more practice with radio procedures and visualizing the traffic pattern entry.

Figure 11-19. Postflight tasks include taxiing to parking, closing your flight plan, and evaluating the flight.

SUMMARY CHECKLIST

✓ Prior to departure, tune comm and nav frequencies, program a flight plan if you are using GPS navigation, set course selectors on the navigation indicators, and set heading and altitude bugs as reminders, if applicable.

✓ Set the emergency frequency of 121.5 on COM 2 to monitor any time that you are not in communication with ATC.

✓ Include any items specific to your cross-country flight in your takeoff briefing, such as the intercept heading for your course, and altitude that you selected for your flight planning.

✓ Departure procedures in the takeoff briefing include a review of the actions to intercept your course, any ATC clearances, and any communication frequency changes that might occur during departure, such a switch to departure control or Flight Service.

✓ As you climb, the actions that you take to intercept your course depend on the takeoff runway, the airspace surrounding the airport, significant terrain and obstacles, and any clearance that you might receive.

✓ Transition to cruise climb after you are above the traffic pattern altitude to improve forward visibility and engine cooling, and increase groundspeed.

✓ Contact Flight Service to activate your VFR flight plan during climb or initial cruise.

✓ Enroute activities include monitoring the flight progress, revising your plan if necessary, and contacting ATC for transition through airspace.

✓ Note the time that you pass over each checkpoint and compare it to the nav log estimate. Keep track of your time enroute and your ETA using the nav log or GPS equipment.

✓ If your flight time is significantly longer than you originally planned, you might need to revise your flight plan, including adding a fuel stop, if necessary.

✓ If you are not already in contact with approach or departure control, you can contact the local ARTCC for flight following.

✓ Conditions that might requiring a diversion are deteriorating weather, turbulence, low fuel, system or equipment malfunctions, pilot or passenger medical problems, and approaching nightfall.

✓ During a diversion, take advantage of all possible shortcuts, such as rule-of-thumb computations, and the Nearest and Direct-To navigation functions of your GPS equipment.

✓ Use all available resources, such as the autopilot, POH, and ATC services, to reduce your workload when diverting to an alternate airport.

✓ The descent activities depend on the time and distance required for the descent.

✓ Determine where to initiate the descent using simple calculations or a vertical navigation (VNAV) feature on your GPS or EFB.

✓ Accomplish as many tasks to prepare for approach and landing at your destination as you can before arriving in the terminal area.

✓ Listen to ATIS or ASOS/AWSS/AWOS for airport and weather information before you communicate with approach or tower control or announce your intentions at an uncontrolled airport.

✓ Visualize your entry into the traffic pattern and perform the before-landing briefing when you have enough information to prepare for traffic pattern entry.

✓ Postflight tasks include taxiing to parking, closing your flight plan, and evaluating the flight.

SECTION B ■ **The Flight**

QUESTIONS

1. List the items included in setting up the communication and navigation equipment for a cross-country flight prior to departure.

2. Which are actions to take during climb and initial cruise?
   A. Set comm and nav frequencies and program a GPS flight plan.
   B. Intercept the course to your first checkpoint and contact Flight Service to activate your flight plan.
   C. Contact Flight Service for updated weather and extend the time enroute, if the time between checkpoints is significantly longer than you originally planned.

3. What is an example of an action that you perform during the enroute phase of flight?
   A. Contact ATC to request a transition through airspace designated by a TFR.
   B. Determine the time enroute between checkpoints and calculate a revised groundspeed.
   C. Contact Flight Service to revise your flight plan if a tailwind decreases your estimated time enroute by five minutes.

4. List six conditions that might require a diversion.

5. List at least three actions to take after you decide to divert to an alternate airport.

6. Your descent rate is 500 feet/min at 110 knots. When should you start a descent from a cruise altitude of a 7,500 feet MSL to reach a traffic pattern altitude of 3,500 feet MSL at your destination?

7. List at least 3 items to include in a before-landing briefing.

8. Select the true statement regarding postflight actions.
   A. Verify that ground control has closed your flight plan.
   B. Evaluate the flight to improve your performance on future flights.
   C. Request a progressive taxi from the control tower prior to landing to ensure you exit the runway at the proper taxiway intersection.

SECTION B ■ **The Flight**

11-25

SECTION B ■ **The Flight**

Answers

CHAPTER 1

SECTION A

1. 16 years of age; be able to read, speak and understand the English language
2. C
3. C
4. A
5. B
6. C
7. C
8. B
9. A
10. C
11. B
12. B

SECTION B

1. C
2. A high performance airplane has an engine with more than 200 horsepower
3. A complex airplane has retractable landing gear, flaps, and a controllable-pitch propeller or FADEC
4. B
5. Instrument rating
6. True
7. B
8. True

SECTION C

1. B
2. The I'M SAFE checklist consists of:
   - Illness
   - Medication
   - Stress
   - Alcohol
   - Fatigue
   - Eating/Emotion
3. The five hazardous attitude and their antidotes are:
   - Anti-authority; antidote—Follow the rules. They are usually right.
   - Impulsivity; antidote—Not so fast. Think first.
   - Invulnerability; antidote—It could happen to me.
   - Macho; antidote—Taking chances is foolish.
   - Resignation; antidote—I'm not helpless. I can make a difference.
4. The risk factors associated with the 5Ps are:
   - Pilot
   - Passengers
   - Plane
   - Programming
   - Plan
5. C
6. The passenger briefing includes:
   - **S**afety belts
   - **A**ir vents
   - **F**ire extinguisher
   - **E**gress and emergency
   - **T**raffic and talking
   - **Y**our questions
7. B
8. C
9. An undesired aircraft state (UAS) is a pilot-induced airplane position/speed deviation, misapplied flight controls, or an incorrect airplane configuration.
10. Threats:
    - Occur outside the influence of, and are not controlled by you as the pilot.
    - Increase the operational complexity of a flight.
    - Can appear suddenly and limit the time available for you to analyze.
    - Must be effectively managed to contain risk within acceptable levels.
11. Errors can be skill-based, decision-based, perceptual, or involve SRM/CRM.
12. Failing to carefully consider choices and reacting quickly and impulsively; failing to respond in a timely manner; reacting with resignation.
13. Adverse weather; Challenging airports surrounded by high mountains; Congested airspace; In-flight mechanical malfunction; Unanticipated runway closure; Communication failure.
14. B
15. Techniques to prevent or alleviate motion sickness are:
    - Avoid lessons in turbulent conditions.
    - Start with shorter flights and graduate to longer instruction periods.
    - Open fresh air vents.
    - Focus on objects outside the airplane.

16. Factors that can impair your fitness for flight are:

    - Stress
    - Fatigue
    - Noise
    - Alcohol
    - Drugs

17. A

18. 8 hours

SECTION D

1. C
2. H
3. G
4. A
5. E
6. B
7. C
8. I
9. D
10. F

CHAPTER 2

SECTION A

1. The major components of the airplane are:

    A. Fuselage

    B. Wings

    C. Powerplant

    D. Empennage

    E. Landing Gear

2. The monocoque design uses the skin to support almost all imposed loads while the semi-monocoque system uses a substructure attached to the airplane's skin to maintain the shape of the airframe and increase its strength.

3. The left aileron moves up and the right aileron moves down.

4. False

5. Trim devices aerodynamically help move a control surface, or maintain the surface in a desired position.

6. Conventional landing gear

7. Accessories mounted on, or connected to the engine generate electrical power, provide vacuum power for some of the flight instruments, and, in most single-engine airplanes, provide a source of heat for the pilot and passengers.

8. Modern airplane engineers favor composite materials for strength and sustainability. Composite materials provide:

    - A longer lifespan with less maintenance because they are less vulnerable to corrosion.
    - A stronger, smoother skin, with a lighter weight structure.
    - A reduction in drag and greater fuel efficiency.

9. C

10. The documents are:

    - Airworthiness certificate.
    - Registration certificate.
    - Radio station class license (if operating outside the U.S.).
    - Operating limitations (contained in the AFM/POH and on placards.
    - Weight and balance data.

11. C

12. A

13. Determine if the inoperative equipment is required by:

    - The VFR-day type certificate requirements prescribed in the airworthiness certification regulations.
    - FAR 91.205 for the specific kind of flight operation (e.g. day or night VFR) or by other flight rules for the specific kind of flight to be conducted.
    - The aircraft's equipment list or the kinds of operations equipment list (KOEL).
    - An airworthiness directive (AD).

SECTION B

1. The four-stroke operating cycle steps are:

    A. Power

    B. Intake

    C. Exhaust

    D. Compression

2. Lean the mixture as the airplane climbs.

3. The first indication of carburetor icing in an airplane with a fixed-pitch propeller is a decrease in engine RPM.

4. A sharp temperature drop occurs in a float-type carburetor due to fuel vaporization and decreasing air pressure in the venturi. If water vapor in the air condenses when the carburetor temperature is at or below freezing, ice can form. Because fuel injection systems eliminate fuel vaporization in a venturi and introduce the fuel directly into the hot engine cylinders, they are relatively free from the formation of induction icing.

5. A magneto is a self-contained, engine-driven unit that supplies electrical current to the spark plugs. It uses a permanent magnet to generate an electrical current completely independent of the aircraft's electrical system.

6. Detonation. Because detonation can occur when the engine overheats, if you suspect detonation while in flight you should attempt to lower the cylinder temperature. Methods include reducing power, enriching the fuel mixture, and lowering the nose to increase airspeed and cooling airflow around the engine.

7. C

8. True

9. The engine oil system performs many functions including lubricating the engine's moving parts, cooling the engine by reducing friction and removing some of the heat from the cylinders, providing a seal between the cylinder walls and pistons, and carrying away contaminants, which are removed as the oil passes through a filter.

10. Low

11. True

12. No. The ammeter should initially show a charge because the battery power used to start the engine is being replace.

SECTION C

1. Standard atmospheric pressure and temperature at sea level equals 29.92 in. Hg and 15°C, respectively.

2. The airspeed indicator is the only instrument which uses pitot pressure.

3. The V-speeds are:

   A. V_{S0}

   B. V_{S1}

   C. V_{FE}

   D. V_{NO}

   E. V_{NE}

4. Maneuvering speed (V_A)

5. C

6. D

7. B

8. A

9. No. If you fly from an area of high pressure to an area of low pressure without resetting your altimeter, the altimeter will sense the decrease in pressure as an increase in altitude. The altitude indicated on the altimeter will be higher than the true altitude of the airplane.

10. At altitudes above the point where the static ports became clogged, the airspeed indicator will indicate slower than actual because the trapped static pressure is higher than normal for that altitude. At altitudes lower than the point where the static ports became clogged, the airspeed will indicate faster than actual since the trapped static pressure is lower than normal for that altitude.

11. Pitch

12. True

13. The Attitude Heading Reference Systems (AHRS) provides attitude, heading, rate of turn, and slip/skid information.

14. C

15. The trend vector helps you make standard-rate turns.

16. B

17. A

18. C

19. If the PFD screen turns black, the PFD instruments should automatically display on the MFD in reversionary mode.

20. False

CHAPTER 3

SECTION A

1. B

2.

   A. Upwash

   B. Leading edge

   C. Trailing edge

   D. Downwash

   E. Flight path

   F. Relative wind

   G. Camber

   H. Chord line

   I. Angle of attack

3. According to Bernoulli's principle, the increase in speed of air on the top of an airfoil produces a drop in pressure and this lowered pressure is a component of total lift. In addition, the decrease in the speed of air on the bottom of an airfoil produces an increase in pressure, and this higher pressure is the other component of total lift. These pressure changes cause the airflow to curve downward, creating downwash behind the airfoil. Newton's third law of motion explains that the action of causing downwash results in a reaction of positive lift.

4. False

5. The aspect ratios are:

   A. 7

   B. 7

   C. 3.4

6. You can control lift by changing airspeed, changing angle of attack, or by employing high-lift devices such as trailing edge flaps.

7. Increase

8. It is preferable for the wing root to stall first. If the wingtips stall before the root, the disrupted airflow near the wingtip can reduce aileron effectiveness to such an extent that it may be impossible to control the airplane about its longitudinal axis.

9. Parasite drag normally is divided into three types; form drag, interference drag, and skin friction drag. Streamlining decreases form drag, and design features, such as wheel fairings and retractable landing gear, can reduce both form and interference drag. Skin friction drag can be minimized by eliminating protruding rivet heads, and employing a glossy, smooth finish to airplane surfaces.

10. Induced drag is caused by the downwash created by wingtip vortices formed when the wing is generating lift. As the air pressure differential between the upper and lower surfaces of the wing becomes greater with an increase in angle of attack, stronger vortices form and induced drag is increased. Because the wing is usually at a low angle of attack at high speeds, and a high angle at low speeds, a relationship between induced drag and speed can be determined. Induced drag is inversely proportional to the square of the speed.

11. One wingspan

SECTION B

1. The three axes of flight and types of movement are:

    A. Lateral

    B. Pitch

    C. Vertical

    D. Yaw

    E. Roll

    F. Longitudinal

2. B

3. A

4. C

5. Forward

6. On many training aircraft, an increase in thrust increases nose heaviness due to the placement of the thrustline above the center of gravity of the aircraft. This helps counteract the nose-up moment caused by the increased downwash resulting from an increase in power.

7. Decreases

8. Spiral instability

9. True

10. The basic guidelines for stall recovery are:

    • Decrease the angle of attack.

    • Smoothly apply maximum power.

    • Adjust the power as required. As the airplane recovers, maintain coordinated flight while adjusting the power to a normal level.

11. The general procedures for recovery from an erect spin are:

    • Move the throttle to idle.

    • Neutralize the ailerons.

    • Determine the direction of rotation.

    • Apply full opposite rudder.

    • Briskly apply elevator/stabilator to neutral or slightly forward of neutral.

    • As rotation stops, neutralize the rudders.

    • Gradually apply aft elevator/stabilator to return to level flight.

SECTION C

1. Thrust

2. Low airspeed, high power settings, and high angles of attack.

3. Left turning tendencies can be reduced, in part, by a rudder trim tab, an offset vertical stabilizer, or a horizontally canted engine.

4. Yes. Variations in weight do not affect the glide ratio of an airplane, but the heavier aircraft will sink faster, and reach the ground sooner. To travel the same distance as the lighter aircraft, a higher airspeed will need to be maintained by the heavier airplane. This higher airspeed, which corresponds to the best glide airspeed at that weight, increases ground speed and enables the heavier aircraft to cover the same horizontal distance even though its rate of descent is higher.

5. When an airplane is banked, lift is comprised of two components. The airplane turns because the horizontal component of lift creates a force directed inward toward the center of rotation.

6. Decrease airspeed

7. 59 knots

8. 66 knots

9. 110 knots

10. False

11. Potential energy from altitude, and kinetic energy from airspeed.

12. Reducing airspeed can demand more power than the engine can supply, or an unplanned reduction in power could result in an involuntary descent.

13. B

14. If you are below the glide path and your airspeed is too fast, the total energy is okay; potential energy is low; and kinetic energy is high.

CHAPTER 4

SECTION A

1. In an emergency situation, as PIC, you have the authority to deviate from any rule as required to handle the emergency.

2. If a deviation is necessary, you are required to submit a written report upon request of the FAA Administrator.

3. The most effective way to scan during daylight is through a series of short, regularly spaced eye movements, focusing for at least one second in each 10 degree sector. This method, which brings successive areas of the sky into the central visual field, is compatible with how the eyes function. Although two normal healthy eyes have a visual field of approximately 200 degrees, the area in which the eye can focus sharply and perceive detail is a relatively narrow cone (usually only about 10 degrees wide) directly in the center of the field of vision. Beyond this area, visual acuity decreases sharply in all directions.

4. False

5. Formation flight must be arranged prior to the flight by the pilot in command of each aircraft in the formation. To safely fly in formation is challenging so you should not attempt this type of flying without comprehensive instruction by a flight instructor with formation training experience.

6. B

7. C

8. B

9. A

10. A

11. A

12. C

13. A

14. B

15. C

16. Flight crewmembers are required to keep their safety belts and shoulder harnesses fastened during takeoffs and landings. Safety belts must stay fastened while enroute.

17. A three step process is recommended when exchanging flight controls. The pilot passing the controls should announce, *"You have the flight controls."* The pilot passing the controls should continue to fly until the pilot taking the controls acknowledges the exchange by saying, *"I have the flight controls."* A visual check is recommended to ensure that the other pilot actually has the controls. The pilot passing should then state, *"You have the flight controls."*

18. A

SECTION B

1. Runway numbers correspond to the runway's magnetic direction, rounded off to the nearest 10 degrees, with the last zero omitted. The number at the approach end of the runway corresponds to your heading when taking off or landing on that runway. For example, a runway that is labeled 9 on one end is labeled 27 on the opposite end.

2. A

3. Usually, a displaced threshold indicates that there are obstructions such as trees, powerlines, or buildings off the end of the runway. This might prevent a normal descent and landing on the beginning portion of the pavement. Although the pavement leading up to a displaced threshold may not be used for landing, it can be available for taxiing, the landing rollout, and takeoffs.

4. A closed runway is designated by a yellow X.

5. A

6. C

7. B

8. A

9. Take these actions to prevent a runway incursion:

   • Review the airport diagram and taxi routes.

   • Complete checklist items before taxi or while holding short.

   • Readback all clearances involving active runway crossing, hold short, or line up and wait instructions.

   • Do not become absorbed in other tasks, or conversation, while your airplane is moving.

   • Stop and ask for assistance or a progressive taxi at controlled airports.

   • In the runup area, position your airplane so you can see landing aircraft.

   • Monitor the appropriate frequencies and be alert for aircraft on other frequencies.

   • After landing, stay on the tower frequency until instructed to change frequencies.

   • During periods of reduced visibility, use your taxi/landing lights.

   • Report deteriorating or confusing airport markings, signs, and lighting.

   • Understand the required procedures if LAHSO is in effect.

10. B

11. A

12. C

13. To conduct LAHSO, you should have at least a private pilot certificate, have the published ALDs and runway slopes available, and know which runway LAHSO combinations provide acceptable landing distances with the existing conditions upon arrival.

14. False

15. C

16. E

17. F

18. B

19. To activate three-step pilot-controlled lighting, key the mic seven times on the specified frequency to turn all the lights on at maximum intensity. Key the mic five times for medium-intensity, and three times for the lowest intensity lighting. The mic must be keyed the required number of times within a period of five seconds.

20. B

SECTION C

1. 40°9 N — 122°15 W

2. 4,000 feet MSL. The contour line indicates that the elevation of the Conboy Lake National Wildlife Refuge is 2,000 feet MSL. To fly 2,000 feet above the surface over this special conservation area, you should fly at least 4,000 feet MSL.

3. D

4. C

5. A

6. E

7. B

8. False

9. 120.1 MHz

10. 544 feet MSL

11. True

12. Pounds Airport has services and fuel available during normal business hours.

13. The airport has a rotating beacon which normally operates from sunset to sunrise.

14. No

15. Fort Worth Flight Service

16. B

17. VFR sectional and terminal area charts are updated every 56 days.

SECTION D

1. Class G—clear of clouds; 1 SM

2. Class E—500 ft below, 1,000 ft above, 2,000 ft horizontal; 3 SM

3. Class E—1,000 ft below, 1,000 ft above, 1 SM horizontal; 5 SM

4. 8 NM wide

5. B

6. B

7. Class E

8. No, the Class D ceiling is 2,600 ft MSL.

9. 500 ft below, 1,000 ft above, 2,000 horizontal; 3 SM,

10. Class C

11. 3 SM, 500 ft below, 1,000 ft above, 2,000 horizontal

12. Student pilot certificate; establish radio contact; Mode C transponder/ADS-B (2020)

13. Class B

14. 3 SM, clear of clouds

15. Private pilot certificate or student pilot with a logbook endorsement (in some cases); ATC clearance; mode C transponder/ADS-B (2020)

16. Class E

17. C

18. B

19. 250 knots

20. A military operations area (MOA) is a block of airspace in which military training and other military maneuvers are conducted. MOAs usually have specified floors and ceilings for containing military activities. VFR aircraft are not prevented from flying through active MOAs, but it is wise to avoid them when possible.

21. A

22. Types of TFRs are:
    • Disaster/hazard TFR
    • Emergency air traffic rules
    • VIP TFR
    • Space flight operations
    • Air shows and sporting events
    • Special security instructions

23. TFRs are issued in NOTAMs that specify the dimensions, restrictions, and effective times. To determine if a TFR affects your flight, obtain NOTAMs from Flight Service during your online or phone briefing. You can also obtain a list of TFR NOTAMs with graphic depictions at tfr.faa.gov.

24. Rock your wings or flash your lights (night) to acknowledge. Then, match the heading and follow the aircraft.

CHAPTER 5

SECTION A

1. B

2. C

3. The limitations of radar include: bending of radar waves (anomalous propagation), which results in ground clutter; reflection of waves by dense objects, such as heavy clouds, precipitation, ground obstacles, or mountains, which can weaken or eliminate the display of targets at a greater range; and screening by high terrain, which obstructs relatively low-altitude aircraft.

4. A

5. VFR—1200; hijacking—7500; two-way communication failure—7600; emergency—7700

6. False

7. ATC asks you to "ident" to cause the transponder return to blossom on the ATC display for a few seconds, enabling the controller to easily identify you.

8. B

9. A

10. C

11. If the sky condition and visibility are absent from the ATIS broadcast, it means that the visibility is 5 miles or more and the cloud ceiling is 5,000 feet or higher.

12. B

13. B

14. A

SECTION B

1. B

2. *Cessna six four niner Sierra Papa*

3. *C*

4. 2100Z = 1300 (T/O CST) + 2:00 (time enroute) − 2:00 (time difference CST to PST) + 8:00 (PST to UTC)

5. 1700Z = 0700 (T/O PDT) + 3:00 (time enroute) + 2:00 (time difference PDT to CDT) + 5:00 (CDT to UTC)

6. A

7. B

8. B

9. Set your transponder code to 7600. Remain outside or above the Class D airspace until you have determined the direction and flow of traffic. Then, join the airport traffic pattern and maintain visual contact with the tower to receive light signals. In the daytime, acknowledge light signals by rocking your wings, and at night, by blinking your landing or navigation lights.

10. Light signals:
    • Steady green: ground—cleared for takeoff; flight—cleared to land
    • Flashing green: ground—cleared to taxi; flight—return for landing
    • Steady red: ground—stop; flight—give way to other aircraft and continue circling
    • Flashing red: ground—taxi clear of landing area (runway) in use; flight—airport unsafe, do not land
    • Flashing white: ground—return to starting point on airport; flight—no assigned meaning
    • Alternating red and green: ground and flight—exercise extreme caution

11. The FARs require that the ELT battery must be replaced, or recharged if the battery is rechargeable, after one-half of the battery's useful life or if the transmitter has been used for more than one cumulative hour.

SECTION C

1. You can find flight information from:
    • Faa.gov
    • Electronic flight bag (EFB)
    • GPS unit
    • Pilot supply stores.

2. B

3. 3 nautical miles east; 2,777 feet MSL

4. Runway 12L/30R is 8,001 feet long and 150 feet wide.

5. Yes, you will be able to obtain fuel. To convert local time to UTC, add 6 hours to obtain 1600Z as your estimated time of arrival. (The identification and location information at the top of the chart indicates that to convert local to UTC, you must subtract 6 hours; UTC -6). The Services section shows that 100LL is available and the Airport Remarks indicate that the airport is attended from 1200 to 0500Z.

6. ASOS—118.425; UNICOM—123.0; LBF VORTAC—117.4

7. C

8. NOTAM(D) (distant NOTAM) information is disseminated for all navigational facilities that are part of the U.S. airspace system, and all public use airports, seaplane bases, and heliports listed in the Chart Supplements. FDC NOTAMs, issued by the National Flight Data Center, contain regulatory information, such as TFRs or amendments to aeronautical charts.

9. C

10. Advisory circulars (ACs) provide nonregulatory guidance and information in a variety of subject areas and explain methods for complying with the FARs.

CHAPTER 6

SECTION A

1. The four layers of the atmosphere are:

   A. Troposphere

   B. Stratosphere

   C. Mesosphere

   D. Thermosphere

2. The troposphere

3. As air is heated, it expands and becomes less dense than the surrounding air. As air cools, its molecules become packed more closely together, making it denser and heavier than warm air. As a result, the cool, heavy air tends to sink and replace warmer, rising air.

4. Unequal heating of the earth's surface

5. False

6. The amount of deflection an object experiences due to Coriolis force is a function of distance traveled, position on the earth's surface (latitude), and speed of the object.

7. In the upper atmosphere, pressure gradient force and Coriolis force cause wind to flow roughly parallel to the isobars. However, within approximately 2,000 feet of the ground, friction caused by the earth's surface slows the movement of air. This reduces the effect of Coriolis force and causes the pressure gradient force to divert the wind toward the lower pressure area.

8. D

9. A

10. B

11. C

SECTION B

1. 2°C (3.5°F) per 1,000 feet

2. When air rises, it expands to a larger volume. As the distance between molecules increases, the temperature of the air lowers. As a result, when a parcel of air rises, pressure decreases, volume increases, and temperature decreases. When air descends, pressure increases, volume decreases, and temperature increases.

3. Evaporation and sublimation

4. 2,000 feet AGL

5. The clouds are:

   A. Cirrus

   B. Altocumulus

   C. Stratus

   D. Towering Cumulus

6. False

7. For precipitation to occur, water or ice particles must grow in size until they can no longer be supported by the atmosphere.

8. Rain generally falls at a relatively steady rate and stops gradually. On the other hand, rain showers generally begin, change intensity, and stop suddenly.

9. A stable air mass generally exhibits widespread stratiform clouds, restricted visibility, smooth air, and steady rain or drizzle.

10. B

11. C

12. A

SECTION C

1. Thunderstorm formation requires unstable conditions, a lifting force, and high moisture levels.

2. As the vertical motions slow near the top of the storm, winds tend to spread the cloud horizontally, forming the anvil shape. Because the anvil's shape is formed by upper-level winds, it points in the approximate direction of the storm's movement.

3. If you encounter turbulence during flight, slow the airplane to maneuvering speed or less, maintain a level flight attitude, and accept variations in altitude. If you expect turbulent or gusty conditions during an approach to a landing, you should consider flying a power-on approach at an airspeed slightly above the normal approach speed.

4. False

5. The most dangerous condition for landing is a light, quartering tailwind because it can move the upwind vortex of a landing aircraft over the runway and forward into the touchdown zone.

6. C

7. Mountain wave turbulence

8. In humid climates where the bases of convective clouds tend to be low, microbursts are associated with a visible rainshaft. In drier climates, the higher thunderstorm cloud bases result in the evaporation of the rainshaft. The only visible indications under these conditions might be virga at the cloud base and a dust ring on the ground.

9. B

10. C

11. A

12. A

CHAPTER 7

SECTION A

1. False

2. Short periods of time

3. Numerical weather prediction uses mathematical equations which relate atmospheric conditions with other variables. The system of equations represents the physical laws that govern the behavior of the atmosphere. The computer applies the model to the current atmospheric conditions to forecast minutes in the future. The process is repeated many times to create a prediction for the next day or two.

4. A

5. Large storms and major heat waves

SECTION B

1. C

2. C

3. A

4. The actual temperature is 6.7°C and dewpoint is 6.1°C.

5. 1008.6 mb (hPa)

6. True

7. "TEMPO" in a TAF means conditions are expected to last less than an hour.

8. A

9. 2SM visibility, light rain, and overcast at 800' AGL.

10. C

11. False

12. Winds are from 250° at 110kts. Temperature -15°C.

13. True

14. A

15. B

SECTION C

1. C

2. C

3. C

4. C

5. A

6. K4V1

7. KC08

8. B

9. Light snow

10. 30.28 in Hg

11. 5,000 broken; 3 SM

12. KALS, KRTN, KSKX, KSAF

13. Weather phenomena—thunderstorms; visibility—below 1 SM to 5 SM; winds—from the east at 10 knots; advisories—AIRMET for IFR conditions

14. New Mexico

15. Snow and snow showers

16. B

17. FL240

18. B

19. B

20. FL160

21. B

SECTION D

1. When you request a briefing, identify yourself as a pilot flying VFR and provide the briefer with your aircraft number or your name, aircraft type, departure airport, route of flight, destination, flight altitude(s), estimated time of departure (ETD), and estimated time enroute (ETE).

2. The three types of briefings are:

   • Standard briefing—you are planning a trip and have not obtained preliminary weather or a previous briefing.

   • Abbreviated briefing—you need only one or two specific items or would like to update weather information from a previous briefing or other weather sources.

   • Outlook briefing—your proposed departure time is six or more hours in the future, to obtain forecast information that helps you make an initial judgment about the feasibility of your flight.

3. A

4. False

5. The NEXRAD radar image that you view on your cockpit display is not real time and can be up to 5 minutes old. Even small-time differences between the age indicator and actual conditions can be important for safety of flight, especially when considering fast-moving weather hazards, quickly developing weather scenarios, and/or fast-moving aircraft. At no time should you use the images as storm penetrating radar nor to navigate through a line of storms.

CHAPTER 8

SECTION A

1. In the Performance section of your airplane's POH

2. As density altitude increases, engine horsepower decreases for normally aspirated engines since the actual amount of air to support combustion has decreased. In addition, both the wing and the propeller lose efficiency in the thinner air.

3. True

4. Headwind—17 knots; crosswind—10 knots

5. B

6. Ground roll—1,665 feet; total distance to clear a 50-foot obstacle—3,300 feet

7. False

8. Time, fuel, and distance

9. B

10. 495 nautical miles with reserves; 555 nautical miles without reserves.

11. False

SECTION B

1. Datum or reference datum
2. Fuel which cannot be drained or safely used in flight
3. A
4. C
5. C
6. Lower
7. False
8. 75,000 inch-pounds
9. 38½ gallons
10. 99.4 inches
11. 102 pounds
12. C
13. A
14. B

SECTION C

1. 88 knots
2. 12.4 gallons per hour
3. 107 knots; Indicated rather than calibrated airspeed was used.
4. 21.8 gallons
5. 129 nautical miles
6. 147°; 90 knots
7. 81 minutes
8. 248° at 27 knots
9. 001°; 107 knots

CHAPTER 9

SECTION A

1. Navigating by reference to landmarks is called pilotage. When flying by pilotage, you match as much as possible of what you see on the ground to the features on your chart to determine the direction to fly. To navigate by dead reckoning, you determine a heading to fly and you calculate time, speed, and distance for your flight. When dead reckoning is combined with pilotage, you select checkpoints based on landmarks and record time, speed, and distance between checkpoints on a nav log. During flight, you refer to the chart and the nav log date to monitor your progress.

2. True

3. The following apply to this example:
   - True Course: 081°
   - True Heading: 086°
   - Magnetic Variation: +4°
   - Magnetic Heading: 090°
   - Compass Heading: 091°

4. 085°

5. Nav Logs typically include:
   - Checkpoints (fixes)
   - Courses
   - Cruising altitudes
   - Climb airspeed
   - True airspeed
   - True course
   - Wind correction angle
   - True heading
   - Magnetic variation
   - Magnetic heading
   - Compass heading
   - Distance between checkpoints and remaining distance
   - Groundspeeds (estimated and actual)
   - ETE and ATE between checkpoints
   - Fuel consumption
   - Communication frequencies

6. Factors when choosing a cruising altitude are:
   - Height above the terrain and obstructions
   - Effect of winds on your groundspeed and the performance of your airplane
   - Visibility of checkpoints
   - Radio range and reception
   - Options in the event of an emergency.
   - VFR cruising altitude rule based on magnetic course

7. C
8. Day—30 minutes; night—45 minutes
9. A
10. C
11. B

SECTION B

1. Course guidance and distance information
2.
   A. Omnibearing selector (OBS) or course selector
   B. Course deviation indicator (CDI)
   C. Course index
   D. TO/FROM indicator
   E. Reciprocal course index.
3. To avoid reverse sensing
4. C
5. True
6. 2 NM
7. C
8. C
9. A
10. C
11. B

SECTION C

1. The three segments are:

   * Space segment—at least 24 GPS satellites in 6 equally-spaced orbital planes that ensure users can view at least four satellites from any point on the planet. Each satellite continuously broadcasts radio signals used by GPS receivers to calculate accurate position information.

   * Control segment—a global network of ground facilities that track the GPS satellites, monitor and analyze their transmissions, and send commands and data to the constellation.

   * User segment—the GPS receivers that receive the signals from the GPS satellites and use the transmitted information to calculate the user's three-dimensional position.

2. B

3. B

4. A

5. Two common errors associated with using a moving map are: using the moving map as a primary navigation instrument and overreliance on the moving map leading to complacency.

6. Types of waypoints include:

   * Waypoints in the GPS database, which normally include VORs, NDBs, intersections, and airports.

   * User waypoints created by entering latitude and longitude coordinates, selecting a position on the map using a cursor or pointer, referencing a bearing/distance from an existing waypoint, or capturing your present position.

   * VFR waypoints (five letters beginning with the letters "VP") shown on sectional and terminal area charts to provide a supplementary navigation tool.

7. B

8. A

9. Auto-sequencing of waypoints occurs when you program a flight plan and the receiver senses when the airplane passes a waypoint and automatically cycles to the next waypoint.

10. Indications of waypoint passage are:

    * A message advises you to begin the turn to the new course so the airplane completes the turn on the course centerline.

    * Navigation information for the next leg is displayed.

    * The moving map shows the active leg in magenta.

    * Navigation indications automatically change to the new course.

    * The flight plan window displays the next leg as the active leg.

CHAPTER 10

SECTION A

1. Cones are concentrated in the center of the retina in a slight depression known as the fovea. The cones are sensitive to bright light and colors. The cones also allow you to see fine detail, but they do not function well in darkness. The rods, which are dispersed outside the foveal area, react to low light and are responsible for most of your night and peripheral vision. The rods do not perceive color or detail well.

2. B

3. Autokinesis

4. Flying too low on an approach.

5. Awareness of your body's position is a result of input from three primary sources: vision, the vestibular system, and your kinesthetic sense.

6. During a prolonged constant-rate turn, your vestibular system may not sense any motion because the fluid in the semicircular canals eventually reaches equilibrium and the hair cells are no longer deflected.

7. A

8. Motion sickness

9. B

10. A

11. D

12. C

13. B

14. Suggest that the passenger slow his or her breathing rate, breathe into a paper bag, or talk aloud.

15. Hypothermia

SECTION B

1. The National Transportation Safety Board (NTSB) is an independent Federal agency responsible for investigating every U.S. civil aviation accident and issuing safety recommendations aimed at preventing future accidents. In addition, the NTSB maintains the government's database on civil aviation accidents and conducts research regarding safety issues of national significance. Examining NTSB accident and incident reports can increase your awareness of the factors which affect a pilot's decision-making ability and help you learn to recognize the chain of events which lead to an accident.

2. Aviation Safety Reporting System (ASRS)

3. General steps in the ADM process include to:

   * Recognize a change.

   * Define the problem.

   * Choose a course of action.

   * Implement your decision.

   * Evaluate the outcome.

4. C

5. You can refer to a personal checklist that specifies limitations such as a maximum amount of time that may pass between flights without refresher training or review of the regulations and POH; your own weather minimums, which may be higher than those listed in the FARs, and the maximum amount of crosswind that you are comfortable with. After you have reviewed your personal limitations, you can use the I'm Safe Checklist to further evaluate your fitness for flight.

6. Operational pitfalls include:
   - Peer pressure.
   - Mind-set.
   - Get-there-itis.
   - Duck under syndrome.
   - Scud running.
   - Continuing VFR flight into instrument conditions.
   - Getting behind the aircraft.
   - Loss of situational awareness.
   - Operating without adequate fuel reserves.
   - Descent below the minimum enroute altitude.
   - Flying outside the envelope.
   - Neglect of flight planning, preflight inspections, and checklists.

7. C

8. B

9. A

10. E

11. D

12. The examples should include risk factors associated with these categories:
   - Pilot—training, experience, fitness.
   - Passengers—experience, fitness, flexibility.
   - Plane—airworthiness, performance, configuration.
   - Programming—avionics airworthiness, operation, and configuration.
   - Plan—airport conditions, terrain, airspace, mission, weather.

13. C

14. The aircraft POH; aeronautical charts; flight computer or GPS flight planning functions; the *Chart Supplement,* an EFB, or airport information pages on the MFD; ATC assistance including Flight Service specialists; another pilot or passenger on board; navigation equipment such as VOR or GPS; an autopilot; and your own ingenuity, knowledge, and skills.

15. Taxi, takeoff, and landing

16. C

17. A

18. Controlled flight into terrain (CFIT)

19. An automation surprise occurs when the system does something you do not expect, or fails to do something you do expect. Prevent an automation surprise by anticipating and monitoring mode indications. Cope with an automation surprise by using a lower equipment operating level, such as by disengaging the autopilot.

20. Root cause analysis (RCA) is the process of examining a problem or why something went wrong to determine how to prevent the issue from reoccurring in the future.

21. Causes of errors include:
   - Insufficient training and experience.
   - Psychological effects, such as stress and emotion, expectation bias, and personality.
   - Inadequate flight planning and preparation.
   - Physiological effects.

22. To manage a startle event:
   - Fly—control your flight path; do not hurry.
   - Focus—control your emotions; observe, identify, and confirm the situation.
   - Act—manage the event by performing the ADM process, implementing your decision, and communicating with ATC and crew.

CHAPTER 11

SECTION A

1. The six primary steps in the flight planning process are:
   - Prepare a flight overview.
   - Develop the route.
   - Obtain a weather briefing.
   - Complete the nav log.
   - File the flight plan.
   - Perform preflight tasks.

2. C

3. C

4. C

5. Specify whether the flight is VFR or IFR. Provide the aircraft registration number and type, the departure airport, the intended flight route, and the destination airport. Include the intended cruise altitude, the estimated time of departure, and estimated time enroute.

6. A

7. 700 feet

8. $072° = 080°$ (TC) $- 4°$(WCA) $- 4°$E (magnetic variation) $± 0°$ (compass deviation)

9. A

10. 1.0 gallon

11. Flight plan information includes:
   - Type of flight: VFR
   - True airspeed: 120
   - Altitude: 5,500
   - Time enroute: 42 minutes (plus any additional time for traffic pattern and landing at the destination)
   - Fuel on board: 5 hours (approximate) 40 gal ÷ 7.5 GPH = 5.3 hrs

SECTION B

1. Tune comm and nav frequencies, program a flight plan if you are using GPS navigation, set course selectors on the navigation indicators, and set heading and altitude bugs as reminders.

2. B

3. B

4. Conditions that might requiring a diversion are:
   - Deteriorating weather.
   - Turbulence.
   - Low fuel.
   - System or equipment malfunctions.
   - Pilot or passenger medical problems.
   - Approaching nightfall.

5. To divert to an alternate airport:
   - Fly the airplane.
   - Turn toward the airport of intended landing.
   - Compute course, time, speed, and distance using rule of thumb calculations or the Nearest and Direct-To functions of your GPS equipment.
   - Prioritize tasks and use all available resources, such as the autopilot and POH.
   - Contact ATC to inform them of your situation and to request assistance.
   - Use the 5P Check to maintain situational awareness and be prepared to adapt to changes required by the conditions.

6. Calculate the distance:
   - 7,500 ft − 3,500 ft = 4,000 ft
   - 4,000 ft ÷ 500 ft/min = 8 min
   - 110 NM/hr × 8 min ÷60 min = 15 NM

7. The Before-Landing briefing includes:
   - Airport information and weather conditions
   - Active runway
   - Terrain and obstacles
   - Airport elevation and pattern altitude
   - Traffic pattern entry
   - Ground operations

8. B

APPENDIX B

ABBREVIATIONS

A

AC — advisory circular

AC — alternating current

AC — convective outlook (weather)

AD — airworthiness directive

ADF — automatic direction finder

ADIZ — air defense identification zone

ADS-B — automatic dependent surveillance-broadcast

ADM — aeronautical decision making

A/FD — airport/facility directory

AFM — aircraft flight manual

AFSS — automated flight service station, now *Flight Service*

AGL — above ground level

AHRS — attitude and heading reference system

AIM — *Aeronautical Information Manual*

AIRMET — airman's meteorological information

ALS — approach light system

AM — amplitude modulation

AME — aviation medical examiner

ARINC — Aeronautical Radio, Incorporated

ARP — airport reference point

ARSR — air route surveillance radar

ARTCC — air route traffic control center

ARTS — automated radar terminal system

ASOS — automated surface observing system

ASR — airport surveillance radar

ATA — actual time of arrival

ATC — air traffic control

ATCRBS — ATC radar beacon system

ATD — actual time of departure

ATD — along track distance

ATD — aviation training device

ATE — actual time enroute

ATIS — automatic terminal information service

ATP — airline transport pilot

ATP CTP — airline transport pilot certification training program.

ATS — air traffic service

AWC — Aviation Weather Center

AWOS — automated weather observing system

AWSS — automated weather sensor system

AWW — alert severe weather watch

B

BFO — beat frequency oscillator

BHP — brake horsepower

BRG — bearing

C

CAP — Civil Air Patrol

CAS — calibrated airspeed

CAT — clear air turbulence

CBTA — competency-based training and assessment

CCFP — Collaborative Convective Forecast Product

CDI — course deviation indicator

CFI — certificated flight instructor

CFIT — controlled flight into terrain

CFR — Code of Federal Regulations

CG — center of gravity

CHT — cylinder head temperature

CL — center of lift

C_L — coefficient of lift

C_{Lmax} — maximum coefficient of lift

CNF — computer navigation fix

CO — carbon monoxide

CO_2 — carbon dioxide

COM — communication competency

CRM — crew resource management

CTAF — common traffic advisory frequency

CW — continuous wave

CWA — center weather advisory

D

DA — density altitude

DALR — dry adiabatic lapse rate

DC — direct current

DCS — decompression sickness

DG — directional gyro

DME — distance measuring equipment

DOD — Department of Defense

DHS — Department of Homeland Security

DVFR — defense visual flight rules

E

EFB — electronic flight bag

ELT — emergency locator transmitter

EGT — exhaust gas temperature

ETA — estimated time of arrival

ETD — estimated time of departure

ETE — estimated time enroute

F

FAA — Federal Aviation Administration

FARs — Federal Aviation Regulations

FBO — fixed base operator

FCC — Federal Communications Commission

FD — winds and temperatures aloft forecast

FDC — Flight Data Center

FIS-B — flight information service-broadcast

FL — flight level

FM — frequency modulation

FMS — flight management system

FPA — airplane flight path management — automation competency

FPA — flight path angle

FPM — airplane flight path management — manual control competency

f.p.m. — feet per minute

FRZ — flight restricted zone

FSDO — Flight Standards District Office

FSS — flight service station or Flight Service

ft/min — feet per minute

G

G — gravity; unit of measure for acceleration

gal/hr — gallons per hour

GCO — ground communication outlet

GLONASS — global navigation satellite system

GNSS — global navigation satellite system

g.p.h. — gallons per hour

GPS — global positioning system

GPWS — ground proximity warning system

GS — groundspeed

H

HF — high frequency

HIRLs — high intensity runway lights

HMR — hazardous materials regulations

hPa — hectoPascal

HSI — horizontal situation indicator

HVOR — high altitude VOR

Hz — hertz

I

IAS — indicated airspeed

ICAO — International Civil Aviation Organization

IFR — instrument flight rules

ILS — instrument landing system

IMC — instrument meteorological conditions

in. Hg — inches of mercury

INS — inertial navigation system

IOAT — indicated outside air temperature

IR — IFR military training route

IR — infrared

ISA — International Standard Atmosphere

IVSI — instantaneous vertical speed indicator

K

kg — kilogram

KCAS — knots calibrated airspeed

kHz — kilohertz

KIAS — knots indicated airspeed

km — kilometer

KNO — application of knowledge competency

KTAS — knots true airspeed

kts — knots

kw — kilowatt

kwh — kilowatt hour

L

L/MF — low/medium frequency

LAA — local airport advisory

LAHSO — land and hold short operations

LAT — latitude

L/D$_{max}$ — maximum lift/drag ratio

LF — low frequency

LIRLs — low intensity runway lights

LLWAS — low level wind shear alert system

LONG — longitude

LOP — line of position

LORAN — long range navigation

LTW — leadership and teamwork competency

LVOR — low altitude VOR

M

M — Mach

mb — millibar

MB — magnetic bearing

MC — magnetic course

MEF — maximum elevation figure

MEL — minimum equipment list

METAR — aviation routine weather report

MH — magnetic heading

MHz — megahertz

MIRLs — medium intensity runway lights

MOA — military operations area

m.p.h. — miles per hour

MSAW — minimum safe altitude warning

MSL — mean sea level

MTR — military training route

MULTICOM — multiple communication frequency used at airports without a tower, FSS, or UNICOM

MVFR — marginal VFR

N

NA — not authorized

NAS — National Airspace System

NASA — National Aeronautics and Space Administration

NAVAID — navigational aid

NDB — nondirectional radio beacon

n.m. — nautical miles

NM — nautical miles

NMC — National Meteorological Center

NOAA — National Oceanic and Atmospheric Administration

NORDO — no radio

NOS — National Ocean Service

NOTAM — notice to air missions

NPRM — notice of proposed rule making

NSA — national security area

NWS — National Weather Service

O

OAT — outside air temperature

OB — observable behavior

OBS — omnibearing selector

OTS — out of service

P

PA — pressure altitude

PAPI — precision approach path indicator

PAVE — pilot, aircraft, environment, external pressures

PCL — pilot controlled lighting

PF — pilot flying

PFD — primary flight display

PIC — pilot in command

PIM — pilot information manual

PIREP — pilot weather report

PLASI — pulse light approach slope indicator

PM — pilot monitoring

POH — pilot's operating handbook

PRO — application of procedures competency

PSD — problem solving — decision making competency

p.s.i. — pounds per square inch

PTS — practical test standards

PVASI — pulsating visual approach slope indicator

R

RA — resolution advisory

RAF — Research Aviation Facility

RAREP — radar weather report

RB — relative bearing

RCA — root cause analysis

RCC — rescue coordination center

RCLS — runway centerline light system
RCO — remote communications outlet
REIL — runway end identifier lights
RMI — radio magnetic indicator
RMK — remarks
RNAV — area navigation
r.p.m. — revolutions per minute
RPM — revolutions per minute
RRL — runway remaining lights
RVR — runway visual range

S

SAR — search and rescue
SAW — situation awareness and management of information competency
SCATANA — Security Control of Air Traffic and Air Navigation Aids
SD — radar weather report
SFRA — special flight rules area
SIGMET — significant meteorological information
SLP — sea level pressure
s.m. — statute miles
SM — statute miles
SPECI — non-routine (special) aviation weather report
SRM — single-pilot resource management
SSV — standard service volume
SVFR — special visual flight rules

T

TA — traffic advisory
TAA — technically advanced aircraft
TACAN — tactical air navigation
TAF — terminal aerodrome forecast
TAS — true airspeed
TC — true course
TCAS — traffic alert and collision avoidance system
TCU — towering cumulonimbus
TDWR — terminal Doppler weather radar
TEM — threat and error management
TFR — temporary flight restriction
TH — true heading
TRACON — terminal radar approach control facilities
TRSA — terminal radar service area
TSA — Transportation Security Administration
TSO — technical standard order
TVOR — terminal VOR

U

UA — pilot report
UAS — undesired aircraft state
UCAR — University Corporation for Atmospheric Research
UHF — ultra high frequency
UNICOM — universal communication frequency for airport communication station
UTC — Coordinated Universal Time (Zulu time)
UUA — urgent pilot report
UWS — urgent weather SIGMET

V

V_A — design maneuvering speed
VAFTAD — volcanic ash transport and dispersion chart
VASI — visual approach slope indicator
V_{FE} — maximum flap extended speed
VFR — visual flight rules
VHF — very high frequency
V_{LE} — maximum landing gear extended speed
V_{LO} — maximum landing gear operating speed
VMC — visual meteorological conditions
VOR — VHF omnidirectional receiver
VOR/DME — collocated VOR and DME
VORTAC — collocated VOR and TACAN
VOT — VOR test facility
V_R — rotation speed
VR — VFR military training route
VSI — vertical speed indicator
V_{S0} — stalling speed or minimum steady flight speed in the landing configuration
V_{S1} — stalling speed or minimum steady flight speed obtained in a specified configuration
V_X — best angle of climb speed
V_Y — best rate of climb speed

W

WA — AIRMET
WAAS — wide area augmentation system
WAC — world aeronautical chart
WCA — wind correction angle
WFO — Weather Forecast Office
WH — hurricane advisory
WLM — workload management competency

WPT — waypoint
WS — SIGMET
WSP — weather systems processor
WST — convective SIGMET
WW — severe weather watch bulletin
wx — weather

Z

Z — Zulu time (UTC)

Numerals

5P — pilot, passengers, plane, programming, plan

APPENDIX B ■ **Abbreviations**

APPENDIX C

GLOSSARY

ABSOLUTE ALTITUDE — Actual height above the surface of the earth, either land or water.

ABSOLUTE CEILING — The altitude where a particular airplane's climb rate reaches zero.

ADIABATIC COOLING — A process of cooling air through expansion, as when air moves up a slope, expands with the reduction of atmospheric pressure and cools as it expands.

ADIABATIC HEATING — A process of heating dry air through compression, as when air moves down a slope and is compressed, which results in an increase in temperature.

ADVECTION FOG — Fog resulting from the movement of warm, humid air over a cold surface.

ADMINISTRATOR — The Federal Aviation Administrator or any person to whom he has delegated his authority in the matter concerned.

AERONAUTICAL DECISION MAKING (ADM) — A systematic approach to the mental process used by aircraft pilots to consistently determine the best course of action in response to circumstances.

AIRPLANE FLIGHT PATH MANAGEMENT — AUTOMATION (FPA) — One of the nine CBTA competencies. To master FPA, you must be able to engage the autopilot to maintain the correct flightpath for the phase of flight by selecting the appropriate level and mode.

AIRPLANE FLIGHT PATH MANAGEMENT — MANUAL CONTROL (FPM) — One of the nine CBTA competencies. To master FPM, you must be able to control the airplane through manual handling, practice basic aircraft control, detect deviations, and take appropriate action with manual control of the aircraft.

AIRPORT/FACILITY DIRECTORY (A/FD) — Airport, communication and NAVAID information from the Chart Supplement.

AGONIC LINE — Line along which the variation between true and magnetic values is zero.

AIR DENSITY — Mass per unit volume of air. Dense air has more molecules per unit volume than less-dense air. Air density decreases with altitude above the surface of the earth and with increasing temperature.

AIR ROUTE TRAFFIC CONTROL CENTER (ARTCC) — A facility established to provide air traffic control service to aircraft operating on IFR flight plans within controlled airspace, principally during the enroute phase of flight.

AIR TRAFFIC CONTROL (ATC) — An FAA service to enable safe, orderly, and expeditious flow of air traffic.

AIR MASS — An extensive body of air having fairly uniform properties of temperature and moisture.

AIRLINE TRANSPORT PILOT CERTIFICATION TRAINING PROGRAM — An advanced-level flight training program that is required by the FAA prior to applying for an FAA Airline Transport Pilot knowledge test.

AIRMET — In-flight weather advisory concerning moderate icing, moderate turbulence, sustained winds of 30 knots or more at the surface, and widespread areas of ceilings less than 1,000 feet and/or visibility less than 3 miles.

ALERT AREA — Special use airspace which may contain a high volume of pilot training activities or an unusual type of aerial activity.

ALTIMETER — A flight instrument that indicates altitude by sensing pressure changes.

ALTIMETER SETTING — Station pressure reduced to sea level. To compensate for variations in atmospheric pressure, you adjust the scale in the barometric pressure setting window to the altimeter setting.

ALTITUDE — Height expressed in units of distance above a reference plane, usually above mean sea level or above ground level.

AMYGDALA HIJACK — A fight or flight response caused by startle events. The amygdala hijacks, or shuts down the pre-frontal cortex, which contains working memory, judgment, planning sequencing or activity, abstract reasoning, and the ability to divide attention.

ANGLE OF ATTACK — Angle between the airfoil chord line and the relative wind.

ANNUAL INSPECTION — Recurring inspection that must occur every 12 calendar months for an aircraft to maintain its airworthiness.

ANGLE OF INCIDENCE — The angle between the chord line of the wing and the longitudinal axis of the airplane.

APPLICATION OF KNOWLEDGE (KNO) — One of the nine CBTA competencies. This competency requires applying knowledge to perform safe, effective ground and flight operations, and to make effective decisions.

APPLICATION OF PROCEDURES AND COMPLIANCE WITH REGULATIONS (PRO) — One of the nine CBTA competencies. PRO promotes safety by ensuring you fly the airplane in accordance with the FARs, aircraft manufacturer recommendations, and flight school guidelines.

ARM — The distance from the reference datum at which a weight may be located. Used in weight and balance calculations to determine moment.

ASPECT RATIO — Span of a wing divided by its average chord.

AUTOMATED SURFACE OBSERVATION SYSTEM (ASOS) — National Weather Service airport reporting system that provides comprehensive surface observations every minute via digitized voice broadcasts and textual reports.

AUTOMATED WEATHER OBSERVING SYSTEM (AWOS) — FAA-funded airport weather system consisting of one or more sensors, that provides surface observations via digitized voice broadcasts. AWOS systems range from providing only an altimeter setting to full automated surface observation system (ASOS) capability.

AUTOMATED WEATHER SENSOR SYSTEM (AWSS) — FAA-funded airport weather observing system with the same capabilities as automated surface observation system (ASOS).

AUTOMATIC DIRECTION FINDER (ADF) — An aircraft radio navigation system which senses and indicates the direction to an L/MF nondirectional radio beacon (NDB) or commercial broadcast station.

AUTOMATIC DEPENDENT SURVEILLANCE-BROADCAST (ADS-B) — A system that incorporates GPS, aircraft transmitters and receivers, and ground stations to provide pilots and ATC with specific data about the position and speed of aircraft. Two forms of ADS-B equipment apply to aircraft—ADS-B Out and ADS-B In.

AUTOMATIC TERMINAL INFORMATION SERVICE (ATIS) — The continuous broadcast of recorded noncontrol information in selected terminal areas. Its purpose is to improve controller effectiveness and to relieve frequency congestion by automating the repetitive transmission of essential but routine information.

AUTOMATION MANAGEMENT — The usage of effective techniques to manage distractions and increase situational awareness when using advanced flight deck avionics.

AUTOMATION SURPRISE — Situation that occurs when the system does something unexpected, or fails to do something that is expected, which can temporarily confuse the pilot.

AVIATE NAVIGATE COMMUNICATE — A reminder that when startled, or faced with a non-normal situation, to Aviate—above all else, fly the airplane and keep it in the air; Navigate—figure out where you are, and where you need to go; and Communicate—call for help, or state your intentions on the radio.

AVIATION WEATHER CENTER (AWC) — The part of the NWS that is designed to provide direct online weather services to pilots.

BASIC RADAR SERVICE — A radar service for VFR aircraft which includes safety alerts, traffic advisories, and limited radar vectoring, as well as aircraft sequencing at some terminal locations.

BEARING — The horizontal direction to or from any point, usually measured clockwise from true north (true bearing), magnetic north (magnetic bearing), or some other reference point, through 360°.

BEST ANGLE-OF-CLIMB AIRSPEED (V_X) — The airspeed that produces the greatest gain in altitude for horizontal distance traveled.

BEST RATE-OF-CLIMB AIRSPEED (V_Y) — The airspeed that produces the greatest gain in altitude per unit of time.

BLAST PAD — An area associated with a runway where propeller blast can dissipate without creating a hazard to others. It cannot be used for landing, takeoffs, or taxiing.

BRACKETING — A navigation technique which uses a series of turns into a crosswind to regain and maintain the desired course.

CALIBRATED AIRSPEED (CAS) — Indicated airspeed of an aircraft, corrected for installation and instrument errors.

CALIBRATED ALTITUDE — Indicated altitude corrected to compensate for instrument error.

CAMBER — The curve of an airfoil section from the leading edge to the trailing edge.

CATEGORY — (1) As used with respect to the certification, ratings, privileges, and limitations of airmen, means a broad classification of aircraft (airplane, rotorcraft, glider, lighter-than-air, and powered-lift). (2) As used with respect to the certification of aircraft, means a grouping of aircraft by intended use or operating limitations (transport, normal, utility, acrobatic, limited, restricted, experimental, and provisional).

CEILING — The height above the earth's surface of the lowest layer of clouds that is reported as broken or overcast or the vertical visibility into an obscuration.

CENTER OF GRAVITY (CG) — The theoretical point where the entire weight of the airplane is considered to be concentrated.

CENTER OF PRESSURE — A point along the wing chord line where lift is considered to be concentrated.

CENTRIFUGAL FORCE — An apparent force, that opposes centripetal force, resulting from the effect of inertia during a turn.

CENTRIPETAL FORCE — A center-seeking force directed inward toward the center of rotation created by the horizontal component of lift in turning flight.

CHORD — An imaginary straight line between the leading and trailing edges of an airfoil section.

CLASS — (1) As used with respect to the certification, ratings, privileges, and limitations of airmen, means a classification of aircraft within a category having similar operating characteristics (single-engine land, multi-engine land, single-engine sea, multi-engine sea, gyroplane, helicopter, airship, and free balloon). (2) As used with respect to certification of aircraft means a broad grouping of aircraft having similar characteristics of propulsion, flight, or landing (airplane, rotorcraft, glider, balloon, landplane, and seaplane).

CLASS A AIRSPACE — Controlled airspace covering the 48 contiguous United States and Alaska, within 12 nautical miles of the coasts, from 18,000 feet MSL up to and including FL600, but not including airspace less than 1,500 feet AGL, in which all flight must be under IFR.

CLASS B AIRSPACE — Controlled airspace designated around certain major airports, extending from the surface or higher to specified altitudes where ATC provides radar separation for all IFR and VFR aircraft. For operations in Class B airspace, all aircraft must receive an ATC clearance to enter, and are subject to the rules and pilot/equipment requirements listed in FAR Part 91.

CLASS C AIRSPACE — Controlled airspace surrounding designated airports where ATC provides radar vectoring and sequencing for all IFR and VFR aircraft. Participation is mandatory, and all aircraft must establish and maintain radio contact with ATC, and are subject to the rules and pilot/equipment requirements listed in FAR Part 91.

CLASS D AIRSPACE — Controlled airspace around at least one primary airport that has an operating control tower. Aircraft are subject to the rules and equipment requirements specified in FAR Part 91.

CLASS E AIRSPACE — Controlled airspace above the United States that typically extends from 1,200 feet AGL to 18,000 feet MSL or extends from 700 feet AGL in conjunction with an airport that has an approved instrument approach procedure.

CLASS G AIRSPACE — Airspace that has not been designated as Class A, B, C, D, or E, and within which air traffic control is not exercised.

CLEAR AIR TURBULENCE (CAT) — Turbulence as a result of two bodies of air meeting at remarkably different speeds. Although CAT is often encountered near the jet stream in clear air, it also can be present at lower altitudes and in non-convective clouds.

CLEARING TURNS — Turns consisting of at least a 180° change in direction, allowing a visual check of the airspace around the airplane to avoid conflicts while maneuvering.

COLD FRONT — The boundary between two air masses where cold air is replacing warm air.

COMMON TRAFFIC ADVISORY FREQUENCY (CTAF) — A frequency designed for the purpose of carrying out airport advisory practices while operating to or from an uncontrolled airport. The CTAF may be a UNICOM, MULTICOM, FSS, or tower frequency and it is identified in appropriate aeronautical publications.

COMMUNICATION (COM) — One of the nine CBTA competencies. Mastering the COM competency will ensure communicating with ATC, conveying information to pilots and passengers, and active listening is effective. You must be proficient in reading, speaking, and comprehending the English language.

COMPASS HEADING — Heading derived by applying correction factors for wind variation, and deviation to true course.

COMPETENCY-BASED TRAINING AND ASSESSMENT (CBTA) — An evolutionary training method originally developed for commercial airline pilots, focusing on nine competencies and observable behaviors, with the goal of developing the behaviors of a resilient pilot, ready to meet the many challenges of flight.

COMPLEX AIRPLANE — An airplane with retractable landing gear, flaps, and a controllable-pitch propeller.

CONDENSATION — A change of state of water from a gas (water vapor) to a liquid.

CONDENSATION NUCLEI — The small particles of solid matter in the air on which water vapor condenses.

CONE OF CONFUSION — The cone-shaped area above a VOR station in which there is no signal and the TO/FROM flag momentarily flickers to OFF (or a similar indication).

CONES — The cells concentrated in the center of the retina that provide color vision and sense fine detail.

CONTROLLED AIRPORT — An airport that has an operating control tower, sometimes called a towered airport.

CONTROLLED AIRSPACE — Airspace designated as Class A, B, C, D, or E, within which some or all aircraft may be subject to air traffic control.

CONTROLLED FLIGHT INTO TERRAIN (CFIT) — Collision with terrain or water with no prior awareness by the pilot that the crash is imminent.

CONVECTION — A circulation process caused by unequal air density that results from heating inequities.

CONVECTIVE SIGMET — A weather advisory concerning convective weather significant to the safety of all aircraft. Convective SIGMETs are issued for tornadoes, lines of thunderstorms, thunderstorms over a wide area, embedded thunderstorms, wind gusts to 50 knots or greater and/or hail 3/4 inch in diameter or greater.

CONVENTIONAL LANDING GEAR — A landing gear configuration with two main wheels located on either side of the fuselage and a third wheel, the tailwheel, positioned at the rear of the airplane.

COORDINATED UNIVERSAL TIME (UTC) — A method of expressing time that places the entire world on one time standard. UTC also is referred to as Zulu time.

CORIOLIS FORCE — A deflective force that is created by the difference in rotational velocity between the equator and the poles of the earth. It deflects air to the right in the northern hemisphere and to the left in the southern hemisphere.

COURSE — The intended or desired direction of flight in the horizontal plane measured in degrees from true or magnetic north.

CREW RESOURCE MANAGEMENT (CRM) — The application of team management concepts in a multi-person flight crew to make effective use of all available resources: human, hardware, and information.

CROSSWIND — A wind that is not parallel to a runway or the path of an aircraft.

CROSSWIND COMPONENT — A wind component that is at a right angle to the runway or the flight path of an aircraft.

DEAD RECKONING — A type of navigation based on the calculations of time, speed, distance, and direction.

DENSITY ALTITUDE — Pressure altitude corrected for nonstandard temperature.

DEPOSITION — The direct transformation of a gas to a solid state, where the liquid state is bypassed.

DEPRESSANTS —Drugs that reduce the body's functioning usually by lowering blood pressure, reducing mental processing, and slowing motor and reaction responses.

DETONATION — An uncontrolled, explosive ignition of the fuel/air mixture within the cylinder's combustion chamber.

DEVIATION — A compass error caused by magnetic disturbances from electrical and metal components in the airplane. The correction for this error is displayed on a compass correction card placed near the magnetic compass in the airplane.

DEWPOINT — The temperature at which air reaches a state where it can hold no more water.

DIHEDRAL — The upward angle of an airplane's wings with respect to the horizontal. Dihedral contributes to the lateral stability of an airplane.

DIRECTIONAL STABILITY — Stability about the vertical axis.

DISPLACED THRESHOLD — When the landing area begins at a point on the runway other than the designated beginning of the runway.

DISTANCE MEASURING EQUIPMENT (DME) — Equipment (airborne and ground) to measure, in nautical miles, the slant range distance of an aircraft from the navigation aid.

DRAG — A backward, or retarding, force that opposes thrust and limits the speed of the airplane.

ELECTRONIC FLIGHT BAG — Electronic information management device that provides items such as airport data, airport diagrams, a moving map, weather displays, flight plans, routes, checklists, aircraft performance charts, and logbooks.

EMERGENCY LOCATOR TRANSMITTER (ELT) — A battery-operated radio transmitter attached to the aircraft structure that transmits on 121.5 MHz and 243.0 MHz. It aids in locating downed aircraft.

EMPENNAGE — The section of the airplane that consists of the vertical stabilizer, the horizontal stabilizer, and the associated control surfaces.

EMPTY FIELD MYOPIA — The normal tendency of the eye to focus at only 10 to 30 feet when looking into a field devoid of objects, contrasting colors, or patterns.

ERRORS — Pilot actions, or inactions that lead to deviations from expected outcomes. Errors absorb attention, increase workload and risk, cause confusion, and reduce safety margins. If a pilot fails to manage, or mismanages an error, the probability of an adverse outcome increases.

EVAPORATION — The transformation of a liquid to a gaseous state, such as the change of water to water vapor.

FLIGHT INFORMATION SERVICE–BROADCAST (FIS-B) — A data service provided through the ADS-B Broadcast Services network that provides pilots of properly-equipped aircraft with a flight deck display of various aviation weather and aeronautical information.

FLIGHT RESTRICTED ZONE (FRZ) — A highly-restricted ring of airspace within 13 to 15 nautical miles of the Washington DC VOR, which is directly over the nation's capital. Flight under VFR and general aviation aircraft operations are prohibited.

FLIGHT SERVICE — A facility that provides various services to pilots, including weather briefings, opening and closing flight plans, and calling search and rescue.

FLY FOCUS ACT — An axiom created to overcome anxious moments, physiological responses, overreactions, or hesitancy. Fly the airplane. Keep it in the air and maintain safety. Do not hurry. Focus. Maintain your composure, observe, identify, and confirm. Act. Manage the event; identify and solve the problem. Communicate.

FREEZING LEVEL — Altitude where the temperature is 32°F (0°C).

FRONT — The boundary between two different air masses.

FUSELAGE — The cabin or flight deck, is located in the fuselage. It also provides room for cargo and attachment points for other major airplane components.

GLOBAL NAVIGATION SATELLITE SYSTEM (GNSS) — The standard generic term for satellite navigation systems that provide autonomous geo-spatial positioning with global coverage.

GLOBAL POSITIONING SYSTEM (GPS) — A United States primary satellite navigation system that is globally available. GPS consists of three segments—space, control, and user.

GREAT CIRCLE — The largest circle that can be drawn on the earth's surface. A great circle's plane passes through the center of the earth dividing it into two equal parts.

GROUND EFFECT — A usually beneficial influence on aircraft performance that occurs while you are flying close to the ground. It results from a reduction in upwash, downwash, and wingtip vortices that provide a corresponding decrease in induced drag.

GROUNDSPEED (GS) — Speed of the aircraft in relation to the ground.

HAZARDS — Factors that could cause, or contribute to, an aircraft incident or accident. Hazards include adverse weather, challenging airports surrounded by high terrain, or congested airspace. Some hazards can be anticipated, and some are unforeseen, such as a sudden in-flight mechanical failure.

HAZARDOUS IN-FLIGHT WEATHER ADVISORY SERVICE (HIWAS) — Continuous recordings of hazardous weather information broadcast over selected VORs.

HEADING — The direction in which the longitudinal axis of the airplane points with respect to true or magnetic north. Heading is equal to course plus or minus any wind correction angle.

HEADWIND COMPONENT — That part of the wind that acts directly on the front of the aircraft and decreases its groundspeed.

HECTOPASCAL (hPa) — The metric equivalent of a millibar (1 hPa = 1 mb).

HIGH-PERFORMANCE AIRPLANE — An airplane that has an engine with more than 200 horsepower.

HOMING — A method of navigating to an NDB by holding a zero relative bearing.

HUMIDITY — The amount of water vapor in the air.

HYPERVENTILATION — The excessive ventilation of the lungs caused by very rapid and deep breathing, which results in an excessive loss of carbon dioxide from the body.

HYPOXIA — The effects on the human body due to an insufficient supply of oxygen.

INDICATED AIRSPEED (IAS) — The speed of an aircraft as shown on the airspeed indicator.

INDICATED ALTITUDE — The altitude shown by an altimeter set to the current altimeter setting.

INDUCED DRAG — That part of total drag that is created by the production of lift. Induced drag increases with a decrease in airspeed.

INSTRUMENT FLIGHT RULES (IFR) — The rules that govern the procedure for conducting flight in weather conditions below VFR weather minimums. The term IFR also is used to define weather conditions and the type of flight plan under which an aircraft is operating.

INTEGRATED AIRMAN CERTIFICATION AND RATING APPLICATION (IACRA) — The web-based application system that guides pilots through the FAA's process to apply for pilot certificates and ratings. IACRA helps ensure that regulatory and policy requirements are met through extensive data validation, uses electronic signatures, eliminates paper forms, and prints temporary certificates.

INTERNATIONAL STANDARD ATMOSPHERE (ISA) — Standard atmospheric conditions consisting of a temperature of 59°F (15°C), and a barometric pressure of 29.92 in. Hg. (1013.2 mb) at sea level. ISA values can be calculated for various altitudes using standard lapse rates.

INVERSION — An increase in temperature with altitude.

ISOBAR — A line that connects points of equal barometric pressure.

ISOGONIC LINES — Lines on charts that connect points of equal magnetic variation.

JET STREAM — A narrow band of winds with speeds of 100 to 200 mph occurring between approximately 32,000 and 49,000 feet.

KATABATIC WIND — Any downslope wind usually stronger than a mountain breeze. A katabatic wind can be either warm or cold.

LAND BREEZE — A coastal breeze blowing from land to sea caused by temperature difference when the sea surface is warmer than the adjacent land. The land breeze usually occurs at night and alternates with a sea breeze, which blows in the opposite direction by day.

LAPSE RATE — The rate of decrease of an atmospheric variable with altitude.

LATERAL STABILITY — Stability about the longitudinal axis.

LATITUDE — Measurement north or south of the equator in degrees, minutes, and seconds. Lines of latitude are also called parallels.

LEADERSHIP AND TEAMWORK (LTW) — One of the nine CBTA competencies. LTW includes team participation and communication, considering input from others, giving and receiving constructive feedback, exercising leadership when needed, and carrying out instructions when directed.

LEARNER-CENTERED GRADING — An evaluation/teaching technique in which a student and instructor independently assess student performance and resolve any differences before creating a plan for improvement.

LIFT — An upward force created by the effect of airflow as it passes over and under the wing.

LOAD FACTOR — The ratio of the load supported by the airplane's wings to the actual weight of the aircraft and its contents.

LOCAL AIRPORT ADVISORY (LAA) — Advisory service provided to pilots by an FSS at airports in Alaska without an operating control tower. Information includes known traffic and weather conditions.

LONGITUDE — Measurement east or west of the Prime Meridian in degrees, minutes, and seconds. Lines of longitude are also called meridians. The Prime Meridian is 0° longitude and runs through Greenwich, England.

LONGITUDINAL STABILITY — Stability about the lateral axis. A desirable characteristic of an airplane whereby it tends to return to its trimmed angle of attack after displacement.

MAGNETIC BEARING — The magnetic course you would fly to go direct to an NDB station.

MAGNETIC COURSE — True course corrected for magnetic variation.

MAGNETO — A self-contained, engine-driven unit that supplies electrical current to the spark plugs, independent of the airplane's electrical system. An engine normally has two magnetos.

MANEUVERING SPEED (V_A) — The maximum speed at which you can use a full, abrupt control movement without overstressing the airframe.

MAYDAY — International radio distress signal. When repeated three times, it indicates imminent and grave danger and that immediate assistance is requested.

MEAN SEA LEVEL (MSL) — The average height of the surface of the sea for all stages of tide.

MESOSPHERE — A layer of the atmosphere above the stratosphere.

MICROBURST — A strong downdraft that normally occurs over horizontal distances of 1 NM or less and vertical distances of less than 1,000 feet. In spite of its small horizontal scale, an intense microburst can induce wind speeds greater than 100 knots and downdrafts as strong as 6,000 feet per minute.

MILITARY OPERATIONS AREA (MOA) — Special use airspace of defined vertical and lateral limits established to help VFR traffic identify locations where military activities are conducted.

MILITARY TRAINING ROUTE (MTR) — Route depicted on an aeronautical chart for the conduct of military flight training at speeds above 250 knots.

MILLIBAR (mb) — A unit of atmospheric pressure equal to a force of 1,000 dynes per square centimeter.

MINIMUM EQUIPMENT LIST — A document provided by an aircraft manufacturer or created by an operator and approved by the FAA, that lists the equipment that may be inoperative for a flight based on the conditions of that flight.

MOMENT — A measurement of the tendency of a weight to cause rotation at the fulcrum.

MOUNTAIN BREEZE — A downslope wind flow at night, caused by the cooling of the air at higher elevations.

MULTICOM — An air-to-air communication frequency (122.9) for pilots to announce their position and intentions to other aircraft in the area.

NONDIRECTIONAL BEACON — An L/MF or UHF radio beacon transmitting nondirectional signals whereby the pilot of an aircraft equipped with direction finding equipment can determine their to or from the radio beacon and "home" on or track too from station.

NOTICE TO AIR MISSIONS (NOTAM) — A notice containing time-critical information that is either of a temporary nature or is not known far enough in advance to permit publication on aeronautical charts or other operational publications.

OBSERVABLE BEHAVIOR (OB) — Each of the nine CBTA competencies include OBs that pilots demonstrate to show proficiency of a particular competency. OBs describe how a pilot demonstrates proficiency of a competency, or in other words, what proficient performance of a competency looks like.

OBSTRUCTION LIGHT — A light, or one of a group of lights, usually red or white, mounted on a surface structure or natural terrain to warn pilots of the presence of a flight hazard.

OCCLUDED FRONT — A frontal occlusion occurs when a fast-moving cold front catches up to a slow-moving warm front. The difference in temperature within each frontal system is a major factor in determining whether a cold or warm front occlusion occurs.

OROGRAPHIC — Associated with or induced by the presence of rising terrain, such as orographic lifting.

PARASITE DRAG — That part of total drag created by the form or shape of airplane parts. Parasite drag increases with an increase in airspeed.

PILOT COMPETENCIES — Nine pilot competencies are part of the competency-based training and assessment (CBTA) process. The pilot competencies describe how to effectively perform as a resilient pilot.

PILOT CONTROLLED LIGHTING (PCL) — Runway lighting systems that are controlled by keying the aircraft's microphone on a specific frequency.

PILOT IN COMMAND (PIC) — The pilot responsible for the operation and safety of an aircraft.

PILOT WEATHER REPORT (PIREP) — A report, generated by pilots, concerning meteorological phenomena encountered in flight.

PILOTAGE — Navigation by visual landmarks.

PRECESSION — The tilting or turning of a gyroscope in response to external forces causing slow drifting and erroneous indications in gyroscopic instruments.

PREIGNITION — A phenomenon when the fuel/air mixture is ignited in advance of the normal timed ignition and is usually caused by a residual hot spot in the cylinder.

PRESSURE ALTITUDE — Height above the standard pressure level of 29.92 in. Hg. Obtained by setting 29.92 in the barometric pressure window and reading the altimeter.

PREVAILING VISIBILITY — The greatest horizontal visibility throughout at least half the horizon.

PREVENTIVE MAINTENANCE — A minor service that pilots without an aviation maintenance technician certificate are allowed to perform on an aircraft.

PROBLEM SOLVING — DECISION MAKING (PSD) — One of the nine CBTA competencies. Being skilled at threat and error management (TEM) and using the aeronautical decision making process (ADM) are key to effectively manage threats and become proficient in PSD.

PROHIBITED AREA — Airspace of defined dimensions within which the flight of aircraft is prohibited.

RADAR CONTACT — Term used by ATC to inform you that your airplane has been identified using an approved ATC surveillance source on a controller's display and that flight following will be provided until radar service is terminated.

APPENDIX C ■ Glossary

RADIAL — A navigational signal generated by a VOR or VORTAC, measured as a magnetic bearing from the station.

REFERENCE DATUM — An imaginary vertical plane from which all horizontal distances are measured for balance purposes.

RELATIVE BEARING — The angular difference between the airplane's longitudinal axis and a straight line drawn from the airplane to a navaid, such as an NDB. It is measured clockwise from the airplane's nose.

RELATIVE HUMIDITY — The actual amount of moisture in the air compared to the total that could be present at that temperature.

RESILIENT PILOTS — The ability to cope with challenges and setbacks, employ stress management strategies, and adapt and quickly recover from adversity by "bouncing back." Many of the qualities that contribute to being a resilient pilot can also be applied to everyday life.

RESTRICTED AREA — Designated special use airspace within which aircraft flight, while not prohibited, is subject to restrictions.

RETINA — The photosensitive portion of the eye which is connected to the optic nerve and contains cells called rods and cones.

RETRACTABLE GEAR — A pilot controllable landing gear system, whereby the gear can be stowed alongside or inside the structure of the airplane during flight.

RIGIDITY IN SPACE — The principle that a wheel with a heavily weighted rim spun rapidly will remain in a fixed position in the plane in which it is spinning.

RISK MANAGEMENT — The process of managing and applying all available resources to ensure a safe flight. Risk management is critical to making effective decisions.

RODS — The cells concentrated outside of the foveal area that are sensitive to low light and not to color.

ROOT CAUSE ANALYSIS — The process of examining a problem, or why something went wrong and determining how to prevent its reoccurrence in the future.

RUNWAY GRADIENT — The amount of change in elevation over the length of the runway.

RUNWAY INCURSION HOT SPOT — A charted location at an airport where a heightened risk of a runway incursion exists.

RUNWAY VISUAL RANGE — An instrumentally derived value representing the horizontal distance a pilot in a moving aircraft should see down the runway.

SAFETY ALERT — An alert issued by an ATC radar facility when an aircraft under its control is in unsafe proximity to terrain, obstruction, or other aircraft.

SATURATED AIR — Air containing the maximum amount of water vapor it can hold at a given temperature (100% relative humidity).

SEA BREEZE — A coastal breeze blowing from sea to land, caused by the temperature difference when the land surface is warmer than the sea surface. The sea breeze usually occurs during the day and alternates with the land breeze, which blows in the opposite direction at night.

SECTIONAL CHART — The most commonly used chart for VFR flight. Each chart covers 6° to 8° of longitude and approximately 4° of latitude and is given the name of a primary city within its coverage.

SEGMENTED CIRCLE — A set of visual indicators that provide traffic pattern information at airports without operating control towers.

SERVICE CEILING — The maximum height above mean sea level, under normal conditions, at which a given airplane is able to maintain a rate of climb of 100 feet per minute.

SIGMET — An in-flight advisory that is considered significant to all aircraft. SIGMET criteria include severe icing, severe and extreme turbulence, duststorms, sandstorms, volcanic eruptions, and volcanic ash lowering visibility to less than three miles.

SINGLE-PILOT RESOURCE MANAGEMENT (SRM) — The skill of managing hardware, information, dispatchers, weather briefers, maintenance personnel, and ATC controllers to gather information, analyze your situation, and make effective decisions about the current and future status of your flight. SRM includes these six concepts: aeronautical decision making, risk management, task management, situational awareness, controlled flight into terrain (CFIT) awareness, and automation management.

SITUATIONAL AWARENESS — The accurate perception of all the operational and environmental factors that affect flight safety before, during, and after a flight.

SITUATION AWARENESS AND MANAGEMENT OF INFORMATION (SAW) — One of the nine CBTA competencies. SAW requires an accurate perception of operational and environmental factors affecting a flight. You must be able to analyze the airplane, the environment, and yourself, to know what is happening, and to anticipate future events.

SKID — Condition in which the rate of turn is too fast for the angle of bank.

SLIP — Condition in which the rate of turn is too slow for the angle of bank.

SPATIAL DISORIENTATION — A feeling of instability caused by a conflict between the information relayed by your central vision, and your peripheral vision.

SPECIAL USE AIRSPACE — Defined airspace areas where aircraft operations may be limited. Examples include: alert area, controlled firing area, military operations area, prohibited area, restricted area, and warning area.

SPECIAL VFR CLEARANCE — An ATC clearance that allows you to operate within the surface areas of Class B, C, D, or E airspace when the ceiling is less than 1,000 feet or visibility is less than 3 statute miles. While operating under special VFR, you must maintain 1 mile of visibility and remain clear of clouds.

SPIN — An aggravated stall that results in the airplane descending in a helical, or corkscrew path.

SQUALL LINE — A continuous line of non-frontal thunderstorms.

STALL — A rapid decrease in lift caused by the separation of airflow from the wing's surface brought on by exceeding the critical angle of attack.

STANDARD LAPSE RATE — The pressure drops by 1.00 in. Hg and the temperature drops by 2°C (3.5°F) per 1,000 feet of altitude increase up to 36,000 feet MSL.

STARTLE EVENT — An unexpected event, such as a sudden engine failure in flight, which could cause you to improperly diagnose and manage a problem. If you do not manage a startle event properly, it could lead to panic, causing an undesired aircraft state (UAS), and potentially result in an incident or accident.

STATIONARY FRONT — A boundary between two air masses that are relatively balanced.

STIMULANTS — Drugs that excite the central nervous system and produce an increase in alertness and activity.

STOPWAY — An area beyond the takeoff runway that is designed to support an airplane during an aborted takeoff without causing structural damage to the airplane. It cannot be used for takeoff, landing or taxiing.

STRATOSPHERE — The first layer above the tropopause extending to a height of approximately 160,000 feet.

SUBLIMATION — Process by which a solid is changed to a gas without going through the liquid state.

SUPERCOOLED WATER DROPLETS — Water droplets that have been cooled below the freezing point, but are still in a liquid state.

TAILWHEEL AIRPLANE — Two main wheels located on either side of the fuselage and a third wheel, the tail wheel, positioned at the rear of the airplane. Also known as a conventional landing gear airplane.

TAILWIND — A wind more than 90° away from the airplane's magnetic heading.

TAILWIND COMPONENT — The effective wind velocity acting along an aircraft's course that increases its groundspeed.

TECHNICALLY ADVANCED AIRCRAFT — An aircraft with a primary flight display (PFD), a multi-function display (MFD) that uses GPS to display the aircraft position, and a two-axis autopilot.

TERMINAL RADAR SERVICE AREA (TRSA) — Airspace surrounding designated airports in which ATC provides radar vectoring, sequencing, and separation for all IFR aircraft and participating VFR aircraft.

TETRAHEDRON — A landing direction indicator, usually at nontowered airports. The small end points into the wind, the general landing direction.

THERMOSPHERE — The area of the atmosphere above the mesosphere.

THREATS — An unexpected risk or hazard. Threats occur outside the influence of and are not controlled by the pilot. Threats increase the operational complexity of a flight; can appear suddenly, and limit time available to analyze; and require a pilot to effectively manage to contain risk within acceptable levels.

THREAT AND ERROR MANAGEMENT (TEM) — A safety management approach of managing threats and errors that can lead to UAS. TEM assumes that as a pilot, you will encounter risk, and make mistakes.

THRESHOLD — The beginning of the landing area of the runway.

THRUST — A forward force that propels the airplane through the air.

TOTAL DRAG — The sum of parasite and induced drag.

TRACK — Or *ground track*; the actual flight path of an aircraft over the ground.

TRACKING — Flying a desired course to or from a station using a sufficient wind correction, if necessary.

TRAFFIC ADVISORIES — ATC advisories to alert a pilot to other air traffic that could be a potential collision threat.

TRAFFIC PATTERN — The traffic flow prescribed for aircraft landing and taking off from an airport, consisting of departure, crosswind, downwind, and base legs; and final approach.

TRANSPONDER — Radio device aboard an aircraft that enhances its identity on an ATC display.

TRICYCLE GEAR — Two main wheels located on either side of the fuselage and a third wheel, the nosewheel, positioned on the nose of the airplane.

TROPOPAUSE — Top of the troposphere, at an average altitude of 36,000 feet, which acts as a lid to most water vapor and associated weather.

TROPOSPHERE — The layer of the atmosphere from the surface to an average altitude of about 36,000 feet.

TRUE AIRSPEED (TAS) — The speed at which an aircraft is moving relative to the surrounding air.

TRUE ALTITUDE — The actual height of an object above mean sea level.

TRUE COURSE (TC) — The intended or desired direction of flight as measured on a chart clockwise from true north.

TRUE HEADING (TH) — The direction the longitudinal axis of the airplane points with respect to true north. True heading is equal to true course plus or minus any wind correction angle.

UNCONTROLLED AIRPORT — An airport where control of VFR traffic is not exercised, also known as a *nontowered airport.*

UNCONTROLLED AIRSPACE — Airspace where air traffic control is not exercised, also known as *Class G airspace.*

UNDESIRED AIRCRAFT STATE (UAS) — An aircraft condition such as pilot-induced airplane flight path/airspeed deviations, misapplied flight controls, an incorrect airplane configuration, taking off or landing on the wrong runway, or unexpectedly approaching an aerodynamic stall. UAS could lead to could lead to an incident or accident.

UNICOM — A privately owned air/ground advisory station that transmits on a limited number of frequencies—122.7, 122.725, 122.8, 122.975, or 123.0.

USABLE FUEL — The amount of fuel available during flight.

USEFUL LOAD — The difference between the basic empty weight of the airplane and the maximum weight allowed by the manufacturer's specification.

VALLEY BREEZE — Upslope wind flow caused by the heating of the mountain slope, which warms the adjacent air.

VAPOR LOCK — Bubbles of vapor in the fuel system that prevent fuel from reaching the cylinders and making restart of a hot engine difficult. It can also occur after running a fuel tank dry, allowing air to enter the fuel system.

VARIATION — The angular difference between true north and magnetic north; indicated on charts by isogonic lines.

VECTOR — A heading issued by ATC to provide a VFR aircraft with navigational guidance on an advisory basis only. ATC might provide vectors for safety reasons or a pilot can request vectors if unfamiliar with the area.

VERY HIGH FREQUENCY — Frequency band from 30 MHz to 300 MHz.

VFR CRUISING ALTITUDE — The rule when flying VFR above 3,000 feet AGL on magnetic headings from 0° to 179° requires odd thousand-foot altitudes plus 500 feet and on headings from 180° to 359° requires even thousand-foot altitudes plus 500 feet.

VFR TERMINAL AREA CHART (TAC) — Aeronautical charts that provide more detail than sectional charts to aid in VFR navigation in and around some of the busiest airports in the country. Most terminal area charts cover airports that have Class B airspace.

VICTOR AIRWAY — An airway system based on the use of VOR facilities.

VISUAL FLIGHT RULES (VFR) — Rules that specify minimum cloud clearance and visibility requirements for flight. The term VFR also defines weather conditions and the type of flight plan under which an aircraft is operating.

VISUAL PURPLE — Another term for *rhodopsin*, the chemical created by the rods for perception of dim light.

VOR — Very high frequency omnidirectional range: ground-based navigational system that provides course guidance.

WARM FRONT — The boundary between two air masses where warm air is replacing cold air.

WARNING AREA — Airspace of defined dimensions, extending from three nautical miles outward from the coast of the United States, which can contain activity that is hazardous to nonparticipating aircraft.

WASHINGTON DC SPECIAL FLIGHT RULES AREA (SFRA) — Airspace where the ready identification, location, and control of aircraft is required in the interests of national security. Depicted on charts, the SFRA includes all airspace within a 30 nautical mile radius of the Washington DC VOR (DCA) from the surface up to but not including flight level 180 (FL180).

WAYPOINT — Predetermined geographical positions for route definition typically in a GPS navigation database.

WEIGHT — A downward force caused by gravity. Weight opposes lift.

WIND CORRECTION ANGLE (WCA) — The angular difference between the heading of the airplane and the course.

WIND SHEAR — A sudden, drastic shift in wind speed or direction in either the vertical or horizontal plane.

WINGS-PILOT PROFICIENCY PROGRAM — An FAA program designed to mitigate the primary factors that cause general aviation accidents by providing ground and flight training opportunities for pilots to apply risk management, enhance their knowledge and skills, and increase proficiency.

WINGTIP VORTICES — Spirals of air created by an airfoil when generating lift. Vortices from medium to heavy aircraft may be extremely hazardous to small aircraft.

WORKLOAD MANAGEMENT (WLM) — One of the nine CBTA competencies. Planning, prioritizing, delegating tasks, and completing tasks during times of low workload are all essential pieces of WLM. Pilots need to use both internal resources, such as checklists, and external resources, such as ATC assistance, to help manage the workload. When overburdened, pilot performance can decline.

ZULU TIME — A term used in aviation for coordinated universal time (UTC), which places the entire world on one time standard.

INDEX